D1159135

McGraw-Hill Ryerson
MATHEMATICS 11

AUTHORS

GEORGE KNILL
B.Sc., M.S.Ed.
Hamilton, Ontario

BARBARA J. CANTON
B.A., B.Ed., M.Ed.
Limestone District School Board

FRED FERNEYHOUGH
B.Math.
Peel District School Board

LYNDA FERNEYHOUGH
B.Math., M.Ed.
Peel District School Board

O. MICHAEL G. HAMILTON
B.Sc., B.Ed., M.Sc.
Ridley College

LOUIS LIM
B.Sc., B.Ed.
Hastings and Prince Edward District School Board

JOHN RODGER
B.Math., B.Ed.
Peel District School Board

MICHAEL WEBB
B.Sc., M.Sc., Ph.D.
Toronto, Ontario

TECHNOLOGY CONSULTANT
Fred Ferneyhough, B.Math.
Peel District School Board

ASSESSMENT CONSULTANT
Lynda Ferneyhough, B.Math., M.Ed.
Peel District School Board

ASSESSMENT CONSULTANT
Chris Dearling, B.Sc., M.Sc.
Burlington, Ontario

WEB CONSULTANT
Frank Maggio, B.A., B.Ed., M.A.
Halton Catholic District School Board

McGraw-Hill
Ryerson

Toronto Montréal Boston Burr Ridge, IL Dubuque, IA Madison, WI New York
San Francisco St. Louis Bangkok Bogotá Caracas Kuala Lumpur Lisbon London
Madrid Mexico City Milan New Delhi Santiago Seoul Singapore Sydney Taipei

McGraw-Hill
Ryerson Limited

A Subsidiary of The McGraw-Hill Companies

COPIES OF THIS BOOK
MAY BE OBTAINED BY
CONTACTING:

McGraw-Hill Ryerson Ltd.

E-MAIL:
orders@mcgrawhill.ca

TOLL FREE FAX:
1-800-463-5885

TOLL FREE CALL:
1-800-565-5758

OR BY MAILING YOUR
ORDER TO:
McGraw-Hill Ryerson
Order Department
300 Water Street
Whitby, ON L1N 9B6

Please quote the ISBN and
title when placing your order.

McGraw-Hill Ryerson MATHEMATICS 11

ISBN 0-07-552910-6

ISBN-13: 978-0-07-552910-1

http://www.mcgrawhill.ca

5 6 7 8 9 0 TRI 0 9 8 7 6

Printed and bound in Canada

Care has been taken to trace ownership of copyright material contained in this text. The publishers will gladly take any information that will enable them to rectify any reference or credit in subsequent printings.

The Geometer's Sketchpad®, Key Curriculum Press, 1150 65th Street, Emeryville, CA 94608 1-800-995-MATH.
CBL™ and CBR™ are trademarks of Texas Instruments.
Zap-a-Graph © Brain Waves Software Inc., 2103 Galetta Sideroad, Fitzroy Harbour, Ontario, K0A 1X0 Tel. 613-623-8686.
Microsoft® Excel are either registered trademarks or trademarks of Microsoft Corporation in the United States and/or other countries.
Screen shot(s) from Quattro® Pro. Corel and Quattro are trademarks or registered trademarks of Corel Corporation or Corel Corporation Limited, reprinted by permission.

National Library of Canada Cataloguing in Publication Data

Main entry under title:

McGraw-Hill Ryerson mathematics 11

Includes index.
ISBN 0-07-552910-6

Mathematics. 2. Functions. I. Knill, George, date. II. Title: Mathematics 11
III. Title: McGraw-Hill Ryerson mathematics eleven.

QA39.2.M225 2001   515'.1   C2001-930279-7

PUBLISHER: Diane Wyman
EDITORIAL CONSULTING: Michael J. Webb Consulting Inc.
DEVELOPMENTAL EDITORS: Jean Ford, Janice Nixon, Lynda Cowan
ASSOCIATE EDITOR: Mary Agnes Challoner
SENIOR SUPERVISING EDITOR: Carol Altilia
PERMISSIONS EDITOR: Ann Ludbrook
EDITORIAL ASSISTANTS: Joanne Murray, Erin Parton
JUNIOR EDITORS: Christopher Cappadocia, Cheryl Stallabrass
ASSISTANT PROJECT COORDINATORS: Melissa Nippard, Janie Reeson
PRODUCTION SUPERVISOR: Yolanda Pigden
COVER AND INTERIOR DESIGN: Matthews Communications Design Inc.
ART DIRECTION: Tom Dart/First Folio Resource Group, Inc.
ELECTRONIC PAGE MAKE-UP: Alana Lai, Claire Milne, Greg Duhaney/First Folio Resource Group, Inc.
COVER IMAGE: Marko Modic/CORBIS/Magma Photo News

Acknowledgements

Reviewers of *McGraw-Hill Ryerson MATHEMATICS 11*

The authors and editors of McGraw-Hill Ryerson MATHEMATICS 11 wish to thank the reviewers listed below for their thoughtful comments and suggestions. Their input has been invaluable in ensuring that this text meets the needs of the students and teachers of Ontario.

Anthony Azzopardi
Toronto Catholic District School Board

Peter Clifford
Toronto District School Board

John Conrad
Waterloo Region District School Board

Chris Dearling
Burlington, ON

Gerry Doerksen
Royal St. George's College

Catherine Dunne
Peel District School Board

Wayne Erdman
Toronto District School Board

Eric Forshaw
Greater Essex County District School Board

Mary-Beth Fortune
Peel District School Board

Jeff Irvine
Peel District School Board

Ann Kajander (Ph.D.)
Lakehead Public Schools

David Kay
Peel District School Board

John Kennedy
District School Board of Niagara

Dianna Knight
Peel District School Board

Sabina Knight
District School Board of Niagara

Mike McGowan
Toronto District School Board

Nick Nolfi
Peel District School Board

Terry Paradellis
Toronto District School Board

Gizele Price
Dufferin Peel Catholic District School Board

Peter Saarimaki
Scarborough, ON

Al Smith
Kawartha Pine Ridge District School Board

Bob Smith
Rainbow District School Board

Susan Smith
Peel District School Board

Gregory Szczachor
Toronto District School Board

Richard Tong
Toronto District School Board

CONTENTS

A Tour of Your Textbook x

CHAPTER ①
Algebraic Tools for Operating With Functions

SPECIFIC EXPECTATIONS xx

MODELLING MATH: Modelling Problems Algebraically 1

GETTING STARTED: Frequency Ranges, Review of Prerequisite Skills . . .2

1.1 Reviewing the Exponent Laws 4

1.2 Rational Exponents .11

1.3 Solving Exponential Equations 19

TECHNOLOGY EXTENSION: Solving Exponential Equations With a
Graphing Calculator 27

1.4 Review: Adding, Subtracting, and Multiplying Polynomials . .28

1.5 Simplifying Rational Expressions 35

1.6 Multiplying and Dividing Rational Expressions 44

1.7 Adding and Subtracting Rational Expressions, I 53

1.8 Adding and Subtracting Rational Expressions, II 62

TECHNOLOGY EXTENSION: Rational Expressions and the Graphing
Calculator 70

1.9 Solving First-Degree Inequalities 72

TECHNOLOGY EXTENSION: Solving Inequalities With a Graphing
Calculator 82

INVESTIGATE & APPLY: Modelling Restrictions Graphically 83

REVIEW OF KEY CONCEPTS 85

CHAPTER TEST 90

CHALLENGE PROBLEMS 92

PROBLEM SOLVING STRATEGY: Model and Communicate Solutions . . .93

PROBLEM SOLVING: Using the Strategies 95

CHAPTER ②

Quadratic Functions and Equations

SPECIFIC EXPECTATIONS 96

MODELLING MATH: Measurements of Lengths and Areas 97

GETTING STARTED: Store Profits, Review of Prerequisite Skills 98

2.1 The Complex Number System 100

2.2 Maximum or Minimum of a Quadratic Function by
Completing the Square 110

2.3 Solving Quadratic Equations 120

TECHNOLOGY EXTENSION: Solving Quadratic Equations 134

2.4 Tools for Operating With Complex Numbers 135

TECHNOLOGY EXTENSION: Radical Expressions and Graphing
Calculators 143

2.5 Operations With Complex Numbers in Rectangular Form . .144

INVESTIGATE & APPLY: Interpreting a Mathematical Model 153

REVIEW OF KEY CONCEPTS 154

CHAPTER TEST 158

CHALLENGE PROBLEMS 160

PROBLEM SOLVING STRATEGY: Look for a Pattern 161

PROBLEM SOLVING: Using the Strategies 164

CUMULATIVE REVIEW: Chapters 1 and 2 165

CHAPTER ③

Transformations of Functions

SPECIFIC EXPECTATIONS 166

MODELLING MATH: Falling Objects .167

GETTING STARTED: Human Physiology, Review of Prerequisite Skills . .168

3.1 Functions 170

3.2 Investigation: Properties of Functions Defined by
$f(x) = \sqrt{x}$ and $f(x) = \frac{1}{x}$ .182

3.3 Horizontal and Vertical Translations of Functions 184

3.4 Reflections of Functions 194

3.5 Inverse Functions 208

3.6 Stretches of Functions 221

3.7 Combinations of Transformations 233

INVESTIGATE & APPLY: Frieze Patterns 244

REVIEW OF KEY CONCEPTS 246

CHAPTER TEST 254

CHALLENGE PROBLEMS 257

PROBLEM SOLVING STRATEGY: Solve Rich Estimation Problems 258

PROBLEM SOLVING: Using the Strategies 261

C H A P T E R   4

Trigonometry

SPECIFIC EXPECTATIONS **262**

MODELLING MATH: Ship Navigation 263

GETTING STARTED: Parallactic Displacement, Review of
     Prerequisite Skills 264

4.1 Reviewing the Trigonometry of Right Triangles 266

4.2 The Sine and the Cosine of Angles Greater Than 90° 276

4.3 The Sine Law and the Cosine Law 283

TECHNOLOGY EXTENSION: Using *The Geometer's Sketchpad®* to
     Explore the SSA Case 296

4.4 The Sine Law: The Ambiguous Case 300

INVESTIGATE & APPLY: The Cosine Law and the Ambiguous Case . . .312

REVIEW OF KEY CONCEPTS 313

CHAPTER TEST 316

CHALLENGE PROBLEMS 318

PROBLEM SOLVING STRATEGY: Use a Diagram 319

PROBLEM SOLVING: Using the Strategies 322

CUMULATIVE REVIEW: Chapters 3 and 4 323

C H A P T E R 5

Trigonometric Functions

SPECIFIC EXPECTATIONS 324

MODELLING MATH: Ocean Cycles 325

GETTING STARTED: Daylight Hours, Review of Prerequisite Skills . . .326

5.1 Radians and Angle Measure 328

5.2 Trigonometric Ratios of Any Angle 341

EXTENSION: Positive and Negative Angles of Rotation 351

5.3 Modelling Periodic Behaviour 355

5.4 Investigation: Sketching the Graphs of $f(x) = \sin x$,
  $f(x) = \cos x$, and $f(x) = \tan x$ 363

5.5 Stretches of Periodic Functions 367

5.6 Translations and Combinations of Transformations 378

TECHNOLOGY EXTENSION: Sinusoidal Regression 392

5.7 Trigonometric Identities 393

5.8 Trigonometric Equations 402

INVESTIGATE & APPLY: Modelling Double Helixes 411

REVIEW OF KEY CONCEPTS 412

CHAPTER TEST 418

CHALLENGE PROBLEMS 420

PROBLEM SOLVING STRATEGY: Solve a Simpler Problem 421

PROBLEM SOLVING: Using the Strategies 423

C H A P T E R 6

Sequences and Series

SPECIFIC EXPECTATIONS 424

MODELLING MATH: The Motion of a Pendulum 425

GETTING STARTED: Exploring Sequences, Review of
  Prerequisite Skills 426

6.1 Sequences 428

6.2 Arithmetic Sequences 436

6.3 Geometric Sequences 447

6.4 Recursion Formulas 457

6.5 Arithmetic Series 465

6.6 Geometric Series 472

INVESTIGATE & APPLY: Relating Sequences and Systems of Equations .479

REVIEW OF KEY CONCEPTS 480

CHAPTER TEST 486

CHALLENGE PROBLEMS 488

PROBLEM SOLVING STRATEGY: Guess and Check 489

PROBLEM SOLVING: Using the Strategies 492

CUMULATIVE REVIEW: Chapters 5 and 6 493

C H A P T E R ⑦

Compound Interest and Annuities

SPECIFIC EXPECTATIONS **494**

MODELLING MATH: Making Financial Decisions 495

GETTING STARTED: Comparing Costs, Review of Prerequisite Skills . .496

7.1 Investigation: Simple Interest, Arithmetic Sequences, and
       Linear Growth .498

7.2 Compound Interest 501

7.3 Investigation: Compound Interest, Geometric Sequences,
       and Exponential Growth 512

7.4 Present Value 516

7.5 Amount of an Ordinary Annuity 526

7.6 Present Value of an Ordinary Annuity 534

7.7 Technology: Amortization Tables and Spreadsheets 544

7.8 Mortgages 559

INVESTIGATE & APPLY: The Cost of Car Ownership 571

REVIEW OF KEY CONCEPTS 572

CHAPTER TEST 577

CHALLENGE PROBLEMS 579

PROBLEM SOLVING STRATEGY: Use Logic 580

PROBLEM SOLVING: Using the Strategies 583

C H A P T E R 8

Loci and Conics

SPECIFIC EXPECTATIONS 584

MODELLING MATH: Motion in Space 585

GETTING STARTED: Communications Satellites, Review of
Prerequisite Skills 586

8.1 Technology: Constructing Loci Using *The Geometer's
Sketchpad®* 588

8.2 Equations of Loci 594

8.3 Technology: Loci and Conics 601

8.4 The Circle 608

8.5 The Ellipse 619

8.6 The Hyperbola 637

8.7 The Parabola 653

8.8 Conic Sections With Equations in the Form
$ax^2 + by^2 + 2gx + 2fy + c = 0$ 665

8.9 Intersections of Lines and Conics 675

INVESTIGATE & APPLY: Confocal Conics 688

REVIEW OF KEY CONCEPTS 689

CHAPTER TEST 694

CHALLENGE PROBLEMS 696

PROBLEM SOLVING STRATEGY: Use a Data Bank 697

PROBLEM SOLVING: Using the Strategies 700

CUMULATIVE REVIEW: Chapters 7 and 8 701

Appendix A Review of Prerequisite Skills 702

Appendix B Graphing Calculator Keystrokes 717

Appendix C Computer Software 743

Answers 774

Glossary 812

Technology Index 822

Index 824

Credits 827

A Tour of Your Textbook

To understand the textbook's structure, begin by taking a brief tour.

CHAPTER INTRODUCTION

SPECIFIC EXPECTATIONS

- The specific expectations listed on the first page of each chapter describe the concepts and skills that you are expected to develop and demonstrate. The table includes references to the sections in which the specific expectations for the Functions and the Functions and Relations courses are covered.

MODELLING MATH

- Each chapter opens with a real-life problem that can be solved using a mathematical model. Examples of mathematical models include graphs, diagrams, formulas, equations, tables of values, and computer models.

- The lessons in the chapter prepare you to solve the problem posed on the opening pages, and related problems found throughout the sections. The related problems are identified with the Modelling Math logo.

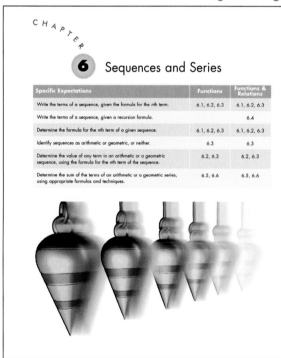

CHAPTER

6 Sequences and Series

Specific Expectations	Functions	Functions & Relations
Write the terms of a sequence, given the formula for the *n*th term.	6.1, 6.2, 6.3	6.1, 6.2, 6.3
Write the terms of a sequence, given a recursion formula.		6.4
Determine the formula for the *n*th term of a given sequence.	6.1, 6.2, 6.3	6.1, 6.2, 6.3
Identify sequences as arithmetic or geometric, or neither.	6.3	6.3
Determine the value of any term in an arithmetic or a geometric sequence, using the formula for the *n*th term of the sequence.	6.2, 6.3	6.2, 6.3
Determine the sum of the terms of an arithmetic or a geometric series, using appropriate formulas and techniques.	6.5, 6.6	6.5, 6.6

The Motion of a Pendulum

Some scientists, including Aristarchus of Samos in the 3rd century B.C. and Copernicus in the 16th century A.D., believed that the Earth rotates. However, no one had been able to demonstrate this rotation scientifically. In 1851, the French astronomer Jean Bernard Léon Foucault (1819-1868) constructed a 67-m long pendulum by suspending a 28-kg iron ball from the dome of the Panthéon in Paris. He used the pendulum to show that the Earth rotates about its axis.

In the Modelling Math questions on pages 445, 455, and 478, you will solve the following problem and other problems that involve the motion of a pendulum.

The period of a pendulum is the time it takes to complete one back-and-forth swing. On the Earth, the period, T seconds, is approximately given by the formula $T = 2\sqrt{l}$, where l metres is the length of the pendulum. If a 1-m pendulum completes its first period at a time of 10:15:30, or 15 min 30 s after 10:00,

a) at what time would it complete 100 periods? 151 periods?
b) how many periods would it have completed by 10:30:00?

Use your research skills to answer the following questions now.

1. Describe how Foucault demonstrated that the Earth rotates about its axis.

2. Describe one of the Foucault pendulums in Ontario. Examples include those at the University of Guelph and at Queen's University.

3. The angle through which the floor under a Foucault pendulum rotates each day depends on the latitude. Describe the relationship between the angle and the latitude.

Web Connection
www.school.mcgrawhill.ca/resources/
To use the Internet for your research on Foucault pendulums, visit the above web site. Go to **Math Resources**, then to *MATHEMATICS 11*, to find out where to go next.

REVIEWING PREREQUISITE SKILLS

GETTING STARTED

- Before the first numbered section in each chapter, a two-page Getting Started section reviews the mathematical skills you will need.

- The first page reviews skills in an interesting context.

- The second page is Review of Prerequisite Skills. Each skill area named in purple on a Review of Prerequisite Skills page is referenced to the alphabetical list of skills in **Appendix A: Review of Prerequisite Skills**, located at the back of the text.

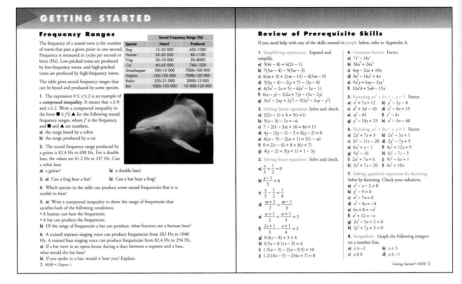

APPENDIX A

- If you need help with any of the skills named in purple on the Review of Prerequisite Skills page at the beginning of each chapter, refer to this alphabetical list.

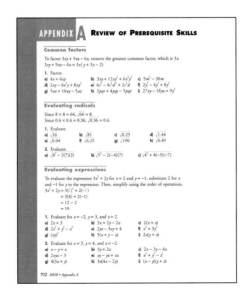

CONCEPT DEVELOPMENT

INVESTIGATE AND INQUIRE

- Most core sections introduce topics in an interesting context. You are then guided through an investigation.

- The investigation uses the process of inquiry, which allows you to discover the new concepts for yourself. This process is an important part of learning mathematics.

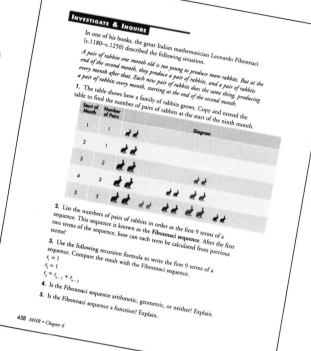

EXAMPLES

- The worked examples demonstrate how to use, and extend, what you have learned. They also provide model solutions to problems.

- The worked examples include graphing calculator solutions where appropriate.

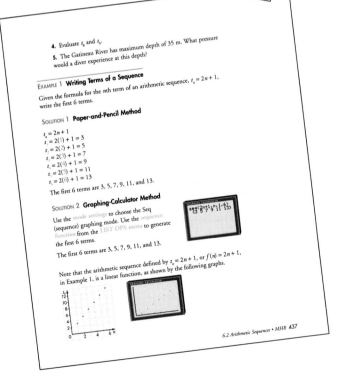

KEY CONCEPTS

- Following the worked examples, the concepts in the section are summarized.

- You can use this summary when you are doing homework or studying.

PRACTISE

- Completing these questions allows you to master essential mathematical skills by practising what you have learned.

APPLY, SOLVE, COMMUNICATE

- Mathematics is powerful when it is applied. The questions in this section allow you to use what you have learned to solve problems, and to apply and extend what you have learned.

- Selected questions are labelled with an Achievement Chart descriptor in red. These questions identify an opportunity for you to improve that particular Achievement Chart skill.

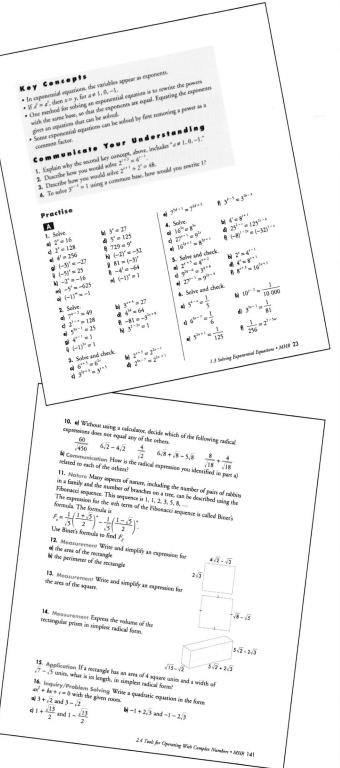

TECHNOLOGY

- Graphing calculators, geometry software, graphing software, and spreadsheets are technology tools that will make you a more powerful learner, by allowing you to explore mathematical concepts more easily and quickly.

- The use of these tools is integrated within the Investigate & Inquire, the worked examples, the Practise questions, and the Apply, Solve, Communicate questions.

- Two appendixes at the back of the text provide specific instructions in the operation of the various technology tools.

- Web Connections appear throughout the textbook providing links to interesting information.

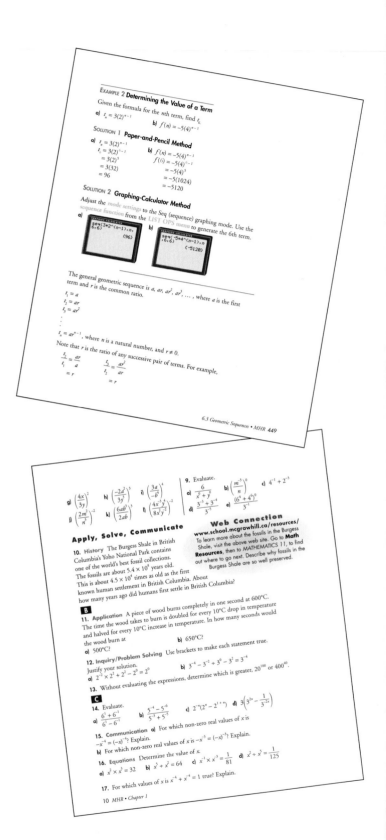

EXAMPLE 2 Determining the Value of a Term

Given the formula for the nth term, find t_6.

a) $t_n = 3(2)^{n-1}$

b) $f(n) = -5(4)^{n-1}$

SOLUTION 1 Paper-and-Pencil Method

a) $t_n = 3(2)^{n-1}$
$t_6 = 3(2)^{6-1}$
$= 3(2)^5$
$= 3(32)$
$= 96$

b) $f(n) = -5(4)^{n-1}$
$f(6) = -5(4)^{6-1}$
$= -5(4)^5$
$= -5(1024)$
$= -5120$

SOLUTION 2 Graphing-Calculator Method

Adjust the mode settings to the Seq (sequence) graphing mode. Use the sequence function from the LIST OPS menu to generate the 6th term.

a) seq(3*2^(n-1),n,6,6)
(96)

b) seq(-5*4^(n-1),n,6,6)
(-5120)

The general geometric sequence is $a, ar, ar^2, ar^3, \ldots$, where a is the first term and r is the common ratio.

$t_1 = a$
$t_2 = ar$
$t_3 = ar^2$
$\vdots$
$t_n = ar^{n-1}$, where n is a natural number, and $r \neq 0$.

Note that r is the ratio of any successive pair of terms. For example,

$\dfrac{t_2}{t_1} = \dfrac{ar}{a}$ $\dfrac{t_3}{t_2} = \dfrac{ar^2}{ar}$
$= r$ $= r$

g) $\left(\dfrac{4x}{3y}\right)^2$ h) $\left(\dfrac{-2d^2}{3y}\right)^3$ i) $\left(\dfrac{3a}{-b^2}\right)^4$

j) $\left(\dfrac{2m^2}{n^3}\right)^{-2}$ k) $\left(\dfrac{6ab^3}{2ab}\right)^3$ l) $\left(\dfrac{4x^{-3}y^4}{8x^3y^{-1}}\right)^{-2}$

9. Evaluate.

a) $\dfrac{6}{x^2+y}$ b) $\left(\dfrac{m^{-3}}{n}\right)^0$ c) $4^{-1}+2^{-3}$

d) $\dfrac{3^{-3}+3^{-4}}{3^{-5}}$ e) $\dfrac{(6^4+4^6)^0}{3^{-1}}$

Apply, Solve, Communicate

10. History The Burgess Shale in British Columbia's Yoho National Park contains one of the world's best fossil collections. The fossils are about 5.4×10^8 years old. This is about 4.5×10^4 times as old as the first known human settlement in British Columbia. About how many years ago did humans first settle in British Columbia?

B

11. Application A piece of wood burns completely in one second at 600°C. The time the wood takes to burn is doubled for every 10°C drop in temperature and halved for every 10°C increase in temperature. In how many seconds would the wood burn at

a) 500°C?

b) 650°C?

12. Inquiry/Problem Solving Use brackets to make each statement true. Justify your solution.

a) $2^{-2} \times 2^2 + 2^1 - 2^0 = 2^0$

b) $3^{-4} - 3^{-2} + 3^0 - 3^2 = 3^{-4}$

13. Without evaluating the expressions, determine which is greater, 20^{100} or 400^{40}.

C

14. Evaluate.

a) $\dfrac{6^1+6^{-1}}{6^1-6^{-1}}$ b) $\dfrac{5^{-4}-5^{-6}}{5^{-5}+5^{-7}}$ c) $2^{-n}(2^n-2^{1+n})$ d) $3\left(3^{2x}-\dfrac{1}{3^{-2x}}\right)$

15. Communication a) For which non-zero real values of x is $-x^{-4} = (-x)^{-4}$? Explain.

b) For which non-zero real values of x is $-x^{-3} = (-x)^{-3}$? Explain.

16. Equations Determine the value of x.

a) $x^2 \times x^3 = 32$ b) $x^2 + x^2 = 64$ c) $x^{-1} \times x^{-3} = \dfrac{1}{81}$ d) $x^2 + x^5 = \dfrac{1}{125}$

17. For which values of x is $x^{-4} + x^{-4} = 1$ true? Explain.

Web Connection
www.school.mcgrawhill.ca/resources/
To learn more about the fossils in the Burgess Shale, visit the above web site. Go to **Math Resources**, then to MATHEMATICS 11, to find out where to go next. Describe why fossils in the Burgess Shale are so well preserved.

Appendix B: Graphing Calculator
Keystrokes includes detailed instructions in an alphabetical listing of functions used in *McGraw-Hill Ryerson MATHEMATICS 11*.

Appendix C: Software Appendix
reviews the essential skills needed to use Microsoft® Excel, Corel® Quattro® Pro, Zap-a-Graph and *The Geometer's Sketchpad*®.

• Technology Extension features provide additional instruction and opportunities to use technology tools in applications related to the chapter.

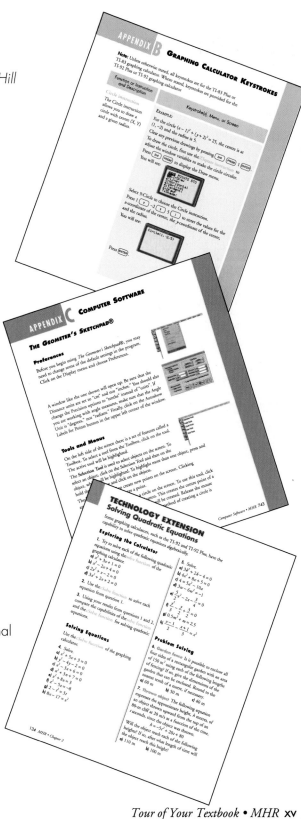

INVESTIGATE & APPLY

- One investigation that uses the process of inquiry is included at the end of each chapter before the Review of Key Concepts.

- These problems explore and extend concepts in a rich problem solving situation.

- Rubrics for these problems are available as blackline masters in the Teacher's Resource.

CAREER CONNECTIONS

- These activities give you the opportunity to investigate mathematics-related careers that make use of the chapter content.

REVIEW AND ASSESSMENT

Several features within the text provide opportunities for review and assessment:

COMMUNICATE YOUR UNDERSTANDING

- Following each summary of Key Concepts are questions designed to help you communicate your understanding of what you have learned.

ACHIEVEMENT CHECK

- This feature provides questions designed to assess your knowledge and understanding, your problem solving skills, your communication skills, and your ability to apply what you have learned.

- Achievement Checks appear throughout the chapter, with one in the chapter test.

- Rubrics for these problems are available as blackline masters in the Teacher's Resource.

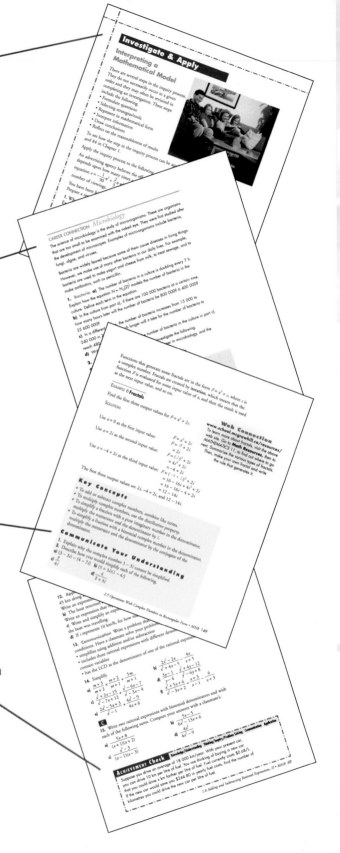

REVIEW OF KEY CONCEPTS

- Near the end of each chapter are questions designed to review the concepts learned in the chapter.

- The review is organized section by section, and refers to each summary of Key Concepts in the chapter.

CHAPTER TEST

- Each chapter includes a test to assess the skills addressed in the chapter.

- Each test includes an Achievement Check that provides an open-ended question designed to assess your knowledge and understanding, your problem solving skills, your communication skills, and your ability to apply what you have learned.

CHALLENGE PROBLEMS

The page of Challenge Problems found near the end of each chapter encourages you to apply the steps of an inquiry/problem solving process to extend your understanding

CUMULATIVE REVIEW

- The cumulative reviews, found at the end of Chapters 2, 4, 6, and 8, review concepts from the two preceding chapters.

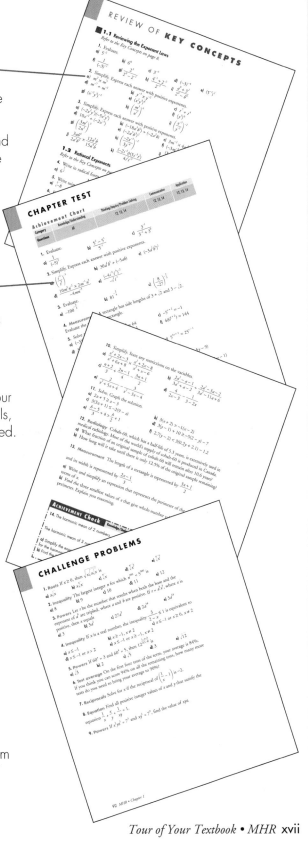

PROBLEM SOLVING

- In the Apply, Solve, Communicate sections, numerous problems allow you to apply your problem solving skills.

- Near the end of each chapter is a Problem Solving page that teaches a specific strategy and provides questions for practice.

- Each chapter ends with a Problem Solving: Using the Strategies page. These pages include a variety of problems that can be solved using different strategies.

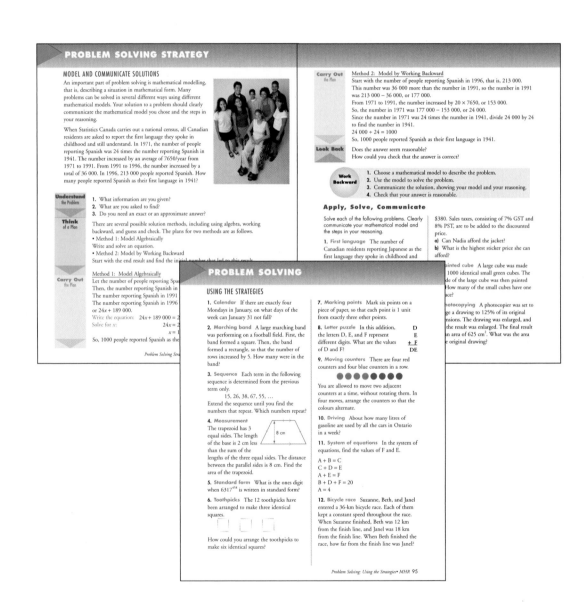

END-OF-TEXT FEATURES

APPENDIX A: A REVIEW OF PREREQUISITE SKILLS

- This review is found on pages 702–716.

APPENDIX B: GRAPHING CALCULATOR KEYSTROKES

- This review is found on pages 717–742.

APPENDIX C: COMPUTER SOFTWARE

- Includes Microsoft® Excel, Corel® Quattro® Pro, Zap-a-Graph and *The Geometer's Sketchpad*®, and is found on pages 743–773.

ANSWERS

- Answers are found on pages 774–811.

GLOSSARY

- Mathematical terms used in the text are listed and defined on pages 812–821.

INDEXES

- The book includes a technology index and a general index on pages 822–826.

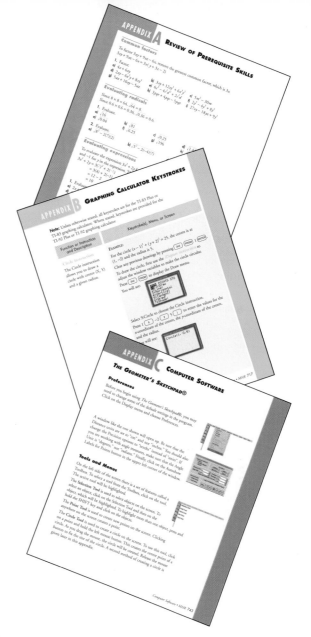

INTERACTIVE E-BOOK

An exciting and interactive e-book in the back of the text will enrich your opportunities for understanding the mathematics you are studying. It contains

- the entire student text in PDFs, including all technical art and most photographs

- a glossary search capability to help you find and learn key terms and definitions

- an index search to link topics and expectations to content

- the ability to highlight key sections to copy to your own study notes

- all answers, including the graphical answers, easily linked from each section

1 Algebraic Tools for Operating With Functions

Specific Expectations	Functions	Functions & Relations
Simplify and evaluate expressions containing integer and rational exponents, using the laws of exponents.	1.1, 1.2	1.1, 1.2
Solve exponential equations.	1.3	1.3
Add, subtract, and multiply polynomials.	1.4	1.4
Add, subtract, multiply, and divide rational expressions, and state the restrictions on the variable values.	1.5, 1.6, 1.7, 1.8	1.5, 1.6, 1.7, 1.8
Solve first-degree inequalities and represent the solutions on number lines.	1.9	1.9
Explain mathematical processes, methods of solution, and concepts clearly to others.	throughout the book	throughout the book
Present problems and solutions to a group and answer questions about the problems and the solutions.	throughout the book	throughout the book
Communicate solutions to problems and to findings of investigations clearly and concisely, orally and in writing, using an effective integration of essay and mathematical forms.	throughout the book	throughout the book
Demonstrate correct use of mathematical language, symbols, visuals, and conventions.	throughout the book	throughout the book
Use graphing technology effectively.	throughout the book	throughout the book

Modelling Problems Algebraically

Medications are not used or eliminated immediately by the body. The biological half-life of a medication is the time taken for the quantity of the medication in a patient to decrease to half the original quantity as a result of biological processes. Different medications have different biological half-lives.

The thyroid gland, located in the front of the neck, produces hormones that we need to live. The two most important thyroid hormones have the abbreviated names T4 and T3. Thyroid gland problems are very common, affecting about 5% of the Canadian population. One type of problem is called hypothyroidism, in which the thyroid gland produces too little hormone. The usual treatment is to take medications that contain synthetic T4 hormone. The body is able to convert some of the T4 to T3.

In the Modelling Math questions on pages 25, 60, and 80, you will solve the following problem and other problems that can be modelled algebraically.

The biological half-life of thyroid hormone T4 is about 6.5 days. If a dose of T4 was not followed by repeat doses,
a) what fraction of the original dose would remain in the body after 19.5 days?
b) how long would it take until only 6.25% of the original dose would remain in the body?

Use your research skills to answer the following questions now.

1. Find the biological half-life of another medication. Compare your findings with your classmates'.

2. Explain how and why the biological half-life of a medication is allowed for when repeated doses of the medication are prescribed.

Frequency Ranges

The frequency of a sound wave is the number of waves that pass a given point in one second. Frequency is measured in cycles per second or hertz (Hz). Low-pitched tones are produced by low-frequency waves, and high-pitched tones are produced by high-frequency waves.

The table gives sound frequency ranges that can be heard and produced by some species.

Species	Sound Frequency Range (Hz)	
	Heard	Produced
Dog	15–50 000	452–1080
Human	20–20 000	80–1100
Frog	50–10 000	50–8000
Cat	60–65 000	760–1520
Grasshopper	100–15 000	7000–100 000
Dolphin	150–150 000	7000–120 000
Robin	250–21 000	2000–13 000
Bat	1000–120 000	10 000–120 000

1. The expression $0 \leq x \leq 2$ is an example of a **compound inequality**. It means that $x \geq 0$ and $x \leq 2$. Write a compound inequality in the form $\blacksquare \leq f \leq \blacktriangle$ for the following sound frequency ranges, where f is the frequency, and $\blacksquare$ and $\blacktriangle$ are numbers.
a) the range heard by a robin
b) the range produced by a cat

2. The sound frequency range produced by a guitar is 82.4 Hz to 698 Hz. For a double bass, the values are 41.2 Hz to 247 Hz. Can a robin hear
a) a guitar?   **b)** a double bass?

3. a) Can a frog hear a bat?   **b)** Can a bat hear a frog?

4. Which species in the table can produce some sound frequencies that it is unable to hear?

5. a) Write a compound inequality to show the range of frequencies that satisfies both of the following conditions.
• A human can hear the frequencies.
• A bat can produce the frequencies.
b) Of the range of frequencies a bat can produce, what fraction can a human hear?

6. A trained soprano singing voice can produce frequencies from 262 Hz to 1046 Hz. A trained bass singing voice can produce frequencies from 82.4 Hz to 294 Hz.
a) If a bat were in an opera house during a duet between a soprano and a bass, what would the bat hear?
b) If you spoke to a bat, would it hear you? Explain.

Review of Prerequisite Skills

If you need help with any of the skills named in purple below, refer to Appendix A.

1. Simplifying expressions Expand and simplify.

a) $3(4t - 8) + 6(2t - 1)$
b) $7(3w - 4) - 5(5w - 3)$
c) $6(m + 3) + 2(m - 11) - 4(3m - 9)$
d) $5(3y - 4) - 2(y + 7) - (3y - 8)$
e) $4(3x^2 - 2x + 5) - 6(x^2 - 2x - 1)$
f) $6(x - y) - 2(2x + 7y) - (3x - 2y)$
g) $3(x^2 - 2xy + 2y^2) - 5(2x^2 - 2xy - y^2)$

2. Solving linear equations Solve and check.

a) $2(2r - 1) + 4 = 5(r + 1)$
b) $5(x - 3) - 2x = -6$
c) $7 - 2(1 - 3x) + 16 = 8x + 11$
d) $4y - (3y - 1) - 3 + 6(y - 2) = 0$
e) $4(w - 5) - 2(w + 1) = 3(1 - w)$
f) $0 = 2(t - 6) + 8 + 4(t + 7)$
g) $4(y - 2) = 3(y + 1) + 1 - 3y$

3. Solving linear equations Solve and check.

a) $\dfrac{x}{3} + \dfrac{1}{2} = 0$

b) $\dfrac{y - 1}{3} = 6$

c) $\dfrac{x}{3} - \dfrac{1}{2} = \dfrac{1}{4}$

d) $\dfrac{m + 2}{2} = \dfrac{m - 1}{3}$

e) $\dfrac{w + 1}{2} + \dfrac{w + 1}{3} = 5$

f) $\dfrac{2x + 1}{3} - \dfrac{x + 1}{4} = 3$

g) $0.4(c - 8) + 3 = 4$
h) $0.5x - 0.1(x - 3) = 4$
i) $1.5(a - 3) - 2(a - 0.5) = 10$
j) $1.2(10x - 5) - 2(4x + 7) = 8$

4. Common factors Factor.

a) $7t^2 - 14t^3$
b) $36x^7 + 24x^5$
c) $4xy - 2xz + 10x$
d) $8x^3 - 16x^2 + 4x$
e) $9x^2y + 6xy - 3xy^2$
f) $10a^2b + 5ab - 15a$

5. Factoring $ax^2 + bx + c$, $a = 1$ Factor.

a) $x^2 + 7x + 12$ b) $y^2 - 2y - 8$
c) $d^2 + 3d - 10$ d) $x^2 - 8x + 15$
e) $w^2 - 81$ f) $t^2 - 4t$
g) $y^2 - 10y + 25$ h) $x^2 - 3x - 40$

6. Factoring $ax^2 + bx + c$, $a \ne 1$ Factor.

a) $2x^2 + 7x + 3$ b) $2x^2 - 3x + 1$
c) $3t^2 - 11t - 20$ d) $2y^2 - 7y + 5$
e) $6x^2 + x - 1$ f) $4x^2 + 12x + 9$
g) $9a^2 - 16$ h) $6s^2 - 7s - 3$
i) $2u^2 + 7u + 6$ j) $9x^2 - 6x + 1$
k) $3x^2 + 7x - 20$ l) $4v^2 + 10v$

7. Solving quadratic equations by factoring Solve by factoring. Check your solutions.

a) $x^2 - x - 2 = 0$
b) $y^2 - 9 = 0$
c) $n^2 - 7n = 0$
d) $x^2 - 4x = -4$
e) $6x + 8 = -x^2$
f) $z^2 + 12 = -z$
g) $2x^2 - 5x + 2 = 0$
h) $2y^2 + 7y + 3 = 0$

8. Inequalities Graph the following integers on a number line.

a) $x > -2$ b) $x < 3$
c) $x \ge 0$ d) $x \le -1$

1.1 Reviewing the Exponent Laws

An order of magnitude is an approximate size of a quantity, expressed as a power of 10.

The table shows some speeds in metres per second, expressed to the nearest order of magnitude.

Entity	Speed (m/s)
Light (in space)	10^8
Sound (in air)	10^2
Horse (galloping)	10^1
Human (walking)	10^0
Garden snail	10^{-3}

1. Express 10^0 metres per second in standard form.

2. Use division to determine, to the nearest order of magnitude, how many times as fast
a) light is as sound   **b)** a horse is as a snail

3. Write the rule you used to divide two powers of 10.

4. To the nearest order of magnitude, the moon orbits the Earth 10^6 times as fast as a snail can travel. Use multiplication to express the speed of the moon in metres per second, to the nearest order of magnitude.

5. Write the rule you used to multiply two powers of 10.

The following summary shows the exponent laws for integral exponents.

Exponent Law for Multiplication

$$3^2 \times 3^4 = (3 \times 3)(3 \times 3 \times 3 \times 3)$$
$$= 3 \times 3 \times 3 \times 3 \times 3 \times 3$$
$$= 3^6$$

$$a^m \times a^n = \underbrace{(a \times a \times \ldots \times a)}_{m \text{ factors}}\underbrace{(a \times a \times \ldots \times a)}_{n \text{ factors}}$$

$$= \underbrace{a \times a \times a \times \ldots \times a}_{m + n \text{ factors}}$$

$$= a^{m+n}$$

Exponent Law for Division

$$\frac{6^5}{6^2} = \frac{6 \times 6 \times 6 \times 6 \times 6}{6 \times 6}$$

$$= 6 \times 6 \times 6$$

$$= 6^3$$

m factors

$$\frac{a^m}{a^n} = \frac{\overbrace{a \times a \times a \times \ldots \times a}}{\underbrace{a \times a \times \ldots \times a}}, \; a \neq 0$$

n factors

$$= \underbrace{a \times a \times a \times \ldots \times a}$$

$m - n$ factors

$$= a^{m-n}$$

Power Law

$$(5^2)^3 = (5 \times 5)^3$$

$$= (5 \times 5)(5 \times 5)(5 \times 5)$$

$$= 5 \times 5 \times 5 \times 5 \times 5 \times 5$$

$$= 5^6$$

$$(a^m)^n = (\underbrace{a \times a \times \ldots \times a})^n$$

m factors

$$= \underbrace{(\underbrace{a \times a \times \ldots \times a})}_{m \text{ factors}} \times \underbrace{(\underbrace{a \times a \times \ldots \times a})}_{m \text{ factors}} \times \ldots \times \underbrace{(\underbrace{a \times a \times \ldots \times a})}_{m \text{ factors}}$$

n times

$$= \underbrace{a \times a \times a \times \ldots \times a}$$

mn factors

$$= a^{mn}$$

Power of a Product

$$(5 \times 2)^3 = (5 \times 2) \times (5 \times 2) \times (5 \times 2)$$

$$= 5 \times 5 \times 5 \times 2 \times 2 \times 2$$

$$= 5^3 \times 2^3$$

$$(ab)^m = \underbrace{(ab) \times (ab) \times \ldots \times (ab)}$$

m factors

$$= \underbrace{(a \times a \times \ldots \times a)}_{m \text{ factors}} \times \underbrace{(b \times b \times \ldots \times b)}_{m \text{ factors}}$$

$$= a^m b^m$$

Power of a Quotient

$$\left(\frac{2}{5}\right)^3 = \left(\frac{2}{5}\right) \times \left(\frac{2}{5}\right) \times \left(\frac{2}{5}\right)$$

$$= \frac{2 \times 2 \times 2}{5 \times 5 \times 5}$$

$$= \frac{2^3}{5^3}$$

$$\left(\frac{a}{b}\right)^m = \underbrace{\left(\frac{a}{b}\right) \times \left(\frac{a}{b}\right) \times \ldots \times \left(\frac{a}{b}\right)}$$

m factors

m factors

$$= \frac{\overbrace{a \times a \times \ldots \times a}}{\underbrace{b \times b \times \ldots \times b}}$$

m factors

$$= \frac{a^m}{b^m}, \; b \neq 0$$

A **power** is an expression in the form a^m. The exponent laws can be used to simplify expressions with powers.

EXAMPLE 1 Simplifying Expressions With Powers

Simplify.

a) $(3a^2b)(-2a^3b^2)$ b) $(m^3)^4$ c) $(-4p^3q^2)^3$

SOLUTION

a) $(3a^2b)(-2a^3b^2) = 3 \times (-2) \times a^2 \times a^3 \times b \times b^2$
$$= -6a^5b^3$$

b) $(m^3)^4 = m^{3 \times 4}$
$$= m^{12}$$

c) $(-4p^3q^2)^3 = (-4)^3 \times (p^3)^3 \times (q^2)^3$
$$= -64p^9q^6$$

EXAMPLE 2 Simplifying a Power of a Quotient

Simplify $\left(\dfrac{6x^5y^3}{8y^4}\right)^2$.

SOLUTION 1

Use the power of a quotient law first.
$$\left(\frac{6x^5y^3}{8y^4}\right)^2 = \frac{(6)^2(x^5)^2(y^3)^2}{(8)^2(y^4)^2}$$
$$= \frac{36x^{10}y^6}{64y^8}$$
$$= \frac{9x^{10}}{16y^2}$$

SOLUTION 2

Simplify the quotient first.
$$\left(\frac{6x^5y^3}{8y^4}\right)^2 = \left(\frac{3x^5}{4y}\right)^2$$
$$= \frac{(3)^2(x^5)^2}{(4)^2(y)^2}$$
$$= \frac{9x^{10}}{16y^2}$$

The following summarizes the rules for zero and negative exponents.

Zero Exponent

$$\frac{2^3}{2^3} = 2^{3-3}$$

$$= 2^0$$

but $\dfrac{2^3}{2^3} = 1$

so $2^0 = 1$

$$\frac{a^m}{a^m} = a^{m-m}$$

$$= a^0$$

but $\dfrac{a^m}{a^m} = 1$

so, if $a \neq 0$, $a^0 = 1$ **Note that 0^0 is not defined.**

Negative Exponents

$$2^3 \times 2^{-3} = 2^{3 + (-3)}$$

$$= 2^0$$

so $2^3 \times 2^{-3} = 1$

$$\frac{2^3 \times 2^{-3}}{2^3} = \frac{1}{2^3}$$ Divide both sides by 2^3.

$$2^{-3} = \frac{1}{2^3}$$

$$a^m \times a^{-m} = a^{m + (-m)}$$

$$= a^0$$

so $a^m \times a^{-m} = 1$

$$\frac{a^m \times a^{-m}}{a^m} = \frac{1}{a^m}$$ Divide both sides by a^m.

so, if $a \neq 0$, $a^{-m} = \dfrac{1}{a^m}$

Similarly, if $a \neq 0$, $\dfrac{1}{a^{-m}} = a^m$

EXAMPLE 3 Simplifying Expressions With Negative Exponents

Simplify $\dfrac{(-6x^{-2}y)(-9x^{-5}y^{-2})}{3x^2y^{-4}}$. Express the answer with positive exponents.

SOLUTION

$$\frac{(-6x^{-2}y)(-9x^{-5}y^{-2})}{3x^2y^{-4}}$$

Multiply: $= \dfrac{54x^{-7}y^{-1}}{3x^2y^{-4}}$

Divide: $= 18x^{-9}y^3$

Rewrite: $= \dfrac{18y^3}{x^9}$

EXAMPLE 4 Evaluating Expressions With Zero and Negative Exponents

Evaluate.

a) $\left(\dfrac{3}{4}\right)^{-2}$

b) $\dfrac{(-6)^0}{2^{-3}}$

c) $\dfrac{2^{-4} + 2^{-6}}{2^{-3}}$

Solution 1 Paper-and-Pencil Method

a) $\left(\dfrac{3}{4}\right)^{-2} = \dfrac{1}{\left(\dfrac{3}{4}\right)^2}$

$= \dfrac{1}{\dfrac{9}{16}}$

$= \dfrac{16}{9}$

b) $\dfrac{(-6)^0}{2^{-3}} = \dfrac{1}{2^{-3}}$

$= \dfrac{1}{\dfrac{1}{2^3}}$

$= \dfrac{1}{\dfrac{1}{8}}$

$= 8$

c) $\dfrac{2^{-4} + 2^{-6}}{2^{-3}} = \dfrac{\dfrac{1}{2^4} + \dfrac{1}{2^6}}{\dfrac{1}{2^3}}$

$= \dfrac{\dfrac{2^2 + 1}{2^6}}{\dfrac{1}{2^3}}$

$= \dfrac{2^2 + 1}{2^6} \times \dfrac{2^3}{1}$

$= \dfrac{2^2 + 1}{2^3}$

$= \dfrac{5}{8}$

Solution 2 Graphing-Calculator Method

The first answer given by a graphing calculator may be a decimal. If necessary, convert the decimal to a fraction using the ▶Frac function.

a)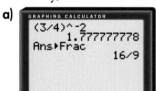

b)

c)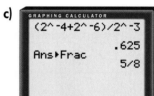

Key Concepts

- Exponent law for multiplication: $a^m \times a^n = a^{m+n}$
- Exponent law for division: $a^m \div a^n = a^{m-n}$
- Power law: $(a^m)^n = a^{mn}$
- Power of a product law: $(ab)^m = a^m b^m$
- Power of a quotient law: $\left(\dfrac{a}{b}\right)^m = \dfrac{a^m}{b^m}$
- Zero exponent property: if $a \neq 0$, $a^0 = 1$
- Negative exponent property: if $a \neq 0$, $a^{-m} = \dfrac{1}{a^m}$ and $\dfrac{1}{a^{-m}} = a^m$

Communicate Your Understanding

1. Describe how you would simplify $(-4x^2y^3)(3xy^4)$.

2. Describe how you would evaluate $\dfrac{3^{-2}}{3^{-1}+3^{-2}}$ using paper and pencil.

3. What is the value of 0^4? Explain.
4. Explain why $a \neq 0$ for the negative exponent property.

Practise

A

1. Express as a power of 2.
   a) $2^4 \times 2^3$ b) $2^6 \div 2^2$ c) $(2^4)^3$
   d) 2×2^7 e) $2^3 \times 2^m$ f) $2^7 \div 2^y$
   g) $2^x \div 2^4$ h) $(2^x)^y$ i) $2^{-3} \times 2^4$
   j) $2^{-2} \div 2^{-5}$ k) $(2^3)^{-1}$ l) $2^{-4} \times 2^0$

2. Evaluate.
   a) 3^{-2} b) 5^0 c) $(-2)^{-4}$
   d) $(2^{-1})^2$ e) $-(-3)^0$ f) $\dfrac{1}{5^{-2}}$
   g) $\dfrac{1}{(-4)^{-1}}$ h) $-(2^3)^{-2}$

3. Simplify. Express each answer with positive exponents.
   a) $a^4 \times a^3$ b) $(m^6)(m^2)$
   c) $b^5 \times b^6 \times b$ d) $a \times b^2 \times a^4$
   e) $(x^3)(y)(y^4)(x^5)$ f) $(x^3)(x^{-5})$
   g) $m^{-4} \times m^{-5}$ h) $y^{-1} \times y^{-3} \times y^2$
   i) $a^5 \times a^0$ j) $(a^{-3})(b^{-2})(a^2)$

4. Simplify. Express each answer with positive exponents.
   a) $x^6 \div x^3$ b) $m^7 \div m$ c) $t^4 \div t^{-2}$
   d) $y^{-5} \div y^{-3}$ e) $m^4 \div m^0$ f) $t^0 \div t^{-5}$

5. Simplify. Express each answer with positive exponents.
   a) $(x^3)^2$ b) $(a^2b^3)^4$ c) $(x^2)^{-1}$
   d) $(t^4)^0$ e) $(a^{-1}b^2)^{-2}$ f) $(x^2y^3)^{-3}$

6. Simplify. Express each answer with positive exponents.
   a) $\left(\dfrac{x}{2}\right)^3$ b) $\left(\dfrac{a}{b}\right)^4$ c) $\left(\dfrac{x^2}{y^3}\right)^5$
   d) $\left(\dfrac{x}{3}\right)^{-1}$ e) $\left(\dfrac{a^{-2}}{b^{-3}}\right)^{-2}$

7. Simplify. Express each answer with positive exponents.
   a) $5m^4 \times 3m^2$ b) $(4ab^4)(-5a^3b^2)$
   c) $5a(-2ab^2)(-3b^3)$ d) $(-6m^3n^2)(-4mn^5)$
   e) $(7x^2)(6x^{-2})$ f) $(3x^{-2}y^2)(-2x^2y^{-3})$
   g) $(-6a^{-1}b^2)(-a^{-3}b^{-4})$ h) $(-10x^4) \div (-2x)$
   i) $\dfrac{45a^2b^4}{9ab^2}$ j) $\dfrac{(4m^2n^4)(7m^3n)}{14mn^5}$
   k) $\dfrac{3ab^3 \times 10a^4b^2}{15a^2b^6}$ l) $\dfrac{4a^4b^3}{a^5b^6} \times \dfrac{-a^3}{-(b^2)}$
   m) $(35x^5) \div (5x^{-3})$ n) $\dfrac{-54a^5b^{-7}}{-6a^{-2}b^{-3}}$
   o) $(-6m^{-4}n^2) \div (2m^{-1}n^{-6})$
   p) $\dfrac{(-2x^{-3}y)(-12x^{-4}y^{-2})}{6xy^{-3}}$

8. Simplify. Express each answer with positive exponents.
   a) $(2m^3)^2$ b) $(-4x^2)^3$
   c) $(-3m^3n^2)^2$ d) $(5c^{-3}d^3)^{-2}$
   e) $(2a^{-3}b^{-2})^{-3}$ f) $(-3x^3y^{-2})^{-4}$

g) $\left(\dfrac{4x}{3y}\right)^2$ h) $\left(\dfrac{-2a^2}{3y^3}\right)^3$ i) $\left(\dfrac{3a}{-b^4}\right)^4$

j) $\left(\dfrac{2m^2}{n^3}\right)^{-2}$ k) $\left(\dfrac{6ab^3}{2ab}\right)^3$ l) $\left(\dfrac{4x^{-3}y^4}{8x^2y^{-2}}\right)^{-2}$

9. Evaluate.

a) $\dfrac{6}{x^0 + y^0}$ b) $\left(\dfrac{m^{-3}}{n}\right)^0$ c) $4^{-1} + 2^{-3}$

d) $\dfrac{3^{-3} + 3^{-4}}{3^{-5}}$ e) $\dfrac{(6^4 + 4^6)^0}{3^{-1}}$

Apply, Solve, Communicate

10. History The Burgess Shale in British Columbia's Yoho National Park contains one of the world's best fossil collections. The fossils are about 5.4×10^8 years old. This is about 4.5×10^4 times as old as the first known human settlement in British Columbia. About how many years ago did humans first settle in British Columbia?

Web Connection
www.school.mcgrawhill.ca/resources/
To learn more about the fossils in the Burgess Shale, visit the above web site. Go to **Math Resources**, then to *MATHEMATICS 11*, to find out where to go next. Describe why fossils in the Burgess Shale are so well preserved.

B

11. Application A piece of wood burns completely in one second at 600°C. The time the wood takes to burn is doubled for every 10°C drop in temperature and halved for every 10°C increase in temperature. In how many seconds would the wood burn at

a) 500°C? b) 650°C?

12. Inquiry/Problem Solving Use brackets to make each statement true. Justify your solution.

a) $2^{-2} \times 2^2 + 2^2 - 2^0 = 2^0$ b) $3^{-4} - 3^{-2} \div 3^0 - 3^2 = 3^{-4}$

13. Without evaluating the expressions, determine which is greater, 20^{100} or 400^{40}.

C

14. Evaluate.

a) $\dfrac{6^1 + 6^{-1}}{6^1 - 6^{-1}}$ b) $\dfrac{5^{-4} - 5^{-6}}{5^{-3} + 5^{-5}}$ c) $2^{-n}(2^n - 2^{1+n})$ d) $3\left(3^{2x} - \dfrac{1}{3^{-2x}}\right)$

15. Communication a) For which non-zero real values of x is $-x^{-4} = (-x)^{-4}$? Explain.
b) For which non-zero real values of x is $-x^{-3} = (-x)^{-3}$? Explain.

16. Equations Determine the value of x.

a) $x^2 \times x^3 = 32$ b) $x^5 \div x^2 = 64$ c) $x^{-1} \times x^{-3} = \dfrac{1}{81}$ d) $x^2 \div x^5 = \dfrac{1}{125}$

17. For which values of x is $x^{-4} \div x^{-4} = 1$ true? Explain.

1.2 Rational Exponents

Most of the power used to move a ship is needed to push along the bow wave that builds up in front of the ship. Ships are designed to use as little power as possible.

To ensure that the design of a ship is energy-efficient, designers test models before the real ship is built. To calculate the speed to use when testing a model the following formula is used.

$$S_m = \frac{S_r \times L_m^{\frac{1}{2}}}{L_r^{\frac{1}{2}}}$$

where S_m is the speed of the model in metres per second, S_r is the speed of the real ship in metres per second, L_m is the length of the model in metres, and L_r is the length of the real ship in metres. This formula includes powers with fractional exponents.

INVESTIGATE & INQUIRE

Using the power law for exponents, 9 can be written as $\left(9^{\frac{1}{2}}\right)^2$, because

$$\left(9^{\frac{1}{2}}\right)^2 = 9^{\frac{1}{2} \times 2}$$
$$= 9^1 \text{ or } 9$$

1. Copy and complete the following statements by replacing each ■ with a natural number. The first statement has been partially completed.

a) $9 = \left(9^{\frac{1}{2}}\right)^2$

but $9 = (3)^2$

so $\left(9^{\frac{1}{2}}\right)^2 = 3^2$

and $9^{\frac{1}{2}} = $ ■

b) $25 = \left(25^{\frac{1}{2}}\right)^2$

but $25 = (■)^2$

so $\left(25^{\frac{1}{2}}\right)^2 = (■)^2$

and $25^{\frac{1}{2}} = $ ■

c) $8 = \left(8^{\frac{1}{3}}\right)^3$

but $8 = (■)^3$

so $\left(8^{\frac{1}{3}}\right)^3 = (■)^3$

and $8^{\frac{1}{3}} = $ ■

d) $16 = \left(16^{\frac{1}{4}}\right)^4$

but $16 = (■)^4$

so $\left(16^{\frac{1}{4}}\right)^4 = (■)^4$

and $16^{\frac{1}{4}} = $ ■

2. Evaluate.

a) $36^{\frac{1}{2}}$ **b)** $27^{\frac{1}{3}}$ **c)** $81^{\frac{1}{4}}$ **d)** $100^{\frac{1}{2}}$

3. A ship is to be built 100 m long and able to travel at 15 m/s. The model of the ship is 4 m long. At what speed should the model be tested?

In the power law for exponents, $(a^m)^n = a^{mn}$, substituting $m = \dfrac{1}{n}$ gives

$$\left(a^{\frac{1}{n}}\right)^n = a^{\frac{1}{n} \times n}$$
$$= a^1 \text{ or } a$$

If $a \geq 0$, we can take the nth root of both sides of the equation $\left(a^{\frac{1}{n}}\right)^n = a$, which gives $a^{\frac{1}{n}} = \sqrt[n]{a}$.

This result suggests the following definition.

$a^{\frac{1}{n}} = \sqrt[n]{a}$, where n is a natural number.

The symbol $\sqrt[n]{}$ indicates an nth root, and $\sqrt[n]{x}$ represents the principal nth root of x. For example, $64^{\frac{1}{3}} = \sqrt[3]{64}$. The expression $\sqrt[3]{64}$ is read as "the cube root of 64."

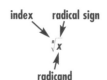

index   radical sign

$\sqrt[n]{x}$

radicand

Finding the cube root of a number is the inverse operation of cubing. To find the cube root of 64, find the number whose cube is 64.

Since $4^3 = 64$, $\sqrt[3]{64} = 4$.

- If n is an even number, then we must have $a \geq 0$ for the nth root to be real. Suppose that n is even and a is negative. For example, if $n = 2$ and $a = -4$, then $(-4)^{\frac{1}{2}}$ becomes $\sqrt{-4}$. There is no real square root of -4.
- If n is an odd number, then a can be any real number. For example, if $n = 3$ and $a = -8$, then $(-8)^{\frac{1}{3}}$ becomes $\sqrt[3]{-8}$, which is -2. In this case, the principal root is negative.

Note how brackets are used with fractional exponents. The expression $\sqrt{-4}$ has no meaning in the real number system, but $-\sqrt{4} = -2$. Similarly, $(-4)^{\frac{1}{2}}$ becomes $\sqrt{-4}$, which has no meaning in the real number system. But $-4^{\frac{1}{2}}$ becomes $-\left(4^{\frac{1}{2}}\right) = -\sqrt{4} = -2$.

EXAMPLE 1 Exponents in the Form $\frac{1}{n}$

Evaluate.

a) $49^{\frac{1}{2}}$

b) $(-27)^{\frac{1}{3}}$

c) $(-8)^{-\frac{1}{3}}$

SOLUTION

a) $49^{\frac{1}{2}} = \sqrt{49}$

$\quad = 7$

b) $(-27)^{\frac{1}{3}} = \sqrt[3]{-27}$

$\quad = -3$

c) $(-8)^{-\frac{1}{3}} = \dfrac{1}{(-8)^{\frac{1}{3}}}$

$\quad = \dfrac{1}{\sqrt[3]{-8}}$

$\quad = -\dfrac{1}{2}$

The following suggests how to evaluate an expression with a fractional exponent in which the numerator is not 1, such as $4^{\frac{3}{2}}$.

The power law $(a^m)^n = a^{mn}$ is used.

Method 1

$4^{\frac{3}{2}} = \left(4^{\frac{1}{2}}\right)^3$

$\quad = \left(\sqrt{4}\right)^3$

$\quad = 2^3$

$\quad = 8$

Method 2

$4^{\frac{3}{2}} = (4^3)^{\frac{1}{2}}$

$\quad = \sqrt{4^3}$

$\quad = \sqrt{64}$

$\quad = 8$

Notice that $\left(\sqrt{4}\right)^3$ and $\sqrt{4^3}$ have the same value.

This result suggests the following definition for rational exponents.

$a^{\frac{m}{n}} = \sqrt[n]{a^m} = \left(\sqrt[n]{a}\right)^m$, where m and n are natural numbers.

If n is an even number, then $a \geq 0$.

If n is an odd number, then a can be any real number.

To calculate $a^{\frac{m}{n}}$

- take the nth root of a, then raise the result to the mth power

$9^{\frac{3}{2}} = \left(\sqrt{9}\right)^3$

$\quad = 3^3$

$\quad = 27$

or

- raise a to the mth power, then take the nth root

$$9^{\frac{3}{2}} = \sqrt{9^3}$$
$$= \sqrt{729}$$
$$= 27$$

It is common practice to take the nth root first.

EXAMPLE 2 Exponents in the Form $\dfrac{m}{n}$

Evaluate.

a) $(-8)^{\frac{4}{3}}$ 　　　　 b) $9^{-2.5}$ 　　　　 c) $\left(\dfrac{25}{4}\right)^{-\frac{3}{2}}$

SOLUTION 1 Paper-and-Pencil Method

a)
$$(-8)^{\frac{4}{3}} = \left(\sqrt[3]{-8}\right)^4$$
$$= (-2)^4$$
$$= 16$$

b)
$$9^{-2.5} = 9^{-\frac{5}{2}}$$
$$= \frac{1}{9^{\frac{5}{2}}}$$
$$= \frac{1}{\left(\sqrt{9}\right)^5}$$
$$= \frac{1}{3^5}$$
$$= \frac{1}{243}$$

c)
$$\left(\frac{25}{4}\right)^{-\frac{3}{2}} = \frac{1}{\left(\dfrac{25}{4}\right)^{\frac{3}{2}}}$$
$$= \frac{1}{\dfrac{\left(\sqrt{25}\right)^3}{\left(\sqrt{4}\right)^3}}$$
$$= \frac{1}{\dfrac{5^3}{2^3}}$$
$$= \frac{1}{\dfrac{125}{8}}$$
$$= \frac{8}{125}$$

SOLUTION 2 Graphing-Calculator Method

The first answer given by a graphing calculator may be a decimal.
If necessary, convert the decimal to a fraction using the ▶Frac function.

a)

b)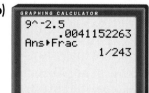

c)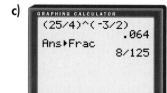

Note the use of brackets on the calculator.

Example 3 Evaluating Approximate Roots

Use a calculator to evaluate the following, to the nearest hundredth.

a) $2^{3.5}$

b) $7^{\frac{2}{3}}$

Solution

a)

GRAPHING CALCULATOR

2^3.5
        11.3137085

Estimate

$2^3 = 8$
$2^4 = 16$
$2^{3.5} \doteq 12$

$2^{3.5} \doteq 11.31$

b)

GRAPHING CALCULATOR

7^(2/3)
        3.65930571

Estimate

$7^{\frac{2}{3}} \doteq 8^{\frac{2}{3}}$
$\doteq 2^2$
$\doteq 4$

$7^{\frac{2}{3}} \doteq 3.66$

Key Concepts

- $a^{\frac{1}{n}} = \sqrt[n]{a}$, where n is a natural number.
- To evaluate $a^{\frac{1}{n}}$ or $\sqrt[n]{a}$ in the real number system,
  if n is even, then $a \geq 0$,
  if n is odd, then a can be any real number.
- $a^{\frac{m}{n}} = \sqrt[n]{a^m} = \left(\sqrt[n]{a}\right)^m$, where m and n are natural numbers.
- To calculate $a^{\frac{m}{n}}$ using paper and pencil, either take the nth root of a, then raise the result to the mth power, or raise a to the mth power, then take the nth root.

Communicate Your Understanding

1. Describe how you would evaluate each of the following using paper and pencil.

a) $27^{\frac{1}{3}}$ **b)** $27^{-\frac{1}{3}}$ **c)** $(-27)^{-\frac{1}{3}}$

2. Describe two ways to evaluate $8^{\frac{2}{3}}$ using paper and pencil.

3. Explain whether it is possible to evaluate each of the following in the real number system.

a) $16^{\frac{1}{4}}$ **b)** $-16^{\frac{1}{4}}$ **c)** $(-16)^{\frac{1}{4}}$

Practise

A

1. Write in radical form.

a) $2^{\frac{1}{3}}$　　　b) $37^{\frac{1}{2}}$　　　c) $x^{\frac{1}{2}}$

d) $a^{\frac{3}{5}}$　　　e) $6^{\frac{4}{3}}$　　　f) $6^{\frac{3}{4}}$

g) $7^{-\frac{1}{2}}$　　　h) $9^{-\frac{1}{5}}$　　　i) $x^{-\frac{3}{7}}$

j) $b^{-\frac{6}{5}}$　　　k) $(3x)^{\frac{1}{2}}$　　　l) $3x^{\frac{1}{2}}$

2. Write using exponents.

a) $\sqrt{7}$　　　b) $\sqrt{34}$　　　c) $\sqrt[3]{-11}$

d) $\sqrt[5]{a^2}$　　　e) $\sqrt[3]{6^4}$　　　f) $(\sqrt[3]{b})^4$

g) $\dfrac{1}{\sqrt{x}}$　　　h) $\dfrac{1}{\sqrt[3]{a}}$　　　i) $\dfrac{1}{\sqrt[5]{x^4}}$

j) $\sqrt[3]{2b^3}$　　　k) $\sqrt{3x^5}$　　　l) $\sqrt[4]{5t^3}$

3. Evaluate.

a) $4^{\frac{1}{2}}$　　　b) $125^{\frac{1}{3}}$　　　c) $16^{-\frac{1}{4}}$

d) $(-32)^{\frac{1}{5}}$　　　e) $25^{0.5}$　　　f) $(-27)^{-\frac{1}{3}}$

g) $64^{-\frac{1}{6}}$　　　h) $0.04^{\frac{1}{2}}$　　　i) $81^{0.25}$

j) $0.001^{\frac{1}{3}}$　　　k) $\left(\dfrac{4}{9}\right)^{\frac{1}{2}}$　　　l) $\left(\dfrac{-27}{-8}\right)^{\frac{1}{3}}$

4. Evaluate.

a) $8^{\frac{2}{3}}$　　　b) $4^{\frac{3}{2}}$　　　c) $9^{2.5}$

d) $81^{\frac{3}{4}}$　　　e) $16^{-\frac{3}{4}}$　　　f) $(-32)^{\frac{2}{5}}$

g) $(-8)^{-\frac{5}{3}}$　　　h) $(-27)^{-\frac{2}{3}}$　　　i) $1^{\frac{5}{3}}$

j) $(-1)^{-\frac{8}{5}}$　　　k) $\left(\dfrac{100}{9}\right)^{\frac{3}{2}}$　　　l) $\left(\dfrac{27}{8}\right)^{-\frac{2}{3}}$

5. Evaluate in the real number system, if possible.

a) $(-9)^{\frac{1}{2}}$　　　　　b) $100\,000^{\frac{3}{5}}$

c) $\left(\dfrac{27}{8}\right)^{\frac{2}{3}}$　　　　d) $3^{\frac{1}{2}} \times 3^{\frac{1}{2}}$

e) $-9^{\frac{1}{2}}$　　　f) $(2^5)^{0.4}$

g) $-8^{\frac{5}{3}}$　　　h) $4^{\frac{3}{2}} \div 16^{\frac{1}{4}}$

i) $(-1)^{-\frac{3}{2}}$　　　j) $(\sqrt[3]{5^2})(\sqrt[3]{5})$

k) $\left(\dfrac{36}{121}\right)^{-\frac{1}{2}}$　　　l) $81^{0.75}$

m) $(-0.0016)^{\frac{1}{4}}$　　　n) $\dfrac{(0.027)^{-\frac{2}{3}}}{(0.25)^{-\frac{1}{2}}}$

o) $(625^{-1})^{-\frac{1}{4}}$　　　p) $9^{\frac{3}{7}} \times 3^{\frac{1}{7}}$

q) $\left[\left(\sqrt{125}\right)^4\right]^{\frac{1}{6}}$　　　r) $\sqrt[3]{\sqrt{64}}$

s) $\sqrt{\sqrt[3]{729}}$　　　t) $\dfrac{(0.09)^{\frac{1}{2}}}{(0.008)^{\frac{1}{3}} \times 2^{-3}}$

6. Communication Write an equivalent expression using exponents.

a) $\sqrt{\sqrt{x^4}}$　　　　　b) $\sqrt[3]{\sqrt{x^6}}$

c) $\sqrt{\sqrt{3x^6}}$　　　　d) $\sqrt{\sqrt[3]{8x^7}}$

e) $\sqrt{\sqrt{81x^8}}$　　　f) $\left(x^{\frac{2}{3}}y^{\frac{1}{3}}\right)^3$

g) $\left(a^{\frac{1}{3}}b^{\frac{1}{4}}\right)^{12}$　　　h) $\sqrt[3]{-27x}$

i) $(81a^8b^4)^{\frac{1}{4}}$　　　j) $(27x^6y^{-9})^{\frac{2}{3}}$

k) $\left(\sqrt{x^3}\right)\left(\sqrt[3]{x}\right)$　　　l) $\left(\sqrt[3]{x^2}\right)\left(\sqrt[4]{x^3}\right)$

m) $\left(\sqrt[5]{x^3}\right)\left(\sqrt[3]{x^2}\right)$　　　n) $\left(\sqrt[3]{a^2b^4}\right)^2$

o) $\left(\sqrt[4]{a^3b^5}\right)^{\frac{1}{2}}$

7. Estimate. Then, find an approximation for each, the nearest hundredth.

a) $6^{0.4}$　　　　　b) $3^{2.8}$

c) $4^{-1.2}$　　　　d) $5^{\frac{1}{3}}$

e) $7^{-\frac{3}{5}}$　　　　f) $10^{\frac{3}{7}}$

Apply, Solve, Communicate

8. Ship building The design of a new ship calls for the ship to be 300 m long and travel at 12 m/s. To test the design, a model 15 m long is used. Using the formula from the beginning of this section, find the speed at which the model should be tested, to the nearest tenth of a metre per second.

B

9. Horizon Because the Earth is curved, it is impossible to see beyond the horizon. The distance, d, to the horizon depends on the observer's height, h, above the ground. The radius of the Earth is r. The formula for the distance to the horizon is

$$d = (2rh + h^2)^{\frac{1}{2}}.$$

a) Use the diagram to show that the formula is valid.
b) Assume that the radius of the Earth is 6370 km. Find the distance to the horizon, to the nearest kilometre, for an observer in an aircraft 10 km above the Earth; in a spacecraft 200 km above the Earth.

10. Weather Meteorologists have determined that violent storms, such as tornadoes and hurricanes, can be described using the formula $D = 9.4t^{\frac{2}{3}}$. In this formula, D kilometres is the diameter of the storm and t hours is the time for which the storm lasts. If a typical hurricane lasts for about 18 h, what is its diameter, to the nearest kilometre?

11. Mining The volume of nickel Canada produces in a year is about 21 000 m^3.
a) If this volume of nickel were made into a single cube, what would be the length of each edge, to the nearest tenth of a metre?
b) How does this volume of nickel compare with the volume of your school gymnasium?

12. Application The frequency of any note on a piano is measured in vibrations per second, or hertz (Hz). The frequency of each of the other notes in the octave above middle C is a multiple of the frequency of middle C. The table shows the approximate frequency of middle C. Copy and complete the table by finding the approximate frequencies of the other notes, to the nearest tenth of a hertz.

Note	Multiple of C	Frequency (Hz)
C	1	261.6
C#	$\sqrt[12]{2}$	
D	$\left(\sqrt[12]{2}\right)^2$	
D#	$\left(\sqrt[12]{2}\right)^3$	
E	$\left(\sqrt[12]{2}\right)^4$	
F	$\left(\sqrt[12]{2}\right)^5$	
F#	$\left(\sqrt[12]{2}\right)^6$	
G	$\left(\sqrt[12]{2}\right)^7$	
G#	$\left(\sqrt[12]{2}\right)^8$	
A	$\left(\sqrt[12]{2}\right)^9$	
A#	$\left(\sqrt[12]{2}\right)^{10}$	
B	$\left(\sqrt[12]{2}\right)^{11}$	
C	$\left(\sqrt[12]{2}\right)^{12}$	

C

13. Equations Evaluate x, where x is a natural number.

a) $2^x = 32$ **b)** $3^{x+1} = 81$ **c)** $(-1)^x = 1$

d) $6^{x-2} = 36$ **e)** $2^{2x} = 16$ **f)** $(-1)^x = -1$

14. Inquiry/Problem Solving a) The diagrams show 2 squares with whole-number areas that can be made on a 4-pin by 4-pin geoboard. If the shortest distance between 2 pins is 1 unit, what is the area of each square?

b) Draw the 3 other different-sized squares with whole-number areas that can be made on the same geoboard.
c) Of the 5 different-sized squares, which ones do not have whole-number side lengths? Express their side lengths using fractional exponents.
d) Draw the 8 different-sized squares with whole-number areas that can be made on a 5-pin by 5-pin geoboard. For the squares that do not have whole-number side lengths, express the side lengths using fractional exponents.
e) Repeat part d) for a 6-pin by 6-pin geoboard, and state how many different-sized squares can be made.
f) Can you make any generalizations or state any conclusions from this investigation?

ACHIEVEMENT Check   Knowledge/Understanding   Thinking/Inquiry/Problem Solving   Communication   Application

If @(a, b, c) means $a^b - b^c + c^a$, what does each of the following equal?

a) @(1, −1, 2)   **b)** @$\left(\dfrac{1}{3}, -1, 8\right)$ **c)** @(−0.5, x, 4)

1.3 Solving Exponential Equations

Radioactive isotopes have many uses, including medical diagnoses and treatments, and the production of nuclear energy.

Different radioactive isotopes decay at different rates. The time it takes for half of any sample of an isotope to decay is called the half-life of the isotope. The decay of radioactive isotopes is described by the following equation.

$A_L = A_O\left(\dfrac{1}{2}\right)^{\frac{t}{t_{\frac{1}{2}}}}$ where A_L is the amount of the isotope left, A_O is the original amount of the isotope, t is the elapsed time, and $t_{\frac{1}{2}}$ is the half-life of the isotope.

The half-life of tungsten-187 is 1 day, so the decay rate for tungsten-187 is described by the following equation.

$A_L = A_O\left(\dfrac{1}{2}\right)^{t}$ where t is the elapsed time in days.

The above equations are examples of **exponential equations**. These are equations in which the variables appear as exponents.

INVESTIGATE & INQUIRE

To solve the equation $8^{x-2} = 2^{x+4}$ means to determine the value of x that makes the equation true, or satisfies the equation.

1. Use a table and guess and check to determine the value of x that satisfies this equation. The first row has been completed for you.

GUESS			CHECK
Value of x	L.S. $= 8^{x-2}$	R.S. $= 2^{x+4}$	Does L.S. $=$ R.S.?
1	$8^{1-2} = 8^{-1} = \dfrac{1}{8}$	$2^{1+4} = 2^5 = 32$	No, R.S. $>$ L.S.

2. a) On the left side of the equation $8^{x-2} = 2^{x+4}$, replace the base 8 with a power of 2.
b) Use an exponent law to simplify the exponent on the left side of the equation.
c) Since the bases on both sides are equal, how must the exponents be related to make the equation true?
d) Write a linear equation using the two exponents, and solve the equation.

3. Is the solution you found in question 2d) the same as the one you found by guess and check?

4. Using your results from questions 2 and 3, write a rule for solving exponential equations.

5. Test your rule by using it to solve each of the following. Use substitution to verify each solution.

a) $2^{x+1} = 4^{x-1}$ **b)** $9^{x+4} = 27^{2x}$ **c)** $8^{2x-3} = 16^{1-x}$
d) $25^x = 5^{3x}$ **e)** $2^{4x-1} = 4^x$

6. Suppose an original sample of tungsten-187 had a mass of 64 mg, and there are 2 mg left.

a) Substitute the given mass values into the equation $A_L = A_O\left(\dfrac{1}{2}\right)^t$ described above.

b) Express $\left(\dfrac{1}{2}\right)^t$ as a power of 2.

c) Express the right side of the equation as a single power of 2.

d) Use your rule from question 4 to solve the equation for t. What was the elapsed time?

e) Check your solution by substitution in the original equation.

One method for solving an exponential equation is to rewrite the powers with the same base, so that the exponents are equal. Equating the exponents gives a linear equation, which can be solved.

This method of solving an exponential equation is based on the property that, if $a^x = a^y$, then $x = y$, for $a \neq 1, 0, -1$.

EXAMPLE 1 Solving Using a Common Base

Solve and check $4^{x+1} = 2^{x-1}$.

SOLUTION

The base 4 on the left side is a power of 2.

$$4^{x+1} = 2^{x-1}$$

Rewrite using base 2: $(2^2)^{x+1} = 2^{x-1}$
Simplify exponents: $2^{2x+2} = 2^{x-1}$
Equate exponents: $2x + 2 = x - 1$
Solve for x: $x = -3$

Check.

L.S. $= 4^{x+1}$ R.S. $= 2^{x-1}$

$\quad = 4^{-3+1}$ $= 2^{-3-1}$

$\quad = 4^{-2}$ $= 2^{-4}$

$\quad = \dfrac{1}{4^2}$ $= \dfrac{1}{2^4}$

$\quad = \dfrac{1}{16}$ $= \dfrac{1}{16}$

$\qquad\qquad$ L.S. = R.S.

The solution is $x = -3$.

Example 2 Rational Solutions

Solve and check $9^{3x+1} = 27^x$.

Solution

Both bases are powers of 3.

$$9^{3x+1} = 27^x$$

Rewrite using base 3: $(3^2)^{3x+1} = (3^3)^x$

Simplify exponents: $3^{6x+2} = 3^{3x}$

Equate exponents: $6x + 2 = 3x$

Solve for x: $3x = -2$

$$x = -\dfrac{2}{3}$$

The solution is $x = -\dfrac{2}{3}$.

Check.

L.S. $= 9^{3x+1}$ R.S. $= 27^x$

$\quad = 9^{3\left(-\frac{2}{3}\right)+1}$ $= 27^{-\frac{2}{3}}$

$\quad = 9^{-2+1}$ $= \dfrac{1}{\left(\sqrt[3]{27}\right)^2}$

$\quad = 9^{-1}$

$\quad = \dfrac{1}{9}$ $= \dfrac{1}{3^2}$

$\qquad\qquad\qquad\qquad$ $= \dfrac{1}{9}$

$\qquad\qquad$ L.S. = R.S.

Example 3 Solving Using a Common Factor

Solve and check $3^{x+2} - 3^x = 216$.

Solution

$$3^{x+2} - 3^x = 216$$

Remove a common factor: $3^x(3^2 - 1) = 216$

Simplify: $3^x(8) = 216$

Divide both sides by 8: $3^x = 27$

Solve for x: $3^x = 3^3$

$$x = 3$$

The solution is $x = 3$.

Check.

L.S. $= 3^{x+2} - 3^x$ R.S. $= 216$

$\quad = 3^{3+2} - 3^3$

$\quad = 3^5 - 27$

$\quad = 243 - 27$

$\quad = 216$

$\qquad\qquad$ L.S. = R.S.

EXAMPLE 4 Modelling Exponential Decay

A radioactive isotope, iodine-131, is used to determine whether a person has a thyroid deficiency. The iodine-131 is injected into the blood stream. A healthy thyroid gland absorbs all the iodine. The half-life of iodine-131 is 8.2 days, so its decay can be modelled by the exponential equation

$$A_L = A_O\left(\frac{1}{2}\right)^{\frac{t}{8.2}}$$

where A_L is the amount of iodine-131 left, A_O is the original amount of iodine-131, and t is the elapsed time, in days. After how long should 25% of the iodine-131 remain in the thyroid gland of a healthy person?

SOLUTION

$$25\% = \frac{1}{4}$$

so $A_L = \frac{1}{4}A_O$

Write the equation: $\qquad A_L = A_O\left(\frac{1}{2}\right)^{\frac{t}{8.2}}$

Substitute $\frac{1}{4}A_O$ for A_L: $\qquad \frac{1}{4}A_O = A_O\left(\frac{1}{2}\right)^{\frac{t}{8.2}}$

Divide both sides by A_O: $\qquad \frac{1}{4} = \left(\frac{1}{2}\right)^{\frac{t}{8.2}}$

Rewrite using base $\frac{1}{2}$: $\qquad \left(\frac{1}{2}\right)^2 = \left(\frac{1}{2}\right)^{\frac{t}{8.2}}$

Equate exponents: $\qquad 2 = \frac{t}{8.2}$

Solve for t: $\qquad 16.4 = t$

Web Connection
www.school.mcgrawhill.ca/resources/
To investigate the modern uses of radioactive isotopes, visit the above web site. Go to **Math Resources**, then to MATHEMATICS 11, to find out where to go next. Write a short report on a beneficial use of radioactive isotopes.

So, 25% of the iodine-131 should remain in the thyroid gland of a healthy person after 16.4 days.

Check.

After one half-life, $\frac{1}{2}$ or 50% is left.

After two half-lives, $\frac{1}{4}$ or 25% is left.

Two half-lives is 2×8.2 days, or 16.4 days.

Key Concepts

- In exponential equations, the variables appear as exponents.
- If $a^x = a^y$, then $x = y$, for $a \neq 1, 0, -1$.
- One method for solving an exponential equation is to rewrite the powers with the same base, so that the exponents are equal. Equating the exponents gives an equation that can be solved.
- Some exponential equations can be solved by first removing a power as a common factor.

Communicate Your Understanding

1. Explain why the second key concept, above, includes "$a \neq 1, 0, -1$."
2. Describe how you would solve $2^{x+3} = 4^{x-1}$.
3. Describe how you would solve $2^{x+1} + 2^x = 48$.
4. To solve $3^{x-3} = 1$ using a common base, how would you rewrite 1?

Practise

A

1. Solve.

a) $2^x = 16$

b) $3^x = 27$

c) $2^x = 128$

d) $5^x = 125$

e) $4^y = 256$

f) $729 = 9^z$

g) $(-3)^x = -27$

h) $(-2)^x = -32$

i) $(-5)^a = 25$

j) $81 = (-3)^x$

k) $-2^x = -16$

l) $-4^y = -64$

m) $-5^x = -625$

n) $(-1)^x = 1$

o) $(-1)^m = -1$

2. Solve.

a) $7^{w-2} = 49$

b) $3^{x+4} = 27$

c) $2^{1-x} = 128$

d) $4^{3k} = 64$

e) $5^{3x-1} = 25$

f) $-81 = -3^{2x+8}$

g) $4^{x-1} = 1$

h) $3^{2-2x} = 1$

i) $(-1)^{2x} = 1$

3. Solve and check.

a) $6^{x+3} = 6^{2x}$

b) $2^{x+3} = 2^{2x-1}$

c) $3^{2y+3} = 3^{y+5}$

d) $2^{4x-7} = 2^{2x+1}$

e) $7^{5d-1} = 7^{2d+5}$

f) $3^{b-5} = 3^{2b-3}$

4. Solve.

a) $16^{2x} = 8^{3x}$

b) $4^t = 8^{t+1}$

c) $27^{x-1} = 9^{2x}$

d) $25^{2-c} = 125^{2c-4}$

e) $16^{2p+1} = 8^{3p+1}$

f) $(-8)^{1-2x} = (-32)^{1-x}$

5. Solve and check.

a) $2^{x+5} = 4^{x+2}$

b) $2^x = 4^{x-1}$

c) $9^{2q-6} = 3^{q+6}$

d) $4^x = 8^{x+1}$

e) $27^{y-1} = 9^{2y-4}$

f) $8^{x+3} = 16^{2x+1}$

6. Solve and check.

a) $5^{4-x} = \dfrac{1}{5}$

b) $10^{y-2} = \dfrac{1}{10\,000}$

c) $6^{3x-7} = \dfrac{1}{6}$

d) $3^{3x-1} = \dfrac{1}{81}$

e) $5^{2n+1} = \dfrac{1}{125}$

f) $\dfrac{1}{256} = 2^{2-5w}$

7. Solve and check.

a) $4^x = 8$ b) $64^z = 16$

c) $(-8)^y = -2$ d) $9^{-x} = 3$

e) $2^{9x} = \dfrac{1}{8}$ f) $9^{6x} = \dfrac{1}{27}$

g) $2^x = 16^4$ h) $2^{-2g} = 32$

i) $9^{2s+1} = 27$

8. Solve and check.

a) $9^{x+1} = 27^{2x}$ b) $16^y = 64^{2y-1}$

c) $36^{t-2} = 216^{-2t}$ d) $8^{2x-1} = 16^{x-1}$

e) $25^{1-3x} = 125^{-x}$ f) $16^{3+k} = 32^{1-2k}$

9. Solve and check.

a) $5 = 25^{\frac{x}{2}}$ b) $8 = 2^{\frac{x}{3}}$

c) $9^{\frac{y}{5}} = 27$ d) $\dfrac{1}{2} = 2^{\frac{a}{3}}$

e) $4^{\frac{x}{4}} = \dfrac{1}{8}$ f) $\left(\dfrac{3}{2}\right)^{\frac{m}{2}} = \dfrac{4}{9}$

10. Solve.

a) $3(5^{x+1}) = 15$

b) $2(3^{y-2}) = 18$

c) $5(4^x) = 10$

d) $2(4^{v+1}) = 1$

e) $2 = 6(3^{4f-2})$

f) $27(3^{3x+1}) = 3$

11. Solve and check.

a) $2^{x+2} - 2^x = 48$

b) $4^{x+3} + 4^x = 260$

c) $2^{a+5} + 2^a = 1056$

d) $6^{x+1} + 6^{x+2} = 7$

e) $3^{x+3} - 3^{x+1} = 648$

f) $10^{z+4} + 10^{z+3} = 11$

g) $2^{x+2} - 2^{x+5} = -7$

h) $3^{m+1} + 3^{m+2} - 972 = 0$

i) $5^{n+2} - 5^{n+3} = -2500$

Apply, Solve, Communicate

12. Communication Solve $4^{3x+3} = 8^{2x+2}$. Explain your answer.

13. Half-life The half-life of ruthenium-106 is 1 year, so the decay of ruthenium-106 is described by the exponential equation

$$A_L = A_O\left(\dfrac{1}{2}\right)^t$$

where t is the elapsed time, in years. If an original sample of ruthenium-106 had a mass of 128 mg, and there are 2 mg left, what is the elapsed time?

B

14. Paper industry Strontium-90 is used in machines that control the thickness of paper during the manufacturing process. Strontium-90 has a half-life of 28 years. Determine how much time has elapsed if the following fraction of a strontium-90 sample remains.

a) $\dfrac{1}{4}$ b) $\dfrac{1}{8}$ c) $\dfrac{1}{32}$

15. Application The biological half-life of thyroid hormone T4 is about 6.5 days. If a dose of T4 was not followed by repeat doses,

a) what fraction of the original dose would remain in the body after 19.5 days?

b) how long would it take until only 6.25% of the original dose would remain in the body?

16. Scuba diving The percent of sunlight, s, that reaches a scuba diver under water can be modelled by the equation

$$s = 0.8^d \times 100\%$$

where d is the depth of the diver, in metres.

a) At what depth does 64% of sunlight reach the diver?

b) What percent of sunlight reaches the diver at a depth of 10 m, to the nearest percent?

17. Application Determine the half-life of each isotope.

a) In 30 h, a sample of plutonium-243 decays to $\dfrac{1}{64}$ of its original amount.

b) In 40.8 years, a sample of lead-210 decays to 25% of its original amount.

c) In 2 min, a sample of radium-221 decays to 6.25% of its original amount.

18. Circulation Sodium-24 is used to diagnose circulatory problems. The half-life of sodium-24 is 14.9 h. A hospital buys a 40-mg sample of sodium-24. After how long will only 2.5 mg remain?

19. Solve.

a) $\dfrac{27^x}{9^{2x-1}} = 3^{x+4}$

b) $27^x(9^{2x-1}) = 3^{x+4}$

c) $27^{x+1} = \left(\dfrac{1}{9}\right)^{2x-5}$

20. Solve.

a) $2^{x^2+2x} = 2^{x+6}$

b) $3^{x^2-2x} = 3^{x-2}$

c) $2^{2x^2-3x} = 2^{x^2-2x+12}$

C

21. Half-life In 8 days, a sample of vanadium-48 decays to $\dfrac{1}{\sqrt{2}}$ of its original amount. Determine the half-life of vanadium-48.

22. Solve and check.

a) $\dfrac{2^{2x+1}}{2^{x-3}} = 4$

b) $\dfrac{9^{x+4}}{27^{x-1}} = 81$

c) $\dfrac{8^{x+2}}{4^{x+3}} = 16^{x-3}$

23. Find x and y if $\dfrac{16^{x+2y}}{8^{x-y}} = 32$ and $\dfrac{32^{x+3y}}{16^{x+2y}} = \dfrac{1}{8}$.

CAREER CONNECTION *Microbiology*

The science of microbiology is the study of micro-organisms. These are organisms that are too small to be examined with the naked eye. They were first studied after the development of microscopes. Examples of micro-organisms include bacteria, fungi, algae, and viruses.

Bacteria are widely feared because some of them cause diseases in living things. However, we make use of many other bacteria in our daily lives. For example, bacteria are used to make yogurt and cheese from milk, to treat sewage, and to make antibiotics, such as penicillin.

1. Bacteria a) The number of bacteria in a culture is doubling every 7 h. Explain how the equation $N = N_0(2)^{\frac{t}{7}}$ models the number of bacteria in the culture. Define each term in the equation.

b) In the culture from part a), if there are 100 000 bacteria at a certain time, how many hours later will the number of bacteria be 800 000? 6 400 000? 25 600 000?

c) In a different culture, the number of bacteria increases from 15 000 to 240 000 in 24 h. How much longer will it take for the number of bacteria to reach 480 000?

d) Write an equation that models the number of bacteria in the culture in part c).

2. Research Use your research skills to investigate the following.

a) the education and training needed for a career in microbiology, and the employers who hire microbiologists

b) an aspect of microbiology that is important in Canada

NUMBER *Power*

You have 1023 coins. How can you place them in 10 bags so that, if you are asked for any number of coins from 1 to 1023, you can provide the number without opening a bag?

TECHNOLOGY EXTENSION
Solving Exponential Equations With a Graphing Calculator

Solving Graphically

One way to solve the equation $2^{3x-2} = 4$ with a graphing calculator is to enter the equations $y = 2^{3x-2}$ and $y = 4$ into the Y= editor, graph the equations, and use the intersect operation to find the x-coordinate of the point of intersection. If necessary, the ▸Frac function can be used to write the solution as a fraction.

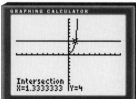 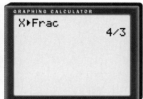

Another method is to rewrite the equation as $2^{3x-2} - 4 = 0$, graph the equation $y = 2^{3x-2} - 4$, and use the zero operation to find the x-intercept of the graph.

 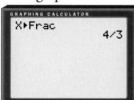

1. Explain why the two methods described above are equivalent.

2. Solve graphically.

a) $2^{x-3} = 8$

b) $5^{2x-1} = \dfrac{1}{25}$

c) $8^{4x+1} = 2$

d) $9^{3x+4} = 3^{2x+1}$

e) $\left(\dfrac{1}{4}\right)^{3x-1} = \left(\dfrac{1}{16}\right)^{x+1}$

f) $3(6^{2x-3}) = 108$

g) $2^{x+1} - 2^{x+3} = -6$

h) $5^{x+1} + 5^{x+2} = 750$

i) $\dfrac{2^{x+1}}{4^{x-1}} = 2^x$

3. Try to solve each of the following graphically. Explain your results.

a) $3^{2x-2} = 9^{x-1}$

b) $2^{3x+2} = -2$

Solving Algebraically

Some graphing calculators, such as the TI-92 and TI-92 Plus, have a solve function with the capability to solve exponential equations algebraically.

4. Solve algebraically using a graphing calculator.

a) $7^{2x+4} = 49$

b) $-8^{2x+3} = -4^{2-x}$

c) $0.001^{\frac{x}{4}} = 0.1$

d) $10(5^{6x+1}) = 6250$

e) $8^{x-1} - 8^{x-2} = 7$

f) $9^{2x} = \dfrac{27^{x+1}}{81^{1-x}}$

5. Try to solve each of the following algebraically using a graphing calculator. Explain your results.

a) $(-4)^{1-2x} = -16$

b) $4^{3x+3} = 8^{2x+2}$

1.4 Review: Adding, Subtracting, and Multiplying Polynomials

Adding and Subtracting Polynomials

Recall that terms such as $8x$ and $5x$, which have the same variable factors, are known as like terms. To simplify an expression containing like terms, add their coefficients.

EXAMPLE 1 **Adding Polynomials**

Simplify $(4x^2 - 7x - 5) + (2x^2 - x + 3)$.

SOLUTION

To add polynomials, collect like terms.
$$(4x^2 - 7x - 5) + (2x^2 - x + 3) = 4x^2 - 7x - 5 + 2x^2 - x + 3$$
$$= 4x^2 + 2x^2 - 7x - x - 5 + 3$$
$$= 6x^2 - 8x - 2$$

EXAMPLE 2 Subtracting Polynomials

Simplify $(4s^2 + 5st - 7t^2) - (6s^2 + 3st - 2t^2)$.

SOLUTION

To subtract, add the opposite.
Multiply each term to be subtracted by -1.

$$
\begin{aligned}
(4s^2 + 5st - 7t^2) - (6s^2 + 3st - 2t^2) &= (4s^2 + 5st - 7t^2) - 1(6s^2 + 3st - 2t^2) \\
&= 4s^2 + 5st - 7t^2 - 6s^2 - 3st + 2t^2 \\
&= 4s^2 - 6s^2 + 5st - 3st - 7t^2 + 2t^2 \\
&= -2s^2 + 2st - 5t^2
\end{aligned}
$$

Practise

1. Add.
a) $(3x^2 - x + 2) + (4x^2 + 3x - 1)$
b) $(2t^2 + 5t - 7) + (3t^2 - 4t + 6)$
c) $(7m^2 - mn - 8n^2) + (6m^2 + 9mn + 11n^2)$
d) $(-4y^2 + 2xy - 6x^2) + (5y^2 - 6xy + 7y^2)$
e) $(3xy - 2x + 7) + (6xy + 5x - 3)$
f) $(5x + 3y - 8xy) + (6xy + 2x - 5y)$

2. Subtract.
a) $(3x^2 - 7x + 3) - (x^2 + 5x - 2)$
b) $(5s^2 + 8s - 12) - (6s^2 - s + 4)$
c) $(9x^2 - 4xy - y^2) - (6y^2 + 3xy + 10x^2)$
d) $(-r^2 + 4rs + s^2) - (6r^2 - rs + 11s^2)$
e) $(3x + 4y - 5z) - (x - y - z)$
f) $(5m - 3n) - (2m - 7n + 4)$

3. Add the sum of $3x^2 - 6x + 5$ and $-3x^2 + 6$ to $-x^2 - x - 1$.

4. Add $4x + 2y - 7$ to the sum of $-2x + 3y - 2$ and $3x + y - 4$.

5. Subtract $3t^2 + 4t - 7$ from the sum of $2t^2 - 5t + 3$ and $4t^2 + 2t + 3$.

6. Subtract the sum of $m^2 + 2m - 3$ and $4m^2 - m + 2$ from $3m^2 + 4m - 1$.

7. Measurement The perimeter of a triangle is $5x - 2y + 3z$. If two sides have lengths $3y + z$ and $4x - y + z$, what is the length of the third side?

Multiplying Polynomials by Monomials

To multiply a polynomial by a monomial, use the distributive property to multiply each term in the polynomial by the monomial.

EXAMPLE 3 Expanding

Expand $3a(2a^2 - 4a - 5)$.

SOLUTION

Use the distributive property.

$$3a(2a^2 - 4a - 5) = 3a(2a^2 - 4a - 5)$$
$$= 6a^3 - 12a^2 - 15a$$

EXAMPLE 4 Expanding and Simplifying

Expand and simplify $2x(3x - 5) - 4x(x - 7) + 3x(x - 1)$.

SOLUTION

Use the distributive property to remove the brackets.
Then, collect like terms.

$$2x(3x - 5) - 4x(x - 7) + 3x(x - 1) = 2x(3x - 5) - 4x(x - 7) + 3x(x - 1)$$
$$= 6x^2 - 10x - 4x^2 + 28x + 3x^2 - 3x$$
$$= 6x^2 - 4x^2 + 3x^2 - 10x + 28x - 3x$$
$$= 5x^2 + 15x$$

When more than one set of brackets is used, simplify to remove the innermost brackets first.

EXAMPLE 5 More Than One Set of Brackets

Expand and simplify $2[3(2x + 3) - 2(x - 1)]$.

SOLUTION

$$2[3(2x + 3) - 2(x - 1)] = 2(6x + 9 - 2x + 2)$$
$$= 2(4x + 11)$$
$$= 8x + 22$$

Practise

8. Expand.

a) $2(3x + 4)$ **b)** $-5(2 - 3x)$

c) $4y(2y - 3)$ **d)** $-3(3m + 2n)$

e) $2t(4s - 5t)$ **f)** $4(2b^2 + b - 1)$

g) $-2(q^2 - 5b - 4)$ **h)** $3p(2p^2 - p + 4)$

i) $-4g(1 + 3g - 3g^2)$

9. Expand and simplify.

a) $2(x - 4) - 3(x - 5)$

b) $3(y^2 - 9x + 5) - 5(y - 4)$

c) $5(3x - 4y) - (2x - 5y) + 7$

d) $4(a - 2b - c) - 6(4a + 2b - 6c)$

e) $3(2x - 9) - 3 - (4x + 1) + 2$

f) $7(3t - 1) - 4(5t + 2) - 6$

g) $2(1 - 3s + 2s^2) - (1 - 4s + 5s^2)$

h) $4x(x - 1) + 6x(x + 3)$

i) $3a(2a + 3) + 5a(a - 4) - a(4a + 1)$

j) $2m(1 - 2m) - (2m - 3) + m$

k) $-4x(2x - 1) - x(1 - 2x) + 2x(x + 4)$

l) $2r(3 - r) + 4r(5r + 3) - r(5 - 2r)$

10. Expand and simplify.

a) $3[5 + 4(x - 7)]$

b) $-3[2(x - 5) - 4(2x - 3)]$

c) $2[3(2t - 4) + 5(t + 3)]$

d) $4[1 - 2(3y - 1) + 2[4(y - 6) - 1]$

e) $2x[x + 2(x - 3)] - x(3x - 4)$

f) $3y[1 - y(y - 3)] - [2 - y(y - 4)]$

Multiplying Polynomials

The distributive property can be used to multiply two binomials.

EXAMPLE 6 Using the Distributive Property

Expand and simplify $(2x + 3)(4x - 5)$.

SOLUTION

$$(2x + 3)(4x - 5) = 2x(4x - 5) + 3(4x - 5)$$
$$= 8x^2 - 10x + 12x - 15$$
$$= 8x^2 + 2x - 15$$

In Example 6, the same result is obtained if each term in the first binomial is multiplied by each term in the second binomial. You can remember this method with the acronym FOIL, which stands for adding the products of First terms, Outside terms, Inside terms, and Last terms.

$$(2x + 3)(4x - 5) = (2x + 3)(4x - 5)$$
$$= 8x^2 - 10x + 12x - 15$$
$$= 8x^2 + 2x - 15$$

EXAMPLE 7 Expanding and Simplifying

Expand and simplify $3(x - 4)(x + 2) - 2(x + 5)(x - 3)$.

SOLUTION

$$3(x - 4)(x + 2) - 2(x + 5)(x - 3) = 3(x - 4)(x + 2) - 2(x + 5)(x - 3)$$
$$= 3(x^2 + 2x - 4x - 8) - 2(x^2 - 3x + 5x - 15)$$
$$= 3(x^2 - 2x - 8) - 2(x^2 + 2x - 15)$$
$$= 3x^2 - 6x - 24 - 2x^2 - 4x + 30$$
$$= x^2 - 10x + 6$$

Recall that some graphing calculators, such as the TI-92 and TI-92 Plus, have the capability to perform operations with polynomials.

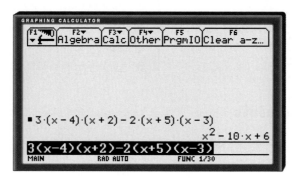

To find the product of any two polynomials, multiply each term of one of the polynomials by each term of the other polynomial. Then, collect like terms.

EXAMPLE 8 Multiplying Trinomials

Expand and simplify $(x^2 - 3x - 1)(2x^2 + x - 2)$.

SOLUTION

$$(x^2 - 3x - 1)(2x^2 + x - 2) = 2x^4 + x^3 - 2x^2 - 6x^3 - 3x^2 + 6x - 2x^2 - x + 2$$
$$= 2x^4 + x^3 - 6x^3 - 2x^2 - 3x^2 - 2x^2 + 6x - x + 2$$
$$= 2x^4 - 5x^3 - 7x^2 + 5x + 2$$

EXAMPLE 9 **Squaring a Trinomial**

Expand and simplify $(y^2 - 2y + 5)^2$.

SOLUTION

$$(y^2 - 2y + 5)^2 = (y^2 - 2y + 5)(y^2 - 2y + 5)$$
$$= y^4 - 2y^3 + 5y^2 - 2y^3 + 4y^2 - 10y + 5y^2 - 10y + 25$$
$$= y^4 - 4y^3 + 14y^2 - 20y + 25$$

Practise

11. Expand and simplify.

a) $(x - 7)(x + 6)$ **b)** $(t - 5)(t + 8)$

c) $(y - 3)(y - 9)$ **d)** $(3y - 1)(4y + 7)$

e) $(4x + 3)(2x + 7)$ **f)** $(5 + 2m)(3 - 4m)$

g) $2(8 - x)(5x + 2)$ **h)** $3(2x - 5)^2$

i) $-(5x - 6)(5x + 6)$

12. Expand and simplify.

a) $(7x + 2y)(8x - 7y)$

b) $(3s + t)(2s - 3t)$

c) $(4x - 5y)(3x - 10y)$

d) $3(6w - 11x)(w + 3x)$

e) $(5x^2 - 4x)(3x^2 + 2x)$

f) $(2m - 3m^2)(m^2 + 2m)$

g) $(3x - 4y)^2$

h) $-2(5x + 6y)(5x - 6y)$

i) $5(1 - xy)(1 + xy)$

13. Expand and simplify.

a) $(x - 7)(x + 1) + (x + 6)(x + 2)$

b) $(2t - 1)(t + 4) - (t + 6)(3t + 2)$

c) $2(x - 4)(x + 3) + 5(2x - 1)(x + 6)$

d) $2(2y - 5)(y - 4) - (5y - 3)(y + 4)$

e) $2(m - 3)(m - 4) - 3(m + 5)^2$

f) $3(2x + 3)^2 - (x - 5)^2 - (3x - 4)(x - 5)$

g) $5(2y - 5)(2y + 5) - 4(y - 2)(y + 3) - (2y + 1)^2$

h) $5t^2 - (t - 3)^2 - 2(t^2 - 5t) + 2(2t + 3)^2$

i) $4(x^2 - 3xy) - (x + y)^2 - 2(x - y)(x + y) + 5$

j) $(2r + 3t)(r - t) - 4(r - 2t)^2 + 5(r^2 - t^2)$

Apply, Solve, Communicate

14. Communication a) Explain how the diagram illustrates the product $(2x + 1)(x + 2y + 3)$.

b) State the product in simplified form.

15. Expand and simplify.

a) $(x+3)(x^2+2x+4)$

b) $(y-2)(y^2-y-5)$

c) $(3m+2)(2m^2+3m-4)$

d) $(t^2-5t-7)(2t+1)$

e) $(x^2+2x-1)(x^2-x-4)$

f) $(y-2)(y^3-2y^2+3y-1)$

g) $(3a^2-4a+2)(a^2-a-5)$

h) $(x^3-7)(3x^3+7)$

i) $(x^2-4x+1)^2$

j) $(2n^2-n-1)^2$

k) $(2a-b+3c)^2$

l) $(2x-1)(x^3-2x^2+5x-3)$

m) $2(x-1)(x^2-3x+2)-(2x^2-3x+4)(2x+3)$

n) $4(x-y+z)(x-2y-3z)-(x+y+z)^2-(x-y-2z)$

o) $(3x-5)[3+(2x+4)(x-1)]$

16. a) Multiply $(x+1)(x+2)$. Then, multiply the result by $x-3$ and simplify.

b) Multiply $(x+1)(x-3)$. Then, multiply the result by $x+2$ and simplify.

c) Multiply $(x-3)(x+2)$. Then, multiply the result by $x+1$ and simplify.

d) Does the order in which you multiply three binomials affect the result?

17. Expand and simplify.

a) $(2x+1)(x-3)(4x-5)$

b) $(x+2y)(x-3y)(2x-y)$

c) $(a+b+c+d)^2$

18. Measurement The dimensions of a rectangular prism are represented by binomials, as shown.

a) Write a simplified expression that represents the surface area of this prism.

b) Write a simplified expression that represents the volume of this prism.

c) If x represents 7 cm, what are the surface area and the volume of the prism?

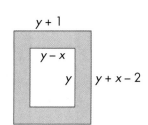

$x+4$ $x-3$ $2x+3$

19. Application Write and simplify an expression to represent the area of the shaded region.

$y+1$ $y-x$ y $y+x-2$

20. Inquiry/Problem Solving Is the product of two binomials always a trinomial? Explain.

21. Expand and simplify.

a) $\left(x+\dfrac{1}{x}\right)\left(x-\dfrac{1}{x}\right)$

b) $\left(y-\dfrac{2}{y}\right)\left(y+\dfrac{3}{y}\right)$

1.5 Simplifying Rational Expressions

Canada officially has two national games, lacrosse and hockey. Lacrosse is thought to have originated with the Algonquin tribes in the St. Lawrence Valley. The game was very popular in the late nineteenth century and was at one time an Olympic sport. Canadian lacrosse teams won gold medals at the Summer Olympics in 1904 and 1908.

There are two forms of lacrosse—box lacrosse, which is played indoors, and field lacrosse. When field lacrosse is played under international rules, the width of the rectangular field can be represented by x and the area of the field by the polynomial $2x^2 - 10x$. Thus, the length of the field can be represented by the expression $\dfrac{2x^2 - 10x}{x}$.

| x | $2x^2 - 10x$ |

This is an example of a **rational expression**, which is a quotient whose numerator and denominator are polynomials. The following are also rational expressions.

$$\frac{3}{x+2} \qquad \frac{y-4}{7} \qquad \frac{x+1}{x+3} \qquad \frac{5y}{y^2 - 1} \qquad \frac{a^2 + b^2}{a^2 - b^2}$$

INVESTIGATE & INQUIRE

1. a) Factor x from the expression for the area of a lacrosse field, $2x^2 - 10x$.
b) Record the other factor and explain why it represents the length of the field.
c) Describe how you could simplify the other expression for the length, $\dfrac{2x^2 - 10x}{x}$, to give the same expression as in part b).

2. The rectangle shown has a width of $2y$ and an area of $2y^2 + 6y$.
a) Factor $2y$ from the expression for the area.
b) Record the other factor and explain why it represents the length.

| $2y$ | $2y^2 + 6y$ |

c) Use the width and the area to write a rational expression that represents the length.

d) Describe how you could simplify the rational expression from part c) to give the same expression as in part b).

3. Use your results from questions 1 and 2 to write a rule for simplifying a rational expression in which the denominator is a monomial factor of the numerator.

4. Use your rule to simplify each of the following.

a) $\dfrac{4t^2 + 8t}{4t}$
b) $\dfrac{10m^3 + 5m^2 + 15m}{5m}$
c) $\dfrac{6r^4 - 3r^3 + 6r^2}{3r^2}$

5. The expressions from question 1 represent the dimensions of a lacrosse field for both the women's and the men's games.

a) For the women's game, played 12-a-side, x represents 60 m. What are the dimensions of the field, in metres?

b) For the men's game, played 10-a-side, x represents 55 m. What are the dimensions of the field, in metres?

EXAMPLE 1 Monomial Denominator

Simplify $\dfrac{24x^3 + 6x^2 + 12x}{6x}$. State the restriction on the variable.

SOLUTION

$$\dfrac{24x^3 + 6x^2 + 12x}{6x}$$

Factor the numerator:
$$= \dfrac{6x(4x^2 + x + 2)}{6x}$$

Divide by the common factor, $6x$:
$$= \dfrac{\overset{1}{\cancel{6x}}(4x^2 + x + 2)}{\underset{1}{\cancel{6x}}}$$

$$= 4x^2 + x + 2$$

Division by 0 is not defined, so exclude values of x for which $6x = 0$.
$6x = 0$ when $x = 0$, so $x \neq 0$

Therefore, $\dfrac{24x^3 + 6x^2 + 12x}{6x} = 4x^2 + x + 2, \ x \neq 0$

Excluded values are known as restrictions on the variable.

The solution to Example 1 could have been found by another method, since the distributive property also applies to division.

For example, $\dfrac{3}{7} = \dfrac{2+1}{7}$

$$= \dfrac{2}{7} + \dfrac{1}{7}$$

So, $\dfrac{24x^3 + 6x^2 + 12x}{6x} = \dfrac{24x^3}{6x} + \dfrac{6x^2}{6x} + \dfrac{12x}{6x}$

$$= 4x^2 + x + 2, \; x \neq 0$$

EXAMPLE 2 Binomial Denominator

Express $\dfrac{x}{2x^2 - 4x}$ in simplest form. State the restrictions on the variable.

SOLUTION

$$\dfrac{x}{2x^2 - 4x}$$

Factor the denominator: $= \dfrac{x}{2x(x-2)}$

Divide by the common factor, x: $= \dfrac{\overset{1}{\cancel{x}}}{2\cancel{x}(x-2)}$

$$= \dfrac{1}{2(x-2)}$$

Exclude values of x for which $2x^2 - 4x = 0$.
$2x^2 - 4x = 2x(x-2)$, so $2x^2 - 4x = 0$ when $2x(x-2) = 0$
$\qquad\qquad 2x = 0$ or $x - 2 = 0$
$\qquad\qquad\;\; x = 0$ or $x = 2$

Therefore, $\dfrac{x}{2x^2 - 4x} = \dfrac{1}{2(x-2)}$, $x \neq 0, 2$.

EXAMPLE 3 Removing a Common Factor of –1

Simplify $\dfrac{3-2x}{4x-6}$. State any restrictions on the variable.

SOLUTION

$$\dfrac{3-2x}{4x-6}$$

Factor the denominator:

$$= \dfrac{3-2x}{2(2x-3)}$$

Factor –1 from the numerator:

$$= \dfrac{-1(2x-3)}{2(2x-3)}$$

Divide by the common factor, $(2x-3)$:

$$= \dfrac{-1(2\cancel{x-3})^{1}}{2(2\cancel{x-3})_{1}}$$

$$= \dfrac{-1}{2} \text{ or } -\dfrac{1}{2}$$

Exclude values of x for which $4x - 6 = 0$.
$4x - 6 = 0$ when $2(2x - 3) = 0$.
$\quad 2x - 3 = 0$

$$x = \dfrac{3}{2}$$

Therefore, $\dfrac{3-2x}{4x-6} = -\dfrac{1}{2}$, $x \neq \dfrac{3}{2}$.

EXAMPLE 4 Trinomial Numerator and Denominator

Express $\dfrac{x^2 + 3x - 10}{x^2 + 8x + 15}$ in simplest form. State the restrictions on the variable.

SOLUTION

$$\dfrac{x^2 + 3x - 10}{x^2 + 8x + 15}$$

Factor the numerator and the denominator:

$$= \dfrac{(x+5)(x-2)}{(x+5)(x+3)}$$

Divide by the common factor, $(x+5)$:

$$= \dfrac{(\cancel{x+5})^{1}(x-2)}{(\cancel{x+5})_{1}(x+3)}$$

$$= \dfrac{x-2}{x+3}$$

Exclude values of x for which $x^2 + 8x + 15 = 0$.

$x^2 + 8x + 15 = (x + 5)(x + 3)$, so $x^2 + 8x + 15 = 0$ when $(x + 5)(x + 3) = 0$

$$x + 5 = 0 \text{ or } x + 3 = 0$$
$$x = -5 \quad \text{or} \quad x = -3$$

Therefore, $\dfrac{x^2 + 3x - 10}{x^2 + 8x + 15} = \dfrac{x - 2}{x + 3}$, $x \neq -5, -3$.

EXAMPLE 5 Trinomial Numerator and Denominator

Simplify $\dfrac{2y^2 - y - 15}{4y^2 - 13y + 3}$. State the restrictions on the variable.

SOLUTION

$$\dfrac{2y^2 - y - 15}{4y^2 - 13y + 3}$$

Factor the numerator and denominator: $\quad = \dfrac{(y - 3)(2y + 5)}{(4y - 1)(y - 3)}$

Divide by the common factor, $(y - 3)$: $\quad = \dfrac{\overset{1}{\cancel{(y - 3)}}(2y + 5)}{(4y - 1)\underset{1}{\cancel{(y - 3)}}}$

$$= \dfrac{2y + 5}{4y - 1}$$

Exclude values of y for which $4y^2 - 13y + 3 = 0$.

$4y^2 - 13y + 3 = (4y - 1)(y - 3)$, so $4y^2 - 13y + 3 = 0$ when $(4y - 1)(y - 3) = 0$

$$4y - 1 = 0 \text{ or } y - 3 = 0$$
$$y = \dfrac{1}{4} \quad \text{or} \quad y = 3$$

Therefore, $\dfrac{2y^2 - y - 15}{4y^2 - 13y + 3} = \dfrac{2y + 5}{4y - 1}$, $y \neq \dfrac{1}{4}, 3$.

Key Concepts

- To simplify rational expressions,
  a) factor the numerator and the denominator
  b) divide by common factors
- To state the restriction(s) on the variable in a rational expression, determine and exclude the value(s) of the variable that make the denominator 0.

Communicate Your Understanding

1. Explain why $x \neq 3$ is a restriction on the variable for the expression $\dfrac{x+4}{x-3}$.

2. Describe how you would simplify $\dfrac{x^2 - x}{x}$.

3. a) Describe how you would simplify $\dfrac{x^2 + 3x + 2}{x^2 - x - 2}$.

b) Describe how you would determine the restrictions on the variable.

4. Write an expression in one variable for the denominator of a rational expression, if the restrictions on the variable are $x \neq 2, -3$.

Practise

In each of the following, state any restrictions on the variables.

A

1. Simplify.

a) $\dfrac{3t^3 + 6t^2 - 15t}{3t}$

b) $\dfrac{6a^2 + 9a}{12a^2}$

c) $\dfrac{10y^4 + 5y^3 - 15y^2}{5y}$

d) $\dfrac{14n^4 - 4n^3 + 6n^2 + 8n}{2n^2}$

e) $\dfrac{4m^2 - 8mn}{4mn}$

f) $\dfrac{-6x^2 y^3}{-18x^3 y}$

g) $\dfrac{16a^2 bc}{4a^2 b^2 c^2}$

h) $\dfrac{-4x^4 y^2 z}{20x^3 y^3 z}$

i) $\dfrac{21m(m-4)}{7m^2}$

2. Express in simplest form.

a) $\dfrac{5x}{5(x+4)}$

b) $\dfrac{8t^2(t+5)}{4t(t-5)}$

c) $\dfrac{7x(x-3)}{14x^2(x-3)}$

d) $\dfrac{(m-1)(m+2)}{(m+4)(m-1)}$

e) $\dfrac{2x}{2x+8}$

f) $\dfrac{y^2}{y^2 + 2y}$

g) $\dfrac{10x}{5x^2 - 15x}$

h) $\dfrac{4x}{16x^3 - 12x}$

i) $\dfrac{3xy}{6x^2 y - 12xy^2}$

3. Simplify.

a) $\dfrac{6t - 36}{t - 6}$

b) $\dfrac{4m + 24}{8m - 24}$

c) $\dfrac{5x - 10}{3x - 6}$

d) $\dfrac{a^2 + 2a}{a^2 - 3a}$

e) $\dfrac{8x^2 + 4x}{6x^2 + 3x}$

f) $\dfrac{2x^2 - 2x}{2x^2 + 2x}$

g) $\dfrac{4x + 4y}{5x + 5y}$

h) $\dfrac{4a^2 b + 8ab}{6a^2 - 6a}$

i) $\dfrac{5xy + 10x}{2y^2 + 4y}$

4. Express in simplest equivalent form.

a) $\dfrac{m - 2}{m^2 - 5m + 6}$

b) $\dfrac{y^2 + 10y + 25}{y + 5}$

c) $\dfrac{2x + 6}{x^2 - 6x - 27}$

d) $\dfrac{r^2 - 4}{5r + 10}$

e) $\dfrac{a^2 + a}{a^2 + 2a + 1}$

f) $\dfrac{x^2 - 9}{2x^2 y - 6xy}$

g) $\dfrac{2w + 2}{2w^2 + 3w + 1}$

h) $\dfrac{3t^2 - 8t + 4}{6t^2 - 4t}$

i) $\dfrac{8z + 6z^2}{9z^2 - 16}$

j) $\dfrac{5x^2 + 3xy - 2y^2}{3x^2 + 3xy}$

5. Simplify.

a) $\dfrac{y - 2}{2 - y}$

b) $\dfrac{3 - x}{x - 3}$

c) $\dfrac{2t - 1}{4 - 8t}$

d) $\dfrac{6 - 10w}{15w - 9}$

e) $\dfrac{x^2 - 1}{1 - x^2}$

f) $\dfrac{1 - 4y^2}{8y^2 - 2}$

6. Simplify.

a) $\dfrac{x^2 + 4x + 4}{x^2 + 5x + 6}$

b) $\dfrac{a^2 - a - 12}{a^2 - 9a + 20}$

c) $\dfrac{m^2 - 5m + 6}{m^2 + 2m - 15}$

d) $\dfrac{y^2 - 8y + 15}{y^2 - 25}$

e) $\dfrac{x^2 - 10x + 24}{x^2 - 12x + 36}$

f) $\dfrac{n^2 - n - 2}{n^2 + n - 6}$

g) $\dfrac{p^2 + 8p + 16}{p^2 - 16}$

h) $\dfrac{2t^2 - t - 1}{t^2 - 3t + 2}$

i) $\dfrac{6v^2 + 11v + 3}{4v^2 + 8v + 3}$

j) $\dfrac{6x^2 - 13x + 6}{8x^2 - 6x - 9}$

k) $\dfrac{3z^2 - 7z + 2}{9z^2 - 6z + 1}$

l) $\dfrac{2m^2 - mn - n^2}{4m^2 - 4mn - 3n^2}$

Apply, Solve, Communicate

7. Saskatchewan flag The area of a Saskatchewan flag can be represented by the polynomial $x^2 + 3x + 2$ and its width by $x + 1$.

a) Write a rational expression that represents the length.

b) Write the expression in simplest form.

c) If x represents 1 unit of length, what is the ratio length:width for a Saskatchewan flag?

B

8. Simplify, if possible.

a) $\dfrac{1 - x}{x - 1}$

b) $\dfrac{x - 1}{x + 1}$

c) $\dfrac{y^2 + 1}{y^2 - 1}$

d) $\dfrac{3t - 7}{3t - 7}$

e) $\dfrac{t^2 - s^2}{(s + t)^2}$

f) $\dfrac{x^3 - 2x^2 + 3x}{2x^2 - 4x + 6}$

9. For which values of x are the following rational expressions not defined?

a) $\dfrac{2x - y}{x - y}$

b) $\dfrac{4x}{3x + y}$

c) $\dfrac{3}{x^3}$

d) $\dfrac{x^2}{x^3 - 8}$

e) $\dfrac{x^2 + 3x - 11}{x^2 - 1}$

f) $\dfrac{3x^2 + 5xy + 2y^2}{4x^2 - 9y^2}$

10. Communication State whether each of the following rational expressions is equivalent to the expression $\dfrac{x+1}{x-1}$. Explain.

a) $\dfrac{x+2}{x-2}$

b) $\dfrac{x^2+x}{x^2-x}$

c) $\dfrac{4+4x}{4x-4}$

d) $\dfrac{3x+1}{3x-1}$

e) $\dfrac{(x+1)^2}{(x-1)^2}$

f) $\dfrac{1+x}{1-x}$

11. Cube For a cube of edge length x, find the ratio of the volume to the surface area. Simplify, if possible.

12. Application For a sphere of radius r, find the ratio of the volume to the surface area. Simplify, if possible.

13. Pattern The first 4 diagrams of two patterns are shown.

Pattern 1

```
                    *
              *     *
         *    *     *
    *    *    *     *
    *    *    *     *
n = 1   2    3     4
```

Pattern 2

```
                              *****
                        ****  *****
                  ***   ****  *****
            **    ***   ****  *****
            **    ***   ****  *****
            **    ***   ****  *****
            **    ***   ****  *****
n = 1       2      3      4
```

a) For pattern 1, express the number of asterisks in the nth diagram in terms of n.

b) For pattern 2, the number of asterisks in the nth diagram is given by the binomial product $(n + \blacktriangle)(n + \blacksquare)$, where $\blacktriangle$ and $\blacksquare$ represent whole numbers. Replace $\blacktriangle$ and $\blacksquare$ in the binomial product by their correct values.

c) Divide your polynomial from part b) by your expression from part a).

d) Use your result from part c) to calculate how many times as many asterisks there are in the 10th diagram of pattern 2 as there are in the 10th diagram of pattern 1.

e) If a diagram in pattern 1 has 20 asterisks, how many asterisks are in the corresponding diagram of pattern 2?

f) If a diagram in pattern 2 has 1295 asterisks, how many asterisks are in the corresponding diagram in pattern 1?

14. Measurement Find the ratio of the area of the square to the area of the trapezoid. Simplify, if possible.

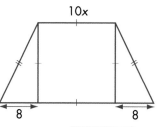

15. Rectangular prism Find the ratio of the volume to the surface area for the rectangular prism shown. Simplify, if possible.

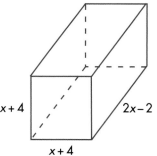

C

16. Write rational expressions in one variable so that the restrictions on the variables are as follows.

a) $x \neq 1$ **b)** $y \neq 0, -3$ **c)** $a \neq \dfrac{1}{2}, -\dfrac{3}{4}$ **d)** $t \neq -1, \pm\sqrt{3}$

17. Technology a) Use a graphing calculator to graph the equations $y = \dfrac{2x^2 + 3x}{x}$ and $y = 2x + 3$ in the same standard viewing window. Explain your observations.

b) Display the tables of values for the two equations. Compare and explain the values of y when $x = 0$.

18. Inquiry/Problem Solving a) For a solid cone with radius r, height h, and slant height s, find the ratio of the volume to the surface area. Simplify, if possible.

b) Determine whole-number values of r, h, and s that give the ratio in part a) a numerical value of 1.

WORD *Power*

Lewis Carroll invented a word game called doublets. The object of the game is to change one word to another by changing one letter at a time. You must form a real word each time you change a letter. The best solution has the fewest steps. Change the word RING to the word BELL by changing one letter at a time.

1.6 Multiplying and Dividing Rational Expressions

The game of badminton originated in England around 1870. Badminton is named after the Duke of Beaufort's home, Badminton House, where the game was first played. The International Badminton Federation now has over 50 member countries, including Canada.

INVESTIGATE & INQUIRE

In a doubles game of badminton, there are four service courts.

The width of each service court is half the width of the whole court. The length of each service court is one third of the distance between the long service lines. The width of the whole court and the distance between the long service lines can be modelled by the expressions shown.

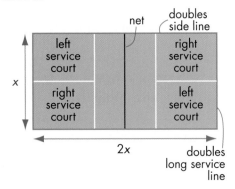

1. a) Write an expression that represents the width of each service court.

b) Write an expression that represents the length of each service court.

c) Write an expression that represents the area of each service court.

Leave your answer in the form $\dfrac{\blacksquare}{\bullet} \times \dfrac{\blacksquare}{\blacktriangle}$, where each numerator is a monomial and ● and ▲ represent whole numbers.

2. a) Use the dimensions x and $2x$ to write an expression that represents the whole area shown.
b) Simplify the expression.

3. a) What fraction of the whole area does each service court cover?
b) Write this fraction of the expression you wrote in question 2b). Do not simplify.

4. How is the expression you wrote in question 3b) related to your expression for the area of each service court in question 1c)? Explain.

5. A large rectangle of width $x - 1$ and length $x + 5$ is divided into 12 small rectangles, as shown.
a) Write an expression that represents the width of each small rectangle.
b) Write an expression that represents the length of each small rectangle.
c) Write an expression that represents the area of each small rectangle.

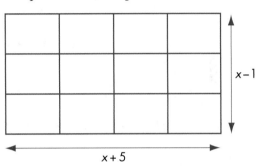

Leave your answer in the form $\dfrac{\blacksquare}{\bullet} \times \dfrac{\blacksquare}{\blacktriangle}$, where each numerator is a binomial, and $\bullet$ and $\blacktriangle$ represent whole numbers.

6. a) Use the dimensions $x - 1$ and $x + 5$ to write an expression that represents the area of the large rectangle.
b) Expand and simplify the expression.

7. a) What fraction of the large rectangle does each small rectangle cover?
b) Write this fraction of the expression you wrote in question 6b).

8. How are the expressions from questions 5c) and 7b) related? Explain.

9. Write a rule for multiplying rational expressions.

10. Multiply. Simplify the product, if possible.

a) $\dfrac{x}{3} \times \dfrac{y}{4}$

b) $\dfrac{3}{a} \times \dfrac{a^2}{b}$

c) $\dfrac{x^2}{2} \times \dfrac{4}{xy}$

d) $\dfrac{x+1}{3} \times \dfrac{x-1}{4}$

e) $\dfrac{3t}{2t+1} \times \dfrac{4}{t+2}$

f) $\dfrac{4x^2}{3y^3} \times \dfrac{9y^4}{8x^4}$

11. In the expressions that model the badminton court, x represents about 6 m. Find the area of each service court, in square metres.

Rational expressions can be multiplied in the same way that fractions are multiplied.

$$\frac{3}{4} \times \frac{5}{6}$$

Multiply the numerators:
Multiply the denominators:
$$= \frac{3 \times 5}{4 \times 6}$$

$$= \frac{15}{24}$$

Divide by the common factor:
$$= \frac{5}{8}$$

For rational expressions $\dfrac{P}{Q}$ and $\dfrac{R}{S}$, $\dfrac{P}{Q} \times \dfrac{R}{S} = \dfrac{PR}{QS}$, $Q, S \neq 0$.

EXAMPLE 1 Multiplying Rational Expressions

Simplify $\dfrac{3a^3}{2b^2} \times \dfrac{10b^3}{9a^2}$. State the restrictions on the variables.

SOLUTION

$$\frac{3a^3}{2b^2} \times \frac{10b^3}{9a^2}$$

Multiply the numerators:
Multiply the denominators:
$$= \frac{30a^3 b^3}{18a^2 b^2}$$

Divide by the common factors:
$$= \frac{5ab}{3}$$

Exclude values for which $2b^2 = 0$ or $9a^2 = 0$.
$$b = 0 \qquad a = 0$$

So, $a \neq 0$, $b \neq 0$.

Therefore, $\dfrac{3a^3}{2b^2} \times \dfrac{10b^3}{9a^2} = \dfrac{5ab}{3}$, $a \neq 0$, $b \neq 0$.

When multiplying some rational expressions, you may find it easier to factor the numerators and the denominators first.

EXAMPLE 2 Multiplying Rational Expressions Involving Polynomials

Simplify $\dfrac{x^2 + x - 6}{x^2 + 2x - 15} \times \dfrac{x - 3}{x - 2}$. State the restrictions on the variable.

SOLUTION

$$\dfrac{x^2 + x - 6}{x^2 + 2x - 15} \times \dfrac{x - 3}{x - 2}$$

Factor:

$$= \dfrac{(x + 3)(x - 2)}{(x + 5)(x - 3)} \times \dfrac{x - 3}{x - 2}$$

Multiply the numerators:
Multiply the denominators:

$$= \dfrac{(x + 3)(x - 2)(x - 3)}{(x + 5)(x - 3)(x - 2)}$$

Divide by the common factors:

$$= \dfrac{(x + 3)(\cancel{x - 2})(\cancel{x - 3})}{(x + 5)(\cancel{x - 3})(\cancel{x - 2})}$$

$$= \dfrac{x + 3}{x + 5}$$

Exclude values for which $(x + 5)(x - 3) = 0$ or $x - 2 = 0$.

$$x + 5 = 0 \quad \text{or} \quad x - 3 = 0 \quad \text{or} \quad x - 2 = 0$$
$$x = -5 \qquad\qquad x = 3 \qquad\qquad x = 2$$

Therefore, $\dfrac{x^2 + x - 6}{x^2 + 2x - 15} \times \dfrac{x - 3}{x - 2} = \dfrac{x + 3}{x + 5}$, $x \neq 2, 3, -5$.

Rational expressions can be divided in the same way fractions are divided.

$$\dfrac{2}{3} \div \dfrac{5}{7}$$

Multiply by the reciprocal:

$$= \dfrac{2}{3} \times \dfrac{7}{5}$$

$$= \dfrac{14}{15}$$

For rational expressions $\dfrac{P}{Q}$ and $\dfrac{R}{S}$, $\dfrac{P}{Q} \div \dfrac{R}{S} = \dfrac{P}{Q} \times \dfrac{S}{R} = \dfrac{PS}{QR}$, $Q, R, S \neq 0$. **Note the restrictions on Q, R, and S.**

EXAMPLE 3 Dividing Rational Expressions

Simplify $\dfrac{2ab}{5c} \div \dfrac{14a^2b^2}{15c^2}$. State the restrictions on the variables.

SOLUTION

$$\dfrac{2ab}{5c} \div \dfrac{14a^2b^2}{15c^2}$$

Multiply by the reciprocal: $= \dfrac{2ab}{5c} \times \dfrac{15c^2}{14a^2b^2}$

Multiply the numerators:
Multiply the denominators: $= \dfrac{30abc^2}{70a^2b^2c}$

Divide by the common factors: $= \dfrac{3c}{7ab}$

Exclude values for which $5c = 0$, $14a^2b^2 = 0$, or $15c^2 = 0$.
$5c = 0$ when $c = 0$.
$14a^2b^2 = 0$ when $a = 0$ or $b = 0$.
$15c^2 = 0$ when $c = 0$.
So, $a \neq 0$, $b \neq 0$, $c \neq 0$.

Therefore, $\dfrac{2ab}{5c} \div \dfrac{14a^2b^2}{15c^2} = \dfrac{3c}{7ab}$, $a \neq 0$, $b \neq 0$, $c \neq 0$.

When dividing some rational expressions, you may find it easier to factor the numerators and the denominators first.

EXAMPLE 4 Dividing Rational Expressions Involving Polynomials

Simplify $\dfrac{x^2 - x - 20}{x^2 - 6x} \div \dfrac{x^2 + 9x + 20}{x^2 - 12x + 36}$. State the restrictions on the variable.

SOLUTION

$$\dfrac{x^2 - x - 20}{x^2 - 6x} \div \dfrac{x^2 + 9x + 20}{x^2 - 12x + 36}$$

Factor: $= \dfrac{(x-5)(x+4)}{x(x-6)} \div \dfrac{(x+4)(x+5)}{(x-6)(x-6)}$

Multiply by the reciprocal:
$$= \frac{(x-5)(x+4)}{x(x-6)} \times \frac{(x-6)(x-6)}{(x+4)(x+5)}$$

Multiply the numerators:
Multiply the denominators:
$$= \frac{(x-5)(x+4)(x-6)(x-6)}{x(x-6)(x+4)(x+5)}$$

Divide by the common factors:
$$= \frac{(x-5)(\overset{1}{\cancel{x+4}})(\overset{1}{\cancel{x-6}})(x-6)}{x(\underset{1}{\cancel{x-6}})(\underset{1}{\cancel{x+4}})(x+5)}$$

$$= \frac{(x-5)(x-6)}{x(x+5)}$$

Exclude values for which $x(x-6) = 0$, $(x+4)(x+5) = 0$, or $(x-6)(x-6) = 0$.

$x = 0$ or $x - 6 = 0$ $x + 4 = 0$ or $x + 5 = 0$ $x - 6 = 0$

         $x = 6$ $x = -4$ $x = -5$ $x = 6$

Therefore, $\dfrac{x^2 - x - 20}{x^2 - 6x} \div \dfrac{x^2 + 9x + 20}{x^2 - 12x + 36} = \dfrac{(x-5)(x-6)}{x(x+5)}$, $x \neq 0, 6, -4, -5$.

Key Concepts

- For rational expressions $\dfrac{P}{Q}$ and $\dfrac{R}{S}$, $\dfrac{P}{Q} \times \dfrac{R}{S} = \dfrac{PR}{QS}$, $Q, S \neq 0$.

- To multiply rational expressions,
  a) factor any binomials and trinomials
  b) multiply the numerators and multiply the denominators
  c) divide by common factors
  d) determine and exclude the values of the variable that make the denominators 0

- For rational expressions $\dfrac{P}{Q}$ and $\dfrac{R}{S}$, $\dfrac{P}{Q} \div \dfrac{R}{S} = \dfrac{P}{Q} \times \dfrac{S}{R} = \dfrac{PS}{QR}$, $Q, R, S \neq 0$.

- To divide rational expressions,
  a) factor any binomials and trinomials
  b) multiply by the reciprocal of the divisor
  c) multiply the numerators and multiply the denominators
  d) divide by common factors
  e) determine and exclude the values of the variable that make the denominators 0

Communicate Your Understanding

1. Write two rational expressions whose quotient is $\dfrac{12a}{5b^2}$.

2. Describe how you would simplify $\dfrac{x^2 + x}{x^2 - 5x - 6} \times \dfrac{x^2 - 9}{x^2 + 2x + 1}$.

3. a) Describe how you would simplify $\dfrac{x+1}{x-2} \div \dfrac{x+1}{x-3}$.

   b) What are the restrictions on the variable?

Practise

In each of the following, state any restrictions on the variables.

A

1. Simplify.

a) $\dfrac{y^2}{3} \times \dfrac{8}{y}$

b) $\dfrac{7}{2x^3} \times \dfrac{-x^4}{14}$

c) $\dfrac{-5n^2}{12} \times \dfrac{4}{-15n^5}$

d) $\dfrac{-4m}{9} \times 6$

2. Simplify.

a) $\dfrac{3}{x} \div \dfrac{12}{x^2}$

b) $\dfrac{y^3}{6} \div \dfrac{y^2}{-3}$

c) $\dfrac{-15}{2m^2} \div \dfrac{10}{3m^4}$

d) $\dfrac{-8t^4}{3} \div \dfrac{-6t^2}{5}$

e) $\dfrac{20}{3x^5} \div \dfrac{-15}{8x^2}$

f) $\dfrac{4r^3}{-3} \div 2r^4$

3. Simplify.

a) $\dfrac{3x^3}{2y} \times \dfrac{8y^2}{9x}$

b) $\dfrac{8m^3}{3n^2} \div \dfrac{5m^2}{6n}$

c) $\dfrac{21xy}{4t^2} \times \dfrac{12}{7x^2y}$

d) $\dfrac{-4a}{7b^3} \div \dfrac{-8a^4}{7}$

e) $\dfrac{12m}{-5t} \div \dfrac{8m^2}{-15}$

f) $\dfrac{15a^2b}{4c} \div \dfrac{8abc}{-3}$

4. Simplify.

a) $\dfrac{16ab}{9x^4y^2} \times \dfrac{3x^5y^4}{8a^2b^2}$

b) $\dfrac{6x^2y}{5mn^3} \div \dfrac{9xy}{10mn^4}$

c) $\dfrac{5xy}{6x^2y} \div \dfrac{10xy^2}{9x^3y^2}$

d) $-12a^2b \times \dfrac{4ab^2}{-3ab^3}$

e) $6x^3y^4 \div \dfrac{2xy}{-3}$

f) $\dfrac{4a^2b^2c}{-3ab} \div 6c^2$

5. Simplify.

a) $\dfrac{3}{x-4} \times \dfrac{x-4}{6}$

b) $\dfrac{m+2}{5} \div \dfrac{y+1}{10}$

c) $\dfrac{5(y-2)}{y+1} \times \dfrac{y+1}{10}$

d) $\dfrac{2(x+1)}{x-2} \div \dfrac{x+1}{x-2}$

e) $\dfrac{4a^2b}{3(a+b)} \div \dfrac{-8ab^2}{a+b}$

f) $\dfrac{3(m+4)}{5m} \times \dfrac{6m^3}{2(m+4)}$

6. Simplify.

a) $\dfrac{4x+4}{3x-3} \times \dfrac{6x-6}{5x+5}$

b) $\dfrac{6m^3}{m+3} \times \dfrac{5m+15}{8m^3}$

c) $\dfrac{3a+6}{9a^2} \div \dfrac{a+2}{-3a}$

d) $\dfrac{x^2-4}{x+3} \div \dfrac{4x-8}{3x+9}$

e) $\dfrac{7y^2}{y^2-9} \times \dfrac{4y+12}{14y^3}$

f) $\dfrac{m^2-25}{m^2-16} \div \dfrac{2m-10}{4m+16}$

g) $\dfrac{4x-6}{8x^2y} \times \dfrac{4xy}{6x-9}$ **h)** $\dfrac{2x^2-8}{6x+3} \div \dfrac{6x-12}{18x+9}$ **f)** $\dfrac{12w^2-5w-2}{8w^2+2w-21} \div \dfrac{12w^2+w-6}{8w^2-2w-15}$

7. Simplify.

a) $\dfrac{x^2+5x+6}{x^2-6x+5} \times \dfrac{x^2+x-30}{x^2+9x+18}$

b) $\dfrac{a^2+7a+12}{a^2+4a+4} \times \dfrac{a^2-a-6}{a^2-9}$

c) $\dfrac{m^2-3m-4}{m^2+5m} \div \dfrac{m^2-7m+12}{m^2+2m-15}$

d) $\dfrac{12a^2-19a+5}{4a^2-9} \times \dfrac{2a-3}{3a-1}$

e) $\dfrac{2x^2-5x-3}{2x^2-11x+15} \times \dfrac{4x^2-8x-5}{4x^2+4x+1}$

8. Simplify.

a) $\dfrac{x^2-xy-20y^2}{x^2-8xy+15y^2} \div \dfrac{x^2+2xy-8y^2}{x^2-xy-6y^2}$

b) $\dfrac{x^2+3xy}{x^2-xy-42y^2} \times \dfrac{x^2-10xy+21y^2}{x^2-9y^2}$

c) $\dfrac{a^2+15ab+56b^2}{a^2-3ab-54b^2} \div \dfrac{a^2+6ab-16b^2}{a^2+4ab-12b^2}$

d) $\dfrac{9s^2+30st+25t^2}{25s^2-25st-6t^2} \times \dfrac{20s^2-49st+30t^2}{12s^2+5st-25t^2}$

Apply, Solve, Communicate

9. Communication On a soccer pitch, the goal area or goal box is inside the penalty area or penalty box and forms part of it. The dimensions of the goal box and the penalty box can be represented as shown.

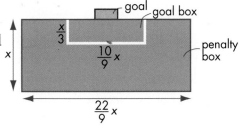

a) Write an expression that represents the area of the goal box.

b) Write an expression that represents the area of the penalty box.

c) Determine how many times as great the area of the penalty box is as the area of the goal box.

d) Does the fact that x represents 16.5 m affect your answer to part c)? Explain.

B

10. Measurement Write the area of the rectangle in simplest form.

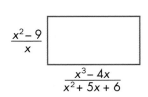

11. Measurement The area of the trapezoid is $6y^2-5y-6$. What is its height?

$$S = \tfrac{1}{2}(a+b) \times H$$

12. Measurement a) Write, but do not simplify, an expression for the area of $\triangle$ABC.
b) Write, but do not simplify, an expression for the area of $\triangle$DEF.
c) Write and simplify an expression that represents the ratio of the area of $\triangle$ABC to the area of $\triangle$DEF.

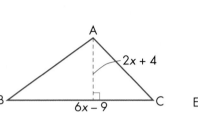
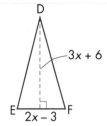

13. Application Write and simplify an expression that represents the fraction of the area of the large rectangle covered by the shaded rectangle.

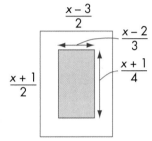

14. In divisions of the form $\dfrac{a}{b} \div \dfrac{c}{d}$, the expressions b, c, and d must all be examined for possible restrictions on the variables. Explain why.

15. Measurement Two rectangles have common sides with a right triangle, as shown. The areas and widths of the rectangles are as indicated. Write and simplify an expression for the area of the triangle.

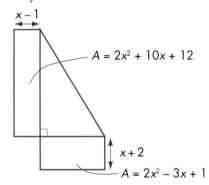

C

16. As the value of y increases, what happens to the value of each of the following expressions? Explain.

a) $\dfrac{15y^2 - 2y - 1}{6y^2 + 7y - 3} \times \dfrac{2y - 1}{10y^2 - 3y - 1}$ **b)** $\dfrac{8y^2 + 2y - 1}{6y^2 - y - 2} \div \dfrac{8y - 2}{9y - 6}$

17. Inquiry/Problem Solving Write two different pairs of rational expressions with a product of $\dfrac{3x^2 + 7xy + 2y^2}{x^2 - y^2}$.

18. Write four different pairs of rational expressions with a product of $\dfrac{4x^2 - 8x + 4}{2x^2 + 5x - 3}$. Compare your expressions with a classmate's.

1.7 Adding and Subtracting Rational Expressions, I

The blimp that provided overhead television coverage of the first World Series played in Canada was based in Miami. The blimp flew 1610 km from Miami to Washington, D.C., and then 634 km to Toronto.

The time taken to fly from Miami to Washington was $\dfrac{1610}{s}$ hours, where s was the average speed in kilometres per hour.

The time taken to fly from Washington to Toronto was $\dfrac{634}{s}$ hours.

The total flying time from Miami to Toronto was $\dfrac{1610}{s} + \dfrac{634}{s}$ hours.

The expression $\dfrac{1610}{s} + \dfrac{634}{s}$ is the sum of two rational expressions with the same denominator.

INVESTIGATE & INQUIRE

Rectangle A and Rectangle B have different areas but the same width.

Rectangle C is formed by placing Rectangles A and B end to end.

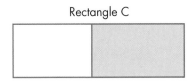

Rectangle A — $x + 2$ — Area $x^2 + 3x + 1$

Rectangle B — $x + 2$ — Area $x^2 + 4x + 2$

Rectangle C

1. a) Using the areas of rectangles A and B, write and simplify an expression that represents the area of rectangle C.
b) What is the width of rectangle C?

2. Write, but do not divide, a rational expression that represents the length of
a) rectangle A
b) rectangle B

3. Using the area and the width of rectangle C, write, but do not divide, a rational expression that represents the length of rectangle C.

4. How does the expression you wrote in question 3 compare with the two expressions you wrote in question 2? Explain.

5. Write a rule for adding two rational expressions with the same denominator.

6. Add.

a) $\dfrac{x}{3} + \dfrac{4x}{3}$

b) $\dfrac{5}{3t} + \dfrac{2}{3t}$

c) $\dfrac{n+1}{n+3} + \dfrac{n-1}{n+3}$

d) $\dfrac{x^2+1}{x^2} + \dfrac{2x^2+1}{x^2}$

7. The flying time of the blimp from Miami to Toronto was $\dfrac{1610}{s} + \dfrac{634}{s}$ hours.

a) Add the rational expressions.
b) If the average speed, s, of the blimp was 85 km/h, what was the total flying time, in hours?

Rational expressions with a common denominator can be added or subtracted in the same way as fractions with a common denominator.

$$\frac{5}{7} + \frac{1}{7} - \frac{2}{7}$$

Write with the common denominator: $\quad = \dfrac{5+1-2}{7}$

Add or subtract the numerators: $\quad = \dfrac{4}{7}$

Example 1 Adding and Subtracting With Common Denominators

Simplify each of the following. State the restriction on the variable.

a) $\dfrac{3}{x^2} + \dfrac{5}{x^2} - \dfrac{2}{x^2}$

b) $\dfrac{4x-1}{x+2} - \dfrac{x+3}{x+2}$

SOLUTION

a)

$$\frac{3}{x^2} + \frac{5}{x^2} - \frac{2}{x^2}$$

Write with the common denominator:

$$= \frac{3 + 5 - 2}{x^2}$$

Add or subtract the numerators:

$$= \frac{6}{x^2}$$

Exclude values for which $x^2 = 0$.

$$x = 0$$

Therefore, $\dfrac{3}{x^2} + \dfrac{5}{x^2} - \dfrac{2}{x^2} = \dfrac{6}{x^2}$, $x \neq 0$.

b)

$$\frac{4x - 1}{x + 2} - \frac{x + 3}{x + 2}$$

Write with the common denominator:

$$= \frac{(4x - 1) - (x + 3)}{x + 2}$$

Subtract the numerators:

$$= \frac{4x - 1 - x - 3}{x + 2}$$

Simplify:

$$= \frac{3x - 4}{x + 2}$$

Exclude values for which $x + 2 = 0$.

$$x = -2$$

Therefore, $\dfrac{4x - 1}{x + 2} - \dfrac{x + 3}{x + 2} = \dfrac{3x - 4}{x + 2}$, $x \neq -2$.

Rational expressions with different denominators can be added or subtracted in the same way as fractions with different denominators.

$$\frac{1}{6} + \frac{3}{4}$$

Rewrite with a common denominator:

$$= \frac{2}{12} + \frac{9}{12}$$

Add the numerators:

$$= \frac{11}{12}$$

$$\frac{3}{5} - \frac{1}{2}$$

Rewrite with a common denominator:

$$= \frac{6}{10} - \frac{5}{10}$$

Subtract the numerators:

$$= \frac{1}{10}$$

Note that the least common denominator (LCD) is normally used but is not necessary. If a greater common denominator is used, the result will reduce to give the same answer.

$$\frac{1}{6} + \frac{3}{4} = \frac{4}{24} + \frac{18}{24}$$
$$= \frac{22}{24}$$
$$= \frac{11}{12}$$

EXAMPLE 2 Adding and Subtracting With Whole-Number Denominators

Simplify $\dfrac{3x+2}{4} + \dfrac{x-4}{8} - \dfrac{2x-1}{6}$.

SOLUTION

To find the LCD, find the least common multiple (LCM) of the denominators 4, 8, and 6. The LCM can be found by factoring. It must contain all the separate factors of 4, 8, and 6.

$4 = 2 \times 2$
$8 = 2 \times 2 \times 2$
$6 = 2 \times 3$
So, the LCD is 24.

The LCM is $2 \times 2 \times 2 \times 3 = 24$

$$\frac{3x+2}{4} + \frac{x-4}{8} - \frac{2x-1}{6}$$

Rewrite with a common denominator: $= \dfrac{6(3x+2)}{6(4)} + \dfrac{3(x-4)}{3(8)} - \dfrac{4(2x-1)}{4(6)}$

$$= \frac{6(3x+2)}{24} + \frac{3(x-4)}{24} - \frac{4(2x-1)}{24}$$

Add or subtract the numerators: $= \dfrac{6(3x+2) + 3(x-4) - 4(2x-1)}{24}$

Expand the numerator: $= \dfrac{18x + 12 + 3x - 12 - 8x + 4}{24}$

Simplify: $= \dfrac{13x+4}{24}$

Therefore, $\dfrac{3x+2}{4} + \dfrac{x-4}{8} - \dfrac{2x-1}{6} = \dfrac{13x+4}{24}$.

Sometimes it is necessary to factor -1 from one of the denominators to recognize the common denominator.

EXAMPLE 3 Factoring −1 From a Denominator

Simplify $\dfrac{5}{x-3} + \dfrac{2}{3-x}$. State the restriction on the variable.

SOLUTION

Factor -1 from the denominator $3 - x$.

$$3 - x = -1(-3 + x)$$
$$= -(x - 3)$$

Rewrite $\dfrac{2}{3-x}$ so that there is a common denominator.

$$\frac{5}{x-3} + \frac{2}{3-x} = \frac{5}{x-3} + \frac{2}{-(x-3)}$$

$$= \frac{5}{x-3} - \frac{2}{x-3}$$

$$= \frac{5-2}{x-3}$$

$$= \frac{3}{x-3}$$

Exclude values for which $x - 3 = 0$ or $3 - x = 0$.

$$x = 3 \qquad 3 = x$$

Therefore, $\dfrac{5}{x-3} + \dfrac{2}{3-x} = \dfrac{3}{x-3}$, $x \neq 3$.

Key Concepts

- To add or subtract rational expressions with a common denominator, write the numerators over the common denominator, and add or subtract the numerators.
- To add or subtract rational expressions with different denominators, rewrite the expressions with a common denominator. Then, write the numerators over the common denominator, and add or subtract the numerators.

Communicate Your Understanding

1. a) Describe how you would simplify $\dfrac{5x}{x+4} - \dfrac{2x}{x+4}$.

   b) What is the restriction on the variable?

2. Describe how you would simplify $\dfrac{x+4}{2} + \dfrac{x-3}{6} - \dfrac{x+5}{4}$.

Practise

In each of the following, state any restrictions on the variables.

A

1. Simplify.

a) $\dfrac{2}{y} + \dfrac{4}{y} - \dfrac{5}{y}$

b) $\dfrac{5}{x^2} - \dfrac{3}{x^2} + \dfrac{6}{x^2}$

c) $\dfrac{4}{x+3} + \dfrac{5}{x+3}$

d) $\dfrac{x}{x-2} - \dfrac{y}{x-2}$

2. Simplify.

a) $\dfrac{x+7}{2} + \dfrac{x+4}{2}$

b) $\dfrac{2y-1}{3} + \dfrac{3y-6}{3}$

c) $\dfrac{3a-1}{a} - \dfrac{4a+2}{a}$

d) $\dfrac{5x-y}{3x} - \dfrac{4x+y}{3x}$

e) $\dfrac{x^2+4}{x+1} + \dfrac{2x^2}{x+1}$

f) $\dfrac{6t-8}{7} + \dfrac{3-5t}{7}$

g) $\dfrac{5z}{2z-1} - \dfrac{z-3}{2z-1}$

h) $\dfrac{2x+3}{x^2-1} + \dfrac{3x-4}{x^2-1}$

i) $\dfrac{4x+1}{x^2+5x+6} + \dfrac{3x+2}{x^2+5x+6}$

j) $\dfrac{1-2y}{2x^2+3x+1} - \dfrac{5y+3}{2x^2+3x+1}$

3. Find the LCM.

a) 4, 5, 6

b) 4, 9, 12

c) 8, 10, 12

d) 20, 15, 10

4. Simplify.

a) $\dfrac{2x}{2} + \dfrac{x}{3}$

b) $\dfrac{3a}{4} + \dfrac{a}{2} - \dfrac{2a}{6}$

c) $\dfrac{x}{5} - \dfrac{y}{2} + \dfrac{7}{10}$

d) $\dfrac{3m}{8} - \dfrac{m}{6} - \dfrac{2m}{3}$

5. Simplify.

a) $\dfrac{2m+3}{2} + \dfrac{3m+4}{7}$

b) $\dfrac{4x-3}{4} + \dfrac{x+2}{3}$

c) $\dfrac{y-5}{6} - \dfrac{2y-3}{4}$

d) $\dfrac{2x+3y}{5} - \dfrac{4x-y}{2}$

e) $\dfrac{4t-1}{6} + \dfrac{3t+2}{2} - \dfrac{2t+1}{3}$

f) $\dfrac{3a-b}{9} - \dfrac{a-2b}{3} - \dfrac{4a-3b}{6}$

g) $\dfrac{5x-1}{5} + 1 - \dfrac{4x-3}{6}$

6. Simplify.

a) $\dfrac{3}{2-x} + \dfrac{-2}{-(x-2)}$

b) $\dfrac{1}{x-1} - \dfrac{1}{1-x}$

c) $\dfrac{a-2}{2a-3} + \dfrac{a+3}{3-2a}$

d) $\dfrac{2y+3}{3-4y} + \dfrac{4+y}{4y-3}$

e) $\dfrac{5x-5}{x^2-9} - \dfrac{3x+1}{9-x^2}$

f) $\dfrac{2x^2+3x+1}{4x^2-9} - \dfrac{x^2-3x-1}{9-4x^2}$

Apply, Solve, Communicate

7. Flying times a) Write an expression that represents the time, in hours, it takes a plane to fly 1191 km from Winnipeg to Calgary at an average speed of s kilometres per hour.
b) Write an expression that represents the time, in hours, it takes a plane to fly 685 km from Calgary to Vancouver at the same speed as in part a).
c) Write and simplify an expression that represents the total flying time for a trip from Winnipeg to Vancouver via Calgary.
d) If the average speed of the plane is 700 km/h, what is the total flying time, in hours, for the trip in part c)?

B

8. Application A backgammon game board consists of two rectangles of the same size, known as tables, separated by a divider, called the bar.
a) The area of each table on a backgammon board can be modelled by the expression $x^2 + 8x$, and the width of each table by x. Write and simplify an expression that represents the width, w, of the whole board in terms of x.
b) If the width of the bar is $\dfrac{x}{5}$, write and simplify an expression that represents the length of the whole board in terms of x.
c) If x represents 15 cm, what are the dimensions of each table? of the whole board?

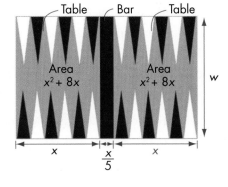

9. Application Two triangles have the same base length, represented by x. The height of one triangle is $x + 1$. The height of the other triangle is $x + 3$. Write and simplify an expression that represents the total area of the two triangles.

10. Communication Rectangle A and rectangle B each have a length of $2x + 1$. Rectangle A has an area of $6x^2 + 5x + 1$, and rectangle B has an area of $4x^2 - 4x - 3$.
a) Write but do not simplify an expression for the width of rectangle A.
b) Write but do not simplify an expression for the width of rectangle B.
c) Subtract the width of rectangle A from the width of rectangle B. Simplify the resulting expression.
d) Subtract the width of rectangle B from the width of rectangle A. Simplify the resulting expression.
e) How do the results of parts c) and d) compare? Explain.

11. Modelling problems algebraically The diameter of the smaller circle is d. The diameter of the larger circle is $d + 1$.
a) Write an expression that represents the area of the smaller circle in terms of d.
b) Write an expression that represents the area of the larger circle in terms of d.
c) Write and simplify an expression that represents the area of the shaded part of the diagram in terms of d.
d) If d represents 10 cm, find the area of the shaded part of the diagram, to the nearest tenth of a square centimetre.

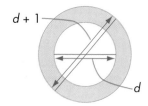

12. Measurement The diagram shows trapezoid ABCD divided into rhombus ABCE and isosceles triangle ADE.
a) Write an expression that represents the area of the triangle in terms of x.
b) Write an expression that represents the area of the rhombus in terms of x.
c) Add and simplify the expressions you wrote in parts a) and b).
d) Write and simplify an expression that represents the longer base of the trapezoid in terms of x.
e) Use the formula for the area of a trapezoid to write and simplify an expression that represents the area of the trapezoid in terms of x.
f) Compare your expressions from parts c) and e).

C

13. Pattern Triangular numbers of objects can be arranged to form triangles. The first four triangular numbers are as shown.

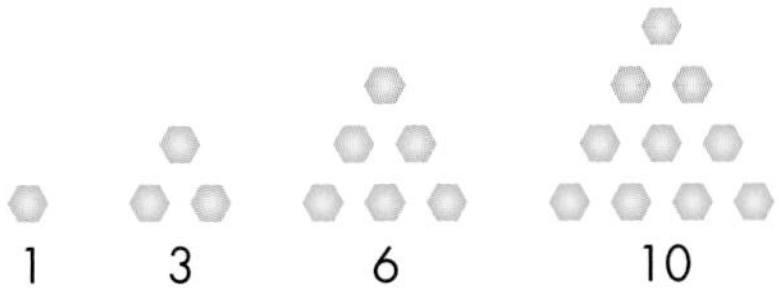

1   3   6   10

a) An expression for finding the nth triangular number can be written in the form $\dfrac{n(n + \blacktriangle)}{\blacksquare}$, where $\blacktriangle$ and $\blacksquare$ represent whole numbers. Copy and complete the expression by finding the numbers represented by $\blacktriangle$ and $\blacksquare$.

b) Write the 5th, 6th, 7th, 8th, and 9th triangular numbers.

c) Add any two consecutive triangular numbers. What kind of number results?

d) Write an expression that represents the $(n + 1)$th triangular number.

e) Add your expressions from parts a) and d). Simplify the result and express it in factored form.

f) How does your result from part e) explain your result from part c)?

ACHIEVEMENT Check   Knowledge/Understanding   Thinking/Inquiry/Problem Solving   Communication   Application

Your company makes fridge magnets. The materials for each magnet cost $0.14. Your company has additional expenses of $27 000 a year. The per-magnet cost is

$\dfrac{\text{total costs per year}}{\text{number produced per year}}$. If your company can make and sell twice as many

magnets next year as this year, the per-magnet cost will be reduced by $0.90. How many magnets is your company making and selling this year?

1.8 Adding and Subtracting Rational Expressions, II

The three-toed sloth of South America moves very slowly. It can travel twice as fast in a tree as it can on the ground. If its speed on the ground is s metres per minute, its speed in a tree is $2s$ metres per minute. The sloth can travel 15 m on the ground in $\dfrac{15}{s}$ minutes and 15 m in a tree in $\dfrac{15}{2s}$ minutes. The total time it takes to travel 15 m on the ground and 15 m in a tree is $\dfrac{15}{s} + \dfrac{15}{2s}$ minutes. The expression $\dfrac{15}{s} + \dfrac{15}{2s}$ is the sum of the two rational expressions with different denominators. Adding these rational expressions involves finding the LCM of two monomials that include variables.

INVESTIGATE & INQUIRE

1. To find the LCM of each group of monomials, copy and complete the table.

	Monomials	Factored Form	LCM
a)	$2a^2b$ $6b^2$	$2 \times a \times a \times b$ $2 \times 3 \times b \times b$	
b)	$10x^3$ $15x^2y^2$		
c)	$3xy$ $6yz$ $9xz$		
d)	$5x^3$ $8x^2y$ $10xy^2$		

2. Describe a method for mentally finding the LCM of monomials that include variables.

3. a) Factor the binomials $2x + 4$ and $3x + 6$.
b) Write the LCM of the binomials in factored form.

4. Write the LCM of each pair of expressions in factored form.

a) $3a - 9, 4a - 12$ **b)** $x^2 + x, 2x^2 + 2x$

c) $2y - 6, y^2 - 9$ **d)** $x^2 + 5x + 6, x^2 + x - 2$

5. The total time a three-toed sloth takes to travel 15 m on the ground and 15 m in a tree is $\dfrac{15}{s} + \dfrac{15}{2s}$ minutes.

a) State the common denominator of the two rational expressions.

b) Add the expressions.

c) If s represents 2.5 m/min, what is the total time, in minutes, the sloth takes to travel 15 m on the ground and 15 m in a tree?

EXAMPLE 1 Adding and Subtracting With Monomial Denominators

Simplify $\dfrac{4}{5a} - \dfrac{3}{2a^2} + \dfrac{1}{a^3}$. State the restriction on the variable.

SOLUTION

Find the LCD.

$5a = 5 \times a$

$2a^2 = 2 \times a \times a$

$a^3 = a \times a \times a$

The LCD is $5 \times 2 \times a \times a \times a$ or $10a^3$.

$$\frac{4}{5a} - \frac{3}{2a^2} + \frac{1}{a^3}$$

Write with the common denominator: $= \dfrac{2a^2(4)}{2a^2(5a)} - \dfrac{5a(3)}{5a(2a^2)} + \dfrac{10(1)}{10(a^3)}$

$$= \frac{8a^2}{10a^3} - \frac{15a}{10a^3} + \frac{10}{10a^3}$$

Add or subtract the numerators: $\quad = \dfrac{8a^2 - 15a + 10}{10a^3}$

Exclude the values for which $5a = 0$ or $2a^2 = 0$, or $a^3 = 0$.

$$a = 0 \qquad a = 0 \qquad a = 0$$

So, $a \neq 0$.

Therefore, $\dfrac{4}{5a} - \dfrac{3}{2a^2} + \dfrac{1}{a^3} = \dfrac{8a^2 - 15a + 10}{10a^3}$, $a \neq 0$.

EXAMPLE 2 Denominators With a Common Binomial Factor

Simplify $\dfrac{m}{2m-4} - \dfrac{3}{3m-6} + 1$. State the restriction on the variable.

SOLUTION

$2m - 4 = 2(m - 2)$
$3m - 6 = 3(m - 2)$
The LCD is $2 \times 3 \times (m - 2)$ or $6(m - 2)$.

$$\dfrac{m}{2m-4} - \dfrac{3}{3m-6} + 1$$

$$= \dfrac{m}{2(m-2)} - \dfrac{3}{3(m-2)} + \dfrac{1}{1}$$

Write with the common denominator: $= \dfrac{3(m)}{3 \times 2(m-2)} - \dfrac{2(3)}{2 \times 3(m-2)} + \dfrac{6(m-2)(1)}{6(m-2)(1)}$

$$= \dfrac{3m}{6(m-2)} - \dfrac{6}{6(m-2)} + \dfrac{6(m-2)}{6(m-2)}$$

Add or subtract the numerators: $= \dfrac{3m - 6 + 6(m-2)}{6(m-2)}$

Expand the numerator: $= \dfrac{3m - 6 + 6m - 12}{6(m-2)}$

Simplify: $= \dfrac{9m - 18}{6(m-2)}$

Factor: $= \dfrac{3 \times 3 \times (m-2)}{3 \times 2 \times (m-2)}$

The numerator and denominator have common factors, so simplify further.

Divide by the common factors: $= \dfrac{\overset{1}{3} \times 3 \times (\overset{1}{m-2})}{\underset{1}{3} \times 2 \times (\underset{1}{m-2})}$

$$= \dfrac{3}{2}$$

Exclude the values for which $2m - 4 = 0$ or $3m - 6 = 0$.
$$m = 2 \qquad\qquad m = 2$$

So, $m \neq 2$.

Therefore, $\dfrac{m}{2m-4} - \dfrac{3}{3m-6} + 1 = \dfrac{3}{2}$, $m \neq 2$.

EXAMPLE 3 Denominators With Different Binomial Factors

Simplify $\dfrac{x}{6x+6} + \dfrac{5}{4x-12}$. State the restrictions on the variable.

SOLUTION

$6x + 6 = 6(x + 1)$
$4x - 12 = 4(x - 3)$
The LCD is $12(x + 1)(x - 3)$.

$$\dfrac{x}{6x+6} + \dfrac{5}{4x-12}$$

$$= \dfrac{x}{6(x+1)} + \dfrac{5}{4(x-3)}$$

Write with the common denominator: $\ = \dfrac{2(x-3)}{2(x-3)} \times \dfrac{x}{6(x+1)} + \dfrac{3(x+1)}{3(x+1)} \times \dfrac{5}{4(x-3)}$

$$= \dfrac{2x(x-3)}{12(x+1)(x-3)} + \dfrac{15(x+1)}{12(x+1)(x-3)}$$

Add the numerators: $\ = \dfrac{2x(x-3) + 15(x+1)}{12(x+1)(x-3)}$

Expand the numerator: $\ = \dfrac{2x^2 - 6x + 15x + 15}{12(x+1)(x-3)}$

Simplify: $\ = \dfrac{2x^2 + 9x + 15}{12(x+1)(x-3)}$

Exclude the values for which $6x + 6 = 0$ or $4x - 12 = 0$.
$$x = -1 \qquad\qquad x = 3$$

So, $x \neq -1, 3$.

Therefore, $\dfrac{x}{6x+6} + \dfrac{5}{4x-12} = \dfrac{2x^2 + 9x + 15}{12(x+1)(x-3)}, \ x \neq -1, 3$.

EXAMPLE 4 Trinomial Denominators

Simplify $\dfrac{4}{y^2 + 5y + 6} - \dfrac{5}{y^2 - y - 12}$. State the restrictions on the variable.

SOLUTION

$y^2 + 5y + 6 = (y + 2)(y + 3)$
$y^2 - y - 12 = (y + 3)(y - 4)$
The LCD is $(y + 2)(y + 3)(y - 4)$.

$$\frac{4}{y^2+5y+6} - \frac{5}{y^2-y-12}$$

$$= \frac{4}{(y+2)(y+3)} - \frac{5}{(y+3)(y-4)}$$

Write with the common denominator: $= \dfrac{y-4}{y-4} \times \dfrac{4}{(y+2)(y+3)} - \dfrac{y+2}{y+2} \times \dfrac{5}{(y+3)(y-4)}$

$$= \frac{4(y-4)}{(y+2)(y+3)(y-4)} - \frac{5(y+2)}{(y+2)(y+3)(y-4)}$$

Subtract the numerators: $= \dfrac{4(y-4)-5(y+2)}{(y+2)(y+3)(y-4)}$

Expand the numerator: $= \dfrac{4y-16-5y-10}{(y+2)(y+3)(y-4)}$

Simplify: $= \dfrac{-y-26}{(y+2)(y+3)(y-4)}$

Exclude the values for which $(y+2)(y+3)=0$ or $(y+3)(y-4)=0$.

$y+2=0$ or $y+3=0$ $y+3=0$ or $y-4=0$

$\qquad y=-2 \qquad y=-3 \qquad\qquad y=-3 \qquad y=4$

So, $y \neq -2, -3, 4$.

Therefore, $\dfrac{4}{y^2+5y+6} - \dfrac{5}{y^2-y-12} = \dfrac{-y-26}{(y+2)(y+3)(y-4)}, \ y \neq -2, -3, 4$.

Key Concepts

• To add or subtract rational expressions with a common polynomial denominator, write the numerators over the common denominator, and add or subtract the numerators.
• To add or subtract rational expressions with different polynomial denominators, rewrite the expressions with a common denominator. Then, write the numerators over the common denominator, and add or subtract the numerators.

Communicate Your Understanding

1. a) Describe how you would simplify $\dfrac{5}{x^3} + \dfrac{3}{2x^2} - \dfrac{7}{3x}$.

b) What is the restriction on the variable?

2. a) Describe how you would simplify $\dfrac{5}{x^2-4} + \dfrac{2}{x^2-x-2}$.

b) What are the restrictions on the variable?

Practise

In each of the following, state any restrictions on the variables.

A

1. Write an equivalent expression with a denominator of $12x^2y^2$.

a) $\dfrac{2}{xy}$

b) $\dfrac{x}{y}$

c) $\dfrac{5}{3xy^2}$

d) $\dfrac{-y}{6x^2}$

2. Find the LCM.

a) $10a^2b,\ 4ab^3$

b) $3m^2n,\ 2mn^2,\ 6mn$

c) $2x^3,\ 6xy^2,\ 4y$

d) $10s^2t^2,\ 20s^2t,\ 15st^2$

3. Simplify.

a) $\dfrac{3}{2x} + \dfrac{4}{5x}$

b) $\dfrac{2}{4y} + \dfrac{3}{3y} - \dfrac{1}{2y}$

c) $\dfrac{1}{2x^2} + \dfrac{3}{3x} - \dfrac{2}{x^3}$

d) $\dfrac{3}{2m^2n} - \dfrac{1}{m^2n^3} + \dfrac{4}{5mn}$

e) $x - \dfrac{2}{x} + 5$

f) $\dfrac{3m+4}{mn} - \dfrac{1}{m} - 2$

g) $\dfrac{4x-1}{3x^2} - \dfrac{2x+3}{x} + \dfrac{5x+2}{5x^2}$

h) $\dfrac{x-2y}{x} - \dfrac{4x+y}{xy} - \dfrac{3x-4y}{y}$

4. Find the LCM of each of the following. Leave answers in factored form.

a) $3m+6,\ 2m+4$

b) $3y-3,\ 5y+10$

c) $4m-8,\ 6m-18$

d) $8x-12,\ 10x-15$

5. Simplify.

a) $\dfrac{4}{x+3} + \dfrac{5}{4x+12}$

b) $\dfrac{1}{3y-15} - \dfrac{2}{y-5}$

c) $\dfrac{t}{t-4} - \dfrac{2t}{3t-12}$

d) $\dfrac{2}{2m+2} + \dfrac{5}{3m+3}$

e) $\dfrac{3}{4y-8} - \dfrac{2}{3y-6}$

f) $\dfrac{1}{4a+2} + \dfrac{4}{6a+3}$

6. Simplify.

a) $\dfrac{2}{x+1} + \dfrac{3}{x+2}$

b) $\dfrac{m}{m-3} - \dfrac{5}{m+2}$

c) $\dfrac{3}{x} + \dfrac{5}{x-1}$

d) $\dfrac{2}{t-1} + \dfrac{1}{5} + 2$

e) $\dfrac{2x}{x-2} - \dfrac{3x}{x+2}$

f) $\dfrac{4}{3n-1} - \dfrac{3}{2n+3}$

g) $\dfrac{1}{2x-2} + \dfrac{3}{4x-8}$

h) $\dfrac{t}{3t+15} - \dfrac{1}{6t-24}$

i) $\dfrac{4}{2s-12} - \dfrac{s}{5s-5}$

j) $\dfrac{2m}{3m-15} + \dfrac{m}{4m-8}$

7. State the LCM in factored form.

a) $x+2,\ x^2+4x+4$

b) $y^2+6y+8,\ y^2-4$

c) $t^2-t-12,\ t^2-3t-4$

d) $2x-4,\ x^2-3x-4$

e) $m^2+6m+9,\ m^2-2m-15$

8. Simplify.

a) $\dfrac{2}{x+3} + \dfrac{3}{x^2+5x+6}$

b) $\dfrac{y}{y^2-16} - \dfrac{4}{y+4}$

c) $\dfrac{3x}{x-5} + \dfrac{2x}{x^2-4x-5}$

d) $\dfrac{a}{a^2 - 7a + 12} - \dfrac{2a}{a - 3}$

e) $\dfrac{4}{2x^2 + 3x + 1} + \dfrac{2}{2x + 1}$

f) $\dfrac{6}{2n - 1} - \dfrac{3}{6n^2 - 5n + 1}$

9. Simplify.

a) $\dfrac{2}{m^2 + 4m + 3} + \dfrac{1}{m^2 + 7m + 12}$

b) $\dfrac{1}{x^2 + 4x + 4} - \dfrac{3}{x^2 - 4}$

c) $\dfrac{a}{a^2 - 25} - \dfrac{2}{a^2 - 9a + 20}$

d) $\dfrac{4m}{m^2 - 9m + 18} + \dfrac{2m}{m^2 - 11m + 30}$

e) $\dfrac{5}{3x^2 + 4x + 1} + \dfrac{2}{3x^2 - 2x - 1}$

f) $\dfrac{3y}{4y^2 - 9} - \dfrac{2y}{4y^2 - 12y + 9}$

10. Simplify.

a) $\dfrac{t + 1}{t - 1} + \dfrac{2}{t^2 - 5t + 4}$

b) $\dfrac{y + 1}{y - 1} + \dfrac{y - 1}{y^2 + y - 2}$

c) $\dfrac{x - 2}{x^2 + 4x + 3} - \dfrac{2x + 1}{x + 3}$

d) $\dfrac{n^2 + 4n - 3}{n^2 - 16} + \dfrac{4 - 3n}{3n - 12}$

e) $\dfrac{m + 4}{m^2 - m - 12} - \dfrac{m}{m^2 - 5m + 4}$

f) $\dfrac{a + 2}{a^2 - 1} - \dfrac{a - 1}{a^2 + 2a + 1}$

g) $\dfrac{3w - 4}{w^2 + 5w + 4} + \dfrac{2w - 3}{w^2 + 2w - 8}$

h) $\dfrac{2x - 1}{2x^2 + 3x + 1} + \dfrac{2x + 1}{3x^2 + 4x + 1}$

i) $\dfrac{2z - 1}{4z^2 - 25} - \dfrac{2z + 5}{4z^2 - 8z - 5}$

Apply, Solve, Communicate

B

11. Inquiry/Problem Solving a) Copy and complete the table. The first line has been completed.

Expressions	Product	LCM	GCF	LCM × GCF
3x, 5x	$15x^2$	15x	x	$15x^2$
12, 8				
$15y^2$, 9y				
a + 1, a − 1				
2t − 2, 3t − 3				

b) How is the product LCM × GCF related to the product of each pair of expressions?

c) Explain why the relationship you found in part b) exists.

12. Application a) An RCMP patrol boat left Goderich and travelled for 45 km along the coast of Lake Huron at a speed of s kilometres per hour. Write an expression that represents the time taken, in hours.
b) The boat returned to Goderich at a speed of $2s$ kilometres per hour. Write an expression that represents the time taken, in hours.
c) Write and simplify an expression that represents the total time, in hours, the boat was travelling.
d) If s represents 10 km/h, for how many hours was the boat travelling?

13. Communication Write a problem that satisfies the following conditions. Have a classmate solve your problem.
• simplifies using addition and/or subtraction
• includes three rational expressions with different denominators that contain variables
• has the LCD as the denominator of one of the rational expressions

14. Simplify.

a) $\dfrac{m+3}{m+2} \times \dfrac{m+2}{m+1} + \dfrac{5m}{m+1}$

b) $\dfrac{2x^2 - 2x}{x^2 + 4x - 5} - \dfrac{4x}{x+5}$

c) $\dfrac{x^2 + 2x - 15}{x^2 - 7x + 12} + \dfrac{x^2 - 6x - 7}{x^2 - 3x - 4}$

d) $\dfrac{3y-1}{y-4} - \dfrac{y^2 + 4y - 12}{y^2 - 6y + 8}$

e) $\dfrac{2z^2 - 5z + 3}{z^2 - 1} + \dfrac{4z^2 - 9}{4z + 6}$

f) $\dfrac{x^2 + 5x + 6}{x^2 - 3x + 2} \div \dfrac{x+3}{x-1} - \dfrac{6}{x+3}$

C

15. Write two rational expressions with binomial denominators and with each of the following sums. Compare your answers with a classmate's.

a) $\dfrac{5x+8}{(x+1)(x+2)}$

b) $\dfrac{5x-5}{6x^2 - 13x + 6}$

c) $\dfrac{x^2 - 3}{(x-1)(x-3)}$

d) $\dfrac{4x^2}{4x^2 - 9}$

ACHIEVEMENT Check   Knowledge/Understanding   Thinking/Inquiry/Problem Solving   Communication   Application

Suppose you drive an average of 18 000 km/year. With your present car, you can drive 10 km per litre of fuel. You are thinking of buying a new car that you could drive x km farther per litre of fuel. Fuel currently costs $0.68/L. If the new car would save you $244.80 in yearly fuel costs, find the number of kilometres you could drive the new car per litre of fuel.

TECHNOLOGY EXTENSION
Rational Expressions and the Graphing Calculator

Complete the following using a graphing calculator, such as a TI-92, TI-92 Plus, or TI-89, with the capability to simplify and perform operations on rational expressions. When you enter a rational expression into the calculator, be careful with your use of brackets. Check that the display shows the intended expression.

Simplifying Rational Expressions

1. Simplify.

a) $\dfrac{3x^2y}{6x^5y^3}$
b) $\dfrac{-42a^9b^2c}{14a^4b^2c^3}$
c) $\dfrac{15p^6q^3rs^2}{-6pq^4r^2s^6}$

To simplify $\dfrac{x^2-x-12}{x^2-10x+24}$ automatically, enter the expression.

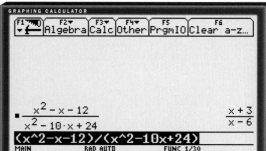

2. Simplify.

a) $\dfrac{4x^3-6x^2+8x}{2x^2}$
b) $\dfrac{3t^3}{12t^4+6t^3-3t^2}$

c) $\dfrac{5m^2+25m}{5m(3-m)}$
d) $\dfrac{8x^3+6x^2-4x}{10x^3+2x^2+4x}$

3. Simplify.

a) $\dfrac{x^2+7x+10}{x^2-3x-10}$
b) $\dfrac{x^2+2xy+y^2}{x^2-y^2}$

c) $\dfrac{12n^2-13n+3}{8n^2+14n-15}$
d) $\dfrac{10m^2-17m+6}{8m^2-14m+5}$

e) $\dfrac{x^4-4}{x^4-4x^2+4}$
f) $\dfrac{(9a^3+4ab^2)(3a^2+ab-2b^2)}{81a^4-16b^4}$

Operations on Rational Expressions

4. Multiply.

a) $\dfrac{-2m^2n^5}{15}\times\dfrac{3}{m^3n^4}$
b) $\dfrac{12x^2y^3}{25ab}\times\dfrac{5ab^2}{6x^3y^3}$

c) $\dfrac{3a}{4a-12}\times\dfrac{2a-6}{9a^2}$
d) $\dfrac{x^2+2x+1}{x^2-5x+6}\times\dfrac{x-3}{x+1}$

e) $\dfrac{9x^2-4}{8x^2-6x-9}\times\dfrac{4x^2-9}{6x^2+13x+6}$

f) $\dfrac{8p^2-22p+5}{15p^2+14p+3}\times\dfrac{10p^2+p-3}{6p^2-11p-10}$

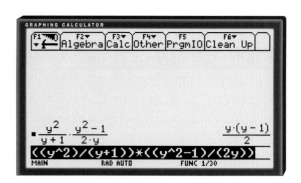

5. Divide.

a) $\dfrac{3s^3t^4}{4} \div \dfrac{9s^2t^5}{2}$

b) $\dfrac{36y^3z^2}{-7mn^2} \div \dfrac{24yz}{21m^2n}$

c) $\dfrac{2x+4}{3x-9} \div \dfrac{6x+12}{4x-12}$

d) $\dfrac{y^2-3y-18}{y^2-9y+14} \div \dfrac{y^2+7y+12}{y^2-2y-35}$

e) $\dfrac{6m^2+13m-5}{8m^2+16m-10} \div \dfrac{9m^2-4}{6m^2+m-2}$

f) $\dfrac{20x^2+17x+3}{18x^2-3x-10} \div \dfrac{25x^2+30x+9}{18x^2-9x-5}$

6. a) Add $\dfrac{5q-1}{6} + \dfrac{2q+1}{4}$.

b) If your calculator gives the answer in the form of two rational expressions with different denominators, use the common denominator function to write the answer as a single rational expression.

7. Add. Use the common denominator function as necessary.

a) $\dfrac{2x+1}{5} + \dfrac{4x-3}{2}$

b) $\dfrac{2}{y} + \dfrac{5}{y^2} + \dfrac{3}{y^3}$

c) $\dfrac{4}{3n-9} + \dfrac{3}{4n-12}$

d) $\dfrac{t}{4-4t} + \dfrac{3t}{4t-4}$

e) $\dfrac{n+1}{n^2+4n+4} + \dfrac{4n}{2^2-4}$

f) $\dfrac{c+1}{4c^2-2c-2} + \dfrac{c-1}{2c^2+3c+1}$

8. a) Subtract $\dfrac{5q-1}{6} - \dfrac{2q+1}{4}$.

b) If necessary, use the common denominator function to write the answer as a single rational expression.

9. Subtract. Use the common denominator function as necessary.

a) $\dfrac{3z+1}{10} - \dfrac{4z-5}{15}$

b) $\dfrac{3}{2x} - \dfrac{1}{3x^2}$

c) $\dfrac{5}{6r+9} - \dfrac{2}{12r+18}$

d) $\dfrac{5x}{3x+4} - \dfrac{3x}{2x-1}$

e) $\dfrac{2y}{y^2-3y+2} - \dfrac{3y}{y^2-4y+3}$

f) $\dfrac{2t-3}{9t^2+6t+1} - \dfrac{t}{6t^2-19t-7}$

10. The side lengths in a rectangle are $\dfrac{x+3}{4}$ and $\dfrac{x-1}{3}$.

a) Write an expression that represents the area.

b) Write and simplify an expression that represents the perimeter.

c) Write and simplify an expression that represents the ratio of the perimeter to the area.

d) Can $x = 1$ in your expression from part c)? Explain.

1.9 Solving First-Degree Inequalities

Canadian long-track speed skater Catriona LeMay
Doan broke world records in both the 500-m and the
1000-m events on the same day in Calgary.

Event	Catriona's Time (s)	Old World Record (s)
500-m	37.90	38.69
1000-m	76.07	77.65

Since her winning times were less than the old world
records, or the old world records were greater than her
winning times, we can describe her achievement in the
form of an **inequality**. A mathematical inequality may
contain a symbol such as $<$, $\leq$, $>$, $\geq$, or $\neq$.

In a **first-degree inequality**, such as $x + 2 > 7$, the
variable has the exponent 1. To solve an inequality,
find values of the variable that make the inequality
true. For example, the inequality $x + 2 > 7$ is true for
$x = 5.1$, $x = 6$, $x = 7.25$, and all other real values of x
greater than 5. These values are said to *satisfy* the
inequality. We write the solution as $x > 5$.

INVESTIGATE & INQUIRE

1. Passenger aircraft land at 240 km/h, but their speeds on their landing
approaches are higher than this. When passenger aircraft descend for
a landing, their speeds during descent are given by the inequalities
$s - 320 \geq 0$ and $s - 320 \leq 80$, where s is the speed in kilometres per hour.
a) Solve the equations $s - 320 = 0$ and $s - 320 = 80$.
b) Solve the inequalities $s - 320 \geq 0$ and $s - 320 \leq 80$ using the same steps
as you used in part a).

2. What is the lowest speed at which a passenger aircraft can descend?

3. What is the highest speed at which a passenger aircraft can descend?

4. a) List the three greatest whole-number solutions for $s - 320 \leq 80$.
b) Use substitution to show that your three values from part a) satisfy both
inequalities.

5. Solve each of the following inequalities using the same rules used to solve equations.

a) $4x + 7 < 15$

b) $7x + 2 > 23$

c) $0.8 + 1.3x > 7.3$

d) $\frac{1}{2}x - 5 < 3$

6. a) The table shows the results of various operations on both sides of the inequality $9 > 6$. Copy and complete the table by replacing each ● with > or <.

Original Inequality	Operation	Resulting Inequality
$9 > 6$	Add 3	$9 + 3 \bullet 6 + 3$
$9 > 6$	Subtract 3	$9 - 3 \bullet 6 - 3$
$9 > 6$	Multiply by 3	$9 \times 3 \bullet 6 \times 3$
$9 > 6$	Multiply by –3	$9 \times (-3) \bullet 6 \times (-3)$
$9 > 6$	Divide by 3	$\frac{9}{3} \bullet \frac{6}{3}$
$9 > 6$	Divide by –3	$\frac{9}{-3} \bullet \frac{6}{-3}$

b) State the operations that reverse the direction of the inequality symbol.

7. Test your statement from question 6b) by determining the results of the following operations.

	Inequality	Operation		Inequality	Operation
a)	$4 > -3$	Add 5	**b)**	$2 < 6$	Add –1
c)	$-3 < -1$	Subtract 2	**d)**	$-1 > -4$	Subtract –2
e)	$2 > -1$	Multiply by 4	**f)**	$-3 < -2$	Multiply by –3
g)	$4 > 3$	Multiply by 2	**h)**	$-4 < -3$	Multiply by –1
i)	$3 < 6$	Divide by 3	**j)**	$2 > -2$	Divide by –2
k)	$-4 < -2$	Divide by 2	**l)**	$-4 > -8$	Divide by –4

8. Use your answers from questions 6 and 7 to solve each of the following. Use substitution to verify your solutions.

a) $-4x > 4$

b) $\frac{x}{-3} < 1$

First-degree inequalities in one variable can be solved by performing the same operations on both sides to isolate the variable. When multiplying or dividing both sides of an inequality by a negative number, reverse the direction of the inequality symbol.

In the following examples and problems, assume that all variables represent real numbers.

EXAMPLE 1 Solving an Inequality

Solve and check $3x - 2 < 13$.

SOLUTION

$$3x - 2 < 13$$

Add 2 to both sides: $3x - 2 \boxed{+ 2} < 13 \boxed{+ 2}$

$$3x < 15$$

Divide both sides by 3: $\dfrac{3x}{3} < \dfrac{15}{3}$

$$x < 5$$

Check.
Try $x = 4$:   **L.S.** $= 3x - 2$ **R.S.** $= 13$
$= 3(4) - 2$
$= 10$
L.S. $<$ R.S.

The solution is any real number less than 5.

EXAMPLE 2 Solving and Graphing

Solve $2(3 - x) - 1 \geq 7$. Graph the solution.

SOLUTION

$$2(3 - x) - 1 \geq 7$$

Expand to remove brackets: $6 - 2x - 1 \geq 7$

$$5 - 2x \geq 7$$

Subtract 5 from both sides: $5 - 2x \boxed{- 5} \geq 7 \boxed{- 5}$

$$-2x \geq 2$$

Divide both sides by -2: $\dfrac{-2x}{-2} \leq \dfrac{2}{-2}$

$$x \leq -1$$

When you multiply or divide by a negative number, reverse the direction of the symbol.

The graph is as shown. The closed dot at $x = -1$ shows that -1 is included in the solution.

EXAMPLE 3 Solving an Inequality Involving Fractions

Solve $\dfrac{3x}{4} + \dfrac{x}{2} > 5$. Graph the solution.

SOLUTION

The LCD is 4.

$$\frac{3x}{4} + \frac{x}{2} > 5$$

Multiply both sides by 4: $\boxed{4 \times}\left(\dfrac{3x}{4} + \dfrac{x}{2}\right) > \boxed{4 \times}\ 5$

$$3x + 2x > 20$$
$$5x > 20$$

Divide both sides by 5: $\dfrac{5x}{\boxed{5}} > \dfrac{20}{\boxed{5}}$

$$x > 4$$

Check.

Try $x = 8$:   L.S. $= \dfrac{3x}{4} + \dfrac{x}{2}$ R.S. $= 5$

$$= \frac{3(8)}{4} + \frac{(8)}{2}$$
$$= 6 + 4$$
$$= 10$$

$$\text{L.S.} > \text{R.S.}$$

The solution is any real number greater than 4.

The graph is as shown. The open dot at $x = 4$ shows that 4 is not included in the solution.

EXAMPLE 4 Solving an Inequality Involving Decimals

Solve $0.5(x+4) - 0.2(x+6) \le 0.5(x+1) - 2.8$. Graph the solution.

SOLUTION

$$0.5(x+4) - 0.2(x+6) \le 0.5(x+1) - 2.8$$

Expand to remove brackets: $0.5x + 2 - 0.2x - 1.2 \le 0.5x + 0.5 - 2.8$

Simplify: $\qquad\qquad\qquad\quad 0.3x + 0.8 \le 0.5x - 2.3$

Subtract 0.8 from both sides: $\quad 0.3x + 0.8 - 0.8 \le 0.5x - 2.3 - 0.8$

$$0.3x \le 0.5x - 3.1$$

Subtract $0.5x$ from both sides: $\quad 0.3x - 0.5x \le 0.5x - 3.1 - 0.5x$

$$-0.2x \le -3.1$$

Divide both sides by -0.2: $\qquad \dfrac{-0.2x}{-0.2} \ge \dfrac{-3.1}{-0.2}$

$$x \ge 15.5$$

The graph is as shown.

Note that Example 4 could also be solved by isolating the variable on the right side.

After expanding and simplifying,

$$0.3x + 0.8 \le 0.5x - 2.3$$

Add 2.3 to both sides: $\qquad\qquad 0.3x + 3.1 \le 0.5x$

Subtract $0.3x$ from both sides: $\qquad 3.1 \le 0.2x$

Divide both sides by 0.2: $\qquad\quad 15.5 \le x$

$$\text{or } x \ge 15.5$$

EXAMPLE 5 Selling Hiking Staffs

Volunteers from a hiking association are selling hiking staffs as a fundraiser. The cost of making the staffs is a fixed overhead of $2000, plus $10 per staff. Each staff is sold for $30. What number of staffs must be sold for the revenue to exceed the cost?

SOLUTION 1 Paper-and-Pencil Method

Let x represent the number of staffs made and sold.
The cost of making the staffs is $C = 2000 + 10x$.
The revenue from selling the staffs is $R = 30x$.

For the revenue to exceed the cost, $R > C$,

$$\text{so } 30x > 2000 + 10x.$$

$$30x > 2000 + 10x$$

Subtract $10x$ from both sides: $\quad 30x - 10x > 2000 + 10x - 10x$

$$20x > 2000$$

Divide both sides by 20: $\quad \dfrac{20x}{20} > \dfrac{2000}{20}$

$$x > 100$$

Over 100 staffs must be sold for the revenue to exceed the cost.

SOLUTION 2 Graphing-Calculator Method

Let x represent the number of staffs made and sold.
The cost of making the staffs is $C = 2000 + 10x$.
The revenue from selling the staffs is $R = 30x$.
Enter the equations $y = 2000 + 10x$ and $y = 30x$ in the Y= editor of a graphing
calculator. Graph the equations using suitable values of the window variables.
Use the intersect operation to find the coordinates of the point of intersection.

For these graphs, the window variables include Xmin = 0, Xmax = 150, Ymin = 0, and Ymax = 5000.

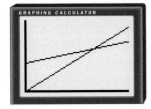
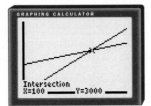

The value of x where the graphs intersect is 100.
For the revenue to exceed the cost, $R > C$,

$$\text{so } 30x > 2000 + 10x.$$

The graph of $y = 30x$ is the graph that starts at the origin. This graph is above
the graph of $y = 2000 + 10x$ when $30x > 2000 + 10x$. So, $R > C$ when $x > 100$.

Over 100 staffs must be sold for the revenue to exceed the cost.

Key Concepts

- The results of performing operations on an inequality are summarized in the table. Similar results are observed for inequalities that include the symbols $<$, $\geq$, and $\leq$. These results and the methods used to solve equations can be used to solve inequalities.

Original Inequality	Operation	Resulting Inequality
$a > b$	Add c	$a + c > b + c$
$a > b$	Subtract c	$a - c > b - c$
$a > b$	Multiply by c, $c > 0$	$ac > bc$, $c > 0$
$a > b$	Multiply by c, $c < 0$	$ac < bc$, $c < 0$
$a > b$	Divide by c, $c > 0$	$\dfrac{a}{c} > \dfrac{b}{c}$, $c > 0$
$a > b$	Divide by c, $c < 0$	$\dfrac{a}{c} < \dfrac{b}{c}$, $c < 0$

- For a graph of an inequality on a number line, a closed dot shows that an endpoint is included, and an open dot shows that an endpoint is not included.

Communicate Your Understanding

1. Describe each step in solving $5x - 7 > 2x + 11$ algebraically.

2. Describe when it is necessary to reverse the direction of the inequality symbol when solving an inequality algebraically.

3. Describe how you would write two inequalities that have the solution $x \geq 3$ and that have variables on both sides of the inequality symbol.

4. Describe how the graphs of $x > -2$ and $x \geq -2$ compare.

Practise

A

1. Solve and check.

a) $y + 9 < 11$ **b)** $2w + 5 > 3$

c) $3x - 4 \geq 5$ **d)** $2z + 9 \leq 3$

e) $-3x < 6$ **f)** $4t > 3t - 4$

g) $2(m - 3) \leq 0$ **h)** $4(n + 2) \geq 8$

2. Solve and check.

a) $2x + 1 > 2$ **b)** $3x + 4 < 2$

c) $6y + 4 \leq 5y + 3$ **d)** $4z - 3 \geq 3z + 2$

e) $7 + 3x < 2x + 9$ **f)** $5(2x - 1) > 5$

g) $2(3x - 2) \leq -4$ **h)** $4(2x + 1) \geq 2$

3. Solve. Graph the solution.

a) $6x + 2 \leq 4x + 8$ **b)** $4x - 1 > x + 5$

c) $2(x + 3) < x + 4$ **d)** $3(x - 2) > x - 4$

e) $3(y + 2) \geq 2(y + 1)$ **f)** $3(2z - 1) \leq 2(1 + z)$

g) $6x - 3(x + 1) > x + 5$

h) $2(x - 2) - 1 < 4(1 - x) + 1$

4. Solve.

a) $6 - 2x > 4$

b) $8 - 3x < 5$

c) $3y - 8 \geq 7y + 8$

d) $6 - 3c \leq 2(c - 2)$

e) $4(1 - x) \geq 3(x - 1)$

f) $-2(3 + x) < 4(x - 2)$

g) $4x - 3(2x + 1) \leq 4(x - 3)$

h) $2(3t - 1) - 5t > -6(1 - t) + 7$

5. Solve. Graph the solution.

a) $\dfrac{y}{3} + 2 < 1$

b) $\dfrac{w}{2} + 2 > 3$

c) $\dfrac{2x}{3} + 1 \geq 2$

d) $\dfrac{3z}{4} + 5 \leq -1$

e) $1.2x - 0.1 > 3.5$

f) $0.8x + 2.5 < -2.3$

g) $1.9 \geq 4.9 - 1.5q$

h) $4.6 - 1.8n \leq -0.8$

6. Solve.

a) $2(1.2a + 2.5) > 0.2$

b) $4(1.8 - 0.5x) \leq 5.2$

c) $0.75y - 2.6 < 0.25y - 3.1$

d) $3(1.3n + 0.3) \geq 3.5n + 0.1$

e) $1.5(x + 2) + 1 > 2.5(1 - x) - 0.5$

f) $2(1.5x + 1) - 1 < 5(0.2x + 0.3) - 0.5$

7. Solve.

a) $\dfrac{x + 1}{2} < \dfrac{x + 2}{3}$

b) $\dfrac{2 - x}{2} \geq \dfrac{2x + 1}{4}$

c) $\dfrac{z + 2}{4} > \dfrac{z - 1}{5} + 1$

d) $\dfrac{2 - 3x}{2} + \dfrac{2}{3} \leq \dfrac{3x - 2}{6}$

Apply, Solve, Communicate

8. Art supplies Katrina has a $50 gift voucher for an arts supply store. She wants to buy a sketch pad and some markers. Including taxes, a sketch pad costs $18 and a marker costs $4. Use the inequality $4m + 18 \leq 50$ to determine the number of markers, m, she can buy.

9. Measurement In $\triangle ABC$, $\angle A$ is obtuse and measures $(5x + 10)°$. Solve the inequalities $5x + 10 > 90$ and $5x + 10 < 180$ to find the possible values of x.

B

10. Application The cost of an extra large tomato and cheese pizza is $12.25, plus $1.55 for each extra topping.

a) Let n represent the number of extra toppings. Write an expression, including n, to represent the total cost of the pizza.

b) Suppose you have $20 you can spend on the pizza. Write and solve an inequality to find the number of extra toppings you can afford.

11. Geometry $\triangle ABC$ is not an acute triangle. $\angle B$ is the greatest angle. The measure of $\angle B$ is $(4x)°$. What are the possible values of x?

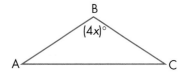

12. Weekly earnings Mario earns $15/h after taxes and other deductions. He spends a total of $75/week on lunches and travel to and from work.
a) Write an expression to represent how much Mario has at the end of a week in which he works t hours.
b) Write and solve an inequality to determine how many hours Mario must work to have at least $450 at the end of the week.

13. Baseball caps A college baseball team raises money by selling baseball caps. The cost of making the caps includes a fixed cost of $500, plus $7 per cap. The caps sell for $15 each. What is the minimum number of caps the team can order in one batch and still raise money?

14. Populations Paris and Aylmer are towns in Ontario. From 1991 to 1996, the population of Paris increased from 8600 to 9000. Over the same period, the population of Aylmer increased from 6200 to 7000. If each population continues to increase at the same rate as it did from 1991 to 1996, over what time period would you expect the population of Aylmer to be greater than the population of Paris?

15. Measurement a) What values of x give this rectangle a perimeter of more than 32 cm?
b) What values of x give the rectangle an area of less than 40 cm^2?
c) Communication In part b), does x have a minimum value? Explain.

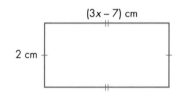

16. Modelling problems algebraically Determine the values of x that give this triangle a perimeter of no more than 15 and no less than 12.

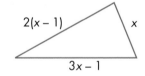

17. Solve $3(x + 2) - 5 \neq 2(1 - x) + 4$. Graph the solution.

18. Charity auction A charity wants to raise at least $90\ 000 for hospital equipment. A car dealer donates a car to be used as a raffle prize. The charity determines that it can sell 1500 to 2000 tickets. If advertising and other costs are estimated to be $4500,
a) what should the charity set as the ticket price?
b) what is the range of profit expected?

19. Inquiry/Problem Solving Write the following inequalities.
Have a classmate solve them.
a) variables on both sides and the solution $x \leq 2$
b) brackets on both sides and the solution $x > -3$
c) denominators of 3 and 2 and the solution $x < 0$

C

20. Driving times Jason left Hamilton at 10:00 and drove 620 km to
Montréal at an average speed of 80 km/h. Hakim left Hamilton an
hour later and drove to Montréal at an average speed of 100 km/h. Between
what times of the day was Hakim further from Hamilton than Jason was?

21. a) Try to solve the equation $4x + 2(x + 1) = 6x - 2$. What is the result?
b) What real values of x satisfy the equation?
c) Try to solve the inequality $4x + 2(x + 1) > 6x - 2$. What is the result?
d) What real values of x satisfy the inequality?

22. Technology a) Predict the graph of the following expression, if it is
drawn using a graphing calculator.
$y = (x - 3)(x < 2)$
Check your prediction using a graphing calculator in the dot mode.
Describe the result.
b) Repeat part a) for the expression $y = (2 - x)(x < 5)$.
c) Repeat part a) for the expression $y = (x + 4)(x > -4)$.

ACHIEVEMENT Check | Knowledge/Understanding Thinking/Inquiry/Problem Solving Communication Application

For what values of x is the triangle possible?

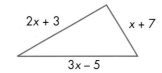

TECHNOLOGY EXTENSION
Solving Inequalities With a Graphing Calculator

Displaying a Solution

Some graphing calculators can be used to display the solution to an inequality in one variable.

To display the solution to $2x + 1 < 3x - 2$, choose the dot mode from the mode settings, and enter the inequality in the Y= editor as follows. You can enter the inequality symbol using the TEST menu.

$$Y_1 = 2X + 1 < 3X - 2$$

Then, graph Y_1 in the standard viewing window.

1. Using paper and pencil, solve the inequality algebraically and graph the solution.

2. Compare your graph from question 1 with the way the calculator displays the solution.

3. Using the TRACE instruction, find the values of x in the solution.

4. Describe how each of the following can be used to give a more accurate answer than you found in question 3.
a) the ZOOM menu
b) the TABLE key and the TABLE SETUP screen
c) changing the window variables

5. a) Using the TRACE instruction, find the value of y for each value of x in the solution.
b) Explain why the graph of the solution does not lie on the x-axis.

6. Graph Y_1 using the connected mode instead of the dot mode. Explain why the dot mode should be used.

7. a) Modify the inequality by changing the $<$ symbol to $\leq$. Display the solution on the calculator.
b) Does the graphing calculator distinguish between $<$ and $\leq$? Explain.

Solving Inequalities

8. Display the solution to each of the following inequalities using a graphing calculator. In your notebook, sketch each display and describe how it should be modified to show the solution fully. State the solution to each inequality.
a) $2x + 3 \geq 7$ **b)** $3x - 1 < 8$
c) $2x - 3 > -x$ **d)** $5x + 8 \leq 4x + 5$
e) $-4x + 2 \leq -2$ **f)** $3x + 2 \geq 5x - 6$
g) $2(x - 3) \leq 4x - 2$
h) $3(x - 1) - x > 4(x + 1) - 1$
i) $\dfrac{2}{3}x \geq x - 1$

j) $\dfrac{3}{4}x + \dfrac{2}{3}x - \dfrac{5}{6} < 2(x - 1)$

k) $\dfrac{x - 1}{2} < \dfrac{x - 2}{3}$

l) $3 + \dfrac{x - 3}{6} \leq x$

m) $x - 2 \geq 3x - 1$
n) $4(x + 2) > 2(5 - x)$

9. Some graphing calculators, such as the TI-92 and TI-92 Plus, have the capability to solve inequalities algebraically. If this type of calculator is available, use the solve function to solve the inequalities in question 8. The inequality symbols are found in the test submenu.

Investigate & Apply

Modelling Restrictions Graphically

There are several steps in the inquiry process. They do not necessarily occur in a given order and they may often be revisited in completing an investigation. These steps include the following.

Formulate questions (FQ)

Select strategies/tools (ST)

Represent in mathematical form (RM)

Interpret information (II)

Draw conclusions (DC)

Reflect on the reasonableness of results (RR)

1. Apply the inquiry process to describe the graph of $y = \dfrac{x}{x^2 - 2x}$.

FQ   Can you identify the shape of the graph from the equation?

FQ   What are the restrictions on x?

ST   • Choose a strategy for finding the restriction.

RM, DC   • Find the restrictions on x.

FQ	How do the restrictions affect the graph?
ST	• Choose a tool to graph the equation.
RM	• Graph the equation.
II, DC	• Describe what you see.

FQ	What is happening around the restriction at $x = 2$?
ST	• Choose a tool to investigate the value of y as x gets closer to 2 from both directions. One way is to complete a table of values.
RM	• Complete a table of values

x	y		x	y
1.5			2.5	
1.7			2.3	
1.9			2.1	
1.99			2.01	
1.999			2.001	

II, DC	• Describe what is happening to y as x gets closer to 2 from each direction.
FQ, ST, RM, II, DC	What is happening around the restriction at $x = 0$?
FQ, DC, RR	Why do the two restrictions have different effects on the graph?
FQ	How does the graph of $y = \dfrac{1}{x - 2}$ compare with the graph of $y = \dfrac{x}{x^2 - 2x}$?
RM, II, DC, RR	• Sketch both graphs.
DC, RR	• Explain the results.

2. Apply your conclusions to describe the effects of the restrictions on the following graphs.

a) $y = \dfrac{x - 3}{x^2 - 4x + 3}$

b) $y = \dfrac{x^2 - 4}{x - 2}$

REVIEW OF **KEY CONCEPTS**

■ **1.1** Reviewing the Exponent Laws

Refer to the Key Concepts on page 8.

1. Evaluate.

a) 5^{-2} b) 6^0 c) 3^{-3} d) $(-3)^{-4}$ e) $(5^{-1})^2$

f) $\dfrac{1}{(-3)^{-1}}$ g) $\dfrac{2^3}{2^0 - 2^{-1}}$ h) $\dfrac{4^{-1} + 2^{-2}}{2^{-3}}$ i) $\dfrac{a^0 + 3^2}{2^4 - b^0}$

2. Simplify. Express each answer with positive exponents.

a) $m^2 \times m^5$ b) $y^{-3} \times y^{-2}$ c) $t^7 \div t^4$

d) $m^{-7} \div m^{-2}$ e) $(x^2 y^3)^4$ f) $(y^3)^0$

g) $(x^{-2} y^3)^{-2}$ h) $\left(\dfrac{m^3}{n^2}\right)^4$ i) $\left(\dfrac{x^{-3}}{y^{-2}}\right)^{-2}$

3. Simplify. Express each answer with positive exponents.

a) $(-2x^2 y^3)(-5x^3 y^4)$ b) $(-18a^3 b^2) \div (-2a^2 b)$ c) $3m^{-2} \times 4m^6$

d) $10x^{-2} \div (-2x^{-3})$ e) $(-2a^5 b^3)^2$ f) $(-3m^{-3} n^{-1})^{-3}$

g) $\left(\dfrac{3m^2}{2n^3}\right)^3$ h) $\left(\dfrac{-2x^{-3}}{3y^{-4}}\right)^{-2}$ i) $\dfrac{(3x^3 y)(6xy^4)}{-9xy^2}$

j) $\dfrac{3ab^4}{2a^3 b^2} \times \dfrac{12a^5 b}{15a^4 b}$ k) $\dfrac{(-2s^{-2} t)(5s^{-3} t^2)}{4s^2 t^3}$ l) $\left(\dfrac{6a^{-2} b^{-3}}{2a^2 b^{-1}}\right)^{-2}$

1.2 Rational Exponents

Refer to the Key Concepts on page 15.

4. Write in radical form.

a) $6^{\frac{1}{2}}$ b) $5^{-\frac{1}{2}}$ c) $7^{\frac{3}{5}}$ d) $10^{-\frac{4}{3}}$

5. Write using exponents.

a) $\sqrt[3]{-8}$ b) $(\sqrt[3]{m})^5$ c) $\sqrt[3]{x^2}$ d) $\sqrt{\sqrt[5]{4a^4}}$

6. Evaluate.

a) $25^{\frac{1}{2}}$ b) $\left(\dfrac{1}{27}\right)^{\frac{1}{3}}$ c) $49^{-\frac{1}{2}}$ d) $1^{-\frac{1}{4}}$

e) $0.09^{0.5}$ f) $(-8)^{-\frac{1}{3}}$ g) $0.008^{-\frac{1}{3}}$ h) $27^{\frac{2}{3}}$

i) $-16^{-\frac{3}{4}}$ j) $\left(\dfrac{81}{16}\right)^{\frac{5}{4}}$ k) $\left(\dfrac{1}{9}\right)^{2.5}$ l) $\left(\dfrac{27}{125}\right)^{-\frac{2}{3}}$

m) $(-32)^{\frac{4}{5}}$ n) $(-8^{-1})^{-\frac{1}{3}}$ o) $\sqrt{\sqrt{16}}$

7. Simplify. Express each answer using exponents, if necessary.

a) $\sqrt[3]{\sqrt{y^4}}$

b) $\sqrt{\sqrt{81m^8}}$

c) $\sqrt[3]{-8x}$

d) $(\sqrt{x^3})(\sqrt{x})$

e) $(\sqrt[3]{-64})x$

f) $\sqrt[3]{-64x}$

8. Measurement The exact edge length of a cube is $\left(\dfrac{5}{2}\right)^{-\frac{2}{3}}$ units. Determine the volume of the cube.

1.3 Solving Exponential Equations
Refer to the Key Concepts on page 23.

9. Solve and check.

a) $2^x = 64$

b) $(-5)^x = -125$

c) $2^{x+3} = 128$

d) $\dfrac{5^{x-1}}{25} = 1$

e) $5^{y+2} = 1$

f) $4^{2x+1} = 8$

g) $2(3^{n+2}) = 18$

h) $4^{x-2} + 1 = 5$

10. Solve.

a) $2^{x+5} = 2^{2x-1}$

b) $27^{x-2} = 3^{x+6}$

c) $8^{2m+2} = 16^{m-2}$

d) $5^{y-1} = 25^{2y-1}$

e) $4^{2t+1} = 8^{2t-1}$

f) $6^{3x+5} = 36^{3x+6}$

11. Solve and check.

a) $2^{x+3} + 2^x = 288$

b) $3^{g+3} - 3^{g+2} = 1458$

c) $-500 = 5^{y+1} - 5^{y+2}$

12. Bacteria The number of bacteria in a culture is doubling every 3.75 h. How long will it take for the number of bacteria to increase from 30 000 to 7 680 000?

1.4 Adding, Subtracting, and Multiplying Polynomials
Refer to the Key Concepts on page 28.

13. Add.

a) $(5x^2 - 4x - 2) + (8x^2 + 3x - 3)$

b) $(2x^2 - 6xy + 7y^2) + (4x^2 + 3xy - 11y^2)$

14. Subtract.

a) $(7y^2 + 4y - 7) - (9y^2 + 3y - 3)$

b) $(3m^2 + mn - 7n^2) - (5m^2 + 3mn - 8n^2)$

15. Expand and simplify.

a) $4(x + 5) + 3(x - 7)$

b) $6(3s - 4t) - (7s - t) + 5$

c) $2x(x + 3) - x(3x + 8)$

d) $3y(y - 2) + 2y(3y + 4) - 4y(2y - 3)$

16. Expand and simplify.

a) $3[4 - 2(y - 3)] + 4[3(2 - y) - 5]$

b) $2x[2 - x(x - 1)] - [3 - x(x + 20)]$

17. Expand and simplify.

a) $(y - 8)(y - 9)$

b) $2(7 - 3x)(4 + x)$

c) $3(3x - 1)^2$

d) $(4x + 3y)(2x - 5y)$

18. Expand and simplify.

a) $(m - 4)(m + 4) + (m - 3)^2$

b) $(x + 6)^2 - (x + 4)(x - 7)$

c) $3(4y + 1)^2 + 2(3y - 4)(2y - 3)$

d) $2(3x - 2y)(x + 3y) - 2(2x - y)^2$

19. Expand and simplify.

a) $(x - 3)(x^2 - 3x + 2)$

b) $(2t + 1)(3t^2 - t - 1)$

c) $(x^2 + 2x + 3)(x^2 - x - 1)$

d) $(3z^2 - 2z + 1)(2z^2 + 2z - 3)$

20. Measurement The dimensions of a rectangle are $2x + 1$ and $x - 1$. If each dimension is increased by 2 units, write and simplify an expression that represents the increase in area.

1.5 Simplifying Rational Expressions

Refer to the Key Concepts on page 39.

21. Simplify. State any restrictions on the variables.

a) $\dfrac{3x}{3x + 9}$

b) $\dfrac{8y^2 - 10xy}{4y}$

c) $\dfrac{5x - 5y}{7x - 7y}$

d) $\dfrac{6x - 10}{5 - 3x}$

e) $\dfrac{3w}{3w^2 - 12w}$

f) $\dfrac{3m^2 - 3m}{4m^2 - 4m}$

g) $\dfrac{t - 2}{t^2 - 3t + 2}$

h) $\dfrac{2a^2 - 7a - 15}{a - 5}$

i) $\dfrac{y^2 - 9}{y^2 + y - 12}$

j) $\dfrac{6n^2 - 7n - 3}{12n^2 + 7n + 1}$

22. Alberta flag The area of an Alberta flag can be represented by the expression $2x^2 + 4x + 2$, and its width by $x + 1$.

a) Write and simplify an expression for the length.

b) Write and simplify an expression that represents the ratio length:width for an Alberta flag.

1.6 Multiplying and Dividing Rational Expressions

Refer to the Key Concepts on page 49.

23. Simplify. State any restrictions on the variables.

a) $\dfrac{5x^3}{2y} \times \dfrac{8y}{15x^2}$

b) $\dfrac{-4a^3}{3b} \div \dfrac{2a}{3b^2}$

c) $\dfrac{3a^2b}{-4xy} \times \dfrac{-5x^2y}{6ab^2}$

d) $\dfrac{b^2}{8x^3y} \div \dfrac{3b}{4xy}$

e) $\dfrac{3x-3}{2x+2} \times \dfrac{5x+5}{6x-6}$

f) $\dfrac{4m+8}{3n-3} \div \dfrac{2m+6}{7n-7}$

g) $\dfrac{t^2+4t+4}{t-2} \div \dfrac{3t+6}{t^2-5t+6}$

h) $\dfrac{2x^2-5x-3}{2x^2-5x+2} \times \dfrac{2x^2+3x-2}{x^2-4x+3}$

i) $\dfrac{6y^2-5y+1}{12y^2-5x-2} \div \dfrac{3y^2-4y+1}{4y^2+3y-1}$

24. Measurement Rectangles B and C are attached to rectangle A, as shown. The area of rectangle B is $2t^2 - 3t + 1$. The area of rectangle C is $3t^2 - 2t - 1$. The lengths of these two rectangles are as shown.

a) Write a rational expression that represents the width of rectangle B.

b) Write a rational expression that represents the width of rectangle C.

c) Write and simplify the product of the expressions you wrote in parts a) and b).

d) What type of rectangle is A? Explain.

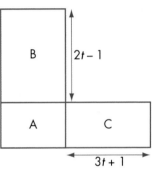

1.7 Adding and Subtracting Rational Expressions, I

Refer to the Key Concepts on page 57.

25. Simplify. State any restrictions on the variables.

a) $\dfrac{5}{x} + \dfrac{1}{x} - \dfrac{8}{x}$

b) $\dfrac{2m+1}{m-2} + \dfrac{3m-5}{m-2}$

c) $\dfrac{4z-3}{z^2} - \dfrac{3z-1}{z^2}$

d) $\dfrac{2t}{3} - \dfrac{3t}{4} + \dfrac{t}{6}$

e) $\dfrac{4x+1}{5} + \dfrac{2x-1}{4}$

f) $\dfrac{2a-3b}{6} - \dfrac{3a-2b}{4}$

g) $\dfrac{4}{2y-5} + \dfrac{2}{5-2y}$

h) $\dfrac{x^2+5}{x^2-4} - \dfrac{x^2-2}{4-x^2}$

i) $\dfrac{2x+5}{x^2+3x+2} - \dfrac{x+4}{x^2+3x+2}$

26. Measurement The perimeter of a triangle is $\dfrac{9x+1}{4}$. If two of the side lengths are $\dfrac{x+1}{2}$ and $\dfrac{2x-1}{2}$, what is the third side length?

1.8 Adding and Subtracting Rational Expressions, II

Refer to the Key Concepts on page 66.

27. Simplify. State any restrictions on the variables.

a) $\dfrac{2}{y} + \dfrac{4}{y^2} - \dfrac{1}{y}$

b) $\dfrac{4}{x^2} - \dfrac{5}{xy} + \dfrac{2}{y^2}$

c) $\dfrac{a}{2a-2} + \dfrac{2}{3a-3}$

d) $\dfrac{2}{x+3} - \dfrac{4}{x+1}$

e) $\dfrac{2}{t^2+3t+2} - \dfrac{1}{t^2+t-2}$

f) $\dfrac{x+1}{3x^2+4x+1} + \dfrac{2x-1}{3x^2-5x-2}$

1.9 Solving First-Degree Inequalities

Refer to the Key Concepts on page 78.

28. Solve and check.

a) $y + 3 < 9$
b) $3w + 4 > 10$
c) $2x - 5 \geq -7$
d) $4z - 5 \leq 3$

e) $-5k < 10$
f) $2t > t - 8$
g) $3(m - 2) \leq 6$
h) $2(n + 4) \geq 0$

29. Solve.

a) $3x + 2 > -10$
b) $5y + 1 < 1$
c) $7m + 3 \leq 6m + 2$

d) $3z - 8 > 2z + 3$
e) $9 + 5b < 6b + 1$
f) $4(2q - 1) > 4$

g) $3(2b - 2) \leq 6$
h) $2(4 - n) \geq 0$

30. Solve. Graph the solution.

a) $5m + 4 \leq 3m + 10$
b) $w + 2 > 6w - 8$

c) $2(x - 7) < x + 3$
d) $4(3z - 1) \leq 2(5 - z)$

e) $2(y - 3) + 1 \geq -4(2 - y) + 7$
f) $5n - 2(n + 3) - 1 < 2(n - 5) + 6$

31. Solve.

a) $\dfrac{x}{4} - 3 > -1$

b) $\dfrac{w}{5} + 5 \leq 2$

c) $1.9m + 2.4 < 6.2$

d) $3.3 - 2.6p \geq 8.5$

e) $\dfrac{x+1}{2} > \dfrac{5-x}{2} - 2$

f) $\dfrac{5-w}{4} \geq \dfrac{5-w}{5}$

g) $1.4(y + 3) + 6.1 > 2.5(1 - y)$

h) $3(1.2k + 2) - 12.6 \leq 5(0.3k + 0.8) - 2.2$

32. Measurement A triangle has side lengths $2x + 1$, $2x + 3$, and $2x - 2$. What values of x give the triangle a perimeter of

a) 44 or more?
b) less than 56?

33. Fundraising The symphony orchestra is holding a fundraising banquet in a hotel. The hotel charges $200 for the banquet hall, plus $60 per person. Tickets are $100 each. How many tickets must be sold to raise more than $10 000?

CHAPTER TEST

Achievement Chart

Category	Knowledge/Understanding	Thinking/Inquiry/Problem Solving	Communication	Application
Questions	All	12, 13, 14	12, 13, 14	12, 13, 14

1. Evaluate.

a) $\dfrac{1}{(-5)^2}$

b) $\dfrac{5^2 - 5^1}{5^{-1}}$

c) $\dfrac{3^{-1}}{3^{-2} + 3^0}$

2. Simplify. Express each answer with positive exponents.

a) $\left(\dfrac{s^{-2}}{t^3}\right)^{-3}$

b) $30a^4b^2 \div (-5ab)$

c) $(-3a^2b^5)^2$

d) $\dfrac{10m^2n^{-2} \times 2m^{-1}n^4}{-4mn^{-3}}$

e) $\dfrac{(-4s^{-2}t^{-3})^{-2}}{-s^2t^{-1}}$

3. Evaluate.

a) $-100^{-\frac{3}{2}}$

b) $81^{-\frac{3}{4}}$

c) $\left(\dfrac{8}{-27}\right)^{-\frac{2}{3}}$

4. Measurement A rectangle has side lengths of $3 + \sqrt{2}$ and $3 - \sqrt{2}$. Evaluate the area of the rectangle.

5. Solve and check.

a) $(-3)^x = 81$

b) $2^{x-3} = 64$

c) $-5^{x+2} = -1$

d) $3^{2y-3} = 9$

e) $2^{3x+2} = \dfrac{1}{16}$

f) $4(6^{g+2}) = 144$

6. Solve and check.

a) $3^{x-2} = 3^{2x+1}$

b) $2^{x+2} = 4^{x+3}$

c) $5^{4x+2} = 25^{x-1}$

d) $6^{x+2} + 6^x = 222$

e) $2^{x+2} - 2^{x+3} = -64$

7. Simplify.

a) $(2x^2 + 3x - 7) + (7x^2 - 6x - 11)$ b) $(4y^2 - 7y - 7) - (8y^2 + 5y - 9)$

8. Expand and simplify.

a) $3t(t - 7) - 2t(4t + 5)$

b) $4w(2w - 3) - 2w(w + 5) - 3w(2w - 1)$

c) $(x - 5)(x + 11)$

d) $3(2x - y)(x - 3y)$

e) $-2(2s + 3t)^2$

f) $2(x - 3)^2 - (2x + 1)(3x + 2)$

g) $3(2x - 3y)(2x + 3y) - (x - y)(3x + y)$

9. Simplify. State any restrictions on the variables.

a) $\dfrac{3x - 3y}{5x - 5y}$

b) $\dfrac{2y^2 + 4y}{3y^2 + 6y}$

c) $\dfrac{t^2 - 16}{t^2 - t - 12}$

d) $\dfrac{2m^2 + m - 3}{3m^2 + 2m - 5}$

10. Simplify. State any restrictions on the variables.

a) $\dfrac{x^2 + 2x - 3}{x^2 + 6x + 8} \times \dfrac{x^2 + 2x - 8}{x^2 + x - 6}$

b) $\dfrac{2a^2 - a - 1}{3a^2 + a - 2} \div \dfrac{2a^2 - 3a - 2}{3a^2 - 11a + 6}$

c) $\dfrac{n + 2}{3} + \dfrac{2n - 1}{4} - \dfrac{3n + 1}{2}$

d) $\dfrac{4}{2x - 3} - \dfrac{1}{3 - 2x}$

e) $\dfrac{2}{x^2 + 5x + 4} - \dfrac{3}{x^2 - 3x - 4}$

11. Solve. Graph the solution.

a) $2z + 5 \geq z - 3$

b) $3(x + 2) > -1(x - 2)$

c) $3(3z + 1) \leq -2(9 - z)$

d) $3(y - 1) + 10 \geq -5(2 - y) - 7$

e) $\dfrac{h - 5}{3} + 4 > \dfrac{h}{2} + 1$

f) $2.7(y - 2) < 3(0.2y + 2.1) - 1.2$

12. Radiology Cobalt-60, which has a half-life of 5.3 years, is extensively used in medical radiology. Most of the world's supply of cobalt-60 is produced in Canada.
a) What fraction of an original sample of cobalt-60 will remain after 10.6 years?
b) How long will it take until there is only 12.5% of the original sample remaining?

13. Measurement The length of a rectangle is represented by $\dfrac{3x + 1}{2}$ and its width is represented by $\dfrac{2x - 1}{3}$.

a) Write and simplify an expression that represents the perimeter of the rectangle in terms of x.
b) Find the three smallest values of x that give whole-number values for the perimeter. Explain your reasoning.

ACHIEVEMENT Check   Knowledge/Understanding   Thinking/Inquiry/Problem Solving   Communication   Application

14. The harmonic mean of 2 numbers, a and b, is $\dfrac{2}{\dfrac{1}{a} + \dfrac{1}{b}}$.

The harmonic mean of 3 numbers, a, b, and c, is $\dfrac{3}{\dfrac{1}{a} + \dfrac{1}{b} + \dfrac{1}{c}}$.

a) Simplify the expressions for 2 and 3 numbers and find a simplified expression for the harmonic mean of 4 numbers.
b) Find the harmonic mean of the numbers 2, 3, 4, 7, and 9.

CHALLENGE PROBLEMS

1. Roots If $x \geq 0$, then $\sqrt{x\sqrt{x\sqrt{x}}}$ is

a) $x\sqrt{x}$ b) $x\sqrt[4]{x}$ c) $\sqrt[8]{x}$ d) $\sqrt[8]{x^3}$ e) $\sqrt[8]{x^7}$

2. Inequality The largest integer n for which $n^{200} < 5^{300}$ is

a) 8 b) 9 c) 10 d) 11 e) 12

3. Powers Let r be the number that results when both the base and the exponent of a^b are tripled, where a and b are positive. If $r = a^b x^b$, where x is positive, then x equals

a) 3 b) $3a^2$ c) $27a^2$ d) $2a^{3b}$ e) $3a^{2b}$

4. Inequality If x is a real number, the inequality $\dfrac{3}{2-x} \leq 1$ is equivalent to

a) $x \leq -1$ b) $x \geq -1$, $x \neq 2$ c) $x \leq -1$ or $x \geq 0$, $x \neq 2$

d) $x \leq -1$ or $x > 2$ e) $x \leq -1$ or $x \geq -1$, $x \neq 2$

5. Powers If $60^a = 3$ and $60^b = 5$, then $12^{\frac{1-a-b}{2(1-b)}}$ is

a) $\sqrt{3}$ b) 2 c) $\sqrt{5}$ d) 3 e) $\sqrt{12}$

6. Test average On the first four tests of the term, your average is 84%. If you think you can score 94% on all the remaining tests, how many more tests do you need to bring your average to 90%?

7. Reciprocals Solve for x if the reciprocal of $\left(\dfrac{1}{x} - 1\right)$ is -2.

8. Equation Find all positive integer values of x and y that satisfy the equation $\dfrac{1}{x} + \dfrac{x}{y} + \dfrac{1}{xy} = 1$.

9. Powers If $x^2 yz^3 = 7^3$ and $xy^2 = 7^9$, find the value of xyz.

PROBLEM SOLVING STRATEGY

MODEL AND COMMUNICATE SOLUTIONS

An important part of problem solving is mathematical modelling, that is, describing a situation in mathematical form. Many problems can be solved in several different ways using different mathematical models. Your solution to a problem should clearly communicate the mathematical model you chose and the steps in your reasoning.

When Statistics Canada carries out a national census, all Canadian residents are asked to report the first language they spoke in childhood and still understand. In 1971, the number of people reporting Spanish was 24 times the number reporting Spanish in 1941. The number increased by an average of 7650/year from 1971 to 1991. From 1991 to 1996, the number increased by a total of 36 000. In 1996, 213 000 people reported Spanish. How many people reported Spanish as their first language in 1941?

Understand
the Problem

1. What information are you given?
2. What are you asked to find?
3. Do you need an exact or an approximate answer?

Think
of a Plan

There are several possible solution methods, including using algebra, working backward, and guess and check. The plans for two methods are as follows.
• Method 1: Model Algebraically
Write and solve an equation.
• Method 2: Model by Working Backward
Start with the end result and find the initial number that led to this result.

Carry Out
the Plan

Method 1: Model Algebraically
Let the number of people reporting Spanish in 1941 be x.
Then, the number reporting Spanish in 1971 was $24x$.
The number reporting Spanish in 1991 was $24x + 20 \times 7650$, or $24x + 153\ 000$.
The number reporting Spanish in 1996 was $24x + 153\ 000 + 36\ 000$, or $24x + 189\ 000$.
Write the equation: $24x + 189\ 000 = 213\ 000$
Solve for x: $24x = 24\ 000$
                                              $x = 1000$
So, 1000 people reported Spanish as their first language in 1941.

Method 2: Model by Working Backward

Start with the number of people reporting Spanish in 1996, that is, 213 000.
This number was 36 000 more than the number in 1991, so the number in 1991 was 213 000 − 36 000, or 177 000.
From 1971 to 1991, the number increased by 20 × 7650, or 153 000.
So, the number in 1971 was 177 000 − 153 000, or 24 000.
Since the number in 1971 was 24 times the number in 1941, divide 24 000 by 24 to find the number in 1941.
24 000 ÷ 24 = 1000
So, 1000 people reported Spanish as their first language in 1941.

Look Back

Does the answer seem reasonable?
How could you check that the answer is correct?

Work Backward

1. Choose a mathematical model to describe the problem.
2. Use the model to solve the problem.
3. Communicate the solution, showing your model and your reasoning.
4. Check that your answer is reasonable.

Apply, Solve, Communicate

Solve each of the following problems. Clearly communicate your mathematical model and the steps in your reasoning.

1. First language The number of Canadian residents reporting Japanese as the first language they spoke in childhood and still understood dropped by an average of 400/year from 1941 to 1951. The number then remained constant from 1951 to 1961, and dropped by a total of 1000 from 1961 to 1971. From 1971 to 1996, the number doubled to 34 000. How many Canadian residents reported Japanese as their first language in 1941?

2. Clothes shopping Nadia has $345 to spend on a jacket. In a store that is selling all jackets at a discount of 25% off the sticker price, she sees a jacket with a sticker price of

$380. Sales taxes, consisting of 7% GST and 8% PST, are to be added to the discounted price.
a) Can Nadia afford the jacket?
b) What is the highest sticker price she can afford?

3. Painted cube A large cube was made from 1000 identical small green cubes. The outside of the large cube was then painted red. How many of the small cubes have one red face?

4. Photocopying A photocopier was set to enlarge a drawing to 125% of its original dimensions. The drawing was enlarged, and then the result was enlarged. The final result had an area of 625 cm^2. What was the area of the original drawing?

PROBLEM SOLVING

USING THE STRATEGIES

1. Calendar If there are exactly four Mondays in January, on what days of the week can January 31 not fall?

2. Marching band A large marching band was performing on a football field. First, the band formed a square. Then, the band formed a rectangle, so that the number of rows increased by 5. How many were in the band?

3. Sequence Each term in the following sequence is determined from the previous term only.

15, 26, 38, 67, 55, …

Extend the sequence until you find the numbers that repeat. Which numbers repeat?

4. Measurement
The trapezoid has 3 equal sides. The length of the base is 2 cm less than the sum of the lengths of the three equal sides. The distance between the parallel sides is 8 cm. Find the area of the trapezoid.

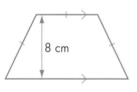
8 cm

5. Standard form What is the ones digit when 6317^{458} is written in standard form?

6. Toothpicks The 12 toothpicks have been arranged to make three identical squares.

How could you arrange the toothpicks to make six identical squares?

7. Marking points Mark six points on a piece of paper, so that each point is 1 unit from exactly three other points.

8. Letter puzzle In this addition, the letters D, E, and F represent different digits. What are the values of D and F?

$$\begin{array}{r} D \\ E \\ + F \\ \hline DE \end{array}$$

9. Moving counters There are four red counters and four blue counters in a row.

●●●●●●●●

You are allowed to move two adjacent counters at a time, without rotating them. In four moves, arrange the counters so that the colours alternate.

10. Driving About how many litres of gasoline are used by all the cars in Ontario in a week?

11. System of equations In the system of equations, find the values of F and E.

$A + B = C$
$C + D = E$
$A + E = F$
$B + D + F = 20$
$A = 4$

12. Bicycle race Suzanne, Beth, and Janel entered a 36-km bicycle race. Each of them kept a constant speed throughout the race. When Suzanne finished, Beth was 12 km from the finish line, and Janel was 18 km from the finish line. When Beth finished the race, how far from the finish line was Janel?

CHAPTER

2 Quadratic Functions and Equations

Specific Expectations	Functions	Functions & Relations
Identify the structure of the complex number system and express complex numbers in the form $a + bi$, where $i^2 = -1$.	2.1	2.1
Determine the maximum or minimum value of a quadratic function whose equation is given in the form $y = ax^2 + bx + c$, using the algebraic method of completing the square.	2.2	2.2
Determine the real or complex roots of quadratic equations, using an appropriate method (factoring, the quadratic formula, completing the square), and relate the roots to the x-intercepts of the graph of the corresponding function.	2.3	2.3
Add, subtract, multiply, and divide complex numbers in rectangular form.		2.4, 2.5

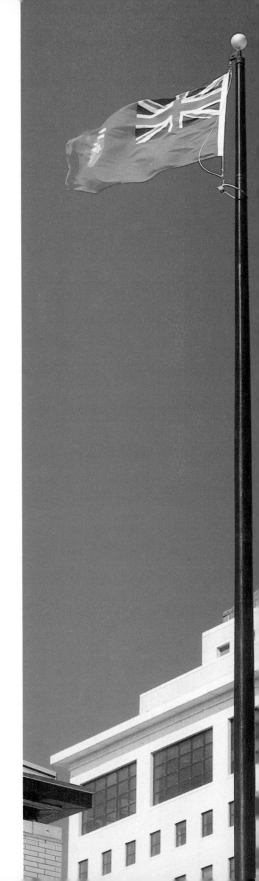

Measurements of Lengths and Areas

In the Modelling Math questions on pages 108, 118, and 132, you will solve the following problem and other problems that involve lengths and areas.

The ratio of the length to the width for the provincial flag of Ontario is always 2:1.

a) Determine the length of the diagonal of each of the following Ontario flags. Write each answer in simplest radical form.

1

2

3

2

4

6

b) Describe the relationship between the length of the diagonal and either dimension of the flag.

c) Use the relationship from part b) to predict the length of the diagonal of a 150 cm by 75 cm Ontario flag. Leave your answer in simplest radical form.

d) Describe the relationship between the length of the diagonal and the area of the Ontario flag.

e) Use the relationship from part d) to predict the length of the diagonal of an Ontario flag with an area of 24 200 cm^2. Leave your answer in simplest radical form.

Use your research skills to answer the following questions now.

1. Determine the ratio of the length to the width for each provincial and territorial flag, and for the Canadian flag.

2. Identify any flags from question 1 for which whole-number values of the dimensions result in whole-number values for the lengths of the diagonals.

3. Why do you think that most official flags are rectangular and are designed with a small whole-number ratio of length to width?

Store Profits

A company is hiring staff for its new mega-bookstore. If there are too few staff members, they will not be able to run the store effectively. If there are too many staff members, their salary costs will be too high.

A consultant has advised the company that the average weekly profit per staff member, P dollars, will be related to the number of staff members, s, by the function

$$P = -s^2 + 50s - 400.$$

1. What is the maximum possible weekly profit per staff member?

2. What number of staff members gives the maximum weekly profit per staff member?

3. What is the least number of staff members that will result in a profit?

4. What is the greatest number of staff members that will result in a profit?

5. What numbers of staff members will result in a weekly profit of at least $200 per staff member?

6. The total weekly profit of the store is the product of the number of staff members and the weekly profit per staff member.
a) Use guess and check to determine whether your answers to questions 1 and 2 result in the greatest total weekly profit for the store.
b) What should the company do to make the greatest total weekly profit?
c) What is the greatest total profit the store could make in a year?

Review of Prerequisite Skills

If you need help with any of the skills named in purple below, refer to Appendix A.

1. Evaluating radicals Evaluate.

a) $\sqrt{121}$ b) $\sqrt{225}$

c) $\sqrt{0.09}$ d) $\sqrt{1.69}$

e) $\sqrt{0.0016}$ f) $\sqrt{5^2 - 3^2}$

g) $\sqrt{5^2 + 12^2}$ h) $\sqrt{2 \times 200}$

i) $\sqrt{9^2 - 4(2)(4)}$ j) $\sqrt{6^2 - 4(8)(-2)}$

2. Graphing quadratic functions Sketch the graph of each function and find the coordinates of the vertex, the equation of the axis of symmetry, the maximum or minimum value, and any intercepts.

a) $y = 2x^2 - 8$ b) $y = -3x^2 + 6$

c) $y = (x - 2)^2 + 3$ d) $y = -2(x + 1)^2 + 8$

3. Solving quadratic equations by graphing Solve by graphing. Check your solutions.

a) $x^2 - x - 6 = 0$ b) $x^2 + 5x + 4 = 0$

c) $x^2 - 4 = 0$ d) $x^2 - 6x = 0$

4. Solving quadratic equations by factoring Solve by factoring. Check your solutions.

a) $x^2 + x - 12 = 0$ b) $x^2 - 10x + 25 = 0$

c) $y^2 + 8y + 15 = 0$ d) $t^2 - 4t - 32 = 0$

e) $2z^2 = 5z + 3$ f) $9s^2 + 6s + 1 = 0$

g) $6w^2 - w = 12$ h) $2x^2 - 12 = 5x$

i) $0 = 5m^2 + 8m + 3$ j) $4x^2 = 15x + 4$

k) $2x^2 - 3x = 0$ l) $9x^2 - 25 = 0$

5. The quadratic formula Solve using the quadratic formula. Check your solutions.

a) $x^2 + 6x + 8 = 0$ b) $y^2 - 2y - 15 = 0$

c) $4x^2 = 3 + x$ d) $2r^2 - r = 3$

e) $6x^2 + 5x = 6$ f) $6x^2 + 7x - 20 = 0$

6. The quadratic formula Solve using the quadratic formula. Express answers as exact roots and as approximate roots, to the nearest hundredth.

a) $x^2 - 3x + 1 = 0$ b) $y^2 + 3y - 3 = 0$

c) $2m^2 - m = 5$ d) $3t^2 = t + 1$

e) $4s^2 + 7s = -1$ f) $x^2 - x - 1 = 0$

g) $2w^2 + 5w + 1 = 0$ h) $3x + 3 = 5x^2$

7. Find the value of c that will make each expression a perfect square trinomial.

a) $x^2 + 10x + c$ b) $x^2 - 12x + c$

c) $x^2 - 2x + c$ d) $x^2 + 8x + c$

e) $x^2 - 14x + c$ f) $x^2 + 4x + c$

g) $x^2 - 30x + c$ h) $x^2 + 18x + c$

8. Rewriting in the form $y = a(x - h)^2 + k$, $a = 1$ Rewrite each of the following in the form $y = a(x - h)^2 + k$, and state the maximum or minimum value of y and the value of x when it occurs.

a) $y = x^2 + 2x - 5$ b) $y = x^2 - 4x + 6$

c) $y = x^2 + 6x + 2$ d) $y = -x^2 + 8x - 6$

e) $y = -x^2 - 6x + 3$ f) $y = -x^2 + 2x - 5$

g) $y = x^2 + 10x$ h) $y = -x^2 + 4x + 1$

9. Rewriting in the form $y = a(x - h)^2 + k$, $a \neq 1$ Rewrite each of the following in the form $y = a(x - h)^2 + k$, and state the maximum or minimum value of y and the value of x when it occurs.

a) $y = 2x^2 - 8x + 3$ b) $y = 3x^2 + 6x - 7$

c) $y = -2x^2 - 12x - 9$ d) $y = -4x^2 + 8x - 2$

e) $y = 2x^2 - 20x + 11$ f) $y = -3x^2 + 18x + 5$

g) $y = 6x^2 - 12x$ h) $y = -5x^2 - 20x + 2$

2.1 The Complex Number System

The approximate speed of a car prior to an accident can be found using the length of the tire marks left by the car after the brakes have been applied. The formula $s = \sqrt{121d}$ gives the speed, s, in kilometres per hour, where d is the length of the tire marks, in metres. Radical expressions like $\sqrt{121d}$ can be simplified.

Copy the table. Complete it by replacing each ■ with a whole number.

$\sqrt{4 \times 9} = \sqrt{\blacksquare} = \blacksquare$	$\sqrt{4} \times \sqrt{9} = \blacksquare \times \blacksquare = \blacksquare$
$\sqrt{9 \times 16} = \sqrt{\blacksquare} = \blacksquare$	$\sqrt{9} \times \sqrt{16} = \blacksquare \times \blacksquare = \blacksquare$
$\sqrt{25 \times 4} = \sqrt{\blacksquare} = \blacksquare$	$\sqrt{25} \times \sqrt{4} = \blacksquare \times \blacksquare = \blacksquare$
$\sqrt{\dfrac{36}{4}} = \sqrt{\blacksquare} = \blacksquare$	$\dfrac{\sqrt{36}}{\sqrt{4}} = \dfrac{\blacksquare}{\blacksquare} = \blacksquare$
$\sqrt{\dfrac{100}{25}} = \sqrt{\blacksquare} = \blacksquare$	$\dfrac{\sqrt{100}}{\sqrt{25}} = \dfrac{\blacksquare}{\blacksquare} = \blacksquare$
$\sqrt{\dfrac{144}{9}} = \sqrt{\blacksquare} = \blacksquare$	$\dfrac{\sqrt{144}}{\sqrt{9}} = \dfrac{\blacksquare}{\blacksquare} = \blacksquare$

1. Compare the two results in each of the first three rows of the table.

2. If a and b are whole numbers, describe how $\sqrt{ab}$ is related to $\sqrt{a} \times \sqrt{b}$.

3. Technology Use a calculator to test your statement from question 2 for each of the following.
a) $\sqrt{5} \times \sqrt{9}$ **b)** $\sqrt{3} \times \sqrt{7}$

4. Compare the two results in each of the last three rows of the table.

5. If a and b are whole numbers, and $b \neq 0$, describe how $\sqrt{\dfrac{a}{b}}$ is related to $\dfrac{\sqrt{a}}{\sqrt{b}}$.

6. Technology Use a calculator to test your statement from question 5 for each of the following.

a) $\sqrt{36} \div \sqrt{2}$ **b)** $\sqrt{15} \div \sqrt{3}$

7. Write the expression $\sqrt{121d}$ in the form $\blacksquare\sqrt{d}$, where $\blacksquare$ represents a whole number.

8. Determine the speed of a car that leaves tire marks of each of the following lengths. Round the speed to the nearest kilometre per hour, if necessary.

a) 64 m   **b)** 100 m   **c)** 15 m

The following properties are used to simplify radicals.

- $\sqrt{ab} = \sqrt{a} \times \sqrt{b},\ a \geq 0,\ b \geq 0$

- $\sqrt{\dfrac{a}{b}} = \dfrac{\sqrt{a}}{\sqrt{b}},\ a \geq 0,\ b > 0$

A radical is in simplest form when

- the radicand has no perfect square factors other than 1 $\sqrt{8} = 2\sqrt{2}$ Recall that the radicand is the expression under the radical sign.

- the radicand does not contain a fraction $\sqrt{\dfrac{1}{4}} = \dfrac{1}{2}$

- no radical appears in the denominator of a fraction $\dfrac{1}{\sqrt{3}} = \dfrac{\sqrt{3}}{3}$

EXAMPLE 1 **Simplifying Radicals**

Simplify.

a) $\sqrt{75}$ **b)** $\dfrac{\sqrt{48}}{\sqrt{6}}$ **c)** $\sqrt{\dfrac{2}{9}}$

SOLUTION

a) $\sqrt{75} = \sqrt{25} \times \sqrt{3}$
$\qquad\quad = 5\sqrt{3}$

b) $\dfrac{\sqrt{48}}{\sqrt{6}} = \sqrt{\dfrac{48}{6}}$
$\qquad\quad = \sqrt{8}$
$\qquad\quad = \sqrt{4} \times \sqrt{2}$
$\qquad\quad = 2\sqrt{2}$

c)

$$\sqrt{\frac{2}{9}} = \frac{\sqrt{2}}{\sqrt{9}}$$

$$= \frac{\sqrt{2}}{3} \text{ or } \frac{1}{3}\sqrt{2}$$

Note that, in Example 1, numbers like $\sqrt{75}$ and $\sqrt{\frac{2}{9}}$ are called **entire radicals**.

Numbers like $5\sqrt{3}$ and $\frac{1}{3}\sqrt{2}$ are called **mixed radicals**.

EXAMPLE 2 **Multiplying Radicals**

Simplify.
a) $9\sqrt{2} \times 4\sqrt{7}$ **b)** $2\sqrt{3} \times 5\sqrt{6}$

SOLUTION

a) $9\sqrt{2} \times 4\sqrt{7} = 9 \times 4 \times \sqrt{2} \times \sqrt{7}$
$$= 36\sqrt{14}$$
b) $2\sqrt{3} \times 5\sqrt{6} = 2 \times 5 \times \sqrt{3} \times \sqrt{6}$
$$= 10\sqrt{18}$$
$$= 10 \times \sqrt{9} \times \sqrt{2}$$
$$= 10 \times 3 \times \sqrt{2}$$
$$= 30\sqrt{2}$$

EXAMPLE 3 **Simplifying Radical Expressions**

Simplify $\dfrac{6 - \sqrt{45}}{3}$.

SOLUTION

$$\frac{6 - \sqrt{45}}{3} = \frac{6 - \sqrt{9} \times \sqrt{5}}{3}$$

$$= \frac{6 - 3\sqrt{5}}{3}$$

$$= 2 - \sqrt{5}$$

In mathematics, we can find the square roots of negative numbers as well as positive numbers. Mathematicians have invented a number defined as the principal square root of negative one. This number, i, is the **imaginary unit**, with the following properties.

$i = \sqrt{-1}$ and $i^2 = -1$

In general, if x is a positive real number, then $\sqrt{-x}$ is a **pure imaginary number**, which can be defined as follows.

$$\sqrt{-x} = \sqrt{-1} \times \sqrt{x}$$
$$= i\sqrt{x}$$

So, $\sqrt{-5} = \sqrt{-1} \times \sqrt{5}$ To ensure that $\sqrt{x}i$ is not read as $\sqrt{xi}$, we write $\sqrt{x}i$ as $i\sqrt{x}$.
$$= i\sqrt{5}$$

Despite their name, pure imaginary numbers are just as real as real numbers.

When the radicand is a negative number, there is an extra rule for expressing a radical in simplest form.
• A radical is in simplest form when the radicand is positive.

Numbers such as i, $i\sqrt{6}$, $2i$, and $-3i$ are examples of pure imaginary numbers in simplest form.

EXAMPLE 4 Simplifying Pure Imaginary Numbers

Simplify.

a) $\sqrt{-25}$ **b)** $\sqrt{-12}$

SOLUTION

a) $\sqrt{-25} = \sqrt{-1} \times \sqrt{25}$
$$= i \times 5$$
$$= 5i$$

b) $\sqrt{-12} = \sqrt{-1} \times \sqrt{12}$
$$= i \times \sqrt{4} \times \sqrt{3}$$
$$= i \times 2 \times \sqrt{3}$$
$$= 2i\sqrt{3}$$

When two pure imaginary numbers are multiplied, the result is a real number.

EXAMPLE 5 Multiplying Pure Imaginary Numbers

Evaluate.

a) $3i \times 4i$ b) $2i \times (-5i)$ c) $\left(3i\sqrt{2}\right)^2$

SOLUTION

a) $3i \times 4i = 3 \times 4 \times i^2$
$\qquad\qquad = 12 \times (-1)$
$\qquad\qquad = -12$

b) $2i \times (-5i) = 2 \times (-5) \times i^2$
$\qquad\qquad\quad = -10 \times (-1)$
$\qquad\qquad\quad = 10$

c) $\left(3i\sqrt{2}\right)^2 = 3i\sqrt{2} \times 3i\sqrt{2}$
$\qquad\qquad = 3^2 \times i^2 \times \left(\sqrt{2}\right)^2$
$\qquad\qquad = 9 \times (-1) \times 2$
$\qquad\qquad = -18$

A **complex number** is a number in the form $a + bi$, where a and b are real numbers and i is the imaginary unit. We call a the **real part** and bi the **imaginary part** of a complex number. Examples of complex numbers include $5 + 2i$ and $4 - 3i$. Complex numbers are used for applications of mathematics in engineering, physics, electronics, and many other areas of science.

Complex Numbers		
a	$+$	bi
↑		↑
real part		imaginary part

If $b = 0$, then $a + bi = a$. So, a real number, such as 5, can be thought of as a complex number, since 5 can be written as $5 + 0i$.

If $a = 0$, then $a + bi = bi$. Numbers of the form bi, such as $7i$, are **pure imaginary numbers**. Complex numbers in which neither $a = 0$ nor $b = 0$ are referred to as **imaginary numbers**.

The following diagram summarizes the complex number system.

Complex Numbers
$a + bi$, where a and b are real numbers and $i = \sqrt{-1}$.
$b = 0$ $a, b \neq 0$

Real Numbers
$$\left(e.g., 5, \sqrt{2}, -7, 3.6, \frac{-2}{3} \right)$$

Imaginary Numbers
(e.g., $4 + 3i$, $3 - 2i$)

$\downarrow a = 0$

Rational Numbers
Can be expressed as
the ratio of two integers.
$$\left(e.g., 3, -\sqrt{4}, 0.\overline{27}, -\frac{6}{7} \right)$$

Irrational Numbers
Cannot be expressed as
the ratio of two integers.
$$\left(e.g., \sqrt{7}, -\sqrt{2}, \pi \right)$$

Pure Imaginary Numbers
$$\left(e.g., 7i, -3i, i\sqrt{2} \right)$$

Integers
Whole numbers and their opposites
(e.g., 4, −4, 0, 9, −9)

Whole Numbers
Positive integers and zero
(e.g., 0, 3, 7, 11)

Natural Numbers
Positive integers
(e.g., 1, 5, 8, 23)

EXAMPLE 6 Simplifying Complex Numbers

Simplify.

a) $3 - \sqrt{-24}$ b) $\dfrac{10 + \sqrt{-32}}{2}$

SOLUTION

a) $3 - \sqrt{-24} = 3 - \sqrt{-1} \times \sqrt{24}$
$\phantom{3 - \sqrt{-24}} = 3 - i \times \sqrt{4} \times \sqrt{6}$
$\phantom{3 - \sqrt{-24}} = 3 - i \times 2\sqrt{6}$
$\phantom{3 - \sqrt{-24}} = 3 - 2i\sqrt{6}$

The expression $3 - 2i\sqrt{6}$ cannot be simplified further, because the
real part and the imaginary part are unlike terms.

b)
$$\frac{10 + \sqrt{-32}}{2} = \frac{10 + \sqrt{-1} \times \sqrt{32}}{2}$$
$$= \frac{10 + i \times \sqrt{16} \times \sqrt{2}}{2}$$
$$= \frac{10 + i \times 4\sqrt{2}}{2}$$
$$= \frac{10 + 4i\sqrt{2}}{2}$$
$$= 5 + 2i\sqrt{2}$$

Key Concepts

- Radicals are simplified using the following properties.

  $\sqrt{ab} = \sqrt{a} \times \sqrt{b},\ a \geq 0,\ b \geq 0$

  $\sqrt{\dfrac{a}{b}} = \dfrac{\sqrt{a}}{\sqrt{b}},\ a \geq 0,\ b > 0$

- The number i is the imaginary unit, where $i^2 = -1$ and $i = \sqrt{-1}$.
- A complex number is a number in the form $a + bi$, where a and b are real numbers and i is the imaginary unit.

Communicate Your Understanding

1. Describe the difference between an entire radical and a mixed radical.

2. Describe how you would simplify

a) $\sqrt{60}$ **b)** $\dfrac{\sqrt{14}}{\sqrt{2}}$ **c)** $\dfrac{10 + \sqrt{20}}{2}$

3. Describe how you would simplify

a) $\sqrt{-28}$ **b)** $\dfrac{9 + \sqrt{-54}}{3}$

4. Describe how you would evaluate $(-3i)^2$.

Practise

A

1. Simplify.

a) $\sqrt{12}$ **b)** $\sqrt{20}$ **c)** $\sqrt{45}$

d) $\sqrt{50}$ **e)** $\sqrt{24}$ **f)** $\sqrt{63}$

g) $\sqrt{200}$ **h)** $\sqrt{32}$ **i)** $\sqrt{44}$

j) $\sqrt{60}$ **k)** $\sqrt{18}$ **l)** $\sqrt{54}$

m) $\sqrt{128}$ **n)** $\sqrt{90}$ **o)** $\sqrt{125}$

2. Simplify.

a) $\dfrac{\sqrt{14}}{\sqrt{7}}$ b) $\dfrac{\sqrt{10}}{\sqrt{2}}$ c) $\dfrac{\sqrt{60}}{\sqrt{3}}$ d) $\dfrac{\sqrt{40}}{\sqrt{5}}$

e) $\dfrac{\sqrt{33}}{\sqrt{3}}$ f) $\dfrac{\sqrt{7}}{\sqrt{4}}$ g) $\dfrac{\sqrt{20}}{\sqrt{9}}$ h) $\dfrac{3\sqrt{8}}{\sqrt{2}}$

i) $\dfrac{27\sqrt{15}}{3\sqrt{5}}$ j) $\dfrac{12\sqrt{75}}{4\sqrt{3}}$ k) $\dfrac{4\sqrt{2}}{\sqrt{8}}$ l) $\dfrac{2\sqrt{2}}{\sqrt{18}}$

3. Simplify.

a) $\sqrt{2} \times \sqrt{10}$ b) $\sqrt{3} \times \sqrt{6}$

c) $\sqrt{15} \times \sqrt{5}$ d) $\sqrt{7} \times \sqrt{11}$

e) $4\sqrt{3} \times \sqrt{7}$ f) $3\sqrt{6} \times 3\sqrt{6}$

g) $2\sqrt{2} \times 3\sqrt{6}$ h) $2\sqrt{5} \times 3\sqrt{10}$

i) $3\sqrt{3} \times 4\sqrt{15}$ j) $4\sqrt{7} \times 2\sqrt{14}$

k) $\sqrt{6} \times \sqrt{3} \times \sqrt{2}$ l) $2\sqrt{7} \times 3\sqrt{1} \times \sqrt{7}$

4. Simplify.

a) $\dfrac{10 + 15\sqrt{5}}{5}$ b) $\dfrac{21 - 7\sqrt{6}}{7}$ c) $\dfrac{6 + \sqrt{8}}{2}$

d) $\dfrac{12 - \sqrt{27}}{3}$ e) $\dfrac{-10 - \sqrt{50}}{5}$ f) $\dfrac{-12 + \sqrt{48}}{4}$

5. Simplify.

a) $\sqrt{-9}$ b) $\sqrt{-25}$ c) $\sqrt{-81}$ d) $\sqrt{-5}$

e) $\sqrt{-13}$ f) $\sqrt{-23}$ g) $\sqrt{-12}$ h) $\sqrt{-40}$

i) $\sqrt{-54}$ j) $-\sqrt{-4}$ k) $-\sqrt{-20}$ l) $-\sqrt{-60}$

6. Evaluate.

a) $5i \times 5i$ b) $2i \times 3i$
c) $(-2i) \times (-2i)$ d) $(-3i) \times (-4i)$
e) $2i \times (-5i)$ f) $(-3i) \times 6i$

7. Simplify.

a) i^3 b) i^4
c) i^5 d) $4i \times 5i$
e) $5i^2$ f) $-i^7$
g) $3(-2i)^2$ h) $i(4i)^3$
i) $(3i)(-6i)$ j) $(i\sqrt{2})^2$
k) $-(i\sqrt{5})^2$ l) $(i\sqrt{6})(-i\sqrt{6})$
m) $(2i\sqrt{3})^2$ n) $(-5i\sqrt{2})^2$
o) $(4i\sqrt{5})(-2i\sqrt{5})$

8. Simplify.

a) $4 + \sqrt{-20}$ b) $7 - \sqrt{-18}$
c) $10 + \sqrt{-75}$ d) $11 - \sqrt{-63}$
e) $-2 - \sqrt{-90}$ f) $-6 - \sqrt{-52}$

9. Simplify.

a) $\dfrac{15 + 20i\sqrt{5}}{5}$ b) $\dfrac{14 - 28i\sqrt{6}}{14}$

c) $\dfrac{10 - \sqrt{-16}}{2}$ d) $\dfrac{12 + \sqrt{-27}}{3}$

e) $\dfrac{-8 + \sqrt{-32}}{4}$ f) $\dfrac{-21 - \sqrt{-98}}{7}$

Apply, Solve, Communicate

10. Express each of the following as an integer.

a) $\sqrt{5^2}$ b) $(\sqrt{5})^2$ c) $\sqrt{(-5)^2}$ d) $-\sqrt{5^2}$ e) $-(\sqrt{5})^2$ f) $-\sqrt{(-5)^2}$

B

11. Communication Classify each of the following numbers into one or more of these sets: real, rational, irrational, complex, imaginary, pure imaginary. Explain your reasoning.

a) $\sqrt{5}$ b) $\sqrt{-4}$ c) $1 + \sqrt{3}$ d) $3 + i\sqrt{6}$

12. Measurement Express the exact area of the triangle in simplest radical form.

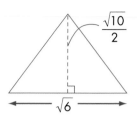

13. Measurement A square has an area of 675 cm². Express the side length in simplest radical form.

14. Application There are many variations on the game of chess. Most are played on square boards that consist of a number of small squares. However, some variations do not use the familiar 64-square board.

a) If each small square on a Grand Chess board is 2 cm by 2 cm, each diagonal of the whole board measures $\sqrt{800}$ cm. How many small squares are on the board?

b) A Japanese variation of chess is called Chu Shogi. If each small square on a Chu Shogi board measures 3 cm by 3 cm, each diagonal of the whole board measures $\sqrt{2592}$ cm. How many small squares are on the board?

Web Connection
www.school.mcgrawhill.ca/resources/
To learn more about variations on the game of chess, visit the above web site. Go to **Math Resources**, then to *MATHEMATICS 11*, to find out where to go next. Summarize how a variation that interests you is played.

15. Pattern a) Simplify i^2, i^3, i^4, i^5, i^6, i^7, i^8, i^9, i^{10}, i^{11}, and i^{12}.

b) Describe the pattern in the values.

c) Describe how to simplify i^n, where n is a whole number.

d) Simplify i^{48}, i^{94}, i^{85}, and i^{99}.

16. Measurements of lengths and areas

a) Determine the length of the diagonal of each of the following Ontario flags. Write each answer in simplest radical form.

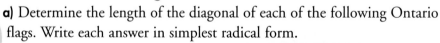

2   4   6

b) Describe the relationship between the length of the diagonal and either dimension of the flag.

c) Use the relationship from part b) to predict the length of the diagonal of a 150 cm by 75 cm Ontario flag. Leave your answer in simplest radical form.

d) Describe the relationship between the length of the diagonal and the area of the Ontario flag.

e) Use the relationship from part d) to predict the length of the diagonal of an Ontario flag with an area of 24 200 cm^2. Leave your answer in simplest radical form.

C

17. Communication Check if $x = -i\sqrt{3}$ is a solution to the equation $x^2 + 3 = 0$. Justify your reasoning.

18. Simplify each of the following by first expressing it as the product of two cube roots.

a) $\sqrt[3]{16}$ **b)** $\sqrt[3]{32}$ **c)** $\sqrt[3]{54}$ **d)** $\sqrt[3]{81}$

19. Equations Solve. Express each answer in simplest radical form.

a) $x\sqrt{2} = \sqrt{14}$ **b)** $5x = \sqrt{50}$ **c)** $\dfrac{x}{\sqrt{3}} = \sqrt{6}$ **d)** $\dfrac{\sqrt{30}}{x} = \sqrt{5}$

20. a) For the property $\sqrt{ab} = \sqrt{a} \times \sqrt{b}$, explain the restrictions $a \geq 0$, $b \geq 0$.

b) For the property $\sqrt{\dfrac{a}{b}} = \dfrac{\sqrt{a}}{\sqrt{b}}$, the restrictions are $a \geq 0$, $b > 0$. Why is the second restriction not $b \geq 0$?

PATTERN *Power*

Subtracting 9 from the two-digit positive integer 21 results in the reversal of the digits to give 12.

1. List all two-digit positive integers for which the digits are reversed when you subtract 9.

2. Find all the two-digit positive integers for which the digits are reversed when you subtract

a) 18   **b)** 27   **c)** 36

3. Describe the pattern in words.

4. Use the pattern to find all the two-digit positive integers for which the digits are reversed when you subtract

a) 54   **b)** 72

2.2 Maximum or Minimum of a Quadratic Function by Completing the Square

To celebrate Canada Day and Independence Day, the International Freedom Festival is held at the end of June in Windsor, Ontario, and Detroit, Michigan. At the end of the festival, there is a fireworks display. The fireworks are launched from barges in the Detroit River.

INVESTIGATE & INQUIRE

The formula for the height, h metres, of an object propelled into the air is

$$h = -\frac{1}{2}gt^2 + v_0 t + h_0$$

where g represents the acceleration due to gravity, which is about 9.8 m/s^2 on Earth, t seconds represents the time, v_0 metres per second represents the initial velocity, and h_0 metres is the initial height of the object.

In fireworks displays, most firework shells are launched from steel tubes. A certain fireworks shell has an initial velocity of 39.2 m/s and is launched 1.5 m above the surface of the water.

1. Substitute known values to write the function, h, for the height of the fireworks shell.

2. Write the function in the form $y = ax^2 + bx + c$.

3. Complete the square to write the function in the form $y = a(x - h)^2 + k$.

4. What is the maximum height reached by the fireworks shell? Justify your reasoning.

5. How many seconds after launch does the fireworks shell reach its maximum height? Justify your reasoning.

6. Describe a way to find the maximum height without completing the square.

To determine the maximum or minimum value of a quadratic function in the form $y = ax^2 + bx + c$ by completing the square, rewrite the function in the form $y = a(x - h)^2 + k$. The maximum or minimum value of the function is k, when $x = h$. If $a > 0$, k is the minimum value of the function. If $a < 0$, k is the maximum value of the function.

EXAMPLE 1 **Finding a Minimum Value**

Find the minimum value of the function $y = 4x^2 - 24x + 31$ by completing the square.

SOLUTION

Factor the coefficient of x^2 from the first two terms. Then, complete the square.

$$y = 4x^2 - 24x + 31$$

Group the terms containing x: $= [4x^2 - 24x] + 31$
Factor the coefficient of x^2 from the first two terms: $= 4[x^2 - 6x] + 31$
Complete the square inside the brackets: $= 4[x^2 - 6x + 9 - 9] + 31$
Write the perfect square trinomial as the square of a binomial: $= 4[(x - 3)^2 - 9] + 31$
Expand to remove the square brackets: $= 4(x - 3)^2 - 36 + 31$
Simplify: $= 4(x - 3)^2 - 5$

The function reaches a minimum value of -5 when $x = 3$.

The solution can be modelled graphically.

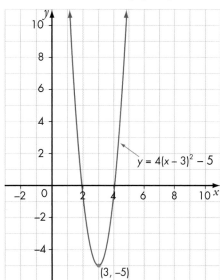

$y = 4(x - 3)^2 - 5$

$(3, -5)$

EXAMPLE 2 **Finding a Maximum Value**

Find the maximum value of the function $y = -0.3x^2 - 2.4x + 7.3$
by completing the square.

SOLUTION

Factor the coefficient of x^2 from the first two terms.
Then, complete the square.

$$y = -0.3x^2 - 2.4x + 7.3$$

Group the terms containing x: $\qquad\qquad = [-0.3x^2 - 2.4x] + 7.3$

Factor the coefficient of x^2 from the first two terms: $\qquad = -0.3[x^2 + 8x] + 7.3$

Complete the square inside the brackets: $\qquad = -0.3[x^2 + 8x + 16 - 16] + 7.3$

Write the perfect square trinomial as the square of a binomial: $= -0.3[(x + 4)^2 - 16] + 7.3$

Expand to remove the square brackets: $\qquad = -0.3(x + 4)^2 + 4.8 + 7.3$

Simplify: $\qquad\qquad\qquad = -0.3(x + 4)^2 + 12.1$

The function reaches a maximum value of
12.1 when $x = -4$.

The solution can be modelled graphically.

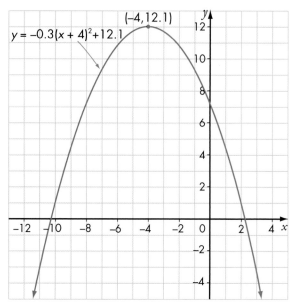

$y = -0.3(x + 4)^2 + 12.1$

$(-4, 12.1)$

In some cases, completing the square requires the use of fractions.

Example 3 Completing the Square Using Fractions

Find the maximum value of the function $y = 5x - 3x^2$ by completing the square.

Solution

Rewrite the equation in the form $y = a(x - h)^2 + k$.

The solution can be modelled graphically.

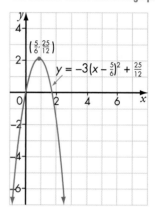

$$y = 5x - 3x^2$$
$$= -3x^2 + 5x$$
$$= [-3x^2 + 5x]$$
$$= -3\left[x^2 - \frac{5}{3}x\right]$$
$$= -3\left[x^2 - \frac{5}{3}x + \frac{25}{36} - \frac{25}{36}\right] \qquad \frac{1}{2} \times \frac{5}{3} = \frac{5}{6} \qquad \left(\frac{5}{6}\right)^2 = \frac{25}{36}$$
$$= -3\left[\left(x - \frac{5}{6}\right)^2 - \frac{25}{36}\right]$$
$$= -3\left(x - \frac{5}{6}\right)^2 + \frac{25}{12}$$

The maximum value is $\dfrac{25}{12}$ when $x = \dfrac{5}{6}$.

Note that a graphing calculator can be used to model the solution to Example 3. On a graphing calculator, the maximum operation or the minimum operation can be used to find the maximum or minimum value of a function. The ▶Frac function can be used to display the value as a fraction, if necessary.

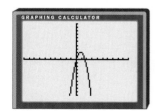
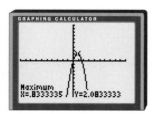

EXAMPLE 4 **Writing a Function**

Find the minimum product of two numbers whose difference is 8.

SOLUTION

The quantity to be minimized is the product, p, of two numbers.
Let one number be n.
Since the difference between the two numbers is 8, the other number can be
represented by $n - 8$ or $n + 8$.
Either expression can be used. We will use $n - 8$.

$p = n \times (n - 8)$
$\quad = n^2 - 8n$

Write the function in the form $y = a(x - h)^2 + k$.

$p = n^2 - 8n$
$\quad = n^2 - 8n + 16 - 16$ Solve the problem using n and $n + 8$ to represent the two
$\quad = (n - 4)^2 - 16$ numbers. Compare the solutions.

The minimum value of the function is -16 when $n = 4$.
So, -16 is the minimum product of two numbers whose difference is 8.

EXAMPLE 5 **Fundraising**

The student council plans to run the annual talent show to raise money for
charity. Last year, tickets sold for $11 each, and 400 people attended. The
student council has decided to raise the ticket price for this year's talent show.
The student council has determined that, for every $1 increase in price, the
talent show attendance would decrease by 20 people. What ticket price would
maximize the revenue?

SOLUTION

Let x represent the number of $1 increases.
The cost of a ticket would be $(11 + x)$.
The number of tickets sold would be $(400 - 20x)$.
The revenue from ticket sales, R, is (number of tickets sold) × (cost per ticket).
So, $R = (11 + x)(400 - 20x)$
Find the maximum value of this function.

$R = (11 + x)(400 - 20x)$
$\quad = 4400 - 220x + 400x - 20x^2$
$\quad = -20x^2 + 180x + 4400$
$\quad = [-20x^2 + 180x] + 4400$
$\quad = -20[x^2 - 9x] + 4400$
$\quad = -20[x^2 - 9x + 20.25 - 20.25] + 4400$
$\quad = -20[(x - 4.5)^2 - 20.25] + 4400$
$\quad = -20(x - 4.5)^2 + 405 + 4400$
$\quad = -20(x - 4.5)^2 + 4805$

If half the coefficient of x is a terminating decimal, it may be more convenient to use decimals than fractions.

In this case, $-\dfrac{9}{2} = -4.5$, and $(-4.5)^2 = 20.25$.

The function reaches a maximum of 4805 when $x = 4.5$.
The ticket price should be increased by \$4.50 to maximize the revenue.
So, a ticket price of \$11 + \$4.50, or \$15.50, would maximize the revenue.

Key Concepts

- To determine the maximum or minimum value of a quadratic function in the form $y = ax^2 + bx + c$, rewrite the function in the form $y = a(x - h)^2 + k$ by completing the square.
- For a quadratic function in the form $y = a(x - h)^2 + k$, the maximum or minimum value is k, when $x = h$. If $a > 0$, k is the minimum value of the function. If $a < 0$, k is the maximum value of the function.

Communicate Your Understanding

1. Describe how you would find the value of c that makes the expression $x^2 + 3x + c$ a perfect square trinomial.

2. Describe the steps you would use to find the maximum or minimum value of the function $y = -2x^2 - 8x + 11$ by completing the square.

3. How can you determine if a quadratic function has a maximum or minimum value without completing the square or graphing the function?

Practise

A

1. Find the value of c that will make each expression a perfect square trinomial.

a) $x^2 + 6x + c$ **b)** $x^2 - 20x + c$

c) $x^2 - 3x + c$ **d)** $x^2 + 5x + c$

e) $x^2 + x + c$ **f)** $x^2 - x + c$

g) $x^2 + 0.8x + c$ **h)** $x^2 - 0.05x + c$

i) $x^2 - 2.4x + c$ **j)** $x^2 + 13.7x + c$

k) $x^2 - \dfrac{2}{3}x + c$ **l)** $x^2 + \dfrac{x}{6} + c$

2. Find the maximum or minimum value of the function and the value of x when it occurs.

a) $y = x^2 + 12x - 7$
b) $y = -x^2 + 6x + 1$
c) $y = 13 + x^2 - 20x$
d) $y = -x^2 - 14x - 5$
e) $y = 10 - 10x - x^2$
f) $y = 2x^2 + 12x$
g) $y = 3x^2 - 12x + 11$
h) $y = -2x^2 - 4x + 1$
i) $y = -36x + 6x^2 - 5$
j) $y = 0.3x^2 + 1.2x - 0.5$
k) $y = 0.2x^2 + 1.6x + 3.1$
l) $y = -0.4x^2 - 2.4x$

3. Find the maximum or minimum value of the function and the value of x when it occurs.

a) $y = x^2 + 3x + 1$
b) $y = x^2 - x - 2$
c) $y = 3x^2 + 2x$
d) $y = -4x^2 + 4x - 9$
e) $y = -\dfrac{1}{3}x^2 + 2x + 4$
f) $y = -2x^2 + 3x - 2$
g) $y = -x^2 - 5x$
h) $y = 3x^2 - 0.6x + 1$
i) $y = -2x^2 - 0.8x - 2$
j) $y = 0.5x^2 + x + 2$
k) $y = -3x^2 + 4x$
l) $y = 0.5x^2 - 0.6x$
m) $y + 4 = -x^2 + 1.8x$
n) $y = -0.2x^2 + 0.5x - 2$

Apply, Solve, Communicate

4. a) Find the minimum product of two numbers whose difference is 12.
b) What are the two numbers?

5. a) Find the minimum product of two numbers whose difference is 9.
b) What are the two numbers?

6. a) Find the maximum product of two numbers whose sum is 23.
b) What are the two numbers?

B

7. Number game A student is asked to do the following: "Choose any number and square it. Then, subtract eight times the original number. Then, add 35. Find the least result and the value of the original number that gives the least result."
a) If x is the original number and y is the result, write an equation that represents the instructions.
b) Find the least result and the value of the original number that gives the least result.

8. Number game A student is asked to do the following: "Choose any number. Subtract ten times the original number and subtract the square of the original number from 375. Find the greatest result and the value of the original number that gives the greatest result."

a) If x is the original number and y is the result, write an equation that represents the instructions.

b) Find the greatest result and the value of the original number that gives the greatest result.

9. Retail sales A sportswear store sells baseball caps with the local baseball team's logo on them. Last year, the store sold 600 caps at $15 each. The store manager is planning to increase the price. A consumer survey shows that for every $1 increase, there will be a drop of 30 sales a year.

a) What should the selling price be to maximize the annual revenue?

b) What is the maximum annual revenue from the caps?

10. Two numbers have a sum of 13.

a) Find the minimum of the sum of their squares.

b) What are the two numbers?

11. Find the maximum or minimum value of y and the value of x when it occurs.

a) $y = x^2 - 9$ **b)** $y = -4x^2 + 25$

12. Measurement Determine the maximum area of a triangle, in square centimetres, if the sum of its base and its height is 13 cm.

13. Measurement A rectangle has dimensions $3x$ and $5 - 2x$.

a) What is the maximum area of the rectangle?

b) What value of x gives the maximum area?

14. Projectile motion The formula for the height, h metres, of an object propelled into the air is

$$h = -\frac{1}{2}gt^2 + v_0 t + h_0$$

where g represents the acceleration due to gravity, which is about 9.8 m/s^2 on the Earth, t seconds represents the time, v_0 metres per second represents the initial velocity, and h_0 metres is the initial height of the object. A projectile has an initial velocity of 34.3 m/s and is launched 2.1 m above the ground.

a) What is the maximum height, in metres, reached by the projectile?

b) How many seconds after launch does the projectile reach its maximum height?

15. Measurement A triangle has a height of $2x$ and a base of $7 - 4x$.

a) What is the maximum area of the triangle?

b) What value of x gives the maximum area?

16. Application The path of a thrown baseball can be modelled by the function
$$h = -0.004d^2 + 0.14d + 2$$
where h is the height of the ball, in metres, and d is the horizontal distance of the ball from the player, in metres.

a) What is the maximum height reached by the ball?

b) What is the horizontal distance of the ball from the player when it reaches its maximum height?

c) How far from the ground is the ball when the player releases it?

17. Application In a nutrient medium, the rate of increase in the surface area of a cell culture can be modelled by the quadratic function
$$S = -0.008t^2 + 0.04t$$
where S is the rate of increase in the surface area, in square millimetres per hour, and t is the time, in hours, since the culture began growing. Find the maximum rate of increase in the surface area and the time taken to reach this maximum.

18. Natural bridge Owachomo Natural Bridge is found in Natural Bridges National Monument in Utah. If the origin is located at one end of the natural arch, the curve can be modelled by the equation
$$h = -0.043d^2 + 2.365d$$
where h metres is the height of the arch, and d metres is the horizontal distance from the origin.

a) What is the maximum height of the arch, to the nearest hundredth of a metre?

b) What is the width of the arch at the base?

19. Measurements of lengths and areas A rectangular corral is to be built using 70 m of fencing.

a) If the fencing has to enclose all four sides of the corral, what is the maximum possible area of the corral, in square metres?

b) In part a), what are the dimensions of the corral with the maximum possible area?

c) If the corral is built with a wall of a large barn as one side and with the fencing enclosing the other three sides, what is the maximum possible area of the corral, in square metres?

d) In part c), what are the dimensions of the corral with the maximum possible area?

C

20. Gravity The table gives the approximate values of the acceleration due to gravity, g, for three planets. Suppose a ball is thrown upward on each planet with an initial velocity of 20 m/s from an initial height of 2 m.

Planet	Value of g (m/s²)
Mercury	4
Neptune	12
Venus	9

a) Write a function for the height of the ball on each planet by substituting known values into the formula $h = -\dfrac{1}{2}gt^2 + v_0t + h_0$, where g represents the acceleration due to gravity, t seconds represents the time, v_0 metres per second represents the initial velocity, and h_0 metres is the initial height of the object.

b) For each planet, determine the maximum height the ball would reach, in metres, and the number of seconds it would take to reach its maximum height.

21. Inquiry/Problem Solving If $y = x^2 + kx + 3$, determine the value(s) of k for which the minimum value of the function is an integer. Explain your reasoning.

22. If $y = -4x^2 + kx - 1$, determine the value(s) of k for which the maximum value of the function is an integer. Explain your reasoning.

ACHIEVEMENT Check   Knowledge/Understanding   Thinking/Inquiry/Problem Solving   Communication   Application

A large dealership has been selling new cars at $6000 over the factory price. Sales have been averaging 80 cars per month. Because of inflation, the $6000 markup is going to be increased. The marketing manager has determined that, for every $100 increase, there will be one less car sold each month. What should the new markup be in order to maximize revenue?

2.3 Solving Quadratic Equations

The process of completing the square, which was used in Section 2.2 to find the maximum or minimum of a quadratic function, is also one of the ways to solve quadratic equations.

In Roman times, a city forum was a large open space surrounded by buildings. It provided a meeting place and a centre for public life.

The Roman city of Pompeii was destroyed by the eruption of Mount Vesuvius in A.D. 79, but a coating of ash protected the city's buildings and streets. The continuing excavation of Pompeii allows visitors to see how citizens of the Roman Empire lived.

INVESTIGATE & INQUIRE

The forum in Pompeii was a rectangle whose length was 120 m greater than the width.

1. a) Let the width of the forum be x. Write an expression in expanded form to represent the area of the forum.
b) The area of the forum was 6400 m². Write an equation with the expression from part a) on the left side and the numerical value of the area on the right side.

2. a) Add a number to the left side of the equation to make the left side a perfect square.
b) Why must this number also be added to the right side of the equation?
c) Write the left side as the square of a binomial, and simplify the right side.
d) Take the square root of both sides.
e) Solve for x.

3. Should either value of x be rejected? Explain.

4. For the forum in Pompeii, what was
a) the width?   **b)** the length?

5. a) Describe two other algebraic methods for solving the same equation.
b) Which of the three methods do you prefer? Explain.

6. Solve each of the following equations using a method of your choice.
a) $x^2 + 4x = 12$ **b)** $x^2 - 2x = 3$ **c)** $x^2 + 6x = -8$

An equation may have a solution in one set of numbers but not in another. For example, the equation $x + 1 = 0$ has no solution in the set of whole numbers. However, it has a solution, $x = -1$, in the set of integers. Similarly, the equation $x^2 - 2 = 0$ has no solution in the set of rational numbers. However, it has two solutions, $x = \pm\sqrt{2}$, in the set of real numbers.

The equation $x^2 + 1 = 0$ has no solution in the set of real numbers. An attempt to solve the equation gives

$$x^2 = -1$$
$$\text{and } x = \pm\sqrt{-1}$$

Since $i = \sqrt{-1}$, the solutions of the equation $x^2 + 1 = 0$ can be written as i and $-i$. These solutions are in the set of pure imaginary numbers. Many quadratic equations have roots that are pure imaginary numbers or imaginary numbers.

All quadratic functions have two zeros, that is, two values of the independent variable that make the value of the function equal to zero. If the graph of a quadratic function intersects the x-axis, the zeros of the function are real zeros, that is, they are real numbers. If there are two real zeros, they can be distinct or equal. If the graph of a quadratic function does not intersect the x-axis, the function has two imaginary zeros. The three possibilities are shown in the graphs.

two distinct real zeros

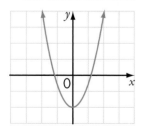

two equal real zeros

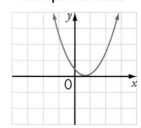

two imaginary zeros

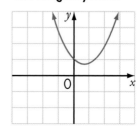

The quadratic equations that correspond to the functions shown in the graphs have two distinct real roots, two equal real roots, and two imaginary roots.

The algebraic method of completing the square gives exact solutions to a quadratic equation. One of the steps in the process involves using the square root principle. For example,

if $(x + 2)^2 = 9$

then $x + 2 = \pm 3$

$\qquad x + 2 = 3 \ \text{ or } \ x + 2 = -3$

$\qquad\qquad x = 1 \quad \text{ or } \quad x = -5$

EXAMPLE 1 Solving by Completing the Square

Solve $x^2 - 6x - 27 = 0$ by completing the square.

SOLUTION

$$x^2 - 6x - 27 = 0$$

Add 27 to both sides: $\qquad\qquad\qquad\qquad\qquad\qquad\qquad x^2 - 6x = 27$

Add the square of half the coefficient of x to both sides: $\ x^2 - 6x + 9 = 27 + 9$

$$x^2 - 6x + 9 = 36$$

Write the left side as the square of a binomial: $\qquad\quad (x - 3)^2 = 36$

Take the square root of both sides: $\qquad\qquad\qquad\quad x - 3 = \pm 6$

Solve for x: $\qquad\qquad\qquad\qquad\qquad x - 3 = 6 \text{ or } x - 3 = -6$

$$x = 9 \quad \text{ or } \quad x = -3$$

The roots are 9 and −3.

The solution to Example 1 can be modelled graphically. The equation has two distinct real roots. The graph of the corresponding function has two distinct real zeros.

Note that the roots of the equation $x^2 - 6x - 27 = 0$ are the x-intercepts of the graph of the corresponding function, $y = x^2 - 6x - 27$.

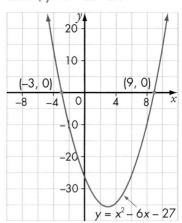

To solve by completing the square when the coefficient of x^2 is not 1, divide each term of the equation by the coefficient of x^2 before completing the square.

EXAMPLE 2 Solving by Completing the Square, $a \neq 1$

Solve $2x^2 - 5x - 1 = 0$ by completing the square. Express answers as exact roots and as approximate roots, to the nearest hundredth.

SOLUTION

$$2x^2 - 5x - 1 = 0$$

Divide both sides by 2:
$$x^2 - \frac{5}{2}x - \frac{1}{2} = 0$$

Add $\frac{1}{2}$ to both sides:
$$x^2 - \frac{5}{2}x = \frac{1}{2}$$

Complete the square:
$$x^2 - \frac{5}{2}x + \frac{25}{16} = \frac{1}{2} + \frac{25}{16}$$

Write the left side as the square of a binomial:
$$\left(x - \frac{5}{4}\right)^2 = \frac{33}{16}$$

Take the square root of both sides:
$$x - \frac{5}{4} = \pm\frac{\sqrt{33}}{4}$$

Solve for x:
$$x = \frac{5}{4} \pm \frac{\sqrt{33}}{4}$$

$$x = \frac{5 \pm \sqrt{33}}{4}$$

Note that the x-intercepts of the graph of $y = 2x^2 - 5x - 1$ are the roots of $2x^2 - 5x - 1 = 0$.

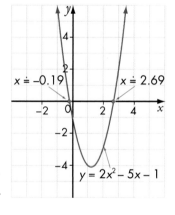

The exact roots are $\dfrac{5 + \sqrt{33}}{4}$ and $\dfrac{5 - \sqrt{33}}{4}$.

Estimate

$$\sqrt{33} \doteq 6$$

$$\frac{5+6}{4} \doteq 3 \qquad \frac{5-6}{4} = -\frac{1}{4}$$

The approximate roots are 2.69 and −0.19, to the nearest hundredth.

EXAMPLE 3 Finding Complex Roots by Completing the Square

Solve $3x^2 + 2x + 6 = 0$ by completing the square.

SOLUTION

$$3x^2 + 2x + 6 = 0$$

Divide both sides by 3:
$$x^2 + \frac{2}{3}x + 2 = 0$$

Subtract 2 from both sides:
$$x^2 + \frac{2}{3}x = -2$$

Add the square of half the coefficient of x to both sides:
$$x^2 + \frac{2}{3}x + \frac{1}{9} = -2 + \frac{1}{9}$$

Write the left side as the square of a binomial:
$$\left(x + \frac{1}{3}\right)^2 = \frac{-17}{9}$$

Take the square root of both sides:
$$x + \frac{1}{3} = \frac{\pm\sqrt{-17}}{3}$$

$$x + \frac{1}{3} = \frac{\pm\sqrt{-1} \times \sqrt{17}}{3}$$

Use the definition of i:
$$x + \frac{1}{3} = \frac{\pm i\sqrt{17}}{3}$$

Solve for x:
$$x = \frac{-1 \pm i\sqrt{17}}{3}$$

The roots are $\dfrac{-1 + i\sqrt{17}}{3}$ and $\dfrac{-1 - i\sqrt{17}}{3}$.

The solution to Example 3 can be modelled graphically. The equation has two imaginary roots. The graph of the corresponding function does not intersect the x-axis and has two imaginary zeros.

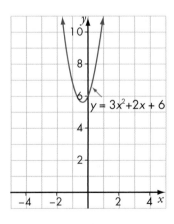

$y = 3x^2 + 2x + 6$

EXAMPLE 4 Solving by Factoring

Solve $4x^2 - 11x = x - 9$ by factoring. Check the solution.

SOLUTION

$$4x^2 - 11x = x - 9$$

Write in the form $ax^2 + bx + c = 0$: $\quad 4x^2 - 12x + 9 = 0$

Factor the left side: $\quad (2x - 3)(2x - 3) = 0$

Use the zero product property: $\quad 2x - 3 = 0 \ \text{ or } \ 2x - 3 = 0$

$$2x = 3 \ \text{ or } \qquad 2x = 3$$
$$x = \frac{3}{2} \qquad\qquad x = \frac{3}{2}$$

Check.

For $x = \dfrac{3}{2}$,

L.S. $= 4x^2 - 11x$ $\qquad\qquad$ R.S. $= x - 9$

$\quad = 4\left(\dfrac{3}{2}\right)^2 - 11\left(\dfrac{3}{2}\right) \qquad\qquad = \dfrac{3}{2} - 9$

$\quad = 4\left(\dfrac{9}{4}\right) - \dfrac{33}{2} \qquad\qquad\quad = \dfrac{3 - 18}{2}$

$\quad = 9 - \dfrac{33}{2} \qquad\qquad\qquad\quad = -\dfrac{15}{2}$

$\quad = \dfrac{18 - 33}{2}$

$\quad = -\dfrac{15}{2}$

$$\text{L.S.} = \text{R.S.}$$

There are two equal roots, $\dfrac{3}{2}$ and $\dfrac{3}{2}$.

The solution to Example 4 can be modelled graphically.
The equation has two equal real roots. The graph of the
corresponding function has two equal real zeros.

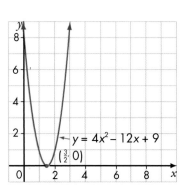

Example 5 Solving Using the Quadratic Formula

Solve $x^2 - 2x + 3 = 0$ using the quadratic formula.

Solution

For $x^2 - 2x + 3 = 0$, $a = 1$, $b = -2$, and $c = 3$.

Use the quadratic formula: $\quad x = \dfrac{-b \pm \sqrt{b^2 - 4ac}}{2a}$

Note that the graph of $y = x^2 - 2x + 3$ does not intersect the x-axis.

$$= \dfrac{-(-2) \pm \sqrt{(-2)^2 - 4(1)(3)}}{2(1)}$$

$$= \dfrac{2 \pm \sqrt{4 - 12}}{2}$$

$$= \dfrac{2 \pm \sqrt{-8}}{2}$$

$$= \dfrac{2 \pm \sqrt{-1} \times \sqrt{8}}{2}$$

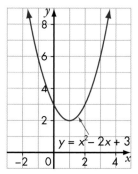

Use the definition of i: $\quad = \dfrac{2 \pm i\sqrt{8}}{2}$

Simplify: $\quad = \dfrac{2 \pm 2i\sqrt{2}}{2}$

$$= 1 \pm i\sqrt{2}$$

The roots are $1 + i\sqrt{2}$ and $1 - i\sqrt{2}$.

Example 6 Checking Imaginary Roots

Solve and check $x^2 + 4 = 0$.

Solution

$$x^2 + 4 = 0$$

Subtract 4 from both sides: $\qquad\qquad x^2 = -4$

Take the square root of both sides: $\qquad x = \pm\sqrt{-4}$

Simplify: $\qquad\qquad\qquad\qquad\qquad x = \pm\sqrt{-1} \times \sqrt{4}$

$$= \pm 2i$$

$$x = 2i \text{ or } x = -2i$$

Check.

For $x = 2i$,

L.S. $= x^2 + 4$ R.S. $= 0$

$\quad = (2i)^2 + 4$

$\quad = 2^2 \times i^2 + 4$

$\quad = 4 \times (-1) + 4$

$\quad = -4 + 4$

$\quad = 0$

$\qquad\qquad$ L.S. = R.S.

For $x = -2i$,

L.S. $= x^2 + 4$ R.S. $= 0$

$\quad = (-2i)^2 + 4$

$\quad = (-2)^2 \times i^2 + 4$

$\quad = 4 \times (-1) + 4$

$\quad = -4 + 4$

$\quad = 0$

$\qquad\qquad$ L.S. = R.S.

The roots are $2i$ and $-2i$.

EXAMPLE 7 Planning a Dining Room

A preliminary floor plan for a new house includes a rectangular dining room that measures 5 m by 4 m. The customers want a larger dining room and agree to an area of 25 m². The architect decides to redesign the floor plan by adding a strip of floor of uniform width to two adjacent sides of the dining room, as shown in the diagram. How wide should the strip be, to the nearest thousandth of a metre?

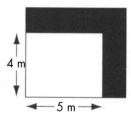

SOLUTION

Let the width of the strip be x metres.
Write and solve an equation to find x.
The dimensions of the new dining room are $(5 + x)$ by $(4 + x)$.
The area of the new dining room is 25 m².

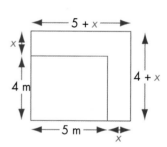

Write the equation: $\qquad\qquad\qquad (5 + x)(4 + x) = 25$

Expand the left side: $\qquad\qquad\qquad 20 + 9x + x^2 = 25$

Write in the form $ax^2 + bx + c = 0$: $\qquad x^2 + 9x - 5 = 0$

For $x^2 + 9x - 5 = 0$, $a = 1$, $b = 9$, and $c = -5$.

Use the quadratic formula.

$$x = \frac{-b \pm \sqrt{b^2 - 4ac}}{2a}$$

$$= \frac{-9 \pm \sqrt{9^2 - 4(1)(-5)}}{2}$$

$$= \frac{-9 \pm \sqrt{81 + 20}}{2}$$

$$= \frac{-9 \pm \sqrt{101}}{2}$$

$$x = \frac{-9 + \sqrt{101}}{2} \text{ or } x = \frac{-9 - \sqrt{101}}{2}$$

$$\doteq 0.525 \qquad \qquad \doteq -9.525$$

Estimate

$$\frac{-9 + 10}{2} = 0.5 \qquad \frac{-9 - 10}{2} = -9.5$$

The width cannot be negative, so the root −9.525 is inadmissible and is rejected.

The strip should be 0.525 m wide, to the nearest thousandth of a metre.

Key Concepts

- To solve a quadratic equation by completing the square, first complete the square, and then take the square root of both sides to find the roots.
- To solve a quadratic equation by factoring, write the equation in the form $ax^2 + bx + c = 0$, factor $ax^2 + bx + c$, use the zero product property, and then solve the two resulting equations to find the roots.
- To solve a quadratic equation using the quadratic formula, write the equation in the form $ax^2 + bx + c = 0$, $a \neq 0$, and substitute the values for a, b, and c into the formula $x = \dfrac{-b \pm \sqrt{b^2 - 4ac}}{2a}$ to find the roots.

Communicate Your Understanding

1. Describe how you would solve $x^2 + 6x + 7 = 0$ by completing the square.
2. Describe how you would solve $2x^2 + 7x = -3$ by factoring.
3. Describe how you would use the quadratic formula to solve $x^2 = 1 - x$.

Practise

A

1. State the value of k that makes each expression a perfect square trinomial. Then, write the trinomial as the square of a binomial.

a) $x^2 + 10x + k$ b) $w^2 - 14w + k$

c) $x^2 + 7x + k$ d) $p^2 - 5p + k$

e) $x^2 + \dfrac{4}{3}x + k$ f) $d^2 - \dfrac{2}{3}d + k$

g) $x^2 + 1.4x + k$ h) $x^2 - 0.06x + k$

2. Solve.

a) $(x + 3)^2 = 9$ b) $(x - 10)^2 - 1 = 0$

c) $(s - 1)^2 = 4$ d) $(y - 4)^2 - 25 = 0$

e) $4 = \left(x + \dfrac{1}{2}\right)^2$ f) $\left(x - \dfrac{1}{3}\right)^2 = \dfrac{1}{9}$

g) $\left(a + \dfrac{3}{4}\right)^2 = \dfrac{9}{16}$ h) $(n - 0.5)^2 = 1.21$

i) $(x + 0.4)^2 - 0.01 = 0$

3. Solve by completing the square. Express solutions in simplest radical form.

a) $x^2 + 6x + 4 = 0$ b) $w^2 - 4w - 11 = 0$

c) $t^2 + 8t - 7 = 0$ d) $x^2 - 10x = 3$

e) $d^2 = 7d - 9$ f) $0 = x^2 - 5x + 2$

g) $x - 3 = -x^2$ h) $4 + y^2 = 20y$

4. Solve by completing the square. Express solutions in simplest radical form.

a) $2x^2 + 8x + 5 = 0$ b) $3x^2 - 6x + 2 = 0$

c) $6x^2 + 3x - 2 = 0$ d) $0 = 3w^2 - 5w - 2$

e) $2x - 6 = -5x^2$ f) $1 - 2z = 5z^2$

g) $\frac{1}{2}x^2 + x - 13 = 0$ h) $0.3y^2 - 0.2y = 0.3$

5. Solve by completing the square. Round solutions to the nearest hundredth.

a) $x^2 + 2x - 1 = 0$ b) $x^2 - 4x + 1 = 0$

c) $d^2 + d = 7$ d) $0 = 2r^2 - 8r + 3$

e) $7x + 4 = -2x^2$ f) $\frac{2}{3}x^2 - 2x - 3 = 0$

g) $\frac{1}{4}n^2 + n = -\frac{1}{8}$ h) $1.2x^2 - 3x - 6 = 0$

6. Solve by completing the square.

a) $x^2 + x + 6 = 0$ b) $y^2 - 2y + 8 = 0$

c) $x^2 + 6x = -17$ d) $2x^2 - 3x + 6 = 0$

e) $2n^2 + 4 = 5n$ f) $3m^2 + 8m + 8 = 0$

g) $\frac{1}{2}x^2 + x + 1 = 0$ h) $0.1g^2 - 0.3g + 0.5 = 0$

7. Solve.

a) $(x - 4)(x + 7) = 0$

b) $(2y + 3)(y - 1) = 0$

c) $(3z + 1)(4z + 3) = 0$

d) $(2x - 5)(2x - 5) = 0$

8. Solve by factoring. Check solutions.

a) $x^2 + 3x - 40 = 0$ b) $x^2 - x = 12$

c) $y^2 = 12y - 36$ d) $z^2 - 30 = -z$

e) $a^2 - 4 = 3a$ f) $b^2 - 5b = 0$

g) $m^2 = 5m + 14$ h) $t^2 = 16$

i) $t^2 + 25 = -10t$ j) $x^2 = 6x + 16$

9. Solve by factoring. Check solutions.

a) $4x^2 - 3 = 11x$ b) $4y^2 - 17y = -4$

c) $9z^2 = -24z - 16$ d) $3x^2 = 4x + 15$

e) $4x^2 = 25$ f) $2m^2 + 9m = 5$

g) $8t^2 = 1 - 2t$ h) $y - 2 = -6y^2$

i) $6x^2 + 7x + 2 = 0$ j) $5z^2 + 44z = 60$

k) $9r^2 - 16 = 0$ l) $2x^2 = -18 - 12x$

10. Solve by factoring.

a) $3p^2 = 15 - 4p$ b) $3x^2 + 7x = 0$

c) $4r^2 + 9 = 12r$ d) $4y^2 - 11y - 3 = 0$

e) $3t^2 + 13t = 10$ f) $2x^2 - 5x = 0$

g) $6n^2 = n + 5$ h) $6t^2 + 7t = 3$

i) $4m^2 - 12 = 13m$ j) $9x^2 - 17x + 8 = 0$

k) $\frac{x^2}{4} + x + 1 = 0$ l) $\frac{x^2}{2} - \frac{x}{3} - \frac{1}{6} = 0$

11. Solve using the quadratic formula.

a) $4x^2 - 12x + 5 = 0$ b) $3y^2 + 5y = 28$

c) $2m^2 = 15 + m$ d) $2z^2 + 3 = 5z$

e) $3r^2 - 20 = 7r$ f) $4x^2 = 11x + 3$

g) $5a^2 - a = 4$ h) $15 + w - 6w^2 = 0$

12. Solve using the quadratic formula. Express answers as exact roots and as approximate roots, to the nearest hundredth.

a) $3x^2 + 6x + 1 = 0$ b) $2t^2 + 6t = -3$

c) $4y^2 + 7 - 12y = 0$ d) $m^2 + 4 = -6m$

e) $2z^2 = 6z - 1$ f) $2x^2 = 11$

g) $3r^2 - 3r = 1$ h) $3n - 1 + n^2 = 0$

i) $3x^2 = 2 - 6x$ j) $5 + 5t - t^2 = 0$

k) $0.1 - 0.3m = 0.2m^2$ l) $\frac{x}{2} + 1 = \frac{7x^2}{2}$

m) $y^2 + \frac{y}{6} - \frac{1}{2} = 0$ n) $\frac{t^2}{5} - \frac{1}{2} = \frac{t}{5}$

13. Solve using the most appropriate method.

a) $x^2 + 2x + 2 = 0$ **b)** $x^2 - 4x + 8 = 0$

c) $z^2 + 5z + 8 = 0$ **d)** $n^2 - 3n + 3 = 0$

e) $x^2 - x + 7 = 0$ **f)** $-y^2 + 3y - 9 = 0$

g) $2x^2 + 3x + 3 = 0$ **h)** $3m^2 - 4m = -2$

i) $5x^2 + 5x + 2 = 0$ **j)** $4y - 1 = 5y^2$

14. Solve and check.

a) $x^2 + 9 = 0$ **b)** $y^2 + 16 = 0$

c) $2k^2 + 50 = 0$ **d)** $5z^2 = -500$

e) $n^2 + 20 = 0$ **f)** $6c^2 + 72 = 0$

g) $-2x^2 - 16 = 0$ **h)** $\dfrac{x^2}{2} - \dfrac{x^2}{3} = -1$

Apply, Solve, Communicate

15. The Olympeion The enormous temple Olympeion was constructed in Athens, Greece, in 174 B.C. The base of the temple is a rectangle with a perimeter of 300 m. The area of the base is 4400 m². What are the dimensions of the base?

16. Gardening A rectangular lawn measures 7 m by 5 m. A uniform border of flowers is to be planted along two adjacent sides of the lawn, as shown in the diagram. If the flowers that have been purchased will cover an area of 6.25 m², how wide is the border?

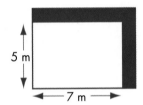

5 m

7 m

B

17. Measurement The base of a triangle is 2 cm more than the height. The area of the triangle is 5 cm². Find the base, to the nearest tenth of a centimetre.

18. Measurement The length of a rectangle is 2 m more than the width. The area of the rectangle is 20 m². Find the dimensions of the rectangle, to the nearest tenth of a metre.

19. Integers The sum of an integer and its square is 210. Find the integer.

20. Measurement A square of side length $x + 1$ has an area of 6 square units. Find the value of x, to the nearest hundredth.

21. Numbers The sum of two numbers is 14, and their product is 37. What are the numbers

a) in simplest radical form?   **b)** to the nearest thousandth?

22. Numbers Subtracting a number from half its square gives a result of 13. Express the possible values of the number in simplest radical form.

23. Application Two positive integers are in the ratio 1:3. If their sum is added to their product, the result is 224. Find the integers.

24. Whole numbers Two whole numbers differ by 3. The sum of their squares is 89. What are the numbers?

25. Architecture Is it possible to design a building so that the rectangular ground floor has a perimeter of 50 m and an area of 160 m^2? Explain.

26. Building a fence Is it possible to build a fence on three sides of a rectangular piece of land with an area of 100 m^2, so that the total length of the fence is as follows? If so, what are the dimensions of the piece of land?
a) 30 m   **b)** 25 m

27. Measurement Is it possible for a rectangle with a perimeter of 44 cm to have each of the following areas? If so, find the dimensions of the rectangle.
a) 125 cm^2 **b)** 121 cm^2 **c)** 117 cm^2

28. Solve. Express each solution in simplest radical form.
a) $x(x+3) = 2x(x+5) + 1$
b) $3(n-1)^2 = (n+1)(2n+1)$
c) $\frac{1}{2}(r+2)^2 = \frac{1}{3}(2r-1)^2$
d) $(4x-1)(3x+7) = (5x-1)(2x+3) - 6$

29. Football The function $h = -5t^2 + 20t + 2$ gives the approximate height, h metres, of a thrown football as a function of the time, t seconds, since it was thrown. The ball hit the ground before a receiver could get near it.
a) How long was the ball in the air, to the nearest tenth of a second?
b) For how many seconds was the height of the ball at least 17 m?

30. Measurement The difference between the length of the hypotenuse and the length of the next longest side of a right triangle is 3 cm. The difference between the lengths of the two perpendicular sides is 3 cm. Find the three side lengths.

31. Numbers Two numbers have a sum of 31. Find the numbers if their product is as follows. Express radical solutions in simplest radical form.
a) 240   **b)** 230   **c)** 250   **d)** $\frac{385}{4}$

32. Technology a) Graph the equation $y = x^2 - 2x + 2$ using a graphing calculator.
b) Describe the results when you try to find the zeros of the function using the Zero operation.
c) Repeat part b) for another quadratic function with imaginary roots.

33. Inquiry/Problem Solving For what values of k is $x^2 + kx + \dfrac{49}{4}$ a perfect square trinomial?

34. Airborne object The height, h metres, of an object fired upward from the ground at 50 m/s is given approximately by the equation $h = -5t^2 + 50t$, where t seconds is the time since the object was launched. Does an object fired upward at 50 m/s reach each of the following heights? If so, after how many seconds is the object at the given height?

a) 45 m   **b)** 125 m   **c)** 150 m

35. Communication Is it possible for a quadratic equation to have one real root and one imaginary root? Explain using examples.

36. Solve and check.
a) $(3t + 2)^2 = (2t - 5)^2$ **b)** $(x - 3)(x + 3) = 7$
c) $(y - 1)^2 = 2y - 3$ **d)** $(m - 2)(m + 2) - 3(m + 5) + 1 = 0$

37. Solve. Express answers as exact roots and as approximate roots, to the nearest hundredth.
a) $(x + 5)(x - 1) = -2$ **b)** $(m + 2)^2 + 7(m + 2) = 3$
c) $2r(r + 2) - 3(1 - r) = 0$ **d)** $(2x - 5)^2 - (x + 1)^2 = 2$

38. Solve.
a) $(x + 6)(2x + 5) + 8 = 0$ **b)** $(2n - 1)^2 = 5(n - 3)$
c) $\dfrac{(w + 1)^2}{3} = -\dfrac{1}{4}$ **d)** $2x(x + 3) - 6 = x(4 - x)$

39. Measurements of lengths and areas A mural is to be painted on a wall that is 15 m long and 12 m high. A border of uniform width is to surround the mural. If the mural is to cover 75% of the area of the wall, how wide must the border be, to the nearest hundredth of a metre?

C

40. Solve each equation for x by completing the square.
a) $x^2 + 2x = k$ **b)** $kx^2 - 2x = k$ **c)** $x^2 = kx + 1$

41. Write a quadratic equation with the given roots in the form $ax^2 + bx + c = 0$.
a) $\sqrt{5}$ and $-\sqrt{5}$ **b)** $2i$ and $-2i$

42. Find the values of k that make $x^2 + (k + 7)x + (7k + 1)$ a perfect square trinomial.

43. Solve $x^2 + bx + c = 0$ for x by completing the square.

44. Solve for x in terms of the other variables.
a) $x^2 - t = 0$ **b)** $ax^2 - b = 0$ **c)** $rx^2 + tx = 0$ **d)** $x^2 - mx - t = 0$

45. Discriminant In the quadratic formula, the discriminant is the quantity $b^2 - 4ac$ under the radical sign. What can you conclude about the value of $b^2 - 4ac$ if the equation has
a) two equal real roots?   **b)** two distinct real roots?
c) two imaginary roots?

ACHIEVEMENT Check   Knowledge/Understanding   Thinking/Inquiry/Problem Solving   Communication   Application

Show that the equation $2x^2 - 5x + 2 = 0$ has roots that are reciprocals of each other. Under what conditions will a quadratic equation in the form $ax^2 + bx + c = 0$ have roots that are reciprocals of each other?

CAREER CONNECTION *Publishing*

Many thousands of publications are produced each year in Canada, including books, magazines, and newspapers, and multimedia and online publications. The publishing industry employs tens of thousands of Canadians and contributes billions of dollars to the economy. Much of the Canadian publishing industry is based in Ontario.

1. Book sales A publishing company expects to sell 5000 copies of a new book from its web site, if the company charges $30 per book. The company expects that 500 more books would be sold for each price reduction of $2.
a) For the company to make revenues of $156 000 from sales of the book, how many books must be sold and at what price?
b) If its assumptions are correct, can the company make $160 000 in revenues from sales of the book? Explain.

2. Research Use your research skills to investigate a publishing career that interests you. Examples might include journalist, editor, designer, or marketer. Describe the education and training needed for the career, and outline the tasks that someone with this background might undertake.

TECHNOLOGY EXTENSION
Solving Quadratic Equations

Some graphing calculators, such as the TI-92 and TI-92 Plus, have the capability to solve quadratic equations algebraically.

Exploring the Calculator

1. Try to solve each of the following quadratic equations using the solve function of the graphing calculator.
a) $x^2 + 3x + 1 = 0$
b) $x^2 - 2x + 4 = 0$
c) $2x^2 + x - 2 = 0$
d) $3x^2 + 2x + 2 = 0$

2. Use the cSolve function to solve each equation from question 1.

3. Using your results from questions 1 and 2, compare the capabilities of the solve function and the cSolve function for solving quadratic equations.

Solving Equations

Use the cSolve function of the graphing calculator.

4. Solve.
a) $x^2 + 5x + 3 = 0$
b) $y^2 - 4y - 2 = 0$
c) $x^2 - 3x + 6 = 0$
d) $n^2 + 3n + 7 = 0$
e) $x^2 + 8x = 3$
f) $z^2 - 5z = -8$
g) $2 - t^2 = 3t$
h) $8x - 17 = x^2$

5. Solve.
a) $3k^2 + 2k - 4 = 0$
b) $4x^2 + 8x + 5 = 0$
c) $4 = 5a^2 - 10a$
d) $3w - 6w^2 = -1$
e) $\dfrac{5}{2}x^2 - 2x - \dfrac{3}{4} = 0$
f) $\dfrac{y^2}{3} - \dfrac{y}{2} + \dfrac{3}{2} = 0$
g) $0.5m^2 + m = 2.5$
h) $\dfrac{x-1}{2} - \dfrac{x+1}{3} = x^2$

Problem Solving

6. Garden fence Is it possible to enclose all four sides of a rectangular garden with an area of 150 m^2 using each of the following lengths of fencing? If so, give the dimensions of the garden that can be enclosed. Round to the nearest tenth of a metre, if necessary.
a) 60 m   b) 50 m   c) 40 m

7. Thrown object The following equation expresses the approximate height, h metres, of an object thrown upward from the top of an 80-m cliff at 20 m/s as a function of the time, t seconds, since the object was thrown.
$$h = -5t^2 + 20t + 80$$
Will the object reach each of the following heights? If so, after what length of time will the object reach this height?
a) 110 m   b) 100 m

2.4 Tools for Operating With Complex Numbers

Solar cells are attached to the surfaces of satellites. The cells convert the energy of sunlight to electrical energy. Solar cells are made in various shapes to cover most of the surface area of satellites.

INVESTIGATE & INQUIRE

The scale drawing shows 6 solar cells. The 3 triangles and 3 rectangles are attached to form one triangular solar panel. The dimensions shown are in centimetres.

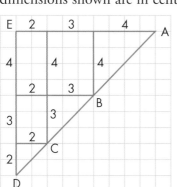

1. Calculate the lengths of AB, BC, and CD. Write your answers as mixed radicals in simplest form.

2. Explain why the three mixed radicals in simplest form from question 1 are called *like radicals*.

3. Use the large right triangle ADE to write an expression for the length of AD. Write your answer as a mixed radical in simplest form.

4. How is the length of AD related to the lengths of AB, BC, and CD?

5. Compare the radical expressions you wrote for the lengths of AB, BC, CD, and AD. Then, write a rule for adding like radicals.

6. Technology Use a calculator to test your rule for each of the following.

a) $3\sqrt{2} + 4\sqrt{2}$

b) $\sqrt{7} + 2\sqrt{7} + 3\sqrt{7}$

7. Simplify.

a) $3\sqrt{5} + 6\sqrt{5}$

b) $4\sqrt{3} + 5\sqrt{3} + \sqrt{3}$

8. a) Describe a method for using information from the same diagram to write a rule for subtracting like radicals.

b) Write and test the rule.

c) Simplify $5\sqrt{6} - 2\sqrt{6}$.

EXAMPLE 1 **Adding and Subtracting Radicals**

Simplify.

a) $\sqrt{12} + \sqrt{18} - \sqrt{27} + \sqrt{8}$ **b)** $4\sqrt{3} + 3\sqrt{20} - \sqrt{12} + 6\sqrt{45}$

SOLUTION

Simplify radicals and combine like radicals.

a) $\sqrt{12} + \sqrt{18} - \sqrt{27} + \sqrt{8} = \sqrt{4} \times \sqrt{3} + \sqrt{9} \times \sqrt{2} - \sqrt{9} \times \sqrt{3} + \sqrt{4} \times \sqrt{2}$

$$= 2\sqrt{3} + 3\sqrt{2} - 3\sqrt{3} + 2\sqrt{2}$$

$$= -\sqrt{3} + 5\sqrt{2}$$

b) $4\sqrt{3} + 3\sqrt{20} - \sqrt{12} + 6\sqrt{45} = 4\sqrt{3} + 3 \times \sqrt{4} \times \sqrt{5} - \sqrt{4} \times \sqrt{3} + 6 \times \sqrt{9} \times \sqrt{5}$

$$= 4\sqrt{3} + 3 \times 2\sqrt{5} - 2\sqrt{3} + 6 \times 3\sqrt{5}$$

$$= 4\sqrt{3} + 6\sqrt{5} - 2\sqrt{3} + 18\sqrt{5}$$

$$= 2\sqrt{3} + 24\sqrt{5}$$

EXAMPLE 2 **Multiplying a Radical by a Binomial**

Expand and simplify $3\sqrt{2}\left(2\sqrt{6} + \sqrt{10}\right)$.

SOLUTION

Use the distributive property.

$3\sqrt{2}\left(2\sqrt{6} + \sqrt{10}\right) = 3\sqrt{2}\left(2\sqrt{6} + \sqrt{10}\right)$

$$= 3\sqrt{2} \times 2\sqrt{6} + 3\sqrt{2} \times \sqrt{10}$$

$$= 6\sqrt{12} + 3\sqrt{20}$$

$$= 6 \times 2\sqrt{3} + 3 \times 2\sqrt{5}$$

$$= 12\sqrt{3} + 6\sqrt{5}$$

EXAMPLE 3 Binomial Multiplication

Simplify $\left(3\sqrt{2} + 4\sqrt{5}\right)\left(4\sqrt{2} - 3\sqrt{5}\right)$.

SOLUTION

Multiply each term in the first binomial by each term in the second binomial.

$$\left(3\sqrt{2} + 4\sqrt{5}\right)\left(4\sqrt{2} - 3\sqrt{5}\right) = \left(3\sqrt{2} + 4\sqrt{5}\right)\left(4\sqrt{2} - 3\sqrt{5}\right)$$

Recall that FOIL means First, Outside, Inside, Last.

$$= 12\sqrt{4} - 9\sqrt{10} + 16\sqrt{10} - 12\sqrt{25}$$
$$= 24 - 9\sqrt{10} + 16\sqrt{10} - 60$$
$$= -36 + 7\sqrt{10}$$

Recall that a radical is in simplest form when no radical appears in the denominator of a fraction.

EXAMPLE 4 Fractions With Radicals in the Denominator

Simplify $\dfrac{1}{3\sqrt{2}}$.

SOLUTION

Multiply the numerator and denominator by $\sqrt{2}$.
This is the same as multiplying the fraction by 1.

$$\frac{1}{3\sqrt{2}} = \frac{1}{3\sqrt{2}} \times \frac{\sqrt{2}}{\sqrt{2}}$$

$$= \frac{1 \times \sqrt{2}}{3\sqrt{2} \times \sqrt{2}}$$

$$= \frac{\sqrt{2}}{3 \times 2}$$

$$= \frac{\sqrt{2}}{6}$$

The process shown in Example 4 is called **rationalizing the denominator**. The denominator has been changed from an irrational number to a rational number.

Binomials of the form $a\sqrt{b} + c\sqrt{d}$ and $a\sqrt{b} - c\sqrt{d}$, where a, b, c, and d are rational numbers, are **conjugates** of each other. The product of conjugates is always a rational number.

Example 5 **Multiplying Conjugate Binomials**

Simplify $\left(\sqrt{7} + 2\sqrt{3}\right)\left(\sqrt{7} - 2\sqrt{3}\right)$.

SOLUTION

$$\left(\sqrt{7} + 2\sqrt{3}\right)\left(\sqrt{7} - 2\sqrt{3}\right) = \left(\sqrt{7} + 2\sqrt{3}\right)\left(\sqrt{7} - 2\sqrt{3}\right)$$

$$= \sqrt{49} - 2\sqrt{21} + 2\sqrt{21} - 4\sqrt{9}$$

$$= 7 - 12$$

$$= -5$$

Conjugate binomials can be used to simplify a fraction with a binomial radical in the denominator.

Example 6 **Rationalizing Binomial Denominators**

Simplify $\dfrac{5}{2\sqrt{6} - \sqrt{3}}$.

SOLUTION

Multiply the numerator and the denominator by the conjugate of $2\sqrt{6} - \sqrt{3}$, which is $2\sqrt{6} + \sqrt{3}$.

$$\frac{5}{2\sqrt{6} - \sqrt{3}} = \frac{5}{2\sqrt{6} - \sqrt{3}} \times \frac{2\sqrt{6} + \sqrt{3}}{2\sqrt{6} + \sqrt{3}}$$

$$= \frac{5\left(2\sqrt{6} + \sqrt{3}\right)}{4\sqrt{36} - \sqrt{9}}$$

$$= \frac{10\sqrt{6} + 5\sqrt{3}}{24 - 3}$$

$$= \frac{10\sqrt{6} + 5\sqrt{3}}{21}$$

Key Concepts

- To simplify radical expressions, express radicals in simplest radical form and add or subtract like radicals.
- To multiply binomial radical expressions, use the distributive property and add or subtract like radicals.
- To simplify a radical expression with a monomial radical in the denominator, multiply the numerator and the denominator by this monomial radical.
- To simplify a radical expression with a binomial radical in the denominator, multiply the numerator and the denominator by the conjugate of the denominator.

Communicate Your Understanding

1. Explain the meaning of the term *like radicals*.
2. Describe how you would simplify each of the following.

a) $\sqrt{24} + \sqrt{54}$

b) $2\sqrt{2}\left(\sqrt{10} - 3\sqrt{2}\right)$

c) $\left(\sqrt{3} + \sqrt{5}\right)\left(2\sqrt{3} - 4\sqrt{5}\right)$

d) $\dfrac{5}{\sqrt{6}}$

e) $\dfrac{2}{\sqrt{5} + \sqrt{3}}$

Practise

A

1. Simplify.

a) $2\sqrt{5} + 3\sqrt{5} + 6\sqrt{5}$

b) $4\sqrt{3} + 2\sqrt{3} - \sqrt{3}$

c) $6\sqrt{2} - \sqrt{2} + 7\sqrt{2} - 3\sqrt{2}$

d) $5\sqrt{7} + 3\sqrt{7} - 2\sqrt{7}$

e) $8\sqrt{10} - 2\sqrt{10} - 7\sqrt{10}$

f) $\sqrt{2} - 3\sqrt{2} - 9\sqrt{2} + 11\sqrt{2}$

g) $\sqrt{5} + \sqrt{5} + \sqrt{5} + \sqrt{5}$

2. Simplify.

a) $5\sqrt{3} + 2\sqrt{6} + 3\sqrt{3}$

b) $8\sqrt{5} - 3\sqrt{7} + 7\sqrt{7} - 4\sqrt{5}$

c) $2\sqrt{2} + 3\sqrt{10} + 5\sqrt{2} - 4\sqrt{10}$

d) $7\sqrt{6} - 4\sqrt{13} - \sqrt{13} + \sqrt{6}$

e) $9\sqrt{11} - \sqrt{11} + 6\sqrt{14} - 3\sqrt{14} - 2\sqrt{11}$

f) $12\sqrt{7} + 9 - 3\sqrt{7} + 4$

g) $8 + 7\sqrt{11} - 9 - 9\sqrt{11}$

3. Simplify.

a) $\sqrt{12} + \sqrt{27}$

b) $\sqrt{20} + \sqrt{45}$

c) $\sqrt{18} - \sqrt{8}$

d) $\sqrt{50} + \sqrt{98} - \sqrt{2}$

e) $\sqrt{75} + \sqrt{48} + \sqrt{27}$

f) $\sqrt{54} + \sqrt{24} + \sqrt{72} - \sqrt{32}$

g) $\sqrt{28} - \sqrt{27} + \sqrt{63} + \sqrt{300}$

4. Simplify.

a) $8\sqrt{7} + 2\sqrt{28}$ b) $3\sqrt{50} - 2\sqrt{32}$

c) $5\sqrt{27} + 4\sqrt{48}$ d) $3\sqrt{8} + \sqrt{18} + 3\sqrt{2}$

e) $\sqrt{5} + 2\sqrt{45} - 3\sqrt{20}$

f) $4\sqrt{3} + 3\sqrt{20} - 2\sqrt{12} + \sqrt{45}$

g) $3\sqrt{48} - 4\sqrt{8} + 4\sqrt{27} - 2\sqrt{72}$

5. Expand and simplify.

a) $\sqrt{2}\left(\sqrt{10} + 4\right)$

b) $\sqrt{3}\left(\sqrt{6} - 1\right)$

c) $\sqrt{6}\left(\sqrt{2} + \sqrt{6}\right)$

d) $2\sqrt{2}\left(3\sqrt{6} - \sqrt{3}\right)$

e) $\sqrt{2}\left(\sqrt{3} + 4\right)$

f) $3\sqrt{2}\left(2\sqrt{6} + \sqrt{10}\right)$

g) $\left(\sqrt{5} + \sqrt{6}\right)\left(\sqrt{5} + 3\sqrt{6}\right)$

h) $\left(2\sqrt{3} - 1\right)\left(3\sqrt{3} + 2\right)$

i) $\left(4\sqrt{7} - 3\sqrt{2}\right)\left(2\sqrt{7} + 5\sqrt{2}\right)$

j) $\left(3\sqrt{3} + 1\right)^2$

k) $\left(2\sqrt{2} - \sqrt{5}\right)^2$

l) $\left(2 + \sqrt{3}\right)\left(2 - \sqrt{3}\right)$

m) $\left(\sqrt{6} - \sqrt{2}\right)\left(\sqrt{6} + \sqrt{2}\right)$

n) $\left(2\sqrt{7} + 3\sqrt{5}\right)\left(2\sqrt{7} - 3\sqrt{5}\right)$

6. Simplify.

a) $\dfrac{1}{\sqrt{3}}$ b) $\dfrac{2}{\sqrt{5}}$ c) $\dfrac{2}{\sqrt{7}}$

d) $\dfrac{\sqrt{1}}{\sqrt{2}}$ e) $\dfrac{5\sqrt{5}}{2\sqrt{3}}$ f) $\dfrac{2\sqrt{2}}{\sqrt{18}}$

g) $\dfrac{4\sqrt{2}}{\sqrt{8}}$ h) $\dfrac{3\sqrt{5}}{\sqrt{3}}$ i) $\dfrac{4\sqrt{7}}{2\sqrt{14}}$

j) $\dfrac{3\sqrt{6}}{4\sqrt{10}}$ k) $\dfrac{7\sqrt{11}}{2\sqrt{3}}$ l) $\dfrac{2\sqrt{5}}{5\sqrt{2}}$

7. Simplify.

a) $\dfrac{1}{\sqrt{2} + 2}$ b) $\dfrac{3}{\sqrt{5} - 1}$

c) $\dfrac{\sqrt{2}}{\sqrt{6} - 3}$ d) $\dfrac{2}{\sqrt{6} + \sqrt{3}}$

e) $\dfrac{3}{\sqrt{5} - \sqrt{2}}$ f) $\dfrac{\sqrt{3}}{\sqrt{3} + \sqrt{2}}$

g) $\dfrac{2\sqrt{6}}{2\sqrt{6} + 1}$ h) $\dfrac{\sqrt{2} - 1}{\sqrt{2} + 1}$

i) $\dfrac{\sqrt{2} + \sqrt{5}}{\sqrt{6} - \sqrt{10}}$ j) $\dfrac{2\sqrt{7} + \sqrt{5}}{3\sqrt{7} - 2\sqrt{5}}$

Apply, Solve, Communicate

8. Measurement Express the perimeter of the quadrilateral in simplest radical form.

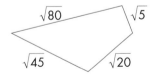

B

9. Without using a calculator, arrange the following expressions in order from greatest to least.

$$\sqrt{3}\left(\sqrt{3} + 1\right),\ \left(\sqrt{3} + 1\right)\left(\sqrt{3} - 1\right),\ \left(1 - \sqrt{3}\right)^2,\ \left(\sqrt{3} + 1\right)^2$$

10. a) Without using a calculator, decide which of the following radical expressions does not equal any of the others.

$$\frac{60}{\sqrt{450}} \qquad 6\sqrt{2} - 4\sqrt{2} \qquad \frac{4}{\sqrt{2}} \qquad 6\sqrt{8} + \sqrt{8} - 5\sqrt{8} \qquad \frac{8}{\sqrt{18}} + \frac{4}{\sqrt{18}}$$

b) Communication How is the radical expression you identified in part a) related to each of the others?

11. Nature Many aspects of nature, including the number of pairs of rabbits in a family and the number of branches on a tree, can be described using the Fibonacci sequence. This sequence is 1, 1, 2, 3, 5, 8, …
The expression for the nth term of the Fibonacci sequence is called Binet's formula. The formula is
$$F_n = \frac{1}{\sqrt{5}}\left(\frac{1 + \sqrt{5}}{2}\right)^n - \frac{1}{\sqrt{5}}\left(\frac{1 - \sqrt{5}}{2}\right)^n.$$
Use Binet's formula to find F_2.

12. Measurement Write and simplify an expression for
a) the area of the rectangle
b) the perimeter of the rectangle

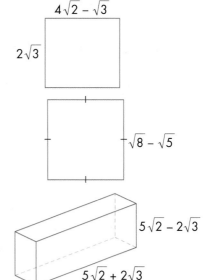

$4\sqrt{2} - \sqrt{3}$

$2\sqrt{3}$

13. Measurement Write and simplify an expression for the area of the square.

$\sqrt{8} - \sqrt{5}$

14. Measurement Express the volume of the rectangular prism in simplest radical form.

$5\sqrt{2} - 2\sqrt{3}$

$5\sqrt{2} + 2\sqrt{3}$

$\sqrt{15} - \sqrt{2}$

15. Application If a rectangle has an area of 4 square units and a width of $\sqrt{7} - \sqrt{5}$ units, what is its length, in simplest radical form?

16. Inquiry/Problem Solving Write a quadratic equation in the form $ax^2 + bx + c = 0$ with the given roots.
a) $3 + \sqrt{2}$ and $3 - \sqrt{2}$ **b)** $-1 + 2\sqrt{3}$ and $-1 - 2\sqrt{3}$
c) $1 + \frac{\sqrt{13}}{2}$ and $1 - \frac{\sqrt{13}}{2}$

C

17. Simplify.

a) $\sqrt[3]{16} + \sqrt[3]{54}$

b) $\sqrt[3]{24} + \sqrt[3]{81}$

c) $2(\sqrt[3]{32}) + 5(\sqrt[3]{108})$

d) $\sqrt[3]{54} + 5(\sqrt[3]{16})$

e) $\sqrt[3]{16} - \sqrt[3]{54}$

f) $\sqrt[3]{108} - \sqrt[3]{32}$

g) $2(\sqrt[3]{40}) - \sqrt[3]{5}$

h) $5(\sqrt[3]{48}) - 2(\sqrt[3]{162})$

18. Measurement Express the ratio of the area of the larger circle to the area of the smaller circle in simplest radical form.

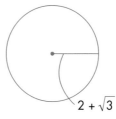

$2 + \sqrt{3}$ $2 - \sqrt{3}$

19. Coordinate geometry State the perimeter of each of the following triangles in simplest radical form.

a)

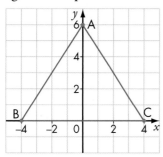

b)

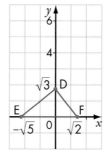

20. Equation Is the statement $\sqrt{a + b} = \sqrt{a} + \sqrt{b}$ always true, sometimes true, or never true? Explain.

LOGIC *Power*

Suppose intercity buses travel from Montréal to Toronto and from Toronto to Montréal, leaving each city on the hour every hour from 06:00 to 20:00. Each trip takes $5\frac{1}{2}$ h. All buses travel at the same speed on the same highways. Your driver waves at each of her colleagues she sees driving an intercity bus in the opposite direction. How many times would she wave during the journey if your bus left Toronto at

a) 14:00?   b) 18:00?   c) 06:00?

TECHNOLOGY EXTENSION
Radical Expressions and Graphing Calculators

Some graphing calculators, such as the TI-92 and TI-92 Plus, have the capability to simplify radical expressions and perform operations on them.

Simplifying Radicals

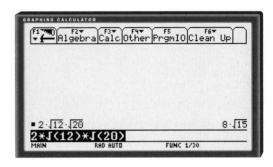

1. Simplify.

a) $\sqrt{425}$ b) $\sqrt{294}$ c) $\sqrt{507}$

d) $\sqrt{8} \times \sqrt{10}$ e) $2\sqrt{21} \times \sqrt{35}$ f) $3\sqrt{15} \times 4\sqrt{20}$

2. Simplify.

a) $\dfrac{3}{\sqrt{27}}$ b) $\dfrac{5}{\sqrt{112}}$ c) $\dfrac{-4}{\sqrt{32}}$

d) $\dfrac{\sqrt{192}}{\sqrt{6}}$ e) $\dfrac{\sqrt{75}}{\sqrt{162}}$ f) $\dfrac{\sqrt{88}}{\sqrt{33}}$

Operations With Radical Expressions

3. Simplify.

a) $\sqrt{20} + \sqrt{45}$ b) $\sqrt{72} + \sqrt{98} + \sqrt{242}$

c) $\sqrt{63} - \sqrt{28}$ d) $\sqrt{54} - \sqrt{96}$

e) $5\sqrt{27} + 3\sqrt{12}$ f) $7\sqrt{90} - 6\sqrt{40}$

g) $2\sqrt{125} + 4\sqrt{5} - 3\sqrt{80}$

h) $3\sqrt{99} - 6\sqrt{44} - 2\sqrt{11}$

4. Expand and simplify.

a) $\sqrt{5}\left(\sqrt{10} + \sqrt{15}\right)$ b) $\sqrt{6}\left(\sqrt{18} - \sqrt{3}\right)$ c) $2\sqrt{3}\left(\sqrt{27} + 5\sqrt{24}\right)$

d) $4\sqrt{7}\left(3\sqrt{21} - 2\sqrt{14}\right)$ e) $\left(2\sqrt{5} - 3\sqrt{3}\right)\left(2\sqrt{5} + 3\sqrt{3}\right)$ f) $\left(4\sqrt{11} + 5\sqrt{2}\right)^2$

g) $\left(3\sqrt{6} - 5\sqrt{10}\right)^2$ h) $\left(4\sqrt{3} - 3\sqrt{2}\right)\left(5\sqrt{2} - 2\sqrt{3}\right)$

5. Simplify.

a) $\dfrac{2}{\sqrt{3} + \sqrt{2}}$ b) $\dfrac{\sqrt{5}}{\sqrt{10} - \sqrt{5}}$ c) $\dfrac{3\sqrt{3}}{2\sqrt{6} - 3\sqrt{2}}$

d) $\dfrac{4 - \sqrt{10}}{7 - 2\sqrt{10}}$ e) $\dfrac{\sqrt{6} + 2\sqrt{3}}{\sqrt{6} - 2\sqrt{3}}$ f) $\dfrac{3\sqrt{7} - 2\sqrt{2}}{3\sqrt{2} - 2\sqrt{7}}$

6. Measurement The area of a triangle is 12 square units, and its base length is $4 + \sqrt{2}$ units. Write a radical expression in simplest form for the height of the triangle.

2.5 Operations With Complex Numbers in Rectangular Form

The computer-generated image shown is called a fractal. Fractals are used in many ways, such as making realistic computer images for movies and squeezing high definition television (HDTV) signals into existing broadcast channels. Meteorologists use fractals to study cloud shapes, and seismologists use fractals to study earthquakes. To understand how fractals are generated, we need to extend our understanding of complex numbers.

INVESTIGATE & INQUIRE

1. Describe each step in the following addition.
$$(2 + 5x) + (4 - 3x) = 2 + 5x + 4 - 3x$$
$$= 2 + 4 + 5x - 3x$$
$$= 6 + 2x$$

2. a) Use the steps from question 1 to simplify $(4 + 3i) + (7 - 2i)$.
b) Explain why the resulting complex number cannot be simplified further.

3. Describe each step in the following subtraction.
$$(5 - 2x) - (3 - 7x) = 5 - 2x - 3 + 7x$$
$$= 5 - 3 - 2x + 7x$$
$$= 2 + 5x$$

4. Use the steps from question 3 to simplify $(3 - 6i) - (7 - 8i)$.

5. What are the results when the following operations are performed on the complex numbers $a + bi$ and $c + di$?
a) $(a + bi) + (c + di)$　　　　　　　**b)** $(a + bi) - (c + di)$

6. Write a rule for adding or subtracting complex numbers.

7. Simplify.
a) $(2 - 5i) + (5 - 4i)$　　　　　　　**b)** $(3 + 2i) - (7 - i)$

8 . Describe each step in the following multiplication.
$$(2 + 3x)(1 - 4x) = 2 - 8x + 3x - 12x^2$$
$$= 2 - 5x - 12x^2$$

9. a) Use the steps from question 8 to simplify $(3 + 4i)(2 - 5i)$.

b) Explain how you can simplify the final term in the resulting expression.

c) Write the expression in simplest form.

d) Write a rule for multiplying complex numbers.

10. Simplify.

a) $(3 + i)(2 + 3i)$

b) $(1 - 2i)(5 - 2i)$

Recall that a complex number is a number in the form $a + bi$, where a is the real part and bi is the imaginary part. Because there are two parts, any complex number can be represented as an ordered pair (a, b). The ordered pair can be graphed using rectangular axes on a plane called the **complex plane**. In the complex plane, the x-axis is referred to as the **real axis**, and the y-axis is referred to as the **imaginary axis**.

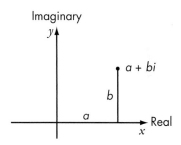

A complex number in the form $a + bi$ is said to be in **rectangular form**, because the ordered pair (a, b) includes the rectangular coordinates of the point $a + bi$ in the complex plane.

To add or subtract complex numbers in rectangular form, combine like terms, that is, combine the real parts and combine the imaginary parts.

EXAMPLE 1 Adding and Subtracting Complex Numbers

Simplify.

a) $(6 - 3i) + (5 + i)$

b) $(2 - 3i) - (4 - 5i)$

SOLUTION

a) $(6 - 3i) + (5 + i) = 6 - 3i + 5 + i$

$= 6 + 5 - 3i + i$

$= 11 - 2i$

b) $(2 - 3i) - (4 - 5i) = 2 - 3i - 4 + 5i$

$= 2 - 4 - 3i + 5i$

$= -2 + 2i$

To perform the operations on a graphing calculator, change the mode settings to the $a + bi$ (rectangular) mode.

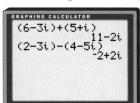

Complex numbers in rectangular form can be multiplied using the distributive property.

EXAMPLE 2 Multiplying Complex Numbers

Simplify.

a) $2i(3 + 4i)$ **b)** $(1 - 2i)(4 + 3i)$ **c)** $(1 - 4i)^2$

SOLUTION

Use the distributive property.

a) $2i(3 + 4i) = 2i(3 + 4i)$
$$= 6i + 8i^2$$
$$= 6i + 8(-1)$$
$$= -8 + 6i$$

Remember to change the mode settings **to the** *a + bi* **(rectangular) mode.**

GRAPHING CALCULATOR
```
2i(3+4i)
                -8+6i
(1-2i)(4+3i)
               10-5i
(1-4i)²
              -15-8i
```

b) $(1 - 2i)(4 + 3i) = (1 - 2i)(4 + 3i)$
$$= 4 + 3i - 8i - 6i^2$$
$$= 4 - 5i - 6i^2$$
$$= 4 - 5i - 6(-1)$$
$$= 4 - 5i + 6$$
$$= 10 - 5i$$

c) $(1 - 4i)^2 = (1 - 4i)(1 - 4i)$
$$= 1 - 4i - 4i + 16i^2$$
$$= 1 - 8i + 16(-1)$$
$$= 1 - 8i - 16$$
$$= -15 - 8i$$

Since i is a radical, $\sqrt{-1}$, any fraction with i in the denominator is not in simplest form. To simplify, rationalize the denominator.

EXAMPLE 3 Rationalizing the Denominator

Simplify $\dfrac{5}{2i}$.

SOLUTION

Multiply the numerator and the denominator by i.
This is the same as multiplying the fraction by 1.

$$\frac{5}{2i} = \frac{5}{2i} \times \frac{i}{i}$$

$$= \frac{5i}{2i^2}$$

$$= \frac{5i}{2(-1)}$$

$$= \frac{5i}{-2}$$

$$= -\frac{5i}{2}$$

Note the use of brackets, because 5/2i means $\frac{5}{2}i$ on the calculator.

Recall that binomials of the form $a\sqrt{b} + c\sqrt{d}$ and $a\sqrt{b} - c\sqrt{d}$ are known as conjugates. Since i represents the radical $\sqrt{-1}$, complex numbers of the form $a + bi$ and $a - bi$ are examples of conjugates and are known as **complex conjugates**.

To simplify a fraction with a binomial complex number in the denominator, multiply the numerator and the denominator by the conjugate of the denominator.

EXAMPLE 4 Rationalizing Binomial Denominators

Simplify $\dfrac{2 + 3i}{1 - 2i}$.

SOLUTION

Multiply the numerator and the denominator by the conjugate of $1 - 2i$, which is $1 + 2i$.

$$\frac{2 + 3i}{1 - 2i} = \frac{2 + 3i}{1 - 2i} \times \frac{1 + 2i}{1 + 2i}$$

$$= \frac{(2 + 3i)(1 + 2i)}{(1 - 2i)(1 + 2i)}$$

$$= \frac{2 + 4i + 3i + 6i^2}{1 + 2i - 2i - 4i^2}$$

$$= \frac{2 + 7i + 6(-1)}{1 - 4(-1)}$$

$$= \frac{2 + 7i - 6}{1 + 4}$$

$$= \frac{-4 + 7i}{5}$$

Use the ▶Frac function to display the decimals as fractions.

Note that 7/5i means $\frac{7}{5}i$ on the calculator.

EXAMPLE 5 Checking Imaginary Roots

Solve and check $x^2 - 4x + 6 = 0$.

SOLUTION

Use the quadratic formula.
For $x^2 - 4x + 6 = 0$, $a = 1$, $b = -4$, and $c = 6$.

$$x = \frac{-b \pm \sqrt{b^2 - 4ac}}{2a}$$

$$= \frac{-(-4) \pm \sqrt{(-4)^2 - 4(1)(6)}}{2}$$

$$= \frac{4 \pm \sqrt{16 - 24}}{2}$$

$$= \frac{4 \pm \sqrt{-8}}{2}$$

$$= \frac{4 \pm 2i\sqrt{2}}{2}$$

$$= 2 \pm i\sqrt{2}$$

$$x = 2 + i\sqrt{2} \quad \text{or} \quad x = 2 - i\sqrt{2}$$

Check.
For $x = 2 + i\sqrt{2}$,

L.S. $= x^2 - 4x + 6$ R.S. $= 0$

$= (2 + i\sqrt{2})^2 - 4(2 + i\sqrt{2}) + 6$

$= (2 + i\sqrt{2})(2 + i\sqrt{2}) - 4(2 + i\sqrt{2}) + 6$

$= 4 + 4i\sqrt{2} - 2 - 8 - 4i\sqrt{2} + 6$

$= 0$

                     L.S. = R.S.

For $x = 2 - i\sqrt{2}$,

L.S. $= x^2 - 4x + 6$ R.S. $= 0$

$= (2 - i\sqrt{2})^2 - 4(2 - i\sqrt{2}) + 6$

$= (2 - i\sqrt{2})(2 - i\sqrt{2}) - 4(2 - i\sqrt{2}) + 6$

$= 4 - 4i\sqrt{2} - 2 - 8 + 4i\sqrt{2} + 6$

$= 0$

                     L.S. = R.S.

The roots are $2 + i\sqrt{2}$ and $2 - i\sqrt{2}$.

Functions that generate some fractals are in the form $F = z^2 + c$, where c is a complex number. Fractals are created by **iteration**, which means that the function F is evaluated for some input value of z, and then the result is used as the next input value, and so on.

_____

EXAMPLE 6 **Fractals**

Find the first three output values for $F = z^2 + 2i$.

SOLUTION

Web Connection
www.school.mcgrawhill.ca/resources/
To learn more about fractals, visit the above web site. Go to **Math Resources,** then to *MATHEMATICS* 11, to find out where to go next. Summarize the various types of fractals. Then, make your own fractal and write the rule that generates it.

$$F = z^2 + 2i$$

Use $z = 0$ as the first input value:
$$F = 0^2 + 2i$$
$$= 2i$$

Use $z = 2i$ as the second input value:
$$F = (2i)^2 + 2i$$
$$= 4i^2 + 2i$$
$$= -4 + 2i$$

Use $z = -4 + 2i$ as the third input value:
$$F = (-4 + 2i)^2 + 2i$$
$$= 16 - 16i + 4i^2 + 2i$$
$$= 16 - 16i - 4 + 2i$$
$$= 12 - 14i$$

The first three output values are $2i$, $-4 + 2i$, and $12 - 14i$.

Key Concepts

- To add or subtract complex numbers, combine like terms.
- To multiply complex numbers, use the distributive property.
- To simplify a fraction with a pure imaginary number in the denominator, multiply the numerator and the denominator by i.
- To simplify a fraction with a binomial complex number in the denominator, multiply the numerator and the denominator by the conjugate of the denominator.

Communicate Your Understanding

1. Explain why the complex number $5 - 3i$ cannot be simplified.
2. Describe how you would simplify each of the following.

a) $(3 - 2i) - (4 - 7i)$ b) $(5 + 3i)(1 - 4i)$

c) $\dfrac{3}{4i}$

d) $\dfrac{4}{2 + 3i}$

Practise

A

1. Simplify.

a) $(4 + 2i) + (3 - 4i)$

b) $(2 - 5i) + (1 - 6i)$

c) $(3 - 2i) - (1 + 3i)$

d) $(6 - i) - (5 - 7i)$

e) $(4 + 6i) + (7i - 6)$

f) $(i - 8) + (4i - 3)$

g) $(9i - 6) - (10i - 3)$

h) $(3i + 11) - (6i - 13)$

i) $2(1 - 7i) + 3(4 - i)$

j) $-3(2i - 4) - (5 + 6i)$

2. Simplify.

a) $2(4 - 3i)$ b) $3i(1 + 2i)$

c) $-4i(3 - 5i)$ d) $2i(3i^2 - 4i + 2)$

e) $(2 - 4i)(1 + 3i)$ f) $(3 + 4i)(3 - 5i)$

g) $(3i - 1)(4i - 5)$ h) $(1 - 5i)(1 + 5i)$

i) $(1 + 2i)^2$ j) $(4i - 3)^2$

k) $(i - 1)^2$ l) $(i^2 - 1)^2$

3. Simplify.

a) $\dfrac{2}{i}$ b) $\dfrac{4}{3i}$ c) $\dfrac{7}{4i}$

d) $\dfrac{-6}{5i}$ e) $\dfrac{5}{-2i}$ f) $-\dfrac{3}{7i}$

4. Simplify.

a) $\dfrac{3 + i}{i}$ b) $\dfrac{2 - 2i}{i}$ c) $\dfrac{5 + 2i}{2i}$

d) $\dfrac{3 - 4i}{-3i}$ e) $-\dfrac{4 + 3i}{2i}$

5. Write the conjugate of each complex number.

a) $3 + 2i$ b) $7 - 3i$

c) $5 - 4i$ d) $6 + 7i$

6. Simplify.

a) $\dfrac{3}{2 - i}$ b) $\dfrac{5}{1 + 2i}$ c) $\dfrac{2i}{3 - 2i}$

d) $\dfrac{i}{4 + 3i}$ e) $\dfrac{4 + i}{3 - i}$ f) $\dfrac{2 - 2i}{3 + i}$

g) $\dfrac{2 + 3i}{2 - 3i}$ h) $\dfrac{-4 - 3i}{-2 + 2i}$

7. Solve and check.

a) $x^2 + 2x + 2 = 0$ b) $y^2 - 4y + 8 = 0$

c) $x^2 - 6x + 10 = 0$ d) $n^2 + 4n + 6 = 0$

e) $z^2 - 2z = -6$ f) $x^2 = 8x - 19$

Apply, Solve, Communicate

8. Fractals Find the first four output values of $F = z^2$ if the first input value is $(1 - i)$.

B

9. Application Imaginary numbers are used in electricity. Three of the basic quantities that can be measured or calculated for an electrical circuit are as follows.

• the electric current, I, measured in amperes (symbol A)

• the resistance or impedance, Z, measured in ohms (symbol Ω)

• the electromotive force, E, sometimes called the voltage and measured in volts (symbol V)

These quantities are related by the formula $E = IZ$. To avoid confusion with the symbol for electric current, I, engineers use j instead of i to represent the imaginary unit.

a) In a circuit, the electric current is $(8 + 3j)$ A and the impedance is $(4 - j)$ Ω. What is the voltage?

b) In a 110-V circuit, the electric current is $(5 + 3j)$ A. What is the impedance?

c) In a 110-V circuit, the impedance is $(6 - 2j)$ Ω. What is the electric current?

10. Communication a) Find the first four output values of $F = iz$ if the first input value is $(1 + i)$.

b) Predict the next four output values of $F = iz$. Explain your reasoning.

11. If $y = x^2 + 4x + 5$, determine the value of y for each of the following values of x.

a) $1 + i$ **b)** $-2 + i$ **c)** $1 - i$

12. Simplify.

a) $(4 + i)^2 + (1 - 3i)^2$ **b)** $(3 - 2i)^2 - (4 + 3i)^2$

c) $2i(6 + 3i) - i(3 - 2i)$ **d)** $3i(-2 + 3i) + 4i(-3 + 2i)$

e) $(3 + i)(2 + i)(1 - i)$ **f)** $(4 - 2i)(-1 + 3i)(3 - i)$

13. Factoring The binomial $a^2 + b^2$ cannot be factored over the real numbers. It can be factored over the complex numbers. Factor $a^2 + b^2$.

14. Reciprocal Write the reciprocal of $a + bi$ in simplest form.

15. Quadratic equations Write a quadratic equation that has each pair of roots.

a) $1 + i$ and $1 - i$ **b)** $\dfrac{3 + 2i}{2}$ and $\dfrac{3 - 2i}{2}$

16. Communication Suppose that the quadratic equation $ax^2 + bx + c = 0$ has real coefficients and complex roots. Explain why the roots must be complex conjugates of each other.

C

17. Determine the values of x and y for which each equation is true.

a) $3x + 4yi = 15 - 16i$ **b)** $2x - 5yi = 6(1 + 5i)$

c) $(x + y) + (x - y)i = 10 + 3i$ **d)** $(x - 2y) - (3x + 4y)i = 4 - 2i$

18. Complex plane If the graph of a complex number $a + bi$ is a point on the imaginary axis in the complex plane, what can you conclude about each of the following? Explain.

a) the value of a **b)** the value of b

19. Transformation Name the transformation that maps the graph of a complex number onto the graph of its conjugate in the complex plane.

20. Quartic equations A quartic equation is a fourth-degree polynomial equation. Quartic equations in the form $ax^4 + bx^2 + c = 0$ can be solved using the same techniques used to solve quadratic equations. To solve $x^4 - x^2 - 12 = 0$, first factor the left side. Then, equate each factor to 0 and solve for x.

$$x^4 - x^2 - 12 = 0$$
$$(x^2 - 4)(x^2 + 3) = 0$$
$$x^2 - 4 = 0 \quad \text{or} \quad x^2 + 3 = 0$$
$$x^2 = 4 \qquad\qquad x^2 = -3$$
$$x = \pm 2 \qquad\qquad x = \pm\sqrt{-3}$$
$$\qquad\qquad\qquad\qquad x = \pm i\sqrt{3}$$

The solutions are 2, -2, $i\sqrt{3}$, and $-i\sqrt{3}$.

Solve the following quartic equations.

a) $x^4 - 8x^2 + 16 = 0$ **b)** $x^4 + 2x^2 + 1 = 0$ **c)** $x^4 + 3x^2 - 4 = 0$
d) $x^4 - 5x^2 + 6 = 0$ **e)** $y^4 - y^2 - 6 = 0$ **f)** $3r^4 - 5r^2 + 2 = 0$
g) $2x^4 + 5x^2 + 3 = 0$ **h)** $2x^4 + x^2 = 6$ **i)** $4a^4 - 1 = 0$
j) $9x^4 - 4x^2 = 0$

21. Quartic equations Is it possible for a quartic equation to have three real roots and one imaginary root? Explain.

22. Fourth roots a) What are the two square roots of 1 and the two square roots of -1 in the complex number system?
b) What are the fourth roots of 1 in the complex number system?
c) What are the fourth roots of -1 in the complex number system?

ACHIEVEMENT Check   Knowledge/Understanding   Thinking/Inquiry/Problem Solving   Communication   Application

a) Express the number 25 as a product of two complex conjugates, $a + bi$ and $a - bi$, in two different ways, with a and b both natural numbers.
b) Find another perfect square that can be expressed as a product of two complex conjugates, $a + bi$ and $a - bi$, in two different ways, with a and b both natural numbers.
c) Describe the most efficient method for finding numbers that satisfy the above relationship.

Investigate & Apply

Interpreting a Mathematical Model

There are several steps in the inquiry process. They do not necessarily occur in a given order and they may often be revisited in completing an investigation. These steps include the following.

- Formulate questions
- Selecting strategies/tools
- Represent in mathematical form
- Interpret information
- Draw conclusions
- Reflect on the reasonableness of results

To see how the step in the inquiry process can be applied, refer to pages 83 and 84 in Chapter 1.

Apply the inquiry process to the following.

An advertising agency believes the effectiveness rating of a TV commercial depends upon how many times a viewer sees it. The agency used the equation $e = -\frac{1}{90}n^2 + \frac{2}{3}n$ where e is the effectiveness rating and n the number of viewings.

You have been hired as mathematical consultant to interpret the model. Prepare a report that includes the following.

1. What is the range of e values and the corresponding domain of n values? Explain and justify your reasoning.

2. Describe the conclusions that can be made from this quadratic equation model.

3. Provide a sketch of an alternative graphical model for showing how the effectiveness rating, e, is related to the number of viewings, n. Justify your model.

REVIEW OF **KEY CONCEPTS**

■ **2.1** The Complex Number System

Refer to the Key Concepts on page 106.

1. Simplify.

a) $\sqrt{18}$ **b)** $\sqrt{32}$ **c)** $\sqrt{500}$

2. Simplify.

a) $\dfrac{\sqrt{40}}{\sqrt{8}}$ **b)** $\dfrac{\sqrt{70}}{\sqrt{10}}$ **c)** $\dfrac{6\sqrt{30}}{\sqrt{5}}$ **d)** $\dfrac{\sqrt{200}}{\sqrt{2}}$

3. Simplify.

a) $\sqrt{10} \times \sqrt{6}$ **b)** $3\sqrt{5} \times 2\sqrt{10}$

4. Simplify.

a) $\dfrac{6 + 9\sqrt{5}}{3}$ **b)** $\dfrac{5 + \sqrt{50}}{5}$ **c)** $\dfrac{4 - \sqrt{20}}{2}$

5. Simplify.

a) $\sqrt{-49}$ **b)** $\sqrt{-18}$ **c)** $\sqrt{-80}$

6. Evaluate.

a) $5i \times 6i$ **b)** $9i \times (-4i)$ **c)** $(-4i)^2$ **d)** $\left(i\sqrt{3}\right)^2$

7. Simplify.

a) $5 - \sqrt{-36}$ **b)** $3 + \sqrt{-20}$ **c)** $\dfrac{6 + \sqrt{-27}}{3}$

8. Measurement A rectangle has side lengths of $2\sqrt{15}$ and $3\sqrt{5}$. Express the area of the rectangle in simplest radical form.

9. Measurement Express the area of the triangle in simplest radical form.

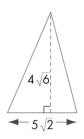

2.2 Maximum or Minimum of a Quadratic Function by Completing the Square

Refer to the Key Concepts on page 115.

10. Find the value of c that will make each expression a perfect square trinomial.

a) $x^2 + 10x + c$ b) $x^2 - 16x + c$ c) $x^2 - 5x + c$ d) $x^2 + 0.6x + c$

11. Find the maximum or minimum value of the function and the value of x when it occurs.

a) $y = x^2 + 6x - 3$ b) $y = x^2 - 12x + 21$ c) $y = -x^2 - 8x + 2$
d) $y = 10x - x^2 - 1$ e) $y = 2x^2 + 6x + 15$ f) $y = -3x^2 - 12x - 9$
g) $y = 3 + x^2 - 5x$ h) $y = x - x^2 - 4$ i) $y = 4x^2 + 2x - 1$
j) $y + 1.5x = -0.5x^2$

12. a) Find the minimum product of two numbers whose difference is 17.
b) What are the two numbers?

13. Measurement The sum of the base and the height of a triangle is 21 cm. Determine the maximum area of the triangle, in square centimetres.

14. Flower bed Reena wants to fence part of her front yard for a flower bed. What is the maximum area she can enclose if she has 30 m of fencing?

15. Basketball The path of a basketball shot can be modelled by the equation $h = -0.125d^2 + 2.5$, where h is the height of the basketball, in metres, and d is the horizontal distance of the ball from the player, in metres.
a) Find the maximum height reached by the ball.
b) What is the horizontal distance of the ball from the player when it reaches its maximum height?
c) How far from the floor is the ball when the player releases it?

2.3 Solving Quadratic Equations

Refer to the Key Concepts on page 128.

16. State the value of k that makes each expression a perfect square trinomial. Then, write the trinomial as the square of a binomial.

a) $x^2 + 8x + k$ b) $y^2 - 18y + k$ c) $m^2 + m + k$

d) $r^2 - 7r + k$ e) $t^2 + \dfrac{3}{2}t + k$ f) $w^2 - 0.04w + k$

17. Solve by completing the square. Express radical answers as exact roots in simplest radical form, and as approximate roots, to the nearest hundredth.

a) $x^2 + 4x + 1 = 0$ b) $t^2 - 6t - 4 = 0$ c) $x^2 + 11 = 8x$

d) $r^2 + 14 = 10r$ e) $y^2 - 3y = 5$ f) $w^2 + 7w - 17 = 0$

g) $x^2 + 5x + 1 = 0$ h) $2a^2 - 6a + 3 = 0$ i) $3z^2 + 5z + 1 = 0$

j) $2x - 5 + 6x^2 = 0$ k) $0.6t^2 - 0.5t = 0.1$ l) $-1 = \dfrac{1}{2}y^2 - 2y$

18. Solve by completing the square.

a) $x^2 + 4x + 5 = 0$ b) $n^2 - n + 1 = 0$ c) $x^2 + 3x = -5$

d) $2x^2 + 2x + 3 = 0$ e) $4g^2 + 3g + 1 = 0$ f) $0.5y^2 + 2.5 = -2y$

19. Solve by factoring. Check solutions.

a) $x^2 + 13x + 36 = 0$ b) $y^2 - y = 56$ c) $m^2 - 36 = 0$

d) $t^2 + 6 = 5t$ e) $w^2 + 24w + 144 = 0$ f) $4x^2 = 9$

g) $2y^2 + 5y = -2$ h) $3x^2 + 3 = -10x$ i) $3t^2 = 5t + 2$

j) $3z^2 - 7z + 2 = 0$ k) $6a^2 - 17a = -12$ l) $9x^2 - 3x - 20 = 0$

20. Solve using the quadratic formula. Express radical answers as exact roots in simplest radical form, and as approximate roots, to the nearest hundredth.

a) $x^2 - 6x - 40 = 0$ b) $t^2 - 2t = 24$ c) $y^2 + 9y + 9 = 0$

d) $k^2 - 10k = 9$ e) $0 = a^2 + 6a + 6$ f) $2x^2 - 7x + 3 = 0$

g) $5w^2 + w - 4 = 0$ h) $3m^2 - 2m = 2$ i) $4b^2 - 2b = 15$

j) $2r^2 = -8r - 3$ k) $8y^2 = 9 + 3y$ l) $4 = 5x + 3x^2$

21. Solve using the quadratic formula.

a) $x^2 - 5x = -9$ b) $0 = 4x^2 + 4x + 3$ c) $3m^2 - 2m + 4 = 0$

d) $2y^2 + 9y + 11 = 0$ e) $0.2k^2 + 1 = 0.1k$ f) $\dfrac{x^2}{2} + \dfrac{x}{5} = -\dfrac{1}{2}$

22. Solve and check.

a) $x^2 + 16 = 0$ b) $-5y^2 - 45 = 0$ c) $9x^2 + 4 = 0$

23. Pool and deck A rectangular swimming pool measuring 10 m by 4 m is surrounded by a deck of uniform width. The combined area of the deck and the pool is 135 m^2. What is the width of the deck?

24. Measurement Is it possible for a rectangle with a perimeter of 46 cm to have an area of 120 cm^2?

25. Real numbers Determine whether two real numbers whose sum is 17 can have each of the following products. If so, find the real numbers.

a) 60   b) 52   c) 80

2.4 Tools for Operating With Complex Numbers

Refer to the Key Concepts on page 139.

26. Simplify.

a) $3\sqrt{2} + 7\sqrt{2} - 5\sqrt{2}$

b) $7\sqrt{3} - 2\sqrt{6} + 5\sqrt{6} - 3\sqrt{3}$

c) $\sqrt{45} + \sqrt{80}$

d) $\sqrt{12} - \sqrt{27}$

e) $\sqrt{18} - \sqrt{50} + \sqrt{32}$

f) $2\sqrt{20} - 3\sqrt{125} + 3\sqrt{80}$

g) $5\sqrt{18} - \sqrt{40} - 2\sqrt{128} + \sqrt{90}$

h) $2\sqrt{27} + 3\sqrt{28} - 5\sqrt{63} - 3\sqrt{12}$

27. Simplify.

a) $\sqrt{3}(\sqrt{2} + 5)$

b) $\sqrt{2}(\sqrt{10} - \sqrt{6})$

c) $(4\sqrt{2} + \sqrt{5})(\sqrt{2} - 3\sqrt{5})$

d) $(2\sqrt{3} + \sqrt{5})^2$

e) $(\sqrt{7} - \sqrt{3})(\sqrt{7} + \sqrt{3})$

f) $(3\sqrt{6} + 5\sqrt{2})(3\sqrt{6} - 5\sqrt{2})$

28. Simplify.

a) $\dfrac{1}{\sqrt{5}}$

b) $\dfrac{\sqrt{2}}{\sqrt{3}}$

c) $\dfrac{4}{3\sqrt{2}}$

d) $\dfrac{\sqrt{3}}{4\sqrt{10}}$

29. Simplify.

a) $\dfrac{2}{\sqrt{3} - 1}$

b) $\dfrac{4}{\sqrt{5} + \sqrt{2}}$

c) $\dfrac{2\sqrt{3}}{\sqrt{2} - 5}$

d) $\dfrac{2\sqrt{7} - \sqrt{3}}{3\sqrt{7} + 2\sqrt{3}}$

30. Measurement A square has a side length of $4 - \sqrt{5}$. Write and simplify an expression for the area of the square.

2.5 Operations With Complex Numbers in Rectangular Form

Refer to the Key Concepts on page 149.

31. Simplify.

a) $(7 + 3i) + (5 - 6i)$

b) $(9 - 2i) - (11 + 4i)$

c) $(5i - 3)(2i + 5)$

d) $(2 - 3i)^2$

32. Simplify.

a) $\dfrac{3}{2i}$

b) $\dfrac{-2}{i}$

c) $\dfrac{4 + 3i}{i}$

d) $\dfrac{5 - 2i}{3i}$

33. Simplify.

a) $\dfrac{4}{3 + i}$

b) $\dfrac{2i}{1 - 3i}$

c) $\dfrac{1 + 2i}{1 - 4i}$

d) $\dfrac{5 - 3i}{2 + 4i}$

34. Solve and check.

a) $x^2 + 2x + 7 = 0$

b) $y^2 - 4y + 11 = 0$

c) $n^2 + 3 = -2n$

35. Fractals Using $z = 0$ as the first input value, find the first three output values for $F = z^2 - 2i$.

CHAPTER TEST

Achievement Chart

Category	Knowledge/Understanding	Thinking/Inquiry/Problem Solving	Communication	Application
Questions	All	8, 10	6, 9, 10	8, 10

1. Simplify.

a) $\sqrt{50}$

b) $\sqrt{44}$

c) $\sqrt{80}$

d) $\sqrt{7} \times \sqrt{5}$

e) $2\sqrt{3} \times \sqrt{6}$

f) $\dfrac{\sqrt{72}}{\sqrt{6}}$

g) $5\sqrt{10} \times 3\sqrt{2}$

h) $\dfrac{8 - \sqrt{40}}{2}$

2. Simplify.

a) $\sqrt{-36}$

b) $-\sqrt{-48}$

c) $5(-3i)^2$

d) $6 - \sqrt{-36}$

e) $5 + \sqrt{-18}$

f) $\dfrac{12 - \sqrt{-27}}{3}$

3. Find the maximum or minimum value of each function and the value of x when it occurs.

a) $y = -4x^2 - 8x + 5$

b) $y = x^2 - 7x + 2$

c) $y = -2x^2 + 5x + 5$

4. Solve by completing the square. Express answers as exact roots in simplest radical form, and as approximate roots, to the nearest hundredth.

a) $2x^2 - 7x + 2 = 0$

b) $3t^2 = 4t + 5$

5. Solve by completing the square.

a) $5x^2 - 2x + 2 = 0$

b) $2x^2 + 4x + 5 = 0$

6. Solve by factoring and check solutions.

a) $2x^2 - 7x = 4$

b) $3x^2 = 6 - 7x$

7. Solve using the quadratic formula. Express answers in simplest radical form.

a) $x^2 + 3x - 5 = 0$

b) $8c = 1 + 5c^2$

c) $n^2 - 5n = -13$

d) $3x^2 = -3x - 7$

8. Picture frame A picture that measures 10 cm by 8 cm is to be surrounded by a mat before being framed. The width of the mat is to be the same on all sides of the picture. The area of the mat is to equal the area of the picture. What is the width of the mat, to the nearest tenth of a centimetre?

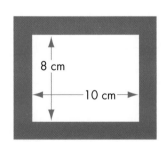

9. State three different methods for solving a quadratic equation. Give an example of each. Explain why each equation is best suited to the solution method chosen.

ACHIEVEMENT Check Knowledge/Understanding Thinking/Inquiry/Problem Solving Communication Application

10. Find k such that the graph of $y = 9x^2 + 3kx + k$
  a) intersects the x-axis in one point only
  b) intersects the x-axis in two points
  c) does not intersect the x-axis

Answer questions 11–15 only if you studied sections 2.4 and 2.5.

11. Simplify.
a) $\sqrt{48} - \sqrt{27} + \sqrt{12}$ **b)** $3\sqrt{40} + 5\sqrt{28} - \sqrt{63} - 2\sqrt{90}$

12. Simplify.
a) $\sqrt{6}\left(3\sqrt{2} + 2\sqrt{8}\right)$ **b)** $\left(2 - \sqrt{3}\right)\left(1 + 3\sqrt{3}\right)$ **c)** $\left(3\sqrt{2} - 2\right)^2$

13. Simplify.
a) $\dfrac{2}{\sqrt{7}}$ **b)** $\dfrac{3}{\sqrt{3} - 4}$ **c)** $\dfrac{5}{2\sqrt{6} + \sqrt{3}}$

14. **Measurement** Find the area of a rectangle with side lengths of $3 + \sqrt{2}$ and $3 - \sqrt{2}$.

15. Simplify.
a) $(8 - 3i) + (5 - 5i)$ **b)** $(7 + 2i) - (9 - 6i)$ **c)** $(6 - 3i)(2 + 5i)$
d) $\dfrac{5}{3i}$ **e)** $\dfrac{5 - 4i}{2i}$ **f)** $\dfrac{3 - 2i}{3 + 4i}$

CHALLENGE PROBLEMS

1. Equation Write an equation in the form $y = ax^2 + bc + c$ for the quadratic function whose graph passes through $(8, 0)$, $(0, 8)$ and $(-2, 0)$.

2. Roots Find the roots of $x^2 + \left(\dfrac{k^2 + 1}{k}\right)x + 1 = 0$.

3. Evaluating If $\left(\dfrac{2}{x} - \dfrac{x}{2}\right)^2 = 0$, evaluate x^6.

4. Real roots Find all values of k that ensure that the roots are real for $x - k(x - 1)(x - 2) = 0$.

5. Factors Find all possible values of k so that $3x^2 + kx + 5$ can be factored as the product of two binomial factors with integer coefficients.

6. Positive integers Show that there are nine pairs of positive integers (m, n) such that $m^2 + 3mn + 2n^2 - 10m - 20n = 0$.

7. Measurement The difference in the length of the hypotenuse of $\triangle ABC$ and the length of the hypotenuse of $\triangle XYZ$ is 3. Hypotenuse $AB = x$, hypotenuse $XY = \sqrt{x - 1}$ and $AB > XY$. Determine the length of each hypotenuse.

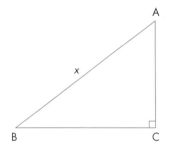
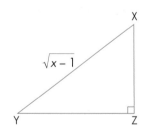

LOOK FOR A PATTERN

The ability to determine patterns is an important aspect of many careers. For example, dynamics, which is the branch of mathematics that deals with forces and their relations to patterns of motion, is applied in meteorology.

Dynamics relates especially to wind and precipitation patterns. To help predict when and where a tornado might occur, meteorologists look for horizontal wind shears, which produce anvil-shaped clouds. Patterns found using Doppler radar, which measures the speed and direction of drops of precipitation, are also used.

Many positive integers can be written as the difference of two squares. For example,

$24 = 7^2 - 5^2$ or $5^2 - 1^2$
$21 = 11^2 - 10^2$ or $5^2 - 2^2$

Which positive integers cannot be expressed as the difference of two squares?

Understand the Problem

1. What information are you given?
2. What are you asked to find?
3. Do you need an exact or an approximate answer?

Think of a Plan

Set up a table and determine which of the first 20 positive integers cannot be expressed as the difference of squares. Determine any pattern and use it to generalize the results.

Carry Out the Plan

Number	Difference of Squares	Number	Difference of Squares
1	$1^2 - 0^2$	11	$6^2 - 5^2$
2	not possible	12	$4^2 - 2^2$
3	$2^2 - 1^2$	13	$7^2 - 6^2$
4	$2^2 - 0^2$	14	not possible
5	$3^2 - 2^2$	15	$8^2 - 7^2$ or $4^2 - 1^2$
6	not possible	16	$5^2 - 3^2$ or $4^2 - 0^2$
7	$4^2 - 3^2$	17	$9^2 - 8^2$
8	$3^2 - 1^2$	18	not possible
9	$5^2 - 4^2$ or $3^2 - 0^2$	19	$10^2 - 9^2$
10	not possible	20	$6^2 - 4^2$

The positive integers 2, 6, 10, 14, and 18 cannot be expressed as the difference of two squares.

Starting with 2, every fourth number in the list cannot be expressed as the difference of two squares.

In general, positive integers represented by the expression $4n - 2$, where n is a positive integer, cannot be expressed as the difference of two squares.

Look Back

How could you check that the answer is correct?

Look for a Pattern

1. Use the given information to find a pattern.
2. Use the pattern to solve the problem.
3. Check that your answer is reasonable.

Apply, Solve, Communicate

1. Products a) Copy and complete the following.

$$143 \times 14 = \blacksquare$$
$$143 \times 21 = \blacksquare$$
$$143 \times 28 = \blacksquare$$
$$143 \times 35 = \blacksquare$$
$$143 \times 42 = \blacksquare$$

b) Explain the pattern.
c) Use the pattern to predict 143×98. Check your prediction.

2. Number puzzle Find the missing number.

43	51	16	24
19	13	45	39
56	61	21	26
24	44	29	■

3. Tables of values Find a rule that relates x and y. Then, copy and complete each table.

a)
x	y
1	10
2	14
3	18
4	22
9	■
■	98
52	■

b)
x	y
0	2
1	3
2	6
3	11
8	■
■	123
15	■

c)
x	y
0	−1
1	1
2	3
3	5
5	■
10	■
■	127

d)
x	y
1	2
3	3
5	4
7	5
11	■
■	20
99	■

4. Asterisks a) How many asterisks are in the next diagram?

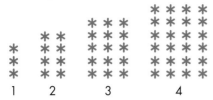

b) Use the pattern to write an expression for the number of asterisks in the *n*th diagram.
c) How many asterisks are in the 30th diagram? the 50th diagram?
d) Which diagram has 483 asterisks?

5. Remainder What is the remainder when 2^{75} is divided by 10?

6. Sequence What are the next two numbers in this sequence?

  9   18   11   16   13   14   15   12

7. Adding digits For each of the following whole numbers, add all the digits. Then, add the digits in the result. Continue, if necessary, until you reach a result that is a single digit.

    128   979   68 576   2 843 976

a) How is the single-digit result related to the middle digit of the original number?
b) For each of the original numbers, add all the digits except the middle digit. What do you notice about the results?
c) Write a 9-digit number, an 11-digit number, and a 15-digit number that fit the same pattern.

8. Number puzzle What number should appear in the final triangle?

9. Using rules According to a certain set of rules, the number 2 can be written as the sum of one or more positive integers in the two ways shown.

$$2 = 2$$
$$2 = 1 + 1$$

Using the same rules, the number 3 can be written as the sum of one or more positive integers in four ways, as follows.

$$3 = 3$$
$$3 = 2 + 1$$
$$3 = 1 + 2$$
$$3 = 1 + 1 + 1$$

Using the same rules, in how many ways can the number 17 be written as the sum of one or more positive integers?

10. Missing letter Find the missing letter.

C	G	N	■	V	W
4	7	3	5	1	6

11. Handball tournament Each school sent one player to a handball tournament. Each player played every other player at the tournament three times. In total, 63 games were played. How many players were in the tournament?

PROBLEM SOLVING

USING THE STRATEGIES

1. Driving Rashad started the 780-km drive from Sault Ste. Marie to Ottawa at 09:00. His average speed was 80 km/h. Three hours later, Tyson left Ottawa for Sault Ste. Marie. He drove at 70 km/h on the same highway. How far from Ottawa, and at what time of day, did they meet?

2. Geometry The side lengths of a triangle are $x + 2$, $8 - x$, and $4x - 1$. What value, or values, of x make the triangle isosceles?

3. Rolling dice You are given a standard die labelled from 1 to 6. How can you label a second die using only the numbers 0, 1, 2, 3, 4, 5, and 6 so that, when you roll both dice, the totals from 1 to 12 are equally likely?

4. Hop, step, and jump Suppose that a hop, a step, and a jump each have a specific length. Suppose p hops equals q steps, r jumps equals s hops, and t jumps equals x metres. How many steps does one metre equal?

5. Measurement The height of a triangle is 12 cm. The side lengths are consecutive whole numbers of centimetres. The area of the triangle is a whole number of square centimetres. What are the side lengths?

6. Rod lengths You have 12 rods, each 13 units long. They are to be cut into pieces measuring 3, 4, and 5 units. The resulting pieces will be assembled into 13 triangles, each with sides of 3, 4, and 5 units. How should the rods be cut?

7. Sequence The ninth term of a geometric sequence is 40. The twelfth term of the sequence is 5. What is the first term of the sequence?

8. Equations The letters R, S, and T represent integers. Find the possible values of R, S, and T.
$$R + S - T = 8$$
$$R \times S \times T = 48$$
$$R - S - T = 0$$

9. Quotients a) What is the quotient when any three-digit number whose digits are the same is divided by the sum of its digits?
b) Explain why the quotient is always the same.
c) Is there a constant quotient for four-digit numbers whose digits are the same? for five-digit numbers? for numbers with any number of digits? Explain.

10. Charity walk Orly and her friends organized a fundraising walk for charity. Without stopping, they walked on a level road, then up a hill, back down the hill, and then back to the start along the level road. The walk took six hours. Their speed was 4 km/h on the level road, 3 km/h up the hill, and 6 km/h down the hill. How far did they walk?

11. Intersecting circles A circle with radius 3 cm intersects a circle with radius 4 cm. At the points of intersection, the radii are perpendicular. What is the difference in the areas of the non-overlapping parts of the circles?

CUMULATIVE REVIEW: CHAPTERS 1 AND 2

Chapter 1

1. Evaluate.

a) $(-4)^{-3}$
b) $\dfrac{3^0 + 4^0}{2^{-1}}$
c) $\dfrac{5^2 - 3^3}{4^{-2} - 4^{-1}}$

d) $27^{-\frac{1}{3}}$
e) $-16^{-\frac{3}{4}}$
f) $\left(\dfrac{-125}{8}\right)^{\frac{2}{3}}$

2. Simplify.

a) $2x^{-3} \times 4x^7$
b) $6a^2 \div (-2a^{-3})$

c) $(-2y^{-1}z^2)^3$
d) $\dfrac{4x^2y^2}{10xy} \times \dfrac{5xy^{-2}}{6x^{-3}y^3}$

3. Solve.

a) $6^{2x-1} = \dfrac{1}{36}$
b) $5^{x+3} - 5^{x+1} = 600$

4. Simplify.

a) $(3x^2 + 4x - 1) - (x^2 - 2x - 2)$
b) $(2a^2 - 3ab + b^2) + (3a^2 + ab - 2b^2)$

5. Expand and simplify.

a) $4y[1 + 2y(1 - y)] + y[2(2y + 1) - 3]$
b) $(3z + 1)(2z^2 - 3z - 4)$
c) $2(a + 2b)^2 - 3(a - b)(2a + b)$

6. Simplify. State any restrictions.

a) $\dfrac{2t + 4}{t^2 + 6t + 8}$
b) $\dfrac{2x^2 - 3x + 1}{3x^2 - 2x - 1}$

c) $\dfrac{t^2 + 6t + 9}{t^2 - 6t + 9} \times \dfrac{3t - 9}{2t + 6}$

d) $\dfrac{2x^2 + 5x + 2}{2x^2 - 3x - 9} \div \dfrac{2x^2 + 3x - 2}{2x^2 + x - 3}$

e) $\dfrac{4}{3y + 1} + \dfrac{5}{1 - 3y}$

f) $\dfrac{4}{2m^2 - m - 1} - \dfrac{2}{m^2 + 2m - 3}$

7. Solve. Graph the solution.

a) $5k - 8 \le 7 + 2k$
b) $2(m - 3) - 5 > 3(4 - m) + 2$
c) $\dfrac{q + 1}{2} \ge \dfrac{5 + q}{3}$

Chapter 2

1. Simplify.

a) $2\sqrt{14} \times 3\sqrt{2}$
b) $\dfrac{\sqrt{80}}{\sqrt{10}}$
c) $\dfrac{9 + \sqrt{45}}{6}$

d) $\sqrt{-54}$
e) $-\sqrt{-24}$
f) $-3(4i)^2$

g) $5i(-2i)$
h) $2 - \sqrt{-27}$
i) $\dfrac{4 + \sqrt{20}}{2}$

2. Find the maximum or minimum value.

a) $y = x^2 - 5x - 1$
b) $y = -3x^2 + 4x + 2$

3. Solve by factoring.

a) $3y^2 + 7y + 4 = 0$
b) $6x^2 - 5x = 4$

4. Solve by completing the square. Express radical answers in simplest radical form.

a) $w^2 + 3w + 4 = 0$
b) $2x^2 - 3x = 3$

c) $\dfrac{1}{2}y^2 - 4y + 2 = 0$

5. Solve using the quadratic formula. Express radical answers in simplest radical form.

a) $x^2 - 4x - 11 = 0$
b) $3x^2 + 4x = 7$

6. Flower garden A rectangular flower garden, 7 m by 6 m, is surrounded by a grass strip of uniform width. If the total area of the garden and the grass strip is 90 m^2, what is the width of the strip?

Answer question 7 only if you studied sections 2.4 and 2.5.

7. Simplify.

a) $2\sqrt{12} + 4\sqrt{20} - 3\sqrt{27} - 5\sqrt{45}$

b) $\sqrt{3}(\sqrt{6} + \sqrt{21})$

c) $(2\sqrt{3} - 3\sqrt{2})(\sqrt{5} + 4\sqrt{3})$
d) $\dfrac{2\sqrt{5}}{3\sqrt{2}}$

e) $\dfrac{\sqrt{2} - \sqrt{3}}{2\sqrt{3} + 3\sqrt{2}}$
f) $(5 - 2i) - (6 + i)$

g) $(3 + 7i)(2 - 3i)$
h) $(5 + 4i)^2$

i) $\dfrac{3i}{2 - 5i}$
j) $\dfrac{4 + 3i}{1 + 2i}$

3 Transformations of Functions

Specific Expectations	Functions	Functions & Relations
Define the term function.	3.1	3.1
Demonstrate facility in the use of function notation for substituting into and evaluating functions.	throughout the chapter	throughout the chapter
Determine, through investigation, the properties of the functions defined by $f(x) = \sqrt{x}$ and $f(x) = \dfrac{1}{x}$.	3.2	3.2
Explain the relationship between a function and its inverse, using examples drawn from linear and quadratic functions, and from the functions $f(x) = \sqrt{x}$ and $f(x) = \dfrac{1}{x}$.	3.5	3.5
Represent inverse functions, using function notation, where appropriate.	3.5	3.5
Represent transformations of the functions defined by $f(x) = x$, $f(x) = x^2$, and $f(x) = \sqrt{x}$, using function notation.	3.3, 3.4, 3.5, 3.6, 3.7	3.3, 3.4, 3.5, 3.6, 3.7
Describe, by interpreting function notation, the relationship between the graph of a function and its image under one or more transformations.	3.3, 3.4, 3.5, 3.6, 3.7	3.3, 3.4, 3.5, 3.6, 3.7
State the domain and range of transformations of the functions defined by $f(x) = x$, $f(x) = x^2$, and $f(x) = \sqrt{x}$.	3.3, 3.4, 3.5, 3.6, 3.7	3.3, 3.4, 3.5, 3.6, 3.7

Falling Objects

A space shuttle depends on gravity to return to the Earth. Below an altitude of about 13 700 m, the descent of the shuttle is not powered. The shuttle's descent toward the runway is at an angle of 19° to the horizontal. This is much steeper than the 3° angle used for the powered descent of a commercial airliner.

In the Modelling Math questions on pages 191, 219, 232, and 242, you will solve the following problem and other problems that involve falling objects.

The approximate height above the ground of a falling object dropped from the top of a building is given by the function
$$h(t) = -5t^2 + d$$
where $h(t)$ metres is the height of the object t seconds after it is dropped, and d metres is the height from which it is dropped. The table shows the heights of three tall buildings in Canada.

Building	Height (m)
Petro-Canada 1, Calgary	210
Two Bloor West, Toronto	148
Complexe G, Québec City	126

a) Write the three functions, f(P-C1), f(TBW), and f(CG), that describe the height of the falling object above the ground t seconds after it is dropped from the top of each building.

b) Graph $h(t)$ versus t for the three functions on the same set of axes or in the same viewing window of a graphing calculator.

c) How could you transform the graph of f(P-C1) onto the graph of f(TBW)?

d) How could you transform the graph of f(CG) onto the graph of f(TBW)?

e) How could you transform the graph of f(P-C1) onto the graph of f(CG)?

Use your research skills to answer the following questions now.

1. The velocity of a falling object increases until the object reaches a terminal velocity. The terminal velocity is not the same for all falling objects. What factors influence terminal velocity?

2. Aristotle believed that the heavier the object, the faster it would fall. Galileo disproved this theory. How did he do it?

Human Physiology

The Heart

1. For the average person at rest, the heart pumps blood at a rate of about 5 L/min. Therefore, the volume, V litres, pumped over a period of time, t minutes, is given by the equation $V = 5t$.

a) Let $V = y$ and $t = x$, complete a table of values, and graph the equation. Use the same scale on each axis.

b) Which quadrant did you use? Explain.

2. For top athletes in competition, the heart can pump blood at 30 L/min, so $V = 30t$.

a) Let $V = y$ and $t = x$, and graph this equation on the same axes as the graph from question 1.

b) For points with the same non-zero x-coordinate, how do the y-coordinates compare for the two graphs?

c) Do the two graphs have any points in common?

3. a) Interchange the x- and y-coordinates in the table from question 1, so that a point (x, y) becomes the point (y, x). Graph the resulting coordinates on the same axes as the graph from question 1.

b) Describe what the graph with the interchanged coordinates represents.

4. a) Graph the line $y = x$ on the same set of axes you used in question 3.

b) Reflect the graph from question 1 in the line $y = x$. How does the result compare with the graph from question 3?

Breathing

5. At rest, the average person breathes about 12 times per minute.

a) Write an equation that expresses the number of breaths in terms of the time, in minutes.

b) Graph the equation.

6. Use the graph from question 5 to graph an equation that expresses the time, in minutes, in terms of the number of breaths. Explain your reasoning.

Review of Prerequisite Skills

If you need help with any of the skills named in purple below, refer to Appendix A.

1. Translations Find the image of the point $(-3, -2)$ under each translation.
a) 4 units upward   **b)** 6 units downward
c) 3 units to the left   **d)** 5 units to the right
e) 2 units downward and 4 units to the right
f) 8 units upward and 9 units to the left
g) 7 units downward and 10 units to the left
h) 12 units upward and 11 units to the right

2. Translations Name the translation that transforms the first point onto the second.
a) $(7, 0)$ onto $(-5, 0)$
b) $(3, 4)$ onto $(3, 12)$
c) $(8, -2)$ onto $(0, -3)$
d) $(-1, -3)$ onto $(-7, -8)$
e) $(-6, 12)$ onto $(5, -9)$
f) $(-11, 3)$ onto $(-4, -16)$

3. Translations Draw $\triangle ABC$ with vertices $A(1, 3)$, $B(-2, -1)$, and $C(4, -3)$ on grid paper. Find the coordinates of the image of $\triangle ABC$ after
a) a translation of 3 units downward and 4 units to the left
b) a translation of 2 units upward and 3 units to the right
c) a translation of 4 units downward and 5 units to the right

4. Reflections Draw $\triangle ABC$ with vertices $A(-2, 4)$, $B(-4, 1)$, and $C(5, -2)$ on grid paper. Find the coordinates of the image of $\triangle ABC$ after
a) a reflection in the x-axis
b) a reflection in the y-axis
c) a reflection in the x-axis followed by a reflection in the y-axis.

5. Stretches and shrinks Describe the stretch or shrink that transforms the function $y = x^2$ onto each of the following functions.
a) $y = 3x^2$ **b)** $y = \frac{1}{3}x^2$

6. Reflections Describe the reflection that transforms the function $y = 2x^2$ onto the function $y = -2x^2$.

7. Translations Describe the translation that transforms the function $y = x^2$ onto
a) $y = x^2 + 3$ **b)** $y = x^2 - 7$
c) $y = (x - 4)^2$ **d)** $y = (x + 6)^2$
e) $y = (x + 3)^2 - 8$ **f)** $y = (x - 7)^2 + 2$

8. Graphing quadratic functions Sketch each parabola. State the direction of the opening, the coordinates of the vertex, the equation of the axis of symmetry, the domain and range, and the maximum or minimum value.
a) $y = x^2 - 4$ **b)** $y = -2x^2 + 5$
c) $y = (x - 2)^2 + 3$ **d)** $y = -(x + 3)^2 - 5$

9. Translations **a)** Translate the graph of $y = x$ upward by 1 unit.
b) Translate the graph of $y = x$ to the left by 1 unit.
c) Compare the results from parts a) and b).
d) Translate the graph of $y = x^2$ upward by 1 unit.
e) Translate the graph of $y = x^2$ to the left by 1 unit.
f) Compare the results from parts d) and e).
g) Explain any differences in your answers to parts c) and f).

3.1 Functions

Cubic packages with edge lengths of 6 cm, 7 cm, and 8 cm have volumes of 6^3 or 216 cm^3, 7^3 or 343 cm^3, and 8^3 or 512 cm^3. These values can be written as a **relation**, which is a set of ordered pairs, (x, y). The relation is $\{(6, 216), (7, 343), (8, 512)\}$. Any set of ordered pairs is a relation.

A function is a special type of relation. A **function** is a set of ordered pairs in which, for every value of x, there is only one value of y.

The relation $\{(1, 5), (2, 6), (3, 7), (4, 8)\}$ is a function because, for each value of x, there is only one value of y.

The relation $\{(2, 7), (3, 8), (3, 9), (4, 10)\}$ is not a function because, when x is 3, y can equal 8 or 9.

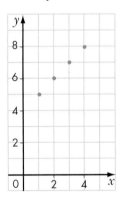

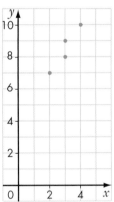

The following graphs model two relations.

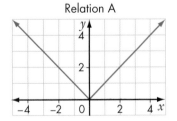

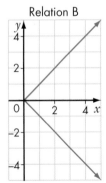

Relation A is a function. For each value of *x*, there is only one value of *y*.
For example, when *x* is −3, *y* is 3.

Relation B is not a function. For each value of *x*, except 0, there are two
values of *y*. For example, when *x* is 2, *y* is 2 or −2.

INVESTIGATE & INQUIRE

The packaging industry is an important application of mathematics.
Packages are made in many shapes and sizes.

An open-topped box can be made from a square sheet of tin by cutting out
smaller squares from the corners and folding up the sides. The diagrams
show that removing squares of side length 1 cm from the corners of an
18 cm by 18 cm sheet of tin results in an open-topped box with dimensions
16 cm by 16 cm by 1 cm and volume $16 \times 16 \times 1$ or 256 cm^3.

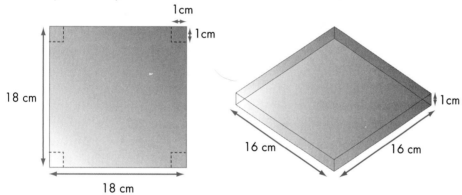

1. Copy and complete a table, like the one shown, or set up a spreadsheet
to determine the following.

Side Length of Removed Squares (cm)	Dimensions of Box (cm)	Volume of Box (cm^3)
1	$16 \times 16 \times 1$	256
2		
3		

a) the maximum volume of an open-topped box that can be made from an
18 cm by 18 cm piece of tin by removing smaller squares with side lengths
that are whole numbers of centimetres
b) the side length of the smaller squares that result in the maximum volume
of the box

2. a) From question 1, list a set of ordered pairs in the form (side length of removed squares, volume of box).
b) Graph the volume of the box versus the side length of the removed squares.
c) Does the graph model a function? Explain.

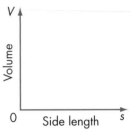

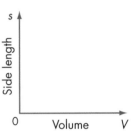

3. a) Graph the side length of the removed squares versus the volume of the box.
b) Does the graph model a function? Explain.

4. Explain why a packaging company might be interested in the answers to question 1.

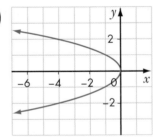

One way to determine if a relation is a function is to graph the relation and then use the **vertical line test**. If any vertical line passes through more than one point on the graph, then the relation is not a function.

EXAMPLE 1 **Vertical Line Test**

Determine if each relation is a function.

a)

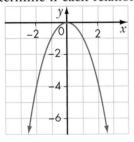

b)

SOLUTION

a) The relation is a function. No vertical line passes through more than one point.

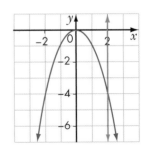

b) The relation is not a function. The vertical line shown passes through two points, $(-4, 2)$ and $(-4, -2)$, so there are two values of y for the same value of x.

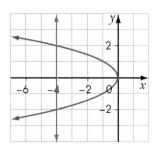

EXAMPLE 2 Falling Object

For an object falling under the effect of gravity, the approximate distance fallen, d metres, is given by the equation $d = 5t^2$, where t seconds is the time since the object was dropped.

a) Graph the equation. **b)** Determine if the relation is a function.

SOLUTION

a) The distance, d, depends on the time, t, so we call d the **dependent variable** and t, the **independent variable**. It is customary to graph the dependent variable versus the independent variable, that is, with the dependent variable on the vertical axis and with the independent variable on the horizontal axis.

Graph d versus t using paper and pencil, a graphing calculator, or graphing software. Note that the graph lies in the first quadrant because the values of d and t cannot be negative.

t	d
0	0
1	5
2	20
3	45
4	80
5	125

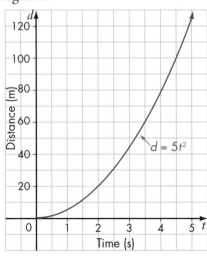

For this graph, the window variables include Xmin = 0, Xmax = 6, Ymin = 0, and Ymax = 130.

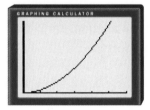

b) Because, for every value of t, there is only one value of d, the relation is a function.

Note that, for a function, the dependent variable is said to be "a function of" the independent variable. In Example 2, the distance, d, depends on, and is a function of, the time, t.

The set of the first elements in a relation is called the **domain** of the relation. For the relation $\{(1, 2), (3, 4), (5, 6), (7, 8)\}$, the domain is $\{1, 3, 5, 7\}$.

The set of the second elements in a relation is called the **range** of the relation. For the relation $\{(1, 2), (3, 4), (5, 6), (7, 8)\}$, the range is $\{2, 4, 6, 8\}$.

A function can be defined as a set of ordered pairs in which, for each element in the domain, there is exactly one element in the range.

EXAMPLE 3 Determining the Domain and Range

State the domain and range of each relation. Determine if each relation is a function.
a) $\{(-2, 2), (-3, 3), (-4, 4), (-3, 5), (-1, 6)\}$
b) $y = x^2$
c) $y = -2x + 3$

SOLUTION

a) The domain is $\{-4, -3, -2, -1\}$.
The range is $\{2, 3, 4, 5, 6\}$.
The relation is not a function because, for the element -3 in the domain, there are two elements, 3 and 5, in the range.

b) Graph the relation $y = x^2$ using paper and pencil, a graphing calculator, or graphing software.

x	y
-3	9
-2	4
-1	1
0	0
1	1
2	4
3	9

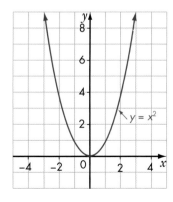

For this graph, the window variables include Xmin = −5, Xmax = 5, Ymin = −2, and Ymax = 9.

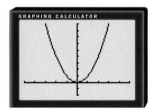

Since x can be any real number, the domain is the set of real numbers.

Since the value of y is always greater than or equal to zero, the range is $y \geq 0$.

The relation is a function because, for each element in the domain, there is exactly one element in the range.

c) Graph the relation $y = -2x + 3$ using paper and pencil, a graphing calculator, or graphing software.

x	y
-2	7
-1	5
0	3
1	1
2	-1
3	-3

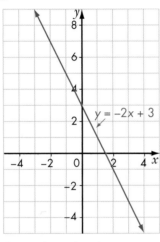

For this graph, the window variables include Xmin = -3, Xmax = 5, Ymin = -4, and Ymax = 8.

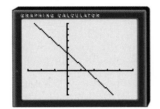

Since x can be any real number, the domain is the set of real numbers. Since y can be any real number, the range is the set of real numbers. The relation is a function because, for each element in the domain, there is exactly one element in the range.

In an equation such as $y = 2x + 3$, y depends on x and is a function of x. An equation that is a function can be named using function notation.

x-y notation	function notation
$y = 2x + 3$	$f(x) = 2x + 3$

In function notation, f names a function. Notice that the symbol $f(x)$ is another name for y. The symbol $f(x)$ is the value of the function f at x. Read $f(x)$ as "the value of f at x" or "f of x."

A function is like a machine. When an x-value in the domain of the function f enters, the machine produces the output $f(x)$. The output $f(x)$ is determined by the rule of the function.

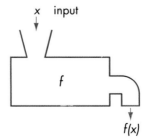

The machine produces exactly one output, f(x), for each x-value.

To find $f(5)$ for the function $f(x) = 2x + 3$, substitute 5 for x in $f(x) = 2x + 3$.
When $x = 5$, the value of y, or $f(5)$, is 13, because $2 \times 5 + 3 = 13$.

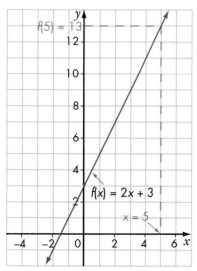

EXAMPLE 4 Evaluating a Function

If $f(x) = 3x + 1$, find

a) $f(6)$ b) $f(-2)$ c) $f(0)$

SOLUTION

a) $f(x) = 3x + 1$
     $f(6) = 3(6) + 1$
         $= 19$

b) $f(x) = 3x + 1$
    $f(-2) = 3(-2) + 1$
         $= -5$

c) $f(x) = 3x + 1$
    $f(0) = 3(0) + 1$
         $= 1$

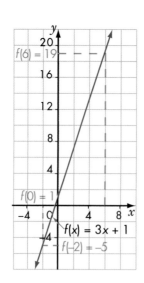

Key Concepts

- A function is a set of ordered pairs in which, for every x, there is only one y.
- If any vertical line passes through more than one point on the graph of a relation, then the relation is not a function.
- The set of the first elements in a relation is called the domain. The set of the second elements in a relation is called the range.
- An equation that is a function can be named using function notation.
- In function notation, the symbol $f(x)$ is another name for y and represents the value of the function f at x.

Communicate Your Understanding

1. a) Is every function a relation? Explain using examples.
b) Is every relation a function? Explain using examples.
2. Describe how you would determine if each of the following relations is a function.
a) $\{(0, 2), (1, 3), (2, 4), (1, 5), (0, 6)\}$
b) $\{(-2, 4), (-1, 1), (0, 0), (1, 1), (2, 4)\}$
3. Describe how you would determine if each graph models a function.

a)

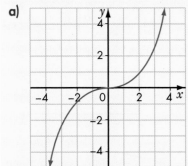

b)

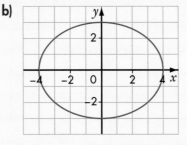

4. Describe how you would determine the domain and range of each of the following relations.
a) $\{(2, -3), (3, -1), (4, 1), (5, 3)\}$ b) $y = 3x + 2$
5. Describe how you would evaluate $f(4)$ for the function $f(x) = 2x - 3$.

Practise

A

1. Determine if each relation is a function.

a)

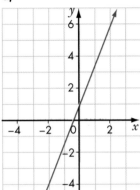

b)

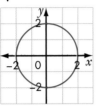

c)

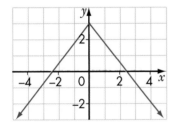

d)

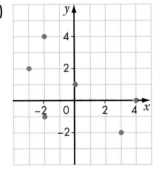

e)

x	y
−2	5
−1	3
0	−1
1	−4
2	−5

f)

x	y
0	−1
0	0
0	2
0	4

2. State the domain and range of each relation.

a) {(0, 5), (1, 6), (2, 7), (3, 8)}

b) {(1, 4), (2, 3), (2, 5)}

c) {(−2, −1), (−1, 0), (0, −1), (1, 2), (2, −1)}

d) {(−2, 1), (0, 1), (3, 1), (4, 1), (7, 1)}

3. State the domain and range of each relation.

a)

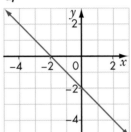

b)

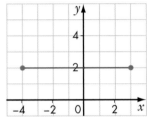

c)

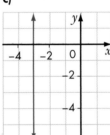

d)

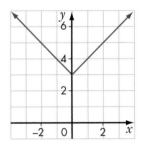

e)

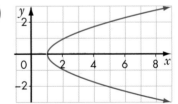

f)

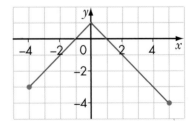

g)

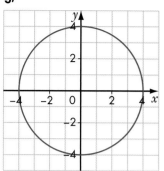

h)

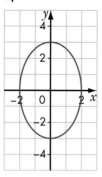

4. a) Graph the equation $y = x^2 - 3$.
b) Is the relation a function?

5. Determine if each relation is a function.
a) $y = 2 - 4x$
b) $y = 2x^2 + 3x - 5$
c) $x^2 + y^2 = 25$

6. If $f(x) = x - 5$, find
a) $f(8)$ **b)** $f(5)$ **c)** $f(1)$
d) $f(0)$ **e)** $f(-2)$

7. If $g(x) = 3x + 4$, find
a) $g(2)$ **b)** $g(0)$ **c)** $g(-1)$
d) $g(-3)$ **e)** $g(0.5)$

8. If $f(x) = x^2 + 2x - 1$, find
a) $f(0)$ **b)** $f(5)$ **c)** $f(-2)$
d) $f(1.5)$ **e)** $f(-0.5)$

9. If $h(x) = 2x^2 - 3x + 6$, find
a) $h(1)$ **b)** $h(10)$ **c)** $h(-3)$
d) $h(0.5)$ **e)** $h(5)$

10. If $f(n) = -4n^2 + 5$, find
a) $f(0)$ **b)** $f(1)$ **c)** $f(4)$
d) $f(-3)$ **e)** $f(-1.5)$

Apply, Solve, Communicate

11. List the ordered pairs of the function $f(x) = 7x - 1$ when the domain is $\{-1, 0, 2, 5\}$.

12. If $f(x) = 4x + 1$, find the value of x when the value of $f(x)$ is
a) 21   **b)** −7   **c)** 53   **d)** −19   **e)** 11

13. If the domain and range of a relation each contain exactly one real number, describe the graph of the relation.

14. Cost The cost, C dollars, of purchasing one type of ballpoint pen is related to the number of pens purchased, p, by the equation $C = 1.25p$.
a) Identify the dependent variable and the independent variable.
b) Is the cost a function of the number of pens purchased? Explain.

B

15. Determine if each of the following relations is a function. If so, state the dependent variable and the independent variable.
a) the time it takes to drive 100 km and the speed of the car
b) the ages of students and the numbers of CDs they own
c) the number of tickets sold for a school play at $8 per ticket and the revenue from ticket sales

16. If a point is on the $f(x)$-axis of a coordinate grid, what is the x-coordinate of the point? Explain.

17. If the graph of a relation is a vertical line, is the relation a function? Explain.

18. Find the range of each function when the domain is $\{-2, 0, 0.5, 3\}$.
a) $f(x) = 2x - 1$ **b)** $f(x) = 8x^2 - 4x + 7$

19. Application Mario sells home theatre systems. He is paid a weekly salary, plus commission on
determined from his
When his weekly sal
When his weekly sal
a) Interpret each fur
b) Identify the depen
c) State the domain
d) What are Mario's

20. Measurement
where $A(r) = \pi r^2$. Sta

21. a) Determine th
b) Is the relation a fu

22. Discount prices
store. All prices are d
a) Write an equation
original price.
b) If the sale price of

23. Measurement
cube as a function of
b) Determine the su

24. Inquiry/Proble
corral with the larges
of the corral will be t
fencing to build the
a) Write an equation
corral as a function o
b) Find the maximum area of the corral and the
dimensions that give this area.

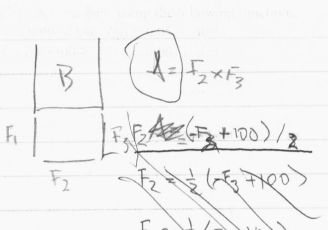

Corral

25. Communication Does the graph of $2x - 3y = 6$ model a function? Explain.

26. Algebra If $f(x) = x^2 + 4x$, find the value(s) of x if the value of $f(x)$ is

a) 5   b) −4   c) 0   d) −3

C

27. a) Sketch a graph of a function that has all real numbers in its domain and all real numbers less than or equal to −3 in its range.
b) Sketch a graph of a relation that is not a function and that has all real numbers in its domain and all real numbers less than or equal to −3 in its range.

28. Jogging Chelsea jogs several laps around a running track at a steady speed.
a) Is the distance she covers a function of the time for which she jogs? Explain.
b) Is her distance from her starting point a function of the time for which she jogs? Explain.

29. Describe how the value of $\dfrac{f(4) - f(1)}{4 - 1}$ is related to the graph of $f(x) = 6x - 5$.

30. Algebra Solve the following system algebraically.
$$f(x) = 2x - 3$$
$$f(x) = 3x + 2$$

31. Measurement Express the area of a circle as a function of its circumference.

32. Algebra a) If $f(x) = 4x + 3$, write and simplify $f(2a)$.
b) If $f(x) = 2 - 3x$, write and simplify $f(n + 1)$.
c) If $g(x) = x^2 + 1$, write and simplify $g(m - 1)$.
d) If $f(x) = 2x^2 - 3$, write and simplify $f(2k + 1)$.
e) If $g(x) = x^2 + 4x - 1$, write and simplify $g(3t - 1)$.
f) If $f(x) = 3x^2 - 2x + 4$, write and simplify $f(3 - 2w)$.

3.2 Investigation: Properties of Functions Defined by $f(x) = \sqrt{x}$ and $f(x) = \dfrac{1}{x}$

Properties of the Function $f(x) = \sqrt{x}$

Recall that $f(x)$ is another name for y.

1. a) Graph the function $y = \sqrt{x}$ manually by copying and completing the table of values, or graph the function using a graphing calculator.

x	0	1	4	9	16
y					

b) Why does the graph of $y = \sqrt{x}$ appear only in the first quadrant?

c) What is the domain of $y = \sqrt{x}$?

d) Does the value of y ever reach a maximum? Explain.

e) What is the range of $y = \sqrt{x}$?

Web Connection
www.school.mcgrawhill.ca/resources/
To investigate the origin of the square root symbol, visit the above web site. Go to **Math Resources**, then to *MATHEMATICS 11*, to find out where to go next. Write a brief report about the origin and history of the square root symbol.

2. a) On the same set of axes as the graph of $y = \sqrt{x}$, graph the function $y = x^2$ manually by copying and completing the table of values, or graph the two functions in the same viewing window of a graphing calculator.

x	3	2	1	0	−1	−2	−3
y							

b) What is the domain of $y = x^2$?

c) What is the range of $y = x^2$?

3. Describe the relationship between the graph of $y = \sqrt{x}$ and the graph of $y = x^2$.

4. a) Graph the function $y = -\sqrt{x}$ on the same set of axes or in the same viewing window as the graphs of $y = \sqrt{x}$ and $y = x^2$.

b) Why does the graph of $y = -\sqrt{x}$ appear only in the fourth quadrant?

c) What is the domain of $y = -\sqrt{x}$?

d) What is the range of $y = -\sqrt{x}$?

e) Describe the relationship between the graphs of $y = \sqrt{x}$, $y = -\sqrt{x}$, and $y = x^2$.

Properties of the Function $f(x) = \dfrac{1}{x}$

5. a) Graph the function $y = \dfrac{1}{x}$ manually by completing the tables of values, or graph the function using a graphing calculator.

x	$\dfrac{1}{4}$	$\dfrac{1}{3}$	$\dfrac{1}{2}$	1	2	3	4
y							

x	$-\dfrac{1}{4}$	$-\dfrac{1}{3}$	$-\dfrac{1}{2}$	−1	−2	−3	−4
y							

b) Why does the graph of $y = \dfrac{1}{x}$ appear only in the first and third quadrants?

6. Copy and complete the following tables of values for the function $y = \dfrac{1}{x}$.

Table 1

x	1	10	100	1000
y				

x	1	0.1	0.01	0.001
y				

Table 2

x	−1	−10	−100	−1000
y				

x	−1	−0.1	−0.01	−0.001
y				

7. For the branch of the graph represented by Table 1 in question 6,
a) as x increases in value, what happens to the value of y?
b) will y ever reach 0? Explain.
c) as x decreases in value, what happens to the value of y?
d) will y ever reach a maximum? Explain.

8. For the branch of the graph represented by Table 2 in question 6,
a) as x decreases in value, what happens to the value of y?
b) will y ever reach 0? Explain.
c) as x increases in value, what happens to the value of y?
d) will y ever reach a minimum? Explain.

9. A line that a curve approaches more and more closely is called an **asymptote**. The word *asymptote* comes from the Greek word *asymptotos*, which means "not meeting." Each example of an asymptote in this book is a line that a curve never touches or crosses. The line $x = 0$ is a vertical asymptote of the graph of $y = \dfrac{1}{x}$. The function $y = \dfrac{1}{x}$ is not defined when $x = 0$, because the denominator cannot equal 0. Since there is a break in the graph of $y = \dfrac{1}{x}$ at $x = 0$, we say that the graph is **discontinuous** at $x = 0$.

a) What is the domain of $y = \dfrac{1}{x}$?
b) What is the range of $y = \dfrac{1}{x}$?

10. a) Graph the function $y = x$ on the same set of axes or in the same viewing window as the graph of $y = \dfrac{1}{x}$.
b) What is the domain of $y = x$?
c) What is the range of $y = x$?

11. Describe the relationship between the graphs of $y = \dfrac{1}{x}$ and $y = x$.

3.3 Horizontal and Vertical Translations of Functions

When an object is dropped from the top of a bridge over a body of water, the approximate height of the falling object above the water is given by the function
$$h(t) = -5t^2 + d$$
where $h(t)$ metres is the height of the object t seconds after it is dropped, and d metres is the height of the bridge.

INVESTIGATE & INQUIRE

1. The table includes the approximate heights, in metres, of three famous Canadian bridges. Write the function that describes the height of a falling object above the water t seconds after it is dropped from the top of each bridge.

Bridge	Height (m)
Ambassador Bridge	45
Confederation Bridge	60
Capilano Canyon Suspension Bridge	70

2. Graph $h(t)$ versus t for the three functions on the same set of axes or in the same viewing window of a graphing calculator.

3. In what quadrant do the three graphs appear? Explain why.

4. For a given t-coordinate, how does the h-coordinate of a point on the graph for the Confederation Bridge compare with the h-coordinate of a point on the graph for the Ambassador Bridge? Explain why.

5. For a given t-coordinate, how does the h-coordinate of a point on the graph for the Capilano Bridge compare with the h-coordinate of a point on the graph for the Ambassador Bridge? Explain why.

6. Graph the three functions $y = -5x^2 + c$ for $c = 45$, $c = 60$, and $c = 70$ on the same set of axes or in the same viewing window of a graphing calculator. If the domain of the three functions is the set of real numbers, how do the three graphs compare with the three graphs of $h(t)$ versus t from question 2? Explain why.

7. How do the three graphs from question 6 compare with the graph of $y = x^2$ over the same domain? Explain why.

EXAMPLE 1 Positive Vertical Translation

a) Graph the functions $y = x^2$ and $y = x^2 + 2$ on the same set of axes.

b) How does the graph of $y = x^2$ compare to the graph of $y = x^2 + 2$?

SOLUTION

a) $y = x^2$

x	y
3	9
2	4
1	1
0	0
−1	1
−2	4
−3	9

$y = x^2 + 2$

x	y
3	11
2	6
1	3
0	2
−1	3
−2	6
−3	11

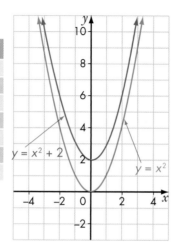

The window variables include Xmin = −4, Xmax = 4, Ymin = 0, and Ymax = 12.

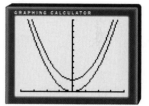

Note that the transformations shown in this chapter can be performed using a graphing software program, such as Zap-a-Graph. For details of how to do this, refer to the Zap-a-Graph section of Appendix C.

b) The graphs of $y = x^2$ and $y = x^2 + 2$ are congruent.

The graph of the function $y = x^2 + 2$ is obtained when the graph of the function $y = x^2$ undergoes a vertical translation of 2 units in the positive direction, that is, upward.

The graph of the function $y = x^2 + 2$ can also be obtained from the graph of $y = x^2$ by adding 2 to each y-value on the graph of $y = x^2$. The point (x, y) on the graph of $y = x^2$ is transformed to become the point $(x, y + 2)$ on the graph of $y = x^2 + 2$.

The graphs show that the functions have the same domain but different ranges. The domain of each function is the set of real numbers. The range of $y = x^2$ is $y \geq 0$. The range of $y = x^2 + 2$ is $y \geq 2$.

EXAMPLE 2 Vertical Translations

How do the graphs of $y = \sqrt{x} + 3$ and $y = \sqrt{x} - 2$ compare with the graph of $y = \sqrt{x}$, where $x \geq 0$.

SOLUTION

$y = \sqrt{x}$

x	y
0	0
1	1
4	2
9	3
16	4

$y = \sqrt{x} + 3$

x	y
0	3
1	4
4	5
9	6
16	7

$y = \sqrt{x} - 2$

x	y
0	-2
1	-1
4	0
9	1
16	2

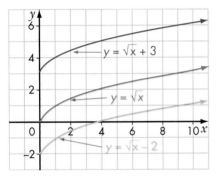

The window variables include Xmin = 0, Xmax = 18, Ymin = -3, and Ymax = 8.

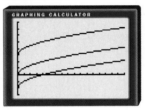

The graph of the function $y = \sqrt{x} + 3$ is the graph of the function $y = \sqrt{x}$ with a vertical translation of 3 units upward. The point (x, y) on the graph of $y = \sqrt{x}$ is transformed to become the point $(x, y + 3)$ on the graph of $y = \sqrt{x} + 3$.

Similarly, the graph of the function $y = \sqrt{x} - 2$ is the graph of the function $y = \sqrt{x}$ with a vertical translation of 2 units downward. The point (x, y) on the graph of $y = \sqrt{x}$ is transformed to become the point $(x, y - 2)$ on the graph of $y = \sqrt{x} - 2$.

All three graphs are congruent and have domain $x \geq 0$. The range of $y = \sqrt{x}$ is $y \geq 0$, of $y = \sqrt{x} + 3$ is $y \geq 3$, and of $y = \sqrt{x} - 2$ is $y \geq -2$.

The results from Examples 1 and 2 can be generalized for all functions as follows.

The graph of $y = f(x) + k$ is congruent to the graph of $y = f(x)$.
If $k > 0$, the graph of $y = f(x) + k$ is the graph of $y = f(x)$ translated upward by k units.
If $k < 0$, the graph of $y = f(x) + k$ is the graph of $y = f(x)$ translated downward by k units.

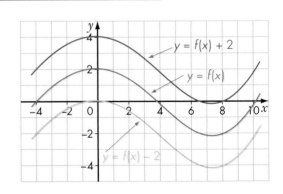

EXAMPLE 3 Horizontal Translations

How do the graphs of $y = \sqrt{x+2}$ and $y = \sqrt{x-3}$ compare to the graph of $y = \sqrt{x}$.

SOLUTION

Complete tables of values using convenient values for x, or use a graphing calculator.

$y = \sqrt{x}$

x	y
0	0
1	1
4	2
9	3
16	4

$y = \sqrt{x+2}$

x	y
-2	1
-1	2
2	3
7	4
14	5

$y = \sqrt{x-3}$

x	y
3	0
4	1
7	2
12	3
19	4

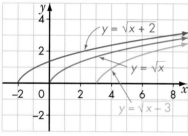

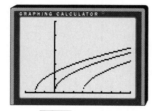

The window variables include Xmin = –3, Xmax = 8, Ymin = 0, and Ymax = 5.

The graphs of $y = \sqrt{x}$, $y = \sqrt{x+2}$, and $y = \sqrt{x-3}$ are congruent.
The graph of $y = \sqrt{x+2}$ is obtained when the graph of $y = \sqrt{x}$ is translated horizontally 2 units to the left.
The graph of $y = \sqrt{x+2}$ is also obtained from the graph of $y = \sqrt{x}$ by subtracting 2 from each x-value on the graph of $y = \sqrt{x}$.
The point (x, y) on the graph of $y = \sqrt{x}$ is transformed to become the point $(x - 2, y)$ on the graph of $y = \sqrt{x+2}$.
Similarly, the graph of $y = \sqrt{x-3}$ is obtained when the graph of $y = \sqrt{x}$ is translated horizontally 3 units to the right.
The graph of $y = \sqrt{x-3}$ is also obtained from the graph of $y = \sqrt{x}$ by adding 3 to each x-value on the graph of $y = \sqrt{x}$.
The point (x, y) on the graph of $y = \sqrt{x}$ is transformed to become the point $(x + 3, y)$ on the graph of $y = \sqrt{x-3}$.
All three graphs have the same range, $y \geq 0$.
The domain of $y = \sqrt{x}$ is $x \geq 0$, of $y = \sqrt{x+2}$ is $x \geq -2$, and of $y = \sqrt{x-3}$ is $x \geq 3$.

The results from Example 3 can be generalized for all functions as follows.

The graph of $y = f(x - h)$ is congruent to the graph of $y = f(x)$.
If $h > 0$, the graph of $y = f(x - h)$ is the graph of $y = f(x)$ translated to the right by h units.
If $h < 0$, the graph of $y = f(x - h)$ is the graph of $y = f(x)$ translated to the left by h units.

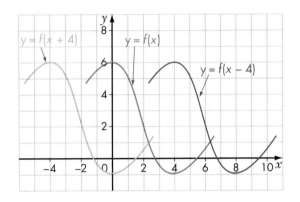

EXAMPLE 4 Horizontal and Vertical Translations

Sketch the graph of $y = (x - 3)^2 + 4$.

SOLUTION

Sketch the graph of $y = x^2$.
Translate the graph of $y = x^2$ three units to the right to obtain the graph of $y = (x - 3)^2$.
Translate the graph of $y = (x - 3)^2$ four units upward to obtain the graph of $y = (x - 3)^2 + 4$.
The point (x, y) on the function $y = x^2$ is transformed to become the point $(x + 3, y + 4)$. For example, $(0, 0)$ becomes $(3, 4)$.

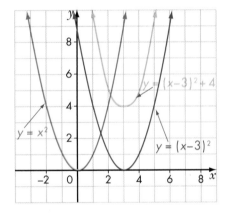

Note that, in Example 4, you could graph the three functions using a graphing calculator. However, it is not necessary to graph $y = x^2$ or $y = (x - 3)^2$ before graphing $y = (x - 3)^2 + 4$.

The window variables include Xmin = −4, Xmax = 7, Ymin = −2, and Ymax = 10.

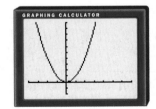
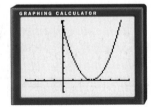
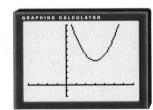

Key Concepts

- A function and its translation image are congruent.
- The table summarizes translations of the function $y = f(x)$.

Translation	Mathematical Form	Effect
Vertical	$y = f(x) + k$	If $k > 0$, then k units upward If $k < 0$, then k units downward
Horizontal	$y = f(x - h)$	If $h > 0$, then h units to the right If $h < 0$, then h units to the left

Communicate Your Understanding

1. Starting with the graph of $y = \sqrt{x}$, describe how you would sketch each of the following graphs.

a) $y = \sqrt{x} + 4$ b) $y = \sqrt{x + 4}$ c) $y = \sqrt{x - 4} - 3$

2. Starting with the graph of $y = x^2$, describe how you would sketch each of the following graphs.

a) $y = x^2 - 5$ b) $y = (x - 5)^2$ c) $y = (x + 5)^2 + 4$

3. Describe how you would find the domain and range of each of the following functions.

a) $y = x^2 + 4$ b) $y = x - 4$ c) $y = \sqrt{x - 5} - 3$

Practise

A

1. The function $y = f(x)$ is given. Describe how the graphs of the following functions can be obtained from the graph of $y = f(x)$.

a) $y = f(x) + 5$ b) $y = f(x) - 6$
c) $y = f(x - 4)$ d) $y = f(x + 8)$
e) $y - 3 = f(x)$ f) $y + 7 = f(x)$
g) $y = f(x + 3) - 5$ h) $y = f(x - 6) + 2$
i) $y = f(x - 5) - 7$ j) $y = f(x + 2) + 9$

2. The function $y = f(x)$ has been transformed to $y = f(x - h) + k$. Determine the values of h and k for each of the following transformations.

a) 6 units upward
b) 8 units downward

c) 3 units to the right
d) 5 units to the left
e) 2 units to the left and 4 units downward
f) 7 units to the right and 7 units upward

3. State the domain and range of each function.

a) $y = x + 2$ b) $y = x - 4$
c) $y = x^2 - 3$ d) $y = (x - 2)^2$
e) $y = (x + 5)^2 - 1$ f) $y = \sqrt{x + 1}$
g) $y = \sqrt{x} - 5$ h) $y = \sqrt{x - 3} + 6$

4. The graph of a function $y = f(x)$ is shown. Sketch the graph of each of the following.

a) $y = f(x) - 4$ b) $y = f(x) + 2$
c) $y = f(x - 4)$ d) $y = f(x + 2)$
e) $y = f(x - 3) - 2$ f) $y = f(x + 4) + 3$

i)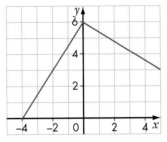

ii)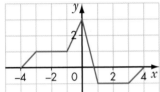

iii)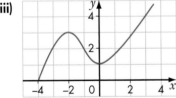

iv)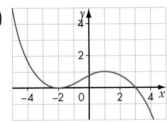

5. The graph of the function drawn in blue is a translation image of the function drawn in red. Write an equation for each function drawn in blue. Check each equation using a graphing calculator.

a)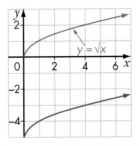

b)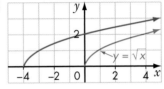

c)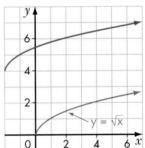

6. Use transformations to sketch the graph of each of the following functions, starting with the graph of $y = x$.

a) $y = x + 4$ b) $y = x - 5$
c) $y = (x - 4)$ d) $y = (x + 2)$
e) $y = (x + 5) - 2$ f) $y = (x - 1) + 6$

7. Use transformations to sketch the graph of each of the following functions, starting with the graph of $y = \sqrt{x}$.

a) $y = \sqrt{x} + 7$ b) $y + 3 = \sqrt{x}$
c) $y = \sqrt{x + 3}$ d) $y = \sqrt{x - 4}$
e) $y = \sqrt{x - 6} + 3$ f) $y = \sqrt{x + 5} + 4$

8. Use transformations to sketch the graph of each of the following functions, starting with the graph of $y = x^2$.

a) $y = x^2 + 3$ b) $y + 2 = x^2$
c) $y = (x - 7)^2$ d) $y = (x + 6)^2$
e) $y = (x + 4)^2 - 3$ f) $y = (x - 5)^2 + 5$

Apply, Solve, Communicate

9. Falling objects The approximate height above the ground of a falling object dropped from the top of a building is given by the function

$$h(t) = -5t^2 + d$$

where $h(t)$ metres is the height of the object t seconds after it is dropped, and d metres is the height from which it is dropped. The table shows the heights of three tall buildings in Canada.

Building	Height (m)
Petro-Canada 1, Calgary	210
Two Bloor West, Toronto	148
Complexe G, Québec City	126

a) Write the three functions, f(P-C1), f(TBW), and f(CG), that describe the height of a falling object above the ground t seconds after it is dropped from the top of each building.

b) Graph $h(t)$ versus t for the three functions on the same set of axes or in the same viewing window of a graphing calculator.

c) How could you transform the graph of f(P-C1) onto the graph of f(TBW)?

d) How could you transform the graph of f(CG) onto the graph of f(TBW)?

e) How could you transform the graph of f(P-C1) onto the graph of f(CG)?

B

10. Service calls Elena and Mario both repair kitchen appliances. Elena charges $45 for a service call, plus $35/h for labour. Mario charges $40 for a service call, plus $35/h for labour. Write an equation for the cost, C dollars, of a service call in terms of the number of hours worked, t

a) for Elena   **b)** for Mario

c) How are the graphs of the two equations related? Explain.

11. Communication When the graph of $y = 2x + 3$ is translated 1 unit to the right and 2 units upward, how is the resulting graph related to the graph of $y = 2x + 3$? Explain.

12. Application Many companies pay their employees using a salary scale that depends on the number of years worked. One salary scale is modelled by the function $S(y) = 25\ 000 + 2250y$, where $S(y)$ dollars is the salary and y is the number of years worked for the company. The employees' union negotiates an increase of $1000 for each employee.

a) How is the graph of $S(y)$ transformed by the increase?

b) Write the function that models the salary scale after the increase.

c) State a reasonable domain and range for the function in part b). Justify your reasoning.

13. Greatest integer function The greatest integer function is defined by $[x]$ = the greatest integer that is less than or equal to x. For example, $[4] = 4$, $[4.83] = 4$, and $[-5.3] = -6$. The graph of $y = [x]$ is shown.

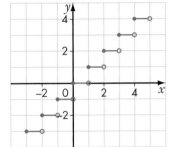

a) Explain the meanings of the open and closed dots on the graph of $y = [x]$.

b) State the domain and range of $y = [x]$.

c) Use transformations to sketch the graph of $y = [x] + 2$; $y = [x - 3]$; $y = [x + 4] - 1$.

14. Parking costs EZ-Park determines its parking charges based on the greatest integer function $y = [x + 2] + 3$, where y is the parking charge, in dollars, and x is the number of hours that a vehicle is in the parking garage.

a) Sketch the graph of the function.

b) How much would a driver pay to park for 30 min? for 1 h? for 1 h 25 min? for 3 h 1 min?

15. What transformation relates the graph of

a) $y = f(x - 4)$ to its image, $y = f(x + 3)$?

b) $y = f(x) + 5$ to its image, $y = f(x) - 7$?

16. Describe a vertical translation that could be applied to the graph of $y = \sqrt{x}$ so that the translation image passes through the point $(4, 0)$.

17. Inquiry/Problem Solving The function $y = x + 3$ could be a vertical translation of $y = x$ three units upward or a horizontal translation of $y = x$ three units to the left. Explain why.

C

18. Chemistry a) One way to describe the concentration of an acid is as a percent by volume. For example, in 40 mL of a 30% acid solution, the volume of pure acid is $40 \times \dfrac{30}{100}$ or 12 mL, and the volume of water is $40 - 12$ or 28 mL. If 5 mL of pure acid is mixed with 20 mL of water to give 25 mL of acid solution, the concentration of the solution is given by $\dfrac{5}{25} \times 100\% = 20\%$. If water is mixed with 50 mL of 40% acid solution, write an equation that describes the acid concentration, $C(x)$, as a function of the volume of water added, x.

b) Graph $C(x)$ versus x.

c) What is the acid concentration after 10 mL of water have been added?

d) Write an equation that describes the acid concentration as a function of the volume of water added to 40 mL of 50% acid solution.

e) Graph $C(x)$ versus x for the function from part d).

f) How could you transform the graph from part e) onto the graph from part b)?

CAREER CONNECTION *Veterinary Medicine*

There are many more domestic animals in Canada than there are people. For example, in addition to the millions of dogs and cats in Canadian homes, there are over 12 000 000 cattle and 10 000 000 pigs on Canadian farms. Medical services for these and other animals are provided by workers in the field of veterinary medicine.

1. Ages of cats and dogs As with humans, the medical needs of domestic animals change as they age. However, humans and domestic animals age differently. For a small dog, aged 3 years or more and with a mass up to about 11 kg, the number of human years equivalent to the age of the dog is given by the formula

$$h(a) = 4a + 20$$

where h is the equivalent number of human years, and a is the age of the dog. For a domestic cat aged 3 years or more, the number of human years equivalent to the age of the cat is given by the formula

$$h(a) = 4a + 15$$

where h is the equivalent number of human years, and a is the age of the cat.

a) Graph h versus a for cats and for small dogs over the domain 3 years to 15 years on the same axes or in the same viewing window of a graphing calculator.

b) Describe how the graphs are related by a transformation.

c) A cat and a small dog were born on the same day and are over 3 years old. How do the numbers of human years equivalent to their ages compare? Explain.

d) If their ages are expressed as equivalent human years, do cats and small dogs age at the same rate after the age of 3? Explain.

e) If their ages are expressed as equivalent human years, do cats and small dogs age at the same rate from birth? How do you know?

2. Research Use your research skills to investigate

a) the training needed to become a veterinarian, also know as a doctor of veterinary medicine

b) the organizations that employ veterinarians

c) other careers that involve animal care

3.4 Reflections of Functions

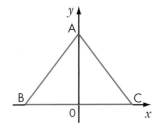

A coordinate grid is superimposed on a cross section of the Great Pyramid, so that the y-axis passes through the vertex of the pyramid. The x-axis bisects two opposite sides of the square base. The two sloping lines in the cross section, AB and AC, are altitudes of two triangular faces of the pyramid.

Web Connection
www.school.mcgrawhill.ca/resources/
To find more information about the Great Pyramid and other Egyptian pyramids, visit the above web site. Go to **Math Resources**, then to *MATHEMATICS 11*, to find out where to go next. Write a brief report about the construction techniques used.

INVESTIGATE & INQUIRE

1. Copy and complete each table of values and sketch each pair of graphs on the same set of axes.

a) $y = \sqrt{x}$ 　　　 $y = -\sqrt{x}$

x	y	x	y
0		0	
1		1	
4		4	
9		9	
16		16	

b) $y = x^2$ 　　　 $y = -x^2$

x	y	x	y
-2		-2	
-1		-1	
0		0	
1		1	
2		2	

2. In question 1a), for equal values of the x-coordinates, how do the y-coordinates of $y = \sqrt{x}$ compare with the y-coordinates of $y = -\sqrt{x}$? How are the graphs the same? How are they different?

3. In question 1b), for equal values of the x-coordinates, how do the y-coordinates of $y = x^2$ compare with the y-coordinates of $y = -x^2$? How are the graphs the same? How are they different?

4. Make a conjecture about the relationship between the graphs of $y = f(x)$ and $y = -f(x)$.

5. Test your conjecture by graphing each of the following pairs of graphs on a graphing calculator.

a) $y = \sqrt{x-1}$ and $y = -\sqrt{x-1}$

b) $y = (x+3)^2$ and $y = -(x+3)^2$

6. Copy and complete each table of values and sketch each pair of graphs on the same set of axes.

a) $y = x + 2$ $\qquad$ $y = (-x) + 2$

x	y		x	y
-2			-2	
-1			-1	
0			0	
1			1	
2			2	

b) $y = 2x - 4$ $\qquad$ $y = 2(-x) - 4$

x	y		x	y
-2			-2	
-1			-1	
0			0	
1			1	
2			2	

7. In question 6a), for equal values of the y-coordinates, how do the x-coordinates of $y = x + 2$ compare with the x-coordinates of $y = (-x) + 2$? How are the graphs the same? How are they different?

8. In question 6b), for equal values of the y-coordinates, how do the x-coordinates of $y = 2x - 4$ compare with the x-coordinates of $y = 2(-x) - 4$? How are the graphs the same? How are they different?

9. Make a conjecture about the relationship between the graphs of $y = f(x)$ and $y = f(-x)$.

10. Test your conjecture from question 9 by graphing each of the following pairs of graphs on a graphing calculator.

a) $y = \sqrt{x}$ and $y = \sqrt{-x}$ $\qquad\qquad$ **b)** $y = (x - 4)^2$ and $y = (-x - 4)^2$

11. In the cross section of the Great Pyramid, one altitude can be modelled by the equation $y = 1.27x + 146$. Which altitude, AB or AC, can be modelled by this equation? Explain.

12. What is the equation of the other altitude? Explain.

EXAMPLE 1 Comparing y = f(x) and y = −f(x)

a) Given the graph of $y = f(x)$, as shown, graph $y = -f(x)$ on the same axes.

b) Describe how the graph of $y = -f(x)$ is related to the graph of $y = f(x)$.

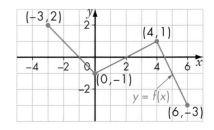

SOLUTION

a) Use the given graph to complete a table of values for the function $y = f(x)$. Then, complete a table of values for the function $y = -f(x)$, and draw the graph.

$y = f(x)$ $y = -f(x)$

x	y
−3	f(−3) = 2
0	f(0) = −1
4	f(4) = 1
6	f(6) = −3

x	y
−3	−f(−3) = −2
0	−f(0) = 1
4	−f(4) = −1
6	−f(6) = 3

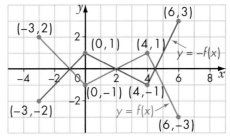

b) The graphs of $y = f(x)$ and $y = -f(x)$ are congruent.
If $y = -f(x)$, then $y = -1f(x)$, so each y-value on the graph of $y = -f(x)$ is the corresponding y-value on the graph of $y = f(x)$ multiplied by −1.
The point (x, y) on the graph of the function $y = f(x)$ becomes the point $(x, -y)$ on the graph of $y = -f(x)$. For example, (−3, 2) becomes (−3, −2), and (6, −3) becomes (6, 3).
The graphs have the same x-intercepts.
The graph of $y = -f(x)$ is a reflection of the graph of $y = f(x)$ in the x-axis.

As noted in Example 1, the two graphs have the same x-intercepts. Points that lie on the x-axis have a y-coordinate of 0, so they are unaltered by the transformation of $y = f(x)$ to $y = -f(x)$. Points that are unaltered by a transformation are said to be **invariant**.

In general, the point (x, y) on the graph of the function $y = f(x)$ becomes the point $(x, -y)$ on the graph of $y = -f(x)$. The graph of $y = -f(x)$ is a reflection of the graph of $y = f(x)$ in the x-axis. Points that lie on the x-axis are invariant, because their y-coordinate is 0.

If, in Example 1, $y = f(x)$ were defined by an equation, such as $f(x) = x - 1$, then $y = -f(x)$ would be defined by the equation $y = -(x - 1)$ or $y = -x + 1$. The graph of $y = -f(x)$ would be a reflection of the graph of $y = f(x)$ in the x-axis. The x-intercepts of the two graphs would be the same. There would be one invariant point, $(1, 0)$.

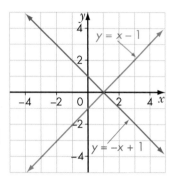

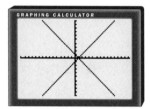

The window can be adjusted using the ZSquare instruction in the ZOOM menu.

EXAMPLE 2 Comparing $y = f(x)$ and $y = f(-x)$

Let $f(x) = 2x + 1$.
a) Write an equation for $f(-x)$.
b) Graph $y = f(x)$ and $y = f(-x)$ on the same axes or in the same viewing window of a graphing calculator.
c) Describe how the graph of $y = f(-x)$ is related to the graph of $y = f(x)$.

SOLUTION

a) Substitute $-x$ for x in $2x + 1$.
$2(-x) + 1 = -2x + 1$
so $f(-x) = -2x + 1$
b) Graph both functions using paper and pencil or a graphing calculator.

$y = 2x + 1$ $y = -2x + 1$

x	y
-2	-3
-1	-1
0	1
1	3
2	5

x	y
2	-3
1	-1
0	1
-1	3
-2	5

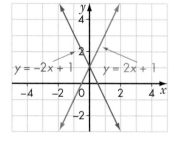

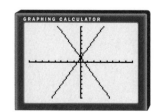

c) The graphs of $y = f(x)$ and $y = f(-x)$ are congruent.
The point (x, y) on the graph of $y = 2x + 1$ becomes the point $(-x, y)$ on the graph of $y = -2x + 1$. For example, the point $(2, 5)$ becomes the point $(-2, 5)$. The point $(-1, -1)$ becomes the point $(1, -1)$.
The graphs have the same y-intercept.
The graph of $y = -2x + 1$ is the graph of $y = 2x + 1$ reflected in the y-axis.

Note that there is one invariant point, (0, 1). For both linear functions, the domain and range are the set of real numbers.

In general, the point (x, y) on the graph of the function $y = f(x)$ becomes the point $(-x, y)$ on the graph of $y = f(-x)$. The graph $y = f(-x)$ is a reflection of the graph of $y = f(x)$ in the y-axis. Points that lie on the y-axis are invariant, because their x-coordinate is 0.

The results for reflecting graphs of functions in the x-axis and the y-axis can be summarized as follows.

Reflection in the x-axis
The graph of $y = -f(x)$ is the graph of $y = f(x)$ reflected in the x-axis.

Reflection in the y-axis
The graph of $y = f(-x)$ is the graph of $y = f(x)$ reflected in the y-axis.

EXAMPLE 3 Reflecting a Radical Function

If $f(x) = \sqrt{x - 2}$, write an equation to represent each of the following functions, describe how the graph of each function is related to the graph of $y = f(x)$, sketch each graph, state the domain and range, and identify any invariant points.

a) $y = -f(x)$ **b)** $y = f(-x)$

SOLUTION

a) For $y = f(x)$ For $y = -f(x)$
$\quad\quad\quad y = \sqrt{x - 2}$ $y = -\sqrt{x - 2}$

The graph of $y = -f(x)$ is the graph of $y = f(x)$ reflected in the x-axis.
Both graphs have the same x-intercept, 2.
For $y = f(x)$, the domain is $x \geq 2$ and the range is $y \geq 0$.
For $y = -f(x)$, the domain is $x \geq 2$ and the range is $y \leq 0$.
There is one invariant point, $(2, 0)$.

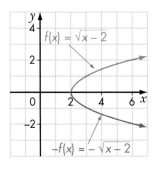

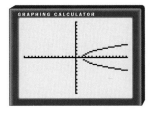

$f(x) = \sqrt{x - 2}$

$-f(x) = -\sqrt{x - 2}$

b) For $y = f(x)$ For $y = f(-x)$

$$y = \sqrt{x - 2} \qquad\qquad y = \sqrt{-x - 2}$$

The graph of $y = f(-x)$ is the graph of $y = f(x)$ reflected in the y-axis.

For $y = f(x)$, the domain is $x \geq 2$ and the range is $y \geq 0$.

For $y = f(-x)$, the domain is $x \leq -2$ and the range is $y \geq 0$.

There are no invariant points.

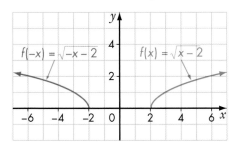

$f(-x) = \sqrt{-x - 2}$ $f(x) = \sqrt{x - 2}$

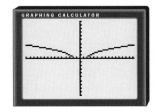

EXAMPLE 4   Reflecting a Quadratic Function

If $f(x) = x^2 + 1$,

a) write an equation for $y = -f(x)$ and $y = f(-x)$

b) describe how the graph of each equation from part a) is related to the graph of $y = f(x)$, and sketch the three graphs on the same axes

SOLUTION

a) For $y = f(x)$ For $y = -f(x)$

$$y = x^2 + 1 \qquad\qquad y = -x^2 - 1$$

     For $y = f(x)$ For $y = f(-x)$

$$y = x^2 + 1 \qquad\qquad y = (-x)^2 + 1$$
$$y = x^2 + 1$$

b) The graph of $y = -f(x)$ is the graph of $y = f(x)$ reflected in the x-axis. Since the graph of $y = x^2 + 1$ is symmetrical about the y-axis, replacing x with $-x$ has no effect on the graph. So, the graphs of $y = f(x)$ and $y = f(-x)$ are the same.

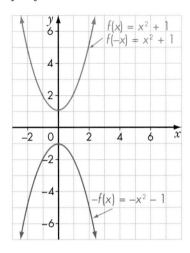 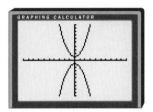

EXAMPLE 5 Reflecting a Quadratic Function

If $f(x) = x^2 + 6x$, write an equation for each of the following functions, describe how the graph of each function is related to the graph of $y = f(x)$, sketch each graph, and identify any invariant points.

a) $y = -f(x)$ **b)** $y = f(-x)$

SOLUTION

a) For $y = f(x)$, $y = x^2 + 6x$.

The graph of $y = x^2 + 6x$ is a parabola that opens up.

Since $x^2 + 6x = 0$ has roots $x = 0$ and $x = -6$, the x-intercepts of $y = x^2 + 6x$ are 0 and -6.

The vertex lies on the axis of symmetry, so the x-coordinate of the vertex is -3.

When $x = -3$, $y = (-3)^2 + 6(-3)$
$$= 9 - 18$$
$$= -9$$

So, the coordinates of the vertex are $(-3, -9)$.

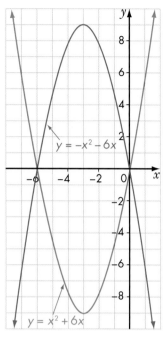

$y = -x^2 - 6x$

$y = x^2 + 6x$

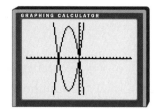

Sketch the graph of $y = f(x)$ using the points $(0, 0)$, $(-6, 0)$, and $(-3, -9)$.

For $y = -f(x)$
$$y = -(x^2 + 6x)$$
$$= -x^2 - 6x$$

The graph of $y = -f(x)$ is the graph of $y = f(x)$ reflected in the x-axis.
For $y = f(x)$, the domain is all real numbers and the range is $y \geq -9$.
For $y = -f(x)$, the domain is all real numbers and the range is $y \leq 9$.
Both graphs have the same x-intercepts, 0 and -6.
There are two invariant points, $(0, 0)$ and $(-6, 0)$.

b) For $y = f(-x)$

$$y = (-x)^2 + 6(-x)$$
$$= x^2 - 6x$$

The graph of $y = f(-x)$ is the graph of $y = f(x)$ reflected in the y-axis. For both $y = f(x)$ and $y = f(-x)$, the domain is all real numbers and the range is $y \geq -9$. The graphs intersect at $(0, 0)$.

There is one invariant point, $(0, 0)$.

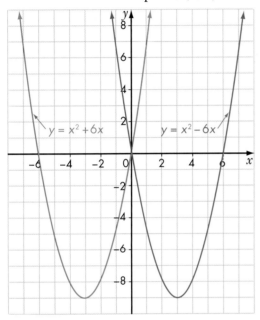

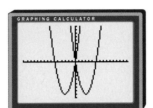

Key Concepts

- A function and its reflection image in the x-axis or y-axis are congruent.
- The graph of $y = -f(x)$ is the graph of $y = f(x)$ reflected in the x-axis.
- The graph of $y = f(-x)$ is the graph of $y = f(x)$ reflected in the y-axis.
- Points that are unaltered by a transformation are known as invariant points.

Communicate Your Understanding

1. If $f(x) = \sqrt{x - 3}$, describe how you would find the equations of $y = -f(x)$ and $y = f(-x)$.

2. If $f(x) = x^2 + 8x$, describe how you would find the equations of $y = -f(x)$ and $y = f(-x)$.

3. If $f(x) = x^2 - 4x$, describe how you would find any points that are invariant under a reflection in

a) the x-axis   **b)** the y-axis

Practise

A

1. Copy the graph of $y = f(x)$ and draw the graphs of $y = -f(x)$ and $y = f(-x)$ on the same axes.

a)

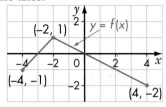

b)

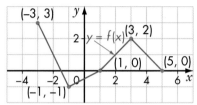

c)

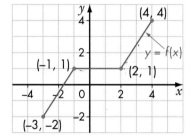

2. The blue graph is a reflection of the red graph in the x-axis. The equation of the red graph is given. Write the equation of the blue graph.

a)

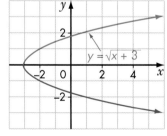

b)

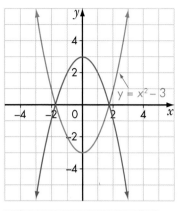

c)

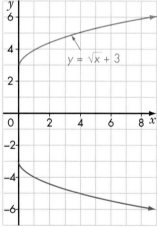

d)

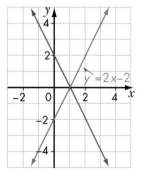

3. The blue graph is a reflection of the red graph in the y-axis. The equation of the red graph is given. Write the equation of the blue graph.

a)

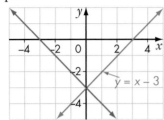

$y = x - 3$

b)

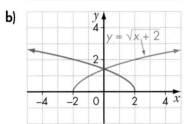

$y = \sqrt{x} + 2$

c)

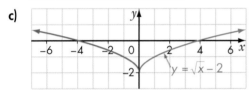

$y = \sqrt{x} - 2$

d)

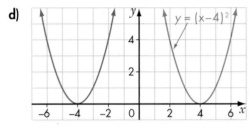

$y = (x-4)^2$

4. Graph $f(x)$ and sketch the specified reflection image.

a) $g(x)$, the reflection of $f(x) = x + 1$ in the x-axis

b) $h(x)$, the reflection of $f(x) = \sqrt{x} + 5$ in the y-axis

c) $g(x)$, the reflection of $f(x) = x^2 - 4$ in the x-axis

d) $h(x)$, the reflection of $f(x) = \sqrt{x+5}$ in the y-axis

5. a) Given $f(x) = 2x - 4$, write equations for $-f(x)$ and $f(-x)$.

b) Sketch the three graphs on the same set of axes.

c) Determine any points that are invariant for each reflection.

d) State the domain and range of each function.

6. a) Given $f(x) = -3x + 2$, write equations for $-f(x)$ and $f(-x)$.

b) Sketch the three graphs on the same set of axes.

c) Determine any points that are invariant for each reflection.

7. a) Given $f(x) = x^2 - 4x$, write equations for $-f(x)$ and $f(-x)$.

b) Sketch the three graphs on the same set of axes.

c) Determine any points that are invariant for each reflection.

8. a) Given $f(x) = x^2 - 9$, write equations for $-f(x)$ and $f(-x)$.

b) Sketch the three graphs on the same set of axes.

c) Determine any points that are invariant for each reflection.

9. a) Given $f(x) = (x + 4)(x - 2)$, write equations for $-f(x)$ and $f(-x)$.

b) Sketch the three graphs on the same set of axes.

c) State the domain and range of each function.

10. a) Given $f(x) = \sqrt{x+4}$, write equations for $-f(x)$ and $f(-x)$.
b) Sketch the three graphs on the same set of axes.
c) State the domain and range of each function.

11. a) Given $f(x) = \sqrt{x} + 4$, write equations for $-f(x)$ and $f(-x)$.
b) Sketch the three graphs on the same set of axes.
c) State the domain and range of each function.

Apply, Solve, Communicate

12. Great Pyramid A coordinate grid is superimposed on a cross section of the Great Pyramid, so that the y-axis passes through the vertex of the pyramid. Two opposite vertices of the square base are on the x-axis. The two sloping lines in the cross section, AD and AE, are opposite edges of the pyramid.
a) In the cross section, one edge can be modelled by the equation $y = 0.9x + 146$. Which edge, AD or AE, can be modelled by this equation? Explain.
b) What is the equation of the other edge? Explain.

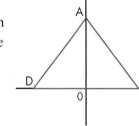

B

13. Communication How are the graphs of $y = x^2 - 2x$ and $y = 2x - x^2$ related. Explain.

14. a) Given $f(x) = x^2 - x - 6$, determine the coordinates of the points where the graph crosses the y-axis and the x-axis.
b) Use the points you found in part a) to sketch the graphs of $y = -f(x)$ and $y = f(-x)$.

15. Sloping roof The diagram shows a set of coordinate axes superimposed on the cross section of a sloping roof of height h metres and width w metres. The equation of one half of the cross section is $y = -0.7x + 1.9$.
a) What is the equation of the other half of the cross section?
b) What is the height of the roof, h?
c) What is the width of the roof, w, to the nearest tenth of a metre?
d) State the domain and range of the function that models each half of the roof.

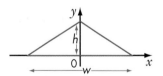

16. Sequencing reflections a) Copy the graph of $y = f(x)$, as shown. Sketch the graph of each relation obtained after a reflection in the y-axis followed by a reflection in the x-axis.
b) Copy the graph of $y = f(x)$, as shown. Sketch the graph of each relation obtained after a reflection in the x-axis followed by a reflection in the y-axis.

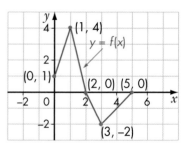

17. Inquiry/Problem Solving When a function $y = f(x)$ is reflected in the x-axis, and its image is reflected in the y-axis, the same point remains invariant for both reflections. Identify the point and explain your reasoning.

18. a) Given $f(x) = \sqrt{x - 3}$, write equations for $-f(x)$, $f(-x)$, and $-f(-x)$.
b) Sketch the four graphs on the same set of axes.
c) State the domain and range of each function.

19. a) Given $f(x) = (x + 1)^2$, write equations for $-f(x)$, $f(-x)$, and $-f(-x)$.
b) Sketch the four graphs on the same set of axes.
c) State the domain and range of each function.

20. Application In the sequence $-7, -4, 1, 8, \ldots$ the number -7 is in position 1, the number -4 is in position 2, and so on.
a) Determine the equation that relates each number, n, to its position, p, in the sequence.
b) Repeat part a) for the sequence $7, 4, -1, -8 \ldots$.
c) How are the graphs of the two sequences related?
d) State the domain and range for each sequence.

C

21. a) Given $f(x) = \sqrt{25 - x^2}$, write the equations for $y = -f(x)$ and $y = f(-x)$.
b) Graph the three functions in the same viewing window of your graphing calculator.
c) Determine any points that are invariant for each reflection.
d) State the domain and range of each function.

22. If the graph of $f(x) = x^2 - 3$ is transformed into the graph of $f(-x)$, how many points are invariant? Explain.

23. If a line is not horizontal or vertical, how is the slope of the line related to the slope of its reflection image in each of the following? Explain.
a) x-axis   **b)** y-axis

A property of a billiard table is that the ball bounces off the side at the same angle that it strikes the side. To find where to hit the ball on the side of the table, you aim for the reflection of the ball you want to hit, as shown.

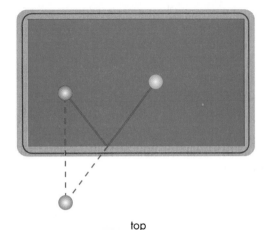

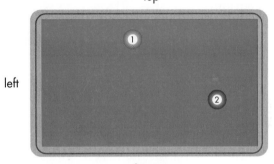

top

left   right

bottom

a) Show on graph paper where ball 1 should be hit so that it bounces off the bottom side and hits ball 2.

b) Show on graph paper where ball 1 should be hit so that it bounces off the left side and hits ball 2.

c) Show on graph paper where ball 1 should be hit so that it bounces once off the top or bottom side and then once off the left or right side and hits ball 2.

d) If the table is placed on a grid with the origin at the bottom left corner, then ball 1 is at the point (5, 6) and ball 2 is at the point (11, 3). Find the coordinates of the point where ball 1 bounces off the bottom side in part a). Label it point H.

e) Find the equation of the path ball 1 takes to the bottom side at H.

f) Find the equation of the path ball 1 takes from point H to ball 2.

g) What is the relationship between the slope of the line from part e) and the slope of the line from part f)? Explain your answer.

3.5 Inverse Functions

Inverse functions are a special class of functions that undo each other. The input and output values for two inverse functions, $f(x) = 2x + 1$ and $g(x) = \dfrac{x - 1}{2}$, are shown.

$$f(x) = 2x + 1 \qquad\qquad g(x) = \dfrac{x - 1}{2}$$

Input, x	Output, f(x)
0	1
1	3
2	5
3	7

Input, x	Output, g(x)
1	0
3	1
5	2
7	3

Notice that the output of the first function, $f(x)$, becomes the input for the second function, $g(x)$. The function $g(x)$ undoes what $f(x)$ does. The ordered pairs of $g(x)$ can be found by switching the coordinates in each ordered pair of $f(x)$.

INVESTIGATE & INQUIRE

The Paralympic Games are the world's second-largest sporting event, after the Olympic Games. At the Paralympics held in Sydney, Australia, over 4000 competitors represented more than 100 countries. Canada finished fifth in the medal standings with 38 gold, 33 silver, and 25 bronze medals. Among the Canadians who won gold medals were wheelchair racers Chantal Petitclerc and Jeff Adams.

1. Wheelchair races are held over various distances, from 100 m to the marathon. The table includes data for some of the events that are held on a 400-m track. Copy and complete the table.

Distance, d (km)	Number of Laps, n
5	
3	
1.5	
	2
	1
	0.5

2. a) Write a function in the form $d(n) = \blacksquare n$, where d is the distance of the race, in kilometres, and n is the number of laps.
b) What operation does the function $d(n)$ perform on each input value?
c) List the ordered pairs of the function $d(n)$.

3. a) Write a function in the form $n(d) = \blacksquare d$, where n is the number of laps and d is the distance of the race, in kilometres.
b) What operation does the function $n(d)$ perform on each input value?
c) List all the ordered pairs of the function $n(d)$.

4. How can the ordered pairs of $n(d)$ be obtained from the ordered pairs of $d(n)$?

5. Are $d(n)$ and $n(d)$ inverse functions? Explain.

6. How does the domain of each function compare with the range of the other?

7. Find the inverse, $g(x)$, of each of the following functions.
a) $f(x) = 2x$ **b)** $f(x) = \dfrac{3x}{4}$ **c)** $f(x) = x + 2$ **d)** $f(x) = x - 4$

Recall that a relation is a set of ordered pairs. The inverse of a relation can be found by interchanging the domain and the range of the relation.

Relation   Inverse Relation
$(-3, 4), (0, 7), (2, 9)$ $(4, -3), (7, 0), (9, 2)$

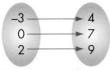   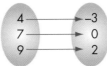

Domain   Range   Domain   Range

Also recall that a function is a special relation. For each element in the domain of a function, there is exactly one element in the range. If the inverse of a function $f(x)$ is also a function, it is called the inverse function of $f(x)$. This inverse function is denoted by $f^{-1}(x)$.

The notation f^{-1} is read as "the inverse of f" or "f inverse."

Note that the -1 in f^{-1} is *not* an exponent, so $f^{-1} \neq \dfrac{1}{f}$.

EXAMPLE 1 Interchanging Coordinates

a) Find the inverse f^{-1} of the function f whose ordered pairs are $\{(-2, -8), (0, -2), (3, 4), (4, 7)\}$.

b) Graph both functions.

SOLUTION

a) Switch the first and second coordinates of each ordered pair.
$f^{-1} = \{(-8, -2), (-2, 0), (4, 3), (7, 4)\}$

b) The graph of f is shown in red and the graph of f^{-1} is shown in blue.

Note that, in Example 1, switching the x- and y-coordinates reflects the function f in the line $y = x$. So, the graph of f^{-1} is the reflection of the graph of f in the line $y = x$. Notice the symmetry of the graphs in the line $y = x$. If the diagram were folded along the line $y = x$, the graph of f and the graph of f^{-1} would exactly match.

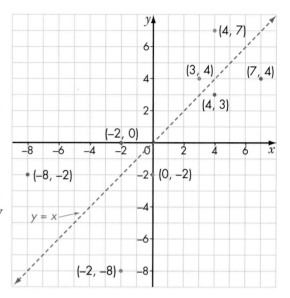

One way to find the inverse of a function is to reverse the operations that the function specifies.

The function $f(x) = 2x + 3$ means: Multiply x by 2, and then add 3.

Reversing the operations means: Subtract 3 from x, and then divide the result by 2.

So, the inverse function of $f(x)$ is $f^{-1} = \dfrac{x - 3}{2}$.

Let $x = 6$ and check that the inverse function undoes the other function.

$$f(x) = 2x + 3 \quad \begin{array}{cccc} \text{Input} & \text{Multiply by 2} & \text{Add 3} & \text{Output} \\ 6 & \longrightarrow & 12 \longrightarrow & 15 \end{array}$$

$$\begin{array}{cccc} \text{Output} & \text{Divide by 2} & \text{Subtract 3} & \text{Input} \\ 6 & \longleftarrow & 12 \longleftarrow & 15 \end{array} \quad f^{-1}(x) = \dfrac{x - 3}{2}$$

The functions $f(x) = 2x + 3$ and $f^{-1} = \dfrac{x - 3}{2}$ are inverses, because one function undoes the other.

One way to reverse the operations that a function specifies is to interchange the variables.

EXAMPLE 2 Inverse of a Linear Function

a) Find the inverse of the function $f(x) = 4x + 3$.

b) Is the inverse of $f(x) = 4x + 3$ a function?

SOLUTION

a)

$$f(x) = 4x + 3$$

Replace $f(x)$ with y: $y = 4x + 3$

Interchange x and y: $x = 4y + 3$

Solve for y: $x - 3 = 4y$

$$\frac{x - 3}{4} = y$$

The inverse of $f(x) = 4x + 3$ is $f^{-1}(x) = \dfrac{x - 3}{4}$.

b) The inverse of $f(x) = 4x + 3$ is a function, since only one y-value can be found for each x-value.

In Example 2, the inverse function can be written in the slope and y-intercept form, $y = mx + b$, as $y = \dfrac{x}{4} - \dfrac{3}{4}$.

The inverse function is linear, with a slope of $\dfrac{1}{4}$ and a y-intercept of $-\dfrac{3}{4}$.

If we graph the functions $f(x) = 4x + 3$ and $f^{-1}(x) = \dfrac{x}{4} - \dfrac{3}{4}$ on the same axes, or in the same viewing window of a graphing calculator, we see two straight lines that are reflections of each other in the line $y = x$.

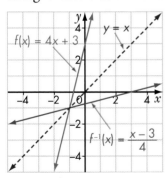

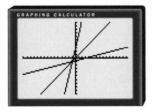

You can set the graph style of $y = x$ to dot style in the Y= editor. The window can be adjusted using the ZSquare instruction in the ZOOM menu.

In general, for the point (x, y) on the graph of $y = f(x)$, there is a corresponding point (y, x) on the graph of its inverse, $x = f(y)$. The graph of $x = f(y)$ is a reflection of the graph of $y = f(x)$ in the line $y = x$. Points that lie on the line $y = x$ are invariant, because their x- and y-coordinates are equal. In Example 2, the point $(-1, -1)$ is invariant.

EXAMPLE 3 Inverse of a Quadratic Function

a) Find the inverse of $f(x) = x^2 - 1$
b) Graph $f(x)$ and its inverse.
c) Is the inverse of $f(x)$ a function?
d) Determine the domain and the range of $f(x)$ and its inverse.

SOLUTION

a)
$$f(x) = x^2 - 1$$
Replace $f(x)$ with y: $\qquad y = x^2 - 1$
Interchange x and y: $\qquad x = y^2 - 1$
Isolate y: $\qquad x + 1 = y^2$
Take the square root of both sides: $\qquad \pm\sqrt{x + 1} = y$
The inverse is $f^{-1}(x) = \pm\sqrt{x + 1}$.

b) The graphs of $f(x)$ and f^{-1} are shown.
To graph $f^{-1}(x) = \pm\sqrt{x + 1}$ manually, graph the two branches $y = \sqrt{x + 1}$ and $y = -\sqrt{x + 1}$.
To graph f and f^{-1} on a graphing calculator, use the DrawInv instruction. You can use DrawInv without finding an equation for f^{-1}.

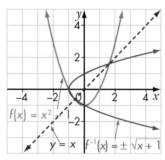
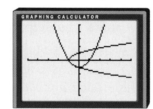

c) For the inverse f^{-1}, there are two values of y for each value of x, except $x = -1$. In other words, f^{-1} does not pass the vertical line test. So, the inverse f^{-1} is not a function.

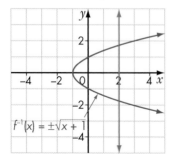

The notation f^{-1} can be used whether or not the inverse is a function.

Recall that, if you can draw a vertical line that intersects a graph in more than one point, the graph does not represent a function.

d) For $f(x) = x^2 - 1$, the domain is the set of real numbers. The range is the set of real numbers, $y \geq -1$.

For $f^{-1}(x) = \pm\sqrt{x+1}$, the domain is the set of real numbers, $x \geq -1$. The range is the set of real numbers.

EXAMPLE 4 Restricting the Domain of f(x)

a) Find the inverse of $f(x) = x^2 + 2$.

b) Graph $f(x)$ and its inverse.

c) Is the inverse of $f(x)$ a function? If not, restrict the domain of $f(x)$ so that its inverse is a function.

SOLUTION

a)
$$f(x) = x^2 + 2$$
Replace $f(x)$ with y:
$$y = x^2 + 2$$
Interchange x and y:
$$x = y^2 + 2$$
Isolate y:
$$x - 2 = y^2$$
Take the square root of both sides: $\pm\sqrt{x-2} = y$

So, $f^{-1}(x) = \pm\sqrt{x-2}$.

b) The graphs of f and f^{-1} are shown.

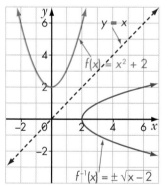

c) The inverse does not pass the vertical line test. So, the inverse f^{-1} is not a function.

The inverse has two branches, $y = \sqrt{x-2}$ and $y = -\sqrt{x-2}$.

The graph of one branch would pass the vertical line test. So, if the domain of f is restricted so that f^{-1} has only one branch, then f^{-1} will be a function.

For example, restricting the domain of $f(x)$ to real values of $x \geq 0$ and reflecting $f(x) = x^2 + 2$, $x \geq 0$, in the line $y = x$ would result in an inverse function $f^{-1}(x) = \sqrt{x-2}$. The graphs are as shown.

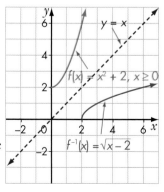

EXAMPLE 5 Car Rental

The cost of renting a car for a day is a flat rate of $40, plus a charge of $0.10 per kilometre driven.

a) Let r dollars be the total rental cost and d kilometres be the distance driven. Write the function $r(d)$ to represent the total cost of a one-day rental.

b) Find the inverse of the function.

c) What does the inverse represent?

d) Give an example of how the inverse could be used.

SOLUTION

a) $r(d) = 0.1d + 40$

b) Substitute y for r and x for d:

$$y = 0.1x + 40$$

Interchange x and y:

$$x = 0.1y + 40$$

Solve for y:

$$x - 40 = 0.1y$$

$$\frac{x - 40}{0.1} = y$$

$$10(x - 40) = y$$

$$10x - 400 = y$$

Because x and y were interchanged, x represents r and y represents d in the inverse.

So, the inverse is $d(r) = 10r - 400$. **Note that solving $r = 0.1d + 40$ for d gives the inverse.**

c) The inverse shows the distance that can be driven for a given rental cost.

d) The distance that can be driven for a total rental cost of $48 for a day is given by

$$d = 10(48) - 400$$
$$= 480 - 400$$
$$= 80$$

So, a distance of 80 km can be driven for a total rental cost of $48.

Note that the use of the inverse cannot include values of $r < 40$, since $40 is the minimum daily rate and the distance driven cannot be negative.

Key Concepts

- A function, $f(x)$, and its inverse function, $f^{-1}(x)$, undo each other.
- The inverse of a function can be found by interchanging the domain and range of the function.
- The inverse of a function can be found by interchanging x and y in the equation of the function.
- The graph of $x = f(y)$ is the graph of $y = f(x)$ reflected in the line $y = x$.
- A function and its inverse are congruent and are symmetric about the line $y = x$.

Communicate Your Understanding

1. Describe how you would find the inverse of $f = \{(-2, 4), (1, 3), (4, 7), (8, 11)\}$.

2. Describe how you would find the inverse of $f(x) = 4x + 7$.

3. Describe how you would find the inverse of $f(x) = x^2 - 5$.

Practise

A

1. Given the ordered pairs of each function, find the inverse, and graph the function and its inverse.

a) $f = \{(0, 2), (1, 3), (2, 4), (3, 5)\}$

b) $g = \{(-1, -3), (1, -2), (3, 4), (5, 0), (6, 1)\}$

2. Given the ordered pairs of each function, find the inverse, and state whether the inverse is a function.

a) $f = \{(-2, 3), (-1, 2), (0, 0), (4, -2)\}$

b) $g = \{(4, -2), (2, 1), (1, 3), (0, -2), (-3, -3)\}$

3. Sketch the inverse of each function.

a)

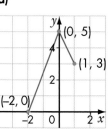

b)

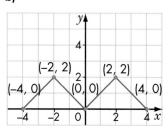

c)

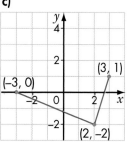

d)

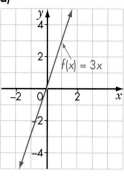

e)

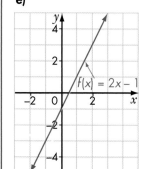

f)

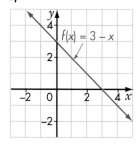

g)

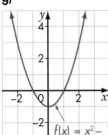

$f(x) = x^2 - 1$

h)

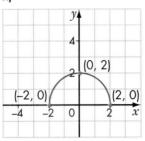

(0, 2)
(-2, 0) (2, 0)

i)

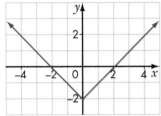

4. Solve each equation for x.

a) $f(x) = 3x + 2$

b) $f(x) = \dfrac{12 - 2x}{3}$

c) $f(x) = 3 - 4x$

d) $f(x) = \dfrac{x + 3}{4}$

e) $f(x) = \dfrac{x}{2} - 5$

f) $f(x) = x^2 + 3$

5. Find the inverse of each function.

a) $f(x) = x - 1$

b) $f(x) = \dfrac{x}{2}$

c) $f(x) = x + 3$

d) $f(x) = \dfrac{4}{3}x$

e) $f(x) = 2x + 1$

f) $f(x) = \dfrac{x + 2}{3}$

g) $g(x) = \dfrac{5}{2}x - 4$

h) $h(x) = 0.2x + 1$

6. Find the inverse of each function. Graph the function and its inverse.

a) $f(x) = x + 2$ **b)** $f(x) = 4x$
c) $f(x) = 3x - 2$ **d)** $f(x) = x$
e) $f(x) = 3 - x$ **f)** $f(x) = \dfrac{x - 2}{3}$

7. Find the inverse of each function and determine whether the inverse is a function.

a) $f(x) = 2x - 5$ **b)** $f(x) = \dfrac{x + 3}{4}$

c) $f(x) = \dfrac{x}{4} + 3$ **d)** $f(x) = 5 - x$

8. Determine if the functions in each pair are inverses of each other.

a) $f(x) = x + 5$ and $g(x) = x - 5$

b) $f(x) = 7x$ and $g(x) = \dfrac{x}{7}$

c) $f(x) = 2x - 1$ and $g(x) = \dfrac{x + 1}{2}$

d) $f(x) = x - 3$ and $g(x) = 3 - x$

e) $f(x) = \dfrac{x}{3} - 4$ and $g(x) = 3x - 4$

f) $g(x) = \dfrac{x}{3} - 5$ and $h(x) = 3x + 5$

g) $h(x) = \dfrac{x - 8}{4}$ and $k(x) = 4(x + 2)$

9. Verify algebraically and graphically that the functions in each pair are inverses of each other.

a) $y = 3x + 4$ and $y = \dfrac{1}{3}(x - 4)$

b) $y = 3 - 2x$ and $y = -\dfrac{1}{2}(x - 3)$

10. For each of the following functions,
a) find the inverse of $f(x)$
b) graph $f(x)$ and its inverse
c) determine the domain and range of $f(x)$ and its inverse
i) $f(x) = x^2 - 3$
ii) $f(x) = x^2 + 1$
iii) $f(x) = -x^2$
iv) $f(x) = -x^2 - 1$
v) $f(x) = (x - 2)^2$
vi) $f(x) = (x + 1)^2$

11. The blue graph is a reflection of the red graph in the line $y = x$. The equation of the red graph is given. Write the equation of the blue graph.

a)

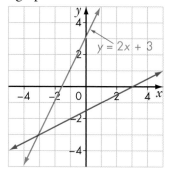

$y = 2x + 3$

b)

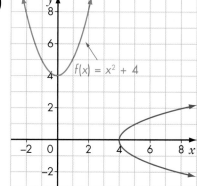

$f(x) = x^2 + 4$

12. Determine if the functions in each pair are inverses of each other.

a) $y = x^2 - 3$ and $y = \sqrt{x + 3}$
b) $y = x^2 + 1$ and $y = \sqrt{x + 1}$

13. Find the inverse of each function. If the inverse is a function, determine the domain and range of the inverse.

a) $y = 2x - 3$ b) $y = 2 - 4x$

c) $y = 3(x - 2)$ d) $y = \dfrac{1}{2}(x - 6)$

e) $y = x^2$ f) $y = x^2 + 2$

g) $y = x^2 - 4$ h) $y = 2x^2 - 1$

i) $y = (x - 3)^2$ j) $y = (x + 2)^2$

14. For each of the following functions,
a) find the inverse of $f(x)$
b) graph $f(x)$ and its inverse
c) determine the domain and range of $f(x)$ and its inverse

i) $f(x) = x^2$, $x \geq 0$
ii) $f(x) = x^2 - 2$, $x \geq 0$
iii) $f(x) = x^2 + 4$, $x \leq 0$
iv) $f(x) = 3 - x^2$, $x \geq 0$
v) $f(x) = (x - 4)^2$, $x \geq 4$
vi) $f(x) = (x + 3)^2$, $x \leq -3$

15. Find the inverse of each of the following functions.

a) $y = \sqrt{x - 2}$
b) $y = \sqrt{3 - x}$
c) $y = \sqrt{x^2 + 9}$

16. For each of the following functions,
a) find the inverse f^{-1}
b) graph $f(x)$ and its inverse
c) restrict the domain of f so that f^{-1} is also a function
d) with the domain of f restricted, sketch a graph of f and f^{-1}

i) $f(x) = x^2 + 3$ ii) $f(x) = 2x^2$
iii) $f(x) = x^2 - 1$ iv) $f(x) = -x^2$
v) $f(x) = 1 - x^2$ vi) $f(x) = (x - 2)^2$
vii) $f(x) = (4 - x)^2$ viii) $f(x) = -(x + 5)^2$

17. a) Find the inverse of $f(x) = \dfrac{1}{x}$.

b) Is the inverse a function? Explain.

18. Communication a) Find the inverse of $f(x) = \sqrt{x}$.

b) Is the inverse a function? Explain.

Apply, Solve, Communicate

19. Van Rental The cost of renting a van for one day is a flat rate of $50, plus a variable rate of $0.15/km.
a) Write a function to express the total cost of a one-day rental, $c(d)$ dollars, in terms of the distance driven, d kilometres.
b) Determine the inverse of the function.
c) What does the inverse represent?
d) What is the domain of the inverse?

B

20. Measurement a) Let x represent the radius of a circle. Write a function $f(x)$ to express the circumference in terms of the radius.
b) Find the inverse of this function.
c) Is the inverse a function?
d) What does the inverse represent?

21. Application a) Let x represent the radius of a sphere. Write a function $f(x)$ to express the surface area in terms of the radius.
b) Find the inverse of this function.
c) Determine the domain and range of the inverse.
d) Is the inverse a function?
e) What does the inverse represent?

22. Retail Sales A sale at an appliance store advertised that all appliances were being sold at 30% off the original selling price.
a) Write a function that gives the sale price as a function of the original selling price.
b) Find the inverse of this function.
c) What does the inverse represent?

23. Foreign currency exchange One day, the Canadian dollar was worth US$0.70.
a) Write a function that expresses the value of the US dollar, u, in terms of the Canadian dollar, c.
b) Find the inverse. Round the coefficient to the nearest hundredth.
c) Use the inverse to convert US$150 to Canadian dollars.

24. Geology The approximate temperature, in degrees Celsius, of rocks beneath the surface of the Earth can be found by multiplying their depth, in kilometres, by 35 and adding 20 to the product.

a) Let d kilometres represent the depth of some rocks. Write a function $T(d)$ that expresses the Celsius temperature of the rocks in terms of their depth.

b) Write the inverse function.

c) At what depth do rocks have a temperature of 90°C?

25. Weekly wages Jana works at a clothing store. She earns $400 a week, plus a commission of 5% of her sales.

a) Write a function that describes Jana's total weekly earnings as a function of her sales.

b) Find the inverse of this function.

c) What does the inverse represent?

d) One week, Jana earned $575. Calculate her sales that week.

26. Measurement The measure of an interior angle, i, of a regular polygon is related to the number of sides n by the function

$$i(n) = 180 - \frac{360}{n}.$$

a) Determine the measure of an interior angle of a regular heptagon.

b) Find the inverse of the function.

c) Use the inverse to identify the regular polygon with interior angles of 144°.

27. Falling objects If an object is dropped from a height of 80 m, its approximate height, $h(t)$ metres, above the ground t seconds after being dropped is given by the function $h(t) = -5t^2 + 80$.

a) Graph the function.   **b)** Find and graph the inverse.

c) Is the inverse a function? Explain.   **d)** What does the inverse represent?

e) After what length of time is the object 35 m above the ground?

f) How long does the object take to reach the ground?

28. a) Given $f(x) = 2x - 4$, write equations for $-f(x)$, $f(-x)$, and $f^{-1}(x)$.

b) Sketch the four graphs on the same set of axes.

c) Determine any points that are invariant for each reflection.

29. a) Given $f(x) = -3x + 2$, write equations for $-f(x)$, $f(-x)$, and $f^{-1}(x)$.

b) Sketch the four graphs on the same set of axes.

c) Determine any points that are invariant for each reflection.

30. a) Given $f(x) = \sqrt{x + 3}$, write equations for $-f(x)$, $f(-x)$, and $f^{-1}(x)$.

b) Sketch the four graphs on the same set of axes.

31. Write an equation for the line obtained by reflecting the line $x = 2$ in the line $y = x$.

32. Sequencing reflections Copy the graph of $y = f(x)$ as shown. Sketch the graph of each relation obtained after a reflection in the y-axis, followed by a reflection in the x-axis, followed by a reflection in the line $y = x$.

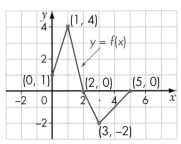

C

33. Inquiry/Problem Solving Write four functions that are their own inverses.

34. The function f includes the ordered pair $(2, 3)$. Can f^{-1} include each of the following ordered pairs? Explain.
a) $(3, 4)$ **b)** $(4, 2)$

35. Analytic geometry Find the area of the figure formed by the intersection of $f(x) = 4 - x$, $g(x) = 12 - 3x$, and g^{-1}.

36. a) Is the relation $y = k$, where k is a constant, a function? Explain.
b) Is the inverse of $y = k$ a function? Explain.

37. What is the inverse of the inverse of a function? Explain.

38. Slope If a line is not horizontal or vertical, how is the slope of the line related to the slope of its inverse?

39. a) In how many different ways could you restrict the domain of $f(x) = x^2 + 3$, so that the inverse is a function? Give examples and use graphs to justify your answer.
b) In how many different ways could you restrict the domain of $f(x) = x^2 + 3$, so that the inverse is not a function? Give examples and use graphs to justify your answer.

ACHIEVEMENT Check   Knowledge/Understanding   Thinking/Inquiry/Problem Solving   Communication   Application

a) Sketch the graph of the function $f(x) = \sqrt{x}$.
b) On the same set of axes, graph $y = f^{-1}(x)$, $y = f^{-1}(-x)$, and $y = -f^{-1}(-x)$.
c) Compare the graphs of $y = f(x)$ and $y = -f^{-1}(-x)$. If the graph of $y = -f^{-1}(-x)$ is drawn from the graph of $y = f(x)$ by a single reflection, what is the equation of the reflection line?

3.6 Stretches of Functions

The clock in the Heritage Hall clock tower, in Vancouver, is known as Little Ben. It was built by the same company that built Big Ben in London, England. Little Ben is a mechanical clock with a pendulum.

On the Earth, the period of a pendulum is approximately represented by the function $T = 2\sqrt{l}$, where T is the period, in seconds, and l is the length of the pendulum, in metres. Since the force of gravity varies from one location to another in the solar system, the function for the period of a pendulum also varies. On the moon, the function is $T = 5\sqrt{l}$. On Pluto, the function is $T = 8\sqrt{l}$.

1. Let $y = T$ and $x = l$, and graph the three functions above, plus the function $y = \sqrt{x}$, on the same set of axes or in the same viewing window of a graphing calculator.

$y = \sqrt{x}$ $\qquad$ $y = 2\sqrt{x}$ $\qquad$ $y = 5\sqrt{x}$ $\qquad$ $y = 8\sqrt{x}$

x	y	x	y	x	y	x	y
0		0		0		0	
1		1		1		1	
4		4		4		4	
9		9		9		9	
16		16		16		16	

2. For the functions $y = \sqrt{x}$ and $y = 2\sqrt{x}$, how do the y-coordinates of any two points that have the same non-zero x-coordinate compare? Explain why.

3. For the functions $y = \sqrt{x}$ and $y = 5\sqrt{x}$, how do the y-coordinates of any two points that have the same non-zero x-coordinate compare? Explain why.

4. For the functions $y = \sqrt{x}$ and $y = 8\sqrt{x}$, how do the y-coordinates of any two points that have the same non-zero x-coordinate compare? Explain why.

5. a) If Little Ben were placed on Pluto, the period of its pendulum would be 9.8 s. What is the length of the pendulum, to the nearest tenth of a metre?
b) If Little Ben were placed on the moon, what would the period of its pendulum be, to the nearest tenth of a second?

EXAMPLE 1 **Vertical Stretching**

a) Copy the graph of $y = f(x)$, as shown. On the same set of axes, graph $y = 3f(x)$ and $y = 0.5f(x)$.
b) Describe how the graphs of $y = 3f(x)$ and $y = 0.5f(x)$ are related to the graph of $y = f(x)$.

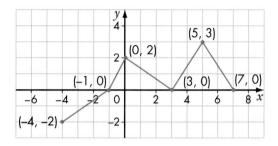

SOLUTION

a) Use the graph to complete a table of values for $y = f(x)$.
Then, complete tables of values for $y = 3f(x)$ and $y = 0.5f(x)$, and draw the graphs.

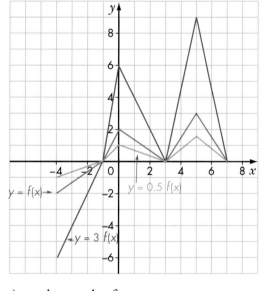

$y = f(x)$

x	y
7	0
5	3
3	0
0	2
−1	0
−4	−2

$y = 3f(x)$

x	y
7	0
5	9
3	0
0	6
−1	0
−4	−6

$y = 0.5f(x)$

x	y
7	0
5	1.5
3	0
0	1
−1	0
−4	−1

b) Comparing $y = f(x)$ and $y = 3f(x)$, the point (x, y) on the graph of $y = f(x)$ is transformed to the point $(x, 3y)$ on the graph of $y = 3f(x)$. The graph of $y = 3f(x)$ is the graph of $y = f(x)$ expanded vertically by a factor of 3.

Comparing $y = f(x)$ and $y = 0.5f(x)$, the point (x, y) on the graph of $y = f(x)$ is transformed to the point $(x, 0.5y)$ on the graph of $y = 0.5f(x)$. The graph of $y = 0.5f(x)$ is the graph of $y = f(x)$ compressed vertically by a factor of 0.5.

The graphs of $y = f(x)$, $y = 3f(x)$, and $y = 0.5f(x)$ are not congruent.

In Example 1, the points $(7, 0)$, $(3, 0)$, and $(-1, 0)$ are points on the graphs of all three functions. Recall that these points are said to be invariant, because they are unaltered by the transformations. The three functions have the same domain, $-4 \le x \le 7$, but different ranges. The range of $y = f(x)$ is $-2 \le y \le 3$, of $y = 3f(x)$ is $-6 \le y \le 9$, and of $y = 0.5f(x)$ is $-1 \le y \le 1.5$.

Note from Example 1 that a stretch may be an expansion or a compression. An **expansion** is a stretch by a factor greater than 1. A **compression** is a stretch by a factor less than 1.

EXAMPLE 2 Vertical Stretching of a Quadratic Function

a) Graph $y = x^2$, $y = 2x^2$, and $y = \dfrac{2}{3}x^2$ on the same grid.

b) Describe how the graphs of $y = 2x^2$ and $y = \dfrac{2}{3}x^2$ are related to the graph of $y = x^2$.

SOLUTION

a) Complete tables of values using convenient values of x, or use a graphing calculator.

$y = x^2$ $y = 2x^2$ $y = \dfrac{2}{3}x^2$

x	y	x	y	x	y
3	9	3	18	6	24
2	4	2	8	3	6
1	1	1	2	0	0
0	0	0	0	-3	6
-1	1	-1	2	-6	24
-2	4	-2	8		
-3	9	-3	18		

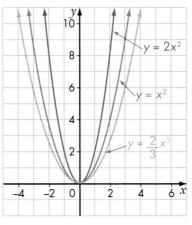

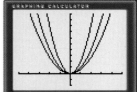

The window variables include Xmin = -5, Xmax = 5, Ymin = -2, and Ymax = 10.

b) Given $y = 2x^2$, then $y = 2(x^2)$. The point (x, y) on the graph of the function $y = x^2$ is transformed to the point $(x, 2y)$ on the graph of $y = 2x^2$. The graph of $y = 2x^2$ is the graph of $y = x^2$ expanded vertically by a factor of 2.

Given $y = \dfrac{2}{3}x^2$, then $y = \dfrac{2}{3}(x^2)$. The point (x, y) on the graph of the function $y = x^2$ is transformed to the point $\left(x, \dfrac{2}{3}y\right)$ on the graph of $y = \dfrac{2}{3}x^2$. The graph of $y = \dfrac{2}{3}x^2$ is the graph of $y = x^2$ compressed vertically by a factor of $\dfrac{2}{3}$.

The three functions have the same domain, all real numbers, and the same range, $y \ge 0$.

In general, for any function $y = f(x)$, the graph of the function $y = af(x)$, where a is any real number, is obtained by multiplying the y-value at each point on the graph of $y = f(x)$ by a.

The point (x, y) on the graph of the function $y = f(x)$ is transformed into the point (x, ay) on the graph of $y = f(x)$.

If $a > 1$, the graph expands vertically by a factor of a.

If $0 < a < 1$, the graph is compressed vertically by a factor of a.

EXAMPLE 3 Horizontal Stretching

Given the graph of $y = f(x)$, compare it to the graphs of

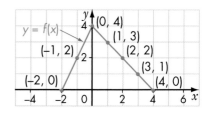

a) $y = f(2x)$

b) $y = f\left(\dfrac{1}{2}x\right)$

SOLUTION

a) Use the given graph to complete a table of values for $y = f(x)$. Then, complete a table of values for $y = f(2x)$. Use convenient values for x.

$y = f(x)$

x	y
-2	0
-1	2
0	4
1	3
2	2
3	1
4	0

$y = f(2x)$

x	y
-1	$f(2 \times (-1)) = f(-2) = 0$
-0.5	$f(2 \times (-0.5)) = f(-1) = 2$
0	$f(2 \times 0) = f(0) = 4$
0.5	$f(2 \times 0.5) = f(1) = 3$
1	$f(2 \times 1) = f(2) = 2$
1.5	$f(2 \times 1.5) = f(3) = 1$
2	$f(2 \times 2) = f(4) = 0$

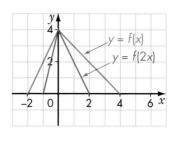

The functions have the same range, $0 \le y \le 4$. The domain of $y = f(x)$ is $-2 \le x \le 4$. The domain of $y = f(2x)$ is $-1 \le x \le 2$.

Note that, in the table for $y = f(2x)$, x-values such as -2 and 4 are not convenient, because $f(2 \times (-2)) = f(-4)$ and $f(2 \times 4) = f(8)$. Both $f(-4)$ and $f(8)$ are not defined for the function $y = f(x)$.

For non-zero values of x, each point on $y = f(2x)$ is half as far from the y-axis as the equivalent point on $y = f(x)$.

The point (x, y) on the graph of the function $y = f(x)$ is transformed to the point $\left(\dfrac{x}{2}, y\right)$ on the graph of $y = f(2x)$. The graph of $y = f(2x)$ is a horizontal compression of the graph of $y = f(x)$ by a factor of $\dfrac{1}{2}$.

b) Use the table of values from part a) for $y = f(x)$.

$y = f(x)$

x	y
-2	0
-1	2
0	4
1	3
2	2
3	1
4	0

Then, complete a table of values for $y = f\left(\frac{1}{2}x\right)$. Use convenient values for x.

$y = f\left(\frac{1}{2}x\right)$

x	y
-4	$f\left(\frac{1}{2} \times (-4)\right) = f(-2) = 0$
-2	$f\left(\frac{1}{2} \times (-2)\right) = f(-1) = 2$
0	$f\left(\frac{1}{2} \times (0)\right) = f(0) = 4$
2	$f\left(\frac{1}{2} \times 2\right) = f(1) = 3$
4	$f\left(\frac{1}{2} \times 4\right) = f(2) = 2$
6	$f\left(\frac{1}{2} \times 6\right) = f(3) = 1$
8	$f\left(\frac{1}{2} \times 8\right) = f(4) = 0$

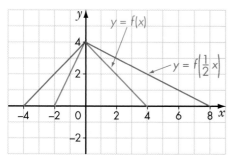

The functions have the same range, $0 \leq y \leq 4$.
The domain of $y = f(x)$ is $-2 \leq x \leq 4$.

The domain of $y = f\left(\frac{1}{2}x\right)$ is $-4 \leq x \leq 8$.

For non-zero values of x, each point on the graph of $y = f\left(\frac{1}{2}x\right)$ is twice as far from the y-axis as the equivalent point on $y = f(x)$. The point (x, y) on the graph of the function $y = f(x)$ is transformed to the point $(2x, y)$ on the graph of $y = f\left(\frac{1}{2}x\right)$.

The graph of $y = f\left(\frac{1}{2}x\right)$ is a horizontal expansion of the graph of $y = f(x)$ by a factor of 2.

Note that the point (0, 4) is invariant under both transformations in Example 3, because this point lies on the y-axis.

EXAMPLE 4 Horizontal Stretching of a Radical Function

Compare the graphs of $y = \sqrt{2x}$ and $y = \sqrt{\frac{1}{2}x}$ to the graph of $y = \sqrt{x}$.

SOLUTION

Complete tables of values using convenient values for x.

$y = \sqrt{x}$ $y = \sqrt{2x}$ $y = \sqrt{\frac{1}{2}x}$

x	y	x	y	x	y
0	0	0	0	0	0
1	1	0.5	1	2	1
4	2	2	2	8	2
9	3	4.5	3	18	3
16	4	8	4	32	4

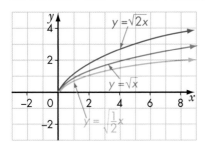

The window variables include Xmin = 0, Xmax = 35, Ymin = 0, and Ymax = 10.

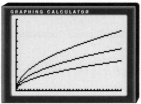

The graph of $y = \sqrt{2x}$ is the graph of $y = \sqrt{x}$ compressed horizontally by a factor of $\frac{1}{2}$.

The graph of $y = \sqrt{\frac{1}{2}x}$ is the graph of $y = \sqrt{x}$ expanded horizontally by a factor of 2.

> The three functions have the same domain, $x \geq 0$, and the same range, $y \geq 0$

In general, for any function $y = f(x)$, the graph of a function $y = f(kx)$, were k is any real number, is obtained by dividing the x-value at each point on the graph of $y = f(x)$ by k.

The point (x, y) on the graph of the function $y = f(x)$ is transformed into the point $\left(\frac{x}{k}, y\right)$ on the graph of $y = f(kx)$.

If $k > 1$, the graph of $y = f(x)$ is compressed horizontally by a factor of $\frac{1}{k}$.

If $k < 1$, the graph of $y = f(x)$ is expanded horizontally by a factor of $\frac{1}{k}$.

EXAMPLE 5 Vertical and Horizontal Stretches

a) Copy the graph of $y = f(x)$, as shown. Compare the graphs of $y = 3f(x)$, and $y = f\left(\frac{1}{2}x\right)$ to the graph of $y = f(x)$.

b) Compare the graph of $y = 3f\left(\frac{1}{2}x\right)$ to the graph of $y = f(x)$.

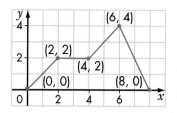

SOLUTION

a) Complete tables of values using convenient values for x.

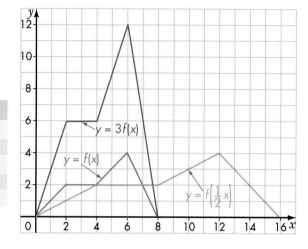

$y = f(x)$

x	y
0	0
2	2
4	2
6	4
8	0

$y = 3f(x)$

x	y
0	0
2	6
4	6
6	12
8	0

$y = f\left(\frac{1}{2}x\right)$

x	y
0	0
4	2
8	2
12	4
16	0

The graph of $y = 3f(x)$ is the graph of $y = f(x)$ expanded vertically by a factor of 3.

The graph of $y = f\left(\frac{1}{2}x\right)$ is the graph of $y = f(x)$ expanded horizontally by a factor of 2.

b) First, apply the transformation $y = 3f(x)$ to the function $y = f(x)$.

Then, apply the transformation $y = f\left(\frac{1}{2}x\right)$ to the function $y = 3f(x)$ to give the transformation $y = 3f\left(\frac{1}{2}x\right)$.

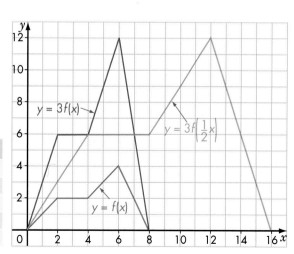

$y = f(x)$

x	y
0	0
2	2
4	2
6	4
8	0

$y = 3f(x)$

x	y
0	0
2	6
4	6
6	12
8	0

$y = 3f\left(\frac{1}{2}x\right)$

x	y
0	0
4	6
8	6
12	12
16	0

The graph of $y = 3f\left(\dfrac{1}{2}x\right)$ is the graph of $y = f(x)$ expanded vertically by a factor of 3 and expanded horizontally by a factor of 2.

In Example 5b), note that the transformation $y = f\left(\dfrac{1}{2}x\right)$ could have been applied to $y = f(x)$ first, followed by the transformation $y = 3f(x)$, to give $y = 3f\left(\dfrac{1}{2}x\right)$.

$y = f(x)$

x	y
0	0
2	2
4	2
6	4
8	0

$y = f\left(\dfrac{1}{2}x\right)$

x	y
0	0
4	2
8	2
12	4
16	0

$y = 3f\left(\dfrac{1}{2}x\right)$

x	y
0	0
4	6
8	6
12	12
16	0

Key Concepts

- The table summarizes stretches of the function $y = f(x)$.

Stretch	Mathematical Form	Effect
Vertical	$y = af(x)$	If $a > 1$, then expand the graph vertically by a factor of a. If $0 < a < 1$, then compress the graph vertically by a factor of a.
Horizontal	$y = f(kx)$	If $k > 1$, then compress the graph horizontally by a factor of $\dfrac{1}{k}$. If $0 < k < 1$, then expand the graph horizontally by a factor of $\dfrac{1}{k}$.

- When a function is stretched horizontally and stretched vertically, the stretches can be performed in either order to give the same image.

Communicate Your Understanding

1. If $y = f(x)$, describe the difference in the graphs of $y = 0.5f(x)$ and $y = 2f(x)$.

2. If $y = f(x)$, describe the difference in the graphs of $y = f(2x)$ and $y = f(0.5x)$.

3. The graph of $y = f(x)$ is shown. Describe how the coordinates of the points on each of the following graphs compare with the coordinates of the points on $y = f(x)$.

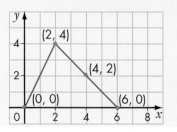

a) $y = 2f(x)$ **b)** $y = f\left(\dfrac{1}{2}x\right)$ **c)** $y = 2f\left(\dfrac{1}{2}x\right)$

Practise

A

1. The graph of function $y = f(x)$ is shown. Sketch the graph of each of the following functions, and state the domain and range.

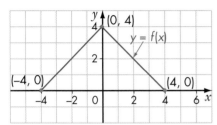

a) $y = 2f(x)$

b) $y = \dfrac{1}{2}f(x)$

c) $y = f(2x)$

d) $y = f\left(\dfrac{1}{2}x\right)$

e) $y = 2f\left(\dfrac{1}{2}x\right)$

f) $y = \dfrac{1}{2}f\left(\dfrac{1}{2}x\right)$

g) $y = 2f(2x)$

h) $y = 0.5f(2x)$

2. The graph of function $y = f(x)$ is shown. Sketch the graph of each of the following functions, and state the domain and range.

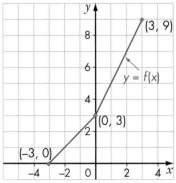

a) $y = 2f(x)$

b) $y = \dfrac{1}{3}f(x)$

c) $y = f(3x)$

d) $y = f\left(\dfrac{1}{3}x\right)$

e) $y = f\left(\dfrac{1}{2}x\right)$

f) $y = \dfrac{1}{3}f(3x)$

3. The graph of $y = f(x)$ is shown.

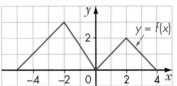

Identify the transformations used to obtain the graphs of $y = g(x)$ and $y = h(x)$.

a)
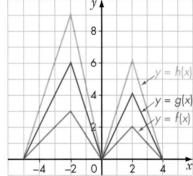

b)
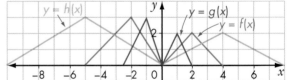

4. For each set of three functions,
a) sketch them on the same grid, or graph them in the same viewing window of a graphing calculator
b) describe how the graphs of the second and third functions are related to the graph of the first function
c) identify any invariant points

i) $y = x$, $y = 2x$, and $y = \dfrac{1}{2}x$

ii) $y = x^2$, $y = 3x^2$, and $y = \dfrac{1}{2}x^2$

iii) $y = \sqrt{x}$, $y = 3\sqrt{x}$, and $y = 1.5\sqrt{x}$

iv) $y = x^2$, $y = (2x)^2$, and $y = \left(\dfrac{1}{2}x\right)^2$

5. Describe how the graphs of the following functions can be obtained from the graph of the function $y = f(x)$.

a) $y = 3f(x)$

b) $y = \frac{1}{2}f(x)$

c) $y = 2f(x)$

d) $y = \frac{1}{3}f(x)$

e) $y = f(2x)$

f) $y = f\left(\frac{1}{2}x\right)$

g) $y = f(4x)$

6. Describe how the graphs of the following functions can be obtained from the graph of the function $y = f(x)$.

a) $y = 3f(2x)$

b) $y = \frac{1}{2}f\left(\frac{1}{3}x\right)$

c) $y = 4f\left(\frac{1}{2}x\right)$

d) $y = \frac{1}{3}f(3x)$

e) $y = 2f(4x)$

f) $y = 5f\left(\frac{1}{2}x\right)$

7. The blue graph is a stretch of the red graph. The equation of the red graph is given. Write an equation for the blue graph. Check each equation using a graphing calculator.

a)
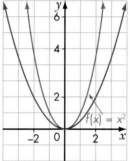

b)
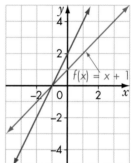

c)
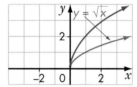

Apply, Solve, Communicate

8. Use transformations and the zeros of the quadratic function $f(x) = (x + 4)(x - 2)$ to determine the zeros of each of the following functions.

a) $y = 3f(x)$

b) $y = f\left(\frac{1}{2}x\right)$

c) $y = f(2x)$

B

9. Stopping distances The distance required to stop a car is directly proportional to the square of the speed of the car. The stopping distance for a car on dry asphalt can be approximated using the function $d(s) = 0.006s^2$, where $d(s)$ is the stopping distance, in metres, and s is the speed of the car, in kilometres per hour. The stopping distance for a car on wet asphalt can be approximated using the function $d(s) = 0.009s^2$. The stopping distance for a car on black ice can be approximated using the function $d(s) = 0.04s^2$.

a) For each of the three surfaces, what is the stopping distance for a car travelling at 80 km/h?

b) Write a reasonable domain and range for each of the functions $d(s) = 0.006s^2$, $d(s) = 0.009s^2$, and $d(s) = 0.04s^2$.

c) Let $y = d(s)$ and $x = s$. Graph the three functions, plus the function $y = x^2$, in the same viewing window of a graphing calculator.

d) Compare the graphs of $y = 0.006x^2$, $y = 0.009x^2$, and $y = 0.04x^2$ to the graph of $y = x^2$.

10. The function $y = f(x)$ has been transformed to $y = af(kx)$. Determine the values of a and k for each of the following transformations.

a) a vertical expansion by a factor of 4

b) a horizontal compression by a factor of $\dfrac{1}{3}$

c) a vertical compression by a factor of $\dfrac{1}{2}$ and a horizontal expansion by a factor of 3

d) a vertical expansion by a factor of 2 and a horizontal compression by a factor of $\dfrac{1}{4}$

11. The graph of $f(x) = \sqrt{16 - x^2}$ is shown. Use transformations to sketch the graph of each of the following functions. State the domain and range of each function.

a) $y = 3f(x)$

b) $y = \dfrac{1}{2}f(x)$

c) $y = f(2x)$

d) $y = f\left(\dfrac{1}{2}x\right)$

e) $y = 2f(x)$

f) $y = f(4x)$

12. Application Use transformations and the zeros of the polynomial function $f(x) = x(x + 3)(x - 6)$ to determine the zeros of each of the following functions.

a) $y = f(3x)$

b) $y = 2f(x)$

c) $y = f\left(\dfrac{1}{2}x\right)$

d) $y = f(2x)$

13. Communication Describe how the graphs of $y = \dfrac{(x-2)(x+2)}{3}$ and $y = x^2 - 4$ are related. Explain.

14. Falling objects If an object is dropped from an initial height of x metres, its approximate height above the ground, h metres, after t seconds is given by $h = -5t^2 + x$ on the Earth and by $h = -0.8t^2 + x$ on the moon. So, for objects dropped from an initial height of 20 m, the functions are $h(t) = -5t^2 + 20$ and $h(t) = -0.8t^2 + 20$.

a) Graph $h(t)$ versus t for $h(t) = -5t^2 + 20$ and $h(t) = -0.8t^2 + 20$ on the same axes or in the same viewing window.

b) Inquiry/Problem Solving Describe how the graph of $h(t) = -0.8t^2 + 20$ can be transformed onto the graph of $h(t) = -5t^2 + 20$. Justify your reasoning.

c) Explain the meaning of the point that the two graphs have in common.

d) State the domain and range of each function.

C

15. Greatest integer function Sketch the graph of $y = [x]$. Then, use transformations to sketch the graphs of the following functions. Check your solutions using a graphing calculator.

a) $y = [2x]$ **b)** $y = [0.5x]$ **c)** $y = 2[x]$

16. a) Describe the horizontal stretch that transforms $y = \sqrt{x}$ to $y = \sqrt{4x}$.

b) Describe the vertical stretch that transforms $y = \sqrt{x}$ to $y = 2\sqrt{x}$.

c) How do the graphs of $y = \sqrt{4x}$ and $y = 2\sqrt{x}$ compare?

d) How do the transformations in parts a) and b) compare? Explain why they compare in this way.

LOGIC *Power*

Copy the diagram. Show two different ways to divide the shape along the lines into four congruent figures.

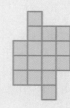

3.7 Combinations of Transformations

Because of its mathematical simplicity, the 3-4-5 right triangle has as much appeal today as it did thousands of years ago. In architecture, a 3-4-5 right triangle, with 4 as the base, has a hypotenuse with the slope of a comfortable stairway.

Some buildings have A-frame roofs. An example can be seen at the Alexander Graham Bell National Historic Site in Baddeck, Nova Scotia. For many A-frame roofs, a cross section is an isosceles triangle.

In some cases, the isosceles triangle can be formed by attaching two 3-4-5 right triangles. Two different isosceles triangles can be formed in this way.

INVESTIGATE & INQUIRE

1. By graphing each function using a table of values, show the right triangle formed by each of the following functions, the x-axis, and the y-axis.

a) $y = -\dfrac{3}{4}x + 3$, $x \geq 0$, $y \geq 0$

x	y
4	
2	
0	

b) $y = -\dfrac{4}{3}x + 4$, $x \geq 0$, $y \geq 0$

x	y
3	
1.5	
0	

2. What are the side lengths of the triangle in question 1a)?

3. What are the side lengths of the triangle in question 1b)?

4. What transformations must be applied to $y = x$ to give the function $y = -\dfrac{3}{4}x + 3$?

5. What transformations must be applied to $y = x$ to give the function $y = -\dfrac{4}{3}x + 4$?

6. In questions 4 and 5, must the transformations be applied in a particular order, or does the order have no effect on the result? Explain.

7. What transformation could be applied to the triangle from question 1a), so that the original triangle and its image form an isosceles triangle? Is there more than one answer? Explain.

8. Repeat question 7 for the triangle from question 1b).

9. State the dimensions and the height of each of the different isosceles triangles that can be formed in questions 7 and 8.

In this section, translations, expansions, compressions, and reflections will be used to perform combinations of transformations on functions. To simplify the procedure and give the desired results, perform the transformations in the following order.
• expansions and compressions
• reflections
• translations

In other words, perform multiplications (expansions, compressions, and reflections) before additions and subtractions (translations).

EXAMPLE 1 Vertical Stretching and Reflecting

a) Copy the graph of $y = f(x)$, as shown.
On the same set of axes, graph $y = -3f(x)$ and $y = -\dfrac{1}{2}f(x)$.

b) Describe how the graphs of $y = -3f(x)$ and $y = -\dfrac{1}{2}f(x)$ are related to the graph of $y = f(x)$.

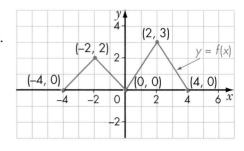

SOLUTION

a) Use the given graph to complete a table of values for the function $y = f(x)$.
Then, complete tables of values for the functions $y = -3f(x)$ and $y = -\frac{1}{2}f(x)$, and draw the graphs.

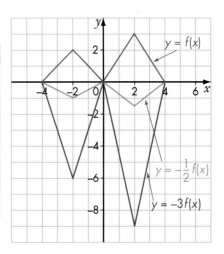

$y = f(x)$

x	y
4	0
2	3
0	0
-2	2
-4	0

$y = -3f(x)$

x	y
4	-3(0) = 0
2	-3(3) = -9
0	-3(0) = 0
-2	-3(2) = -6
-4	-3(0) = 0

$y = -\frac{1}{2}f(x)$

x	y
4	$-\frac{1}{2}(0) = 0$
2	$-\frac{1}{2}(3) = -1.5$
0	$-\frac{1}{2}(0) = 0$
-2	$-\frac{1}{2}(2) = -1$
-4	$-\frac{1}{2}(0) = 0$

b) The point (x, y) on the graph of the function $y = f(x)$ becomes the point $(x, -3y)$ on the graph of $y = -3f(x)$.

The graph of $y = -3f(x)$ is the graph of $y = f(x)$ expanded vertically by a factor of 3 and reflected in the x-axis.

The point (x, y) on the graph of the function $y = f(x)$ becomes the point $\left(x, -\frac{1}{2}y\right)$ on the graph of $y = -\frac{1}{2}f(x)$.

The graph of $y = -\frac{1}{2}f(x)$ is the graph of $y = f(x)$ compressed vertically by a factor of $\frac{1}{2}$ and reflected in the x-axis.

Note that, in Example 1, the three functions have the same domain, $-4 \le x \le 4$. The range of $y = f(x)$ is $0 \le y \le 3$, of $y = -3f(x)$ is $-9 \le y \le 0$, and of $y = -\frac{1}{2}f(x)$ is $-1.5 \le y \le 0$. The points $(4, 0)$, $(0, 0)$, and $(-4, 0)$ are invariant.

EXAMPLE 2 Transforming Quadratic Functions

Sketch the graph of $y = x^2$ and the graph of $y = \frac{1}{2}(x+4)^2 - 5$.

SOLUTION

Sketch the graph of $y = x^2$.

To sketch the graph of $y = \frac{1}{2}(x+4)^2 - 5$, first sketch the graph of $y = \frac{1}{2}x^2$. This graph is a vertical compression of $y = x^2$ by a factor of $\frac{1}{2}$.

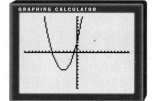

The transformed function can be graphed directly using a graphing calculator.

GRAPHING CALCULATOR

Then, apply the horizontal translation of 4 units to the left and the vertical translation of 5 units downward.

The result is the graph of $y = \frac{1}{2}(x+4)^2 - 5$.

For such functions as $y = (3x + 6)^2$ and of $y = \sqrt{-x+5}$, factor the coefficient of the x-term to identify the characteristics of the function more easily.

The function $y = (3x + 6)^2$ becomes $y = (3(x + 2))^2$. Therefore, the graph of $y = (3x + 6)^2$ is the graph of $y = x^2$ compressed horizontally by a factor of $\frac{1}{3}$ and translated 2 units to the left.

The function $y = \sqrt{-x+5}$ becomes $y = \sqrt{-(x-5)}$. Therefore, the graph of $y = \sqrt{-x+5}$ is the graph of $y = \sqrt{x}$ reflected in the y-axis and translated 5 units to the right.

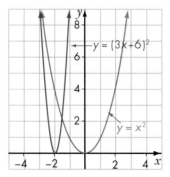

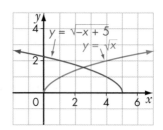

EXAMPLE 3 Transforming Radical Functions

Given $f(x) = \sqrt{x}$, sketch the graph of $y = f(x)$ and the graph of $y = 2f(-x - 3) + 4$.

SOLUTION

Sketch the graph of $y = \sqrt{x}$.
The graph of $y = 2f(-x - 3) + 4$ is the graph
of $y = 2\sqrt{-x - 3} + 4$.
Rewrite $y = 2\sqrt{-x - 3} + 4$ as $y = 2\sqrt{-(x + 3)} + 4$.
To sketch the graph of $y = 2\sqrt{-(x + 3)} + 4$, first
sketch the graph of $y = 2\sqrt{x}$. This graph is a
vertical expansion of the graph of $y = \sqrt{x}$ by a
factor of 2.

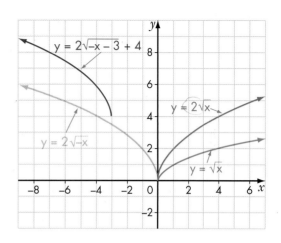

Then, sketch the graph of $y = 2\sqrt{-x}$, which is a
reflection of the graph of $y = 2\sqrt{x}$ in the y-axis.
Then, apply the horizontal translation of 3 units to the left and the
vertical translation of 4 units upward.
The result is the graph of $y = 2f(-x - 3)$ or $y = 2\sqrt{-x - 3} + 4$.

For the function
$y = 2\sqrt{-x - 3} + 4$, the
domain is $x \leq -3$ and the
range is $y \geq 4$.

EXAMPLE 4 Horizontal Stretching and Reflecting of Radical Functions

Compare the graphs of $y = \sqrt{-x}$, $y = \sqrt{-2x}$, and $y = \sqrt{-\dfrac{1}{2}x}$ to the
graph of $y = \sqrt{x}$.

SOLUTION

Complete tables of values using convenient values for x, or use a graphing
calculator.

$y = \sqrt{x}$

x	y
0	0
1	1
4	2
9	3
16	4

$y = \sqrt{-x}$

x	y
0	0
-1	1
-4	2
-9	3
-16	4

$y = \sqrt{-2x}$

x	y
0	0
$-\dfrac{1}{2}$	1
-2	2
$-\dfrac{9}{2}$	3
-8	4

$y = \sqrt{-\dfrac{1}{2}x}$

x	y
0	0
-2	1
-8	2
-18	3
-32	4

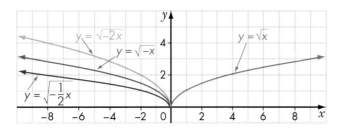

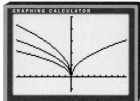

The window variables include Xmin = −9, Xmax = 9, Ymin = −1, and Ymax = 5.

The graph of $y = \sqrt{-x}$ is the graph of $y = \sqrt{x}$ reflected in the y-axis.

The graph of $y = \sqrt{-2x}$ is the graph of $y = \sqrt{x}$ compressed horizontally by a factor of $\frac{1}{2}$ and reflected in the y-axis.

The graph of $y = \sqrt{-\frac{1}{2}x}$ is the graph of $y = \sqrt{x}$ expanded horizontally by a factor of 2 and reflected in the y-axis.

EXAMPLE 5 Writing Equations

The graph of $y = \sqrt{x}$ is expanded vertically by a factor of 5, reflected in the x-axis, and translated 6 units to the right and 3 units downward. Write the equation of the transformed function.

SOLUTION

When $y = \sqrt{x}$ is expanded vertically by a factor of 5, the function becomes $y = 5\sqrt{x}$. This function is reflected in the x-axis, becoming $y = -5\sqrt{x}$.

The function $y = -5\sqrt{x}$ is then translated 6 units to the right and 3 units downward to give $y = -5\sqrt{x - 6} - 3$.

The equation of the transformed function is $y = -5\sqrt{x - 6} - 3$.

The solution can be modelled graphically.

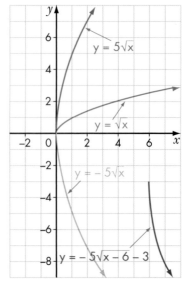

EXAMPLE 6 Transforming Quadratic Functions

Given $f(x) = x^2$, sketch the graph of $y = f(x)$ and the graph of
$y = -f(2(x - 5)) + 6$.

SOLUTION

Sketch the graph of $y = x^2$.
The graph of $y = -f(2(x - 5)) + 6$ is the graph
of $y = -(2(x - 5))^2 + 6$.
To sketch the graph of $y = -(2(x - 5))^2 + 6$, first
sketch the graph of $y = (2x)^2$. This graph is a
horizontal compression of the graph of $y = x^2$ by
a factor of $\dfrac{1}{2}$.
Then, sketch the graph of $y = -(2x)^2$, which is a
reflection of the graph of $y = (2x)^2$ in the x-axis.

Then, apply the horizontal translation of 5 units to
the right and the vertical translation of 6 units upward.
The result is the graph of $y = -f(2(x - 5)) + 6$ or
$y = -(2(x - 5))^2 + 6$.

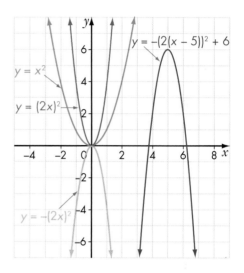

Key Concepts

- Perform combinations of transformations in the following order.
  - * expansions and compressions
  - * reflections
  - * translations
- If necessary, factor the coefficient of the x-term to identify the characteristics
  of a function more easily.

Communicate Your Understanding

1. Identify the combination of transformations on $y = x^2$ that results in the
given function.
a) $y = -x^2 + 4$ b) $y = -3(x - 1)^2 + 5$
2. Describe how you would graph the function $y = 2\sqrt{x + 3} - 7$.
3. Describe how you would graph the function $y = \sqrt{-x - 4}$.

Practise

A

1. Describe how the graph of each of the following functions can be obtained from the graph of $y = f(x)$.

a) $y = 2f(x) + 3$

b) $y = \dfrac{1}{2}f(x) - 2$

c) $y = f(x + 4) + 1$

d) $y = 3f(x - 5)$

e) $y = f\left(\dfrac{1}{2}x\right) - 6$

f) $y = f(2x) + 1$

2. Describe how the graph of each of the following functions can be obtained from the graph of $y = f(x)$.

a) $y = -2f(x)$

b) $y = -\dfrac{1}{3}f(x)$

c) $y = f(-4x)$

d) $y = f\left(-\dfrac{1}{2}x\right)$

3. Describe how the graph of each of the following functions can be obtained from the graph of $y = f(x)$.

a) $y = -f(2x)$

b) $y = 3f(-2x)$

c) $y = -\dfrac{1}{2}f\left(\dfrac{1}{3}x\right)$

d) $y = 4f(x - 6) + 2$

e) $y = -2f(x) - 3$

f) $y = -f(x - 3) + 1$

g) $y = 3f(2x) - 6$

h) $y = \dfrac{1}{2}f\left(\dfrac{1}{2}x\right) - 4$

4. The graph of $y = f(x)$ is shown.

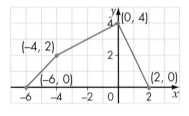

Sketch the graph of each of the following functions, state its domain and range, and identify any invariant points.

a) $y = f(x - 4) + 2$

b) $y = f(x + 2) - 4$

c) $y = \dfrac{1}{2}f(x) - 3$

d) $y = f(2x) + 3$

e) $y = -2f(x)$

f) $y = f(-x) - 2$

g) $y = f\left(-\dfrac{1}{2}x\right)$

h) $y = -\dfrac{1}{2}f(-2x)$

5. The graph of $y = f(x)$ is shown. Sketch the graph of each of the following functions, and state its domain and range.

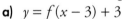

a) $y = f(x - 3) + 3$

b) $y = -f(x) + 1$

c) $y = f(2x) - 3$

d) $y = 3f(x) - 2$

e) $y = f(-x) + 2$

f) $y = 2f(-x)$

6. Sketch each set of functions on the same set of axes in the given order.

a) $y = x$
$y = 3x$
$y = 3(x - 2) + 10$

b) $y = x^2$
$y = 2x^2$
$y = 2(x + 2)^2 - 3$

c) $y = x$
$y = 0.5x$
$y = -0.5(x - 4) + 2$

d) $y = x^2$
$y = \left(\dfrac{1}{2}x^2\right)$
$y = \dfrac{1}{2}(x + 3)^2 + 3$

e) $y = \sqrt{x}$
$y = \sqrt{2x}$
$y = -\sqrt{2x}$
$y = -\sqrt{2(x - 1)} - 3$

f) $y = \sqrt{x}$
$y = 2\sqrt{x}$
$y = 2\sqrt{-x}$
$y = 2\sqrt{-(x - 3)} + 5$

7. Describe how the graph of each of the following functions can be obtained from the graph of $y = f(x)$.

a) $y = f(2(x - 4))$ **b)** $y = f(-(x + 1)) - 1$
c) $y = f(3(x + 4)) + 5$ **d)** $y = -2f(4(x - 2))$
e) $y = f(-x + 2)$ **f)** $y = f(2x + 8) - 4$
g) $y = f(4 - x) + 5$ **h)** $y = f(3x - 6) + 8$

8. Given $f(x) = x^2$, sketch the graph of each of the following, and state the domain and range.

a) $y = f(x - 3) + 1$ **b)** $y = 2f(x + 5) - 4$
c) $y = \frac{1}{2}f\left(\frac{1}{2}x\right) + 3$ **d)** $y = -f(2(x - 2)) - 3$
e) $y = f(3 - x) + 2$ **f)** $y = -\frac{1}{2}f(2x + 6) - 2$

9. Given $f(x) = \sqrt{x}$, sketch the graph of each of the following, and state the domain and range.

a) $y = f(x - 5) - 4$ **b)** $y = 3f(x + 3) + 2$
c) $y = \frac{1}{2}f(2(x - 1)) - 2$ **d)** $y = 2f(3x - 9) + 1$
e) $y = -f(-x) + 5$ **f)** $y = -2f(4 - x) - 3$

10. Technology The calculator display shows the graph of $y = (x + 2)^2 + 3$ and its image after a reflection in the x-axis and a reflection in the y-axis. Write the equation of the image.

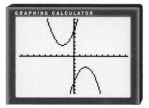

11. Technology The graph of $y = x^2$ is expanded vertically by a factor of 3, translated 4 units to the right, and translated 2 units downward. Write the equation of the transformed function. Check your solution using a graphing calculator.

12. Technology The graph $y = \sqrt{x}$ is expanded horizontally by a factor of 2, reflected in the y-axis, and translated 6 units to the right. Write the equation of the transformed function. Check your solution using a graphing calculator.

Apply, Solve, Communicate

13. Ski chalet A cross section of the roof of a ski chalet can be modelled by the following two functions and the x-axis.

$$y = \frac{5}{3}x + 10, -6 \le x \le 0, 0 \le y \le 10$$

$$y = -\frac{5}{3}x + 10, 0 \le x \le 6, 0 \le y \le 10$$

a) Graph both functions on the same axes or in the same viewing window.
b) Find the side lengths and the height of the isosceles triangle formed by the two functions and the x-axis. Round to the nearest tenth of a unit, if necessary.
c) What transformations must be applied to $y = x$ to give the function
$y = \frac{5}{3}x + 10$? to give the function $y = -\frac{5}{3}x + 10$?

B

14. Application The height, y metres, of an emergency flare fired upward from a small boat can be modelled by the function
$$y = -5(x - 4)^2 + 80$$
where x seconds is the time since the flare was fired.
a) Describe how the graph of $y = -5(x - 4)^2 + 80$ can be obtained by transforming the graph of $y = x^2$.
b) Interpret the equation of the transformed function to find the maximum height reached by the flare and the time it takes to reach this height.

15. Use transformations and the zeros of the quadratic function $f(x) = (x + 2)(x - 6)$ to determine the zeros of each of the following functions.
a) $y = -2f(x)$ **b)** $y = f(-2x)$ **c)** $y = 2f(-x)$

d) $y = -\frac{1}{2}f\left(\frac{1}{2}x\right)$ **e)** $y = -f(x + 1)$ **f)** $y = f(-x - 2)$

16. Communication The graph of $f(x) = \sqrt{16 - x^2}$ is shown. Sketch the graph of each of the following functions, and state its domain and range.

a) $y = -\frac{1}{2}f(x)$ **b)** $y = f(-2x)$

c) $y = \frac{1}{2}f(x) + 5$ **d)** $y = -\frac{1}{2}f(x) - 3$

e) $y = (f(x) - 3) - 2$ **f)** $y = -2f(x + 6) + 5$

17. The function $y = f(x)$ has been transformed to $y = af(k(x - h)) + q$. Determine the values of a, k, h, and q for each of the following transformations.
a) a vertical expansion by a factor of 3 and a reflection in the y-axis

b) a vertical compression by a factor of $\frac{1}{3}$, a horizontal expansion by a factor of 2, a translation of 6 units to the right, and a translation of 1 unit downward

c) a vertical expansion by a factor of 2, a horizontal compression by a factor of $\frac{1}{2}$, a reflection in the x-axis, a reflection in the y-axis, a translation of 7 units to the left, and a translation of 4 units upward

18. Falling objects On the Earth, if an object is dropped from an initial height of x metres, its approximate height above the ground, h metres, after t seconds, is given by $h(t) = -5t^2 + x$. On the moon, the approximate height is given by $h(t) = -0.8t^2 + x$. For objects dropped from an initial height of 20 m, the functions are $h(t) = -5t^2 + 20$ and $h(t) = -0.8t^2 + 20$.

a) Translate the graph of $h(t) = -5t^2 + 20$ upward by 105 units. Explain the meaning of the point that the resulting graph and the graph of $h(t) = -0.8t^2 + 20$ have in common.

b) The approximate height of an object dropped on Jupiter is given by $h(t) = -12.8t^2 + x$. So, for an initial height of 320 m, the function is $h(t) = -12.8t^2 + 320$. Graph $h(t)$ versus t for $h(t) = -12.8t^2 + 320$ and $h(t) = -0.8t^2 + 20$ on the same axes or in the same viewing window.

c) **Inquiry/Problem Solving** Describe how the graph of $h(t) = -0.8t^2 + 20$ can be transformed onto the graph of $h(t) = -12.8t^2 + 320$. Justify your reasoning.

d) Explain the meaning of the point that the two graphs in part c) have in common.

e) State the domain and range of each function in part c).

C

19. a) Describe how the graph of $y = 2\sqrt{x+1}$ can be obtained from the graph of $y = \sqrt{x}$.

b) Describe how the graph of $y = \sqrt{4x+4}$ can be obtained from the graph of $y = \sqrt{x}$.

c) How are the graphs of $y = 2\sqrt{x+1}$ and $y = \sqrt{4x+4}$ related? Explain.

20. Describe how the graph of the function $y = -\sqrt{x+2} - 3$ can be obtained from the graph of $y = x^2$, $x \le 0$.

21. a) Expand the graph of $y = x$ vertically by a factor of 2. Then, translate the result 3 units to the left and 5 units downward. Compare the result to the graph of $y = 2x + 1$. Explain your findings.

b) If the same transformations are performed on the graph of $y = x^2$, instead of $y = x$, is the result the same as the graph of $y = 2x^2 + 1$? Explain.

ACHIEVEMENT Check   Knowledge/Understanding Thinking/Inquiry/Problem Solving Communication Application

Given: a horizontal line segment from (0, 0) to (2, 0)
a vertical line segment from (0, 0) to (0, 2)
a semi-circle centred at (1, 0) with endpoints (0, 0) and (0, 2) and passing through point (1, 1), radius 1 unit

a) Sketch the lower case letters of the alphabet that are possible to print using transformations of the given figures.

b) What other types of transformations would be needed to print the other letters?

Investigate & Apply

Frieze Patterns

A frieze pattern is a pattern that repeats in one direction. The patterns depend on the use of transformations. Many cultures have used frieze patterns to make decorative designs on buildings, textiles, pottery, and so on.

One way of creating a frieze pattern is to make a basic design on a grid and then to transform the design. An example of a basic design is shown.

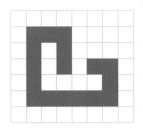

1. Describe how transformations have been used to create each of the following frieze patterns from the basic design.

a)

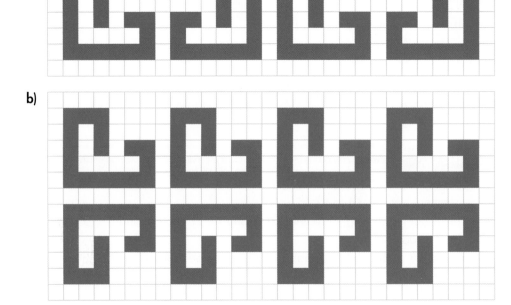

b)

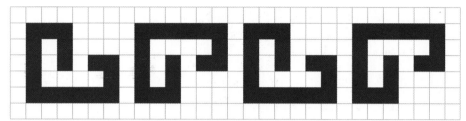

c)

2. The frieze pattern shown was created from equilateral triangles of side length 2 units.

Suppose a coordinate grid is superimposed on the pattern, as shown.

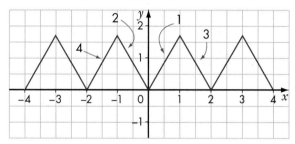

a) What is the exact slope of line segment 1?

b) Write an equation for line segment 1.

3. Use transformations to write an equation for
a) line segment 2   **b)** line segment 3   **c)** line segment 4

4. a) Design your own frieze pattern from a basic design of your choice.

b) Describe how you used transformations to create the pattern.

c) Suggest a possible use for your frieze pattern, and explain why you chose this use.

5. Research Use your research skills to determine how frieze patterns are classified and how many types of frieze patterns there are.

REVIEW OF **KEY CONCEPTS**

3.1 Functions

Refer to the Key Concepts on page 177.

1. Determine if each relation is a function.

a)
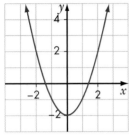

b)
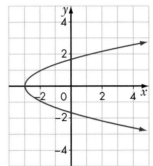

c)
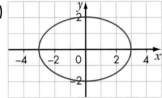

d)
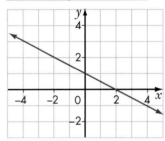

e)
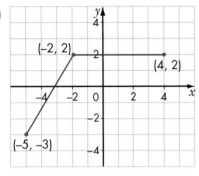

f)
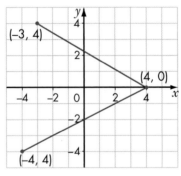

2. State the domain and range of each function.

a)
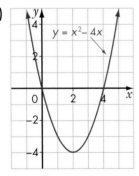

b)
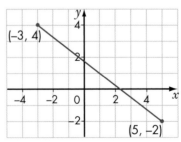

c)

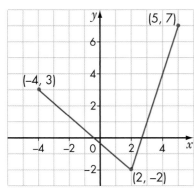

d)

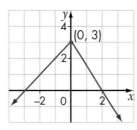

3. If $f(x) = 7 - 3x$, find

a) $f(4)$ **b)** $f(8)$ **c)** $f(0)$ **d)** $f(-1)$ **e)** $f(-4)$
f) $f(0.5)$ **g)** $f(-0.1)$ **h)** $f(100)$ **i)** $f(5000)$

4. If $f(x) = 2x^2 + x - 8$, find

a) $f(1)$ **b)** $f(0)$ **c)** $f(-2)$ **d)** $f(-4)$ **e)** $f(0.5)$
f) $f(-1.5)$ **g)** $f(-0.2)$ **h)** $f(60)$ **i)** $f(-100)$

5. Find the range of each function if the domain is $\{-3, -1, 2, 4\}$.

a) $f(x) = 4x - 5$ **b)** $f(x) = -3x^2 + 2x - 1$

6. Algebra a) If $f(x) = 5x + 2$, write and simplify $f(3a)$.
b) If $f(x) = 4 - 7x$, write and simplify $f(n - 4)$.
c) If $f(x) = 3x^2 + 2x - 5$, write and simplify $f(2k + 3)$.

7. Cost The cost, C dollars, of one type of bottled water is related to the number of bottles purchased, b, by the equation $C = 2.50b$.
a) Identify the dependent variable and the independent variable.
b) Is the cost a function of the number of bottles purchased? Explain.

8. a) Write an equation that expresses the surface area of a sphere as a function of its radius.
b) State the domain and range of the function.

3.2 Investigation: Properties of Functions Defined by $f(x) = \sqrt{x}$ and $f(x) = \dfrac{1}{x}$

9. For $y = \sqrt{x}$, what is
a) the domain? **b)** the range?

10. Describe the relationship between the graph of $y = x^2$ and the graph of $y = \sqrt{x}$.

11. For $y = \dfrac{1}{x}$, what is

a) the domain? 　　　　　　　　　　 **b)** the range?

12. Describe the relationship between the graph of $y = x$ and the graph of $y = \dfrac{1}{x}$.

3.3 Horizontal and Vertical Translations of Functions

Refer to the Key Concepts on page 189.

13. The function $y = f(x)$ is given. Describe how the graphs of the following functions can be obtained from the graph of $y = f(x)$.

a) $y = f(x) - 3$ 　　　　　　　　　 **b)** $y = f(x + 6)$
c) $y = f(x - 4) - 5$ 　　　　　　　 **d)** $y + 5 = f(x)$

14. The graph of the function drawn in blue is a translation of the function drawn in red. Write an equation for each function drawn in blue. Check each equation using a graphing calculator.

a)
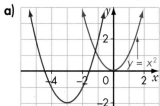

b)
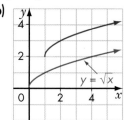

15. The graphs of two functions $y = f(x)$ are shown.

a)
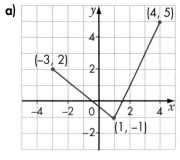

b)
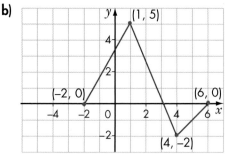

For each function, sketch the graphs of each of the following.

i) $y = f(x) - 3$ 　　　　 **ii)** $y = f(x) + 4$ 　　　　 **iii)** $y = f(x - 3)$
iv) $y = f(x + 4)$ 　　　 **v)** $y = f(x + 3) - 2$ 　　 **vi)** $y = f(x - 2) + 3$

16. Use transformations to sketch the graphs of each of the following functions, starting with the graph of $y = x$.

a) $y = (x - 5)$ 　　　　　　　　　 **b)** $y = (x + 6)$
c) $y = (x + 2) - 3$ 　　　　　　　 **d)** $y = (x - 3) + 5$

17. Use transformations to sketch the graphs of each of the following functions, starting with the graph of $y = \sqrt{x}$. State the domain and range of each function.

a) $y = \sqrt{x} - 4$

b) $y = \sqrt{x+5}$

c) $y = \sqrt{x-2} + 3$

d) $y = \sqrt{x+4} - 3$

18. Use transformations to sketch the graphs of each of the following functions, starting with the graph of $y = x^2$. State the domain and range of each function.

a) $y = (x-5)^2$

b) $y = (x+7)^2$

c) $y = (x+2)^2 - 6$

d) $y = (x-3)^2 + 4$

19. Houseboat rental The cost in dollars, $C(d)$, to rent a houseboat during July and August from a certain company is given by $C(d) = 120d + 100$, where d is the number of rental days.

a) What does the number 100 represent in this function?

b) State the domain and range of the function.

c) How could you transform the graph of the function $C(d) = 120d$ onto the graph of $C(d) = 120d + 100$?

d) During June and September, the company reduces the total cost of each houseboat rental by $40. Write the cost function for June and September.

3.4 Reflections of Functions

Refer to the Key Concepts on page 201.

20. The graphs of two functions are shown.

a)

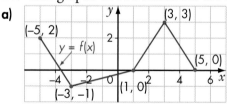

b)

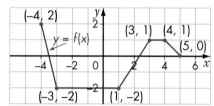

For each graph, draw and label the following pairs of graphs.

i) $y = f(x)$ and $y = -f(x)$

ii) $y = f(x)$ and $y = f(-x)$

21. Graph $f(x)$ and sketch the specified reflection image. State the domain and range of each function and its reflection image.

a) the reflection of $f(x) = \sqrt{x+2}$ in the y-axis

b) the reflection of $f(x) = -2x + 4$ in the x-axis

c) the reflection of $f(x) = x^2 + 6$ in the y-axis

22. a) Given $f(x) = 3 - 2x$, write equations for $-f(x)$ and $f(-x)$.
b) Sketch the three graphs on the same set of axes.
c) Determine any points that are invariant for each reflection.

23. a) Given $f(x) = x^2 + 6x$, write equations for $-f(x)$ and $f(-x)$.
b) Sketch the three graphs on the same set of axes.
c) Determine any points that are invariant for each reflection.

24. a) Given $f(x) = x^2 - 4$, write equations for $-f(x)$ and $f(-x)$.
b) Sketch the three graphs on the same set of axes.
c) Determine any points that are invariant for each reflection.

25. a) Given $f(x) = \sqrt{x} - 4$, write equations for $-f(x)$ and $f(-x)$.
b) Sketch the three graphs on the same set of axes.
c) Determine any points that are invariant for each reflection.

26. a) Given $f(x) = \sqrt{x-5}$, write equations for $-f(x)$ and $f(-x)$.
b) Sketch the three graphs on the same set of axes.
c) Determine any points that are invariant for each reflection.

27. A-frame roof The diagram shows a set of coordinate axes superimposed on the cross section of an A-frame roof. The height of the roof is h metres and the width is w metres. The equation of one half of the cross section is $y = -1.6x + 4$.
a) What is the equation of the other half of the cross section?
b) What is the height of the roof, h?
c) What is the width of the roof, w?
d) State the domain and range of the function that models each half of the roof.

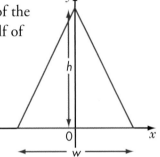

3.5 Inverse Functions

Refer to the Key Concepts on page 215.

28. $f = \{(-1, 5), (3, -2), (2, 2), (0, 3)\}$
a) Graph the function and its inverse.
b) Is the inverse a function? Explain.

29. Find the inverse of each function. State the domain and range of each function and its inverse.

a) $f(x) = x + 7$ **b)** $f(x) = \dfrac{x-4}{3}$ **c)** $f(x) = 3x - 1$

d) $f(x) = x^2 - 5$ **e)** $f(x) = (x+2)^2$ **f)** $f(x) = \sqrt{x-3}$

30. Sketch the graph of the inverse of each function.

a)
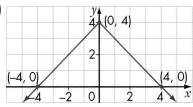

b)
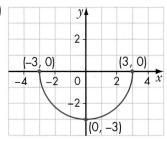

31. Determine if the functions in each pair are inverses of each other.

a) $y = 5x - 1$

$y = \dfrac{x + 1}{5}$

b) $y = 2x + 8$

$y = 8 - \dfrac{1}{2}x$

32. a) Find the inverse of $f(x) = 3 - x^2$.

b) Graph $f(x)$ and its inverse.

c) Restrict the domain of f so that f^{-1} is also a function.

d) With the domain of f restricted, sketch a graph of each function.

e) State the domain and range of each function.

33. Identify any invariant points for each function and its inverse.

a) $f(x) = 3x + 2$

b) $f(x) = 5x - 8$

34. Retail sales A sale at a furniture store advertised that all furniture was being sold at 40% off the original selling price.

a) Write a function that gives the sale price as a function of the original selling price.

b) Find the inverse of this function.

c) What does the inverse represent?

3.6 Stretches of Functions

Refer to the Key Concepts on page 228.

35. The graph of $y = f(x)$ is shown. Sketch the graphs of the following functions, and state the domain and range.

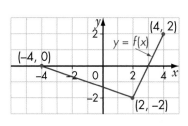

a) $y = 3f(x)$

b) $y = f\left(\dfrac{1}{2}x\right)$

c) $y = 2f\left(\dfrac{1}{2}x\right)$

36. For each of the following sets of three functions,

a) sketch them on the same grid or in the same viewing window of a graphing calculator

b) describe how the graphs of the second and third functions are related to the graph of the first function

c) identify any invariant points

i) $y = x$, $y = 4x$, $y = 0.5x$

ii) $y = x^2$, $y = 2x^2$, $y = 0.5x^2$

iii) $y = \sqrt{x + 3}$, $y = 2\sqrt{x + 3}$, $y = \frac{1}{4}\sqrt{x + 3}$

37. Use transformations and the zeros of the quadratic function $f(x) = (x - 6)(x + 4)$ to determine the zeros of each of the following functions.

a) $y = 2f(x)$ **b)** $y = f(2x)$ **c)** $y = f(0.5x)$

38. The graph of $f(x) = \sqrt{36 - x^2}$ is shown. Use transformations to sketch the graphs of each of the following, and state the domain and range.

a) $y = 2f(x)$

b) $y = 0.5f(x)$

c) $y = f(3x)$

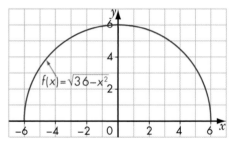

39. Describe how the graphs of the following functions can be obtained from the graph of $y = f(x)$.

a) $y = 4f(x)$ **b)** $y = f(3x)$

c) $y = \frac{1}{2}f(2x)$ **d)** $y = 3f\left(\frac{1}{2}x\right)$

3.7 Combinations of Transformations
Refer to the Key Concepts on page 239.

40. Use transformations to sketch the graphs of the following pairs of functions.

a) $y = x - 4$
$y = -(x - 4)$
c) $y = (x - 3)^2$
$y = -\frac{1}{4}(x - 3)^2$

b) $y = \sqrt{x + 2}$
$y = -3\sqrt{x + 2}$
d) $y = x^2 + 4$
$y = -2x^2 + 4$

41. Describe how the graph of each of the following functions can be obtained from the graph of $y = f(x)$.

a) $y = 3f(x + 2) - 4$

b) $y = -2f(x) + 5$

c) $y = 0.5f(4x) + 2$

d) $y = f(2(x + 1))$

e) $y = -f(0.5(x - 3)) + 1$

f) $y = 3f(4 - x) - 7$

42. Given $f(x) = x^2$, sketch the graph of each of the following, and state the domain and range.

a) $y = 2f(x + 3) - 4$

b) $y = -f(x - 5) + 6$

c) $y = 0.5f(0.5(x + 4)) - 3$

d) $y = -2f(2x - 4) + 3$

43. Given $f(x) = \sqrt{x}$, sketch the graph of each of the following, and state the domain and range.

a) $y = f(x - 3) + 4$

b) $y = -2f(x + 4) - 5$

c) $y = -3f(2 - x) - 2$

d) $y = f(-0.5x + 3) + 2$

44. The graph of $y = \sqrt{x}$ is expanded vertically by a factor of 4, reflected in the y-axis, translated 5 units to the left, and translated 4 units upward. Write the equation of the transformed function.

45. The graph of $f(x) = \sqrt{16 - x^2}$ is shown. Sketch the graph of each of the following functions.

a) $y = 3f(x) - 4$

b) $y = -0.5f(-0.5x) + 6$

c) $y = 2f(x - 3) + 4$

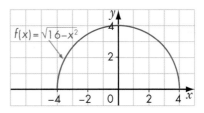

46. Salary raise The annual salary of each employee at an automobile plant was increased by a 6% cost-of-living raise and then a $2000 productivity raise.

a) Write a function that transforms the old annual salary, S, into the new annual salary, N.

b) Does it make any difference in which order the raises were given? Explain.

CHAPTER TEST

1. State whether each set of ordered pairs is a function.

a) {(2, 4), (3, 5), (7, 9), (2, −5), (3, −7)}

b) {(5, 4), (4, 3), (3, 2), (2, 1), (1, 0)}

c) {(−1, 6), (0, −6), (1, −6), (2, −6)}

2. State the domain and range of each relation.

a)

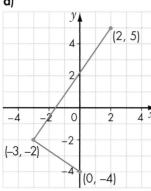

b)

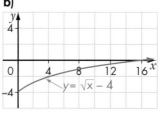

c)

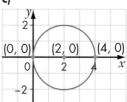

3. If $f(x) = 3 - 2x^2$, find

a) $f(5)$ **b)** $f(-2)$ **c)** $f(-5)$ **d)** $f(0)$

e) $f(7)$ **f)** $f(-0.5)$ **g)** $f(3a)$ **h)** $f(2a - 1)$

4. Describe the graph of each of the following functions.

a) $y = \sqrt{x}$ **b)** $y = \dfrac{1}{x}$

5. The graph of $y = f(x)$ is shown. Sketch the following functions.

a) $y = f(x) + 3$ **b)** $y = -0.5f(x)$

c) $y = f\left(-\dfrac{1}{2}x\right)$ **d)** $y = -2f(-x)$

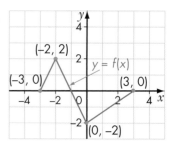

6. Describe how the graph of each of the following functions can be obtained from the graph of $y = f(x)$.

a) $y = f(x) + 4$ **b)** $y = f(x - 2) - 3$ **c)** $y = -f(x + 5) - 1$

d) $y = \frac{1}{3}f(-3x) + 5$ **e)** $y = -2f(2(x + 3)) + 6$ **f)** $y = 2f(-x + 2) - 4$

7. The blue graph is a transformation of the red graph. The equation of the red graph is given. Find the equation of the blue graph, and state its domain and range.

a)

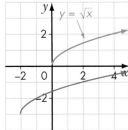

b)

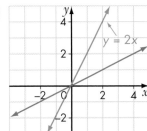

c)

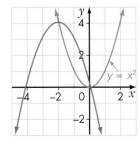

8. Find the inverse of each function. Is the inverse a function? Explain.

a) $y = 3x - 5$ **b)** $y = x^2 - 7$

9. Sketch the graphs of each of the following pairs of functions on the same set of axes.

a) $y = x + 3$ and $y = -2(x + 3)$
b) $y = \sqrt{x}$ and $y = \sqrt{x} - 4$
c) $y = \sqrt{x}$ and $y = -\sqrt{x + 4} + 3$
d) $y = (x + 1)^2$ and $y = -\frac{1}{2}(x + 1)^2 - 3$

10. The graph of $y = x^2$ is expanded vertically by a factor of 2, translated 3 units to the left, and translated 4 units upward. Write the equation of the transformed function, and state its domain and range.

11. The graph of $f(x) = \sqrt{x}$ is compressed horizontally by a factor of 0.5, reflected in the x-axis, and translated 4 units to the left. Write the equation of the transformed function, and state its domain and range.

12. a) Given $f(x) = x^2 - 10x$, write equations for $-f(x)$ and $f(-x)$.
b) Sketch the three graphs on the same set of axes.
c) Determine any points that are invariant for each reflection.
d) State the domain and range of each function.

13. a) Given $f(x) = \sqrt{x} - 3$, write equations for $-f(x)$ and $f(-x)$.
b) Sketch the three graphs on the same set of axes.
c) Determine any points that are invariant for each reflection.
d) State the domain and range of each function.

14. Manufacturing A manufacturing company produces garage doors. The number of garage doors, g, produced per week is related to the number of hours of labour, h, required per week to produce them by the function $g = 1.8\sqrt{h}$.
a) How many doors can be produced per week using 500 h of labour?
b) Write the inverse function and explain its meaning.
c) How many hours of labour are needed each week to keep production at or above 25 doors a week?

ACHIEVEMENT Check   Knowledge/Understanding   Thinking/Inquiry/Problem Solving   Communication   Application

15. Many companies pay their employees using a salary scale that depends on the number of years worked. One salary scale is modelled by the function $S(t) = 25\,000 + 3000t - 150t^2$ for the first ten years of employment and $S(t) = 500\sqrt{t - 10} + 40\,000$ for additional years, where $S(t)$ dollars is the annual salary and t is the number of years worked for the company. The employees' union negotiators are asking for an increase of $1000 for each employee. The company's negotiators are offering a 2.75% raise.
a) Describe how the graph of $S(t)$ is transformed by each type of increase.
b) Write the functions that model the new salary scale for each type of increase.
c) State a reasonable domain and range for each function in part b).
d) If you have worked for the company for 5 years, which model of the new salary scale would be better for you? Justify your answer.
e) If you have worked for the company for 19 years, which model of the new salary scale would be better for you? Justify your answer.

CHALLENGE PROBLEMS

1. Even or odd An even function satisfies $f(x) = f(-x)$ for all values of x in its domain. An odd function satisfies $f(-x) = -f(x)$ for all values of x in its domain. Classify each of the following as even, odd, or neither.

a) $f(x) = -2x + 1$ **b)** $f(x) = 3x^2 - 4$ **c)** $f(x) = \sqrt{x}$

d) $f(x) = \dfrac{1}{x}$ **e)** $f(x) = x^2 - x$

2. Reflection image The straight line $2x - 3y + 6 = 0$ is reflected in the line $y = -x$. Find the equation of the reflection image.

3. Power function A function of the form $f(x) = ax^n$, where $a \neq 0$ and n is a positive integer, is called a power function. How is the exponent in the equation of a power function related to the symmetry of its graph?

4. Inverse function If $f(x) = \dfrac{ax + b}{cx + d}$, $x \neq -\dfrac{d}{c}, \dfrac{a}{c}$, find an equation for $f^{-1}(x)$.

5. Regular polygon A regular polygon with x sides is completely enclosed by x regular polygons with n sides each. When $x = 4$, $n = 8$, as shown in the diagram. If $x = 10$, what is the value of n?

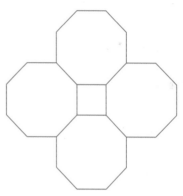

6. Reflection Let G be the graph of $y = \dfrac{1}{x}$. H is the reflection of G in the line $y = 2x$. The equation of H can be written as $12x^2 + bxy + cy^2 + d = 0$. Find the values of b, c, and d.

SOLVE RICH ESTIMATION PROBLEMS

How many words does your daily newspaper print in a year? Estimation problems like this one are known as Fermi problems. They get their name from the physicist Enrico Fermi, who liked to pose them to his students at the University of Chicago. Solving a Fermi problem may require several estimates. Specify any assumptions you make to arrive at the estimates.

About how many soccer balls would be needed to fill the SkyDome with the roof closed?

Understand the Problem

1. What information are you given?
2. What are you asked to find?
3. Do you need an exact or an approximate answer?

Think of a Plan

The problem is an example of a volume problem.

The task is to estimate how many small objects are needed to fill a large object. The number of small objects, n, can be found using

$$n = (\text{large volume}) \div (\text{small volume})$$

Some information is missing from the problem. You can use your research skills to find that the volume inside the SkyDome with the roof closed is about 1.6 million cubic metres. A soccer ball has a diameter of about 22 cm.

Assume that the soccer ball approximates a cube, with each edge 22 cm or 0.22 m. The volume of the cube is 0.22^3, or about 0.01 m^3.

So, $n = \dfrac{1\ 600\ 000}{0.01}$

$= 160\ 000\ 000$

About 160 000 000 soccer balls would be needed to fill the SkyDome with the roof closed.

Look Back

Does the answer seem reasonable?

Is there a way to improve the estimate?

Solve Fermi Problems

1. Locate the information you need.
2. Decide what assumption(s) to make.
3. Estimate the solution to the problem.
4. Check that your estimate is reasonable.

Many Fermi problems require you to use your research skills to locate missing information. You may use the Internet, or you may obtain the information in another way, such as looking it up in a reference book, measuring it, or asking an expert. The following Fermi problem is missing some information.

How many litres of gasoline does a family car use in a year?

The fuel efficiency of a car depends on several things, including the size of the car, where most of the driving is done—city or highway—and whether the transmission is manual or automatic.

Assume that the car is mid-sized, with an automatic transmission, that it is used mainly for city driving, and that it is driven about 25 000 km in a year. Research indicates that the car's fuel efficiency is about 12 L/100 km.

The car will use about $25\ 000 \times \dfrac{12}{100}$, or 3000 L of gasoline in a year.

Apply, Solve, Communicate

Use your research skills to locate any missing information. Then, solve each problem.

1. Dimes About how many dimes would it take to cover the floors of the hallways in your school?

2. Loonies Estimate the volume of one million loonies.

3. Drinking fountain Estimate the number of litres of water that would be wasted if a drinking fountain ran continuously for a school year.

4. Basketballs About how many basketballs would it take to fill your school?

5. Buildings About how many buildings of all types are there in Ontario?

6. Internet About how many families in Ontario are connected to the Internet?

7. Milk Estimate the total number of litres of milk Canadian students in grades 1 to 12 drink in a year.

8. Newspapers Estimate the number of words your daily newspaper prints in a year.

9. Sidewalks About how many kilometres of sidewalk are there in Ontario?

10. Swimming pool About how long would it take to fill a community swimming pool with a garden hose?

11. Variety stores Estimate the number of variety stores in Canada.

12. Sun and Earth If the sun were hollow, about how many spheres the size of the Earth would fill it?

13. Movie theatres About how many movie theatres are there in Canada?

14. Tennis balls Estimate the number of tennis balls it would take to replace all the water in Lake Superior.

15. Television About how many televisions are there in Canada?

16. Flyers Estimate the number of advertising flyers delivered to homes in Ontario in a year.

17. Students If all the grade 11 students in Canada stood shoulder to shoulder, about how far would the line stretch?

18. Driving About how many high school students in Ontario have a driver's licence?

19. Formulating problems Write a problem similar to question 1. Have a classmate solve your problem.

PROBLEM SOLVING

USING THE STRATEGIES

1. Basketball Five basketball players, A, B, C, D, and E, run onto the court in order from shortest to tallest. Each player after the first has a mass of 1 kg more than the player in front. The total of their masses is 310 kg.

E has a mass of 1 kg more than A.
B has a mass of 2 kg more than A.
C has a mass of 1 kg more than D.
E has a mass of 3 kg less than C.

What is the mass of each player?

2. Geometry Given that AC = BC, CD = CE, and ∠ACD = 50°, find the measure of ∠BDE.

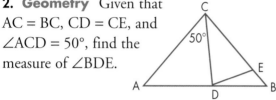

3. Subtractions The two subtractions have the same answer.

$$\begin{array}{r} 3\ 0\ 0 \\ -\ X\ Y\ 3 \\ \hline \end{array} \qquad \begin{array}{r} 3\ X\ Y \\ -\ 3\ 0\ 0 \\ \hline \end{array}$$

Find the values of X and Y.

4. Stamps You have an unlimited number of 3¢ stamps and 5¢ stamps. What amounts of postage can you not make?

5. Boxes Some books are stored in 5 boxes, labelled A, B, C, D, and E. The total mass of A and B is 24 kg, of B and C is 27 kg, of C and D is 23 kg, of D and E is 16 kg, and of A, C, and E is 32 kg What is the mass of each box?

6. Atlantic crossings Every day at noon, one ship leaves Halifax for Lisbon. At exactly the same time every day, one ship leaves Lisbon for Halifax. Each crossing takes exactly 6 days. The ships send each other a radio message when they meet. In one crossing, how many radio messages will a ship from Halifax send to ships from Lisbon?

7. Cycling Damon rode his bicycle to school from his home at 9 km/h and arrived 10 min late. The next day, he left home at the same time, but he rode at 12 km/h and arrived 10 min early. What is the distance from Damon's home to the school?

8. Magic square
Copy the diagram.
Complete it by placing four odd numbers in the empty squares so that the sum of each row, column, and diagonal is 48.

	22	
4	16	28
	10	

9. Standard form What is the last digit when the following product is written in standard form?

$$(3^1)(3^2)(3^3) \ldots (3^{298})(3^{299})(3^{300})$$

10. Marching band The leader of a marching band has decided that the band must be able to form a rectangle. The leader has also decided that the number of band members marching on the outside edges of the rectangle must be the same as the number of band members marching in the interior.

a) What are the possible numbers of band members?

b) Show algebraically that you have found the only solutions.

4 Trigonometry

Specific Expectations	Functions	Functions & Relations
Determine the sine, cosine, and tangent of angles greater than 90°, using a suitable technique, and determine two angles that correspond to a given single trigonometric function value.	4.2, 4.4	4.2, 4.4
Solve problems in two dimensions and three dimensions involving right triangles and oblique triangles, using the primary trigonometric ratios, the cosine law, and the sine law (including the ambiguous case).	4.1, 4.3, 4.4	4.1, 4.3, 4.4

Ship Navigation

In the Modelling Math questions on pages 274, 293, 310, and 311, you will solve the following problem and other problems that involve ship navigation.

From 1857 to 1860, Great Britain financed the construction of ten lighthouses in British North America. They were built because obsolete navigational aids were hindering economic growth. The lighthouses are called the *Imperial Lights*. Four of them were built along the approaches to the Saint Lawrence, and six were built on the eastern shore of Lake Huron. The Point Clark lighthouse, on Lake Huron, is 28.3 m tall. From the top of the lighthouse, the angle of depression of a ship is 3.3°. How far is the ship from the lighthouse, to the nearest metre?

Use your research skills to investigate the following now.

1. Describe the connection between the history of ship navigation and the development of accurate ways of measuring the time.

2. Describe how the following are used in modern methods of ship navigation.
a) radio   **b)** satellites

Parallactic Displacement

Hold a pencil at arm's length and look at it with both eyes open. Now close one eye, open it, and then close the other eye. You will notice that the position of the pencil, relative to the background, appears to change. This effect is known as parallactic displacement, or parallax. Since parallactic displacement can be measured in degrees, it can be used to determine the distance to an object using trigonometry.

Web Connection
www.school.mcgrawhill.ca/resources/
To learn more about parallax, visit the above web site. Go to **Math Resources**, then to *MATHEMATICS 11*, to find out where to go next. Write a brief report on your findings.

The first use of parallactic displacement to determine the distance from the Earth to a star is usually credited to the German astronomer Friedrich Wilhelm Bessell (1784–1846). He studied 61 Cygni, a star in the constellation Cygnus, the Swan. He compared the positions of the star, relative to its background, on different nights half a year apart. He announced his results in 1838.

In the diagram, S represents the sun, C represents the star 61 Cygni, and E_1 and E_2 represent the Earth at its opposite positions around the sun.

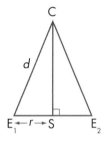

Bessell measured the angular shift in the star's position due to the Earth's motion around the sun, $\angle E_1CE_2$, as 1.7×10^{-4} degrees.

1. Annual parallax is defined as half the angular shift, that is, as $\angle SCE_1$ or $\angle SCE_2$ in the diagram. Determine the degree measure of the annual parallax for 61 Cygni in standard scientific notation.

2. If r, the average radius of the Earth's orbit around the sun, is 1.5×10^8 km, calculate the distance d, in kilometres. Express your answer in standard scientific notation, with the decimal part rounded to the nearest tenth.

3. Stellar distances are usually expressed in light-years, where one light-year is the distance light travels through space in one year. The speed of light is 3×10^5 km/s. Express d to the nearest tenth of a light-year.

4. The accepted value for the distance from the Earth to 61 Cygni is now 11.1 light-years. Using this value, work backward and calculate a more accurate value for the angular shift.

Review of Prerequisite Skills

If you need help with any of the skills named in purple below, refer to Appendix A.

1. Trigonometric ratios For each right triangle, use the Pythagorean theorem to calculate the length of the third side. Then, find the sine, cosine, and tangent of each acute angle. Express answers as fractions in lowest terms.

a)
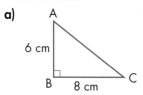

b)
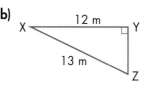

c)
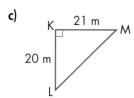

d)
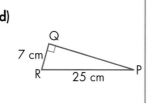

2. Find each of the following, to three decimal places.

a) sin 27°   **b)** cos 56°   **c)** tan 78°
d) cos 7°   **e)** tan 40°   **f)** sin 62°

3. Find the size of each angle, to the nearest degree.

a) sin D = 0.602   **b)** cos A = 0.309
c) tan C = 0.445   **d)** tan R = 2.246
e) sin X = 0.978   **f)** cos W = 0.951

4. Finding a side length in a right triangle Calculate x, to the nearest tenth of a unit.

a)
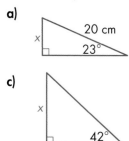

b)

c)

d)
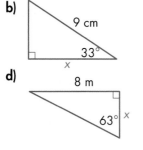

e)
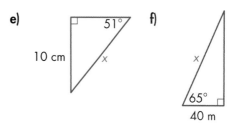

f)

5. Finding an angle in a right triangle Find ∠x, to the nearest degree.

a)   **b)**

c)   **d)**

e)   **f)**
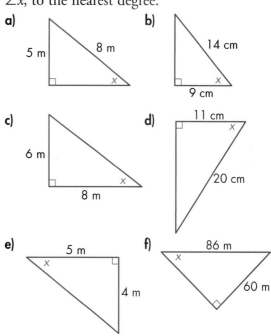

6. Solving proportions Solve for x. Express answers as decimals. Round to the nearest hundredth, if necessary.

a) $\dfrac{x}{3.15} = 11.8$ **b)** $\dfrac{x}{9.73} = \dfrac{7.65}{0.46}$

c) $\dfrac{91.24}{83.56} = \dfrac{x}{71.77}$ **d)** $\dfrac{12.56}{x} = \dfrac{19.83}{27.77}$

e) $\dfrac{0.57}{0.81} = \dfrac{1.52}{x}$ **f)** $\dfrac{1.38}{x} = \dfrac{5.72}{4.11}$

4.1 Reviewing the Trigonometry of Right Triangles

In the short story *The Musgrave Ritual*, Sherlock Holmes found the solution to a mystery at a certain point. To find the point, he had to start near the stump of an elm tree and take 20 paces north, then 10 paces east, then 4 paces south, and finally 2 paces west.

1. Let E be the point where Holmes started pacing and S be the point where he stopped. Draw a diagram of his path. Join ES. Draw another line segment so that ES is the hypotenuse of a right triangle.

2. What are the lengths of the perpendicular sides of the right triangle?

3. What trigonometric ratio can you use to find ∠E from the lengths of the perpendicular sides?

4. Find ∠E, to the nearest degree.

5. What methods could you use to find the length of ES, in paces?

6. What is the length of ES, to the nearest pace?

7. In what direction, and for how many paces, could Holmes have walked in order to go directly from E to S?

8. After Holmes arrived at S, he looked back at the location of point E on the ground. What angle did his line of sight make with the ground, to the nearest degree? Assume that the ground was level and that the height of his eyes above the ground was approximately equal to the length of 2 paces.

The primary trigonometric ratios are

$$\text{sine } \theta = \frac{\text{opposite}}{\text{hypotenuse}}$$

$$\text{cosine } \theta = \frac{\text{adjacent}}{\text{hypotenuse}}$$

$$\text{tangent } \theta = \frac{\text{opposite}}{\text{adjacent}}$$

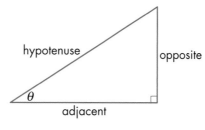

To solve a right triangle means to find all the unknown sides and the unknown angles.

EXAMPLE 1 Solving a Right Triangle, Given a Side and an Angle

In $\triangle ABC$, $\angle B = 90°$, $\angle A = 18.6°$, and $b = 11.3$ cm. Solve the triangle by finding
a) the unknown angle
b) the unknown sides, to the nearest tenth of a centimetre

SOLUTION

a) Using the given information,
$$\angle C = 90° - 18.6°$$
$$= 71.4°$$

b) From the diagram,

$$\frac{a}{11.3} = \sin 18.6°$$
$$a = 11.3 \times \sin 18.6°$$
$$\doteq 3.6$$

$$\frac{c}{11.3} = \cos 18.6°$$
$$c = 11.3 \times \cos 18.6°$$
$$\doteq 10.7$$

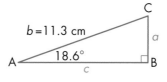

Using the mode settings, ensure that the calculator is in degree mode.

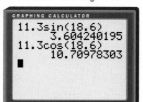

In $\triangle ABC$, $\angle C = 71.4°$, $a = 3.6$ cm, and $c = 10.7$ cm.

EXAMPLE 2 Solving a Right Triangle, Given Two Sides

In △DEF, $\angle E = 90°$, $d = 7.4$ m, and $f = 6.5$ m. Solve the triangle by finding

a) the unknown angles, to the nearest tenth of a degree
b) the unknown side, to the nearest tenth of a metre

SOLUTION

a) From the diagram,

$$\tan D = \frac{7.4}{6.5}$$

$$\angle D \doteq 48.7°$$

$$\angle F = 90° - 48.7°$$

$$= 41.3°$$

b) From the diagram,

$$\sin 48.7° = \frac{7.4}{e}$$

$$e \times \sin 48.7° = 7.4$$

$$e = \frac{7.4}{\sin 48.7°}$$

$$\doteq 9.9$$

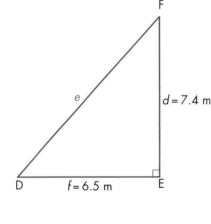

In △DEF, $\angle D = 48.7°$, $\angle F = 41.3°$, and $e = 9.9$ m.

If you are standing on a cliff beside a river, and you look down at a boat, the angle that your line of sight makes with the horizontal is called the **angle of depression**. If you look up at a helicopter, the angle that your line of sight makes with the horizontal is called the **angle of elevation**.

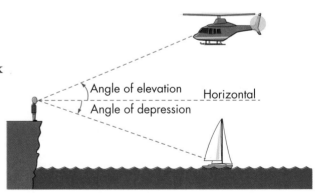

EXAMPLE 3 Western Red Cedars

Cathedral Grove, on Vancouver Island, is a rain forest of firs and western red cedars. From a point 40 m from the foot of one cedar, the angle of elevation of the top is 65°. Find the height of the cedar, to the nearest metre.

SOLUTION

Draw and label a diagram.
Let h represent the height of the cedar.

$$\frac{h}{40} = \tan 65°$$
$$h = 40\tan 65°$$
$$\doteq 86$$

The cedar is 86 m tall, to the nearest metre.

On the Earth, a parallel of latitude is a circle parallel to the equator.

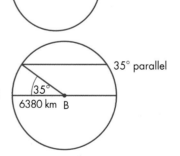

EXAMPLE 4 Parallel of Latitude

Find the length of the 35° parallel of latitude, to the nearest 10 km. Assume that the radius of the Earth is 6380 km.

SOLUTION

In the diagram, B is the centre of the Earth, and A is a point on the equator. D is the centre of the circle defined by the 35° parallel, and E is a point on its circumference. DE is the radius, r, of the 35° parallel.

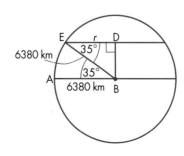

∠BDE is a right angle.
BA = BE, because both are radii of the Earth.
∠DEB = ∠ABE (alternate angles)

In △DEB,

$$\frac{r}{6380} = \cos 35°$$

$$r = 6380\cos 35°$$

$$\doteq 5226$$

The length of the 35° parallel of latitude is its circumference, C.

$$C = 2\pi r$$

$$\doteq 2\pi(5226)$$

$$\doteq 32\ 840$$

To copy the previous answer to the cursor location, enter Ans by pressing 2nd (−).

Estimate

$$2 \times 3 \times 5000 = 30\ 000$$

The length of the 35° parallel of latitude is 32 840 km, to the nearest 10 km.

EXAMPLE 5 Rock Pillars

Rock pillars are interesting geological features found in several national parks in Ontario. Rock pillars, found in rivers and lakes, have been sculpted by wind and water. A geologist wanted to determine the height of a rock pillar in a river. The geologist set up a theodolite at C and measured ∠ACB to be 28.5°. A baseline CD was marked off, perpendicular to BC. The length of CD is 10 m, and ∠CDB = 56.4°. If the height of the theodolite is 1.6 m, what is the height of the rock pillar, to the nearest tenth of a metre?

SOLUTION

Calculate the length of AB, and then add the height of the theodolite to determine the height of the rock pillar.

In △BCD,

$$\frac{BC}{CD} = \tan \angle BCD$$

$$\frac{BC}{10} = \tan 56.4°$$

$$BC = 10 \tan 56.4°$$

$$BC \doteq 15.1$$

In △ABC,

$$\frac{AB}{BC} = \tan \angle ACB$$

$$\frac{AB}{15.1} = \tan 28.5°$$

$$AB = 15.1 \tan 28.5°$$

$$AB \doteq 8.2$$

$$8.2 + 1.6 = 9.8$$

So, the height of the rock pillar is 9.8 m, to the nearest tenth of a metre.

Key Concepts

- For any acute angle θ in a right triangle,

$$\sin \theta = \frac{\text{opposite}}{\text{hypotenuse}} \qquad \cos \theta = \frac{\text{adjacent}}{\text{hypotenuse}}$$

$$\tan \theta = \frac{\text{opposite}}{\text{adjacent}}$$

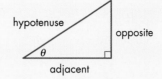

- To use trigonometry to solve a right triangle, given the measure of one acute angle and the length of one side, find
  a) the measure of the third angle using the angle sum in the triangle
  b) the measure of the other two sides using the sine, cosine, or tangent ratios

- To use trigonometry to solve a right triangle, given the lengths of two sides, find
  a) the measure of one angle using its sine, cosine, or tangent ratio
  b) the measure of the third angle using the angle sum in the triangle
  c) the measure of the third side using a sine, cosine, or tangent ratio

- To use trigonometry to solve a problem involving two right triangles,
  a) use a diagram showing the given information and the unknown side length(s) or angle measure(s)
  b) identify the two triangles that can be used to solve the problem, and plan how to use each triangle
  c) carry out the plan

Communicate Your Understanding

1. Describe how you would solve △ABC, given ∠B = 90°, ∠A = 36°, and $c = 12$ cm.

2. Describe how you would solve △RST, given ∠S = 90°, $s = 22$ cm, and $t = 15$ cm.

3. Describe the difference between an angle of elevation and an angle of depression.

4. Describe how you would find the measure of ∠A.

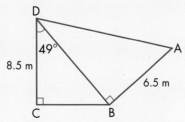

Practise

A

1. Solve each triangle. Round each side length to the nearest unit and each angle to the nearest degree.

a)

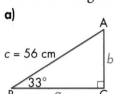

$c = 56$ cm, $33°$ at B, b, a, right angle at C, A

b)

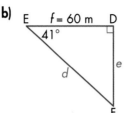

E, $f = 60$ m, D, $41°$, d, e, F

c)

U, $s = 15$ m, $t = 10$ m, S, u, T

d)

Q, $r = 13$ cm, P, $p = 8$ cm, q, R

2. Solve each triangle. Round each side length to the nearest tenth of a unit and each angle to the nearest tenth of a degree.

a)

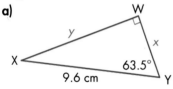

W, y, x, X, $63.5°$, 9.6 cm, Y

b)

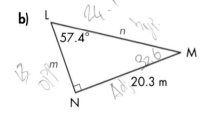

L, $57.4°$, n, M, 20.3 m, N

c)

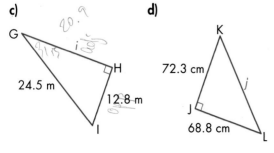

G, i, H, 24.5 m, 12.8 m, I

d)

K, 72.3 cm, j, J, 68.8 cm, L

3. Solve each triangle. Round answers to the nearest tenth, if necessary.

a) In $\triangle XYZ$, $\angle X = 90°$, $x = 9.5$ cm, $z = 4.2$ cm

b) In $\triangle KLM$, $\angle M = 90°$, $\angle K = 37°$, $m = 12.3$ cm

c) In $\triangle ABC$, $\angle A = 90°$, $\angle B = 55.1°$, $b = 4.8$ m

d) In $\triangle DEF$, $\angle E = 90°$, $d = 18.2$ cm, $f = 14.9$ cm

4. Find the measure of $\angle\theta$, to the nearest tenth of a degree.

a)

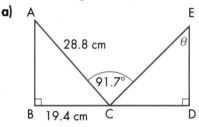

A, 28.8 cm, E, θ, $91.7°$, B, 19.4 cm, C, D

b)

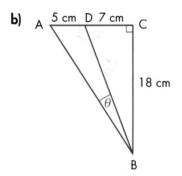

A, 5 cm, D, 7 cm, C, 18 cm, θ, B

c)

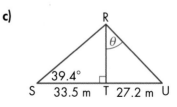

R, θ, $39.4°$, S, 33.5 m, T, 27.2 m, U

d)

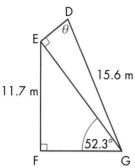

5. Find AB, to the nearest tenth of a metre.

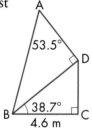

6. Find RS, to the nearest tenth of a centimetre.

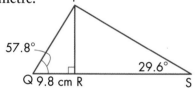

7. Find FH, to the nearest tenth of a metre.

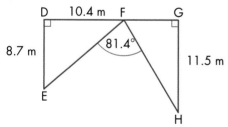

8. Find RT, to the nearest centimetre.

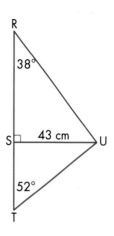

9. Find MN, to the nearest tenth of a centimetre.

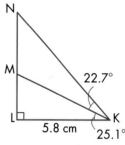

10. Find AB, to the nearest metre.

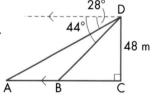

Apply, Solve, Communicate

11. Mica Dam The highest dam in Canada is the Mica Dam, one of three dams on the Columbia River in British Columbia. From a point 600 m from the foot of the dam, the angle of elevation of the top of the dam is 22°. What is the height of the dam, to the nearest metre?

12. Arctic Circle Find the length of the Arctic Circle, which is 66.55° north, to the nearest 10 km. Assume that the radius of the Earth is 6380 km.

13. Surveying A surveyor measured the height of a vertical rock face by determining the measurements shown. If the surveyor's theodolite had a height of 1.7 m, find the height of the rock face, AB, to the nearest tenth of a metre.

14. Communication a) Find the area of △DEF, to the nearest tenth of a square metre.
b) What other minimum sets of conditions (sides, angles) would allow you to calculate the area of △DEF?

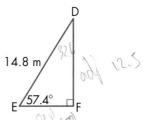

15. Latitude a) Find the length of the 20° parallel of latitude, to the nearest 10 km. Assume that the radius of the Earth is 6380 km.
b) Find the length of the parallel of latitude where you live, to the nearest 10 km.

16. Application From 1857 to 1860, Great Britain financed the construction of ten lighthouses in British North America. They were built because obsolete navigational aids were hindering economic growth. The lighthouses are called the *Imperial Lights.* Four of them were built along the approaches to the Saint Lawrence, and six were built on the eastern shore of Lake Huron. The Point Clark lighthouse, on Lake Huron, is 28.3 m tall. From the top of the lighthouse, the angle of depression of a ship is 3.3°. How far is the ship from the lighthouse, to the nearest metre?

17. Inquiry/Problem Solving Show that the length of any parallel of latitude is equal to the length of the equator times the cosine of the latitude angle.

18. Measurement Find the volume of the triangular prism, to the nearest cubic centimetre.

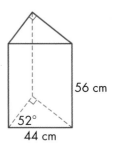

C

19. Great Pyramid The Great Pyramid of Khufu has a
square base with a side length of about 230 m. The four
triangular faces of the pyramid are congruent and
isosceles. The altitude of each triangular face makes an
angle of 52° with the base. Find the measure of each base
angle of the triangular faces, to the nearest degree.

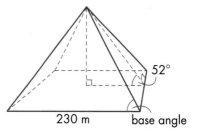

230 m   base angle

20. Geometry a) △ABC is an acute triangle. Show that the area, A,
of △ABC can be found using the formula $A = 0.5ac\sin B$.
b) Show that the area, A, of △ABC can also be found using the
formula $A = 0.5ab\sin C$.
c) Find the other formula, similar to those in parts a) and b),
for the area of △ABC.
d) Describe the pattern in the formulas.
e) What given information is sufficient to find the area of an acute triangle?
Explain.
f) Find the area of the triangle to the right, to the nearest tenth of a
square metre.

g) Find the area of the triangle to the right, to the nearest tenth of
a square centimetre.
h) Do the formulas from parts a), b), and c) apply to obtuse
triangles? Explain and justify your reasoning using diagrams.

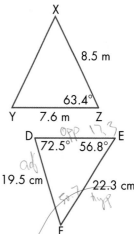

On a wall, a spider is 100 cm above a fly. The fly starts moving horizontally
at the speed of 10 cm/s. After 1 s, the spider begins moving at twice the
speed of the fly, in such a way as to intercept the fly by taking a straight line
path. In what direction does the spider move, and how far has the fly moved
when they meet?

4.2 The Sine and the Cosine of Angles Greater Than 90°

The trigonometric ratios have been defined in terms of sides and acute angles of right triangles. Trigonometric ratios can also be defined for angles in **standard position** on a coordinate grid.

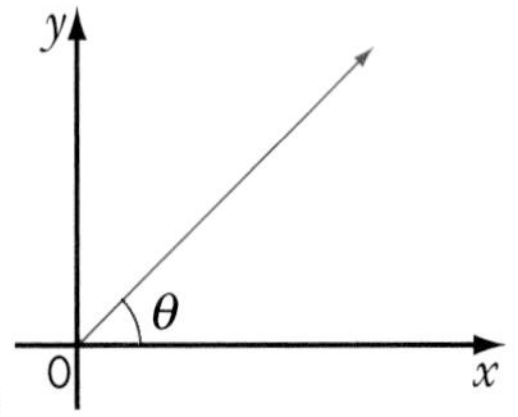

In the diagram, angle θ is in standard position.

The vertex of angle θ is at the origin. One ray, called the **initial arm**, is fixed on the positive x-axis. The other ray, called the **terminal arm**, rotates about the origin. The measure of the angle is the amount of rotation from the initial arm to the terminal arm.

The clapper board used in TV or movie production makes angles that look like angles in standard position. The lower part of the board is used to record such information as the name of the movie or TV program, the director, and the scene and take numbers. At the start of each take, the hinged arm is clapped onto the lower part of the board. In the editing stage, the loud sound and the visual image of the arm hitting the board are used to synchronize the moving pictures with the sound track.

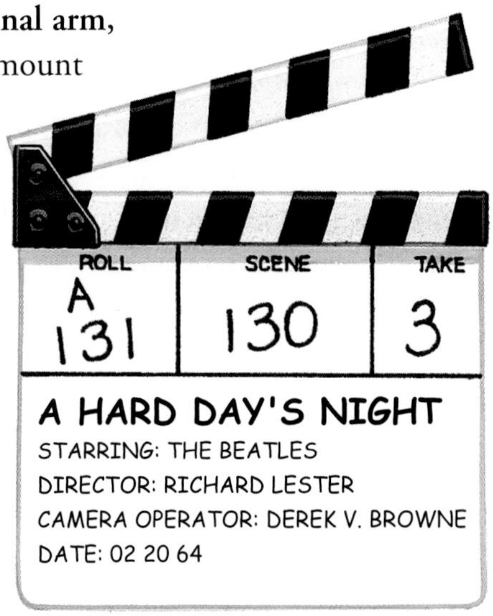

ROLL	SCENE	TAKE
A 131	130	3

A HARD DAY'S NIGHT
STARRING: THE BEATLES
DIRECTOR: RICHARD LESTER
CAMERA OPERATOR: DEREK V. BROWNE
DATE: 02 20 64

INVESTIGATE & INQUIRE

1. Copy the table. Complete it using a calculator. Round each calculated value to four decimal places.

2. How does the value of sin A change as the measure of $\angle A$ goes from 0° to 90°? from 90° to 180°?

3. a) What pairs of angles have equal sine values?
b) How are the angles in each pair related?

Angle A	sin A	cos A
0°		
15°		
30°		
45°		
60°		
75°		
90°		
105°		
120°		
135°		
150°		
165°		
180°		

4. If sin A = sin 40°, and ∠A is between 0° and 180°, what are the possible measures of ∠A?

5. If sin B = sin 130°, and ∠B is between 0° and 180°, what are the possible measures of ∠B?

6. How does the value of cos A change as the measure of ∠A goes from 0° to 90°? from 90° to 180°?

7. a) How are cos 30° and cos 150° the same? How are they different?
b) How are cos 165° and cos 15° the same? How are they different?

8. The measure of the angle made by the clapper board in the photo is 15°.
a) Is there an obtuse angle whose sine equals the sine of the 15° angle? Explain.
b) Is there an obtuse angle whose cosine equals the cosine of the 15° angle? Explain.

Let (x, y) be a point on the terminal arm of an angle θ in standard position.

The side opposite θ is y and the side adjacent to θ is x. The hypotenuse, r, can be found using the Pythagorean Theorem.

When the amount of rotation of an angle in standard position exceeds 90°, the angle is drawn on a coordinate grid as shown.

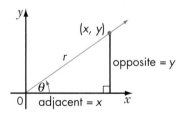 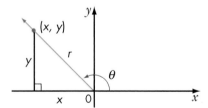

$$r = \sqrt{x^2 + y^2}$$

The trigonometric ratios are expressed in terms of x, y, and r.

$$\sin \theta = \frac{y}{r} \qquad \cos \theta = \frac{x}{r}$$

EXAMPLE 1 The Sine and the Cosine of Angles Less Than 90°

The point (5, 12) is on the terminal arm of an angle θ in standard position.
Find $\sin \theta$ and $\cos \theta$.

SOLUTION

Sketch the angle in standard position. Make a triangle by
drawing a perpendicular from (5, 12) to the x-axis.

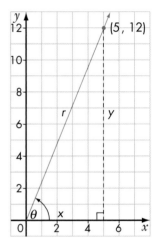

$$r^2 = x^2 + y^2$$
$$r = \sqrt{x^2 + y^2}$$
$$= \sqrt{5^2 + 12^2}$$
$$= \sqrt{169}$$
$$= 13$$

$$\sin \theta = \frac{y}{r} \qquad \cos \theta = \frac{x}{r}$$
$$= \frac{12}{13} \qquad = \frac{5}{13}$$

EXAMPLE 2 The Sine and the Cosine of Angles Greater Than 90°

The point (−4, 3) is on the terminal arm of an angle θ in standard position.
Find $\sin \theta$ and $\cos \theta$.

SOLUTION

Sketch the angle in standard position. Make a triangle by
drawing a perpendicular from the point (−4, 3) to the x-axis.

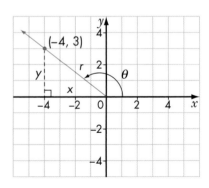

$$r^2 = x^2 + y^2$$
$$r = \sqrt{x^2 + y^2}$$
$$= \sqrt{(-4)^2 + 3^2}$$
$$= \sqrt{16 + 9}$$
$$= \sqrt{25}$$
$$= 5$$

$$\sin \theta = \frac{y}{r} \qquad \cos \theta = \frac{x}{r}$$
$$= \frac{3}{5} \qquad = \frac{-4}{5} \text{ or } -\frac{4}{5}$$

P(a, b) is on the terminal arm of angle θ in standard position.

Then, P'($-a$, b) is on the terminal arm of angle $(180° - \theta)$, following a reflection in the y-axis.

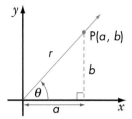 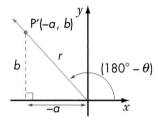

In both cases, $r = \sqrt{a^2 + b^2}$

$$\sin \theta = \frac{b}{r} \qquad\qquad \sin (180° - \theta) = \frac{b}{r}$$

$$= \frac{b}{\sqrt{a^2 + b^2}} \qquad\qquad = \frac{b}{\sqrt{a^2 + b^2}}$$

So, $\sin \theta = \sin (180° - \theta)$
or $\sin (180° - \theta) = \sin \theta$

$$\cos \theta = \frac{a}{r} \qquad\qquad \cos (180° - \theta) = \frac{-a}{r}$$

$$= \frac{a}{\sqrt{a^2 + b^2}} \qquad\qquad = \frac{-a}{\sqrt{a^2 + b^2}}$$

So, $\cos \theta = -\cos (180° - \theta)$
or $\cos (180° - \theta) = -\cos \theta$

For example,
$\sin 123° = \sin (180° - 57°)$
$\qquad = \sin 57°$
$\qquad \doteq 0.8387$
and $\cos 123° = \cos (180° - 57°)$
$\qquad\qquad = -\cos 57°$
$\qquad\qquad \doteq -0.5446$

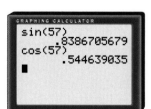

EXAMPLE 3 Evaluating Sines and Cosines Using a Calculator

Evaluate, to four decimal places.

a) sin 133°

b) cos 119.7°

SOLUTION

a)

sin 133° ≐ 0.7314

b)

cos 119.7° ≐ −0.4955

EXAMPLE 4 Finding Angle Measures Using a Calculator

Find ∠A, to the nearest tenth of a degree, if $0° \leq A \leq 180°$.

a) sin A = 0.3214

b) cos A = −0.5804

SOLUTION

a)

The sine of an angle is positive when the terminal arm is in the first or second quadrants.

So, ∠A = 18.7° or ∠A = 180° − 18.7°

= 161.3°

b)

GRAPHING CALCULATOR
cos⁻¹(-.5804)
        125.4786814
■

The cosine of an angle is negative when the terminal arm is in the second quadrant.

So, ∠A = 125.5°

Key Concepts

- $\sin(180° - \theta) = \sin \theta$
- $\cos(180° - \theta) = -\cos \theta$

Communicate Your Understanding

1. The point $(-12, 5)$ is on the terminal arm of an angle θ in standard position, $0° \leq \theta \leq 180°$. Describe how you would find $\sin \theta$ and $\cos \theta$.

2. Describe how $\sin \theta$ changes as the value of θ increases from $0°$ to $180°$.

3. Describe how $\cos \theta$ changes as the value of θ increases from $0°$ to $180°$.

Practise

A

1. If the given point is on the terminal arm of an angle A in standard position, $0° \leq A \leq 180°$, find sin A and cos A.

a) $(6, 8)$ **b)** $(-5, 12)$

2. Evaluate, to four decimal places.

a) $\cos 144°$ **b)** $\sin 105°$ **c)** $\sin 167°$
d) $\cos 92°$ **e)** $\cos 134.7°$ **f)** $\sin 121.3°$
g) $\sin 178.8°$ **h)** $\cos 113.1°$ **i)** $\sin 156.4°$

3. Find $\angle A$, if $0° \leq A \leq 180°$. Round to the nearest tenth of a degree, if necessary.

a) $\sin A = 0.5$ **b)** $\cos A = 0.5$
c) $\cos A = -0.5$ **d)** $\sin A = 0.2568$
e) $\cos A = 0.4561$ **f)** $\cos A = -0.5603$
g) $\sin A = 0.5736$ **h)** $\cos A = -0.0876$
i) $\sin A = 1$ **j)** $\cos A = -1$

Apply, Solve, Communicate

4. If the point $(0, 2)$ is on the terminal arm of $\angle A$ in standard position, $0° \leq A \leq 180°$, what is the measure of $\angle A$?

B

5. If two angles in standard position are supplementary, and have terminal arms that are perpendicular, what are the measures of the angles?

6. Algebra The point $P(k, 24)$ is 25 units from the origin. If P is on the terminal arm of an angle in standard position, find, for each value of k,
a) the sine of the angle   **b)** the cosine of the angle

7. Communication If you assume that the pages of this book are flat, the bottom edges of the two pages facing you form a 180° angle. Describe the changes in the sine of the angle and the cosine of the angle as you close the book.

8. Application The amplitude, *A*, of a pendulum is the maximum horizontal distance of the bob from its central position. The displacement, *x*, is the actual horizontal distance of the bob from its central position at a certain time. The displacement is related to the amplitude by the equation $x = A \sin \frac{360t}{T}$. In this equation, *t* is the time since the bob passed through the central position, and *T* is the time the pendulum takes for a complete back-and-forth swing.

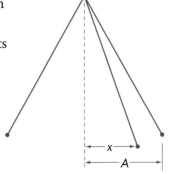

a) A pendulum has an amplitude of 10 cm and takes 2 s to complete one back-and-forth swing. To the nearest centimetre, how far is the bob from the central position when *t* is 0 s, 0.25 s, 0.5 s, 0.75 s, and 1 s?

b) Use your results from part a) to describe the motion of a pendulum from *t* = 0 to *t* = 1.

9. Inquiry/Problem Solving a) Find three pairs of acute angles such that the sine of the first angle equals the cosine of the second angle.

b) How are the angles in each pair from part a) related?

c) Find three pairs of angles, each including one acute angle and one obtuse angle, such that the sine of the first angle equals the cosine of the second angle.

d) How are the angles in each pair from part c) related?

10. If the given point is on the terminal arm of an angle A in standard position, 0° ≤ A ≤ 180°, find the measure of ∠A, to the nearest tenth of a degree.

a) (4, 3)　　　　**b)** (12, 5)　　　　**c)** (−6, 8)　　　　**d)** (−24, 7)

11. Obtuse angles A and B are in standard position. Point P(−3, 4) is on the terminal arm of angle A. Point Q(−9, 12) is on the terminal arm of angle B. How are the measures of angles A and B related?

12. Show that, for any acute angle θ,

a) sin (90° − θ) = cos θ　　　　　　**b)** cos (90° − θ) = sin θ

4.3 The Sine Law and the Cosine Law

The Peace Tower is the tallest part of Canada's Parliament Buildings. A bronze mast, which flies the Canadian flag, stands on top of the Peace Tower.

From a point 25 m from the foot of the tower, the angle of elevation of the top of the tower is 74.8°. From the same point, the angle of elevation of the top of the mast is 76.3°.

INVESTIGATE & INQUIRE

To find the height of the mast, use the diagram shown. The given information is marked on the diagram. DC is the height of the mast, and BC is the height of the tower.

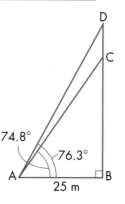

1. Find the height of the mast, to the nearest tenth of a metre, using only right triangles.

2. a) List measurements you would use to find the height of the mast using the cosine law in △ACD.
b) Find these measurements.
c) Use the cosine law to find the height of the mast, to the nearest tenth of a metre.

3. a) List measurements you would use to find the height of the mast using the sine law in △ACD.
b) Find these measurements.
c) Use the sine law to find the height of the mast, to the nearest tenth of a metre.

4. Compare your answers from questions 1, 2, and 3. Which method did you prefer? Explain.

You have previously applied the sine law and the cosine law to acute triangles. You have seen that the sine law and the cosine law also apply to obtuse triangles.

The sine law for acute and obtuse triangles can be developed as follows.

In △ABC, draw AD perpendicular to BC, or to BC extended. AD is the altitude or height, h, of △ABC.

Acute Triangle

Obtuse Triangle

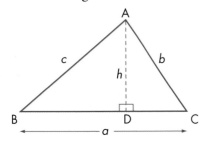
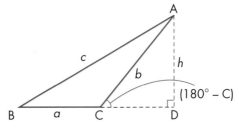

In △ACD, $\dfrac{h}{b} = \sin C$

$h = b\sin C$

In △ACD, $\dfrac{h}{b} = \sin(180° - C)$ **Recall that sin (180° − θ) = sin θ.**

$\dfrac{h}{b} = \sin C$

$h = b\sin C$

In △ABD, $\dfrac{h}{c} = \sin B$

$h = c\sin B$

In △ABD, $\dfrac{h}{c} = \sin B$

$h = c\sin B$

For both the acute and the obtuse triangles,

$$c\sin B = b\sin C$$

Divide both sides by bc:

$$\dfrac{c\sin B}{bc} = \dfrac{b\sin C}{bc}$$

Simplify:

$$\dfrac{\sin B}{b} = \dfrac{\sin C}{c}$$

By drawing the altitude from C, we can similarly show that

$$\dfrac{\sin A}{a} = \dfrac{\sin B}{b}$$

Combining the results gives the following forms of the sine law.

$$\dfrac{\sin A}{a} = \dfrac{\sin B}{b} = \dfrac{\sin C}{c} \qquad \dfrac{a}{\sin A} = \dfrac{b}{\sin B} = \dfrac{c}{\sin C}$$

EXAMPLE 1 The Sine Law, Given Two Angles and a Side

In △RST, ∠S = 40°, ∠T = 21°, and r = 46 cm. Find t, to the nearest centimetre.

SOLUTION

Draw a diagram.

Find the measure of ∠R.
∠R = 180° − 40° − 21°
     = 119°

Use the sine law to find t.

$$\frac{t}{\sin T} = \frac{r}{\sin R}$$

$$\frac{t}{\sin 21°} = \frac{46}{\sin 119°}$$

$$t = \frac{46\sin 21°}{\sin 119°}$$

$$t \doteq 19$$

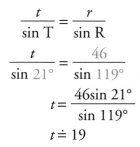

So, t = 19 cm, to the nearest centimetre.

EXAMPLE 2 The Sine Law, Given Two Sides and the Angle Opposite One of Them

In △PQR, ∠P = 105.2°, p = 23.2 cm, and r = 18.5 cm. Solve the triangle, rounding the side length to the nearest tenth of a centimetre and the angles to the nearest tenth of a degree, if necessary.

SOLUTION

Draw a diagram.

Use the sine law to find the measure of ∠R.

$$\frac{\sin R}{r} = \frac{\sin P}{p}$$

$$\frac{\sin R}{18.5} = \frac{\sin 105.2°}{23.2}$$

$$\sin R = \frac{18.5\sin 105.2°}{23.2}$$

$$\angle R \doteq 50.3°$$

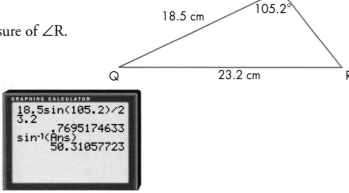

Find the measure of $\angle Q$.

$\angle Q = 180° - 105.2° - 50.3°$
$ = 24.5°$

Use the sine law to find q.

$$\frac{q}{\sin Q} = \frac{p}{\sin P}$$

$$\frac{q}{\sin 24.5°} = \frac{23.2}{\sin 105.2°}$$

$$q = \frac{23.2\sin 24.5°}{\sin 105.2°}$$

$$\doteq 10.0$$

In $\triangle PQR$, $\angle R = 50.3°$, $\angle Q = 24.5°$, and $q = 10.0$ cm.

The cosine law for acute and obtuse triangles can be developed as follows.

In $\triangle ABC$, draw AD perpendicular to BC, or to BC extended.

AD is the altitude or height, h, of $\triangle ABC$.

Acute Triangle

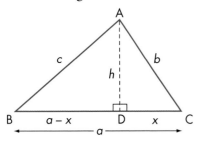

In $\triangle ADC$, $\dfrac{x}{b} = \cos C$

$ x = b\cos C$

and $b^2 = h^2 + x^2$

In $\triangle ABD$, $c^2 = h^2 + (a - x)^2$
$ = h^2 + a^2 - 2ax + x^2$
$ = a^2 + (h^2 + x^2) - 2ax$
$ c^2 = a^2 + b^2 - 2ab\cos C$

Obtuse Triangle

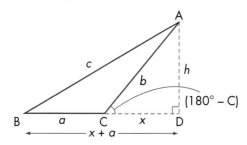

In $\triangle$ADC, $\dfrac{x}{b} = \cos(180° - C)$

$\qquad\qquad x = b\cos(180° - C)$ **Recall that $\cos(180° - \theta) = -\cos\theta$.**

$\qquad\qquad\quad = -b\cos C$

and $b^2 = h^2 + x^2$

In $\triangle$ABD, $c^2 = h^2 + (a + x)^2$

$\qquad\qquad\quad = h^2 + a^2 + 2ax + x^2$

$\qquad\qquad\quad = a^2 + (h^2 + x^2) + 2ax$

$\qquad\qquad\quad = a^2 + b^2 + 2a(-b\cos C)$

$\qquad\qquad c^2 = a^2 + b^2 - 2ab\cos C$

The forms of the cosine law are as follows.

$a^2 = b^2 + c^2 - 2bc\cos A \qquad\qquad \cos A = \dfrac{b^2 + c^2 - a^2}{2bc}$

$b^2 = a^2 + c^2 - 2ac\cos B \qquad\qquad \cos B = \dfrac{a^2 + c^2 - b^2}{2ac}$

$c^2 = a^2 + b^2 - 2ab\cos C \qquad\qquad \cos C = \dfrac{a^2 + b^2 - c^2}{2ab}$

EXAMPLE 3 The Cosine Law, Given Three Sides

In $\triangle$ABC, $a = 9.6$ m, $b = 20.6$ m, and $c = 14.7$ m. Solve the triangle.
Round each angle measure to the nearest tenth of a degree.

SOLUTION

Draw a diagram.

Use the cosine law to find the measure of an angle.

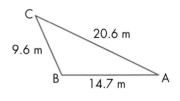

$$\cos B = \frac{a^2 + c^2 - b^2}{2ac}$$

$$= \frac{9.6^2 + 14.7^2 - 20.6^2}{2(9.6)(14.7)}$$

$$\angle B \doteq 114.3°$$

Use the sine law to find the measure of $\angle A$.

$$\frac{\sin A}{a} = \frac{\sin B}{b}$$

$$\frac{\sin A}{9.6} = \frac{\sin 114.3°}{20.6}$$

$$\sin A = \frac{9.6 \sin 114.3°}{20.6}$$

$$\angle A \doteq 25.1°$$

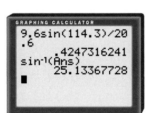

Find $\angle C$.

$$\angle C = 180° - 114.3° - 25.1°$$

$$= 40.6°$$

In $\triangle ABC$, $\angle A = 25.1°$, $\angle B = 114.3°$, and $\angle C = 40.6°$.

EXAMPLE 4 The Cosine Law, Given Two Sides and the Contained Angle

Find the length of CD, to the nearest tenth of a metre.

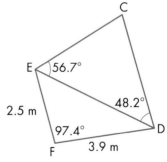

SOLUTION

Use △DEF and the cosine law to find the length of DE.

$$DE^2 = DF^2 + EF^2 - 2(DF)(EF)\cos F$$
$$= 3.9^2 + 2.5^2 - 2(3.9)(2.5)\cos 97.4°$$
$$DE \doteq 4.9$$

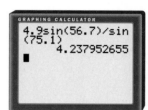

Use △CED to find the measure of ∠C.

$$\angle C = 180° - 56.7° - 48.2°$$
$$\angle C = 75.1°$$

Use △CED and the sine law to find the length of CD.

$$\frac{CD}{\sin 56.7°} = \frac{4.9}{\sin 75.1°}$$

$$CD = \frac{4.9\sin 56.7°}{\sin 75.1°}$$

$$CD \doteq 4.2$$

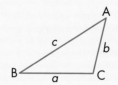

CD = 4.2 m, to the nearest tenth of a metre.

Key Concepts

- The forms of the sine law are

$$\frac{a}{\sin A} = \frac{b}{\sin B} = \frac{c}{\sin C}$$

$$\frac{\sin A}{a} = \frac{\sin B}{b} = \frac{\sin C}{c}$$

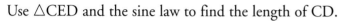

- The sine law can be used to solve any triangle when given
  a) the measures of two angles and any side
  b) the measures of two sides and the angle opposite one of these sides
- The forms of the cosine law are

$$a^2 = b^2 + c^2 - 2bc\cos A \qquad \cos A = \frac{b^2 + c^2 - a^2}{2bc}$$

$$b^2 = a^2 + c^2 - 2ac\cos B \qquad \cos B = \frac{a^2 + c^2 - b^2}{2ac}$$

$$c^2 = a^2 + b^2 - 2ab\cos C \qquad \cos C = \frac{a^2 + b^2 - c^2}{2ab}$$

- The cosine law can be used to solve any triangle when given
  a) the measures of two sides and the contained angle
  b) the measures of three sides

Communicate Your Understanding

1. Describe how you would solve each of the following triangles. Justify your chosen method.

a)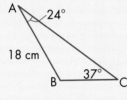

b)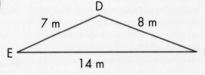

c)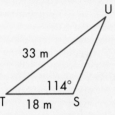

d)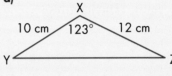

2. Explain why you cannot start with the sine law to solve △XYZ.

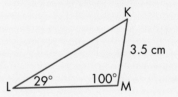

3. Explain why you cannot start with the cosine law to solve △KLM.

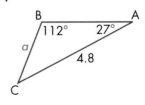

Practise

A

1. Find the length of the indicated side, to the nearest tenth.

a)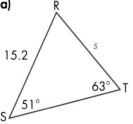

b)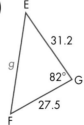

c)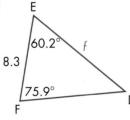

d)

e)

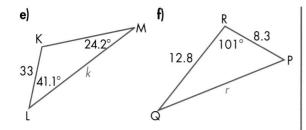

f)

2. Find the measure of the indicated angle, to the nearest tenth of a degree.

a)

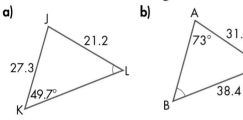

b)

c)

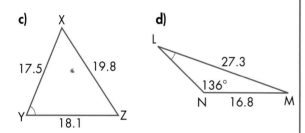

d)

e)

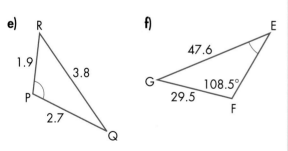

f)

3. Solve each triangle. Round answers to the nearest tenth, if necessary.

a) In $\triangle ABC$, $\angle A = 84°$, $\angle C = 40°$, $a = 5.6$ m.

b) In $\triangle PQR$, $\angle R = 28.5°$, $p = 10.4$ cm, $r = 6.3$ cm.

c) In $\triangle LMN$, $\angle M = 62°$, $l = 16.9$ m, $n = 15.1$ m.

d) In $\triangle UVW$, $\angle W = 123.9°$, $\angle V = 22.2°$, $v = 27.5$ km.

e) In $\triangle XYZ$, $\angle X = 92.3°$, $y = 3.1$ cm, $z = 2.8$ cm.

f) In $\triangle FGH$, $f = 12.6$ m, $g = 8.5$ m, $h = 6.3$ m.

4. Find the length of the indicated side, to the nearest tenth.

a)

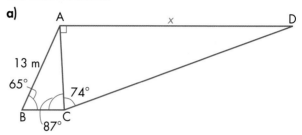

b)

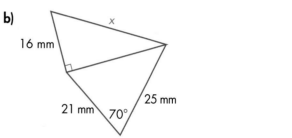

c)

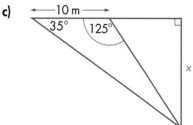

d)

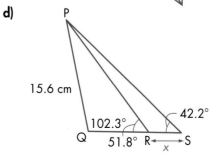

e)

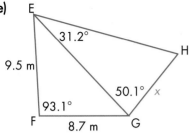

f)

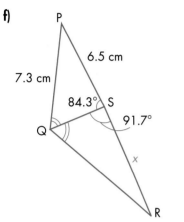

5. Find the measure of the indicated angle, to the nearest tenth of a degree.

a)

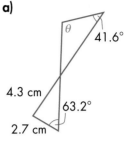

b)

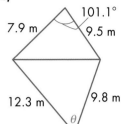

c)

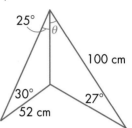

d)

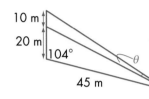

Apply, Solve, Communicate

6. Solve △ABC. Round answers to the nearest tenth.

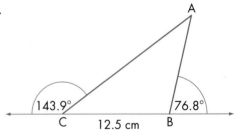

7. Measurement An isosceles triangle has two 5.5-cm sides and two 32.4° angles. Find
a) the perimeter of the triangle, to the nearest tenth of a centimetre
b) the area of the triangle, to the nearest tenth of a square centimetre

B

8. Inquiry/Problem Solving Airport X is 150 km east of airport Y. An aircraft is 240 km from airport Y, and 23° north of due west from airport Y. How far is the aircraft from airport X, to the nearest kilometre?

9. Application To determine the height of the Peace Tower on Parliament Hill in Ottawa, measurements were taken from a baseline AB. It was found that AB = 50 m, $\angle XAY = 42.6°$, $\angle XAB = 60°$, and $\angle ABX = 81.65°$. Calculate the height of the Peace Tower, to the nearest metre.

Web Connection
www.school.mcgrawhill.ca/resources/
To learn more about the history and
construction of the Parliament Buildings, visit
the above web site. Go to **Math Resources**,
then to *MATHEMATICS 11*, to find out where
to go next. Write a brief report.

10. Ship navigation Two ships left Port Hope on Lake Ontario at the same time. One travelled at 12 km/h on a course of 235°. The other travelled at 15 km/h on a course of 105°. How far apart were the ships after four hours, to the nearest kilometre?

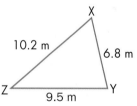

11. Measurement Find the area of $\triangle XYZ$, to the nearest tenth of a square metre.

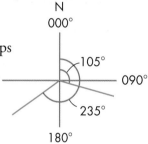

12. Communication a) Use the cosine law to find x, to the nearest tenth.
b) Use the Pythagorean theorem to find x, to the nearest tenth.
c) Explain why the two methods give the same results in a right triangle.

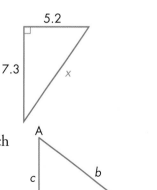

13. Sine law in right triangles Right $\triangle ABC$ is shown. Write each

of the ratios $\dfrac{a}{\sin A}$, $\dfrac{b}{\sin B}$, and $\dfrac{c}{\sin C}$ in terms of a, b, or c, and

verify that $\dfrac{a}{\sin A} = \dfrac{b}{\sin B} = \dfrac{c}{\sin C}$ for a right triangle.

C

14. Stikine Canyon The Stikine Canyon in central British Columbia is often referred to as Canada's Grand Canyon. Two points X and Y are sighted from a baseline AB of length 30 m on the opposite side of the canyon. The angle measurements recorded from positions A and B were ∠XAY = 31.3°, ∠XBY = 18.5°, ∠ABX = 25.6°, and ∠BAY = 27.9°. Find the distance from X to Y, to the nearest metre.

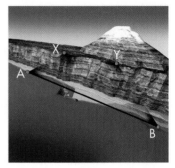

15. Geometry Use the cosine law to show that opposite angles in a parallelogram are congruent.

16. Measurement In △RST, RS = 4.9 m, ST = 3.7 m, and RT = 8.1 m. Find the area of △RST, to the nearest tenth of a square metre.

17. Measurement In △ABC, BC = 46 m, ∠A = 42.2°, and ∠B = 39.5°. Find the area of △ABC, to the nearest tenth of a square metre.

18. Measurement Find the volume of the right prism, to the nearest cubic centimetre.

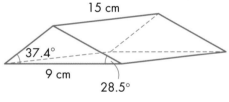

19. Measurement Find the volume of the right prism, to the nearest cubic metre.

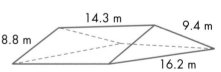

20. Analytic geometry △PQR has vertices P(1, 5), Q(6, −7), and R(−2, 1). Find the angle measures, to the nearest tenth of a degree.

ACHIEVEMENT Check Knowledge/Understanding Thinking/Inquiry/Problem Solving Communication Application

An equilateral triangle ABC has been creased and folded so that its vertex A now rests on BC at D, such that BD = 1 and DC = 2. Find the length of

a) AP   **b)** AQ   **c)** PQ

Surveying is the scientific measurement of natural or artificial features on the Earth's surface. Surveyors are involved in a wide variety of tasks that require very accurate knowledge of locations. The distances and angles determined by surveyors are used in many ways, including drawing maps, positioning buildings and other structures correctly, and defining the property lines that separate one piece of land from another.

Because Canada is the world's second-largest country, surveying Canada has been an enormous task. For example, it took almost 60 years to complete a survey of the Canada-U.S border, part of which runs through four of the Great Lakes. As a result of over 150 years of surveying work, detailed maps now exist for all parts of Canada.

1. From point P, the distance to one end of a pond is 450 m and the distance to the other end is 520 m. The angle formed by the lines of sight is 115°. Find the length of the pond, to the nearest ten metres.

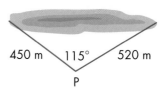

450 m   115°   520 m

P

2. Research Use your research skills to investigate the following.
a) the education and training required to become a surveyor, and the organizations that employ surveyors
b) the use of different types of surveying equipment, including manually controlled, electronic, and photographic instruments, and the use of satellite technology
c) the work of the Geological Survey of Canada in exploring and mapping the country

PATTERN *Power*

a) Copy and complete the pattern.
$11^2 =$ ▇▇▇▇▇
$101^2 =$ ▇▇▇▇▇
$1001^2 =$ ▇▇▇▇▇
b) Describe the pattern in words.
c) Explain why the pattern works.
d) Write the next 2 lines of the pattern.
e) Use the pattern to find $\sqrt{100\ 000\ 020\ 000\ 001}$.

TECHNOLOGY EXTENSION
Using *The Geometer's Sketchpad®* to Explore the SSA Case

Exploring the SSA Case

1. Use the following steps to construct △ABC where ∠A = 50°, AC = 6 cm, and CB = 5 cm.

a) Construct a base segment PQ. From the **Construct** menu, choose the **Point On Object** command. Highlight points A and Q, and measure the distance between them. If necessary, drag point A away from point Q, so that this distance is more than 8 cm.

b) Construct line segment AC. Measure the line segment AC and ∠CAQ. Drag point C so that AC = 6 cm and ∠CAQ = 50°.

c) In an open area, construct line segment EF. Measure the length of EF and drag one endpoint so that EF = 5 cm.

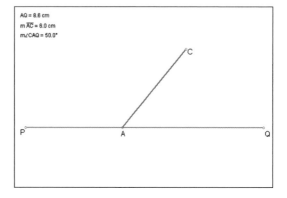

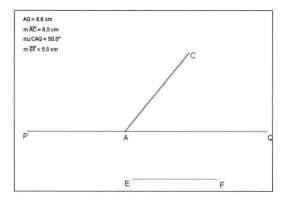

d) Highlight point C and line segment EF. From the **Construct** menu, choose the **Circle By Center And Radius** command. This will create a circle that intersects line segment PQ twice. Highlight line segment PQ and the circle. From the **Construct** menu, choose the **Point At Intersection** command. Label the two points of intersection B1 and B2.

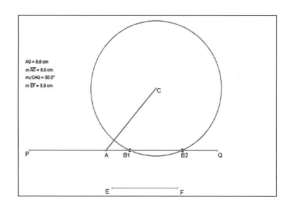

e) Construct a line segment joining C to B1, and construct a line segment joining C to B2. Measure each segment. How many different possible triangles satisfy the given conditions that ∠A = 50°, AC = 6 cm, and CB = 5 cm?

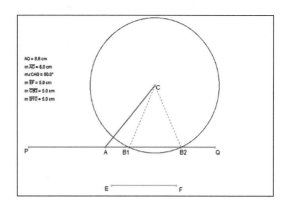

2. Use the following set of steps to construct △ABC where ∠A = 120° and AC = 6 cm.

a) Start with the base line segment with endpoints labelled P and Q. From the **Construct** menu, choose the **Point On Object** command. Label this point A.

b) Using the **Point Tool,** construct a point in the open area above line segment PQ. Label this point C. Construct line segment AC. Measure line segment AC and ∠QAC. Drag point C to a location where AC = 6 cm and ∠QAC = 120°.

c) To complete the triangle, a third point B must be placed on line segment AQ. In an open area of the screen, construct a line segment EF. Measure the length of EF. Highlight point C and line segment EF. From the **Construct** menu, choose the **Circle By Center And Radius** command. Dragging endpoint F, will change the radius of the circle.

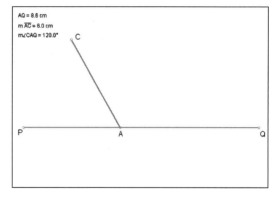

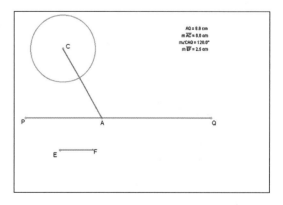

d) Drag point F until the circle with centre C intersects line segment PQ in a point to the right of point A. Choose the **Point At Intersection** command from the **Construct** menu to create point B. Construct line segment CB. Measure the length of CB.

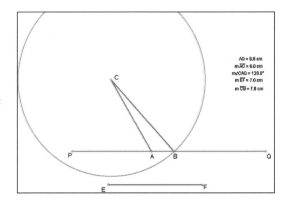

e) △ABC now satisfies the given conditions that ∠A = 120° and AC = 6 cm. Drag point F closer to point E to change the lengths of both EF and CB.
i) For what values of CB will no triangle exist?
ii) For what values of CB will one triangle exist?
iii) Are there any values of CB for which two triangles will exist?

Making Generalizations

1. To make generalizations, investigate constructions that give no triangles, exactly one triangle, or two triangles.
a) Start with a point P on line segment PQ and a point C above the line segment. Measure the length of AC and ∠A. Show the label for line segment AC, and re-label this line segment as b. Construct a line segment EF, and measure the length of this line segment. Construct a circle with centre C and radius EF.

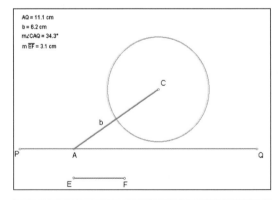

b) Drag point F to determine a range of values for the length of EF such that the circle does not intersect line segment PQ. Record the range of values in your notebook. Drag point F to a point that allows the circle to intersect line segment PQ at two points. Label these points B1 and B2, as shown. Measure line segment CB1 and label it a1. Measure CB2 and label it a2. Measure the

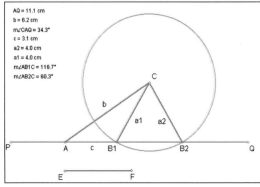

angles ∠AB1C and ∠AB2C. Drag point F and record the range of lengths that produces two triangles.

c) State the condition that must exist so that there is exactly one triangle in the construction. What angle is created at point B?

d) Use an appropriate trigonometric ratio to calculate the exact value for side CB in terms of ∠A and side b so that there is exactly one triangle. How is this value related to the ranges of values you recorded for no triangles and two triangles in part b)?

e) Make a generalization about the conditions on the length of side BC that will give no triangles, one triangle, and two triangles.

f) If you were given the measure of ∠A and the length of side AC, how would you determine if the length of a side BC would produce no triangles, one triangle, or two triangles.

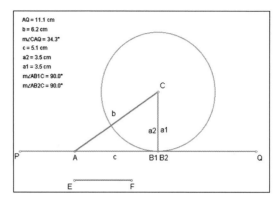

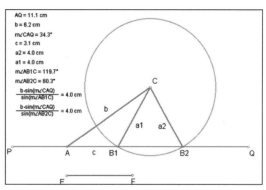

2. As shown in the diagram, use the calculator built into *The Geometer's Sketchpad®* to complete the Sine Law calculation for your construction.

a) What is the relationship between the two angles ∠AB1C and ∠AB2C?

b) Explain why the values from the two calculations are equal.

4.4 The Sine Law: The Ambiguous Case

Canada has 1.3 million square kilometres of wetlands, or almost 25% of all the wetlands in the world. These marshes and swamps help to prevent flooding and act as natural water purifiers.

Wetlands provide a habitat for many species of animals and plants. Millions of waterfowl use wetlands as migration stops or nesting sites.

INVESTIGATE & INQUIRE

A surveyor is preparing a map of a marsh that is to be protected as a conservation area. She places flags at four points around the edge of the marsh and uses them in making various measurements. The flags at points B, C, and D lie in a line. ∠ADC measures 34°. The straight-line distance from A to D is 2 km, and from A to B is 1.3 km. The distance from A to C is also 1.3 km. The diagram shows all the given information.

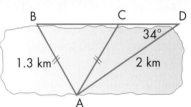

1. Can ∠ABD equal ∠ACD? Explain.

2. In △ABD, write an equation of the form sin ∠ABD = ■ to express sin ∠ABD in terms of AD, AB, and sin ∠ADB.

3. In △ACD, write an equation of the form sin ∠ACD = ■ to express sin ∠ACD in terms of AD, AC, and sin ∠ADB.

4. Using the equations from questions 2 and 3, what can you conclude about how the values of sin ∠ABD and sin ∠ACD compare?

5. If you substituted known values into the equations and used a calculator to find the measures of ∠ABD and ∠ACD, how would the calculated results compare? Explain.

6. Without using the sine law, show why your answer to question 5 is true.

7. Complete the calculation from question 5 and interpret the result.

8. The diagram you used shows an example of the ambiguous case of the sine law. Explain why the word *ambiguous* is appropriate.

When two sides and the non-included angle of a triangle are given, the triangle may not be unique. It is possible that no triangle, one triangle, or two triangles exist with the given measurements.

Suppose that, in $\triangle ABC$, you are given the side lengths a and b, and the measure of $\angle A$, and suppose that $\angle A < 90°$. If $a \geq b$, there is one triangle.

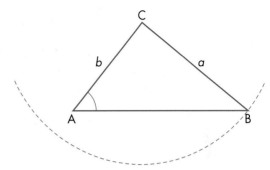

If $a < b$, there are three possibilities.

If $a = b\sin A$, there is one triangle and $\angle B = 90°$.

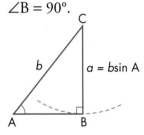

If $a < b\sin A$, there is no triangle.

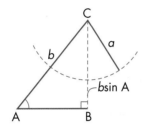

If $a > b\sin A$, there are two triangles.

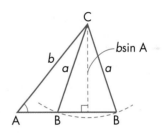

The situation in which there are two triangles is shown in the following diagrams. In the isosceles triangle, the angles opposite the equal sides are equal.

Because they lie on a straight line, the adjacent angles x and y are supplementary. Thus, in the ambiguous case, the two possible values of $\angle ABC$ are supplementary.

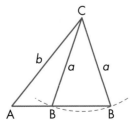

Recall that the sines of supplementary angles are equal. For example, $\sin 30° = 0.5$ and $\sin 150° = 0.5$.

If $\angle A \geq 90°$, there are two possibilities.

If $a \leq b$, there is no triangle.   If $a > b$, there is one triangle.

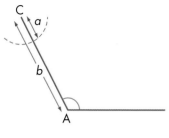

The cases for solving triangles, given the lengths of two sides and the measure of an angle opposite one of the sides, can be summarized as follows.

- In $\triangle ABC$, you are given the measure of acute $\angle A$ and the lengths of sides a and b.
  If $a \geq b$, there is one solution.
  If $a < b$, and $b\sin A$ is the altitude from C to AB, there are the following three possibilities.
  If $a < b\sin A$, there is no solution.
  If $a = b\sin A$, there is one solution, and $\triangle ABC$ is a right triangle.
  If $a > b\sin A$, there are two solutions.

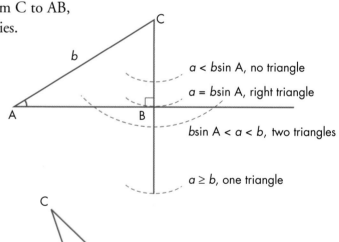

a < b sin A, no triangle

a = b sin A, right triangle

b sin A < a < b, two triangles

a ≥ b, one triangle

- In $\triangle ABC$, you are given the measure of right or obtuse $\angle A$ and the lengths of sides a and b.
  If $a \leq b$, there is no solution.
  If $a > b$, there is one solution.

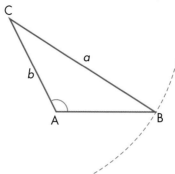

EXAMPLE 1 **One Solution**

Solve △ABC, if ∠A = 44.3°, a = 11.5 m, and b = 7.7 m. Round the side
length to the nearest tenth of a metre and the angles to the nearest tenth
of a degree, if necessary.

SOLUTION

Draw a diagram.

Since ∠A is given, and a > b, there is one solution.

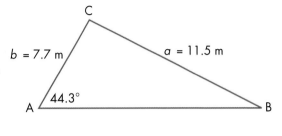

Use the sine law to find the measure of ∠B.

$$\frac{\sin B}{b} = \frac{\sin A}{a}$$

$$\frac{\sin B}{7.7} = \frac{\sin 44.3°}{11.5}$$

$$\sin B = \frac{7.7 \sin 44.3°}{11.5}$$

$$\angle B \doteq 27.9°$$

Find the measure of ∠C.

$$\angle C = 180° - 44.3° - 27.9°$$
$$= 107.8°$$

Use the sine law to find c.

$$\frac{c}{\sin C} = \frac{a}{\sin A}$$

$$\frac{c}{\sin 107.8°} = \frac{11.5}{\sin 44.3°}$$

$$c = \frac{11.5 \sin 107.8°}{\sin 44.3°}$$

$$c \doteq 15.7$$

So, ∠B = 27.9°, ∠C = 107.8°, and c = 15.7 m.

EXAMPLE 2 **Two Solutions**

Solve △ABC, if ∠A = 29.3°, b = 20.5 cm, and a = 12.8 cm. Round the side
length to the nearest tenth of a centimetre and the angles to the nearest
tenth of a degree, if necessary.

SOLUTION

Draw a diagram.

Then, $b\sin A = 20.5\sin 29.3°$

$\doteq 10$

Because $b\sin A < a < b$, there are two locations for point B and two solutions.

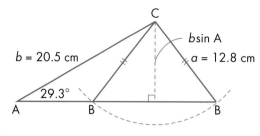

First, use the sine law to find the two possible measures of $\angle ABC$.

$$\frac{\sin B}{b} = \frac{\sin A}{a}$$

$$\frac{\sin B}{20.5} = \frac{\sin 29.3°}{12.8}$$

$$\sin B = \frac{20.5\sin 29.3°}{12.8}$$

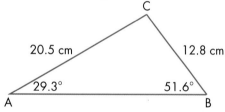

So, $\angle B \doteq 51.6°$ or $\angle B \doteq 180° - 51.6°$

$\doteq 128.4°$

Solve both triangles ABC.

$\angle C = 180° - 29.3° - 51.6°$, or $99.1°$

Use the sine law or the cosine law to find c.

Using the sine law,

$$\frac{c}{\sin C} = \frac{a}{\sin A}$$

$$\frac{c}{\sin 99.1°} = \frac{12.8}{\sin 29.3°}$$

$$c = \frac{12.8\sin 99.1°}{\sin 29.3°}$$

$\doteq 25.8$

$\angle C = 180° - 29.3° - 128.4°$, or $22.3°$

Use the sine law or the cosine law to find c.

Using the sine law,

$$\frac{c}{\sin C} = \frac{a}{\sin A}$$

$$\frac{c}{\sin 22.3°} = \frac{12.8}{\sin 29.3°}$$

$$c = \frac{12.8\sin 22.3°}{\sin 29.3°}$$

$\doteq 9.9$

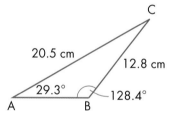

So, there are two solutions: $\angle B = 51.6°$, $\angle C = 99.1°$, and $c = 25.8$ cm

or $\angle B = 128.4°$, $\angle C = 22.3°$, and $c = 9.9$ cm.

Example 3 **Chord of a Circle**

A 15-cm line segment, PC, is drawn at an angle of 47° to a horizontal line segment, PT. A circle with centre C and radius 12 cm intersects PT at R and S. Calculate the length of the chord RS, to the nearest tenth of a centimetre.

Solution

Draw a diagram.

In △PCS, use the sine law to find the measure of ∠PSC.

$$\frac{\sin \angle PSC}{PC} = \frac{\sin \angle CPS}{CS}$$

$$\frac{\sin \angle PSC}{15} = \frac{\sin 47°}{12}$$

$$\sin \angle PSC = \frac{15\sin 47°}{12}$$

$$\angle PSC \doteq 66.1°$$

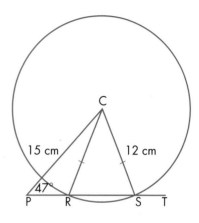

In △RCS, CR = CS, so ∠RSC = ∠SRC = 66.1°.

$$\angle RCS = 180° - 66.1° - 66.1°$$
$$= 47.8°$$

Use the sine law or the cosine law to find the length of RS.
Using the cosine law,

$$RS^2 = CR^2 + CS^2 - 2(CR)(CS)\cos \angle RCS$$
$$= 12^2 + 12^2 - 2(12)(12)\cos 47.8°$$
$$RS \doteq 9.7$$

The length of the chord RS is 9.7 cm, to the nearest tenth of a centimetre.

Example 4 **Offshore Beacon**

The light from a rotating offshore beacon can illuminate effectively up to a distance of 250 m. A point on the shore is 500 m from the beacon. From this point, the sight line to the beacon makes an angle of 20° with the shoreline. What length of shoreline, to the nearest metre, is effectively illuminated by the beacon?

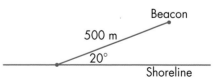

Solution

Complete the diagram using the given information. Label the given point on the shoreline A, and label the beacon B.

Consider the light from the beacon. If the beacon effectively illuminates a length of shoreline, then two points on the shoreline are 250 m from the beacon. Label point C, closer to point A. Label point D, farther from point A. CD is the length of shoreline illuminated by the beacon.

The beacon effectively illuminates a circular area of radius 250 m. CD is a chord of the circle.

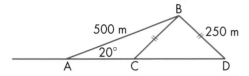

Use △ABD and the sine law to find the measure of ∠D.

$$\frac{\sin D}{AB} = \frac{\sin A}{BD}$$

$$\frac{\sin D}{500} = \frac{\sin 20°}{250}$$

$$\sin D = \frac{500 \sin 20°}{250}$$

$$\angle D \doteq 43.2°$$

In △BCD, find the measure of ∠CBD.
Since △BCD is isosceles, ∠BCD = 43.2°.
∠CBD = 180° − 43.2° − 43.2°
= 93.6°

Use △BCD and the sine law to find the length of CD.

$$\frac{CD}{\sin \angle CBD} = \frac{BC}{\sin \angle D}$$

$$\frac{CD}{\sin 93.6°} = \frac{250}{\sin 43.2°}$$

$$CD = \frac{250 \sin 93.6°}{\sin 43.2°}$$

$$CD \doteq 364$$

So, the length of shoreline effectively illuminated by the beacon is 364 m, to the nearest metre.

Key Concepts

- In △ABC, ∠A < 90°, and the side lengths *a* and *b* are given.
  a) If *a* ≥ *b*, there is one solution.

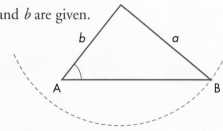

b) If *a* < *b*, there are three possibilities.

a = *b*sin A; there is one solution.	*a* < *b*sin A; there is no solution.	*a* > *b*sin A; there are two solutions.

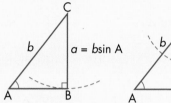 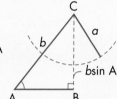 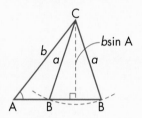

- In △ABC, ∠A ≥ 90°, and the side lengths *a* and *b* are given.
  a) If *a* ≤ *b*, there is no solution.
  b) If *a* > *b*, there is one solution.

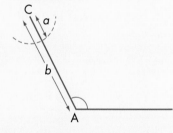 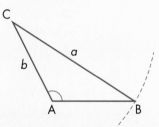

Communicate Your Understanding

1. In △DEF, ∠D = 61°, *d* = 9 cm, and *f* = 8 cm. Describe how you would determine the number of solutions for the triangle.

2. In △RST, ∠S = 35°, *t* = 7 cm, and *s* = 5 cm. From which vertex would you calculate an altitude of the triangle to determine the number of solutions? Justify your reasoning.

3. In △WXY, ∠W = 110° and *x* = 10 cm. Describe how you would find the value(s) for the length *w* that would result in

a) one solution   **b)** no solution

Practise

A

1. Find the measures of the angles x and y, to the nearest tenth of a degree.

a)

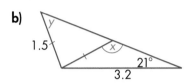

b)

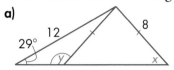

c)

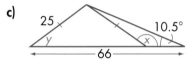

d)

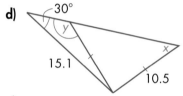

e)

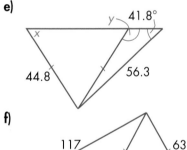

f)

2. Determine the number of possible triangles that could be drawn with the given measures. Then, find the measures of the other angles in each possible triangle. Round to the nearest tenth of a degree, if necessary.

a) $\triangle$ABC, where $\angle A = 42°$, $a = 30$ cm, and $b = 25$ cm

b) $\triangle$ABC, where $\angle B = 27°$, $b = 25$ cm, and $c = 30$ cm

c) $\triangle$PQR, where $\angle P = 30°$, $p = 24$ cm, and $q = 48$ cm

d) $\triangle$KLM, where $\angle M = 37.3°$, $m = 85$ cm, and $l = 90$ cm

e) $\triangle$UVW, where $\angle W = 38.7°$, $w = 10$ cm, and $v = 25$ cm

f) $\triangle$ABC, where $\angle B = 48°$, $c = 15.6$ m, and $b = 12.6$ m

g) $\triangle$XYZ, where $\angle X = 120°$, $x = 40$ cm, and $z = 20$ cm

h) $\triangle$DEF, where $\angle E = 144°$, $e = 10.5$ m, and $f = 12.5$ m

3. Solve each triangle. Round the angle measures to the nearest tenth of a degree and the side length to the nearest tenth of a centimetre, if necessary.

a) $\triangle$ABC, where $\angle A = 45°$, $a = 30$ cm, and $b = 24$ cm

b) $\triangle$XYZ, where $\angle Y = 32.7°$, $y = 54$ cm, and $x = 25$ cm

c) $\triangle$PQR, where $\angle R = 40.3°$, $r = 35.2$ cm, and $q = 40.5$ cm

d) $\triangle$FGH, where $\angle G = 105°$, $f = 3.5$ cm, and $g = 6.1$ cm

e) $\triangle$RST, where $\angle T = 50.2°$, $s = 10.5$ cm, and $t = 7.1$ cm

f) $\triangle$DEF, where $\angle E = 71.2°$, $e = 29.5$ cm, and $f = 30.3$ cm

g) $\triangle$BCD, where $\angle C = 143°$, $d = 12.5$ cm, and $c = 8.9$ cm

h) $\triangle$LMN, where $\angle L = 42.8°$, $l = 15.8$ cm, and $n = 18.5$ cm

4. Calculate
a) the measures of ∠BCD and ∠BDA, to the nearest tenth of a degree
b) the length of CD, to the nearest tenth of a centimetre

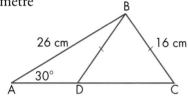

5. Find the length of KL, to the nearest tenth of a metre.

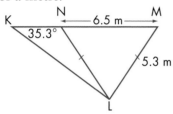

6. A circle with centre C and radius 5.5 cm intersects a line segment AB at points D and E. If ∠CAB = 48.9° and AC = 6.4 cm, what is the length of chord DE, to the nearest tenth of a centimetre?

7. Determine the length of the chord PQ in each circle, to the nearest tenth of a metre.
a)

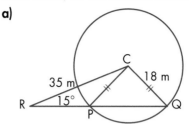

b)

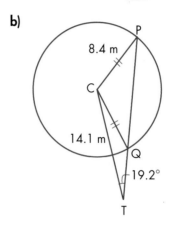

Apply, Solve, Communicate

8. Park light A light in a park can illuminate effectively up to a distance of 100 m. A point on a bike path is 150 m from the light. The sight line to the light makes an angle of 23° with the bike path. What length of the bike path, to the nearest metre, is effectively illuminated by the light?

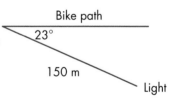

B

9. Application A triangular garden is enclosed by a fence. A dog is on a 5-m leash tethered to the fence at point P. One corner, B, of the fence is 6.5 m from P and forms a 41° angle, as shown in the diagram. Determine the total length of fence that the dog can reach, to the nearest tenth of a metre, if the dog cannot reach side AC.

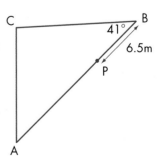

10. Analytic geometry Determine the length of the chord AB in each circle, to the nearest hundredth.

a)

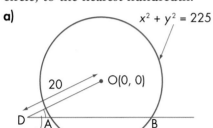

$x^2 + y^2 = 225$

20   O(0, 0)

D A
25°   B

b)

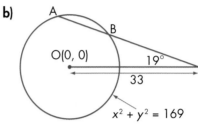

A
B
O(0, 0)
19°
33
$x^2 + y^2 = 169$

11. Forest fires A forest ranger spots a fire on a bearing of 050° from her station. She estimates that the fire is about 10 km away. A second station is due east of the first. A ranger in the second station thinks that the fire is about 8 km away from him. How far apart are the two stations, to the nearest kilometre?

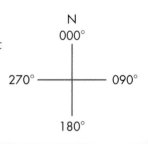

N
000°

270° ——————— 090°

180°

12. Road work A mechanical digger is being used to dig a ditch for laying pipes under a road. The arm of the digger is articulated at A, and the two sections of the arm have lengths of 3.8 m and 2.7 m, as shown. If the operator keeps the first section fixed at 43° to the horizontal, and then moves the second section to scoop out earth, what length of ditch will be worked in one swing of the second section, to the nearest tenth of a metre?

13. Ship navigation From a position 110 km northwest of a coast guard station, an oil tanker makes radio contact with the coast guard. The tanker is travelling due south at 25 km/h. The radar unit at the coast guard station has a range of 90 km. For what length of time can the coast guard expect the tanker to be visible on the radar screen, to the nearest tenth of an hour?

14. Application A satellite is in an orbit 1000 km above the surface of the Earth. A receiving dish is located so that the directions from the satellite to the dish and from the satellite to the centre of the Earth make an angle of 27°, as shown. If a signal from the satellite travels at 3×10^8 m/s, how long does it take to reach the dish, to the nearest thousandth of a second? Assume that the radius of the Earth is 6370 km.

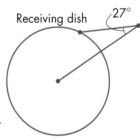

Receiving dish   27°

15. Gardening An underground sprinkler system is laid at an angle of 34.5° to a fence. The sprinkler jets are 10 m apart and have a range of 12 m. Determine the length of the fence that gets wet from the sprinklers, to the nearest tenth of a metre.

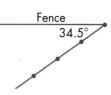

16. Analytic geometry Line segment PQ has endpoints P(0, 0) and Q(5, 12). Point R must be vertically above P. Find the coordinates of R, if they exist, for each of the following lengths of QR. Round to the nearest hundredth, if necessary.

a) 5   **b)** 8   **c)** 4   **d)** 15

17. Inquiry/Problem Solving Triangles are formed by the intersection of the lines $y = x$, $y = 2x$, $y = -2x$, and $y = -4$, as shown. Solve $\triangle ABC$ and $\triangle ABD$.

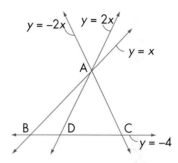

18. Measurement Show that the ratio of the areas of $\triangle ACD$ and $\triangle ACB$ is as follows.

$$\frac{\text{area } \triangle ACD}{\text{area } \triangle ACB} = \frac{\sin \angle ACD}{\sin \angle ACB}$$

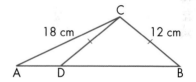

19. Ship navigation Three ships, A, B, and C, are sailing along the same straight line on a course of 240°. Ship A is at the front, and ship C brings up the rear. From ship A, the bearing of a navigation buoy is 025°. The buoy is 2.5 km from ship A and 2 km from ships B and C. To the nearest degree, what is the bearing of the buoy from ship C? from ship B?

ACHIEVEMENT Check   Knowledge/Understanding Thinking/Inquiry/Problem Solving Communication Application

Low-voltage lighting is a popular feature of landscaping. Two pathways meet at 30° to each other. One pathway has the lighting, and the first light is placed where the two paths meet, as shown. The distance between successive lights is 5 m, and each has a range of effective illumination of 6 m. For the second pathway, find, to the nearest tenth of a metre, the length that is
a) effectively illuminated
b) effectively illuminated by both light B and light C
c) effectively illuminated by light C or light D

The Cosine Law and the Ambiguous Case

The ambiguous case may occur when two sides and a non-included angle (SSA) of a triangle are given. The number of triangles that can be constructed or solved may be two, one, or zero.

1. Using the cosine law, find the length of side c. How many triangles are possible? Explain.

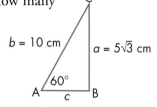

$b = 10$ cm $a = 5\sqrt{3}$ cm $60°$

2. Suppose that the length of side a is changed to each of the following values. By using the cosine law to find the length of c, determine how many triangles are possible. Explain your reasoning.

a) 9 cm   **b)** 12 cm   **c)** 8 cm

3. a) Determine the length of chord LM, to the nearest tenth of a centimetre, using the sine law and then using the cosine law.

b) Which method did you prefer? Justify your choice.

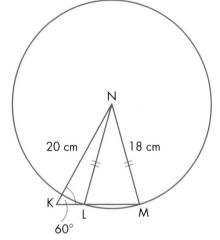

20 cm   18 cm   60°

4. Use the cosine law to determine whether each set of measurements will give one triangle, two triangles, or no triangle. Solve the triangles that exist. Round the side length to the nearest tenth of a centimetre and the angle measures to the nearest tenth of a degree, if necessary.

a) In $\triangle ABC$, $\angle A = 37°$, $a = 3$ cm, and $c = 4$ cm.

b) In $\triangle LMN$, $\angle M = 41.2°$, $m = 6.5$ cm, and $n = 8.9$ cm.

c) In $\triangle ABC$, $\angle A = 30°$, $a = 5.7$ cm, and $b = 11.4$ cm.

d) In $\triangle FGH$, $\angle G = 102.5°$, $g = 9.8$ cm, and $f = 10.9$ cm.

e) In △ABC, $a = 38.4$ m, $b = 25.2$ m, and $c = 19.3$ m.

f) In △GHI, $\angle G = 98.8°$, $g = 42.7$ cm, and $h = 30.1$ cm.

9. Inaccessible cliff To measure the height, XY, of an inaccessible cliff, a surveyor recorded the data shown. If the height of the theodolite used was 1.7 m, find the height of the cliff, to the nearest tenth of a metre.

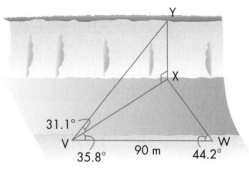

10. Determine the number of triangles that could be drawn with the given measures. Then, find the measures of the other angles and the other side in each possible triangle. Round side lengths to the nearest tenth of a unit and angle measures to the nearest tenth of a degree, where necessary.

a) △ABC, where $\angle A = 125°$, $a = 3$ m, $b = 5$ m

b) △STU, where $\angle S = 29°$, $s = 3.5$ cm, $t = 6$ cm

c) △XYZ, where $\angle X = 96.3°$, $x = 2.5$ m, $y = 0.8$ m

d) △FGH, where $\angle G = 41.7°$, $g = 7.2$ cm, $h = 9.9$ cm

11. The distance between the centres of the circles, A and B, is 60 m. Each circle has a radius of 40 m. Point C lies on the circle with centre B. $\angle BAC = 41°$. What are the possible lengths of AC, to the nearest metre?

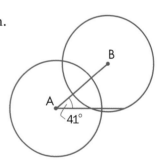

12. A plane leaves an aircraft carrier and flies due north at 500 km/h. The aircraft carrier proceeds 30° west of south at 35 km/h. If the plane has enough fuel for 4 h of flying, what is the maximum distance north it can fly, so that the fuel remaining will allow a safe return to the aircraft carrier?

CHALLENGE PROBLEMS

1. Perpendicular lines In $\triangle PQR$, $\angle PQR = 120°$, $PQ = 3$, and $QR = 4$. If perpendicular lines are constructed to PQ at P and to QR at R and are extended to meet at T, then the measure of RT is

a) 3   b) $\dfrac{8}{\sqrt{3}}$ c) 5   d) $\dfrac{11}{2}$ e) $\dfrac{10}{\sqrt{3}}$

2. Overlap Two strips of ribbon, each 1 cm wide, overlap at an angle θ, as shown. The area of the overlap (shaded), in square centimetres, is

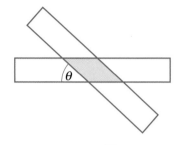

a) $\sin \theta$ b) $\dfrac{1}{\sin \theta}$ c) $\dfrac{1}{1 - \cos \theta}$

d) $\dfrac{1}{\tan \theta}$ e) $\dfrac{1}{1 - \sin \theta}$

3. Unknown length $\triangle LMP$ has $\angle M = 30°$, $LM = 150$, and $LP = 50\sqrt{3}$. Determine the length of MP.

4. Median The sides AB and AC of $\triangle ABC$ are of lengths 4 and 7, respectively. The median AM is of length 3.5. Find the length of BC.

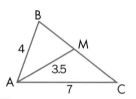

5. Isosceles triangle In an isosceles triangle ABC, $a = b = \sqrt{3}$, and $c \geq 3$. Find the smallest possible measure for $\angle C$.

6. Regular dodecagon A regular dodecagon is inscribed in a circle with radius r cm. Find the area of the dodecagon.

7. Regular polygon A regular polygon of n sides is inscribed in a circle of radius r. If the area of the polygon is $2r^2\sqrt{2}$, how many sides does it have?

PROBLEM SOLVING STRATEGY

USE A DIAGRAM

Diagrams provide insights that help you solve many problems. The following problem has been adapted from a two-thousand-year-old Chinese mathematics book called the *Chiu chang suan-shu*.

A tree trunk has a height of 20 m and a circumference of 3 m. An arrowroot vine winds seven times around the tree and reaches the top of the trunk. What is the length of the vine?

Understand
the Problem

1. What information are you given?
2. What are you asked to find?
3. Do you need an exact or an approximate answer?

Think
of a Plan

The vine winds seven times around the tree.

Let the foot of the vine be fixed, and imagine that the tree is rotated seven times to the right along the ground. As the tree rotates, the vine unwinds.

If the tree is perpendicular to the ground, the vine becomes the hypotenuse of a right triangle formed with the ground and the tree trunk.

Find the length of the hypotenuse of the triangle.

Carry Out
the Plan

Let v metres represent the length of the vine.

The circumference of the tree is 3 m. Each time the tree rotates, it moves 3 m along the ground. In seven rotations, the tree will move 7×3 or 21 m.

Use the Pythagorean theorem in the right triangle.

$$v^2 = 21^2 + 20^2$$
$$= 441 + 400$$
$$= 841$$
$$v = \sqrt{841}$$
$$= 29$$

Estimate

$\sqrt{900} = 30$

20 m

21 m

v

The length of the vine is 29 m.

Look Back

Does the answer seem reasonable?
Is there another way to find the length v in the triangle?

Use a Diagram

1. Draw a diagram to represent the situation.
2. Use the diagram to solve the problem.
3. Check that the answer is reasonable.

Apply, Solve, Communicate

1. **Winding staircase** A staircase winds around the outside of a fire observation tower that is 60 m high. The diameter of the tower is 1.6 m, and the stairs wind around it 6 times. What is the length of the stairs, to the nearest metre?

2. **Vine** If a 30 m long vine winds 5 times around a 22 m high tree trunk to reach the top, what is the radius of the trunk, to the nearest centimetre?

3. **Rail tunnel** A train 100 m long travels at a speed of 90 km/h through a tunnel that is 400 m long. How much time, in seconds, does the train take to pass completely through the tunnel?

4. **River crossing** Three members of the Bold political party and three members of the Timid political party must cross a river in a boat that holds only two people at a time. Only one of the Timids and one of the Bolds know how to swim. For safety reasons, at least one of the swimmers must be in the boat each time it crosses the river. When there are Bolds and Timids on the same side of the river, the Bolds can never outnumber the Timids. How do the six people get across the river?

5. Passing trains Two trains, each having an engine and 40 cars, are approaching each other on a single track from opposite directions. The track has a siding, as shown in the diagram. The siding can hold 20 cars and one engine. Engines can push or pull cars from either end. How can the trains pass each other?

6. Spider and fly The length and width of the floor of a room are each 10 m. The height of the room is 3 m. A spider, on the floor in a corner, sees a fly on the ceiling in the corner diagonally opposite. If the fly does not move, what is the shortest distance the spider can travel to reach the fly, to the nearest tenth of a metre?

7. Measurement A point P is inside rectangle ABCD. The length of PA is 5 cm, PB is 4 cm, and PC is 3 cm. Find the length of PD.

8. Analytic geometry The coordinates of the endpoints of one diagonal of a square are (8, 11) and (4, 5). What are the coordinates of the endpoints of the other diagonal?

9. Coloured grid Each small square on a 2 by 2 grid must be painted red, black, or white. How many different ways of colouring are there, if rotations are not allowed?

10. Measurement A point P, in the plane of an equilateral triangle, is 4 cm, 6 cm, and 9 cm from the vertices of the triangle. What is the side length of the triangle?

11. Perspective drawing The cube was made from 8 identical smaller cubes. One cube has been removed from one corner. Sketch the shape you would see if you looked at the shape along the diagonal from the missing corner D toward the opposite corner E.

12. Toothpicks The toothpicks are arranged to make 5 identical squares. Move 3 toothpicks to make 4 identical squares.

13. Patrol boats One patrol boat leaves the north shore of a lake at the same time as another patrol boat leaves the south shore of the lake. Each travels at a constant speed. The boats meet 500 m from the north shore. Each boat continues to the shore opposite where it started, and then turns around and heads back. The boats meet for the second time 200 m from the south shore. Find the distance across the lake from the north shore to the south shore.

14. Formulating problems Write a problem that can be solved using a diagram. Have a classmate solve your problem.

USING THE STRATEGIES

1. Truck tires You have a four-wheel drive all-terrain truck, and you need to make a 27 000-km trip. Each tire can be used for 12 000 km. The four tires on the truck are new, and you have five new tires in the back of the truck. How can you use the nine tires to complete the trip?

2. Marbles There are twenty marbles in a bag. There are eight yellow marbles, seven purple marbles, and five green marbles. If your eyes are closed, what is the maximum number of marbles you can take from the bag to be certain that you will leave in the bag at least four marbles of one colour and at least three marbles of a second colour?

3. Equilateral triangles The length of each side of an equilateral triangle is 2 cm. The midpoints of the sides are joined to form an inscribed equilateral triangle. If this process is continued without end, find the sum of the perimeters of the triangles.

4. Prime numbers The number 13 is a prime number in which the two digits are different. Reversing the digits gives 31, another prime number. How many other pairs of numbers containing two different digits, like 13 and 31, are both prime?

5. Drops of milk About how many drops of milk are there in a one-litre carton of milk?

6. Cycling speed A cyclist rode up a hill at 4 km/h and back down the same route at 12 km/h. If she made the ride without stopping, what was her average speed for the entire ride?

7. Measurement The sum of the lengths of the sides of a right triangle is 18 cm. The sum of the squares of the lengths of the sides is 128 cm^2. What is the area of the triangle?

8. Cube net The net shown is folded to make a cube. Name the face that will be opposite the face labelled A.

9. Books There were five books in a series. The books were published at six-year intervals. The year the fifth book was published, the sum of the five publication years was 9970. In what year was the first book published?

10. Perfect squares The number 144 is a perfect square. The sum of the digits of 144 is 9, which is also a perfect square. How many other three-digit numbers have this property?

11. Card sales A gift shop began selling cards on May 1. On May 2, it sold 3 more cards than it did on May 1. This pattern continued, with the shop selling 3 more cards each day than on the previous day. The shop was open every day for 9 straight days. At the end of the day on May 9, it had sold a total of 171 cards in the 9 days. How many cards did the shop sell on May 7?

Chapter 3

1. If $f(x) = 2x^2 - 5x + 1$, find
a) $f(0)$ **b)** $f(3)$ **c)** $f(-2)$ **d)** $f(0.5)$

2. Given the graph of $y = f(x)$, sketch each of the following.
a) $y = f(x) + 4$
b) $y = f(x - 3)$
c) $y = f(x + 2) - 1$
d) $y = f(-x)$
e) $y = -f(x)$ **f)** $y = -f(-x)$

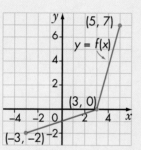

3. Find the inverse of each function. State if the inverse is a function.
a) $f(x) = 2x + 5$ **b)** $f(x) = x^2 - 4$

4. The graph of $y = f(x)$ is shown. Sketch the graphs of the following functions.
a) $2f(x)$ **b)** $\frac{1}{2}f(x)$
c) $f(3x)$ **d)** $\frac{1}{2}f\left(\frac{1}{2}x\right)$

5. Sketch the graphs of each of the following pairs of functions on the same set of axes. State the domain and range of each function.
a) $y = \sqrt{x}$ and $y = -\sqrt{x - 2} + 4$
b) $y = x^2$ and $y = 2(x + 3)^2 - 5$

6. Describe how the graph of each of the following functions can be obtained from the graph of $y = f(x)$.
a) $y = -5f(x) + 3$ **b)** $y = 4f\left(\frac{1}{2}x\right) + 2$
c) $y = 2f(x + 3) - 4$ **d)** $y = f(7(x - 5)) + 1$
e) $y = -f(8(x + 7))$ **f)** $y = \frac{1}{3}f(2x - 6) - 9$

Chapter 4

1. Find the measure of $\angle\theta$, to the nearest tenth of a degree.

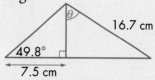

2. If $0° \leq A \leq 180°$, find $\angle A$, to the nearest tenth of a degree.
a) $\cos A = -0.7732$ **b)** $\sin A = 0.2853$

3. The point $(-20, 21)$ is on the terminal arm of an angle θ in standard position. Find $\sin \theta$ and $\cos \theta$.

4. Solve each triangle. Round answers to the nearest tenth, if necessary.
a) In $\triangle JKL$, $\angle K = 125°$, $\angle L = 21°$, $l = 8$ cm
b) In $\triangle ABC$, $\angle A = 43.5°$, $b = 12.5$ cm, $c = 10.8$ cm
c) In $\triangle DEF$, $\angle D = 32.1°$, $e = 8.1$ m, $d = 3.8$ m

5. Determine the length of the chord AB, to the nearest tenth of a centimetre.

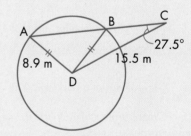

6. **Navigation** Two boats left Providence Bay on Manitoulin Island at the same time. One boat travelled at 10 km/h on a course of 280°. The other boat travelled at 13 km/h on a course of 165°. How far apart were the boats after 2 h, to the nearest kilometre?

5 TRIGONOMETRIC FUNCTIONS

Specific Expectations	Functions	Functions & Relations
Determine the sine, cosine, and tangent of angles greater than 90°, using a suitable technique, and determine two angles that correspond to a given single trigonometric function value.	5.2	5.2
Represent transformations of the functions defined by $f(x) = \sin x$ and $f(x) = \cos x$, using function notation.	5.5, 5.6	5.5, 5.6
State the domain and range of the functions defined by $f(x) = \sin x$ and $f(x) = \cos x$.	5.4	5.4
Define the term *radian measure*.	5.1	5.1
Describe the relationship between radian measure and degree measure.	5.1	5.1
Represent, in applications, radian measure in exact form as an expression involving π and in approximate form as a real number.	5.1	5.1
Determine the exact values of the sine, cosine, and tangent of the special angles $0, \frac{\pi}{6}, \frac{\pi}{4}, \frac{\pi}{3}, \frac{\pi}{2}$ and their multiples less than or equal to 2π.	5.2	5.2
Prove simple identities, using the Pythagorean identity, $\sin^2 x + \cos^2 x = 1$, and the quotient relation, $\tan x = \dfrac{\sin x}{\cos x}$.	5.7	5.7
Solve linear and quadratic trigonometric equations on the interval $0 \le x \le 2\pi$.	5.8	5.8
Demonstrate facility in the use of radian measure in solving equations and in graphing.	5.5, 5.6, 5.8	5.5, 5.6, 5.8
Sketch the graphs of $y = \sin x$ and $y = \cos x$, and describe their periodic properties.	5.4	5.4
Determine through investigation, using graphing calculators or graphing software, the effect of simple transformations on the graphs and equations of $y = \sin x$ and $y = \cos x$.	5.5, 5.6	5.5, 5.6
Determine the amplitude, period, phase shift, domain, and range of sinusoidal functions whose equations are given in the form $y = a\sin (kx + d) + c$ or $y = a\cos (kx + d) + c$.	5.6	5.6
Sketch the graphs of simple sinusoidal functions.	5.4, 5.5, 5.6	5.4, 5.5, 5.6
Write the equation of a sinusoidal function, given its graph and given its properties.	5.5, 5.6	5.5, 5.6

Specific Expectations	Functions	Functions & Relations
Sketch the graph of $y = \tan x$; identify the period, domain, and range of the function; and explain the occurrence of asymptotes.	5.4	5.4
Determine, through investigation, the periodic properties of various models of sinusoidal functions drawn from a variety of applications.	5.3, 5.5, 5.6	5.3, 5.5, 5.6
Explain the relationship between the properties of a sinusoidal function and the parameters of its equation, within the context of an application, and over a restricted domain.	5.5, 5.6	5.5, 5.6
Predict the effects on the mathematical model of an application involving sinusoidal functions when the conditions in the application are varied.	5.5, 5.6	5.5, 5.6
Pose and solve problems related to models of sinusoidal functions drawn from a variety of applications, and communicate the solutions with clarity and justification, using appropriate mathematical forms.	5.5, 5.6	5.5, 5.6

Ocean Cycles

People who work at sea need to be familiar with ocean cycles, which follow regular patterns. One pattern is the wave motion of the ocean. Another pattern is in the times of high and low tides.

In the Modelling Math questions on pages 362, 377, 388, 389, and 390, you will solve the following problem and other problems that involve ocean cycles.

A point on the ocean rises and falls as waves pass. Suppose that a wave passes every 4 s, and the height of each wave from the crest to the trough is 0.5 m.
a) Sketch a graph to model the height of the point relative to its average height for a complete cycle, starting at the crest of a wave.
b) Use exact values to write an equation of the form $h(t) = a\cos kt$ to model the height of the point, $h(t)$ metres, relative to its average height,

as a function of the time, t seconds.
c) If the times on the t-axis were changed from seconds to minutes, what would be the transformational effect on the graph, and what would be the new equation?

Use your research skills to answer the following questions now.

1. What forces on the ocean cause
a) tides?   **b)** waves?

2. Why are there two high tides and two low tides daily?

3. If there is a high tide on one side of the Earth, is there a high tide or a low tide on the opposite side of the Earth? Explain.

Daylight Hours

A *photoperiod* is the number of daylight hours in a given day. *Photoperiodism* is the response to changes in the photoperiod.

Photoperiodism is important in biology, because different species react to changes in the photoperiod in different ways. For example, different species of plants flower at different times of the year.

Photoperiodism is also important in geology and meteorology, because changes in the photoperiod determine the heating and cooling cycles of the Earth, and affect the weather.

The graph shows the approximate number of hours of daylight per day for Niagara Falls, Ontario, for one year.

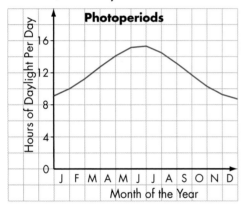

1. Sketch the graph and extend it for a second year. Explain and justify your reasoning.

2. Estimate the greatest number of daylight hours per day in Niagara Falls.

3. Estimate the least number of daylight hours per day in Niagara Falls.

4. Estimate the percent of days in the year with a photoperiod of at least
a) 11 h   b) 14 h

5. How would the graph for Kapuskasing, Ontario, compare with the graph for Niagara Falls? Explain and justify your reasoning.

6. Make a conjecture about how a graph of the non-daylight hours per day for Niagara Falls would compare with the graph of daylight hours.

7. Test your conjecture by sketching the graph of non-daylight hours per day for two years on the same grid as in question 1.

8. How are the graphs from question 7 related?

Review of Prerequisite Skills

If you need help with any of the skills named in purple below, refer to Appendix A.

1. Trigonometric ratios For each of the following right triangles, find the sine, cosine, and tangent of each acute angle, to the nearest thousandth.

a)

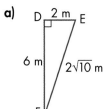

b)

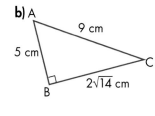

2. The given point lies on the terminal arm of an angle θ in standard position. Find $\sin \theta$ and $\cos \theta$. Round answers to the nearest thousandth, if necessary.

a) (7, 6)   b) (−5, 4)   c) (−8, 1)
d) (0, 6)   e) (−9, 0)

3. The graph of $y = f(x)$ is shown.

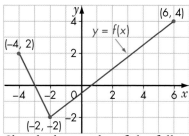

Sketch the graphs of the following functions.

a) $y = 2f(x)$ b) $y = f(2x)$

c) $y = \frac{1}{2}f\left(\frac{1}{2}x\right)$ d) $y = 3f\left(\frac{1}{3}x\right)$

4. The function $y = f(x)$ is given. Describe how the graphs of the following functions can be obtained from the graph of $y = f(x)$.

a) $y = f(x) - 8$ b) $y = f(x - 7)$
c) $y = f(x + 5) + 9$ d) $y = f(x - 6) - 3$

5. The graph of $y = f(x)$ is shown.

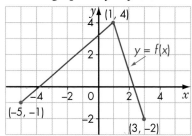

Draw and label the following pairs of graphs.

a) $y = f(x)$ and $y = -f(x)$
b) $y = f(x)$ and $y = f(-x)$

6. Describe how the graph of each of the following functions can be obtained from the graph of $y = f(x)$.

a) $y = 2f(x) + 5$ b) $y = f(2(x + 1)) - 2$

c) $y = \frac{1}{3}f\left(\frac{1}{2}(x - 4)\right)$ d) $y = -3f(x - 1) + 5$

7. Sketch the graph of each of the following functions.

a) $y = (x - 2)^2 + 3$
b) $y = -2(x + 4)^2 - 5$

c) $y = \frac{1}{2}(x + 3)^2$

d) $y = \frac{1}{2}\sqrt{x} + 1$

e) $y = 2\sqrt{x - 1} - 3$
f) $y = -3\sqrt{2(x - 5)} - 2$

8. Solving quadratic equations by factoring
Solve by factoring. Check your solutions.

a) $t^2 - t - 42 = 0$ b) $m^2 + 13m = -40$
c) $w^2 + 80 = 18w$ d) $r^2 - 81 = 0$
e) $3n^2 + 15 = 14n$ f) $15x^2 + 22x + 8 = 0$
g) $8z^2 + 18z - 5 = 0$ h) $3k^2 - 2k = 0$
i) $4d^2 - 20d + 25 = 0$ j) $16x^2 - 9 = 0$

5.1 Radians and Angle Measure

To describe the positions of places on the Earth, cartographers use a grid of circles that are north-south and east-west. The circles through the poles are called meridians of longitude. The circles parallel to the equator are called parallels of latitude. The meridians of longitude each have a radius that equals the radius of the Earth. The radii of the parallels of latitude vary with the latitude.

INVESTIGATE & INQUIRE

1. The angle θ is said to be **subtended** at the centre of the circle by the arc AB. Suppose that the length of the arc AB equals the radius of the circle, r. Estimate the measure of $\angle\theta$. Justify your estimate.

2. In Ontario, the city of London and the town of Moosonee have approximately the same longitude, 81°W. The diagram shows a circular cross section of the Earth through 81°W and 99°E. Determine the measure of each of the following angles.

a) $\angle ECN$ **b)** $\angle ECW$ **c)** reflex $\angle ECS$

3. What fraction of the circumference of the circle is the length of the arc that subtends each angle in question 2?

4. If the radius of the circle is r, what is the circumference of the circle in terms of r and π?

5. What is the length, in terms of π and r, of the arc that subtends each angle in question 2?

6. The radius of the Earth is approximately 6400 km. What is the length of the three arcs in terms of π?

7. Describe a method for finding the length of an arc of a circle using the radius and the fraction of the circumference that equals the arc length.

8. The latitude of London is 43°N, and the latitude of Moosonee is 51.4°N. Determine the distance between London and Moosonee along meridian 81°W, to the nearest kilometre.

9. If the length of an arc on a meridian of longitude is equal to the radius of the Earth, about 6400 km, express the measure of the angle subtended at the centre in terms of π.

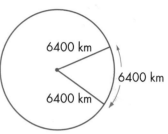

6400 km

6400 km

6400 km

10. If the radius of a circle is *r* and the length of an arc is *r*, express the measure of the angle subtended at the centre
a) in terms of π   **b)** to the nearest tenth of a degree

11. Compare your answer in question 10b) with your estimate in question 1. Explain any difference between the two values.

Recall that an angle is in standard position when its vertex is at the origin and the initial arm is fixed on the positive *x*-axis. The terminal arm rotates about the origin. If the direction of rotation is counterclockwise, the measure of the angle is positive.

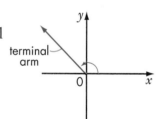

terminal arm

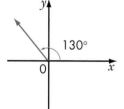

130°

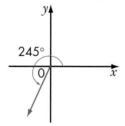

245°

When the terminal arm rotates, it may make one or more revolutions. When the terminal arm makes exactly one revolution counterclockwise, the angle has a measure of 360°.

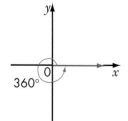

360°

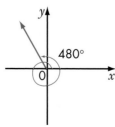

480°

The degree is a commonly used unit for the measures of angles in trigonometric applications. In the late 1800s, mathematicians and physicists saw the need for another unit, called a **radian**, that would simplify some calculations.

One radian is the measure of the angle subtended at the centre of a circle by an arc equal in length to the radius of the circle.

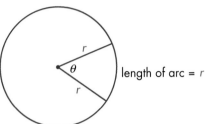

length of arc = r

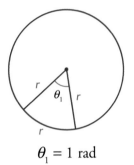

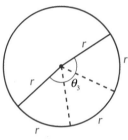

$\theta_1 = 1$ rad $\qquad$ $\theta_2 = 2$ rad $\qquad$ $\theta_3 = 3$ rad $\qquad$ The abbreviation rad means radians.

The arc lengths in the diagrams are r, $2r$, and $3r$. $\qquad$ Note that the arc of length $3r$ is not a semicircle.

So, $\theta_1 = 1$ $\qquad$ $\theta_2 = 2$ $\qquad$ $\theta_3 = 3$

$\quad = \dfrac{r}{r}$ $\qquad\quad = \dfrac{2r}{r}$ $\qquad\quad = \dfrac{3r}{r}$

These relationships lead to the following generalization.

$$\theta = \frac{a}{r} \quad \text{or} \quad \text{number of radians} = \frac{\text{arc length}}{\text{radius}}$$

Since $\theta = \dfrac{a}{r}$, then $a = r\theta$, $\theta > 0$

where a is the arc length, r is the radius, and θ is the measure of the angle, in radians.

Note that, when angle measures are written in radians, the unit rad is often omitted. For example, you should assume that an angle measure written as 2 is 2 rad, and that an angle measure written as π is π rad.

EXAMPLE 1 Using a = rθ

Find the indicated quantity in each diagram.

a)   b)   c)

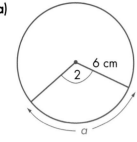

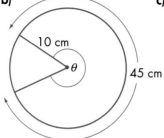

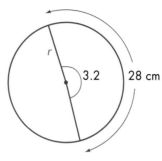

SOLUTION

a) $a = r\theta$

$a = 6 \times 2$

$\quad = 12$

So, $a = 12$ cm.

b) $\theta = \dfrac{a}{r}$

$\theta = \dfrac{45}{10}$

$\quad = 4.5$

So, $\theta = 4.5$ rad.

c) $r = \dfrac{a}{\theta}$

$r = \dfrac{28}{3.2}$

$\quad = 8.75$

So, $r = 8.75$ cm.

In order to convert from degree measure to radian measure, the relationship between degrees and radians must be established.

In degree measure, one revolution is 360°.
In radian measure, one revolution is given by

$$\frac{\text{arc length}}{\text{radius}} = \frac{2\pi r}{r}$$

$$= 2\pi \text{ rad}$$

So, the relationship between degrees and radians is

$$2\pi \text{ rad} = 360°$$

which simplifies to π rad $= 180°$ or $1 \text{ rad} = \left(\dfrac{180}{\pi}\right)°$.

Also, since $180° = \pi$ rad,

then $1° = \left(\dfrac{\pi}{180}\right) \text{ rad}$.

EXAMPLE 2 Radian Measure to Degree Measure

Change each radian measure to degree measure. Round to the nearest tenth of a degree, if necessary.

a) $\dfrac{\pi}{6}$ **b)** $\dfrac{5\pi}{4}$ **c)** 2.2

SOLUTION

To change radian measure to degree measure, multiply the number of radians by $\left(\dfrac{180}{\pi}\right)^{\circ}$.

a) $\dfrac{\pi}{6}\,\text{rad} = \dfrac{\pi}{6}\left(\dfrac{180}{\pi}\right)^{\circ}$

$\quad\quad = 30°$

b) $\dfrac{5\pi}{4}\,\text{rad} = \dfrac{5\pi}{4}\left(\dfrac{180}{\pi}\right)^{\circ}$

$\quad\quad = 225°$

c) $2.2\,\text{rad} = 2.2\left(\dfrac{180}{\pi}\right)^{\circ}$

$\quad\quad \doteq 126.1°$

```
GRAPHING CALCULATOR
2.2(180/π)
            126.0507149
■
```

Estimate

$\dfrac{2}{3} \times 180 = 120$

EXAMPLE 3 Degree Measure to Radian Measure in Exact Form

Find the exact radian measure, in terms of π, for each of the following.

a) 45°   **b)** 60°   **c)** 210°

SOLUTION

To change degree measure to radian measure, multiply the number of degrees by $\left(\dfrac{\pi}{180}\right)$ radians.

a) $45° = 45\left(\dfrac{\pi}{180}\right)$

$\quad = \dfrac{45\pi}{180}$

$\quad = \dfrac{\pi}{4}\,\text{rad}$

b) $60° = 60\left(\dfrac{\pi}{180}\right)$

$\quad = \dfrac{60\pi}{180}$

$\quad = \dfrac{\pi}{3}\,\text{rad}$

c) $210° = 210\left(\dfrac{\pi}{180}\right)$

$\quad = \dfrac{210\pi}{180}$

$\quad = \dfrac{7\pi}{6}\,\text{rad}$

EXAMPLE 4 Degree Measure to Radian Measure in Approximate Form

Change each degree measure to radian measure, to the nearest hundredth.
a) 30°   **b)** 240°

SOLUTION

a) $30° = 30\left(\dfrac{\pi}{180}\right)$
$\doteq 0.52$ rad

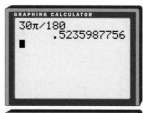

$\dfrac{30 \times 3}{180} = \dfrac{90}{180} = 0.5$

b) $240° = 240\left(\dfrac{\pi}{180}\right)$
$\doteq 4.19$ rad

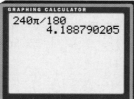

$\dfrac{240 \times 3}{180} = \dfrac{720}{180} = 4$

When an object rotates around a centre point or about an axis, like the one shown, its angular velocity, w, is the rate at which the central angle, θ, changes. Angular velocity is usually expressed in radians per second.

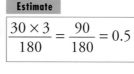

EXAMPLE 5 Angular Velocity

A small electric motor turns at 2000 r/min. Express the angular velocity, in radians per second, as an exact answer and as an approximate answer, to the nearest hundredth.

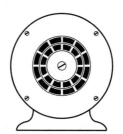

The abbreviation r means revolution.

SOLUTION

To complete one revolution, the motor turns through 2π radians.
2000 r/min = $2000 \times 2\pi$ rad/min

$= \dfrac{2000 \times 2\pi}{60}$ rad/s

$\dfrac{2000 \times 2\pi}{60} = \dfrac{200\pi}{3}$

$\doteq 209.44$

$\dfrac{200 \times 3}{3} = 200$

The angular velocity is exactly $\dfrac{200\pi}{3}$ rad/s, or 209.44 rad/s, to the nearest hundredth.

Key Concepts

- For an angle θ subtended at the centre of a circle of radius r by an arc of length a, number of radians $= \dfrac{\text{arc length}}{\text{radius}}$, or $\theta = \dfrac{a}{r}$.

- π rad $= 180°$ or 1 rad $= \left(\dfrac{180}{\pi}\right)°$

  To change radian measure to degree measure, multiply the number of radians by $\left(\dfrac{180}{\pi}\right)°$.

- Since $180° = \pi$ rad, $1° = \dfrac{\pi}{180}$ rad.

  To change degree measure to radian measure, multiply the number of degrees by $\left(\dfrac{\pi}{180}\right)$ radians.

Communicate Your Understanding

1. Describe how you would find the length of arc a in the diagram.

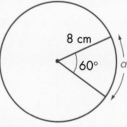

2. Describe how you would change $\dfrac{9}{5}\pi$ rad to degree measure.

3. Describe how you would change $25°$ to radian measure.

Practise

A

1. Find the exact number of degrees in the angles with the following radian measures.

a) $\dfrac{\pi}{3}$

b) $\dfrac{\pi}{4}$

c) 2π

d) $\dfrac{\pi}{2}$

e) $\dfrac{3}{4}\pi$

f) $\dfrac{3}{2}\pi$

g) 4π

h) $\dfrac{5}{6}\pi$

i) $\dfrac{\pi}{18}$

j) $\dfrac{11}{3}\pi$

k) $\dfrac{7}{6}\pi$

l) 5π

2. Find the exact radian measure in terms of π for each of the following angles.

a) 40°   **b)** 75°   **c)** 10°   **d)** 120°
e) 225°   **f)** 315°   **g)** 330°   **h)** 240°
i) 540°   **j)** 1080°

3. Find the approximate number of degrees, to the nearest tenth, in the angles with the following radian measures.

a) 2.5   **b)** 1.75   **c)** 0.35   **d)** 1.25
e) $\frac{5}{7}\pi$ **f)** $\frac{3}{11}\pi$ **g)** $\frac{17}{13}\pi$ **h)** 0.5
i) 3.14   **j)** 1.21

4. Find the approximate number of radians, to the nearest hundredth, in the angles with the following degree measures

a) 60°   **b)** 150°   **c)** 80°   **d)** 145°
e) 230°   **f)** 325°   **g)** 56.4°   **h)** 128.5°
i) 255.4°   **j)** 310.5°

5. Determine the length of each arc, *a*, to the nearest tenth.

a)

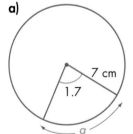

b)

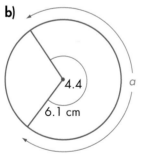

c)

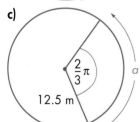

6. Determine the approximate measure of ∠θ, to the nearest hundredth of a radian and to the nearest tenth of a degree.

a)

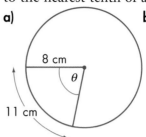

b)

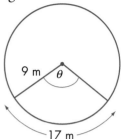

c)

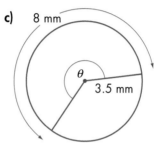

7. Determine the length of each radius, *r*, to the nearest tenth.

a)

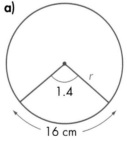

b)

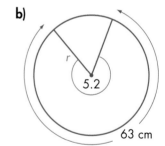

c)

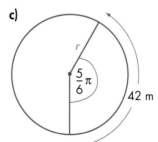

Apply, Solve, Communicate

8. Find the indicated quantity in each diagram. Round lengths to the nearest tenth of a metre and the angle measure to the nearest hundredth of a radian.

a)

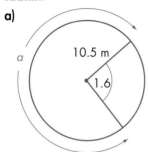

b)

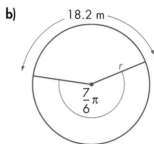

c)

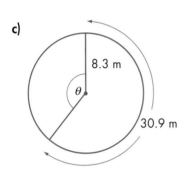

9. Pendulum The end of a 100-cm long pendulum swings through an arc of length 5 cm. Find the radian measure of the angle through which the pendulum swings. Express the radian measure in exact form and in approximate form, to the nearest hundredth.

10. Geometry Express the radian measure of each angle in an equilateral triangle in exact form and in approximate form, to the nearest hundredth.

11. Electric motor An electric motor turns at 3000 r/min. Determine the angular velocity in radians per second in exact form and in approximate form, to the nearest hundredth.

B

12. Discus throw A discus thrower completes 2.5 r before releasing the discus. Express the angle of rotation of the thrower in radian measure in exact form and in approximate form, to the nearest hundredth.

13. An angle has a radian measure of $\frac{\pi}{8}$. Find the exact radian measure of

a) its supplement
b) its complement

14. Geometry The measures of the acute angles in a right triangle are in the ratio 2:3. Write the radian measures of the acute angles in exact form and in approximate form, to the nearest hundredth.

15. Inquiry/Problem Solving How many times as large, to the nearest tenth, is an angle measuring 2 rad as an angle measuring 2°?

16. Geography a) Through how many radians does the Earth turn on its axis in a leap year? Express your answer in exact form.
b) How many hours does it take the Earth to rotate through an angle of 150°?
c) How many hours does it take the Earth to rotate through an angle of $\frac{7}{3}\pi$ rad?

17. Application The revolving restaurant in the CN Tower completes $\frac{5}{6}$ of a revolution every hour. If Shani and Laszlo ate dinner at the restaurant from 19:15 to 21:35, through what angle did their table rotate during the meal? Express your answer in radian measure in exact form and in approximate form, to the nearest tenth.

18. Propeller The propeller of a small plane has an angular velocity of 1100 r/min. Express the angle through which the propeller turns in 1 s
a) in degree measure
b) in radian measure in exact form and in approximate form, to the nearest radian.

19. Watch Express the angle that the second hand on a watch turns through in 1s
a) in degree measure
b) in radian measure in exact form and in approximate form, to the nearest hundredth

20. Clock a) Express the angular velocity of the hour hand on a clock in revolutions per day.
b) Determine the radian measure of the angle that the hour hand turns through from 15:00 on Friday until 09:00 the following Monday. Express your answer in exact form and in approximate form, to the nearest hundredth.

21. Compact discs A music CD rotates inside an audio player at different rates. The angular velocity is 500 r/min when music is read near the centre, decreasing to 200 r/min when music is read near the outer edge.

a) When music is read near the centre, through how many radians does the CD turn each second? Express your answer in exact form and in approximate form, to the nearest hundredth.

b) When music is read near the outer edge, through how many degrees does the CD turn each second?

c) Computer CD-ROM drives rotate CDs at multiples of the values used in music CD players. A 1X drive uses the same angular velocities as a music player, a 2X drive is twice as fast, a 4X drive is four times as fast, and so on. When data are being read near the centre, through how many degrees does a CD turn each second in a 12X drive? a 40X drive? a 50X drive?

d) When data are read near the outer edge, through how many radians does a CD turn each second in a 24X drive? Express your answer in exact form and in approximate form, to the nearest radian.

22. Ferris wheel A Ferris wheel with a radius of 32 m makes two revolutions every minute.

a) Find the exact angular velocity, in radians per second.

b) If the ride lasts 3 min, how far does a rider travel, to the nearest metre?

23. Weather satellite A weather satellite completes a circular orbit around the Earth every 3 h. The satellite is 2500 km above the Earth, and the radius of the Earth is approximately 6400 km. How far does the satellite travel in 1 min, to the nearest kilometre?

24. Rolling wheel A wheel turns with an angular velocity of 8 rad/s.

a) Express the angular velocity to the nearest tenth of a revolution per minute.

b) If the radius of the wheel is 20 cm, how far will it roll in 10 s, to the nearest centimetre?

25. Satellite orbit A satellite with a circular orbit has an angular velocity of 0.002 rad/s.

a) How long will it take the satellite to complete one orbit, to the nearest second?

b) What is the speed of the satellite, to the nearest tenth of a kilometre per second, if it is orbiting 800 km above the Earth, and the radius of the Earth is approximately 6400 km?

26. Navigation satellite The signals from a marine navigation satellite, N, can be received by any ship on arc AB, as shown in the diagram. The signal angle is 15°. What is the length of arc AB, to the nearest 100 km?

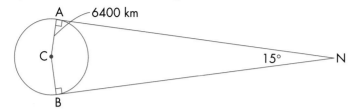

C

27. Trigonometric ratios State the exact value of each of the following. Explain your reasoning.

a) $\sin\left(\dfrac{\pi}{2}\ \text{rad}\right)$

b) $\cos\left(\dfrac{\pi}{3}\ \text{rad}\right)$

c) $\sin\left(\dfrac{5\pi}{6}\right)$

d) $\cos\left(\dfrac{2\pi}{3}\right)$

28. Tire velocity A car is travelling at 100 km/h. The radius of a tire is 36 cm.

a) Find the angular velocity of a tire, to the nearest tenth of a revolution per second.

b) Through how many radians will the tire turn in 30 s, to the nearest radian?

29. Communication The equator and the Arctic Circle have different latitudes. In relation to the Earth's axis of rotation, how do the following compare for a person standing on the equator and a person standing on the Arctic Circle? Explain your reasoning.

a) the angular velocity, in radians per hour

b) the speed, in kilometres per hour

30. Gradians A system of measuring angles uses units called gradians. The circle is divided into 400 equal parts. Each part is called a grad. So one complete revolution measures 400 grad.

a) What are the sine and the cosine of 100 grad?

b) Express 100 grad in degree measure and in exact radian measure.

c) Express one radian in gradian measure in exact form and in approximate form, to the nearest hundredth.

The orbit of the Earth around the sun can be approximately modelled by a circle of radius 150 million kilometres. The Earth completes one revolution every 365 days, and it rotates on its axis once every 24 h. The Earth revolves on its axis in a west to east direction. Assume that the Earth is a sphere with radius 6370 km.

a) Find the speed of the Earth, as it orbits the sun, in kilometres per hour.

b) Find the angular velocity of the Earth, as it orbits the sun, in radians per hour.

c) Find the speed, relative to the Earth's axis, of a point on the equator, in kilometres per hour.

d) Find the speed, relative to the Earth's axis, of the North Pole.

e) Find the speed, relative to the Earth's axis, of Vancouver, which is 49° north of the equator.

f) Find the speed, relative to the Earth's axis, of the top of the CN tower, which is 533 m high and 44° north of the equator.

CAREER CONNECTION *Crafts*

Thousands of Canadians make their living from crafts, such as pottery, glassblowing, textile printing, and making jewelry. Surveys suggest that about 5% of Canadians spend over seven hours a week each on a craft. In some cases, the craft is a hobby, but many people make or supplement their income by working at a craft.

1. Pottery wheel A pottery wheel of radius 16 cm makes 30 r in 10 s.

a) Determine the angular velocity of the wheel, in radians per second.

b) In 0.1 s, what distance does a point on the edge of the wheel travel, to the nearest tenth of a centimetre.

2. Use your research skills to investigate the following for a craft that interests you.

a) how people learn the craft

b) the skills and equipment needed

c) how math is used in the craft

5.2 Trigonometric Ratios of Any Angle

The use of cranes to lift heavy objects is an essential part of the construction and shipping industries. There are many different designs of crane, but they usually include some kind of winding mechanism, called a winch, which pulls or releases a cable.

In one crane design, the cable passes along a straight tube, called a boom or jib, which can be raised or lowered by changing its angle of inclination. The cable passes over a pulley at the upper end of the boom. When the hook at the end of the cable is attached to a heavy object, the winch is used to raise or lower it.

Operating a large crane is a highly skilled occupation. In adjusting the position of the boom, the operator uses the concepts of trigonometry, as will be shown in Example 5.

INVESTIGATE & INQUIRE

Angles that measure 30°, 45°, and 60° occur often in trigonometry. They are sometimes called *special angles*. Right triangles in which they are found are sometimes called *special triangles*.

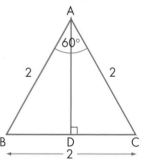

1. △ABC is an equilateral triangle with side lengths of 2 units. AD bisects ∠BAC to form two congruent triangles, △ABD and △ACD. △ABD and △ACD are called 30°-60°-90° triangles. What is the length of BD?

2. Determine the length of AD. Express the answer in radical form.

3. Repeat questions 1 and 2, beginning with equilateral triangles with the following side lengths. Express the length of AD as a mixed radical.
a) 4 units   **b)** 10 units

4. Describe the relationship between the side lengths in any 30°-60°-90° triangle.

5. Use one of the 30°-60°-90° triangles to determine the exact values of the sine, cosine, and tangent of 30° and 60°.

6. △DEF is an isosceles right triangle whose equal sides, DE and DF, have a length of 1 unit. △DEF is called a 45°-45°-90° triangle. Determine the length of EF. Express the answer in radical form.

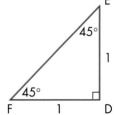

7. Repeat question 6, beginning with isosceles right triangles whose equal sides have the following lengths. Express the length of EF as a mixed radical.
a) 2 units   **b)** 6 units

8. Describe the relationship between the lengths of the sides in any 45°-45°-90° triangle.

9. Use one of the 45°-45°-90° triangles to determine the exact values of the sine, cosine, and tangent of 45°.

_____

The angle θ is shown in standard position.
The point $P(x, y)$ is any point on the terminal arm.
By the Pythagorean theorem, $r = \sqrt{x^2 + y^2}$.
Recall that you have found the sine and cosine ratios of an angle θ in standard position, where $0° \leq \theta \leq 180°$. We will now extend the domain of θ to include angles of any measure, and will include the tangent ratio of any angle in standard position.

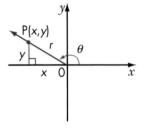

For any angle θ in standard position, with point $P(x, y)$ on the terminal arm, and $r = \sqrt{x^2 + y^2}$, the three primary trigonometric ratios are defined in terms of x, y, and r as follows.

$$\sin \theta = \frac{y}{r} \qquad \cos \theta = \frac{x}{r} \qquad \tan \theta = \frac{y}{x}$$

_____

EXAMPLE 1 Sine, Cosine, and Tangent of Any Angle

The point $P(-3, -6)$ lies on the terminal arm of an angle θ in standard position. Determine the exact values of $\sin \theta$, $\cos \theta$, and $\tan \theta$.

SOLUTION

Sketch the angle in standard position. Construct $\triangle PAO$ by drawing
a line segment perpendicular to the x-axis from the point $P(-3, -6)$.

$r^2 = x^2 + y^2$

$r = \sqrt{x^2 + y^2}$

$\quad = \sqrt{(-3)^2 + (-6)^2}$

$\quad = \sqrt{9 + 36}$

$\quad = \sqrt{45}$

$\quad = 3\sqrt{5}$

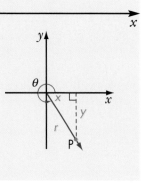

$\sin \theta = \dfrac{y}{r}$ $\cos \theta = \dfrac{x}{r}$ $\tan \theta = \dfrac{y}{x}$

$\quad = \dfrac{-6}{3\sqrt{5}}$ $\quad = \dfrac{-3}{3\sqrt{5}}$ $\quad = \dfrac{-6}{-3}$

$\quad = \dfrac{-2}{\sqrt{5}}$ or $-\dfrac{2}{\sqrt{5}}$ $\quad = \dfrac{-1}{\sqrt{5}}$ or $-\dfrac{1}{\sqrt{5}}$ $\quad = 2$

So, $\sin \theta = -\dfrac{2}{\sqrt{5}}$, $\cos \theta = -\dfrac{1}{\sqrt{5}}$, and $\tan \theta = 2$.

This chart summarizes the signs of trigonometric ratios.

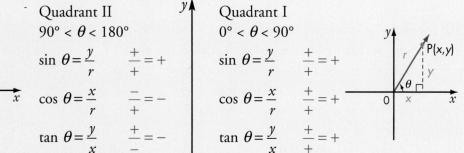

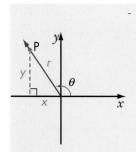

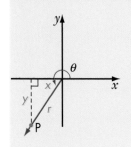

	Quadrant II $90° < \theta < 180°$	Quadrant I $0° < \theta < 90°$	
	$\sin \theta = \dfrac{y}{r}$ $\dfrac{+}{+} = +$	$\sin \theta = \dfrac{y}{r}$ $\dfrac{+}{+} = +$	
	$\cos \theta = \dfrac{x}{r}$ $\dfrac{-}{+} = -$	$\cos \theta = \dfrac{x}{r}$ $\dfrac{+}{+} = +$	
	$\tan \theta = \dfrac{y}{x}$ $\dfrac{+}{-} = -$	$\tan \theta = \dfrac{y}{x}$ $\dfrac{+}{+} = +$	
	Quadrant III $180° < \theta < 270°$	Quadrant IV $270° < \theta < 360°$	
	$\sin \theta = \dfrac{y}{r}$ $\dfrac{-}{+} = -$	$\sin \theta = \dfrac{y}{r}$ $\dfrac{-}{+} = -$	
	$\cos \theta = \dfrac{x}{r}$ $\dfrac{-}{+} = -$	$\cos \theta = \dfrac{x}{r}$ $\dfrac{+}{+} = +$	
	$\tan \theta = \dfrac{y}{x}$ $\dfrac{-}{-} = +$	$\tan \theta = \dfrac{y}{x}$ $\dfrac{-}{+} = -$	

The memory device CAST shows which trigonometric ratios are positive in each quadrant.

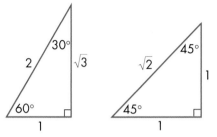

The Pythagorean theorem shows that the side lengths of a 30°-60°-90° triangle have a ratio of 1 to $\sqrt{3}$ to 2, or $1:\sqrt{3}:2$. The side lengths of a 45°-45°-90° triangle have a ratio of 1 to 1 to $\sqrt{2}$, or $1:1:\sqrt{2}$. These special triangles are used to find the exact trigonometric ratios of special angles.

θ in Degrees	θ in Radians	sin θ	cos θ	tan θ
30°	$\dfrac{\pi}{6}$	$\dfrac{1}{2}$	$\dfrac{\sqrt{3}}{2}$	$\dfrac{1}{\sqrt{3}}$
45°	$\dfrac{\pi}{4}$	$\dfrac{1}{\sqrt{2}}$	$\dfrac{1}{\sqrt{2}}$	1
60°	$\dfrac{\pi}{3}$	$\dfrac{\sqrt{3}}{2}$	$\dfrac{1}{2}$	$\sqrt{3}$

EXAMPLE 2 Exact Trigonometric Ratios from Degree Measures

Find the exact values of the sine, cosine, and tangent of 120°.

SOLUTION

Sketch the angle in standard position. Construct a triangle by drawing a perpendicular from the terminal arm to the x-axis. The angle between the terminal arm and the x-axis is 60°. So, the triangle formed is a 30°-60°-90° triangle.
Choose point P on the terminal arm so that $r = 2$.
It follows that $x = -1$ and $y = \sqrt{3}$.

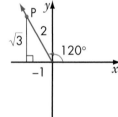

$$\sin \theta = \frac{y}{r} \qquad \cos \theta = \frac{x}{r} \qquad \tan \theta = \frac{y}{x}$$

$$\sin 120° = \frac{\sqrt{3}}{2} \qquad \cos 120° = \frac{-1}{2} \text{ or } -\frac{1}{2} \qquad \tan 120° = \frac{\sqrt{3}}{-1} \text{ or } -\sqrt{3}$$

So, $\sin 120° = \dfrac{\sqrt{3}}{2}$, $\cos 120° = -\dfrac{1}{2}$, and $\tan 120° = -\sqrt{3}$.

EXAMPLE 3 Exact Trigonometric Ratios from Radian Measures

Find the exact values of

a) $\sin \dfrac{7}{4}\pi$

b) $\cos \dfrac{5}{6}\pi$

SOLUTION

a) Sketch the angle in standard position. Construct a triangle by drawing a perpendicular from the terminal arm to the x-axis.

$$\dfrac{7}{4}\pi \text{ rad} = \dfrac{7}{4}\pi\left(\dfrac{180}{\pi}\right)^{\circ}$$
$$= 315^{\circ}$$

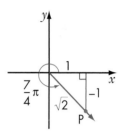

The angle between the terminal arm and the x-axis is 45°.
So, the triangle formed is a 45°-45°-90° triangle.
Choose point P on the terminal arm so that $r = \sqrt{2}$.
It follows that $x = 1$ and $y = -1$.

$$\sin \theta = \dfrac{y}{r}$$

$$\sin \dfrac{7}{4}\pi = \dfrac{-1}{\sqrt{2}} \text{ or } -\dfrac{1}{\sqrt{2}}$$

So, $\sin \dfrac{7}{4}\pi = -\dfrac{1}{\sqrt{2}}$.

b) Sketch the angle in standard position. Construct a triangle by drawing a perpendicular from the terminal arm to the x-axis.

$$\dfrac{5}{6}\pi \text{ rad} = \dfrac{5}{6}\pi\left(\dfrac{180}{\pi}\right)^{\circ}$$
$$= 150^{\circ}$$

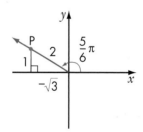

The angle between the terminal arm and the x-axis is 30°.
So, the triangle formed is a 30°-60°-90° triangle.
Choose point P on the terminal arm so that $r = 2$.
It follows that $x = -\sqrt{3}$ and $y = 1$.

$$\cos \theta = \dfrac{x}{r}$$

$$\cos \dfrac{5}{6}\pi = \dfrac{-\sqrt{3}}{2} \text{ or } -\dfrac{\sqrt{3}}{2}$$

So, $\cos \dfrac{5}{6}\pi = -\dfrac{\sqrt{3}}{2}$.

EXAMPLE 4 Trigonometric Ratios of a Right Angle

Find the values of the sine, cosine, and tangent of an angle that measures 90°.

SOLUTION

Sketch the angle in standard position.
Choose P(0, 1) on the terminal arm of the angle.
Therefore, $x = 0$, $y = 1$, and $r = 1$.

$$\sin \theta = \frac{y}{r} \qquad \cos \theta = \frac{x}{r} \qquad \tan \theta = \frac{y}{x}$$

$$\sin 90° = \frac{1}{1} \qquad \cos 90° = \frac{0}{1} \qquad \tan 90° = \frac{1}{0}$$

$$= 1 \qquad\qquad = 0 \qquad\qquad$$

Since division by 0 is not defined, tan 90° is not defined.

So, $\sin 90° = 1$, $\cos 90° = 0$, and tan 90° is not defined.

EXAMPLE 5 Operating a Crane

The boom of a crane can be moved so that its angle of inclination changes. In one location, close to some buildings and some overhead power cables, the minimum value of the angle of inclination is 30°, and the maximum value is 60°. If the boom is 10 m long, find the vertical displacement of the end of the boom when the angle of inclination increases from its minimum value to its maximum value. Express your answer in exact form and in approximate form, to the nearest tenth of a metre.

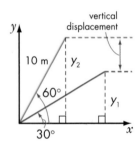

SOLUTION

a) In the diagram, the vertical displacement is $y_2 - y_1$.
Using the exact values of the trigonometric ratios for 30° and 60°, write expressions for y_2 and y_1.

$$\sin 30° = \frac{y_1}{10} \qquad\qquad \sin 60° = \frac{y_2}{10}$$

$$\frac{1}{2} = \frac{y_1}{10} \qquad\qquad \frac{\sqrt{3}}{2} = \frac{y_2}{10}$$

$$\frac{10}{2} = y_1 \qquad\qquad \frac{10\sqrt{3}}{2} = y_2$$

$$5 = y_1 \qquad\qquad 5\sqrt{3} = y_2$$

The vertical displacement is

$y_2 - y_1 = 5\sqrt{3} - 5$

$\doteq 3.7$

The vertical displacement is $5\sqrt{3} - 5$ m in exact form, or 3.7 m in approximate form, to the nearest tenth of a metre.

Key Concepts

- For any angle θ in standard position, with point P(x, y) on the terminal arm, and $r = \sqrt{x^2 + y^2}$, the three primary trigonometric ratios are defined in terms of x, y, and r as follows.

$\sin \theta = \dfrac{y}{r}$ $\cos \theta = \dfrac{x}{r}$ $\tan \theta = \dfrac{y}{x}$

- The memory device CAST shows which trigonometric ratios are positive in each quadrant.

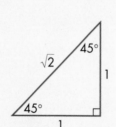

- The two special triangles used to find the exact values of the sine, cosine, and tangent of special angles are as shown.

30°-60°-90° triangle   45°-45°-90° triangle

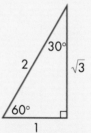

Communicate Your Understanding

1. The point P(3, −4) lies on the terminal arm of an angle θ in standard position. Describe how you would determine the exact values of sin θ, cos θ, and tan θ.

2. Describe how you would find the exact value of $\tan \dfrac{2}{3}\pi$.

3. Describe how you would find the exact value of sin 135°.
4. Describe how you would find the exact value of cos 270°.

Practise

A

1. The coordinates of a point P on the terminal arm of each $\angle\theta$ in standard position are shown, where $0 \leq \theta \leq 2\pi$. Determine the exact values of sin θ, cos θ, and tan θ.

a)
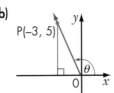

b)
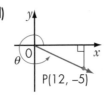

c)
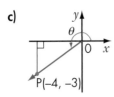

d)
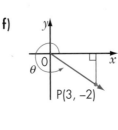

e)
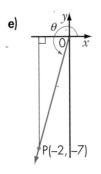

f)
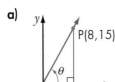

2. The coordinates of a point P on a terminal arm of an $\angle\theta$ in standard position are given, where $0 \leq \theta \leq 2\pi$. Determine the exact values of sin θ, cos θ, and tan θ.
a) P(6, 5)
b) P(−1, 8)
c) P(−2, −5)
d) P(6, −1)
e) P(2, −4)
f) P(−3, −9)
g) P(3, 3)
h) P(−2, 6)

3. Find the exact value of each trigonometric ratio.
a) sin 30°
b) tan 315°
c) cos 240°
d) tan 150°
e) cos 225°
f) sin 45°
g) cos 330°
h) sin 300°

4. Find the exact value of each trigonometric ratio.
a) $\sin \frac{5}{4}\pi$
b) $\tan \frac{11}{6}\pi$
c) $\cos \frac{\pi}{6}$
d) $\cos \frac{7}{4}\pi$
e) $\tan \frac{4}{3}\pi$
f) $\cos \frac{7}{6}\pi$
g) $\sin \frac{5}{6}\pi$
h) $\cos \frac{3}{4}\pi$

Apply, Solve, Communicate

5. Boom crane The arm of a boom crane is 12 m long. Because of the location of the construction site, the angle of inclination of the boom of the crane has a minimum value of 30° and a maximum value of 45°. Find the vertical displacement of the end of the boom as
a) an exact value
b) an approximate value, to the nearest tenth of a metre

6. $\angle\theta$ is in standard position with its terminal arm in the stated quadrant, and $0 \le \theta \le 2\pi$. A trigonometric ratio is given. Find the exact values of the other two trigonometric ratios.

a) $\sin \theta = \dfrac{4}{5}$, quadrant II

b) $\cos \theta = -\dfrac{2}{3}$, quadrant III

c) $\tan \theta = -\dfrac{5}{2}$, quadrant IV

d) $\sin \theta = -\dfrac{3}{7}$, quadrant III

7. $\angle\theta$ is in standard position, and $0 \le \theta \le 2\pi$. A trigonometric ratio is given. Find the exact values of the other two trigonometric ratios.

a) $\sin \theta = \dfrac{1}{3}$

b) $\cos \theta = \dfrac{3}{5}$

c) $\tan \theta = \dfrac{1}{4}$

d) $\cos \theta = -\dfrac{1}{2}$

e) $\tan \theta = -\dfrac{8}{5}$

f) $\sin \theta = -\dfrac{5}{6}$

8. Application A cargo ship is tied up at a dock. At low tide, a 12-m long unloading ramp slopes down from the ship to the dock and makes an angle of 30° to the horizontal. At high tide, the ship is closer to the dock, and the unloading ramp makes an angle of 45° to the horizontal.
a) Determine the change in the horizontal distance from the ship to the dock from low tide to high tide. Express the distance as an exact value and as an approximate value, to the nearest hundredth of a metre.
b) Determine the change in the height of the upper end of the unloading ramp above the dock from low tide to high tide. Express the change in height as an exact value and as an approximate value, to the nearest hundredth of a metre.

9. Bascule bridges Bridges that are hinged so that they can be raised are known as bascule bridges. They are commonly built over water, where they can be opened to allow large boats to pass through. Some bascule bridges are hinged at one end, but others are built in two halves that separate when they are raised. A bascule bridge built in two halves is 80 m long. When the bridge is opened fully, each half of the bridge makes an angle of 60° with the horizontal. When the bridge is opened fully, how far apart are the top ends of the two halves of the bridge?

10. Flagpole A flagpole is anchored by two pairs of guy wires attached 8 m above the ground. One pair is anchored to the ground at 45° angles. The other pair is anchored to the ground at 60° angles. Which pair of guy wires has the greater total length and by how much, to the nearest tenth of a metre?

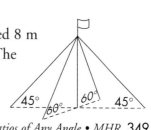

11. If $0° \le A \le 360°$, find the possible measures of $\angle A$.

a) $\sin A = \frac{1}{\sqrt{2}}$ **b)** $\cos A = \frac{1}{2}$ **c)** $\tan A = 1$ **d)** $\sin A = \frac{\sqrt{3}}{2}$

e) $\cos A = -\frac{1}{\sqrt{2}}$ **f)** $\tan A = -\sqrt{3}$ **g)** $\cos A = -\frac{\sqrt{3}}{2}$ **h)** $\sin A = -\frac{1}{2}$

i) $\tan A = -1$ **j)** $\tan A = -\frac{1}{\sqrt{3}}$ **k)** $\cos A = -\frac{1}{2}$

12. The table shows the trigonometric ratios of a right angle, from Example 4. Copy the table and complete it by finding the trigonometric ratios of 0°, 180°, 270°, and 360° angles.

		Angle Measure				
		0°	90°	180°	270°	360°
Ratio	sin θ		1			
	cos θ		0			
	tan θ		n.d.			

13. If $0 \le A \le 2\pi$, and $\cos A = \frac{1}{\sqrt{2}}$, find the values of

a) sin A **b)** tan A

14. If $0 \le A \le 2\pi$, and $\tan A = -\sqrt{3}$, find the values of

a) sin A **b)** cos A

C

15. Inquiry/Problem Solving If the shortest side of a 30°-60°-90° triangle is equal in length to the shortest side of a 45°-45°-90° triangle, what is the ratio of the perimeter of the first triangle to the perimeter of the second triangle, to the nearest hundredth?

16. If the longest side of a 30°-60°-90° triangle is equal in length to the shortest side of a 45°-45°-90° triangle, what is the ratio of the area of the first triangle to the area of the second triangle, to the nearest hundredth?

17. Communication The terminal arm of angle θ in standard position has a slope of -1. If $0 \le \theta \le 2\pi$, what are the possible values of the following? Explain.

a) $\sin \theta$ **b)** $\cos \theta$ **c)** $\tan \theta$

18. Find the exact value of each of the following.

a) $\sin 30° \sin 45° \sin 60°$ **b)** $\sin 30° \cos 30° + \sin 60° \cos 60°$

c) $\sin 60° \cos 30° + \sin 30° \cos 60°$ **d)** $2 \sin 30° \cos 30°$

e) $\frac{5 \tan 60°}{\cos 30°}$

EXTENSION
Positive and Negative Angles of Rotation

The angles you have worked with so far have had measures ranging from 0° to 360°, or 0 to 2π. Other measures are possible.

The measure of an angle in standard position is determined by the amount of rotation from the initial arm to the terminal arm. If the rotation is counterclockwise, the angle is positive. If the rotation is clockwise, the angle is negative.

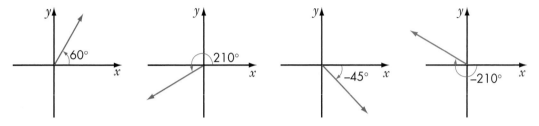

If the terminal arm makes exactly one counterclockwise revolution, the angle has a measure of 360°. If the terminal arm makes exactly one clockwise revolution, the angle has a measure of −360°.

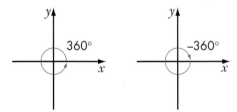

When the terminal arm rotates, it may make one or more revolutions.

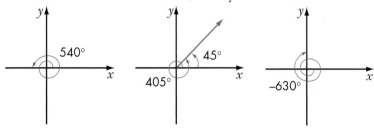

If you draw a 420° angle and a 60° angle in standard position on the same set of axes, the terminal arm of the 420° angle and the terminal arm of the 60° angle are the same. When two angles in standard position have the same terminal arm, they are called **coterminal angles**.

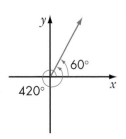

Sine, Cosine, and Tangent of Positive Angles

To find the exact values of the sine, cosine, and tangent of 135°, first draw the angle in standard position. Then, construct a triangle by drawing a perpendicular from the terminal arm to the x-axis. The angle between the terminal arm and the x-axis is 45°.

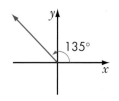

The triangle formed is a 45°-45°-90° triangle.
Choose point P on the terminal arm so that $r = \sqrt{2}$.
It follows that $x = -1$ and $y = 1$.

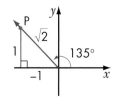

$$\sin \theta = \frac{y}{r} \qquad \cos \theta = \frac{x}{r} \qquad \tan \theta = \frac{y}{x}$$

$$\sin 135° = \frac{1}{\sqrt{2}} \qquad \cos 135° = \frac{-1}{\sqrt{2}} \text{ or } -\frac{1}{\sqrt{2}} \qquad \tan 135° = \frac{1}{-1}$$
$$= -1$$

The same method can be used to find the exact values of the sine, cosine, and tangent of 510°. In this case, the angle between the terminal arm and the x-axis is 30°.

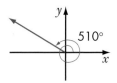

The triangle formed is a 30°-60°-90° triangle.
Choose point P on the terminal arm so that $r = 2$.
It follows that $x = -\sqrt{3}$ and $y = 1$.

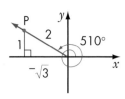

$$\sin \theta = \frac{y}{r} \qquad \cos \theta = \frac{x}{r} \qquad \tan \theta = \frac{y}{x}$$

$$\sin 510° = \frac{1}{2} \qquad \cos 510° = \frac{-\sqrt{3}}{2} \text{ or } -\frac{\sqrt{3}}{2} \qquad \tan 510° = \frac{1}{-\sqrt{3}} \text{ or } -\frac{1}{\sqrt{3}}$$

To find the values of the sine, cosine, and tangent of 630°, first draw the angle in standard position. Choose P(0, −1) on the terminal arm of the angle.

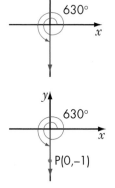

Therefore, $x = 0$, $y = -1$, and $r = 1$.

$$\sin \theta = \frac{y}{r} \qquad \cos \theta = \frac{x}{r} \qquad \tan \theta = \frac{y}{x}$$

$$\sin 630° = \frac{-1}{1} \qquad \cos 630° = \frac{0}{1} \qquad \tan 630° = \frac{-1}{0}$$
$$= -1 \qquad\qquad\qquad = 0 \qquad\qquad \text{Since division by 0 is}$$
$$\text{not defined, } \tan 630° \text{ is}$$
$$\text{not defined.}$$

1. Find the exact values of the sine, cosine, and tangent for each angle.
   a) 390°
   b) 405°
   c) 690°
   d) 450°
   e) $\dfrac{7}{3}\pi$
   f) $\dfrac{13}{4}\pi$
   g) 3π
   h) $\dfrac{10}{3}\pi$

2. Find the exact value of each trigonometric ratio.
   a) cos 495°
   b) tan 480°
   c) sin 660°
   d) cos 720°
   e) tan 570°
   f) sin 390°

Sine, Cosine, and Tangent of Negative Angles

To find the exact values of the sine, cosine, and tangent of −135°, first draw the angle in standard position.

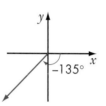

Then, construct a triangle by drawing a perpendicular from the terminal arm to the x-axis.

The angle between the terminal arm and the x-axis is 45°.
The triangle formed is a 45°-45°-90° triangle.
Choose point P on the terminal arm so that $r = \sqrt{2}$.
It follows that $x = -1$ and $y = -1$.

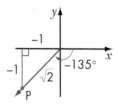

$$\sin\theta = \dfrac{y}{r} \qquad\qquad \cos\theta = \dfrac{x}{r} \qquad\qquad \tan\theta = \dfrac{y}{x}$$

$$\sin(-135°) = \dfrac{-1}{\sqrt{2}} \text{ or } -\dfrac{1}{\sqrt{2}} \quad \cos(-135°) = \dfrac{-1}{\sqrt{2}} \text{ or } -\dfrac{1}{\sqrt{2}} \quad \tan(-135°) = \dfrac{-1}{-1}$$
$$= 1$$

The same method can be used to find the exact values of the sine, cosine, and tangent of −420°. In this case, the angle between the terminal arm and the x-axis is 60°.

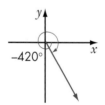

The triangle formed is a 30°-60°-90° triangle.
Choose point P on the terminal arm so that $r = 2$.
It follows that $x = 1$ and $y = -\sqrt{3}$.

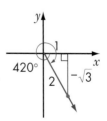

$$\sin\theta = \dfrac{y}{r} \qquad\qquad \cos\theta = \dfrac{x}{r} \qquad \tan\theta = \dfrac{y}{x}$$

$$\sin(-420°) = \dfrac{-\sqrt{3}}{2} \text{ or } -\dfrac{\sqrt{3}}{2} \quad \cos(-420°) = \dfrac{1}{2} \quad \tan(-420°) = \dfrac{-\sqrt{3}}{1} \text{ or } -\sqrt{3}$$

To find the values of the sine, cosine, and tangent for an angle that measures $-180°$, first draw the angle in standard position.
Choose P(-1, 0) on the terminal arm of the angle.
Therefore $x = -1$, $y = 0$, and $r = 1$.

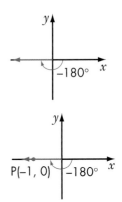

$$\sin \theta = \frac{y}{r} \qquad \cos \theta = \frac{x}{r} \qquad \tan \theta = \frac{y}{x}$$

$$\sin (-180°) = \frac{0}{-1} \qquad \cos (-180°) = \frac{-1}{1} \qquad \tan (-180°) = \frac{0}{-1}$$

$$= 0 \qquad\qquad\qquad = -1 \qquad\qquad\qquad = 0$$

3. Find the exact values of the sine, cosine, and tangent for each angle.
a) $-60°$ b) $-225°$ c) $-90°$ d) $-150°$

e) $-\dfrac{3}{2}\pi$ f) $-\dfrac{11}{6}\pi$ g) $-\dfrac{3}{4}\pi$ h) $-\pi$

4. Find the exact value of each trigonometric ratio.
a) $\sin (-540°)$ b) $\cos (-450°)$
c) $\sin (-630°)$ d) $\cos (-510°)$
e) $\sin (-405°)$ f) $\cos (-675°)$

5.3 Modelling Periodic Behaviour

There are many examples of periodic behaviour in nature. Familiar examples include the rising and setting of the sun, and the rise and fall of tides. The rhythm of the human heartbeat also follows a periodic pattern. Less obvious examples include the motion of sound waves and light waves. Even the populations of some animal species show a periodic pattern in the way they increase and decrease over time.

Web Connection

www.school.mcgrawhill.ca/resources/
To investigate the world's largest Ferris wheel, visit the above web site. Go to **Math Resources**, then to *MATHEMATICS 11*, to find out where to go next. Compare the largest Ferris wheel with Cosmoclock 21.

INVESTIGATE & INQUIRE

One of the world's largest Ferris wheels, Cosmoclock 21, is located in Yokohama City, Japan. The wheel has a diameter of about 100 m.

Suppose that you and a group of friends are riding the Ferris wheel. As shown in the diagram, you are at point A(50, 0) when the last of the 60 gondolas is loaded at point D. The ride then begins with you at point A. The Ferris wheel turns counterclockwise at a constant speed. The wheel takes one minute to complete one revolution.

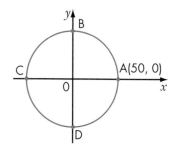

1. Identify the coordinates of the points B, C, and D.

2. Copy and complete the following table by giving your height relative to the *x*-axis after the given rotations.

Rotation of Wheel (degrees)	0	90°	180°	270°	360°	450°	540°	630°	720°
Height, relative to x-axis (metres)	0								

3. Plot the points on a grid like the one shown. Join the points with a smooth curve to show the relationship between your height, *h*, relative to the *x*-axis, and the angle of rotation, *r*.

4. a) Does the graph appear to be linear, quadratic, or neither?
b) Describe the graph.

5. a) Predict the graph of your height, relative to the x-axis, versus the time, in seconds, for two complete revolutions of the wheel. Justify your reasoning.
b) Complete a table of values like the one in question 2, but replace the rotation of the wheel, in degrees, with the time, in seconds.
c) Draw the graph of your height, relative to the x-axis, versus time.
d) Compare your graph with your prediction from part a). Explain any differences.

A function is **periodic** if it has a pattern of y-values that repeats at regular intervals. One complete pattern is called a **cycle**. A cycle may begin at any point on the graph. The horizontal length of one cycle is called the **period** of the function. The period of the function in the graph shown is 4 units.

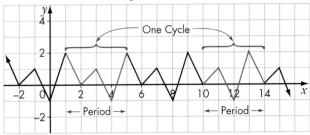

EXAMPLE 1 **Functions**

Determine whether each function is periodic. If it is, state the period.

a)

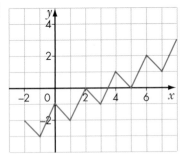

b)

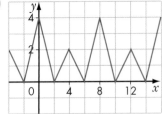

SOLUTION

a) The graph shows a pattern of similar curves, but the pattern of y-values in one section of the graph does not repeat in the other sections of the graph. So, the function is not periodic.

b) The pattern of y-values in one section of the graph repeats at regular intervals in other sections of the graph. This function is periodic.

To determine the period of this function, identify the coordinates of the point at the beginning of one cycle. Next, identify the coordinates of the point at the beginning of the next cycle.

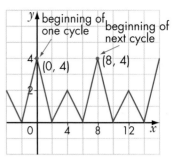

The coordinates of the two points are (0, 4) and (8, 4). Subtract the x-coordinates.

$8 - 0 = 8$

The cycle repeats every 8 units, so the period of this function is 8.

EXAMPLE 2 Finding Function Values

The graph has a period of 7. Find the value of
a) $f(6)$ **b)** $f(20)$

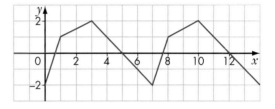

SOLUTION

a) From the graph, $f(6) = -1$.
b) Since the period of this function is 7, then
$$f(6) = f(6 + 7)$$
$$= f(6 + 7 + 7)$$
$$= f(20)$$
Therefore, $f(20) = -1$.

In general, a function f is periodic if there exists a positive number p such that $f(x + p) = f(x)$ for every x in the domain of f. The smallest positive value of p is the period of the function.

In any periodic function, the **amplitude** of the function is defined as half the difference between the maximum value of the function and the minimum value of the function.

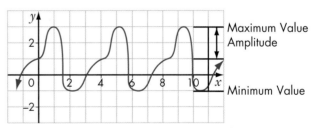

For the function shown, the maximum value is 3 and the minimum value is −1.

$$\text{Amplitude} = \frac{1}{2}(3 - (-1))$$

Note that the amplitude is always positive.

$$= \frac{1}{2}(4)$$
$$= 2$$

The amplitude of this function is 2.

EXAMPLE 3 **Jogging**

André is jogging on a straight boardwalk beside the ocean. The boardwalk is
800 m long, and André begins jogging from one end at a steady pace of
8 km/h.

a) Graph André's distance, in metres, from his starting point versus the
time, in minutes, for four lengths of the boardwalk.

b) Determine the period and amplitude of the function.

c) State the domain and range.

SOLUTION

a) $\text{speed} = \dfrac{\text{distance}}{\text{time}}$

so, $\text{time} = \dfrac{\text{distance}}{\text{speed}}$

The boardwalk is 800 m, or 0.8 km, long.

André jogs one length of the boardwalk in $\dfrac{0.8}{8}$, or 0.1 h, which is 6 min.

After 6 min, André is 800 m from his starting point. After another 6 min,
for a total of 12 min, he is back at his starting point.

Therefore, three points on the graph of distance versus time are (0, 0),
(6, 800), and (12, 0).

These points are joined with straight segments, because André jogs at a
steady pace.

The pattern continues until André completes four lengths of the boardwalk.
The graph is as shown.

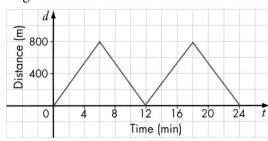

b) The period of the function is 12 min.

The amplitude of the function is given by $\dfrac{1}{2}(800 - 0) = 400.$

So, the amplitude is 400 m.

c) The domain is $0 \le t \le 24$.

The range is $0 \le d \le 800$.

Key Concepts

- A function is periodic if it has a pattern of y-values that repeats at regular intervals.
- One complete pattern of a periodic function is called a cycle.
- The horizontal distance from the beginning of one cycle to the beginning of the next cycle is called the period.
- A function f is periodic if there is a positive number p such that $f(x + p) = f(x)$ for every x in the domain of f. The length of the period is the smallest positive value of p.
- The amplitude of a periodic function is half the difference between the maximum value of the function and the minimum value of the function. The amplitude is always positive.

Communicate Your Understanding

The function shown is periodic. Answer questions 1–3 for this function.

1. Describe how you would determine the period of the function.

2. Describe how you would find

a) $f(4)$ **b)** $f(5)$ **c)** $f(8)$ **d)** $f(13)$

3. Describe how you would determine the amplitude of the function.

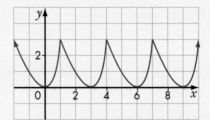

Practise

A

1. Classify the following graphs as periodic or not periodic. Justify your decisions.

a)

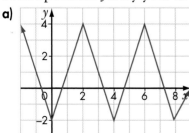

b)

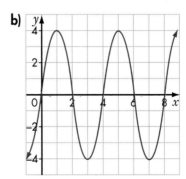

c)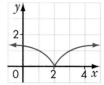

d)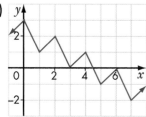

e)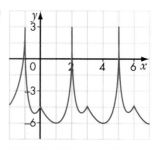

f)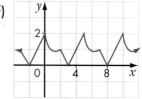

2. Determine the period and amplitude of each of the following functions.

a)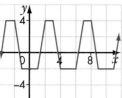

b)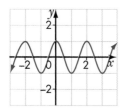

c)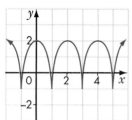

d)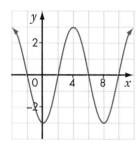

3. Communication Sketch a graph of a periodic function with the given period and amplitude. Compare your graphs with a classmate's.

a) a period of 6 and an amplitude of 4

b) a period of 3 and an amplitude of 5

c) a period of 2 and an amplitude of 2

4. A periodic function f has a period of 12. If $f(7) = -2$ and $f(11) = 9$, determine the value of

a) $f(43)$ **b)** $f(79)$ **c)** $f(95)$ **d)** $f(-1)$

5. Determine the maximum value, the minimum value, the amplitude, the period, the domain, and the range of each function.

a)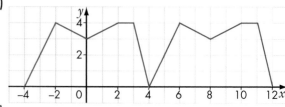

b)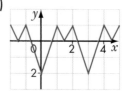

Apply, Solve, Communicate

B

6. Application Anh is training by swimming lengths of a 50-m pool at a constant speed. She swims one length every minute.
a) Graph her distance, in metres, from her starting point versus the time, in minutes, for six lengths of the pool.
b) Determine the period and amplitude of the function.
c) State the domain and range.
d) After a month of training, Anh increases her speed by 10%. If she swims six lengths of the pool, how do the period and amplitude compare with the values from part b)? Explain.

7. Inquiry/Problem Solving Is it possible to model each situation described below with a periodic function? Explain.
a) the sunrise times for Lively, Ontario, recorded daily for three years
b) the number of passenger cars purchased annually in Ontario from 1990 to 2000
c) the average monthly temperature in your community, recorded each month for two years

8. Ferris wheel At a county fair, the Ferris wheel has a diameter of 32 m, and its centre is 18 m above the ground. The wheel completes one revolution every 30 s.
a) Graph a rider's height above the ground, in metres, versus the time, in seconds, during a 2-min ride. The rider begins at the lowest position on the wheel.
b) Determine the period and amplitude of this function.
c) State the domain and range of the function.

9. Climate The mid-season high temperatures for Dorset, Ontario, were recorded over a three-year period. The results are shown in the table.
a) Plot a graph of the temperature versus the date. Draw a periodic function that models the data as accurately as possible.
b) Use the graph to estimate the approximate period and amplitude of the function.

Season	Date	Temperature (°C)
Winter	Feb. 5, 1998	−9
Spring	May 2, 1998	16
Summer	Aug. 3, 1998	25
Fall	Nov. 2, 1998	3
Winter	Feb. 5, 1999	−10
Spring	May 2, 1999	17
Summer	Aug. 3, 1999	27
Fall	Nov. 2, 1999	3
Winter	Feb. 5, 2000	−10
Spring	May 2, 2000	16
Summer	Aug. 3, 2000	26
Fall	Nov. 2, 2000	3

10. **Ocean cycles** The cycle of ocean tides represents periodic behaviour and can be modelled with a periodic function. Each day, at various locations around the world, the height of the tide above the mean low-water level is recorded. Data for low tide and high tide from one location are shown in the table. Estimate the period and amplitude of the function that you would use to model the tide cycle at this location.

Time	Tide Height (m)
10:40	0.3
16:52	2.7
23:05	0.3
05:17	2.7

11. **Research** Use your research skills to describe one example of a periodic function not mentioned in this section. Sketch the function, and describe its period and amplitude.

12. **Pose and solve problems** Pose a problem related to each of the following. Check that you are able to solve each problem. Then, have a classmate solve it.
a) ocean cycles
b) Ferris wheels

C

13. If a periodic function has a whole number of cycles, what is the relationship between the period and the domain? Explain.

14. How is the amplitude of a periodic function related to its range? Explain.

15. **Earth's rotation** A point on the Earth's equator rotates about the Earth's axis. The distance of the point from its starting position is a periodic function of the time.
a) What is the period of the function? Explain.
b) What is the amplitude of the function? Explain.

NUMBER *Power*

If $\dfrac{9^{28} - 9^{27}}{8} = 3^x$, what is the value of x?

5.4 Investigation: Sketching the Graphs of $f(x) = \sin x$, $f(x) = \cos x$, and $f(x) = \tan x$

Graphing $y = \sin x$

1. Copy and complete the following table of values for $y = \sin x$, $0° \le x \le 360°$. Include the exact value of $\sin x$ for each value of x. Also include the decimal value of $\sin x$, rounded to the nearest tenth, if necessary.

Value of x (radians)	0	$\frac{\pi}{6}$	$\frac{\pi}{3}$	$\frac{\pi}{2}$	$\frac{2\pi}{3}$	$\frac{5\pi}{6}$	π	$\frac{7\pi}{6}$	$\frac{4\pi}{3}$	$\frac{3\pi}{2}$	$\frac{5\pi}{3}$	$\frac{11\pi}{6}$	2π
Value of x (degrees)	0	30	60	90	120	150	180	210	240	270	300	330	360
Exact Value of sin x	0	$\frac{1}{2}$	$\frac{\sqrt{3}}{2}$										
Decimal Value of sin x	0	0.5	0.9										

2. Use the decimal values of $\sin x$ and plot the ordered pairs $(x, \sin x)$ on a grid, like the one shown. Join the points with a smooth, continuous curve.

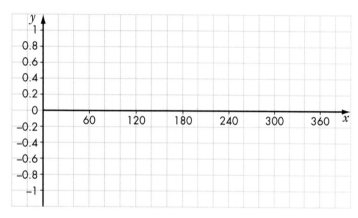

3. a) What is the maximum value of $y = \sin x$?
b) For what value of x does the maximum value of y occur?

4. a) What is the minimum value of $y = \sin x$?
b) For what value of x does the minimum value of y occur?

5. What is the amplitude of $y = \sin x$?

6. a) Make a conjecture about the appearance of the graph of $y = \sin x$ for the domain $0° \le x \le 720°$.
b) Use your conjecture to sketch the graph of $y = \sin x$ for the domain $0° \le x \le 720°$.

7. Test your conjecture from question 6a) by graphing $y = \sin x$ on a graphing calculator. Using the mode settings, select the degree mode. Adjust the window variables to include Xmin = 0, Xmax = 720, Ymin = −1.5, and Ymax = 1.5. Display the graph. Compare the result with your sketch from question 6b).

8. a) Make a conjecture about the appearance of the graph of $y = \sin x$ for the domain $-360° \le x \le 720°$.
b) Use your conjecture to sketch the graph of $y = \sin x$ for the domain $-360° \le x \le 720°$.

9. Test your conjecture from question 8a) by graphing $y = \sin x$ on a graphing calculator. In degree mode, adjust the window variables to include Xmin = −360, Xmax = 720, Ymin = −1.5, and Ymax = 1.5. Display the graph. Compare the result with your sketch from question 8b).

10. a) Is the graph of $y = \sin x$ periodic? Explain.
b) If so, what is the period of the graph of $y = \sin x$?

11. How can you verify that $y = \sin x$ is a function?

12. For the function $y = \sin x$, what is
a) the domain?   **b)** the range?

Graphing y = cos x

13. Copy and complete the following table of values for $y = \cos x$, $0° \le x \le 360°$. Include the exact value of $\cos x$ for each value of x. Also include the decimal value of $\cos x$, rounded to the nearest tenth, if necessary.

Value of x (radians)	0	$\frac{\pi}{6}$	$\frac{\pi}{3}$	$\frac{\pi}{2}$	$\frac{2\pi}{3}$	$\frac{5\pi}{6}$	π	$\frac{7\pi}{6}$	$\frac{4\pi}{3}$	$\frac{3\pi}{2}$	$\frac{5\pi}{3}$	$\frac{11\pi}{6}$	2π
Value of x (degrees)	0	30	60	90	120	150	180	210	240	270	300	330	360
Exact Value of cos x													
Decimal Value of cos x													

14. Use the decimal values of $\cos x$ and graph $y = \cos x$, $0° \le x \le 360°$, on the same set of axes as $y = \sin x$.

15. What is the amplitude of $y = \cos x$?

16. a) Make a conjecture about the appearance of the graph of $y = \cos x$ for the domain $0° \le x \le 720°$.

b) Use your conjecture to sketch the graph of $y = \cos x$ for the domain $0° \le x \le 720°$.

17. Test your conjecture from question 16a) by graphing $y = \cos x$ on a graphing calculator. Compare the result with your sketch from question 16b).

18. a) Make a conjecture about the appearance of the graph of $y = \cos x$ for the domain $-360° \le x \le 720°$.

b) Use your conjecture to sketch the graph of $y = \cos x$ for the domain $-360° \le x \le 720°$.

19. Test your conjecture from question 18a) by graphing $y = \cos x$, $-360° \le x \le 720°$, on a graphing calculator. Compare the result with your sketch from question 18b).

20. a) Is the graph of $y = \cos x$ periodic? Explain.
b) If so, what is the period of the graph of $y = \cos x$?

21. How can you verify that $y = \cos x$ is a function?

22. For the function $y = \cos x$, what is
a) the domain?   **b)** the range?

23. Compare the graphs of $y = \sin x$ and $y = \cos x$. How are they the same? How are they different?

Graphing $y = \tan x$

24. Copy and complete the following table of values for $y = \tan x$. Round decimal values of $\tan x$ to the nearest tenth, if necessary.

Value of x (degrees)	0	45	60	70	80	90	100	110	120	135	150	180
Decimal Value of tan x												
Value of x (degrees)	225	240	250	260	270	280	290	300	315	330	360	
Decimal Value of tan x												

25. Graph $y = \tan x$ on a grid like the one shown.

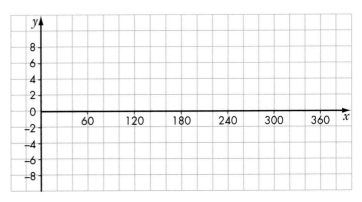

26. What happens to the value of tan x as x increases from 0° to 90°?

27. a) What is the value of tan x when $x = 90°$?
b) How is this value of tan x shown on the graph?

28. What happens to the value of tan x as x increases from 90° to 270°?

29. a) What is the value of tan x when $x = 270°$?
b) How is this value of tan x shown on the graph?

30. a) Make a conjecture about the appearance of the graph of $y = \tan x$ for the domain $0° \le x \le 720°$.
b) Use your conjecture to sketch the graph of $y = \tan x$ for the domain $0° \le x \le 720°$.

31. Test your conjecture from question 30a) by graphing $y = \tan x$, $0° \le x \le 720°$, on a graphing calculator. Compare the result with your sketch from question 30b).

32. a) Make a conjecture about the appearance of the graph of $y = \tan x$ for the domain $-360° \le x \le 720°$.
b) Use your conjecture to sketch the graph of $y = \tan x$ for the domain $-360° \le x \le 720°$.

33. Test your conjecture from question 32a) by graphing $y = \tan x$, $-360° \le x \le 720°$, on a graphing calculator. Compare the result with your sketch from question 32b).

34. a) Is the graph of $y = \tan x$ periodic? Explain.
b) If so, what is the period of the graph of $y = \tan x$?

35. Does $y = \tan x$ have a maximum value? a minimum value? Explain.

36. How can you verify that $y = \tan x$ is a function?

37. For the function $y = \tan x$, what is
a) the domain?   **b)** the range?

5.5 Stretches of Periodic Functions

Light waves can be modelled by sine functions. The
graphs model waves of red and blue light. The units
of the scale on the x-axis are nanometres, where one
nanometre (1 nm) is a billionth of a metre, or 10^{-9} m.

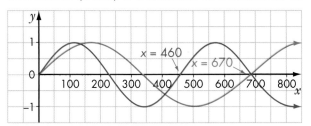

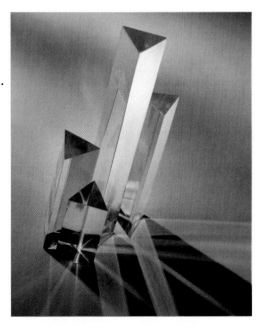

Notice that the graphs for red and blue light are
transformations of $y = \sin x$. These graphs will be
used in Example 5.

INVESTIGATE & INQUIRE

1. Copy and complete the table by finding decimal values for sin x and 2sin x.
Round values to the nearest tenth, if necessary.

x (degrees)	0	30	60	90	120	150	180	210	240	270	300	330	360
sin x	0	0.5											
2sin x	0	1											

2. Sketch the graphs of the
functions $y = \sin x$ and
$y = 2\sin x$ on the same grid
like the one shown.

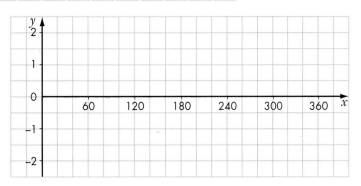

3. For the graphs from question 2, find
a) the amplitudes   b) the periods   c) any invariant points

4. What transformation can be applied to $y = \sin x$ to give $y = 2\sin x$?

5. a) Make a conjecture about how the graph of $y = \dfrac{1}{2}\sin x$ compares with the graph of $y = \sin x$.

b) Use your conjecture to sketch the graphs of $y = \sin x$ and $y = \dfrac{1}{2}\sin x$ on the same grid for the domain $0° \le x \le 360°$.

6. Test your conjecture from question 5a) by graphing $y = \sin x$ and $y = \dfrac{1}{2}\sin x$, $0° \le x \le 360°$, on a graphing calculator. Compare the result with your graph from question 5b).

7. For the graphs from question 6, find
a) the amplitudes   **b)** the periods   **c)** any invariant points

8. What transformation can be applied to $y = \sin x$ to give $y = \dfrac{1}{2}\sin x$?

9. If $a > 0$, write a statement describing the transformational effect of a on the graph of $y = a\sin x$.

10. If $a > 0$, write a conjecture about the transformational effect of a on the graph of $y = a\cos x$.

11. Test your conjecture from question 10 by graphing $y = \cos x$, $y = 2\cos x$, and $y = \dfrac{1}{2}\cos x$, $0° \le x \le 360°$, on a graphing calculator.

12. For the graphs from question 11, find
a) the amplitudes   **b)** the periods   **c)** any invariant points

INVESTIGATE & INQUIRE

1. Copy and complete the tables of values for $y = \sin x$ and $y = \sin 2x$. Round values to the nearest tenth, if necessary.

x (degrees)	0	30	60	90	120	150	180	210	240	270	300	330	360
sin x	0	0.5											
sin 2x	0	0.9											

2. Sketch the graphs of the functions $y = \sin x$ and $y = \sin 2x$ on the same grid.

3. For the graphs from question 2, find
a) the amplitudes   **b)** the periods   **c)** any invariant points

4. What transformation can be applied to $y = \sin x$ to give $y = \sin 2x$?

5. a) Make a conjecture about how the graph of $y = \sin \frac{1}{2}x$ compares with the graph of $y = \sin x$.

b) Use your conjecture to sketch the graphs of $y = \sin x$ and $y = \sin \frac{1}{2}x$ on the same grid for the domain $0° \le x \le 720°$.

6. Test your conjecture from question 5a) by graphing $y = \sin x$ and $y = \sin \frac{1}{2}x$, $0° \le x \le 720°$, on a graphing calculator. Compare the result with your graph from question 5b).

7. For the graphs from question 6, find
a) the amplitudes   **b)** the periods   **c)** any invariant points

8. What transformation can be applied to $y = \sin x$ to give $y = \sin \frac{1}{2}x$?

9. If $k > 0$, write a statement about the transformational effect of k on the graph of $y = \sin kx$.

10. If $k > 0$, write a conjecture about the transformational effect of k on the graph of $y = \cos kx$.

11. Test your conjecture from question 10 by graphing $y = \cos x$, $y = \cos 2x$, and $y = \cos \frac{1}{2}x$, $0° \le x \le 720°$, on a graphing calculator.

12. For the graphs from question 11, find
a) the amplitudes   **b)** the periods   **c)** any invariant points

---

The transformations that apply to algebraic functions also apply to trigonometric functions.

If $a > 1$, the graphs of $y = a\sin x$ and $y = a\cos x$ are stretched vertically by a factor of a. The amplitude of each function is a.

If $0 < a < 1$, the graphs of $y = a\sin x$ and $y = a\cos x$ are compressed vertically by a factor of a. The amplitude of each function is a.

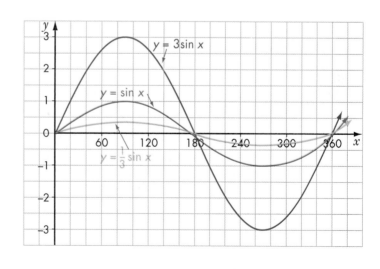

The five-point method is a convenient way to sketch the graph of a sine or cosine function using its amplitude and period. This method uses the fact that one cycle of a sine or cosine function includes a maximum, a minimum, and three zeros. These five key points are equally spaced along the x-axis, so they divide the period into quarters.

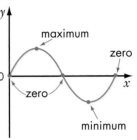

Suppose the graph of a sine function has an amplitude of 3 and a period of 2π. Because the five key points divide the period into quarters, the coordinates of the five key points are $(0, 0)$, $\left(\dfrac{\pi}{2}, 3\right)$, $(\pi, 0)$, $\left(\dfrac{3\pi}{2}, -3\right)$, and $(2\pi, 0)$.

To sketch the graph, use scales on the y-axis and x-axis that are about equal. Use $\pi \doteq 3$. Plot the five key points in the cycle. Draw a smooth curve through the points.

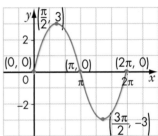

EXAMPLE 1 Sketching $y = a\cos x$

Sketch one cycle of the graph of $y = 4\cos x$, starting at $(0, 0)$, $x \geq 0$. State the domain and range of the cycle.

Note that the transformations shown in this chapter can be performed using a graphing software program, such as Zap-a-Graph. For details of how to do this, refer to the Zap-a-Graph section of Appendix C.

SOLUTION

The graph of $y = 4\cos x$ is the graph of $y = \cos x$ expanded vertically by a factor of 4.
The amplitude is 4, so the maximum value is 4 and the minimum value is -4.
The period of the function $y = 4\cos x$ is 2π.

Use the five-point method to sketch the graph.
The five key points divide the period into quarters.
Therefore, the coordinates of the five key points are

$(0, 4)$, $\left(\dfrac{\pi}{2}, 0\right)$, $(\pi, -4)$, $\left(\dfrac{3\pi}{2}, 0\right)$, and $(2\pi, 4)$.

Plot the 5 key points in the cycle. Draw a smooth curve through the points. Label the graph.

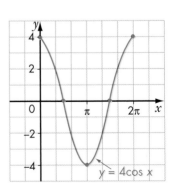

The domain of the cycle is $0 \leq x \leq 2\pi$.
The range of the cycle is $-4 \leq y \leq 4$.

If $k > 1$, the graphs of $y = \sin kx$ and $y = \cos kx$ are compressed horizontally by a factor of $\dfrac{1}{k}$.

The period of each function is $\dfrac{2\pi}{k}$ or $\dfrac{360°}{k}$.

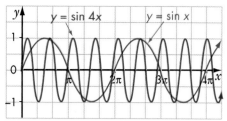

If $0 < k < 1$, the graphs of $y = \sin kx$ and $y = \cos kx$ are expanded horizontally by a factor of k. The period of each function is $\dfrac{2\pi}{k}$ or $\dfrac{360°}{k}$.

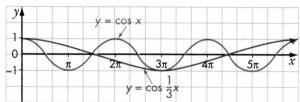

Example 2 Sketching $y = \sin kx$

Sketch one cycle of the graph of $y = \sin 3x$, starting at $(0, 0)$, $x \geq 0$.
State the domain and range of the cycle.

SOLUTION

The graph of $y = \sin 3x$ is the graph of $y = \sin x$ compressed horizontally by a factor of $\dfrac{1}{3}$.

The amplitude is 1, so the maximum value is 1 and the minimum value is -1.

The period of the function is $\dfrac{2\pi}{3}$.

Use the five-point method to sketch the graph.
The five key points divide the period into quarters.
Therefore, the coordinates of the five key points are

$(0, 0)$, $\left(\dfrac{\pi}{6}, 1\right)$, $\left(\dfrac{\pi}{3}, 0\right)$, $\left(\dfrac{\pi}{2}, -1\right)$, and $\left(\dfrac{2\pi}{3}, 0\right)$.

Plot the 5 key points in the cycle. Draw a smooth curve through the points. Label the graph.

The domain of the cycle is $0 \leq x \leq \dfrac{2\pi}{3}$.

The range of the cycle is $-1 \leq y \leq 1$.

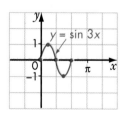

EXAMPLE 3 Sketching a Sine Function

A sine function has an amplitude of 4 and a period of 6π. If one cycle of the graph begins at the origin, and $x \geq 0$, sketch one cycle of this sine function.

SOLUTION

The cycle begins at the origin and the amplitude is 4, so the maximum value is 4 and the minimum value is −4.
The period of the function is 6π.

Use the five-point method to sketch the graph. The five key points divide the period into quarters. Therefore, the coordinates of the five key points are

$(0, 0)$, $\left(\dfrac{3\pi}{2}, 4\right)$, $(3\pi, 0)$, $\left(\dfrac{9\pi}{2}, -4\right)$, and $(6\pi, 0)$.

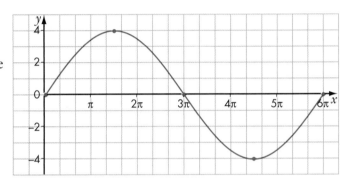

Plot the 5 key points in the cycle. Draw a smooth curve through the points.

EXAMPLE 4 Graphing $y = a\cos kx$, $-\pi \leq x \leq \pi$

Sketch the graph of $y = 3\cos 2x$ over the domain $-\pi \leq x \leq \pi$.

SOLUTION

The graph of $y = 3\cos 2x$ is the graph of $y = \cos x$ expanded vertically by a factor of 3 and compressed horizontally by a factor of $\dfrac{1}{2}$.

The amplitude is 3, so the maximum value is 3 and the minimum value is −3.
The period of the function is $\dfrac{2\pi}{2}$, or π.

Use the five-point method to sketch the graph for $0 \leq x \leq \pi$.
The five key points divide the period into quarters.
Therefore, the coordinates of the five key points are

$(0, 3)$, $\left(\dfrac{\pi}{4}, 0\right)$, $\left(\dfrac{\pi}{2}, -3\right)$, $\left(\dfrac{3\pi}{4}, 0\right)$, and $(\pi, 3)$.

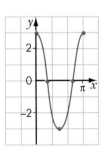

Plot the 5 key points in the cycle. Draw a smooth curve through the points.

Use the pattern to sketch the graph over required domain. Label the graph.

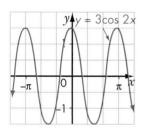

You can verify the graph using a graphing calculator in radian mode, with Xmin = 0 and Xmax = π.

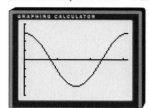

EXAMPLE 5 Writing the Equations of Light Waves

The sine graphs model waves of red light and blue light, where the units of x are nanometres. Write equations for red light waves and blue light waves
a) using exact values
b) using approximate values, to the nearest thousandth

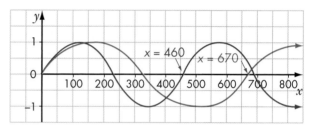

SOLUTION

a) The amplitude for both colours is 1.

From the graph for red light, one cycle has a length of 670 nm, so the period is 670.

$$\text{Period} = \frac{2\pi}{k}$$

$$670 = \frac{2\pi}{k}$$

$$k = \frac{2\pi}{670} \text{ or } \frac{\pi}{335}$$

The equation for red light waves is $y = \sin\left(\frac{\pi x}{335}\right)$, using exact values.

From the graph for blue light, one cycle has a length of 460 nm, so the period is 460.

$$\text{Period} = \frac{2\pi}{k}$$

$$460 = \frac{2\pi}{k}$$

$$k = \frac{2\pi}{460} \text{ or } \frac{\pi}{230}$$

The equation for blue light waves is $y = \sin\left(\frac{\pi x}{230}\right)$.

b) In the equation for red light,

$$k = \frac{\pi}{335}$$

$$\doteq 0.009$$

The approximate equation for red light waves is $y = \sin 0.009x$.

b) In the equation for blue light,

$$k = \frac{\pi}{230}$$

$$\doteq 0.014$$

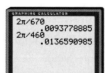

The approximate equation for blue light waves is $y = \sin 0.014x$.

Key Concepts

- The vertical and horizontal stretches of sine and cosine functions can be summarized as follows.

Stretch	Mathematical Form	Effect
Vertical	$y = a\sin x$ $y = a\cos x$	If $a > 1$, the graph is expanded vertically by a factor of a. If $0 < a < 1$, the graph is compressed vertically by a factor of a.
Horizontal	$y = \sin kx$ $y = \cos kx$	If $k > 1$, the graph is compressed horizontally by a factor of $\frac{1}{k}$. If $0 < k < 1$, the graph is expanded horizontally by a factor of $\frac{1}{k}$.

- For $y = a\sin x$ or $y = a\cos x$, $a > 0$, the amplitude is a.
- For $y = \sin kx$ or $y = \cos kx$, $k > 0$, the period is $\dfrac{2\pi}{k}$ or $\dfrac{360°}{k}$.

Communicate Your Understanding

1. Describe how you would find the amplitude and period of the function $y = 2\cos 4x$.

2. Describe how you would find the amplitude and period of the graph shown.

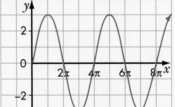

3. Describe how you would sketch one cycle of the graph of $y = \dfrac{1}{2}\sin 3x$, starting at $(0, 0)$, $x \geq 0$.

4. Describe how you would sketch one cycle of the graph of $y = 5 \cos \dfrac{1}{2}x$, $-\pi \leq x \leq \pi$.

Practise

A

1. Sketch one cycle of the graph of each of the following. State the domain and range of the cycle.

a) $y = 3\sin x$

b) $y = 5\cos x$

c) $y = 1.5\sin x$

d) $y = \dfrac{2}{3}\cos x$

2. Find the period, in degrees and radians, for each of the following.

a) $y = \sin 6x$

b) $y = \cos 4x$

c) $y = \cos \dfrac{2}{3}x$

d) $y = 8\sin \dfrac{2}{3}x$

e) $y = 5\sin \dfrac{1}{6}x$

f) $y = 7\cos 8x$

3. Sketch one cycle of the graph of each of the following. State the domain and range of the cycle.

a) $y = \sin 2x$

b) $y = \cos 3x$

c) $y = \sin 6x$

d) $y = \cos \dfrac{1}{4}x$

e) $y = \sin \dfrac{3}{4}x$

f) $y = \cos \dfrac{1}{3}x$

4. Write the equation and sketch one complete cycle for each sine function described. Each graph begins at the origin, $x \geq 0$.

a) amplitude 6, period 180°

b) amplitude 1.5, period 240°

c) amplitude 0.8, period 3π

d) amplitude 4, period 6π

5. Write the equation and sketch one cycle for each cosine function described.

a) amplitude 3, period 180°

b) amplitude 0.5, period 720°

c) amplitude 4, period 4π

d) amplitude 2.5, period 5π

6. Sketch one cycle of the graph of each of the following.

a) $y = 3\sin 2x$

b) $y = 4\cos 3x$

c) $y = \dfrac{1}{2}\sin \dfrac{1}{2}x$

d) $y = \dfrac{5}{3}\sin \dfrac{2}{3}x$

e) $y = 2\sin 4x$

f) $y = 2.5\cos \dfrac{4}{5}x$

7. Determine the equation for each sine function.

a)

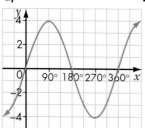

b)

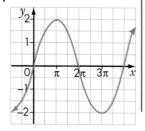

c)

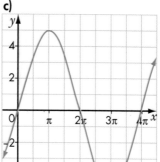

d)

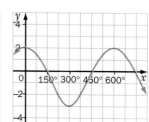

8. Determine the equation for each cosine function.

a)

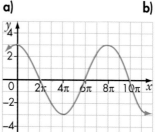

b)

c)

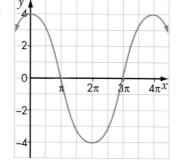

d)

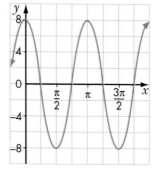

9. Sketch the graphs of the following functions.
a) $y = 4\sin x, -2\pi \le x \le 2\pi$

b) $y = \dfrac{1}{2}\cos 3x, -\pi \le x \le \pi$

c) $y = 3\cos 4x, -2\pi \le x \le \pi$

d) $y = 2\sin \dfrac{1}{5}x, -5\pi \le x \le 5\pi$

10. Identify any invariant points when $y = \sin x$ is transformed onto each of the following.
a) $y = 5\sin x$
b) $y = \sin 4x$
c) $y = 2\sin 3x$
d) $y = 0.5\sin 0.5x$

11. Identify any invariant points when $y = \cos x$ is transformed onto each of the following.
a) $y = 2.5\cos x$
b) $y = \cos 3x$
c) $y = 3\cos 2x$
d) $y = 4\cos 0.5x$

12. State the domain, range, amplitude, and period of each of the following functions.
a) $y = 3.5\sin x$
b) $y = \cos 2.5x$
c) $y = 2\sin \dfrac{1}{6}x$
d) $y = \dfrac{1}{4}\cos \dfrac{1}{2}x$

Apply, Solve, Communicate

13. **Colours** The sine graphs model waves of yellow, green, and violet light. The units of the scale on the x-axis are nanometres. Write equations for the waves
a) using exact values
b) using approximate values, to the nearest thousandth

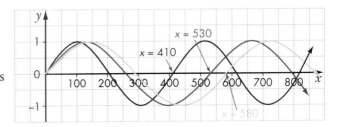

B

14. **Communication** Graph $y = \sin x$ and $y = \cos x$, for $0 \le x \le 360°$.
a) For what values of x is $\sin x > \cos x$?
b) For what values of x is $\sin x < \cos x$?
c) For what values of x is $\sin x = \cos x$?

15. **Sound** A pure tone made by a tuning fork can be observed as a sine curve on a piece of scientific equipment called an oscilloscope. For the note A, above middle C, the equation of the curve is $y = 10\sin 880\pi x$.
a) Determine the amplitude and the period.
b) Describe how $y = \sin x$ could be transformed onto $y = 10\sin 880\pi x$.

16. **Application** The frequency of a periodic function is defined as the number of cycles completed in a second. This is the same as the reciprocal of the period of a periodic function and is typically measured in Hertz (Hz). If t is measured in seconds, what is the frequency of
a) $y = \sin t$?
b) $y = \cos t$?
c) $y = \sin 2t$?
d) $y = 4\cos \dfrac{t}{2}$?

17. Ocean cycles A point on the ocean rises and falls as waves pass. Suppose that a wave passes every 4 s, and the height of each wave from the crest to the trough is 0.5 m.

a) Sketch a graph to model the height of the point relative to its average height for a complete cycle, starting at the crest of a wave.

b) Use exact values to write an equation of the form $h(t) = a\cos kt$ to model the height of the point, $h(t)$ metres, relative to its average height, as a function of time, t seconds.

c) If the times on the t-axis were changed from seconds to minutes, what would be the transformational effect on the graph, and what would be the new equation?

18. Biorhythms According to biorhythm theory, three cycles affect people's lives, giving them favourable and non-favourable days. The physical cycle has a 23-day period, the emotional cycle has a 28-day period, and the intellectual cycle has a 33-day period. The cycles can be shown graphically as sine curves with amplitude 1 and with the person's date of birth being considered as the start of each type of cycle.

a) Write a sine function to represent each type of cycle.

b) Graph the cycles for the first 100 days of someone's life.

c) How old is the person the first time that all three types of cycle reach a maximum on the same day?

19. Electricity The voltage, V, in volts, of a household alternating current circuit is given by $V(t) = 170\sin 120\pi t$, where t is the time in seconds.

a) Determine the amplitude and the period for this function.

b) The number of cycles completed in 1 s is the frequency of the current. Determine the frequency.

c) Graph two cycles of the function.

20. Electricity A generator produces a voltage, V, in volts, given by $V(t) = 120\cos 30\pi t$, where t is the time in seconds. Graph the function for $0 \le t \le 0.5$.

C

21. Inquiry/Problem Solving a) Use sketches to predict the graphs of
$y = \dfrac{1}{\sin x}$ and $y = \dfrac{1}{\cos x}$, $0 \le x \le 4\pi$.

b) Use a graphing calculator to check your predictions.

5.6 Translations and Combinations of Transformations

The highest tides in the world are found in the Bay of Fundy. Tides in one area of the bay cause the water level to rise to 6 m above average sea level and to drop to 6 m below average sea level. The tide completes one cycle approximately every 12 h. The depth of the water can be modelled by a sine function. This function will be modelled in Example 6.

INVESTIGATE & INQUIRE

1. Copy and complete the table by finding decimal values for sin x and sin $x + 2$. Round values to the nearest tenth, if necessary.

x (degrees)	0	45	90	135	180	225	270	315	360
sin x	0	0.7							
sin x + 2	0	2.7							

2. Sketch the graphs of the functions $y = \sin x$ and $y = \sin x + 2$ on the same grid, like the one shown.

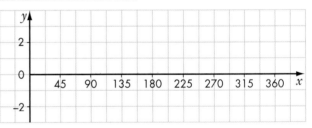

3. For the graphs from question 2, find
a) the amplitudes **b)** the periods

4. What transformation can been applied to $y = \sin x$ to give $y = \sin x + 2$?

5. a) Make a conjecture about how the graph of $y = \sin x - 1$ compares with the graph of $y = \sin x$.
b) Use your conjecture to sketch the graphs of $y = \sin x$ and $y = \sin x - 1$ on the same grid for the domain $0° \le x \le 360°$.

6. Test your conjecture from question 5a) by graphing $y = \sin x$ and $y = \sin x - 1$, $0° \le x \le 360°$, on a graphing calculator. Compare the result with your graph from question 5b).

7. For the graphs from question 6, find
a) the amplitudes   **b)** the periods

8. What transformation can been applied to $y = \sin x$ to give $y = \sin x - 1$?

9. Write a statement about the transformational effect of c on the graph of $y = \sin x + c$.

10. Write a conjecture about the transformational effect of c on the graph of $y = \cos x + c$.

11. Test your conjecture from question 10 by graphing $y = \cos x$, $y = \cos x + 2$, and $y = \cos x - 1$, $0° \le x \le 360°$, on a graphing calculator.

12. For the graphs from question 11, find
a) the amplitudes   **b)** the periods

INVESTIGATE & INQUIRE

1. Copy and complete the tables by finding values of $\sin x$ and $\sin (x + 45°)$. Round values to the nearest tenth, if necessary.

x (degrees)	0	45	90	135	180	225	270	315	360
sin x	0	0.7							
sin (x + 45)	0.7	1							

2. Sketch the graphs of the functions $y = \sin x$ and $y = \sin (x + 45°)$ on the same grid, like the one shown.

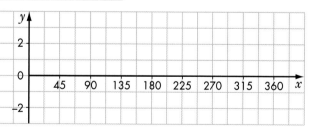

3. For the graphs from question 2, find
a) the amplitudes   **b)** the periods

4. What transformation can been applied to $y = \sin x$ to give $y = \sin (x + 45°)$?

5. a) Make a conjecture about how the graph of $y = \sin (x - 45°)$ compares with the graph of $y = \sin x$.
b) Use your conjecture to sketch the graph of $y = \sin x$ and $y = \sin (x - 45°)$ on the same grid for the domain $0° \le x \le 360°$.

6. Test your conjecture from question 5a) by graphing $y = \sin x$ and $y = \sin (x - 45°)$, $0° \le x \le 360°$, on a graphing calculator. Compare the result with your graph from question 5b).

7. For the graphs from question 6, find
a) the amplitudes   **b)** the periods

8. What transformation can been applied to $y = \sin x$ to give $y = \sin (x - 45°)$?

9. Write a statement about the transformational effect of d on the graph of $y = \sin (x - d)$.

10. Write a conjecture about the transformational effect of d on the graph of $y = \cos (x - d)$.

11. Test your conjecture from question 10 by graphing $y = \cos x$, $y = \cos (x + 45°)$, $y = \cos (x - 45°)$, $0° \leq x \leq 360°$, on a graphing calculator.

12. For the graphs from question 11, find
a) the amplitudes   **b)** the periods

The vertical translations that apply to algebraic functions also apply to trigonometric functions.

If $c > 0$, the graphs of $y = \sin x + c$ and $y = \cos x + c$ are translated upward by c units.

If $c < 0$, the graphs of $y = \sin x + c$ and $y = \cos x + c$ are translated downward by c units.

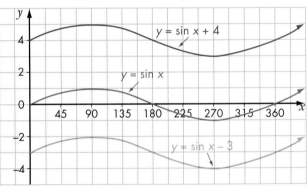

As with algebraic transformations, combinations of trigonometric transformations are performed in the following order.
• expansions and compressions
• reflections
• translations

EXAMPLE 1   Sketching $y = a\sin x + c$

Sketch one cycle of the graph of $y = 2\sin x + 3$.
State the domain and range of the cycle.

SOLUTION

First, sketch the graph of $y = 2\sin x$.

The graph of $y = 2\sin x$ is the graph of $y = \sin x$ expanded vertically by a factor of 2.

The amplitude is 2, so the maximum value is 2 and the minimum value is -2.

The period of the function $y = 2\sin x$ is 2π.

Use the five-point method to sketch the graph.
The five key points divide the period into quarters.
Therefore, the coordinates of the five key points are

$(0, 0)$, $\left(\dfrac{\pi}{2}, 2\right)$, $(\pi, 0)$, $\left(\dfrac{3\pi}{2}, -2\right)$, and $(2\pi, 0)$.

Plot the 5 key points in the cycle. Draw a smooth curve through the points.

Translate the graph three units upward to obtain the graph of $y = 2\sin x + 3$.
Label the graph.

The domain of the cycle is $0 \le x \le 2\pi$.
The range is $1 \le y \le 5$.

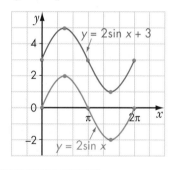

The horizontal translations that apply to algebraic functions also apply to trigonometric functions.

If $d > 0$, the graphs of $y = \sin(x - d)$ and $y = \cos(x - d)$ are translated d units to the right.

If $d < 0$, the graphs of $y = \sin(x - d)$ and $y = \cos(x - d)$ are translated d units to the left.

For trigonometric functions, a horizontal translation is often called the **phase shift** or **phase angle**.

EXAMPLE 2 Sketching $y = a\cos(x - d)$

Sketch one cycle of the graph of $y = 0.5\cos\left(x + \dfrac{\pi}{2}\right)$.

State the domain, range, and phase shift of the cycle.

SOLUTION

First, sketch the graph of $y = 0.5\cos x$.

The graph of $y = 0.5\cos x$ is the graph of $y = \cos x$ compressed vertically by a factor of 0.5.

The amplitude is 0.5, so the maximum value is 0.5 and the minimum value is -0.5.

The period of the function $y = 0.5\cos x$ is 2π.

Use the five-point method to sketch the graph.
The five key points divide the period into quarters.
Therefore, the coordinates of the five key points are

$(0, 0.5)$, $\left(\dfrac{\pi}{2}, 0\right)$, $(\pi, -0.5)$, $\left(\dfrac{3\pi}{2}, 0\right)$, and $(2\pi, 0.5)$.

Plot the 5 key points in the cycle. Draw a smooth curve through the points.

Translate the graph $\dfrac{\pi}{2}$ units to the left to obtain the graph

of $y = 0.5\cos\left(x + \dfrac{\pi}{2}\right)$.

Label the graph.

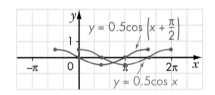

The domain of the cycle is $-\dfrac{\pi}{2} \le x \le \dfrac{3\pi}{2}$.

The range is $-0.5 \le y \le 0.5$.

The phase shift is $\dfrac{\pi}{2}$ units to the left.

EXAMPLE 3 Sketching $y = a\sin k(x - d)$

Sketch one cycle of the graph of $y = 3\sin 2\left(x - \dfrac{\pi}{4}\right)$.

State the domain, range, and phase shift of the cycle.

SOLUTION

First sketch the graph of $y = 3\sin 2x$.

The graph of $y = 3\sin 2x$ is the graph of $y = \sin x$ expanded vertically by a factor of 3 and compressed horizontally by a factor of $\dfrac{1}{2}$.

The amplitude is 3, so the maximum value is 3 and the minimum value is -3.

The period of the function $y = 3\sin 2x$ is $\dfrac{2\pi}{2}$, or π.

Use the five-point method to sketch the graph.
The five key points divide the period into quarters.
Therefore, the coordinates of the five key points are

$(0, 0)$, $\left(\dfrac{\pi}{4}, 3\right)$, $\left(\dfrac{\pi}{2}, 0\right)$, $\left(\dfrac{3\pi}{4}, -3\right)$, and $(\pi, 0)$.

Plot the 5 key points in the cycle. Draw a smooth curve
through the points.

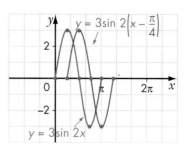

Translate the graph $\dfrac{\pi}{4}$ units to the right to obtain the graph

of $y = 3\sin 2\left(x - \dfrac{\pi}{4}\right)$. Label the graph.

The domain of the cycle is $\dfrac{\pi}{4} \le x \le \dfrac{5\pi}{4}$.

The range is $-3 \le y \le 3$.

The phase shift is $\dfrac{\pi}{4}$ units to the right.

If necessary, factor the coefficient of the x-term to identify the characteristics
of a function more easily.

EXAMPLE 4 Sketching $y = a\cos k(x - d) + c$

Sketch the graph of $y = 4\cos\left(\dfrac{1}{2}x + \dfrac{\pi}{2}\right) - 1$, $-4\pi \le x \le 4\pi$.

SOLUTION

Factor the coefficient of the x-term.

$y = 4\cos\left(\dfrac{1}{2}x + \dfrac{\pi}{2}\right) - 1$ becomes $y = 4\cos\dfrac{1}{2}(x + \pi) - 1$.

Now, sketch the graph of $y = 4\cos\dfrac{1}{2}x$.

The graph of $y = 4\cos\dfrac{1}{2}x$ is the graph of $y = \cos x$ expanded vertically by
a factor of 4 and expanded horizontally by a factor of 2.
The amplitude is 4, so the maximum value is 4 and the minimum value is -4.

The period of the function $y = 4\cos\dfrac{1}{2}x$ is $\dfrac{2\pi}{\dfrac{1}{2}}$, or 4π.

Use the five-point method to sketch the graph.
The five key points divide the period into quarters.
Therefore, the coordinates of the five key points are
$(0, 4)$, $(\pi, 0)$, $(2\pi, -4)$, $(3\pi, 0)$, and $(4\pi, 4)$.
Plot the 5 key points in the cycle. Draw a smooth curve through the points.

Translate the graph π units to the left and one unit downward to obtain

the graph of $y = 4\cos \dfrac{1}{2}(x + \pi) - 1$, $-\pi \le x \le 3\pi$.

Use the pattern to sketch the graph over the domain $-4\pi \le x \le 4\pi$.
Label the graph.

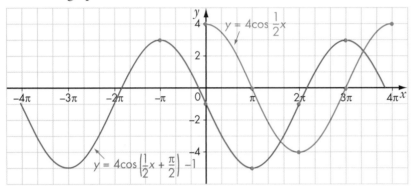

EXAMPLE 5 **Sketching for $a < 1$**

Sketch the graph of $y = -4\sin\left(x - \dfrac{\pi}{2}\right)$, $0 \le x \le 4\pi$.

SOLUTION

First sketch the graph of $y = 4\sin x$.
The graph of $y = 4\sin x$ is the graph of $y = \sin x$ expanded vertically by a factor of 4.
The amplitude is 4, so the maximum value is 4 and the minimum value is -4.
The period of the function $y = 4\sin x$ is 2π.

Use the five-point method to sketch the graph.
The five key points divide the period into quarters.
Therefore, the coordinates of the five key points are

$(0, 0)$, $\left(\dfrac{\pi}{2}, 4\right)$, $(\pi, 0)$, $\left(\dfrac{3\pi}{2}, -4\right)$, and $(2\pi, 0)$.

Plot the 5 key points in the cycle. Draw a smooth curve through the points.

Recall that the graph of $y = -f(x)$ is the graph of $y = f(x)$ reflected in the x-axis. So, reflect the graph of $y = 4\sin x$ in the x-axis to obtain the graph of $y = -4\sin x$.

Translate the reflected graph $\dfrac{\pi}{2}$ units to the right to obtain the graph of

$y = -4\sin\left(x - \dfrac{\pi}{2}\right)$, $\dfrac{\pi}{2} \le x \le \dfrac{5\pi}{2}$.

Use the pattern to sketch the graph over the domain $0 \le x \le 4\pi$. Label the graph.

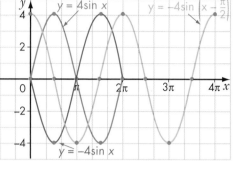

Note that all of the graphs required in Examples 1–5 can be drawn directly using a graphing calculator. The graph shown is $y = -4\sin\left(x - \dfrac{\pi}{2}\right)$, $0 \le x \le 4\pi$, from Example 5. The calculator is in radian mode, and the window variables include $Xmin = 0$, $Xmax = 4\pi$, $Ymin = -5$, and $Ymax = 5$.

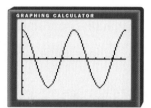

Example 6 Bay of Fundy Tides

In one area of the Bay of Fundy, the tides cause the water level to rise to 6 m above average sea level and to drop to 6 m below average sea level. One cycle is completed approximately every 12 h. Assume that changes in the depth of the water over time can be modelled by a sine function.

a) If the water is at average sea level at midnight and the tide is coming in, draw a graph to show how the depth of the water changes over the next 24 h. Assume that at low tide the depth of the water is 2 m.

b) Write an equation for the graph.

c) If the water is at average sea level at 02:00, and the tide is coming in, write an equation for the graph that shows how the depth changes over the next 24 h.

Solution

a) The depth of water at low tide is 2 m. At low tide, the water level is 6 m below average sea level. So, the depth of water for average sea level is 8 m. This is the depth at midnight. The maximum depth of the water is $8 + 6$, or 14 m. Use the known values to sketch a 12-h cycle of depth versus time. Use the pattern to show the changes over 24 h.

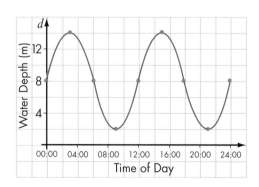

b) The amplitude, a, is 6 m.

The period is 12 h.

$$12 = \frac{2\pi}{k}$$

$$k = \frac{\pi}{6}$$

The graph has been translated 8 units upward, so $c = 8$.

The equation that shows how the depth of the water changes over time is

$$h = 6\sin\frac{\pi t}{6} + 8.$$

c) When the water is at average sea level at 02:00, the depth is 8 m at 02:00.
The graph is translated 2 h to the right.

The equation is $h = 6\sin\frac{\pi}{6}(t - 2) + 8.$

Key Concepts

- Perform combinations of transformations in the following order.
  - * expansions and compressions
  - * reflections
  - * translations
- For trigonometric functions, a horizontal translation is called the phase shift or phase angle.
- If necessary, factor the coefficient of the x-term to identify the characteristics of a function more easily.

Communicate Your Understanding

1. Describe how you would identify the transformations on $y = \sin x$ that result in each of the following functions.

a) $y = 3\sin\frac{1}{2}x - 3$ **b)** $y = 6\sin 3(x - 2\pi)$ **c)** $y = -2\sin (x + \pi) + 2$

2. Describe how you would identify the transformations on $y = \cos x$ that result in the function $y = 3\cos (4x - \pi)$.

3. Describe how you would sketch the graph of one cycle of the function

$$y = 2\sin 2\left(x + \frac{\pi}{2}\right) - 3.$$

Practise

A

1. Determine the vertical translation and the phase shift of each function with respect to $y = \sin x$.

a) $y = \sin x + 3$

b) $y = \sin x - 1$

c) $y = \sin (x - 45°)$

d) $y = \sin \left(x - \dfrac{3\pi}{4} \right)$

e) $y = \sin (x - 60°) + 1$

f) $y = \sin \left(x + \dfrac{\pi}{3} \right) + 4$

g) $y = \sin \left(x + \dfrac{3\pi}{8} \right) - 0.5$

h) $y = \sin (x - 15°) - 4.5$

2. Determine the vertical translation and the phase shift of each function with respect to $y = \cos x$.

a) $y = \cos x + 6$

b) $y = \cos x - 3$

c) $y = \cos \left(x + \dfrac{\pi}{2} \right)$

d) $y = \cos (x + 72°)$

e) $y = \cos (x - 30°) - 2$

f) $y = \cos \left(x + \dfrac{\pi}{6} \right) + 1.5$

g) $y = \cos (x + 110°) + 25$

h) $y = \cos \left(x - \dfrac{5\pi}{12} \right) - 3.8$

3. Sketch one cycle of the graph of each of the following. State the amplitude, period, domain, and range of the cycle.

a) $y = 3\sin x + 2$

b) $y = 2\cos x - 2$

c) $y = 1.5\sin x - 1$

d) $y = \dfrac{1}{2}\cos x + 1$

4. Sketch one cycle of the graph of each of the following. State the amplitude, period, domain, range, and phase shift of the cycle.

a) $y = 2\sin (x - \pi)$

b) $y = \cos \left(x - \dfrac{\pi}{2} \right)$

c) $y = \dfrac{1}{2}\sin \left(x + \dfrac{\pi}{2} \right)$

d) $y = 3\cos \left(x + \dfrac{\pi}{4} \right)$

e) $y = -\cos \left(x + \dfrac{\pi}{2} \right)$

5. Determine the amplitude, period, vertical translation, and phase shift for each function with respect to $y = \sin x$.

a) $y = 2\sin x - 3$

b) $y = 0.5\sin (2x) - 1$

c) $y = 6\sin 3(x - 20°)$

d) $y = -5\sin 2\left(x - \dfrac{\pi}{6} \right) + 1$

6. Determine the amplitude, period, vertical translation, and phase shift for each function with respect to $y = \cos x$.

a) $y = \cos x + 3$

b) $y = \cos 3(x - 90°)$

c) $y = -3\cos 4\left(x - \dfrac{\pi}{4} \right) + 5$

d) $y = 0.8\cos \dfrac{2}{3}\left(x - \dfrac{\pi}{3} \right) - 7$

7. Sketch one cycle of the graph of each of the following. State the amplitude, period, domain, range, and phase shift of the cycle.

a) $y = \sin 2\left(x + \dfrac{\pi}{4} \right)$

b) $y = 2\cos 2\left(x - \dfrac{\pi}{4} \right) + 1$

c) $y = 3\sin \dfrac{1}{2}(x - \pi) - 2$

d) $y = 4\cos \dfrac{1}{3}(x + 2\pi) - 4$

e) $y = -3\sin 2\left(x - \dfrac{\pi}{4} \right) + 2$

8. Communication Sketch one cycle of the graph of each of the following. State the amplitude, period, domain, range, and phase shift of the cycle.

a) $y = \sin\left(2x - \dfrac{\pi}{2}\right)$ **b)** $y = \cos\left(\dfrac{1}{2}x - \pi\right) - 2$

c) $y = 2\sin(3x - \pi) + 2$

d) $y = -3\cos(2x - 4\pi) - 1$

9. Write an equation for the function with the given characteristics, where T is the type, A is the amplitude, P is the period, V is the vertical shift, and H is the horizontal shift.

	T	A	P	V	H
a)	sine	8	2π	–6	none
b)	cosine	7	π	2	none
c)	sine	1	4π	3	π right
d)	cosine	10	$\dfrac{\pi}{2}$	none	$\dfrac{\pi}{2}$ left

10. Sketch the graph of each of the following. State the range.

a) $y = 2\sin x + 2,\ 0 \le x < 3\pi$

b) $y = -\cos 3x - 2,\ -\pi \le x \le \pi$

c) $y = 3\cos\left(x - \dfrac{\pi}{6}\right),\ -2\pi \le x \le 2\pi$

d) $y = 4\sin 2\left(x + \dfrac{\pi}{4}\right) - 1,\ -\pi \le x \le \pi$

e) $y = -2\sin\left(2x - \dfrac{\pi}{3}\right) + 1,\ -\pi \le x \le \pi$

f) $y = 5\cos\left(\dfrac{1}{3}x - \dfrac{\pi}{3}\right) + 2,\ -\pi \le x \le 3\pi$

g) $y = 2\sin\left(2x + \dfrac{\pi}{8}\right),\ -2\pi \le x \le 2\pi$

11. Each graph shows part of the sine function of the form $y = a\sin k(x - d) + c$. Determine the values of a, k, d, and c for each graph. Check by graphing.

a)

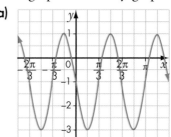

b)

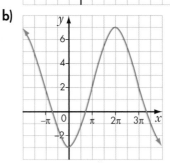

Apply, Solve, Communicate

B

12. Ocean cycles The water depth in a harbour is 21 m at high tide and 11 m at low tide. One cycle is completed approximately every 12 h.

a) Find an equation for the water depth as a function of the time, t hours, after low tide.

b) Draw a graph of the function for 48 h after low tide, which occurred at 14:00.

c) State the times at which the water depth was

i) a maximum   ii) a minimum   iii) at its average value

d) Estimate the depth of the water at

i) 17:00   ii) 21:00

e) Estimate the times at which the depth of the water was

i) 14 m   ii) 20 m   iii) at least 18 m

13. Application An object attached to the end of a spring is oscillating up and down. The displacement of the object, y, in centimetres, is a function of the time, t, in seconds, and is given by $y = 2.4\cos\left(12t + \dfrac{\pi}{6}\right)$.

a) Sketch two cycles of the function.

b) What is the maximum distance through which the object oscillates?

c) What is the period of the function? Give your answer as an exact number of seconds, in terms of π, and as an approximate number of seconds, to the nearest hundredth.

14. Ocean cycles On a certain day, the depth of water off a pier at high tide was 6 m. After 6 h, the depth of the water was 3 m. Assume a 12-h cycle.

a) Find an equation for the depth of water, with respect to its average depth, in terms of the time, t hours, since high tide.

b) Draw a graph of the depth of water versus time for 48 h after high tide.

c) Find the depth of water at $t = 8$ h, 15 h, 20 h, and 30 h.

d) Predict how the equation will change if the first period begins at low tide.

e) Test your prediction from part d) by drawing the graph and finding the equation.

15. Spring An object suspended from a spring is oscillating up and down. The distance from the high point to the low point is 30 cm, and the object takes 4 s to complete 5 cycles. For the first few cycles, the distance from the mean position, $d(t)$ centimetres, with respect to the time, t seconds, is modelled by a sine function.

a) Sketch a graph of this function for two cycles.

b) Write an equation that describes the distance of the object from its mean position as a function of time.

16. Ferris wheel A carnival Ferris wheel with a radius of 7 m makes one complete revolution every 16 s. The bottom of the wheel is 1.5 m above the ground.

a) Draw a graph to show how a person's height above the ground varies with time for three revolutions, starting when the person gets onto the Ferris wheel at its lowest point.

b) Find an equation for the graph.

c) Predict how the graph and the equation will change if the Ferris wheel turns more slowly.

d) Test your predictions from part c) by drawing the graph for three revolutions and finding an equation, if the wheel completes one revolution every 20 s.

17. Ferris wheel A Ferris wheel has a radius of 10 m and makes one revolution every 12 s. Draw a graph and find an equation to show a person's height above or below the centre of rotation for two counterclockwise revolutions starting at

a) point A   **b)** point B   **c)** point C

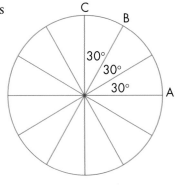

18. For the sine function expressed in the form $y = a\sin k(x - d) + c$, does the value of c affect each of the following? Explain.

a) the amplitude?

b) the period?

c) the maximum and minimum values of the function?

d) the phase shift?

19. Ocean cycles The depth of water, $d(t)$ metres, in a seaport can be approximated by the sine function $d(t) = 2.5\sin 0.164\pi(t - 1.5) + 13.4$, where t is the time in hours.

a) Graph the function for $0 \leq t \leq 24$ using a graphing calculator.

b) Find the period, to the nearest tenth of an hour.

c) A cruise ship needs a depth of at least 12 m of water to dock safely. For how many hours in each period can the ship dock safely? Round your answer to the nearest tenth of an hour.

20. Pose and solve problems Pose a problem related to each of the following. Check that you are able to solve each problem. Then, have a classmate solve it.

a) the vertical motion of a spring   **b)** the motion of a wheel

C

21. The equation of a sine function can be expressed in the form
$y = a\sin k(x - d) + c.$
Describe what you know about a, k, d, and c for each of the following statements to be true.

a) The period is greater than 2π.
b) The amplitude is less than one unit.
c) The graph passes through the origin.
d) The graph has no x-intercepts.

22. a) Predict how the graphs of $y = \sin(x + 45°)$ and $y = \sin(x - 315°)$ are related.
b) Test your prediction using a graphing calculator, and explain your observations.

23. a) Predict how the graphs of $y = \sin x$ and $y = \cos\left(x - \dfrac{\pi}{2}\right)$ are related.
b) Test your prediction using a graphing calculator, and explain your observations.

ACHIEVEMENT Check   Knowledge/Understanding Thinking/Inquiry/Problem Solving Communication Application

The rodent population in a particular region varies with the number of predators that inhabit the region. At any time, you could predict the rodent population, $r(t)$, using the function
$r(t) = 2500 + 1500\sin\dfrac{\pi t}{4}$, where t is the number of years that have passed since 1976.

Web Connection
www.school.mcgrawhill.ca/resources/
To investigate a simulation of a predator-prey relationship, visit the above web site. Go to **Math Resources**, then to *MATHEMATICS 11*, to find out where to go next. Report on your findings.

a) In the first cycle of this function, what was the maximum number of rodents and in which year did this occur?
b) What was the minimum number of rodents in a cycle?
c) What is the period of this function?
d) How many rodents would you predict in the year 2014?
e) Change the function to model a rodent population cycle that lasts 5 years.

TECHNOLOGY EXTENSION
Sinusoidal Regression

The equation of a sine curve that models a set of data can be found using a graphing calculator. Use the STAT EDIT menu to enter the data as two lists. Draw the scatter plot using the STAT PLOTS menu. Choose suitable window variables, or use the ZoomStat instruction to set them automatically. Find the equation of the curve of best fit using the SinReg (sinusoidal regression) instruction. Note that, for an equation found using this instruction, the domain is expressed in radians.

1. The graphing calculator expresses the equation of a sine curve of best fit in the form $y = a\sin(bx + c) + d$. Explain the meanings of a, b, c, and d in this form of the equation.

2. Graph the sinusoidal curve of best fit and find its equation. Write the equation in the form $y = a\sin k(x - d) + c$. Round decimal values to the nearest hundredth.

a)

x	y
0	0
0.5π	1
π	0
1.5π	-1
2π	0

b)

x	y
0	1
0.5	0
1	-1
1.5	0
2	1

c)

x	y
1	4
2	2.5
3	1
4	2.5
5	4

d)

x	y
2	-5
4	-2
6	1
8	-2
10	-5

3. What are the exact equations in the form $y = a\sin k(x - d) + c$ for the curves in question 2? Explain.

4. Weather The temperature was recorded at 4-h intervals on one summer day.

Time	Temperature (°C)
00:00	18.1
04:00	15.6
08:00	20.5
12:00	25.4
16:00	28.1
20:00	24.7

a) Graph the sinusoidal curve of best fit and find its equation. Round decimal values to the nearest hundredth.

b) Use the graph to estimate the temperature at 06:30, to the nearest tenth of a degree.

c) For what length of time was the temperature at least 25°C, to the nearest tenth of an hour?

5. Daylight The table shows the numbers of hours of daylight per day on different days of the year in Thunder Bay.

Day of the Year	Daylight Hours
16	8.72
75	11.82
136	15.18
197	15.83
259	12.68
320	9.18

a) Graph the sinusoidal curve of best fit and finds its equation. Round decimal values to the nearest thousandth.

b) Find the percent of days in the year with less than 10 h of daylight. Round to the nearest percent.

5.7 Trigonometric Identities

When a ball is kicked from the ground, the maximum height the ball will reach can be determined by the formula

$$h = \frac{v_0^2 \sin^2 x}{2g}.$$

In this formula, h metres is the maximum height that the ball will reach, v_0 metres per second is the initial velocity of the ball, x is the angle that the path of the ball makes with the ground when the ball is kicked, and g is the acceleration due to gravity.

Another formula that determines the maximum height that a ball will reach is

$$h = \frac{v_0^2 \cos^2 x \tan^2 x}{2g}.$$

Equating the two expressions for maximum height results in the following equation, which will be proved in Example 5.

$$\frac{v_0^2 \sin^2 x}{2g} = \frac{v_0^2 \cos^2 x \tan^2 x}{2g}$$

This equation is called a **trigonometric identity**. A trigonometric identity is an equation that is true for all values of the variable for which the expressions on both sides of the equation are defined.

INVESTIGATE & INQUIRE

1. Copy and complete the following table, using exact values for sin x, cos x, and tan x.

x	sin x	cos x	tan x
30°			
45°			
60°			
120°			
135°			
150°			

2. Determine the exact value of $\dfrac{\sin x}{\cos x}$ for each angle in the table.

3. Make a conjecture about the relationship between $\dfrac{\sin x}{\cos x}$ and tan x.

4. Test your conjecture from question 3 for other values of x for which both expressions are defined. Can you find a counterexample?

5. The terms $(\sin x)^2$ and $(\cos x)^2$ are usually written as $\sin^2 x$ and $\cos^2 x$. Determine the value of $\sin^2 x + \cos^2 x$ for each angle in the table.

6. Make a conjecture about the value of $\sin^2 x + \cos^2 x$.

7. Test your conjecture from question 6 for other values of x for which the expression is defined. Can you find a counterexample?

In this section, we will derive two trigonometric identities from their definitions.

First consider the ratio $\dfrac{\sin \theta}{\cos \theta}$.

$$\sin \theta = \frac{y}{r}$$

$$\cos \theta = \frac{x}{r}$$

$$\frac{\sin \theta}{\cos \theta} = \frac{y}{r} \times \frac{r}{x}$$

$$= \frac{y}{x}$$

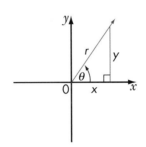

By definition, $\tan \theta = \dfrac{y}{x}$.

Therefore, $\dfrac{\sin \theta}{\cos \theta} = \tan \theta$, or $\dfrac{\sin x}{\cos x} = \tan x$.

Recall that θ and x can be used interchangeably as the angle variable.

This identity is called the **quotient identity** or **quotient relation**.

Another form of this identity is

$$\frac{\sin^2 \theta}{\cos^2 \theta} = \tan^2 \theta \text{ or } \frac{\sin^2 x}{\cos^2 x} = \tan^2 x$$

Now consider $\sin^2 \theta + \cos^2 \theta$.

$$\sin^2 \theta + \cos^2 \theta = \frac{y^2}{r^2} + \frac{x^2}{r^2}$$

($\sin \theta)^2$ and $(\cos \theta)^2$ are usually written as $\sin^2 \theta$ and $\cos^2 \theta$.

$$= \frac{y^2 + x^2}{r^2}$$

$$= \frac{r^2}{r^2}$$

$$= 1$$

So, the identity is $\sin^2 \theta + \cos^2 \theta = 1$, or $\sin^2 x + \cos^2 x = 1$.

This identity is called the **Pythagorean identity**.

Other forms of the Pythagorean identity are
$$\sin^2 \theta = 1 - \cos^2 \theta \text{ or } \sin^2 x = 1 - \cos^2 x$$
$$\text{and } \cos^2 \theta = 1 - \sin^2 \theta \text{ or } \cos^2 x = 1 - \sin^2 x$$
To prove that an equation is an identity, show that one side of the equation is equivalent to the other side. Treat each side of the equation independently. The best approach is to start with the more complex side of the equation and transform the expression on this side into the exact form of the expression on the other side.

EXAMPLE 1 Proving That an Equation is an Identity

Prove that $\dfrac{\cos x \tan x}{\sin x} = 1$.

SOLUTION

$$\text{L.S.} = \frac{\cos x \tan x}{\sin x}$$

Use the quotient identity:
$$= \frac{\cos x \sin x}{\sin x \cos x}$$

Divide by common factors: $= 1$

$$\text{R.S.} = 1$$

L.S. = R.S., so $\dfrac{\cos x \tan x}{\sin x} = 1$.

EXAMPLE 2 Proving That an Equation is an Identity

Prove that $\cos x = \dfrac{1}{\cos x} - \sin x \tan x$.

SOLUTION

$$\text{R.S.} = \frac{1}{\cos x} - \sin x \tan x$$

Use the quotient identity:
$$= \frac{1}{\cos x} - \frac{\sin x \sin x}{\cos x}$$

Simplify:
$$= \frac{1 - \sin^2 x}{\cos x}$$

Use the Pythagorean identity:
$$= \frac{\cos^2 x}{\cos x}$$

Divide by the common factor: $= \cos x$

$$\text{L.S.} = \cos x$$

L.S. = R.S., so $\cos x = \dfrac{1}{\cos x} - \sin x \tan x$.

EXAMPLE 3 Graphing and Proving an Identity

For the equation $\dfrac{\sin^2 x}{1 - \cos x} = 1 + \cos x$,

a) use a graph to show that the equation appears to be an identity

b) prove the equation is an identity

SOLUTION

a) Use a graphing calculator.

Using the Y= editor, enter $y = \dfrac{\sin^2 x}{1 - \cos x}$ as Y1, and $y = 1 + \cos x$ as Y2.

The window variables include Xmin = −20, Xmax = 20, Ymin = −3, and Ymax = 3, with the calculator in radian mode.

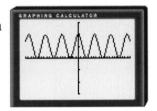

Adjust the mode settings and window variables as necessary, and display both graphs in the same viewing window. In the Y= editor, you may find it helpful to set the graph style of the second graph to thick. You will then see the second graph appear on top of the first.

Since the graphs appear to be identical, the equation appears to be an identity.

b)

$$\text{L.S.} = \dfrac{\sin^2 x}{1 - \cos x}$$

Use the Pythagorean identity:
$$= \dfrac{1 - \cos^2 x}{1 - \cos x}$$

Factor the difference of squares:
$$= \dfrac{(1 - \cos x)(1 + \cos x)}{1 - \cos x}$$

Divide by the common factor:
$$= 1 + \cos x$$

$$\text{R.S.} = 1 + \cos x$$

L.S. = R.S., so $\dfrac{\sin^2 x}{1 - \cos x} = 1 + \cos x.$

EXAMPLE 4 Proving an Identity Using a Common Denominator

Prove the identity $\dfrac{1}{1 - \sin x} - \dfrac{1}{1 + \sin x} = \dfrac{2\tan x}{\cos x}.$

SOLUTION

The more complex side is the left side.

The common denominator of the left side is $(1 - \sin x)(1 + \sin x)$.

$$\text{L.S.} = \frac{1}{1 - \sin x} - \frac{1}{1 + \sin x}$$

Write with a common denominator: $$= \frac{(1 + \sin x) - (1 - \sin x)}{(1 - \sin x)(1 + \sin x)}$$

Simplify: $$= \frac{1 + \sin x - 1 + \sin x}{(1 - \sin x)(1 + \sin x)}$$

$$= \frac{2\sin x}{1 - \sin^2 x}$$

Use the Pythagorean identity: $$= \frac{2\sin x}{\cos^2 x}$$

$$= \frac{2\sin x}{\cos x \cos x}$$

Use the quotient identity: $$= \frac{2\tan x}{\cos x}$$

$$\text{R.S.} = \frac{2\tan x}{\cos x}$$

L.S. = R.S., so $\dfrac{1}{1 - \sin x} - \dfrac{1}{1 + \sin x} = \dfrac{2\tan x}{\cos x}$.

EXAMPLE 5 Kicking a Ball

As shown at the beginning of this section, the equation
$\dfrac{v_0^2 \sin^2 x}{2g} = \dfrac{v_0^2 \cos^2 x \tan^2 x}{2g}$ includes two expressions for finding the
maximum height that a ball kicked from the ground will reach.
Prove that $\dfrac{v_0^2 \sin^2 x}{2g} = \dfrac{v_0^2 \cos^2 x \tan^2 x}{2g}$ is an identity.

SOLUTION

$$\text{R.S.} = \frac{v_0^2 \cos^2 x \tan^2 x}{2g} \qquad \text{L.S.} = \frac{v_0^2 \sin^2 x}{2g}$$

$$= \frac{v_0^2 \cos^2 x \sin^2 x}{2g \cos^2 x}$$

$$= \frac{v_0^2 \sin^2 x}{2g}$$

L.S. = R.S., so $\dfrac{v_0^2 \sin^2 x}{2g} = \dfrac{v_0^2 \cos^2 x \tan^2 x}{2g}$.

Key Concepts

- A trigonometric identity is an equation that is true for all values of the variable for which the expressions on both sides of the equation are defined.
- To prove that an equation is an identity, start with the more complex side of the identity and transform the expression into the exact form of the expression on the other side.
- The quotient identity is $\dfrac{\sin x}{\cos x} = \tan x.$
- The three forms of the Pythagorean identity are
$$\sin^2 x + \cos^2 x = 1$$
$$\sin^2 x = 1 - \cos^2 x$$
$$\cos^2 x = 1 - \sin^2 x$$

Communicate Your Understanding

1. Describe how you would prove that the equation $\sin x \tan x = \cos x \tan^2 x$ is an identity.

2. What would be your first step in proving that the equation $(1 + \sin x)(1 - \sin x) = \cos^2 x$ is an identity? Explain your choice.

3. Describe how you would use a graphing calculator to show that the equation $\tan^2 x \cos^2 x = 1 - \cos^2 x$ appears to be an identity.

Practise

A

1. State an equivalent expression for each of the following.

a) $\cos x \tan x$ **b)** $\sin^2 x$

c) $\cos^2 x$ **d)** $\tan^2 x$

e) $\tan x \sin x$ **f)** $1 - \sin^2 x$

g) $\sin x \tan x \cos x$ **h)** $1 - \cos^2 x$

i) $\sin^2 x + \cos^2 x$

2. Prove each identity.

a) $\dfrac{\sin x}{\tan x} = \cos x$

b) $\sin x \cos x \tan x = 1 - \cos^2 x$

c) $\dfrac{1 - \cos^2 x}{\sin x} = \sin x$

d) $\sin^2 x + \dfrac{\sin x \cos x}{\tan x} = 1$

e) $1 + \dfrac{1}{\tan^2 x} = \dfrac{1}{\sin^2 x}$

f) $2\sin^2 x - 1 = \sin^2 x - \cos^2 x$

g) $\dfrac{1}{\cos x} - \cos x = \sin x \tan x$

h) $\sin x + \tan x = \tan x\, (1 + \cos x)$

i) $\dfrac{1}{1 - \sin^2 x} = 1 + \tan^2 x$

j) $\cos^2 x - \sin^2 x = 2\cos^2 x - 1$

k) $\sin^2 x + \cos^2 x + \tan^2 x = \dfrac{1}{\cos^2 x}$

l) $\dfrac{\sin x}{\sin x + \cos x} = \dfrac{\tan x}{1 + \tan x}$

m) $\dfrac{1 + \tan^2 x}{1 - \tan^2 x} = \dfrac{1}{\cos^2 x - \sin^2 x}$

3. Use a graphing calculator to show that each equation appears to be an identity. Then, prove that the equation is an identity.

a) $\cos x \tan x = \sin x$

b) $\sin x + \tan x = \tan x\,(1 + \cos x)$

c) $1 + \tan^2 x = \dfrac{1}{\cos^2 x}$

d) $\cos^2 x = \sin^2 x + 2\cos^2 x - 1$

e) $\dfrac{1}{1 + \sin x} + \dfrac{1}{1 - \sin x} = \dfrac{2}{\cos^2 x}$

f) $\tan^2 x - \sin^2 x = \sin^2 x \tan^2 x$

4. Prove each identity.

a) $\dfrac{1}{\sin^2 x} + \dfrac{1}{\cos^2 x} = \dfrac{1}{\sin^2 x \cos^2 x}$

b) $\tan x + \dfrac{1}{\tan x} = \dfrac{1}{\sin x \cos x}$

c) $\dfrac{1}{1 - \cos x} + \dfrac{1}{1 + \cos x} = \dfrac{2}{\sin^2 x}$

d) $(\sin x + \cos x)^2 = 1 + 2\sin x \cos x$

e) $(1 - \cos^2 x)\left(1 + \dfrac{1}{\tan^2 x}\right) = 1$

f) $\dfrac{1 + 2\sin x \cos x}{\sin x + \cos x} = \sin x + \cos x$

g) $\dfrac{\sin x}{1 - \cos x} - \dfrac{1 + \cos x}{\sin x} = 0$

h) $\sin^2 x - \sin^4 x = \cos^2 x - \cos^4 x$

i) $(1 + \tan^2 x)(1 - \cos^2 x) = \tan^2 x$

j) $\dfrac{\sin x - 1}{\sin x + 1} = \dfrac{-\cos^2 x}{(\sin x + 1)^2}$

Apply, Solve, Communicate

5. Conical pendulum A conical pendulum is so named because of the cone-shaped path traced by the bob and the wire. The length of the pendulum wire, L, is related to the angle, x, that the wire makes with the vertical by the formula $L = \dfrac{g}{\omega^2 \cos x}$, where g is the acceleration due to gravity and ω is the angular velocity of the bob about the vertical, in radians per second. Another way of expressing the relationship is the formula $L = \dfrac{g\tan x}{\omega^2 \sin x}$.

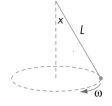

a) Verify that the two formulas are equivalent when $x = \dfrac{\pi}{6}$.

b) Prove that $\dfrac{g}{\omega^2 \cos x} = \dfrac{g\tan x}{\omega^2 \sin x}$ is an identity.

6. Kicking a ball When a ball is kicked from the ground, the time of flight of the ball can be determined by the formula

$$t = \frac{2v_0 \sin x}{g}.$$

In this formula, t seconds is the time of flight, v_0 metres per second is the initial velocity of the ball, x is the angle that the path of the ball makes with the ground when the ball is kicked, and g is the acceleration due to gravity.

a) Write another formula that determines the time of flight of the ball.

b) Equate the trigonometric expressions from the given formula and the formula you found in part a) to write an equation.

c) Use a graphing calculator to check if the equation appears to be an identity.

d) If the equation appears to be an identity, prove that it is an identity.

e) The formula for the horizontal distance, d metres, travelled by a ball kicked from the ground is $d = \dfrac{2v_0^2 \sin x \cos x}{g}$.

Write another formula for the horizontal distance.

f) Equate the trigonometric expressions from the two formulas in part e) to write an equation. Use a graphing calculator to check if the equation appears to be an identity.

g) If the equation appears to be an identity, prove that it is an identity.

B

7. Prove each identity.

a) $\sin^4 x - \cos^4 x = \sin^2 x - \cos^2 x$

b) $\sin^4 x + 2\sin^2 x \cos^2 x + \cos^4 x = 1$

c) $\dfrac{4}{\cos^2 x} - 5 = 4 \tan^2 x - 1$

d) $\dfrac{\cos x - \sin x - \cos^3 x}{\cos x} = \sin^2 x - \tan x$

e) $\dfrac{\sin^2 x - 6 \sin x + 9}{\sin^2 x - 9} = \dfrac{\sin x - 3}{\sin x + 3}$

8. Find a counterexample to show that each equation is not an identity.

a) $\sin x = \sqrt{\sin^2 x}$ b) $\cos x = \sqrt{\cos^2 x}$

9. Technology Use radian measure for the following.

a) In the same viewing window, graph $y = \sin x$ and $y = x$ for $-0.2 \le x \le 0.2$ and $-0.2 \le y \le 0.2$. Do the graphs suggest that $\sin x = x$ is an identity?

b) Repeat part a) for $-2 \le x \le 2$ and $-2 \le y \le 2$.

c) Write a conclusion about verifying identities graphically.

10. Application Use the x, y, and r definitions of sin x and cos x to prove the following identity.

$$\frac{\cos \theta}{1 - \sin \theta} = \frac{1 + \sin \theta}{\cos \theta}$$

11. Inquiry/Problem Solving Determine if each of the following equations is an identity or not.

a) $\dfrac{1}{\tan x} + \cos x = \dfrac{\cos x(1 + \sin x)}{\sin x}$ b) $\dfrac{1}{\tan x} + \cos x = \tan x + \sin x$

c) $\dfrac{1}{\tan x} + \cos x = \dfrac{2 \cos x}{\sin x}$

12. Communication Explain why you think that the equation $(a + b)^2 = a^2 + 2ab + b^2$ can be called an algebraic identity.

13. Algebra If $x = a\cos \theta - b\sin \theta$ and $y = a\sin \theta + b\cos \theta$, show that $x^2 + y^2 = a^2 + b^2$.

14. Since $1 - \cos^2 x = \sin^2 x$ is an identity, is $\sqrt{1 - \cos^2 x} = \sin x$ also an identity? Explain how you know.

15. a) Show graphically that $\sin^2 x + \cos^2 x = (\sin x + \cos x)^2$ is not an identity. Explain your reasoning.
b) Explain how the graph shows if there any values of x for which the equation is true.

16. Write a list of helpful strategies for proving trigonometric identities, and describe situations in which you would try each strategy. Compare your list with your classmates'.

C

17. Formulating problems a) Create a trigonometric identity that has not appeared in this section.
b) Have a classmate check graphically that your equation may be an identity. If so, have your classmate prove your identity.

18. Technology a) Use a graph to show that the equation
$$\frac{\cos^2 x - 1}{\cos x + 1} = \cos x - 1$$ appears to be an identity.
b) Compare the functions defined by each side of the equation by displaying a table of values. Find a value of x for which the values of the two functions are not the same. Have you shown that the equation is not an identity? Explain.

20. Prove that $\dfrac{\tan x \sin x}{\tan x + \sin x} = \dfrac{\tan x - \sin x}{\tan x \sin x}$.

5.8 Trigonometric Equations

To calculate the angle at which a curved section of highway should be banked, an engineer uses the equation $\tan x = \dfrac{v^2}{224\ 000}$, where x is the angle of the bank and v is the speed limit on the curve, in kilometres per hour.

The equation $\tan x = \dfrac{v^2}{224\ 000}$ is an example of a trigonometric equation. It will be used in Example 6.

A **trigonometric equation** is an equation that contains one or more trigonometric functions. Other examples of trigonometric equations include the following.
$$\sin^2 x + \cos^2 x = 1 \qquad 2\cos x - 1 = 0$$

The equation $\sin^2 x + \cos^2 x = 1$ is an identity. Recall that a trigonometric identity is an equation that is true for all values of the variable for which the expressions on both sides of the equation are defined.

The equation $2\cos x - 1 = 0$ is not an identity and is only true for certain values of x. To solve a trigonometric equation that is not an identity means to find all values of the variable, x, that make the equation true.

INVESTIGATE & INQUIRE

1. To graph the system $y = 2x + 3$ and $y = 9 - x$, use the Y= editor to enter Y1 = 2X + 3 and Y2 = 9 − X. Graph the equations in the standard viewing window.

2. Use the intersect operation to find the coordinates of the point of intersection.

3. Equate the right sides of the two equations from question 1 and solve the resulting equation for x algebraically.

4. Compare the value of x you found in question 3 to the value of x you found in question 2. Explain your findings.

5. To graph the equations $y = 4\cos x$ and $y = 1 + 2\cos x$, use the Y= editor to enter $Y_1 = 4\cos(X)$ and $Y_2 = 1 + 2\cos(X)$. Using the mode settings, select the degree mode. Using the window variables, adjust the window to $0° \le x \le 360°$ and $-5 \le y \le 5$ by setting $Xmin = 0$, $Xmax = 360$, $Ymin = -5$, and $Ymax = 5$. Display the graph.

6. Use the intersect operation to find the coordinates of the points of intersection.

7. Equate the right sides of the two equations from question 5, and solve the resulting equation for $\cos x$ algebraically.

8. If $0° \le x \le 360°$, what values of x, in degrees, give the value of $\cos x$ from question 7?

9. Compare the values of x you found in question 8 to the values of x you found in question 6. Explain your findings.

EXAMPLE 1 Solving a Trigonometric Equation

Solve $\sqrt{2}\sin\theta - 1 = 0$ on the interval $0° \le \theta \le 360°$.

SOLUTION 1 Paper-and-Pencil Method

$$\sqrt{2}\sin\theta - 1 = 0$$

Add 1 to both sides: $\sqrt{2}\sin\theta = 1$

Divide both sides by $\sqrt{2}$: $\sin\theta = \dfrac{1}{\sqrt{2}}$

The sine of an angle is positive in the first and second quadrants.

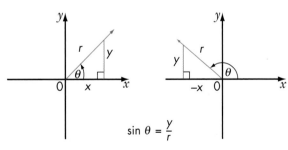

$\sin\theta = \dfrac{y}{r}$

The solutions are $45°$ and $135°$.

In radians, the solutions are $\dfrac{\pi}{4}$ and $\dfrac{3\pi}{4}$.

Use the memory aid CAST to remember which trigonometric ratios are positive in each quadrant.

SOLUTION 2 Graphing-Calculator Method

Graph the related trigonometric function
$y = \sqrt{2}\sin x - 1$ for $0° \leq x \leq 360°$.
Using the mode settings, select the degree mode.
Find the x-intercepts using the zero operation.
The graph intersects the x-axis at $45°$ and $135°$.

The solutions are $45°$ and $135°$.

In radians, the solutions are $\theta = \dfrac{\pi}{4}$ and $\dfrac{3\pi}{4}$.

The window variables include Xmin = 0, Xmax = 360,
Ymin = –3, Ymax = 1.

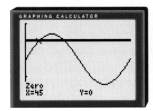
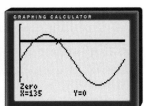

EXAMPLE 2 Solving a Trigonometric Equation

Find the exact solutions for $3\cos \theta = \cos \theta + 1$, if $0° \leq \theta \leq 360°$.

SOLUTION 1 Paper-and-Pencil Method

$$3\cos \theta = \cos \theta + 1$$

Subtract $\cos \theta$ from both sides: $2\cos \theta = 1$

Divide both sides by 2: $\cos \theta = \dfrac{1}{2}$

The cosine of an angle is positive in the first
and fourth quadrants.

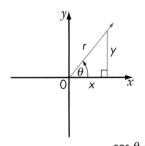
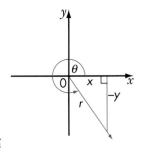

$\cos \theta = \frac{x}{r}$

The exact solutions are $60°$ and $300°$.

In radians, the exact solutions are $\dfrac{\pi}{3}$ and $\dfrac{5\pi}{3}$.

Solution 2 Graphing-Calculator Method

Graph $y = 3\cos x$ and $y = \cos x + 1$ in the same viewing window for $0° \le x \le 360°$. Use the intersect operation to determine the coordinates of the points of intersection.

The window variables include Xmin = 0, Xmax = 360, Ymin = −4, Ymax = 4.

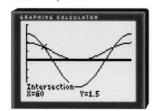 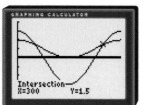

The graphs intersect at $x = 60°$ and $x = 300°$.

The solutions are 60° and 300°.

In radians, the solutions are $\dfrac{\pi}{3}$ and $\dfrac{5\pi}{3}$.

Note that an alternative graphing-calculator method for solving Example 2 involves rewriting $3\cos x = \cos x + 1$ as $2\cos x - 1 = 0$. The equation $2\cos x - 1 = 0$ can be solved by graphing $y = 2\cos x - 1$ and using the zero operation to find the x-intercepts.

The window variables include Xmin = 0, Xmax = 360, Ymin = −4, Ymax = 2.

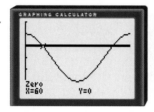 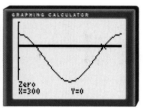

Example 3 Solving by Factoring

Solve the equation $2\cos^2 x - \cos x - 1 = 0$ on the interval $0 \le x \le 2\pi$.

Solution

$$2\cos^2 x - \cos x - 1 = 0$$

Factor the left side:
$$(2\cos x + 1)(\cos x - 1) = 0$$

Use the zero product property:
$$2\cos x + 1 = 0 \quad \text{or} \quad \cos x - 1 = 0$$
$$2\cos x = -1 \qquad\qquad \cos x = 1$$
$$\cos x = -\frac{1}{2} \qquad\qquad x = 0 \text{ or } 2\pi$$
$$x = \frac{2\pi}{3} \text{ or } \frac{4\pi}{3}$$

The solutions are 0, $\dfrac{2\pi}{3}$, $\dfrac{4\pi}{3}$, and 2π.

The solution can be modelled graphically. The window variables include Xmin = 0, Xmax = 360, Ymin = −3, Ymax = 3.

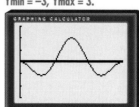

EXAMPLE 4 Solving by Factoring

Solve the equation $2\sin^2 x - 7\sin x + 3 = 0$ for $0 \leq x \leq 2\pi$. Express answers as exact solutions and as approximate solutions, to the nearest hundredth of a radian.

SOLUTION

$$2\sin^2 x - 7\sin x + 3 = 0$$

Factor the left side: $\quad\quad\quad (\sin x - 3)(2\sin x - 1) = 0$

Use the zero product property: $\quad \sin x - 3 = 0 \quad$ or $\quad 2\sin x - 1 = 0$

$$\sin x = 3 \quad\quad\quad\quad 2\sin x = 1$$

$$\sin x = \frac{1}{2}$$

$$x = \frac{\pi}{6} \text{ or } \frac{5\pi}{6}$$

There is no solution to $\sin x = 3$, since all values of $\sin x$ are ≥ -1 or ≤ 1.

The exact solutions are $\dfrac{\pi}{6}$ and $\dfrac{5\pi}{6}$.

The approximate solutions are 0.52 and 2.62.

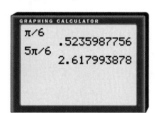

EXAMPLE 5 Using a Trigonometric Identity

Solve $6\cos^2 x - \sin x - 5 = 0$ for $0° \leq x \leq 360°$. Round approximate solutions to the nearest tenth of a degree.

SOLUTION

Use the Pythagorean identity $\cos^2 x = 1 - \sin^2 x$ to write an equivalent equation that involves only the sine function.

$$6\cos^2 x - \sin x - 5 = 0$$

Substitute $1 - \sin^2 x$ for $\cos^2 x$: $\quad 6(1 - \sin^2 x) - \sin x - 5 = 0$

Expand: $\quad\quad\quad\quad\quad\quad\quad 6 - 6\sin^2 x - \sin x - 5 = 0$

Simplify: $\quad\quad\quad\quad\quad\quad\quad\quad -6\sin^2 x - \sin x + 1 = 0$

Multiply both sides by -1: $\quad\quad\quad 6\sin^2 x + \sin x - 1 = 0$

Factor the left side: $\quad\quad\quad\quad (2\sin x + 1)(3\sin x - 1) = 0$

Use the zero product property: $\quad 2\sin x + 1 = 0 \quad$ or $\quad 3\sin x - 1 = 0$

$$2\sin x = -1 \quad\quad \text{or} \quad\quad 3\sin x = 1$$

$$\sin x = -\frac{1}{2} \quad\quad \text{or} \quad\quad \sin x = \frac{1}{3}$$

$$x = 210° \text{ or } 330° \quad\quad x \doteq 19.5° \text{ or } 160.5°$$

The solutions are 19.5°, 160.5°, 210°, and 330°.

EXAMPLE 6 Bank Angles

Engineers use the equation $\tan x = \dfrac{v^2}{224\,000}$ to calculate the angle at which a curved section of highway should be banked. In the equation, x is the angle of the bank and v is the speed limit on the curve, in kilometres per hour.

a) Calculate the angle of the bank, to the nearest tenth of a degree, if the speed limit is 100 km/h.

b) The four turns at the Indianapolis Motor Speedway are banked at an angle of 9.2°. What is the maximum speed through these turns, to the nearest kilometre per hour?

SOLUTION

a)
$$\tan x = \frac{v^2}{224\,000}$$
$$= \frac{100^2}{224\,000}$$
$$x \doteq 2.6°$$

The angle of the bank is 2.6°, to the nearest tenth of a degree.

b)
$$\frac{v^2}{224\,000} = \tan x$$
$$v^2 = 224\,000\tan x$$
$$v = \sqrt{224\,000\tan x}$$
$$v = \sqrt{224\,000\tan 9.2°}$$
$$v \doteq 190$$

The maximum speed is 190 km/h, to the nearest kilometre per hour.

Key Concepts

- To solve a trigonometric equation that is not an identity, find all values of the variable that make the equation true.
- Trigonometric equations can be solved
  a) with paper and pencil using the methods used to solve algebraic equations
  b) graphically using a graphing calculator
- Answers can be expressed in degrees or radians.

Communicate Your Understanding

1. Describe how you would solve $2\sin x - \sqrt{3} = 0$, $0° \le x \le 360°$.
2. Describe how you would solve $2\cos^2 x - 3\cos x + 1 = 0$, $0 \le x \le 2\pi$. Justify your method.
3. Explain why the equation $\cos x - 2 = 0$ has no solutions.

Practise

A

1. Solve each equation for $0 \le x \le 2\pi$.
a) $\sin x = 0$
b) $2\cos x + 1 = 0$
c) $\tan x = 1$
d) $\sqrt{2}\sin x + 1 = 0$
e) $2\cos x - \sqrt{3} = 0$
f) $2\sin x + \sqrt{3} = 0$

2. Solve each equation for $0 \le x \le 360°$.
a) $\sin x + 1 = 0$
b) $\sqrt{2}\cos x - 1 = 0$
c) $2\sin x - \sqrt{3} = 0$
d) $\sqrt{2}\cos x + 1 = 0$
e) $2\sin x + 1 = 0$
f) $\tan x = -1$

3. Solve each equation for $0° \le x \le 360°$.
a) $2\cos^2 x - 7\cos x + 3 = 0$
b) $3\sin x = 2\cos^2 x$
c) $2\sin^2 x - 3\sin x - 2 = 0$
d) $\sin^2 x - 1 = \cos^2 x$
e) $\tan^2 x - 1 = 0$
f) $2\sin^2 x + 3\sin x + 1 = 0$
g) $2\cos^2 x + 3\sin x - 3 = 0$

4. Solve each equation for $0 \le x \le 2\pi$. Express answers as exact solutions and as approximate solutions, to the nearest hundredth of a radian.
a) $\sin^2 x - 2\sin x - 3 = 0$
b) $2\cos^2 x = \sin x + 1$
c) $2\sin x \cos x + \sin x = 0$
d) $\sin^2 x = 6\sin x - 9$
e) $\sin^2 x + \sin x = 0$
f) $\cos x = 2\sin x \cos x$

5. Solve for x on the interval $0° \le x \le 360°$. Round approximate solutions to the nearest tenth of a degree.
a) $4\cos x - 3 = 0$
b) $1 + \sin x = 4\sin x$
c) $6\cos^2 x - \cos x - 1 = 0$
d) $9\sin^2 x - 6\sin x + 1 = 0$
e) $16\cos^2 x - 4\cos x + 1 = 0$
f) $6\cos^2 x + \sin x - 4 = 0$

6. Solve for x on the interval $0 \le x \le 2\pi$. Give the exact solution, where possible. Otherwise, round to the nearest hundredth of a radian.
a) $\sin x - 3\sin x \cos x = 0$
b) $6\cos^2 x + \cos x - 1 = 0$
c) $8\cos^2 x + 14\cos x = -3$
d) $8\sin^2 x - 10\cos x - 11 = 0$

Apply, Solve, Communicate

7. Refraction of light The diagram shows how a light ray bends as it travels from air into water. The bending of the ray is a process known as refraction. In the diagram, i is the angle of incidence and r is the angle of refraction. For water, the index of refraction, n, is defined as follows.

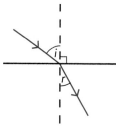

$$n = \frac{\sin i}{\sin r}$$

a) The index of refraction of water has a value of 1.33. For a ray with an angle of incidence of 40°, find the angle of refraction, to the nearest tenth of a degree.

b) State any restrictions on the value of r in the above equation. Explain.

B

8. Isosceles triangle a) Show that the area, A, of an isosceles triangle with base b and equal angles x, can be found using the equation

$$A = \frac{b^2 \tan x}{4}.$$

b) For an isosceles triangle with $A = 40$ cm^2 and $b = 8$ cm, use the equation to find x, to the nearest tenth of a degree.

9. Application In right triangle ABC, BD $\perp$ AC, AC = 4, and BD = 1.

a) Show that $\sin A \cos A = \dfrac{1}{4}$.

b) Describe how you would use a graphing calculator to find the measure of $\angle A$, in degrees.

c) Find the exact measure of $\angle A$, in degrees.

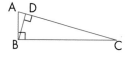

10. Solve for x on the interval $0 \le x \le 2\pi$. Give the exact solution, if possible. Otherwise, round to the nearest hundredth of a radian.

a) $\tan^2 x - 4\tan x = 0$ **b)** $\tan^2 x - 5\tan x + 6 = 0$

c) $\tan^2 x - 4\tan x + 4 = 0$ **d)** $6\tan^2 x - 7\tan x + 2 = 0$

11. Solve for x on the interval $0° \le x \le 360°$. Check your solutions.

a) $2\sin x \tan x - \tan x - 2\sin x + 1 = 0$

b) $\cos x \tan x - 1 + \tan x - \cos x = 0$

12. a) Graph $y = \sin x$ and $y = 0.5$ for $0° \le x \le 360°$.

b) For what values of x is $\sin x \ge 0.5$?

c) For what values of x is $\sin x \le 0.5$?

13. a) Communication Write a short paragraph, including examples, to distinguish the terms *trigonometric identity* and *trigonometric equation*.

b) Inquiry/Problem Solving If $(\cos x + \sin x)^2 + (\cos x - \sin x)^2 = k$, for what value(s) of k is the equation an identity?

C

14. Solve each equation on the interval $0° \le x \le 360°$.

a) $\sin 2x = 1$ **b)** $\cos 2x = -1$ **c)** $2\sin 2x = 1$

d) $2\cos 2x = 1$ **e)** $\sqrt{2}\cos 2x = 1$ **f)** $2\sin 2x = \sqrt{3}$

g) $2\cos 2x = -\sqrt{3}$ **h)** $\sin 2x = 0$ **i)** $2\sin 0.5x = 1$

15. Technology Use a graphing calculator to find the exact solutions, in degrees, for $0° \le x \le 360°$.

a) $\cos 2x = \cos x$ **b)** $\sin 2x = 2\cos x$ **c)** $\tan x = \sin 2x$

16. Technology a) Describe how you would use a graphing calculator to solve the equation $\sin (\cos x) = 0$, $0° \le x \le 360°$.

b) Find the exact solutions to this equation.

ACHIEVEMENT Check   Knowledge/Understanding   Thinking/Inquiry/Problem Solving   Communication   Application

A storm drain has a cross section in the shape of an isosceles trapezoid. The shorter base and each of the two equal sides measure 2 m, and x is the angle formed by the longer base and each of the equal sides.

a) Write an expression for the area, A, of the cross section in terms of sin x and cos x.

b) Describe how you would use a graphing calculator to find the value of x, in degrees, if the area of the cross section is 5 m².

c) Find the value of x, to the nearest degree.

NUMBER *Power*

Place the digits from 1 to 9 in the boxes to make the statements true.
Use the order of operations.

$$■ × ■ - ■ = 2$$

$$(■ + ■) ÷ ■ = 2$$

$$■ + ■ - ■ = 2$$

Investigate & Apply

Modelling Double Helixes

1. Spiral staircases The up and down handrails of two spiral staircases inside a building are in the shape of a double helix. The staircases are 25 m high and there are 12 turns (half periods) of the handrails from the bottom to the top of the staircases. Each staircase is 2.2 m wide. The projection (side view) of the down handrail, as shown in the diagram, can be approximately modelled in the coordinate plane by $y = 1.1\sin 1.5x$, where x is in radians and y is in metres. Find at least two models that represent the projection of the up handrail. Explain your models.

2. DNA A strand of DNA is also in the shape of a double helix. The projection of the "down" side makes 5 turns (half periods) in a distance of 17 nm, where 1 nm = 10^{-9} m.
a) If the projection is 2 nm wide, find a model to represent the projection of the "down" side of a strand of DNA. Justify your model.
b) Find a model to represent the projection of the "up" side of a strand of DNA. Explain your model.

■ Down handrail
■ Up handrail

REVIEW OF **KEY CONCEPTS**

■ **5.1** Radians and Angle Measure

Refer to the Key Concepts on page 328.

1. Find the exact number of degrees in the angles with the following radian measures.

a) $\dfrac{2\pi}{3}$ b) $\dfrac{5\pi}{2}$ c) $\dfrac{3\pi}{5}$ d) $\dfrac{2\pi}{9}$ e) $\dfrac{8\pi}{3}$ f) 4π

2. Find the exact radian measure in terms of π for each of the following angles.

a) $50°$ b) $270°$ c) $135°$ d) $210°$ e) $225°$ f) $720°$

3. Find the approximate number of degrees, to the nearest tenth, in the angles with the following radian measures.

a) 3.5 b) $\dfrac{\pi}{7}$ c) 0.75 d) $\dfrac{5\pi}{11}$ e) 1.45 f) $\dfrac{7\pi}{13}$

4. Find the approximate number of radians, to the nearest hundredth, in the angles with the following degree measures

a) $30°$ b) $120°$ c) $70°$ d) $46.4°$ e) $231.6°$ f) $315.1°$

5. Determine the length of each arc a, to the nearest tenth.

a)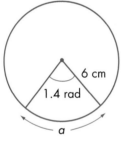

b)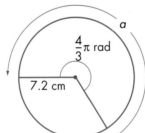

6. Determine the approximate measure of $\angle\theta$, to the nearest hundredth of a radian and to the nearest tenth of a degree.

a)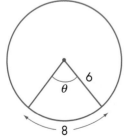

b)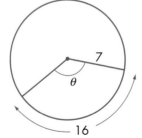

7. Determine the length of each radius, r, to the nearest tenth.

a)

b)

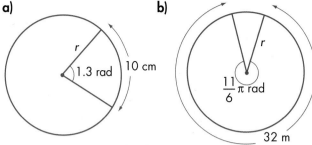

8. Electric motor An electric motor turns at 2400 revolutions per minute, or 2400 r/min. Find the angular velocity, in radians per second, in exact form and in approximate form, to the nearest hundredth.

9. Geometry The radian measure of one acute angle in a right triangle is $\frac{\pi}{5}$.

a) What is the exact radian measure of the other acute angle?

b) What is the exact degree measure of each acute angle?

5.2 Trigonometric Ratios of Any Angle

Refer to the Key Concepts on page 341.

10. The coordinates of a point P on the terminal arm of an $\angle\theta$ in standard position are given, where $0 \le \theta \le 2\pi$. Determine the exact values of $\sin\theta$, $\cos\theta$, and $\tan\theta$.

a) P(4, 5)   **b)** P(−2, 7)   **c)** P(−3, −6)   **d)** P(7, −4)

11. Find the exact value of each trigonometric ratio.

a) $\cos 30°$ **b)** $\tan 225°$ **c)** $\sin 210°$ **d)** $\cos 150°$

12. Find the exact value of each trigonometric ratio.

a) $\sin\dfrac{\pi}{6}$ **b)** $\cos\dfrac{4\pi}{3}$ **c)** $\tan\dfrac{7\pi}{4}$ **d)** $\cos\dfrac{5\pi}{6}$

13. $\angle\theta$ is in standard position with its terminal arm in the stated quadrant, and $0 \le \theta \le 2\pi$. Find the exact values of the other two trigonometric ratios.

a) $\sin\theta = \dfrac{2}{5}$, quadrant II   **b)** $\cos\theta = -\dfrac{4}{7}$, quadrant III

c) $\tan\theta = -\dfrac{5}{6}$, quadrant IV

14. If $0° \le A \le 360°$, find the possible measures of $\angle A$.

a) $\sin A = \dfrac{1}{2}$ **b)** $\cos A = \dfrac{1}{\sqrt{2}}$ **c)** $\tan A = -\sqrt{3}$ **d)** $\cos A = \dfrac{\sqrt{3}}{2}$

5.3 Modelling Periodic Behaviour

Refer to the Key Concepts on page 355.

15. The graph of $y = f(x)$ is shown.

a) Explain why this function is periodic.

b) Find the period and amplitude.

c) Find the value of

i) $f(2\pi)$ **ii)** $f(3\pi)$ **iii)** $f(9\pi)$

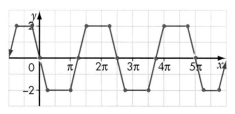

16. Sketch the graph of a periodic function with a period of 18 and an amplitude of 5.

5.4 Investigation: Sketching the Graphs of $f(x) = \sin x$, $f(x) = \cos x$, and $f(x) = \tan x$

Refer to the Key Concepts on page 363.

17. a) Sketch the graphs of $y = \sin x$ and $y = \cos x$ for the domain $-180° \leq x \leq 450°$.

b) Identify the period, amplitude, and range for each function.

18. a) Sketch the graph of $y = \tan x$, $-180° \leq x \leq 450°$.

b) Identify the period, domain, and range.

5.5 Stretches of Periodic Functions

Refer to the Key Concepts on page 367.

19. Sketch one cycle of the graph of each of the following. State the domain and range of the cycle.

a) $y = 4\sin x$ **b)** $y = \dfrac{1}{2}\cos x$ **c)** $y = \sin 3x$

d) $y = \cos 2x$ **e)** $y = \sin \dfrac{1}{3}x$ **f)** $y = \cos \dfrac{3}{4}x$

20. Find the period, in degrees and radians, for each of the following.

a) $y = \sin 2x$ **b)** $y = \cos \dfrac{3}{2}x$

c) $y = 2\sin \dfrac{1}{2}x$ **d)** $y = 1.5\cos 0.25x$

21. Write an equation and sketch one complete cycle for each sine function described. Each cycle begins at the origin, $x \geq 0$.

a) amplitude 4, period 3π **b)** amplitude 2, period π

22. Sketch one cycle of the graph of each of the following.

a) $y = 2\sin 2x$

b) $y = 5\cos 3x$

c) $y = 4\sin \dfrac{1}{2}x$

d) $y = \dfrac{2}{3}\cos \dfrac{1}{3}x$

23. Sketch the graphs of the following functions. State the range, amplitude, and period of each function.

a) $y = 5\sin 4x,\ -\pi \le x \le \pi$

b) $y = 0.5\cos 3x,\ -\pi \le x \le \pi$

c) $y = 3\cos \dfrac{1}{3}x,\ -6\pi \le x \le 6\pi$

d) $y = 2\sin \dfrac{2}{3}x,\ -6\pi \le x \le 6\pi$

24. Clock A clock pendulum completes one back-and-forth cycle every 2 s. The maximum measure of the angle of the pendulum with the vertical is 12°.

a) Express the measure of the angle between the pendulum and the vertical as a function of time. Assume the relationship is a sine function.

b) Draw a graph of the function for $0 \le t \le 6$, starting from the vertical position.

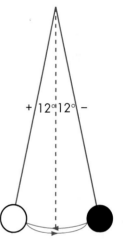

5.6 Translations and Combinations of Transformations

Refer to the Key Concepts on page 378.

25. Sketch one cycle of the graph of each of the following. State the domain, range, amplitude, period, and phase shift of the cycle.

a) $y = 4\sin x - 3$

b) $y = 3\cos x + 2$

c) $y = 2\sin\left(x - \dfrac{\pi}{2}\right)$

d) $y = \dfrac{1}{2}\cos\left(x + \dfrac{\pi}{2}\right)$

26. Sketch one cycle of the graph of each of the following. State the domain, range, amplitude, period, and phase shift of the cycle.

a) $y = 2\sin \dfrac{3}{2}x$

b) $y = -3\cos \dfrac{1}{4}x$

27. Sketch one cycle of the graph of each of the following. State the domain, range, amplitude, period, and phase shift of the cycle.

a) $y = 2\sin 2\left(x - \dfrac{\pi}{4}\right)$

b) $y = 3\cos 2\left(x + \dfrac{\pi}{4}\right) - 1$

c) $y = -\sin \dfrac{1}{3}(x - 2\pi) + 3$

d) $y = 2\cos \dfrac{1}{2}(x - \pi) + 1$

e) $y = \dfrac{1}{2}\sin\left(\dfrac{1}{2}x - \pi\right) - 2$

f) $y = -2\cos (3x - \pi) + 2$

28. Write an equation for the function with the given characteristics.

	Type	Amplitude	Period	Vertical Translation	Horizontal Translation
a)	sine	3	2π	-4	π right
b)	cosine	5	π	1	$\frac{\pi}{2}$ left

29. Sketch the graph of each of the following. State the range.

a) $y = 3\cos x - 2,\ 0 \le x < 3\pi$

b) $y = -2\sin 2x + 2,\ -\pi \le x \le \pi$

c) $y = \frac{1}{2}\sin\left(x + \frac{\pi}{4}\right),\ -2\pi \le x \le \pi$

d) $y = 5\cos 2\left(x - \frac{\pi}{2}\right) + 1,\ -2\pi \le x \le 2\pi$

e) $y = -2\cos\left(2x + \frac{\pi}{3}\right) - 1,\ -\pi \le x \le \pi$

f) $y = -4\sin\left(\frac{1}{3}x - \pi\right),\ -\pi \le x \le 3\pi$

30. Ferris wheel A carnival Ferris wheel with a radius of 9.5 m rotates once every 10 s. The bottom of the wheel is 1.2 m above the ground.

a) Find the equation of the sine function that gives a rider's height above the ground, in metres, as a function of the time, in seconds, with the rider starting at the bottom of the wheel.

b) Sketch the graph showing two complete cycles.

31. Tides The water depth in a harbour is 8 m at low tide and 20 m at high tide. One cycle is completed approximately every 12 h.

a) Find an equation for the water depth, $d(t)$ metres, as a function of the time, t hours, after high tide, which occurred at 03:00.

b) Draw a graph of the function for 48 h after high tide.

5.7 Trigonometric Identities

Refer to the Key Concepts on page 393.

32. Prove each identity.

a) $\dfrac{1 - \sin^2 x}{\cos x} = \cos x$

b) $\dfrac{\tan x}{\sin x} = \dfrac{1}{\cos x}$

c) $\dfrac{\sin x \cos x}{\tan x} = 1 - \sin^2 x$

d) $\cos^2 x + \dfrac{\sin x \cos x}{\tan x} = 2\cos^2 x$

e) $1 + \tan^2 x = \dfrac{1}{\cos^2 x}$

f) $\cos^2 x - \sin^2 x = 2\cos^2 x - 1$

g) $\dfrac{1}{\sin x} - \sin x = \dfrac{\cos x}{\tan x}$

h) $\dfrac{1 - \tan^2 x}{1 + \tan^2 x} = \cos^2 x - \sin^2 x$

33. Use a graphing calculator to show that each equation appears to be an identity. Then, prove the equation is an identity.

a) $\dfrac{1}{1 + \cos x} + \dfrac{1}{1 - \cos x} = \dfrac{2}{\sin^2 x}$

b) $\dfrac{1 + \cos x}{\sin x} - \dfrac{\sin x}{1 - \cos x} = 0$

34. Prove each identity.

a) $(\sin x - \cos x)(\sin x + \cos x) = 2\sin^2 x - 1$

b) $(\sin x - \cos x)^2 = 1 - 2\sin x \cos x$

c) $1 + \tan^2 x = \dfrac{1}{\cos^2 x}$

d) $\cos^2 x - \cos^4 x = \cos^2 x \sin^2 x$

e) $(1 - \cos^2 x)(1 + \tan^2 x) = \tan^2 x$

f) $\dfrac{\sin x}{1 - \cos x} - \dfrac{1 + \cos x}{\sin x} = 0$

5.8 Trigonometric Equations

Refer to the Key Concepts on page 402.

35. Solve each equation for $0 \le x \le 2\pi$.

a) $\cos x = 0$

b) $2\sin x - 1 = 0$

c) $\tan x = -1$

d) $\sqrt{2}\sin x = 1$

e) $2\cos x - 3 = 0$

f) $2\sin x + \sqrt{3} = 0$

g) $\sqrt{2}\cos x + 1 = 0$

h) $\cos x - 1 = 0$

i) $\tan x = \sqrt{3}$

36. Solve each equation for $0 \le x \le 2\pi$.

a) $\cos^2 x - 1 = \sin^2 x$

b) $2\cos^2 x + 3\cos x = -1$

c) $\sin^2 x + \sin x - 2 = 0$

d) $\sin x \cos x - \sin x = 0$

e) $2\sin^2 x + 1 = 3\sin x$

f) $\cos x + 1 = 2\sin^2 x$

g) $2\cos^2 x - 5\cos x + 3 = 0$

h) $2\sin^2 x - 7\sin x = 4$

i) $\cos^2 x + 4 = 4\cos x$

37. Solve on the interval $0° \le x \le 360°$. Give the exact solution in degrees and radians. Round approximate solutions to the nearest tenth of a degree and the nearest hundredth of a radian.

a) $5\cos x - 4 = 0$

b) $3\sin x = 3 - \sin x$

c) $9\cos^2 x + 6\cos x + 1 = 0$

d) $6\sin^2 x - 7\sin x + 2 = 0$

CHAPTER TEST

Achievement Chart

Category	Knowledge/Understanding	Thinking/Inquiry/Problem Solving	Communication	Application
Questions	All	11, 12, 13	7, 8, 9, 11, 13	9, 10, 11, 12,13

1. Change each radian measure to degree measure. Round to the nearest tenth of a degree, if necessary.

   a) $\frac{5}{3}\pi$
   b) $\frac{7}{6}\pi$
   c) 1.7

2. Find the approximate number of radians, to the nearest hundredth, in the angles with the following degree measures.

   a) 60°
   b) 205°
   c) 312.6°

3. The coordinates of a point P on the terminal arm of an $\angle\theta$ in standard position are given, where $0° \le \theta \le 2\pi$. Determine the exact values of sin θ, cos θ, and tan θ.

   a) P(−4, 6)
   b) P(5, −3)

4. Find the exact value of each trigonometric ratio.

   a) $\tan \frac{5}{4}\pi$
   b) $\sin \frac{4}{3}\pi$
   c) cos 120°
   d) sin 330°

5. If $0° \le A \le 360°$, find the possible measures of $\angle A$.

   a) $\sin A = -\frac{1}{\sqrt{2}}$
   b) $\cos A = \frac{\sqrt{3}}{2}$
   c) $\tan A = -1$

6. a) Sketch the graph of $y = \tan x$, $-2\pi \le x \le \pi$.
   b) State the period, domain, and range.

7. Sketch one cycle of the graph of each of the following.
   State the domain, range, amplitude, period, and phase shift of the cycle.

   a) $y = 4\sin 3x$
   b) $y = -3\cos \frac{1}{2}x$
   c) $y = 2\cos 2\left(x - \frac{\pi}{4}\right)$
   d) $y = -4\sin \frac{1}{2}(x + \pi) + 2$
   e) $y = 3\sin (3x - \pi) - 2$

8. Sketch the graph of each of the following. State the range, amplitude, period, and phase shift of each function.

   a) $y = 4\sin 2x - 3$, $0 \le x \le \pi$
   b) $y = -2\cos \frac{1}{2}(x - \pi)$, $0 \le x \le 4\pi$
   c) $y = -3\sin (2x + \pi) - 1$, $-\pi \le x \le 2\pi$

9. Tides The average depth of the water in a port on a tidal river is 4 m. At low tide, the depth of the water is 2 m. One cycle is completed approximately every 12 h.
a) Find an equation of the depth, $d(t)$ metres, with respect to the average depth, as a function of the time, t hours, after low tide, which occurred at 15:00.
b) Draw a graph of the function for 48 h after low tide.

10. Blood pressure a) The function $p(t) = 20\sin 2\pi t + 100$ models a person's blood pressure while resting, where $p(t)$ represents the blood pressure, in millimetres of mercury, and t is the time, in seconds. Sketch a graph of this function for $0 \le t \le 3$.
b) Find the period of the function in part a), and describe what the period represents.
c) The function $p(t) = 20\sin 4\pi t + 100$ models a person's blood pressure while exercising. Sketch a graph of this function for $0 \le t \le 3$.
d) Find the period of the function in part c), and describe what the period represents.
e) Find the amplitude of each function, and describe what the amplitude represents.

11. Prove each identity.
a) $\dfrac{\cos^2 x}{1 + 2\sin x - 3\sin^2 x} = \dfrac{1 + \sin x}{1 + 3\sin x}$
b) $\dfrac{\sin x + \tan x}{\cos x + 1} = \tan x$
c) $\dfrac{\tan x}{1 + \tan x} = \dfrac{\sin x}{\sin x + \cos x}$
d) $(2\sin x + 3\cos x)^2 + (3\sin x - 2\cos x)^2 = 13$

12. Solve each equation for $0 \le x \le 2\pi$.
a) $\cos x = \sqrt{3} - \cos x$
b) $\cos x \sin x - \cos x = 0$
c) $2\sin^2 x + 5\sin x + 3 = 0$
d) $4\sin^2 x - 1 = 0$

ACHIEVEMENT Check Knowledge/Understanding Thinking/Inquiry/Problem Solving Communication Application

13. In a resort town, the number of people employed during any given month, $f(x)$, in thousands, can be modelled by the function

$f(x) = 2.3\sin \dfrac{\pi}{6}(x + 1) + 5.5$, where x represents the month, with

January = 1, February = 2, and so on.
a) Approximately how many people are employed in August?
b) During which month is employment at its highest level? Use the function model to justify your answer.
c) In which two months are about 5000 people employed in the town?

CHALLENGE PROBLEMS

1. Intersection How many times can an arbitrary circle intersect the graph of $y = \sin x$?

a) at most 2 times   **b)** at most 4 times   **c)** at most 6 times
d) at most 8 times   **e)** more than 16 times

2. Sum of roots Find the sum of the roots of $\tan^2 x - 9\tan x + 1 = 0$, $0 \le x \le 2\pi$.

3. If ... , then ... If $f\left(\dfrac{x}{x+1}\right) = \dfrac{1}{x}$ for all $x \ne 0$, 1 and $0 < \theta < \dfrac{\pi}{2}$, then simplify $f\left(\dfrac{1}{\cos^2 \theta}\right)$.

4. Sin x cos x If $\sin x = 3\cos x$, what is the value of $\sin x \cos x$?

5. Sine function The graph represents the function $y = a\sin b\theta$. Find $\tan k$.

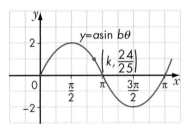

6. Number of roots If $0 < x < 2\pi$, determine the number of roots of $2 \sin^3 x - 5 \sin^2 x + 2 \sin x = 0$.

7. Sin x If $8\tan x = 3\cos x$, $0 < x < \pi$, find the value of $\sin x$.

8. Proof Prove that $1 - \dfrac{\sin^2 x}{1 + \cot x} - \dfrac{\cos^2 x}{1 + \tan x} = \sin x \cos x$.

PROBLEM SOLVING STRATEGY

SOLVE A SIMPLER PROBLEM

For some problems, a simpler or related problem may be easier to solve than the original problem. The solution to the simpler or related problem may give clues for solving the original problem.

At the end of a play, the 13 actors are introduced to the audience. The actors wait behind the curtain until their names are announced. The first actor introduced takes a spot on the stage facing the audience. The second actor introduced could stand on either side of the first actor. The third actor introduced could stand on either side of the two already on stage. The process continues for the remaining 10 actors. When introduced, each actor takes a position at one end of the line of actors. In how many different ways can the actors be lined up on the stage after all 13 have been introduced.

Understand the Problem

1. What information are you given?
2. What are you asked to find?
3. Do you need an exact or an approximate answer?

Think of a Plan

List the possible ways the first few actors could line up on the stage. Then, see if there is a pattern.

Carry Out the Plan

Actors	Possible Ways to Line Up							Number of Ways
A	A							1
A, B	AB			BA				2
A, B, C	CAB	ABC		CBA		BAC		4
A, B, C, D	DCAB	CABD	DABC	ABCD	DCBA	CBAD	DBAC	BACD	8

The numbers of ways are all powers of 2.
$1 = 2^0$, $2 = 2^1$, $4 = 2^2$, and $8 = 2^3$.
In each case, the exponent in the power of 2 is one less than the number of actors on stage.
For n actors, the number of ways is 2^{n-1}.
For 13 actors, the number of ways is 2^{13-1} or 2^{12} or 4096.

The actors can be lined up in 4096 different ways.

Look Back

Is there another way to solve the problem?

Solve a Simpler Problem

1. Break the problem into smaller parts.
2. Solve the problem.
3. Check that your answer is reasonable.

Apply, Solve, Communicate

1. Operations Evaluate.
$1 \times 2 - 2 \times 2 + 3 \times 2 - 4 \times 2 \ldots - 100 \times 2$

2. Whole numbers For how many whole numbers between 0 and 1000 does the sum of the digits equal 9?

3. Differences Find the exact value of this difference.
$(5\ 555\ 555\ 555)^2 - (4\ 444\ 444\ 445)^2$

4. Perimeters The first figure is made from one n-shape, the second from two n-shapes, the third from three n-shapes, and so on.

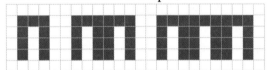

If the smallest squares on the grid measure 1 unit by 1 unit, what is the perimeter of the 60th figure?

5. Number cards Suppose you have a set of cards numbered consecutively from 1 to 100. You try to put the cards in pairs so that the sum of the numbers on the cards in each pair is 94. How many pairs will have a sum of 94?

6. Transformations Consider the following transformations on the number 119.

119	$1^2 + 1^2 + 9^2 = 83$
83	$8^2 + 3^2 = 73$
73	$7^2 + 3^2 = 58$
58	$5^2 + 8^2 = 89$

and so on

If this transformation is carried out on the number 42, what number results after 500 transformations?

7. Money How much more does a million dollars in toonies weigh than a million dollars in ten-dollar bills?

8. Digits Without using a calculator, find the number of digits in the number n, if $n = 2^{23} + 5^{19}$.

9. Number table The whole numbers are arranged as follows.

Row	Number				
1	1				
2	2	3			
3	4	5	6		
4	7	8	9	10	
5	11	12	13	14	15
and so on					

a) Write an expression for the last number in each row in terms of the row number.
b) What is the last number in the 38th row?
c) What number is in the 22nd column of the 56th row?

10. Sum Evaluate.
$1 + 2 + 3 + \ldots + 1000 + \ldots + 3 + 2 + 1$

11. Triangles a) Find the number of triangles in each figure.
b) How many triangles are in the 17th figure? the 99th figure?

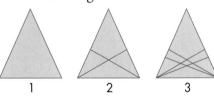

12. Formulating problems Write a problem that can be solved by solving a simpler problem. Have a classmate solve your problem.

PROBLEM SOLVING

USING THE STRATEGIES

1. Natural numbers There are three natural numbers that give the same result when they are added and when they are multiplied.

$1 + 2 + 3 = 1 \times 2 \times 3$

There is only one set of four natural numbers that has this property.

$1 + 1 + 2 + 4 = 1 \times 1 \times 2 \times 4$

There are three sets of five natural numbers, all less than 6, that have this property. What are they?

2. a) If $9x + 24 = A[x + B(x + C)]$ and A, B, and C are integers, find the values of A, B, and C.

b) If $-4x + 40 = A[x + B(x + C)]$ and A, B, and C are integers, find the values of A, B, and C.

3. Coins Among nine similar coins, one is counterfeit and lighter than the others. Using a balance only twice, how can the counterfeit coin be identified?

4. Baseball a) Use the following information to sketch a graph of distance from home plate versus time for the baseball. *The pitcher throws the baseball. The batter hits a long fly ball to right field. The ball hits the fence and is caught by the right fielder. The right fielder throws the ball to second base where it is caught by the shortstop.*

b) Use the same information to sketch a graph of the height of the ball versus time.

5. Numbers The sum of two numbers is 3. The product of the numbers is 2.

a) Find the numbers.

b) Find the sum of the reciprocals of the numbers.

6. Floor tiles A rectangular floor is made up of identical square tiles. The floor is 40 tiles long and 30 tiles wide. If a straight line is drawn diagonally across the floor from corner to corner, how many tiles will it cross?

7. Test marks On the six tests that she wrote, Kelly had an average of 87%. Her highest mark was 95%, which she achieved on only one of the tests. What is the lowest percent she could have achieved on one of the tests?

8. Stationery Tom and Alicia bought identical boxes of stationery. Tom used his to write 1-page letters. Alicia used hers to write 3-page letters. Tom used all the envelopes and had 50 sheets of paper left over. Alicia used all the paper and had 50 envelopes left over. How many sheets of paper were in each box?

9. Intersecting circles A circle with radius 3 cm intersects a circle with radius 4 cm. At the points of intersection, the radii are perpendicular. What is the difference in the areas of the non-overlapping parts of the circles?

10. Team logos About what percent of the people in your city or town own at least one piece of clothing with a professional sports team name or logo on it?

CHAPTER 6 Sequences and Series

Specific Expectations	Functions	Functions & Relations
Write the terms of a sequence, given the formula for the *n*th term.	6.1, 6.2, 6.3	6.1, 6.2, 6.3
Write the terms of a sequence, given a recursion formula.		6.4
Determine the formula for the *n*th term of a given sequence.	6.1, 6.2, 6.3	6.1, 6.2, 6.3
Identify sequences as arithmetic or geometric, or neither.	6.3	6.3
Determine the value of any term in an arithmetic or a geometric sequence, using the formula for the *n*th term of the sequence.	6.2, 6.3	6.2, 6.3
Determine the sum of the terms of an arithmetic or a geometric series, using appropriate formulas and techniques.	6.5, 6.6	6.5, 6.6

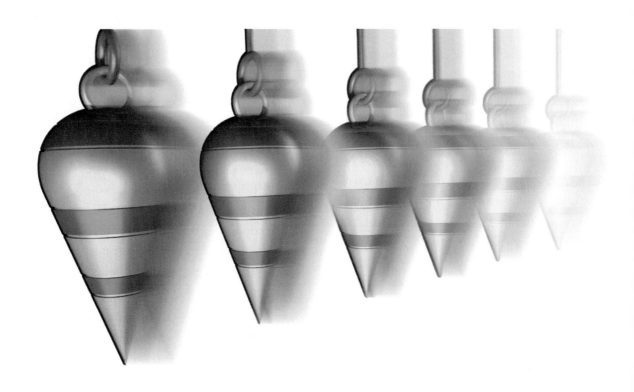

The Motion of a Pendulum

Some scientists, including Aristarchus of Samos in the 3rd century B.C. and Copernicus in the 16th century A.D., believed that the Earth rotates. However, no one had been able to demonstrate this rotation scientifically. In 1851, the French astronomer Jean Bernard Léon Foucault (1819-1868) constructed a 67-m long pendulum by suspending a 28-kg iron ball from the dome of the Panthéon in Paris. He used the pendulum to show that the Earth rotates about its axis.

In the Modelling Math questions on pages 445, 455, and 478, you will solve the following problem and other problems that involve the motion of a pendulum.

The period of a pendulum is the time it takes to complete one back-and-forth swing. On the Earth, the period, T seconds, is approximately given by the formula $T = 2\sqrt{l}$, where l metres is the length of the pendulum. If a 1-m pendulum completes its first period at a time of 10:15:30, or 15 min 30 s after 10:00,

a) at what time would it complete 100 periods? 151 periods?

b) how many periods would it have completed by 10:30:00?

Use your research skills to answer the following questions now.

1. Describe how Foucault demonstrated that the Earth rotates about its axis.

2. Describe one of the Foucault pendulums in Ontario. Examples include those at the University of Guelph and at Queen's University.

3. The angle through which the floor under a Foucault pendulum rotates each day depends on the latitude. Describe the relationship between the angle and the latitude.

Web Connection

www.school.mcgrawhill.ca/resources/

To use the Internet for your research on Foucault pendulums, visit the above web site. Go to **Math Resources**, then to *MATHEMATICS 11*, to find out where to go next.

Exploring Sequences

Patterns in Numbers and Letters

1. Describe the pattern and write the next three terms.

a) 4, 7, 10, 13, …
b) 3, 6, 12, 24, …
c) 1, 2, 4, 7, 11, 16, …
d) 4, 12, 6, 18, 9, …
e) 2, 3, 9, 10, 30, …
f) 3, 4, 7, 11, 18, 29, …
g) 9, 27, 45, 63, …
h) 2, 5, 10, 17, 26, …
i) 11, 8, 3, 5, −2, 7, …

2. Describe the pattern and write the next three terms.

a) A, D, G, J, …
b) A, D, B, E, C, F, D, …
c) A, C, D, F, G, I, J, …
d) 1, Z, 4, Y, 8, X, 13, …

Patterns in Diagrams

3. The I-Shapes are made from asterisks.
a) How many asterisks are in the fourth diagram? the fifth diagram?
b) Describe the pattern in words.
c) Write an expression that represents the number of asterisks in the nth I-shape in terms of n.
d) Using the expression from question 3, how many asterisks are there in the 65th I-shape? the 100th I-shape?

4. Diagram 1 has 1 triangle. If you count only the small triangles, diagram 2 has 4 triangles. Diagram 3 has 16 small triangles.

a) How many small triangles are there in the fourth diagram? the fifth diagram?
b) Describe the pattern in words.
c) How many small triangles are there in the 10th diagram?
d) Write an expression that represents the number of small triangles in the nth diagram in terms of n.

Review of Prerequisite Skills

If you need help with any of the skills named in purple below, refer to Appendix A.

1. Exponent rules Evaluate.
a) 2^7 b) $(-3)^4$ c) $3^4 \times 3^3$
d) $4^8 \div 4^5$ e) $(2^3)^4$ f) $3(2)^5$
g) $2(-5)^3$ h) $-3(-2)^7$ i) $-8(-1)^6$

2. Exponent rules Simplify.
a) $(3x^4)(5x^2)$ b) $16y^4 \div 2y^2$
c) $-24x^7 \div 6x^3$ d) $-36a^6 \div (-3a^6)$
e) $(-7t^6)(-4t^2)$ f) $(11g^5)(-4g^3)$

3. Evaluating expressions Evaluate for $x = 2$.
a) $(-x)^2$ b) 5^x
c) $(x-4)^3$ d) $(1-2x)^5$
e) $\dfrac{x(3^x-1)}{x-1}$ f) $\dfrac{3(x^6-1)}{x-1}$

4. Evaluating expressions Evaluate for $y = -3$.
a) $2y^2$ b) $-2y^3$
c) $(-y)^2$ d) $-y^2$
e) 2^y f) $(y-3)^2$
g) $\dfrac{y(2+y)^7}{y-1}$ h) $\dfrac{(y+2)(y-1)}{y+4}$

5. Simplifying expressions Simplify.
a) $4(x-3) - 2(3x+5)$
b) $2y(y-4) + y^2 - 7(y+2)$
c) $4(z-5) - (z-3) - 2z + 1$
d) $6(3t+4) - 2(t-6) - (t+5)$

6. Evaluating expressions Evaluate for
$x = 3$, $y = 2$, and $z = -1$.
a) $3xy + 4xz + 2yz$
b) $2x^2y - xz - 4z^2$
c) $3x(5-z) - (3-y) + 4(z-x)$
d) $zx^2 - 4yz^2 - 3xy^2$
e) $\dfrac{y}{2}[2z + x(y-1)]$
f) $\dfrac{z(x^y-1)}{x-1}$

7. Solving linear equations Solve and check.
a) $2x + 7 = 6$
b) $\dfrac{2}{3}n - 11 = -5$
c) $5y - 4 = 26 + 3y$
d) $0.2x + 3 = x + 7$
e) $2(d+1) - 4 = -18$
f) $15 = 3(z-5) + 6$

8. Solving linear equations Solve.
a) $5(a-7) - 3(a+5) = 10$
b) $3(2d-1) = 4(d-6) - 5$
c) $5(n+3) - 4 + 3(n-1) = 8$
d) $3(y+1) - 4(y-3) - (y+5) = -5$
e) $5 = 2(3d-1) - 4(d+2) - 11$

9. Exponent rules Use guess and check to find the value of x.
a) $4^x = 256$ b) $2^x = 128$
c) $3^{x-1} = 81$ d) $2^{x+1} = 256$
e) $7^{x-1} = 343$ f) $10^{x+1} = 1\ 000\ 000$

10. Graphing equations Graph each equation. The domain is the set of real numbers.
a) $y = x + 4$ b) $y = 2x - 3$
c) $y = x^2$ d) $y = -x^2$

11. Solving linear systems Solve and check.
a) $x + 3y = 16$ b) $3x + 2y = -7$
    $x + 8y = 31$ $4x - 5y = 6$
c) $8x + 3y = 4$ d) $y = 3x - 7$
    $2x + 7y = 1$ $y = -2x - 2$
e) $\dfrac{x}{2} + \dfrac{y}{3} = 7$ f) $\dfrac{x}{3} + \dfrac{y}{5} = 0$
    $\dfrac{x}{4} - \dfrac{y}{9} = 1$ $\dfrac{x}{6} - \dfrac{y}{2} = 6$

6.1 Sequences

A number **sequence** is a set of numbers, usually separated by commas, arranged in an order. The first term is referred to as t_1, the second term as t_2, the third term as t_3, and so on. The nth term is referred to as t_n.

A sequence may stop at some number, or continue indefinitely. The following are examples of sequences.

5, 7, 9, 11
2, 6, 18, 54
80, 40, 20, 10, ...   The three dots, called an **ellipsis**, indicate that the sequence continues indefinitely.

In the first sequence above, $t_1 = 5$, $t_2 = 7$, $t_3 = 9$, and $t_4 = 11$.

INVESTIGATE & INQUIRE

Spirals can be found in many places. The most sensational spirals are galaxies in the universe.

Examples of two common spirals are shown. The distances of successive loops from the centre of each spiral are marked on the diagrams.

Diagram 1   Diagram 2

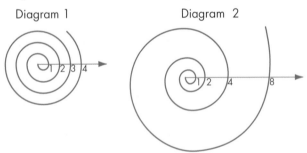

The spiral in diagram 1 was defined by Archimedes and is called an Archimedean spiral. The spiral in diagram 2 is called an exponential spiral.

For each spiral, the distances from the centre form a sequence.

In the first sequence   In the second sequence

$t_1 = 1$ $t_1 = 1$

$t_2 = 2$ $t_2 = 2$

$t_3 = 3$ $t_3 = 4$

$t_4 = 4$ $t_4 = 8$

1. How are the terms in the first sequence related?

2. Find t_5, t_6, and t_7 for the first sequence.

3. How are the terms in the second sequence related?

4. Find t_5, t_6, and t_7 for the second sequence.

5. A roll of tape is an Archimedean spiral. State two other examples.

6. Research Our galaxy is an example of an exponential spiral. Use your research skills to find another example.

7. Explain why spirals in which the distances from the centre follow such patterns as 1, 2, 4, 8, ... or 1, 3, 9, 27, ... are called exponential spirals.

Web Connection
www.school.mcgrawhill.ca/resources/
To learn more about galaxies, visit the above web site. Go to **Math Resources**, then to *MATHEMATICS 11*, to find out where to go next. Name some galaxies that are spirals, and describe the shapes of some galaxies that are not spirals.

Sometimes a pattern can lead to a general rule for finding the terms of a sequence. The rule is called the formula for the nth term, t_n.

For the sequence 1, 3, 5, 7, ...

$t_n = 2n - 1$
$t_1 = 2(1) - 1 = 1$
$t_2 = 2(2) - 1 = 3$
$t_3 = 2(3) - 1 = 5$
$t_4 = 2(4) - 1 = 7$

For the sequence 1, 3, 9, 27, ...

$t_n = 3^{n-1}$
$t_1 = 3^{1-1} = 3^0 = 1$
$t_2 = 3^{2-1} = 3^1 = 3$
$t_3 = 3^{3-1} = 3^2 = 9$
$t_4 = 3^{4-1} = 3^3 = 27$

In both sequences, the value of a term depends on the value of n.

EXAMPLE 1 Writing a Sequence Given the Formula for the nth Term

Given the formula for the nth term, write the first five terms of each sequence, and graph t_n versus n.

a) $t_n = 3n - 2$ **b)** $t_n = n^2 + 1$

SOLUTION 1 Paper-and-Pencil Method

a) $t_n = 3n - 2$
$t_1 = 3(1) - 2 = 1$
$t_2 = 3(2) - 2 = 4$
$t_3 = 3(3) - 2 = 7$
$t_4 = 3(4) - 2 = 10$
$t_5 = 3(5) - 2 = 13$

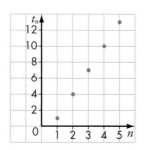

b) $t_n = n^2 + 1$

$t_1 = (1)^2 + 1 = 2$

$t_2 = (2)^2 + 1 = 5$

$t_3 = (3)^2 + 1 = 10$

$t_4 = (4)^2 + 1 = 17$

$t_5 = (5)^2 + 1 = 26$

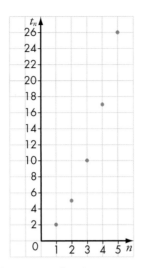

Solution 2 Graphing-Calculator Method

Use the mode settings to choose the Seq (sequence) graphing mode. Use the sequence function from the LIST OPS menu to generate the first 5 terms of the sequence. Store the terms in list L2. Also generate the first 5 natural numbers and store them in list L1. Using the STAT PLOTS menu and the ZoomStat instruction, draw the scatter plot of L2 versus L1.

The first display in part a) could be as follows. If the increment in n is not given, the default value is 1.

a)

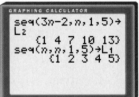

b)

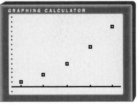

In the graphs in Example 1, there is exactly one value of t_n for each value of n. Therefore, each graph models a function. In general, a sequence is a function whose domain is the set, or a subset, of the natural numbers 1, 2, 3, 4, ... The values in the range of the function are the terms of the sequence. Because a sequence is a function, the formula for the nth term can be written using function notation.

EXAMPLE 2 Formula for the *n*th Term in Function Notation

Given the formula for the *n*th term in function notation, find t_8.

a) $f(n) = 5 - 2n$ b) $f(n) = 2n^2 - 1$

SOLUTION

a) $f(n) = 5 - 2n$
$f(8) = 5 - 2(8)$
$ = 5 - 16$
$ = -11$

b) $f(n) = 2n^2 - 1$
$f(8) = 2(8)^2 - 1$
$ = 2(64) - 1$
$ = 127$

A sequence can be entered into a graphing
calculator as a function of the independent
variable *n*. To input the function from
Example 2b), use the mode settings to choose
the Seq (sequence) graphing mode and the
dot mode. Display the sequence Y= editor
and enter $2n^2 - 1$ for the function u(*n*).

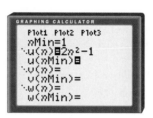

With the function u(*n*) defined, the graphing calculator can be used to find
a set of sequence terms or a specific term of the sequence. The TABLE key
can be used to display all the terms of the sequence.

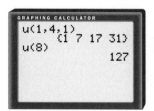
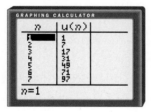

Note that, in the first
display, we could enter
u(1,4) instead of u(1,4,1).
The default increment is 1.

To graph the first 4 terms of the sequence, set the window variables to
*n*Min = 1 and *n*Max = 4, and choose suitable values for Xmin, Xmax, Ymin,
and Ymax. Press the GRAPH key to display the graph.

For this graph, Xmin = 0, Xmax = 5,
Ymin = 0, Ymax = 35.

EXAMPLE 3 Finding the Formula for the *n*th Term

a) Find the formula for the *n*th term that determines the sequence 4, 7, 10, 13, ...
b) Use the formula for the *n*th term to find t_{35} and t_{50}.

SOLUTION

a) Find a pattern and use it to write an expression for each term as a function of its term number.

$t_1 = 4$ or $3(1) + 1$
$t_2 = 7$ or $3(2) + 1$
$t_3 = 10$ or $3(3) + 1$
$t_4 = 13$ or $3(4) + 1$
So, $t_n = 3(n) + 1$ or $3n + 1$

The formula for the *n*th term is $t_n = 3n + 1$.

b) $t_{35} = 3(35) + 1$
$\quad\quad = 106$

$t_{50} = 3(50) + 1$
$\quad\quad = 151$

Key Concepts

- A number sequence is a set of numbers arranged in an order.
- Given the formula for the *n*th term, t_n or $f(n)$, the terms of a sequence can be written by substituting term numbers for *n*.
- To determine the formula for the *n*th term of a sequence, find a pattern and use it to write an expression for each term as a function of its term number.

Communicate Your Understanding

1. Given the formula for the *n*th term, $t_n = 3n + 5$, describe how you would write the first four terms of the sequence.
2. Given the formula for the *n*th term, $f(n) = 7 - 4n$, describe how you would find the 10th term of the sequence.
3. a) Describe how you would find the formula for the *n*th term that determines the sequence 4, 8, 12, 16, ...
b) Describe how you would use the formula for the *n*th term from part a) to find t_{44}.

Practise

A

1. Given the formula for the nth term, state the first five terms of the sequence and graph the solution.

a) $t_n = 3n$

b) $t_n = 2n + 4$

c) $t_n = 5 - 2n$

d) $f(n) = 10 - n$

e) $t_n = 2^n$

f) $f(n) = n^2 - 1$

2. Find a formula for the nth term that determines each sequence. Then, list the next three terms.

a) 5, 10, 15, 20, ...

b) 2, 3, 4, 5, ...

c) 6, 5, 4, 3, ...

d) 1, 4, 9, 16, ...

e) 2, 4, 6, 8, ...

f) $-3, -6, -12, -24, ...$

g) $-1, 0, 1, 2, ...$

h) 0.1, 0.2, 0.3, 0.4, ...

i) $\dfrac{1}{2}, \dfrac{2}{3}, \dfrac{3}{4}, \dfrac{4}{5}, ...$

j) $x, 2x, 3x, 4x, ...$

k) $1, 1 + d, 1 + 2d, 1 + 3d, ...$

3. List the first four terms of the sequence determined by each of the following.

a) $t_n = 3(n - 1)$

b) $t_n = (n - 1)^2$

c) $f(n) = \dfrac{1}{n}$

d) $f(n) = \dfrac{n + 1}{n}$

e) $t_n = (n + 1)(n - 1)$ **f)** $t_n = (-1)^n$

g) $t_n = 2^{n-1}$

h) $f(n) = 2^n - 1$

i) $f(n) = \dfrac{n - 1}{n + 1}$

j) $t_n = (-1)^{n-1}$

k) $f(n) = \dfrac{1}{3^n}$

l) $f(n) = \dfrac{1}{2^n} + 1$

4. Find the indicated terms.

a) $t_n = 2n + 7$; t_6 and t_{15}

b) $t_n = 8n - 5$; t_9 and t_{11}

c) $f(n) = 12 + 5n$; $f(3)$ and $f(10)$

d) $f(n) = 9 - 4n$; $f(4)$ and $f(8)$

e) $t_n = n^2 + 4$; t_3 and t_7

f) $t_n = \dfrac{n - 2}{2}$; t_4 and t_{12}

g) $f(n) = (n - 3)^2$; $f(2)$ and $f(15)$

Apply, Solve, Communicate

5. List the first four terms of each sequence. Then, find the formula for the nth term for each.

a)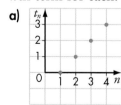

b)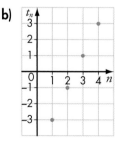

c)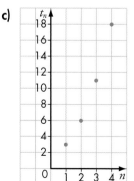

d)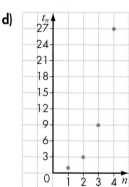

6. Whooping cranes A Canadian zoo started with 16 whooping cranes and planned to hatch 20 chicks a year. A few birds were kept for breeding and the rest released into the wild. At the end of last year, a total of 136 cranes had spent some time in the zoo. If the pattern continues, how many cranes will have spent time in the zoo by the end of this year? by the end of next year?

B

7. Gold production Canada ranks fifth in the world in the production of gold. By the end of one year, one of Canada's mines had produced about 198 t of gold since it opened. If the annual production averaged 4.5 t, predict the total production by the end of

a) the next year

b) the year after that

c) another 18 years after that

8. Hairstyles Human scalp hair grows at a rate of about 0.25 cm/week. Tania has decided to wear her hair longer. It is now 10 cm long.

a) How long will her hair be in 1 week? 2 weeks? 6 weeks?

b) How many weeks will it take for Tania's hair to grow from 10 cm to 15 cm in length?

9. Communication
The Earth is about 150 000 000 km from the sun. Astronomers call this distance one astronomical unit (1 AU). The distances of other planets from the sun can be expressed in astronomical units. For example, if a planet were twice as far from the sun as the Earth, the planet's distance from the sun would be 2 AU. In 1776, the astronomer Johann Bode

Planet	Bode's Distance (AU)	Actual Distance (AU)
Mercury	$\frac{(0 + 4)}{10}$	0.387
Venus	$\frac{(3 + 4)}{10}$	0.723
Earth	$\frac{(6 + 4)}{10}$	1
Mars	$\frac{(12 + 4)}{10}$	1.524
Minor Planets	$\frac{(24 + 4)}{10}$	2.3 to 3.3
Jupiter		5.203
Saturn		9.555
Uranus		19.22
Neptune		30.11
Pluto		39.84

found a sequence that he thought could determine each planet's distance from the sun in astronomical units.

a) Describe Bode's sequence in words.
b) Copy Bode's sequence and continue the sequence for the last five planets.
c) Calculate each planet's distance from the sun using Bode's sequence.
d) Compare the results from part c) with the actual distances. Name the first planet for which Bode's distance is not close to the actual distance.

10. **Application** Joshua exercised by cycling for 30 min, then taking a brisk 15-min walk. The bike ride used 1000 kJ of energy. The walk used energy at a rate of 20 kJ/min.
a) What was the total amount of energy used for cycling and walking after Joshua had walked for 1 min? 5 min? 15 min?
b) Write the formula for the nth term that determines the sequence.

11. **Stadium seating** In one section of a stadium, there are 30 seats in the first row, 32 in the second row, 34 in the third row, and so on. There are 60 rows.
a) Write the first 5 terms of the sequence.
b) Write the general term that determines the sequence.
c) How many seats are there in the 60th row?

12. **Inquiry/Problem Solving** A sequence begins 1, 2, 4, ... Can you predict the fourth term? Explain.

C

13. a) Write three different sequences of your own.
b) Describe each sequence in words.
c) Write the formula for the nth term of each sequence.

14. **Car depreciation** A car dealership has determined that a car costing $60 000.00 new depreciates in value by 20% in the first year. In the second year, the car depreciates by 20% of the value it had at the end of the first year. The car continues to depreciate each year by 20% of the value it had at the end of the previous year.
a) What is the value of the car at the end of the first year? the second year?
b) Write a formula to determine the value of the car at the end of the nth year.
c) After how many years is the car worth about $10 000.00?

NUMBER *Power*

Without using a calculator, decide which is greater, 3^{99} or $3^{97} + 3^{97} + 3^{97}$.
Show your reasoning.

6.2 Arithmetic Sequences

A sequence like 2, 5, 8, 11, … , where the difference between consecutive terms is a constant, is called an **arithmetic sequence**. In an arithmetic sequence, the first term, t_1, is denoted by the letter a. Each term after the first is found by adding a constant, called the **common difference**, d, to the preceding term.

INVESTIGATE & INQUIRE

For about 200 years, the Gatineau River was used as a highway by logging companies. Logs from the Canadian Shield were floated down the river to the Ottawa River. It has been estimated that 2% of the hundreds of millions of logs that floated down the Gatineau sank. Those that sank below the oxygen level are perfectly preserved and are now being harvested by water loggers, who wear scuba-diving gear.

The pressure that a diver experiences is the sum of the pressure of the atmosphere and the pressure of the water. The increase in pressure with depth under water follows an arithmetic sequence. If a diver enters the water when the atmospheric pressure is 100 kPa (kilopascals), the pressure at a depth of 1 m is about 110 kPa. At a depth of 2 m, the pressure is about 120 kPa, and so on.

1. Copy and complete the table for this sequence.

	Term				
	t_1	t_2	t_3	t_4	t_5
Pressure (kPa)	100	110	120	130	140
Pressure (kPa) Expressed Using 100 and 10	100	100 + 1(10)			
Pressure Expressed Using a and d	a	$a +$			

2. What are the values of a and d for this sequence?

3. When you write an expression for a term using the letters a and d, you are writing a formula for the term. What is the formula for t_6? t_8? t_9?

4. Evaluate t_8 and t_9.

5. The Gatineau River has maximum depth of 35 m. What pressure would a diver experience at this depth?

EXAMPLE 1 Writing Terms of a Sequence

Given the formula for the nth term of an arithmetic sequence, $t_n = 2n + 1$, write the first 6 terms.

SOLUTION 1 Paper-and-Pencil Method

$t_n = 2n + 1$
$t_1 = 2(1) + 1 = 3$
$t_2 = 2(2) + 1 = 5$
$t_3 = 2(3) + 1 = 7$
$t_4 = 2(4) + 1 = 9$
$t_5 = 2(5) + 1 = 11$
$t_6 = 2(6) + 1 = 13$

The first 6 terms are 3, 5, 7, 9, 11, and 13.

SOLUTION 2 Graphing-Calculator Method

Use the mode settings to choose the Seq (sequence) graphing mode. Use the sequence function from the LIST OPS menu to generate the first 6 terms.

The first 6 terms are 3, 5, 7, 9, 11, and 13.

Note that the arithmetic sequence defined by $t_n = 2n + 1$, or $f(n) = 2n + 1$, in Example 1, is a linear function, as shown by the following graphs.

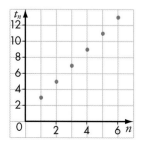

EXAMPLE 2 Determining the Value of a Term

Given the formula for the nth term, find t_{10}.

a) $t_n = 7 + 4n$ **b)** $f(n) = 5n - 8$

SOLUTION 1 Pencil-and-Paper Method

a) $t_n = 7 + 4n$
$t_{10} = 7 + 4(10)$
$= 7 + 40$
$= 47$

b) $f(n) = 5n - 8$
$f(10) = 5(10) - 8$
$= 50 - 8$
$= 42$

SOLUTION 2 Graphing-Calculator Method

Use the mode settings to choose the Seq (sequence) graphing mode. Use the sequence function from the LIST OPS menu to generate the 10th term.

a)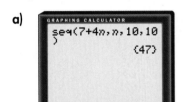

b)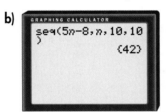

Note that the general arithmetic sequence is
$a, a + d, a + 2d, a + 3d, \ldots$
where a is the first term and d is the common difference.
$t_1 = a$
$t_2 = a + d$
$t_3 = a + 2d$
$\vdots$

$t_n = a + (n - 1)d$, where n is a natural number.

Note that d is the difference between any successive pair of terms.
For example,
$t_2 - t_1 = (a + d) - a$
$= d$
$t_3 - t_2 = (a + 2d) - (a + d)$
$= a + 2d - a - d$
$= d$

EXAMPLE 3 Finding the Formula for the *n*th Term

Find the formula for the *n*th term, t_n, and find t_{19}, for the arithmetic sequence 8, 12, 16, …

SOLUTION

For the given sequence, $a = 8$ and $d = 4$.

$$t_n = a + (n - 1)d$$

Substitute known values: $\quad = 8 + (n - 1)4$

Expand: $\quad = 8 + 4n - 4$

Simplify: $\quad = 4n + 4$

Three ways to find t_{19} are as follows.

Method 1	*Method 2*	*Method 3*
$t_n = a + (n - 1)d$	$t_n = 4n + 4$	Use a graphing calculator.
$t_{19} = a + (19 - 1)d$	$t_{19} = 4(19) + 4$	
$= a + 18d$	$= 76 + 4$	
$= 8 + 18(4)$	$= 80$	
$= 8 + 72$		
$= 80$		

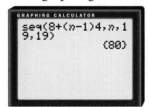

GRAPHING CALCULATOR
seq(8+(n-1)4,n,1
9,19)
{80}

So, $t_n = 4n + 4$ and $t_{19} = 80$.

EXAMPLE 4 Finding the Number of Terms

How many terms are there in the following sequence?
−3, 2, 7, … , 152

SOLUTION

For the given sequence, $a = -3$, $d = 5$, and $t_n = 152$.

Substitute the known values in the formula for the general term and solve for *n*.

$$t_n = a + (n-1)d$$

Substitute known values:	$152 = -3 + (n-1)5$
Expand:	$152 = -3 + 5n - 5$
Simplify:	$152 = 5n - 8$
Solve for n:	$152 + 8 = 5n - 8 + 8$

$$160 = 5n$$
$$\frac{160}{5} = \frac{5n}{5}$$
$$32 = n$$

The sequence has 32 terms.

EXAMPLE 5 Finding t_n Given Two Terms

In an arithmetic sequence, $t_7 = 121$ and $t_{15} = 193$. Find the first three terms of the sequence and t_n.

SOLUTION

Substitute known values in the formula for the nth term to write a system of equations. Then, solve the system.

	$t_n = a + (n-1)d$	
Write an equation for t_7:	$121 = a + (7-1)d$	
	$121 = a + 6d$	(1) **The (1) shows that we are naming the equation as "equation one."**
Write an equation for t_{15}:	$193 = a + (15-1)d$	
	$193 = a + 14d$	(2)
Subtract (1) from (2):	$72 = 8d$	
Solve for d:	$9 = d$	
Substitute 9 for d in (1):	$121 = a + 6(9)$	
Solve for a:	$121 = a + 54$	
	$67 = a$	**You can check by substituting 67 for a and 9 for d in (2).**

Since $a = 67$ and $d = 9$, the first three terms of the sequence are 67, 76, and 85.

To find t_n, substitute 67 for a and 9 for d in the formula for the nth term.

	$t_n = a + (n-1)d$
	$t_n = 67 + (n-1)9$
Simplify:	$t_n = 67 + 9n - 9$
	$t_n = 9n + 58$

So, the first three terms are 67, 76, and 85, and $t_n = 9n + 58$.

Key Concepts

- The general arithmetic sequence is a, $a + d$, $a + 2d$, $a + 3d$, ... , where a is the first term and d is the common difference.
- The formula for the nth term, t_n or $f(n)$, of an arithmetic sequence is $t_n = a + (n - 1)d$, where n is a natural number.

Communicate Your Understanding

1. Given the formula for the nth term of an arithmetic sequence, $t_n = 4n - 3$, describe how you would find the first 6 terms.

2. a) Describe how you find the formula for the nth term of the arithmetic sequence 3, 8, 13, 18, ...

b) Describe how you would find t_{46} for this sequence.

3. Describe how you would find the number of terms in the sequence 5, 10, 15, ... , 235.

4. Given that $t_5 = 11$ and $t_{12} = 25$ for an arithmetic sequence, describe how you would find t_n for the sequence.

Practise

A

1. Find the next three terms of each arithmetic sequence.

a) 3, 7, 11, ...
b) 33, 27, 21, ...
c) −23, −18, −13, ...
d) 25, 18, 11, ...
e) 5.8, 7.2, 8.6, ...
f) $\dfrac{3}{4}, \dfrac{5}{4}, \dfrac{7}{4}$, ...

2. Given the formula for the nth term of an arithmetic sequence, write the first 4 terms.

a) $t_n = 3n + 5$
b) $f(n) = 2n - 7$
c) $t_n = 4n - 1$
d) $f(n) = 6 - n$
e) $t_n = -5n - 2$
f) $f(n) = \dfrac{n + 3}{2}$

3. Given the formula for the nth term of an arithmetic sequence, write the indicated term.

a) $t_n = 2n - 5$; t_{11}
b) $t_n = 4 + 3n$; t_{15}
c) $f(n) = -4n + 5$; t_{10}
d) $f(n) = 0.1n - 1$; t_{25}
e) $t_n = 2.5n + 3.5$; t_{30}
f) $f(n) = \dfrac{2n - 1}{3}$; t_{12}

4. Determine which of the following sequences are arithmetic. If a sequence is arithmetic, write the values of a and d.

a) 5, 9, 13, 17, ...
b) 1, 6, 10, 15, 19, ...
c) 2, 4, 8, 16, 32, ...
d) −1, −4, −7, −10, ...
e) 1, −1, 1, −1, 1, ...
f) $\dfrac{1}{2}, \dfrac{2}{3}, \dfrac{3}{4}, \dfrac{4}{5}$, ...
g) −4, −2.5, −1, 0.5, ...
h) y, y^2, y^3, y^4, ...
i) x, $2x$, $3x$, $4x$, ...
j) c, $c + 2d$, $c + 4d$, $c + 6d$, ...

5. Given the values of a and d, write the first five terms of each arithmetic sequence.

a) $a = 7, d = 2$ b) $a = 3, d = 4$

c) $a = -4, d = 6$ d) $a = 2, d = -3$

e) $a = -5, d = -8$ f) $a = \frac{5}{2}, d = \frac{1}{2}$

g) $a = 0, d = -0.25$ h) $a = 8, d = x$

i) $t_1 = 6, d = y + 1$ j) $t_1 = 3m, d = 1 - m$

6. Find the formula for the nth term and find the indicated terms for each arithmetic sequence.

a) $6, 8, 10, \dots ; t_{10}$ and t_{34}

b) $12, 16, 20, \dots ; t_{18}$ and t_{41}

c) $9, 16, 23, \dots ; t_9$ and t_{100}

d) $-10, -7, -4, \dots ; t_{11}$ and t_{22}

e) $-4, -9, -14, \dots ; t_{18}$ and t_{66}

f) $\frac{1}{2}, \frac{3}{2}, \frac{5}{2}, \dots ; t_{12}$ and t_{21}

g) $5, -1, -7, \dots ; t_8$ and t_{14}

h) $7, 10, 13, \dots ; t_{15}$ and t_{30}

i) $10, 8, 6, \dots ; t_{13}$ and t_{22}

j) $x, x + 4, x + 8, \dots ; t_{14}$ and t_{45}

7. Find the number of terms in each of the following arithmetic sequences.

a) $10, 15, 20, \dots , 250$

b) $1, 4, 7, \dots , 121$

c) $40, 38, 36, \dots , -30$

d) $-11, -7, -3, \dots , 153$

e) $-2, -8, -14, \dots , -206$

f) $-6, -\frac{7}{2}, -1, \dots , 104$

g) $x + 2, x + 9, x + 16, \dots , x + 303$

8. Find a and d, and then write the formula for the nth term, t_n, of arithmetic sequences with the following terms.

a) $t_5 = 16$ and $t_8 = 25$

b) $t_{12} = 52$ and $t_{22} = 102$

c) $t_{50} = 140$ and $t_{70} = 180$

d) $t_2 = -12$ and $t_5 = 9$

e) $t_7 = -37$ and $t_{10} = -121$

f) $t_8 = 166$ and $t_{12} = 130$

g) $t_4 = 2.5$ and $t_{15} = 6.9$

h) $t_3 = 4$ and $t_{21} = -5$

9. The third term of an arithmetic sequence is 24 and the ninth term is 54.

a) What is the first term?

b) What is the formula for the nth term?

10. The fourth term of an arithmetic sequence is 14 and the eleventh term is -35.

a) What are the first four terms?

b) What is the formula for the nth term?

Apply, Solve, Communicate

11. The graph of an arithmetic sequence is shown.

a) What are the first five terms of the sequence?

b) What is t_{50}? t_{200}?

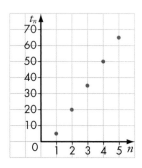

12. Find the common difference of the sequence whose formula for the nth term is $t_n = 2n - 3$.

13. Copy and complete each arithmetic sequence. Graph t_n versus n for each sequence.

a) ■, ■, 14, ■, 26   b) ■, 3, ■, ■, −18

14. **Olympic Games** The modern summer Olympic Games were first held in Athens, Greece, in 1896. The games were to be held every four years, so the years of the games form an arithmetic sequence.
a) What are the values of a and d for this sequence?
b) **Research** In what years were the games cancelled and why?
c) What are the term numbers for the years the games were cancelled?
d) What is the term number for the next summer games?

15. **Multiples** How many multiples of 5 are there from 15 to 450, inclusive?

16. The 18th term of an arithmetic sequence is 262. The common difference is 15. What is the first term of the sequence?

17. **Driving** Barrie is 60 km north of Toronto by road. If you drive north from Barrie at 80 km/h, how far are you from Toronto by road after
a) 1 h?   b) 2 h?   c) t hours?

18. **Inquiry/Problem Solving** Comets approach the Earth at regular intervals. For example, Halley's Comet reaches its closest point to the Earth about every 76 years. Comet Finlay is expected to reach its closest point to the Earth in 2009, 2037, and three times between these years. In which years between 2009 and 2037 will Comet Finlay reach its closest point to the Earth?

19. **Electrician** Amber works as an electrician. She charges $60 for each service call, plus an hourly rate. If she charges $420 for an 8-h service call
a) what is her hourly rate?
b) how much would she charge for a 5-h service call?

20. **Salary** Franco is the manager of a health club. He earns a salary of $25 000 a year, plus $200 for every membership he sells. What will he earn in a year if he sells 71 memberships? 88 memberships? 104 memberships?

21. Ring sizes A ring size indicates a standardized inside diameter of a ring. The table gives the inside diameters for 5 ring sizes.

Ring Size	Inside Diameter (mm)
1	12.37
2	13.2
3	14.03
4	14.86
5	15.69

a) Determine the formula for the nth term of the sequence of inside diameters.
b) Use the formula to find the inside diameter of a size 13 ring.

22. Displaying merchandise Boxes are stacked in a store display in the shape of a triangle. The numbers of boxes in the rows form an arithmetic sequence. There are 41 boxes in the 3rd row from the bottom. There are 23 boxes in the 12th row from the bottom.
a) How many boxes are there in the first (bottom) row?
b) What is the formula for the nth term of the sequence?
c) What is the maximum possible number of rows of boxes?

23. Application On the first day of practice, the soccer team ran eight 40-m wind sprints. On each day after the first, the number of wind sprints was increased by two from the day before.
a) What are the values of a and d for this sequence?
b) Write the formula for the nth term of the sequence.
c) How many wind sprints did the team run on the 15th day of practice? How many metres was this?

24. Pattern How many dots are in the 51st figure?

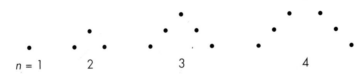

$n = 1$ 2   3   4

25. Pattern The U-shapes are made from asterisks.
a) How many asterisks are in the 4th diagram?
b) What is the formula for the nth term of the sequence in the numbers of asterisks?
c) How many asterisks are in the 25th diagram?
d) Which diagram contains 139 asterisks?

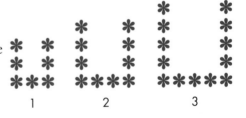

26. Astronomy The time from one full moon to the next is 29.53 days. The first full moon of a year occurred 12.31 days into the year.
a) How many days into the year did the 9th full moon occur?
b) At what time of day did the 9th full moon occur?

27. The eighth term of an arithmetic sequence is 5.3 and the fourteenth term is 8.3. What is the fifth term?

28. Communication a) Use finite differences to explain why the graph of t_n versus n for an arithmetic sequence is linear.
b) Explain why the points on a graph of t_n versus n for an arithmetic sequence are not joined by a straight line.

29. Motion of a pendulum The period of a pendulum is the time it takes to complete one back-and-forth swing. On the Earth, the period, T seconds, is approximately given by the formula $T = 2\sqrt{l}$, where l metres is the length of the pendulum. If a 1-m pendulum completes its first period at a time of 10:15:30, or 15 min 30 s after 10:00,
a) at what time would it complete 100 periods? 151 periods?
b) how many periods would it have completed by 10:30:00?

30. Motion of a pendulum Repeat question 29 for a 9-m pendulum on the Earth.

31. Motion of a pendulum The period of a pendulum depends on the acceleration due to gravity, so the period would be different on the moon than on the Earth. On the moon, the period, T seconds, would be given approximately by the formula $T = 5\sqrt{l}$, where l metres is the length of the pendulum. Repeat question 29 for a 1-m pendulum on the moon.

C

32. Measurement The side lengths in a right triangle form an arithmetic sequence with a common difference of 2. What are the side lengths?

33. The sum of the first two terms of an arithmetic sequence is 16. The sum of the second and third terms is 28. What are the first three terms of the sequence?

34. a) How does the sum of the first and fourth terms of an arithmetic sequence compare with the sum of the second and third terms? Explain.
b) Find two other pairs of terms whose sums compare in the same way as the two pairs of terms in part a).

35. The first four terms of an arithmetic sequence are 4, 13, 22, and 31. Which of the following is a term of the sequence?
      316   317   318   319   320

36. Algebra The first term of an arithmetic sequence is represented by $3x + 2y$. The third term is represented by $7x$. Write the expression that represents the second term.

37. Algebra Determine the value of x that makes each sequence arithmetic.
a) $2, 8, 14, 4x, \ldots$
b) $1, 3, 5, 2x - 1, \ldots$
c) $x - 2, x + 2, 5, 9, \ldots$
d) $x - 4, 6, x, \ldots$
e) $x + 8, 2x + 8, -x, \ldots$

38. Algebra Find the value of x so that the three given terms are consecutive terms of an arithmetic sequence.
a) $2x - 1, 4x,$ and $5x + 3$
b) $x, 0.5x + 7,$ and $3x - 1$
c) $2x, 3x + 1,$ and $x^2 + 2$

39. Algebra Find a, d, and t_n for the arithmetic sequence with the terms $t_7 = 3 + 5x$ and $t_{11} = 3 + 23x$.

40. Algebra Show that $t_n - t_{n-1} = d$ for any arithmetic sequence.

ACHIEVEMENT Check   Knowledge/Understanding   Thinking/Inquiry/Problem Solving   Communication   Application

3, 14, 25, ... and 2, 9, 16, ... are two arithmetic sequences. Find the first ten terms common to both sequences.

LOGIC *Power*

A box contains 5 coloured cubes and an empty space the size of a cube.

Use moves like those in checkers. In one move, one cube can slide to an empty space or jump over one cube to an empty space. Find the smallest number of moves needed to reverse the order of the cubes.

6.3 Geometric Sequences

In the sequence 2, 10, 50, 250, … , each term after the first is found by multiplying the preceding term by 5. Therefore, the ratio of consecutive terms is a constant.

$$\frac{10}{2} = 5 \qquad \frac{50}{10} = 5 \qquad \frac{250}{50} = 5$$

This type of sequence is called a **geometric sequence**. The ratio of consecutive terms is called the **common ratio**. For the sequence 2, 10, 50, 250, … , the common ratio is 5.

In a geometric sequence, the first term, t_1, is denoted by the letter a. Each term after the first is found by multiplying the previous term by the common ratio, r.

INVESTIGATE & INQUIRE

An acoustic piano has 88 keys. Each key plays a note with a different frequency. The frequency of a note is measured in hertz, symbol Hz. One hertz is one vibration per second. The first and lowest note is assigned the letter A. It has a frequency of 27.5 Hz.

There are eight As on a piano. The second A has a frequency two times the frequency of the first A, or 55 Hz. The third A has a frequency two times the frequency of the second A, or 110 Hz. Each subsequent A has a frequency two times the frequency of the A that precedes it.

The first four As have the frequencies 27.5 Hz, 55 Hz, 110 Hz, and 220 Hz. The numbers 27.5, 55, 110, and 220 form a geometric sequence.

1. Copy and complete the table for this sequence.

	Note							
	A_1	A_2	A_3	A_4	A_5	A_6	A_7	A_8
Frequency (Hz)	27.5	55	110	220				
Frequency (Hz) Expressed Using 27.5 and Powers of 2	27.5	27.5×2^1						
Frequency Expressed Using a and r	a	$a \times$						

2. What are the values of a and r for this sequence?

3. When you write an expression for a term using the letters a and r, you are writing a formula for the term. What is the formula for t_6? t_7? t_8?

4. Evaluate t_6, t_7, and t_8.

5. Write the formula for the nth term of this geometric sequence.

6. There are seven Gs on a piano. The frequency doubles from one G to the next. The lowest G has a frequency of about 49 Hz. Find the frequency of the highest G.

7. Describe the similarities and differences in the formulas for the nth term of an arithmetic sequence and the nth term of a geometric sequence.

8. Write the formula for the nth term of the following sequence.
2, −10, 50, −250, ...

EXAMPLE 1 Writing Terms of a Sequence

Given the formula for the nth term of a geometric sequence, $t_n = 5(-2)^{n-1}$, write the first 5 terms.

SOLUTION 1 Paper-and-Pencil Method

$$t_n = 5(-2)^{n-1}$$

$t_1 = 5(-2)^{1-1}$	$t_2 = 5(-2)^{2-1}$	$t_3 = 5(-2)^{3-1}$	$t_4 = 5(-2)^{4-1}$	$t_5 = 5(-2)^{5-1}$
$= 5(-2)^0$	$= 5(-2)^1$	$= 5(-2)^2$	$= 5(-2)^3$	$= 5(-2)^4$
$= 5(1)$	$= 5(-2)$	$= 5(4)$	$= 5(-8)$	$= 5(16)$
$= 5$	$= -10$	$= 20$	$= -40$	$= 80$

The first 5 terms are 5, −10, 20, −40, and 80.

SOLUTION 2 Graphing-Calculator Method

Adjust the mode settings to the Seq (sequence) graphing mode. Use the sequence function from the LIST OPS menu to generate the first 5 terms.

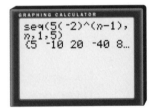

The first 5 terms are 5, −10, 20, −40, and 80.

EXAMPLE 2 Determining the Value of a Term

Given the formula for the nth term, find t_6.

a) $t_n = 3(2)^{n-1}$ b) $f(n) = -5(4)^{n-1}$

SOLUTION 1 Paper-and-Pencil Method

a) $t_n = 3(2)^{n-1}$

$t_6 = 3(2)^{6-1}$

$= 3(2)^5$

$= 3(32)$

$= 96$

b) $f(n) = -5(4)^{n-1}$

$f(6) = -5(4)^{6-1}$

$= -5(4)^5$

$= -5(1024)$

$= -5120$

SOLUTION 2 Graphing-Calculator Method

Adjust the mode settings to the Seq (sequence) graphing mode. Use the sequence function from the LIST OPS menu to generate the 6th term.

a)
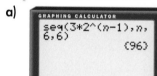

b)
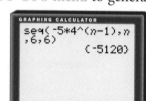

The general geometric sequence is a, ar, ar^2, ar^3, ... , where a is the first term and r is the common ratio.

$t_1 = a$

$t_2 = ar$

$t_3 = ar^2$

.
.
.

$t_n = ar^{n-1}$, where n is a natural number, and $r \neq 0$.

Note that r is the ratio of any successive pair of terms. For example,

$\dfrac{t_2}{t_1} = \dfrac{ar}{a}$

$= r$

$\dfrac{t_3}{t_2} = \dfrac{ar^2}{ar}$

$= r$

EXAMPLE 3 Finding the Formula for the *n*th Term

Find the formula for the *n*th term, t_n, and find t_6 for the geometric sequence 2, 6, 18, …

SOLUTION

For the given sequence, $a = 2$ and $r = 3$.

$$t_n = ar^{n-1}$$

Substitute known values: $\quad = 2(3)^{n-1}$

The formula for the *n*th term is $t_n = 2(3)^{n-1}$.

Three ways to find t_6 are as follows.

Method 1

$t_n = ar^{n-1}$
$t_6 = 2(3)^{6-1}$
$\quad = 2(3)^5$
$\quad = 2(243)$
$\quad = 486$

Method 2

$t_n = 2(3)^{n-1}$
$t_6 = 2(3)^{6-1}$
$\quad = 2(3)^5$
$\quad = 2(243)$
$\quad = 486$

Method 3

Use a graphing calculator.

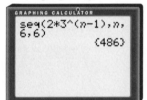

So, $t_n = 2(3)^{n-1}$ and $t_6 = 486$.

EXAMPLE 4 Finding the Number of Terms

Find the number of terms in the geometric sequence 3, 6, 12, … , 384.

SOLUTION

For the given sequence, $a = 3$, $r = 2$, and $t_n = 384$.

Substitute the known values in the formula for the general term and solve for *n*.

$$t_n = ar^{n-1}$$

Substitute known values: $\quad 384 = 3(2)^{n-1}$

Divide both sides by 3: $\quad \dfrac{384}{3} = \dfrac{3(2)^{n-1}}{3}$

Simplify: $\quad 128 = 2^{n-1}$

Write 128 as a power of 2: $\quad 2^7 = 2^{n-1}$

Equate the exponents: $\quad 7 = n - 1$

Solve for *n*: $\quad 8 = n$

There are 8 terms in the sequence.

EXAMPLE 5 Finding t_n Given Two Terms

In a geometric sequence, $t_5 = 1875$ and $t_7 = 46\,875$. Find the first three terms of the sequence and t_n.

SOLUTION

Substitute known values in the formula for the nth term to write a system of equations. Then, solve the system.

$$t_n = ar^{n-1}$$

Write an equation for t_5:
$$1875 = ar^{5-1}$$
$$1875 = ar^4 \qquad (1)$$

Write an equation for t_7:
$$46\,875 = ar^{7-1}$$
$$46\,875 = ar^6 \qquad (2)$$

Divide (2) by (1):
$$\frac{46\,875}{1875} = \frac{ar^6}{ar^4}$$
$$25 = r^2$$

Solve for r:
$$\pm 5 = r$$

Since $r = 5$ or $r = -5$, there are two possible solutions.

Substitute 5 for r in (1): $1875 = a(5)^4$
$$1875 = 625a$$
Solve for a: $3 = a$

Substitute -5 for r in (1): $1875 = a(-5)^4$
$$1875 = 625a$$
Solve for a: $3 = a$

Since $a = 3$ and $r = 5$, the first three terms are 3, 15, and 75.

Since $a = 3$ and $r = -5$, the first three terms are 3, -15, and 75.

Substitute 3 for a and 5 for r in the formula for the nth term.
$$t_n = ar^{n-1}$$
$$t_n = 3(5)^{n-1}$$

Substitute 3 for a and -5 for r in the formula for the nth term.
$$t_n = ar^{n-1}$$
$$t_n = 3(-5)^{n-1}$$

So, the first three terms are 3, 15, and 75, and $t_n = 3(5)^{n-1}$, or the first three terms are 3, -15, and 75, and $t_n = 3(-5)^{n-1}$.

Key Concepts

- The general geometric sequence is a, ar, ar^2, ar^3, ... , where a is the first term and r is the common ratio.
- The formula for the nth term, t_n or $f(n)$, of a geometric sequence is $t_n = ar^{n-1}$, where n is a natural number.

Communicate Your Understanding

1. Given a sequence of numbers written in order from least to greatest, explain how you would determine if the sequence is arithmetic, geometric, or neither.

2. Given the formula for the nth term of a geometric sequence, $t_n = 4(3)^{n-1}$, describe how you would find the first 5 terms.

3. a) Describe how you would find the formula for the nth term of the geometric sequence 4, 8, 16, 32, ...

b) Describe how you would find t_{12} for this sequence.

4. Describe how you would find the number of terms in the sequence 5, 10, 20, ... , 1280.

5. Given that $t_3 = 28$ and $t_4 = 56$ for a geometric sequence, describe how you would find t_n for the sequence.

Practise

A

1. Determine whether each sequence is arithmetic, geometric, or neither. Then, find the next two terms of each sequence.

a) 1, 4, 9, 16, ...   **b)** 1, 2, 4, 8, ...
c) 7, 14, 21, 28, ...   **d)** 1, 2, 4, 7, 11, ...
e) 20, 16, 12, 8, ...   **f)** 32, 16, 8, 4, ...
g) $\dfrac{11}{3}, \dfrac{10}{3}, \dfrac{8}{3}, \dfrac{5}{3}, \ldots$ **h)** 0.5, 1.5, 4.5, 13.5, ...

2. State the common ratio and write the next three terms of each geometric sequence.

a) 1, 3, 9, 27, ...   **b)** 5, 10, 20, 40, ...
c) 2, −8, 32, −128, ...   **d)** 7, −7, 7, −7, ...
e) 0.5, 5, 50, 500, ...   **f)** $\dfrac{1}{3}, \dfrac{2}{3}, \dfrac{4}{3}, \dfrac{8}{3}, \ldots$

g) 64, 32, 16, 8, ...
h) 800, −400, 200, −100, ...

3. Write the first 5 terms of each geometric sequence.

a) $a = 4$ and $r = 3$
b) $a = 20$ and $r = 4$
c) $a = 1024$ and $r = 0.5$
d) $a = 0.043$ and $r = 10$
e) $a = 8$ and $r = -1$
f) $a = -10$ and $r = -5$

4. Given the formula for the *n*th term of a geometric sequence, write the first 4 terms.

a) $t_n = 4(2)^{n-1}$

b) $t_n = 10(3)^{n-1}$

c) $t_n = 2(-2)^{n-1}$

d) $f(n) = 5(-3)^{n-1}$

e) $t_n = -3(2)^{n-1}$

f) $t_n = -2(-3)^{n-1}$

g) $f(n) = 0.5(4)^{n-1}$

h) $t_n = -(-1)^{n-1}$

i) $f(n) = 200(0.5)^{n-1}$

j) $f(n) = -1000(-0.1)^{n-1}$

5. Find the formula for the *n*th term and find the indicated terms for each of the following geometric sequences.

a) 2, 4, 8, ... ; t_7 and t_{12}

b) 1, 5, 25, ... ; t_6 and t_9

c) 4, 12, 36, ... ; t_8 and t_{10}

d) 64, 32, 16, ... ; t_7 and t_{10}

e) 6, 0.6, 0.06, ... ; t_6 and t_8

f) -3, 6, -12, ... ; t_7 and t_9

g) 729, -243, 81, ... ; t_6 and t_{10}

h) 4, -40, 400, ... ; t_8 and t_{12}

6. Find the number of terms in each of the following geometric sequences.

a) 4, 12, 36, ... , 2916

b) 3, 6, 12, ... , 1536

c) 2, -4, 8, ... , -1024

d) 4374, 1458, 486, ... , 2

e) $\dfrac{1}{2}, \dfrac{1}{4}, \dfrac{1}{8}, \ldots , \dfrac{1}{1024}$

f) $\dfrac{1}{25}, \dfrac{1}{5}, 1, \ldots , 625$

g) $\dfrac{2}{81}, \dfrac{4}{27}, \dfrac{8}{9}, \ldots , 6912$

h) -409.6, 102.4, -25.6, ... , 0.025

7. Given two terms of each geometric sequence, find t_n for the sequence.

a) $t_3 = 36$, $t_4 = 108$

b) $t_2 = 6$, $t_3 = -12$

c) $t_4 = 64$, $t_5 = 32$

d) $t_2 = 4$, $t_4 = 64$

e) $t_5 = 80$, $t_7 = 320$

f) $t_3 = 99$, $t_5 = 11$

Apply, Solve, Communicate

8. Measurement The diagrams show the side lengths of three 30°-60°-90° triangles. Find the side lengths of the next triangle in the sequence.

 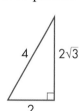 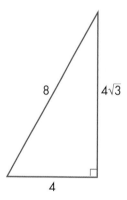

B

9. Helium balloon A balloon filled with helium has a volume of 20 000 cm³. The balloon loses one fifth of its helium every 24 h.

a) Write the sequence giving the volume of helium in the balloon at the beginning of each day for 5 days, including the first day.

b) What is the common ratio for this sequence?

c) What volume of helium will be in the balloon at the start of the sixth day? the seventh day?

10. Demographics Each year for 10 years, the population of a city increased by 5% of its value in the previous year. If the initial population was 200 000, what was the population after 10 years?

11. Vacuum pump Each stroke of a vacuum pump removes one third of the air remaining in a container. What percent of the original quantity of air remains in the container after 10 strokes, to the nearest percent?

12. Photocopying Many photocopiers can reduce the dimensions of the image of an original. Usually the maximum reduction capability is to 64% of the original dimensions. How many reductions, at the maximum setting, would it take to reduce an image to less than 10% of its original dimensions?

13. Biology A single bacterium divides into two bacteria every 10 min. If the same rate of division continues for 2 h, how many bacteria will there be?

14. The first two terms of a sequence are 3 and 6.
a) Write the first 5 terms of the sequence if the sequence is arithmetic.
b) Write the first 5 terms of the sequence if the sequence is geometric.
c) Graph each sequence and compare the graphs.

15. Which term of the geometric sequence 4, 12, 36, … , is 2916?

16. Application The rate of decay for a radioactive isotope varies from one isotope to another. The time it takes for half of any sample to decay is called the *half-life*. Barium-123 has a half-life of 2 min.
a) Write an equation to determine the quantity of barium-123 remaining in a sample after n half-lives.
b) A fresh sample of 80 mg of barium-123 was obtained for an experiment. It took 10 min to set up the experiment. What mass of the barium-123 was left when the experiment began?

17. Communication If each term of a geometric sequence is multiplied by the same number, is the resulting sequence a geometric sequence? Explain.

18. Rare stamp The value of a rare stamp is expected to follow a geometric sequence from year to year. The stamp is now worth $800 and is expected to be worth $1250 two years from now. How much is it expected to be worth
a) one year from now?   **b)** three years from now?

19. a) Is the graph of t_n versus n for a geometric sequence linear or non-linear?
b) Use finite differences to explain your answer to part a).

20. Motion of a pendulum On the first swing, a pendulum swings through an arc of 50 cm. On each successive swing, the length of the arc is 0.97 of the previous length. What is the length of the arc, to the nearest hundredth of a centimetre, on

a) the 10th swing?   **b)** the 15th swing?

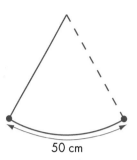

50 cm

C

21. Algebra Show that $t_n \div t_{n-1} = r$ for any geometric sequence.

22. Algebra Determine the value of x that makes each sequence geometric.

a) 4, 8, 16, $3x + 2$, ...   **b)** 2, 6, $5x - 2$, ...

23. Inquiry/Problem Solving Can the side lengths of a triangle be consecutive terms of a geometric sequence? Explain your reasoning.

24. Algebra The first three terms of a geometric sequence are w, x, and y. Express y in terms of w and x. Check your solution using the first three terms of any geometric sequence.

25. Geometric mean If a, x, and b are consecutive terms of a geometric sequence, then x is called the geometric mean of a and b. Find the geometric mean of

a) 2 and 8   **b)** 5 and 180   **c)** m and n

26. Find a, r, and t_n for the following geometric sequences.

a) $t_5 = 48$ and $t_8 = 384$ **b)** $t_3 = 24$ and $t_6 = -192$

27. Algebra Find a, r, and $f(n)$ for the following geometric sequences.

a) $f(3) = 5x^6$ and $f(10) = 5x^{20}$ **b)** $f(4) = 8x^3$ and $f(9) = 256x^8$

28. Algebra Find the indicated terms for each of the following geometric sequences.

a) t_n and t_{10} for $2x$, $4x^2$, $8x^3$, ...   **b)** t_n and t_6 for $\dfrac{1}{2}, \dfrac{x}{4}, \dfrac{x^2}{8}$, ...

c) t_n and t_{25} for $\dfrac{1}{x^4}, \dfrac{1}{x^2}, 1$, ...   **d)** t_n and t_{20} for $3x^{10}$, $-3x^9$, $3x^8$, ...

CAREER CONNECTION *Accounting*

All organizations need to monitor and record their financial activities. Accountants are the professionals who measure and report on the financial activities of organizations. The history of accounting can be traced back to ancient times. The first official meeting of accountants in North America was held in Montréal in 1879.

Accounting reports are used within an organization to plan for the future. They are also made available to selected outside groups, such as banks and government tax departments, that need to be aware of the organization's finances. These outside groups employ their own accountants to interpret the reports they receive and to monitor their own financial affairs.

1. Depreciation The Canada Customs and Revenue Agency administers Canada's tax laws. The agency allows tax deductions based on the depreciation of equipment used for business purposes. The agency allows a 30% annual depreciation rate for computer equipment. Suppose that you are an accountant working for a company that purchased $60 000 worth of computer equipment today.

a) Calculate the value that the Canada Customs and Revenue Agency will assign to the computer equipment four years from now.

b) In the company's tax return for any year, you can claim a tax deduction equal to the decrease in value of the computer equipment that year. Calculate the deduction you will claim for the fourth year from now.

2. Research Use your research skills to determine the meanings of the accounting qualifications CA, CMA, and CGA. Compare the education and training required for these qualifications, and describe some possible career paths open to people who obtain them.

6.4 Recursion Formulas

In previous sections, you have determined sequences using a formula for the nth term. An example is the formula $t_n = 2n + 3$ or $f(n) = 2n + 3$, which determines the arithmetic sequence 5, 7, 9, 11, 13, ...

Another example is the formula $t_n = 2^{n-1}$ or $f(n) = 2^{n-1}$, which determines the geometric sequence 1, 2, 4, 8, 16, ...

Such formulas are known as **explicit formulas**. They can be used to calculate any term in a sequence without knowing the previous term. For example, the tenth term in the sequence determined by the formula $t_n = 2n + 3$ is $2(10) + 3$, or 23.

It is sometimes more convenient to calculate a term in a sequence from one or more previous terms in the sequence. Formulas that can be used to do this are called **recursion formulas**.

A recursion formula consists of at least two parts. The parts give the value(s) of the first term(s) in the sequence, and an equation that can be used to calculate each of the other terms from the term(s) before it. An example is the formula
$t_1 = 5$
$t_n = t_{n-1} + 2$
The first part of the formula shows that the first term is 5. The second part of the formula shows that each term after the first term is found by adding 2 to the previous term. The sequence is 5, 7, 9, 11, 13, ... Thus, this recursion formula determines the same sequence as the explicit formula $t_n = 2n + 3$.

In one of his books, the great Italian mathematician Leonardo Fibonnaci (c.1180–c.1250) described the following situation.

A pair of rabbits one month old is too young to produce more rabbits. But at the end of the second month, they produce a pair of rabbits, and a pair of rabbits every month after that. Each new pair of rabbits does the same thing, producing a pair of rabbits every month, starting at the end of the second month.

1. The table shows how a family of rabbits grows. Copy and extend the table to find the number of pairs of rabbits at the start of the ninth month.

Start of Month	Number of Pairs	Diagram
1	1	
2	1	
3	2	
4	3	
5	5	

2. List the numbers of pairs of rabbits in order as the first 9 terms of a sequence. This sequence is known as the **Fibonnaci sequence.** After the first two terms of the sequence, how can each term be calculated from previous terms?

3. Use the following recursion formula to write the first 9 terms of a sequence. Compare the result with the Fibonnaci sequence.

$t_1 = 1$

$t_2 = 1$

$t_n = t_{n-1} + t_{n-2}$

4. Is the Fibonnaci sequence arithmetic, geometric, or neither? Explain.

5. Is the Fibonnaci sequence a function? Explain.

EXAMPLE 1 Writing an Arithmetic Sequence From a Recursion Formula

Write the first 5 terms of the sequence determined by the recursion formula.

$t_1 = 11$

$t_n = t_{n-1} - 4$

SOLUTION

From the first part of the recursion formula, the first term is 11.

From the second part of the recursion formula,

After finding the second term, as shown on the screen, press the ENTER key repeatedly to find more terms.

$$t_2 = t_1 - 4 \qquad t_3 = t_2 - 4$$
$$\quad = 11 - 4 \qquad \quad = 7 - 4$$
$$\quad = 7 \qquad \qquad \quad = 3$$
$$t_4 = t_3 - 4 \qquad t_5 = t_4 - 4$$
$$\quad = 3 - 4 \qquad \quad = -1 - 4$$
$$\quad = -1 \qquad \qquad \quad = -5$$

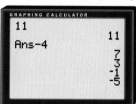

The first 5 terms of the sequence are 11, 7, 3, −1, and −5.

Note that the sequence in Example 1 is an arithmetic sequence with a first term of 11 and a common difference of −4. The sequence could be determined by the explicit formula $t_n = 15 - 4n$ or $f(n) = 15 - 4n$.

EXAMPLE 2 Writing a Geometric Sequence From a Recursion Formula

Write the first 5 terms of the sequence determined by the recursion formula.

$t_1 = 2$

$t_n = -3t_{n-1}$

SOLUTION

From the first part of the recursion formula, the first term is 2.

From the second part of the recursion formula,

$$t_2 = -3 \times t_1 \qquad t_3 = -3 \times t_2$$
$$\quad = -3 \times 2 \qquad \quad = -3 \times (-6)$$
$$\quad = -6 \qquad \qquad \quad = 18$$
$$t_4 = -3 \times t_3 \qquad t_5 = -3 \times t_4$$
$$\quad = -3 \times 18 \qquad \quad = -3 \times (-54)$$
$$\quad = -54 \qquad \qquad \quad = 162$$

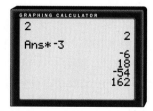

The first 5 terms of the sequence are 2, −6, 18, −54, and 162.

Note that the sequence in Example 2 is a geometric sequence with a first term of 2 and a common ratio of −3. The sequence could be determined by the explicit formula $t_n = 2(-3)^{n-1}$ or $f(n) = 2(-3)^{n-1}$.

The graphing calculator screens in the examples show how the ENTER key can be used to write terms of a sequence recursively. However, this method is laborious if you want to write many terms. Another method is to use the mode settings to choose the Seq(sequence) graphing mode and the dot mode, and then to use the sequence Y= editor to enter the sequence as a function. The TABLE key can be used to display the terms of the sequence.

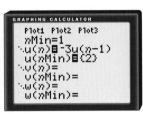
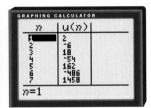

Key Concepts

- A formula for calculating the nth term of a sequence without knowing any previous terms is called an explicit formula.
- A formula used to calculate a term in a sequence from one or more previous terms is called a recursion formula.
- A recursion formula consists of at least two parts. The first part(s) give the value(s) of the first term(s) in the sequence. The last part is an equation that can be used to calculate each of the other terms from the term(s) before it.

Communicate Your Understanding

1. Describe the difference between a recursion formula and an explicit formula.

2. Describe how finding the 15th term of a sequence using an explicit formula is different from finding the 15th term using a recursion formula.

3. Explain why a recursion formula must have at least two parts.

4. Describe how you would write an explicit formula for the sequence determined by the recursion formula $t_1 = 2; t_n = t_{n-1} + 4$.

5. Explain why the recursion formula $t_1 = 1; t_n = 2t_{n-1}$ determines the same sequence as the recursion formula $f(1) = 1; f(n) = 2f(n-1)$.

Practise

A

1. Use the recursion formula to write the first 5 terms of each sequence.

a) $t_1 = 4; t_n = t_{n-1} + 3$

b) $t_1 = 3; t_n = t_{n-1} - 2$

c) $t_1 = -1; t_n = 2t_{n-1}$

d) $t_1 = 48; t_n = 0.5t_{n-1}$

e) $t_1 = 6; t_n = t_{n-1} + 2n$

f) $t_1 = -2; t_n = 4t_{n-1} + n^2$

g) $t_1 = 2; t_n = t_{n-1} - 2n + 1$

h) $t_1 = -3; t_n = 4 - 2t_{n-1}$

2. Use the recursion formula to write the first 5 terms of each sequence.

a) $t_1 = 3; t_2 = 5; t_n = t_{n-2} - t_{n-1}$

b) $t_1 = -2; t_2 = 3; t_n = 2t_{n-2} + t_{n-1}$

c) $t_1 = 2; t_2 = 1; t_n = t_{n-2} + t_{n-1} - 2n$

d) $t_1 = 1; t_2 = -2; t_n = t_{n-2} \times t_{n-1}$

e) $t_1 = -1; t_2 = 2; t_3 = 4; t_n = 2t_{n-3} - t_{n-2} + 3t_{n-1}$

f) $t_1 = 1; t_2 = 1; t_n = (t_{n-2})^2 + (t_{n-1})^2$

3. Use the recursion formula to write the first 6 terms of each sequence.

a) $f(1) = 12; f(n) = f(n-1) + 6$

b) $f(1) = 4; f(n) = 3f(n-1)$

c) $f(1) = 1.5; f(2) = 2.5;$
$f(n) = f(n-1) + f(n-2)$

d) $f(1) = -1; f(2) = 1;$
$f(n) = f(n-1) \div f(n-2)$

4. Use the recursion formula to write the first 6 terms of each sequence.

a) $t_1 = 5; t_{n+1} = t_n + 1$

b) $t_1 = 80; t_{n+1} = \dfrac{t_n}{2}$

c) $t_1 = -1; t_{n+1} = 3(t_n + 1)$

d) $t_1 = 1; t_2 = 1; t_{n+2} = t_{n+1} - t_n$

5. Write an explicit formula for the sequence determined by each recursion formula.

a) $t_1 = 1; t_n = t_{n-1} - 4$

b) $t_1 = 2; t_n = 3t_{n-1}$

c) $t_1 = 20; t_n = -\dfrac{1}{2}t_{n-1}$

d) $t_1 = \dfrac{1}{2}; t_n = \dfrac{1}{2}t_{n-1}$

Apply, Solve, Communicate

6. Verify that the sequence determined by the recursion formula $t_1 = 10; t_n = t_{n-1} - 2$ is arithmetic.

7. Verify that the sequence determined by the recursion formula $t_1 = 20; t_n = 0.5t_{n-1}$ is geometric.

8. Is the sequence determined by the recursion formula $t_1 = 5; t_n = t_{n-1} + n - 4$ arithmetic, geometric, or neither? Explain.

9. **Shading pattern** The first 4 diagrams in a pattern are shown.

a) Write the numbers of shaded squares in the first 6 diagrams as a sequence.

b) Verify that the recursion formula $t_1 = 1$; $t_n = t_{n-1} + 4n - 3$ determines the sequence.

c) Use the recursion formula to predict the number of shaded squares in the seventh diagram; the eighth diagram.

10. **Theatre seats** The numbers of seats in the rows of a theatre are represented by the recursion formula $t_1 = 20$; $t_n = t_{n-1} + 2$.

a) Determine the number of seats in each of the first 8 rows.

b) Write an explicit formula to represent the number of seats in the nth row.

11. **Application** Hans bought a computer for \$3000. The value of the computer at the start of each year, beginning from the date of purchase, is determined by the recursion formula $t_1 = 3000$; $t_n = 0.4t_{n-1}$.

a) Determine the value of the computer at the beginning of each year for the first six years.

b) What is the annual rate of depreciation of the computer?

c) Write an explicit formula for the nth term of the sequence.

12. a) Write the first 6 terms of the sequence determined by the recursion formula $t_1 = 1$; $t_n = t_{n-1} + 2n - 1$.

b) What do all the numbers in the sequence have in common?

c) Write an explicit formula for the nth term of the sequence.

13. Write an explicit formula for the sequence determined by each of the following recursion formulas.

a) $t_1 = \dfrac{1}{2}$; $t_n = t_{n-1} + \dfrac{1}{n(n+1)}$

b) $t_1 = 0$; $t_n = t_{n-1} + \dfrac{2}{n(n+1)}$

c) $t_1 = 2$; $t_n = t_{n-1} - \dfrac{1}{n(n-1)}$

14. **Communication** Use the explicit formula $t_n = n^2 - 5n + 2$ and the recursion formula $t_1 = -2$; $t_n = t_{n-1} + 2(n-3)$ to find the eighth term of a sequence.

a) Which method do you prefer if you use paper and pencil? Explain.

b) Which method do you prefer if you use a graphing calculator? Explain.

6.5 Arithmetic Series

The following is an arithmetic sequence.
1, 3, 5, 7, 9, …
A **series** is the sum of the terms of a sequence.
An **arithmetic series** is the sum of the terms of an
arithmetic sequence.
The series that corresponds to the sequence above is
$1 + 3 + 5 + 7 + 9 + …$
For this series, S_5 means the sum of the first 5 terms, so
$S_5 = 25$.

 **INVESTIGATE & INQUIRE**

In the version of pool known as "rotation," the aim
is to pocket the balls in order from 1 to 15. Players
receive points equal to the numbers on the
pocketed balls. The number of points on the table at the start of the game
is the sum of the numbers from 1 to 15, that is, the sum of the following
arithmetic series.
$1 + 2 + 3 + … + 15$

One way to find this sum is to add the numbers directly. Another way is
to look for a pattern in the sums of the following simpler arithmetic series.

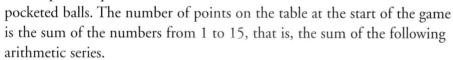
a) $1 + 2 + 3$ **b)** $1 + 2 + 3 + 4$ **c)** $1 + 2 + 3 + 4 + 5$

1. What is the sum of each of the series a) to c)?

2. What is the average of all the terms in each series?

3. How many terms are there in each series?

4. For each series, how is the sum related to the number of terms and the
average of all the terms?

5. For each series, what is the average of the first and last terms?

6. For each series, how is the average of the first and last terms related to
the average of all the terms?

7. Write a rule for finding the sum of an arithmetic series from the number
of terms and the average of the first and last terms.

8. Use your rule to find the sum of the 15 pool ball numbers.

9. Describe a situation in which there would be an advantage to using your rule, rather than adding all the terms of an arithmetic series directly.

10. Find the sum of each of the following arithmetic series.
a) $1 + 3 + \ldots + 29$ **b)** $2 + 4 + \ldots + 40$ **c)** $350 + 345 + \ldots + 5$

When the mathematician Karl Gauss was eight years old, he used the following method to find the sum of the natural numbers from 1 to 100.

Let S_{100} represent the sum of the first 100 natural numbers. Write out the series in order and then in reverse order.

$$S_{100} = \quad 1 + \quad 2 + \quad 3 + \ldots + \; 98 + \; 99 + 100$$

Reverse: $\quad S_{100} = 100 + \; 99 + \; 98 + \ldots + \quad 3 + \quad 2 + \quad 1$

Add: $\quad 2S_{100} = 101 + 101 + 101 + \ldots + 101 + 101 + 101$
$$= 100(101)$$

Divide both sides by 2: $\quad \dfrac{2S_{100}}{2} = \dfrac{100(101)}{2}$

$$S_{100} = 5050$$

The same method can be used to derive the formula for the sum of the general arithmetic series.

The general arithmetic sequence
$a, (a + d), (a + 2d), \ldots , (t_n - d), t_n$
has n terms, with the first term a and the last term t_n.

The corresponding series is

$$S_n = \quad a \quad + (a + d) + (a + 2d) + \ldots + (t_n - d) + \quad t_n$$

Reverse: $\quad S_n = \quad t_n \quad + (t_n - d) + (t_n - 2d) + \ldots + (a + d) + \quad a$

Add: $\quad 2S_n = (a + t_n) + (a + t_n) + (a + t_n) \; + \ldots + (a + t_n) + (a + t_n)$
$$2S_n = n(a + t_n)$$

Divide both sides by 2: $\quad S_n = \dfrac{n}{2}(a + t_n) \quad (1)$

The (1) shows that we are naming the formula as "formula one."

Since $t_n = a + (n - 1)d$, substitute for t_n in formula (1).

$$S_n = \dfrac{n}{2}[a + a + (n - 1)d]$$

$$S_n = \dfrac{n}{2}[2a + (n - 1)d] \quad (2)$$

EXAMPLE 1 Sum of a Series Given First Terms

Find the sum of the first 60 terms of each series.

a) $5 + 8 + 11 + ...$ b) $-6 - 8 - 10 - ...$

SOLUTION

a) $a = 5$, $d = 3$, and $n = 60$

$$S_n = \frac{n}{2}[2a + (n-1)d]$$

$$S_{60} = \frac{60}{2}[2(5) + (60-1)3]$$

$$= 30(10 + 177)$$

$$= 30(187)$$

$$= 5610$$

Estimate
$30 \times 200 = 6000$

b) $a = -6$, $d = -2$, and $n = 60$

$$S_n = \frac{n}{2}[2a + (n-1)d]$$

$$S_{60} = \frac{60}{2}[2(-6) + (60-1)(-2)]$$

$$= 30(-12 - 118)$$

$$= 30(-130)$$

$$= -3900$$

The sum of the series in Example 1 could be found using a graphing calculator. The following method requires the formula for the nth term. Adjust the mode settings to the Seq (sequence) graphing mode. Use the sequence function from the LIST OPS menu to list the first 60 terms of the series. Use the (STO▸) key to store the list in L₁. Then, use the sum function from the LIST MATH menu to find the sum.

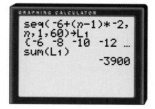

EXAMPLE 2 Sum of a Series Given First and Last Terms

Find the sum of the arithmetic series.

$5 + 9 + 13 + ... + 201$

SOLUTION 1 Paper-and-Pencil Method

To use either formula (1), or formula (2), the number of terms is needed.
For the given series, $a = 5$, $d = 4$, and $t_n = 201$.

$$t_n = a + (n-1)d$$

Substitute known values: $201 = 5 + (n-1)4$

Expand: $201 = 5 + 4n - 4$

Simplify: $201 = 4n + 1$

Solve for n: $200 = 4n$

$$50 = n$$

Two ways to find the sum are as follows.

Using formula (1)

$$S_n = \frac{n}{2}(a + t_n)$$

$$S_{50} = \frac{50}{2}(5 + 201)$$

$$= 25(206)$$

$$= 5150$$

Using formula (2)

$$S_n = \frac{n}{2}[2a + (n-1)d]$$

$$S_{50} = \frac{50}{2}[2(5) + (50-1)4]$$

$$= 25(206)$$

$$= 5150$$

Estimate

$25 \times 200 = 5000$

SOLUTION 2 Graphing-Calculator Method

Adjust the mode settings to the Seq(sequence) graphing mode. Use
the sequence function from the LIST OPS menu to list the terms
of the series. Use the (STO▸) key to store the list in L1. Then, use
the sum function from the LIST MATH menu to find the sum.

Key Concepts

- Given the first terms of an arithmetic series, the sum of the first n terms can
  be found using the formula $S_n = \frac{n}{2}[2a + (n-1)d]$

- Given the first and last terms of an arithmetic series, the sum of the series can
  be found using the formula $S_n = \frac{n}{2}(a + t_n)$ or the formula $S_n = \frac{n}{2}[2a + (n-1)d]$.

Communicate Your Understanding

1. Describe the difference between an arithmetic sequence and an arithmetic series.
2. Describe how you would find S_{30} for the series $4 + 6 + 8 + \ldots$
3. Describe how you would find the sum of the series $5 + 10 + 15 + \ldots + 105$.

Practise

A

1. Find the sum of the first 100 terms of each arithmetic series.
a) $1 + 5 + 9 + ...$
b) $2 + 5 + 8 + ...$
c) $10 + 8 + 6 + ...$
d) $0 - 3 - 6 - ...$

2. Find the indicated sum for each arithmetic series.
a) S_{10} for $2 + 4 + 6 + ...$
b) S_{20} for $10 + 15 + 20 + ...$
c) S_{30} for $-2 + 4 + 10 + ...$
d) S_{18} for $40 + 38 + 36 + ...$
e) S_{15} for $80 + 76 + 72 + ...$
f) S_{18} for $1.5 + 2.5 + 3.5 + ...$

3. Find the sum of each arithmetic series.
a) $4 + 6 + 8 + ... + 200$
b) $3 + 7 + 11 + ... + 479$
c) $100 + 90 + 80 + ... - 50$
d) $-8 - 5 - 2 + ... + 139$

e) $18 + 12 + 6 + ... - 216$
f) $-7 - 11 - 15 - ... - 171$
g) $\dfrac{5}{2} + 4 + \dfrac{11}{2} + ... + 100$
h) $6 - \dfrac{16}{3} - \dfrac{14}{3} - ... - 12$

4. Given the first and last terms, find the sum of each arithmetic series.
a) $a = 6, t_9 = 24$
b) $a = 3, t_{12} = 36$
c) $f(1) = 5, f(10) = -13$
d) $a = -4, t_{22} = -46$
e) $a = 4.5, t_{11} = 19.5$
f) $f(1) = 20, f(31) = 110$
g) $a = -5, t_{45} = 17$
h) $a = -0.3, t_{51} = -10.3$

Apply, Solve, Communicate

5. Find the sum of
a) the first 50 positive integers
b) the first 100 odd positive integers
c) the first 75 positive multiples of 3
d) the first 100 even positive integers

6. **Theatre seats** A theatre has 30 seats in the front row, 31 seats in the second row, 32 seats in the third row, and so on. If there are 20 rows of seats, how many seats are there?

B

7. Find the sum of
a) the positive multiples of 5 up to 500
b) the positive multiples of 7 up to 245

8. **Communication** Show that the right-hand side of the formula $S_n = \dfrac{n}{2}[2a + (n-1)d]$ represents the product of the number of terms and the average of the first and last terms of an arithmetic series.

9. Pattern The first four diagrams in a pattern are shown. If the pattern continues, what is the total number of squares in the first 50 diagrams?

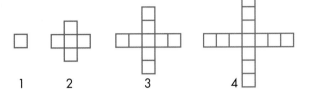

1   2   3   4

10. Marching band A marching band has 8 musicians in the first row, 10 musicians in the second row, 12 musicians in the third row, and so on. If there are 12 rows, how many musicians are in the band?

11. Salary increases Michelle is a marine biologist. She accepted a job that pays $46 850 in the first year and $56 650 in the eighth year. Her salary is an arithmetic sequence with eight terms.
a) What raise can Michelle expect each year?
b) What will her salary be in the sixth year?
c) In what year will her salary first exceed $50 000?
d) What is the total amount that Michelle will earn in the eight years?

12. Measurement The perimeter of a triangle is 30 units. The side lengths form an arithmetic sequence. If each side length is a whole number, what are the possible sets of side lengths?

13. Falling object An object dropped from the same height as the Fairbanks building in Toronto takes about 5 s to hit the ground. The object falls 4.9 m in the first second, 14.7 m in the second second, 24.5 m in the third second, and so on. How tall is the building?

14. Store display The top three layers of boxes in a store display are arranged as shown. If the pattern continues, and there are 12 layers in the display, what is the total number of boxes in the display?

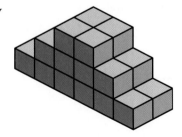

15. Rocket flight An accelerating rocket rises 10 m in the first second, 40 m in the second second, and 70 m in the third second. If the arithmetic sequence continues, how high will the rocket be after 20 s?

16. Application A clothing store ordered 300 sweaters and sold 20 of them at $100 each in the first week. In the second week, the selling price was lowered by $10, and 40 sweaters were sold. In the third week, the selling price was lowered by another $10, and 60 sweaters were sold. If the pattern continued,
a) how many weeks did it take to sell all the sweaters?
b) what was the selling price in the final week?

17. Pipe stacks Pipes are stored in stacks, as shown in the diagram. Find the number of pipes in a stack of
a) 6 rows, if the bottom row has 12 pipes
b) 10 rows, if the bottom row has 15 pipes

18. Construction A company building a new library was required to pay a penalty of $1000 for the first day the completion was late, $1500 for the second day, $2000 for the third day, and so on. If the company paid a penalty of $115 000, how many days late was the completion of the library?

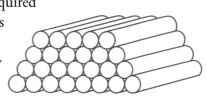

19. In an arithmetic series, $t_2 = 10$ and $t_5 = 31$. Find the sum of the first 16 terms.

20. In an arithmetic series, $f(3) = 11$ and $f(7) = 13$. Find the sum of the first 20 terms.

21. Inquiry/Problem Solving The interior angles of a hexagon are in an arithmetic sequence. The largest angle is 130°. What are the other angles?

C

22. Write four different arithmetic series, each with five terms and each with a sum of 100. Compare your series with a classmate's.

23. The first eight terms of an arithmetic series have a sum of 148. The common difference is 3. What are the first three terms of the series?

24. Determine the first and last terms of an arithmetic series with 50 terms, a common difference of 6, and a sum of 7850.

25. The first three terms of an arithmetic series have a sum of 24 and a product of 312. What is the fourth term of the series?

26. Find the first five terms of the arithmetic series with $t_{12} = 35$ and $S_{20} = 610$.

27. Algebra Find the sum of each arithmetic series.
a) $4x + 6x + \ldots + 22x$ **b)** $(x + y) + (x + 2y) + \ldots + (x + 10y)$

6.6 Geometric Series

The following is a geometric sequence.
3, 6, 12, 24, ... , where $a = 3$ and $r = 2$.

A **geometric series** is the sum of the terms of a geometric sequence.
The geometric series that corresponds to the geometric sequence above is
$3 + 6 + 12 + 24 + \dots$, where $a = 3$ and $r = 2$.
For this series, S_3 means the sum of the first 3 terms, so
$S_3 = 3 + 6 + 12$
$\quad = 21$

INVESTIGATE & INQUIRE

Rosa decided to research her ancestry for the last 6 generations, which included her 2 parents, 4 grandparents, 8 great-grandparents, and so on. To go back 6 generations, the total number of people Rosa needed to research was the sum of the following geometric series.

$$S_6 = 2 + 4 + 8 + 16 + 32 + 64$$

1. a) Use addition to find the total number of people Rosa needed to research.
b) To develop a second method for finding the sum of the series, first write the equation that represents the sum of the first 6 terms and label it equation (1).

$$S_6 = 2 + 4 + 8 + 16 + 32 + 64 \quad (1)$$

Then, multiply both sides of equation (1) by the common ratio, 2, and label the result equation (2).

$$2S_6 = 4 + 8 + 16 + 32 + 64 + 128 \quad (2)$$

Write equations (1) and (2) so that the equal terms on the right side line up as shown.

$$S_6 = 2 + 4 + 8 + 16 + 32 + 64 \qquad (1)$$
$$2S_6 = \quad\;\; 4 + 8 + 16 + 32 + 64 + 128 \quad (2)$$

Subtract equation (1) from equation (2). Do not simplify the right side.
c) The right side shows the difference between two terms of the series
$2 + 4 + 8 + 16 + \dots$ Which two terms?

d) Simplify the right side. What is the value of S_6?

e) What two numbers would you subtract to find S_8 for this series?

2. a) If $S_5 = 1 + 3 + 9 + 27 + 81$, find S_5 by addition.

b) Predict the next term in the series, and subtract the first term from it.

c) Compare your answer from part b) with your answer from part a).

d) By what factor would you divide your answer from part b) to give the correct sum?

e) What is the difference between the factor you found in part d) and the value of the common ratio for this series?

3. Repeat question 2 for the series $S_4 = 3 + 12 + 48 + 192$.

4. Explain why no division step was necessary in question 1 parts d) and e) to find the correct sum.

5. a) Using your results from questions 1 to 4, describe two different methods for finding the sum of a geometric series.

b) In what situations would it be better to use each method?

The method developed in the Investigate & Inquire can be used to write a formula for finding the sum, S_n, of the general geometric series.

For the general geometric series,
$S_n = a + ar + ar^2 + \ldots + ar^{n-1}$,
where S_n represents the sum of n terms.

Write the series:	$S_n = a + ar + ar^2 + \ldots + ar^{n-1}$	(1)
Multiply both sides by r:	$rS_n = \quad ar + ar^2 + \ldots + ar^{n-1} + ar^n$	(2)
Subtract (1) from (2):	$rS_n - S_n = -a \qquad\qquad\qquad + ar^n$	
Rearrange the right side:	$rS_n - S_n = ar^n - a$	
Factor the left side:	$S_n(r - 1) = ar^n - a$	

Divide both sides by $(r - 1)$: $\quad S_n = \dfrac{ar^n - a}{r - 1}, \ r \neq 1$ The sum is the next term minus the first term divided by the common ratio minus 1.

$$\text{or } S_n = \frac{a(r^n - 1)}{r - 1}, \ r \neq 1$$

So, the sum, S_n, of the first n terms of a geometric series can be found using the formula

$$S_n = \frac{a(r^n - 1)}{r - 1}, \ r \neq 1$$

EXAMPLE 1 Sum of a Geometric Series When $r > 0$

Find S_8 for the series $2 + 8 + 32 + \ldots$

SOLUTION

$a = 2$, $r = 4$, and $n = 8$

$$S_n = \frac{a(r^n - 1)}{r - 1}$$

$$S_8 = \frac{2(4^8 - 1)}{4 - 1}$$

$$= \frac{2(4^8 - 1)}{3}$$

$$= 43\,690$$

```
GRAPHING CALCULATOR
2(4^8-1)/3
                43690
```

The sum of the first 8 terms is $43\,690$.

EXAMPLE 2 Sum of a Geometric Series When $r < 0$

Find S_9 for the series $3 - 9 + 27 \ldots$

SOLUTION

$a = 3$, $r = -3$, $n = 9$

$$S_n = \frac{a(r^n - 1)}{r - 1}$$

$$S_9 = \frac{3((-3)^9 - 1)}{-3 - 1}$$

$$= \frac{3((-3)^9 - 1)}{-4}$$

$$= 14\,763$$

```
GRAPHING CALCULATOR
3((-3)^9-1)/-4
                14763
```

The sum of the first 9 terms is $14\,763$.

EXAMPLE 3 Sum of a Geometric Series Given First and Last Terms

Find the sum of the series $4 + 12 + 36 + \ldots + 2916$.

SOLUTION

To use the formula $S_n = \dfrac{a(r^n - 1)}{r - 1}$, the number of terms is needed.

For the given series $a = 4$, $r = 3$, and $t_n = 2916$.

$$t_n = ar^{n-1}$$

Substitute known values: $\quad 2916 = 4(3)^{n-1}$

Divide both sides by 4: $\quad \dfrac{2916}{4} = \dfrac{4(3)^{n-1}}{4}$

Simplify: $\quad 729 = 3^{n-1}$
Write 729 as a power of 3: $\quad 3^6 = 3^{n-1}$
Equate the exponents: $\quad 6 = n - 1$
Solve for n: $\quad 7 = n$

The series has 7 terms.

$$S_n = \frac{a(r^n - 1)}{r - 1}$$

$$S_7 = \frac{4(3^7 - 1)}{3 - 1}$$

$$= \frac{4(3^7 - 1)}{2}$$

$$= 4372$$

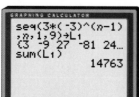

```
GRAPHING CALCULATOR
4(3^7-1)/2
                  4372
```

The sum of the series is 4372.

The sums in Examples 1–3 could be found using a graphing calculator. The following method requires the formula for the nth term.

Use the sequence function from the LIST OPS menu to list the terms to be added. Use the $\boxed{\text{STO·}}$ key to store the list in L_1. Then, use the sum function from the LIST MATH menu to find the sum.

```
GRAPHING CALCULATOR
seq(2*4^(n-1),n,
1,8)→L1
{2 8 32 128 512…
sum(L1)
             43690
```

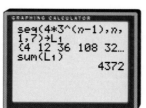

```
GRAPHING CALCULATOR
seq(3*(-3)^(n-1)
,n,1,9)→L1
{3 -9 27 -81 24…
sum(L1)
             14763
```

```
GRAPHING CALCULATOR
seq(4*3^(n-1),n,
1,7)→L1
{4 12 36 108 32…
sum(L1)
             4372
```

Key Concepts

- The sum of the first n terms of a geometric series can be found using the formula $S_n = \dfrac{a(r^n - 1)}{r - 1}$.

- Given at least the first two terms and the last term of a geometric series, the number of terms can be found by substituting known values in $t_n = ar^{n-1}$ and solving for n.

Communicate Your Understanding

1. Describe the similarities and the differences between a geometric series and an arithmetic series.

2. Describe how you would find S_{15} for the series $4 + 8 + 16 + \dots$

3. Describe how you would find the sum of the series $5 + 15 + 45 + \dots + 10\ 935$.

Practise

A

1. Find the indicated sum for each geometric series.

a) S_{12} for $1 + 2 + 4 + \dots$

b) S_7 for $1 + 4 + 16 + \dots$

c) S_6 for $3 + 15 + 75 + \dots$

d) S_8 for $2 - 6 + 18 - \dots$

e) S_9 for $3 - 6 + 12 - \dots$

f) S_6 for $256 + 128 + 64 + \dots$

g) S_7 for $972 + 324 + 108 + \dots$

h) S_6 for $1 - \dfrac{1}{2} + \dfrac{1}{4} - \dots$

2. Find S_n for each geometric series.

a) $a = 5$, $r = 3$, $n = 8$

b) $a = 4$, $r = -3$, $n = 10$

c) $a = 625$, $r = 0.6$, $n = 5$

d) $f(1) = 4$, $r = 0.5$, $n = 7$

e) $a = 100\ 000$, $r = -0.1$, $n = 5$

f) $f(1) = \dfrac{1}{2}$, $r = -5$, $n = 6$

3. Find the sum of each geometric series.

a) $1 + 2 + 4 + \dots + 256$

b) $1 + 3 + 9 + \dots + 2187$

c) $2 - 4 + 8 - \dots + 512$

d) $5 - 15 + 45 - \dots + 3645$

e) $729 + 243 + 81 + \dots + 1$

f) $1200 + 120 + 12 + \dots + 0.0012$

4. If $f(1) = 0.8$ and $f(2) = 1.6$ for a geometric series, find S_{10}.

5. If $f(1) = 2$ and $f(2) = -8$ for a geometric series, find S_{15}.

Apply, Solve, Communicate

6. Genealogy Suppose you researched your ancestors back ten generations. How many people would you research?

B

7. Pattern The first four rectangles in a sequence have dimensions of 2 cm by 3 cm, 2 cm by 9 cm, 2 cm by 27 cm, and 2 cm by 81 cm. If the pattern continues, what is the total area of the first 10 rectangles?

8. Measurement Larger and larger squares are formed consecutively, as shown in the diagram. The side lengths of the squares form a geometric sequence, starting at 10 cm. For the first 10 squares in the sequence, find
a) the sum of the perimeters   **b)** the sum of the areas

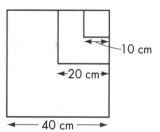

9. Communication Some companies use a telephone chain to notify employees that the company is closing because of bad weather. Suppose that, in the first round of calls, the first person in the chain calls four people. Each person called then makes four calls, and so on. What is the total number of people called in the first six rounds of calls?

10. Lottery The first prize in a lottery is $100 000. Each succeeding winning number pays 40% as much as the winning number before it. How much is paid out in prizes if 6 numbers are drawn?

11. Application The air in a hot-air balloon cools as the balloon rises. If the air is not reheated, the balloon rises more slowly every minute. Suppose that a hot-air balloon rises 50 m in the first minute. In each succeeding minute, the balloon rises 70% as far as it did in the previous minute. How far does the balloon rise in 7 min, to the nearest metre?

12. Find the value of a for each geometric series.
a) $S_7 = 70\ 993$ and $r = 4$ **b)** $S_6 = -364$ and $r = -3$
c) $S_5 = 310$ and $r = 0.5$

13. Billionaire's club Frank had a plan to become a billionaire. He would put aside 1 cent on the first day, 2 cents on the second day, 4 cents on the third day, and so on, doubling the number of cents each day.
a) How much money would he have after 20 days?
b) How many days would it take Frank to become a billionaire?
c) Do you see any problems with Frank's plan? Explain.

14. Motion of a pendulum On the first swing, a pendulum swings through an arc of 40 cm. On each successive swing, the length of the arc is 0.98 of the previous length. In the first 20 swings, what is the total distance that the lower end of the pendulum swings, to the nearest hundredth of a centimetre?

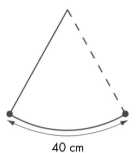

40 cm

C

15. Inquiry/Problem Solving The sides of the large square are 16 cm. The midpoints of the sides are joined to form a new square. Find the sum of the areas of all the squares.

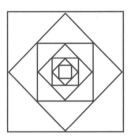

16. Bouncing ball A ball is thrown 16 m into the air. The ball falls, rebounds to half of its previous height, and falls again. If the ball continues to rebound and fall in this manner, find the total distance the ball travels until it hits the ground for the sixth time.

17. A geometric series has three terms. The sum of the three terms is 42. The third term is 3.2 times the sum of the other two. What are the terms?

18. If a sequence is defined by $f(x) = 4^{x-1}$, where x is a natural number, find the sum of the first 8 terms.

19. Algebra Determine S_{15} for the series $3 + 3x^2 + 3x^4 + \ldots$

20. Chess According to an old tale, the inventor of chess, Sissa Dahir, was granted anything he wished by the Indian king, Shirham. Sissa asked for one grain of wheat for the first square on the chess board, two grains for the second square, four grains for the third, eight grains for the fourth, and so on, for all 64 squares.
a) How many grains did Sissa ask for?
b) Research If one grain of wheat has a mass of 65 mg, how did the mass of wheat that Sissa asked for compare with the world's annual wheat production?

ACHIEVEMENT Check Knowledge/Understanding   Thinking/Inquiry/Problem Solving   Communication   Application

In a geometric series, $t_2 = 36$ and $t_6 = 2016$. Find the sum of the first 8 terms.

Relating Sequences and Systems of Equations

1. Solve the following system of equations.

$$2x + 5y = 8$$
$$7x + 3y = -1$$

Note that in each equation, the coefficients, including the constant term, form an arithmetic sequence.

2. For the system of equations

$$Ax + By = C$$
$$Px + Qy = R$$

the coefficients A, B, C form an arithmetic sequence and the coefficients P, Q, R form another arithmetic sequence. Prove that $(x, y) = (-1, 2)$ is always an exact solution to this system.

3. What if the coefficients form a geometric sequence? Is there an exact solution? Is there a general solution? If there is a solution, develop it. If not, show that it does not exist.

4. Extend the above ideas to at least one other type of sequence. Develop a feasible solution or show that none exists.

REVIEW OF **KEY CONCEPTS**

■ **6.1** Sequences

Refer to the Key Concepts on page 432.

1. Given the formula for the nth term, write the first 5 terms of each sequence, and graph t_n or $f(n)$ versus n.

a) $t_n = 2n + 1$

b) $t_n = n^2 - 3$

c) $f(n) = 7 - 2n$

d) $t_n = 3^n - 1$

2. Given the formula for the nth term, find the tenth term of each sequence.

a) $t_n = \dfrac{3n + 2}{n - 2}$

b) $f(n) = 2n^2 - 5$

3. Find the indicated terms.

a) $t_n = 5n - 4$; t_6 and t_{14}

b) $f(n) = 7 - 4n$; t_5 and t_9

c) $t_n = n^2 - 5$; t_7 and t_{10}

d) $f(n) = \dfrac{n + 2}{2}$; t_8 and t_{30}

4. Find the formula for the nth term that determines each sequence. Then, find t_{12}.

a) 4, 8, 12, 16, …

b) 1, 3, 5, 7, …

c) 2, 5, 10, 17, …

d) −6, −11, −16, −21, …

5. Fingernails A human fingernail grows about 0.6 mm a week. If the visible part of a fingernail is 15 mm long, how long will the nail be in

a) 1 week?

b) 2 weeks?

c) 4 weeks?

6.2 Arithmetic Sequences

Refer to the Key Concepts on page 441.

6. Find the next 3 terms of each arithmetic sequence.

a) 9, 15, 21, …

b) 6, 1, −4, …

c) −3.5, −1, 1.5, …

d) 1, $\dfrac{1}{2}$, 0, …

7. Given the formula for the nth term of each arithmetic sequence, write the first 4 terms.

a) $t_n = 5n + 2$

b) $t_n = 4n - 1$

c) $f(n) = 6 - 3n$

d) $f(n) = -5n + 3$

e) $t_n = \dfrac{2n - 1}{3}$

f) $f(n) = 0.2n + 4$

8. Given the values of a and d, write the first 5 terms of each arithmetic sequence.

a) $a = 3, d = 5$
b) $a = -5, d = 2$
c) $a = 4, d = -3$
d) $a = 0, d = -2.3$

9. Find the formula for the nth term, and find the indicated term for each arithmetic sequence.

a) $3, 5, 7, \ldots ; t_{30}$
b) $-2, -6, -10, \ldots ; t_{25}$
c) $-4, 3, 10, \ldots ; t_{18}$

10. Find the number of terms in each arithmetic sequence.

a) $4, 9, 14, \ldots , 169$
b) $19, 11, 3, \ldots , -229$

11. Find a and d, and then write the formula for the nth term, t_n, of arithmetic sequences with the following terms.

a) $t_7 = 9$ and $t_{12} = 29$
b) $t_4 = 12$ and $t_{11} = -2$

12. Pattern The rectangular shapes are made from asterisks.

a) How many asterisks are used for the 4th rectangle? the 5th rectangle?

b) Write the formula for the nth term for the sequence in the numbers of asterisks.

c) Predict the number of asterisks used for the 25th rectangle.

d) Which rectangle is made from 92 asterisks?

13. Soccer The Women's World Cup of Soccer tournament was first held in 1991. The next two tournaments were held in 1995 and 1999.

a) Write a formula for finding the year in which the nth tournament will be held.

b) Predict the year of the 35th tournament.

6.3 Geometric Sequences

Refer to the Key Concepts on page 452.

14. Determine whether each sequence is arithmetic, geometric, or neither. Then, find the next 2 terms.

a) $1, 8, 27, 64, \ldots$
b) $1, 3, 9, 27, \ldots$
c) $6, 12, 18, 24, \ldots$
d) $64, -32, 16, -8, \ldots$
e) $15, 14, 12, 9, 5, \ldots$
f) $3, 2.7, 2.4, 2.1, \ldots$

15. Given the values of a and r, write the first 5 terms of each geometric sequence.

a) $a = 6$ and $r = 4$ b) $a = 5$ and $r = -2$

c) $a = -3$ and $r = -5$ d) $a = 10$ and $r = 0.1$

16. Given the formula for the nth term of each geometric sequence, write the first 5 terms.

a) $t_n = 3(2)^{n-1}$ b) $t_n = 2(-3)^{n-1}$

c) $f(n) = 4(-2)^{n-1}$ d) $f(n) = -(4)^{n-1}$

e) $t_n = -2(-2)^{n-1}$ f) $t_n = -1000(0.5)^{n-1}$

17. Find the formula for the nth term and find the indicated term for each geometric sequence.

a) $3, 6, 12, \ldots ; t_{10}$ b) $2, 8, 32, \ldots ; t_8$

c) $27, 9, 3, \ldots ; t_6$ d) $1, -3, 9, \ldots ; t_7$

18. Find the number of terms in each geometric sequence.

a) $1, 2, 4, \ldots , 1024$ b) $16\ 384, 4096, \ldots , 1$

19. Find a, r, and t_n for each geometric sequence.

a) $t_4 = 24$ and $t_6 = 96$ b) $t_2 = -6$ and $t_5 = -162$

20. Population Approximate values for Canada's population were 3.5 million in 1870, 7 million in 1910, 14 million in 1950, and 28 million in 1990. If the pattern continues, predict the population in 2070.

21. Nuclear chemistry The time it takes for half of a sample of a radioactive isotope to decay is called the half-life. Iodine-131 has a half-life of 8 days.

a) Write an equation to determine the amount of iodine-131 remaining in a sample after n half-lives.

b) If a hospital receives a shipment of 200 mg of iodine-131, how much iodine-131 would remain after 32 days?

6.4 Recursion Formulas

Refer to the Key Concepts on page 460.

22. Use the recursion formula to write the first 5 terms of each sequence.

a) $t_1 = 19; t_n = t_{n-1} - 8$

b) $f(1) = -5; f(n) = f(n-1) + 3$

c) $t_1 = -1; t_n = -2t_{n-1}$

d) $f(1) = 8; f(n) = 0.5f(n-1)$

e) $t_1 = 3; t_2 = 3; t_n = t_{n-1} + t_{n-2}$

f) $t_1 = -12; t_n = t_{n-1} + 3n$

g) $t_1 = 11; t_n = 2t_{n-1} - n^2$

h) $t_1 = -1; t_2 = 1; t_n = t_{n-1} \div t_{n-2}$

23. Write the first 5 terms of the sequence defined by each of the following recursion formulas. State whether each sequence is arithmetic, geometric, or neither.

a) $t_1 = 2; t_n = 4t_{n-1}$

b) $t_1 = 0; t_n = 3t_{n-1} + n - 1$

c) $t_1 = -3; t_n = t_{n-1} - 4$

24. Write an explicit formula for the sequence determined by each recursion formula.

a) $t_1 = -5; t_n = t_{n-1} + 3$

b) $t_1 = 3; t_n = 4t_{n-1}$

c) $t_1 = 40; t_n = 0.5t_{n-1} + n$

25. Pattern The first four diagrams in a pattern are shown. The shaded part of the first diagram has an area of 3 square units.

a) Determine which of the following recursion formulas determines the sequence in the areas of the shaded parts of the diagrams.

1   2   3   4

$t_1 = 3$ $t_1 = 3$ $t_1 = 3$
$t_n = t_{n-1} + 3(n-1)$ $t_n = t_{n-1} + 2n - 1$ $t_n = t_{n-1} + n + 1$

b) Use the correct recursion formula to predict the area of the shaded region in the 5th diagram; the 6th diagram.

c) Write an explicit formula that determines the sequence in the areas of the shaded parts of the diagrams.

26. Taxi fare In one city, a taxi fare includes an initial fixed charge, plus another charge based on each tenth of a kilometre travelled. The possible fares charged, in dollars, in this city, can be represented by the recursion formula

$$t_1 = 3; \ t_n = t_{n-1} + 0.2$$

a) What is the initial fixed charge?

b) What is the charge per tenth of a kilometre travelled?

c) Write an explicit formula to model the possible taxi fares in this city. Define each variable in your formula.

d) Calculate the fare for a 4.5-km trip.

6.5 Arithmetic Series

Refer to the Key Concepts on page 468.

27. Find the indicated sum for each arithmetic series.

a) S_{20} for $3 + 8 + 13 + \ldots$

b) S_{25} for $-20 - 18 - 16 \ldots$

c) S_{27} for $-2.5 + 1 + 4.5 + \ldots$

28. Find the sum of each arithmetic series.

a) $7 + 10 + 13 + \ldots + 70$

b) $65 + 59 + 53 + \ldots - 85$

c) $1 + \dfrac{5}{4} + \dfrac{3}{2} + \ldots + 20$

29. Find S_n for each arithmetic series.

a) $a = 3, \ t_n = 147, \ n = 19$

b) $a = -1, \ t_n = -37, \ n = 10$

30. Integers Find the sum of the 50 greatest negative integers.

31. Measurement The side lengths in a quadrilateral form an arithmetic sequence. The perimeter is 38 cm, and the shortest side measures 5 cm. What are the other side lengths?

32. Piling logs In a triangular pile of logs, the top row contains one log. Each row below the top row contains one log more than the row above. If the bottom row contains 21 logs, how many logs are in the pile?

6.6 Geometric Series

Refer to the Key Concepts on page 476.

33. Find the indicated sum for each geometric series.
a) S_{10} for $2 + 4 + 8 + 16 + \ldots$
b) S_9 for $1 - 3 + 9 - 27 + \ldots$
c) S_6 for $1024 + 512 + 256 + \ldots$
d) S_{12} for $4 - 8 + 16 - 32 + \ldots$

34. Find S_n for each geometric series.
a) $a = 1$, $r = 3$, $n = 9$
b) $a = 6$, $r = -2$, $n = 7$
c) $a = 8$, $r = 0.5$, $n = 6$
d) $a = -3$, $r = -1$, $n = 10$

35. If $f(1) = 1\ 000\ 000$ and $f(2) = -500\ 000$ for a geometric series, find S_7.

36. Find the sum of each geometric series.
a) $5 + 10 + 20 + \ldots + 1280$
b) $2 + 6 + 18 + \ldots + 4374$
c) $-4 + 12 - 36 + \ldots + 972$
d) $3645 - 1215 + 405 - \ldots + 5$

37. Pattern The first four diagrams in a pattern are shown. Each shape is made from small squares of area 1 cm^2. Find the total area of the first 9 diagrams in the pattern.

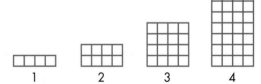

1 2 3 4

38. Bouncing ball A ball is thrown 6.4 m into the air. The ball falls, rebounds to 60% of its previous height, and falls again. If the ball continues to rebound and fall in this manner, find the total distance the ball travels until it hits the ground for the fifth time.

CHAPTER TEST

Achievement Chart

Category	Knowledge/Understanding	Thinking/Inquiry/Problem Solving	Communication	Application
Questions	All	11, 12, 13	5, 6, 13, 15	11, 12, 13

1. Given the formula for the nth term, state the first 5 terms of each sequence. Then, graph t_n or $f(n)$ versus n.
   a) $t_n = 2n - 3$
   b) $f(n) = n^2 + 3$

2. Given the formula for the nth term, find the twelfth term of each sequence.
   a) $t_n = 4 + 3n$
   b) $f(n) = (n + 1)^2$

3. Find the indicated terms.
   a) $t_n = 6n - 1$; t_8 and t_{24}
   b) $f(n) = 8 + 3n^2$; t_6 and t_{20}

4. Given the formula for the nth term of each sequence, write the first 5 terms.
   a) $t_n = 0.5n + 3$
   b) $f(n) = 5 - 3n$
   c) $t_n = 6(2)^{n-1}$
   d) $f(n) = 10(-2)^{n-1}$

5. Find the formula for the nth term that determines each sequence. Then, find $t_{21.}$
   a) 6, 10, 14, 18, ...
   b) −5, −11, −17, −23, ...

6. Find the formula for the nth term that determines each sequence. Then, find $t_{8.}$
   a) 1, 4, 16, 64, ...
   b) 10 000, −5000, 2500, −1250, ...

7. Find the indicated sum for each arithmetic series.
   a) S_{15} for $4 + 11 + 18 + \dots$
   b) S_{20} for $99 + 88 + 77 + \dots$

8. Find the sum of the arithmetic series $-12 - 9 - 6 - \dots + 39$.

9. Find the indicated sum for each geometric series.
   a) S_9 for $7 + 14 + 28 + 56 + \dots$
   b) S_6 for $2000 - 400 + 80 - \dots$

10. Find the sum of the geometric series $7 - 21 + 63 + \dots - 1701$.

11. Space shuttle When Dr. Roberta Bondar flew on the space shuttle, it lifted off at about 10:00 and then orbited Earth once every 90 min. At what time of day did it complete its 9th orbit?

12. Bacterial culture The number of bacteria in a culture is doubling every 30 min. If there are 10 000 bacteria at 16:00, how many will there be at 20:00 on the same day?

13. a) If a number is added to each term of a geometric sequence, is the resulting sequence still geometric? Is it arithmetic? Explain.
b) If a number is multiplied by each term of an arithmetic sequence, is the resulting sequence still arithmetic? Is it geometric? Explain.
c) If an arithmetic sequence is added to a geometric sequence, term by term, is the resulting sequence arithmetic or geometric? Explain.

Answer questions 14 and 15 only if you studied section 6.4.

14. Write the first 4 terms determined by each recursion formula.

a) $t_1 = 7$; $t_n = t_{n-1} - 3$ **b)** $t_1 = -2$; $t_n = t_{n-1} + n$

c) $t_1 = 2000$; $t_n = -0.4t_{n-1}$ **d)** $t_1 = 2$; $t_2 = 3$; $t_n = t_{n-1} - t_{n-2}$

15. Write an explicit formula for the sequence determined by each recursion formula.

a) $t_1 = 2$; $t_n = 5t_{n-1}$ **b)** $t_1 = -7$; $t_n = t_{n-1} + 8$

CHALLENGE PROBLEMS

1. Counting backward Starting at 888 and counting backward by 7, a student counts 888, 881, 874, and so on. Which of the following numbers will be included?

a) 35   **b)** 34   **c)** 33   **d)** 32   **e)** 31

2. Positive integers The sum of the first n even positive integers is h. The sum of the first n odd positive integers is k. Then $h - k$ is equal to:

a) $\dfrac{n}{2}$ **b)** $\dfrac{n}{2} - 1$ **c)** n **d)** $-n$ **e)** $n - 1$

3. Arithmetic sequence The largest four-digit number to be found in the arithmetic sequence 1, 6, 11, 16, 21, ... is:

a) 9995   **b)** 9996   **c)** 9997   **d)** 9998   **e)** 9999

4. Even integers The sum of 50 consecutive even integers is 3250. The largest of these integers is:

a) 64   **b)** 66   **c)** 112   **d)** 114   **e)** 116

5. Calendar January 1, 1986 was a Wednesday. January 1, 1992, was what day of the week?

6. Integer table The integers greater than 1 are arranged in a table, with three numbers in each row and with five columns, as shown. If the pattern continues, in which column will the number 1000 appear?

A	B	C	D	E
2	3	4		
		7	6	5
8	9	10		
		13	12	11
14	15	16		
		19	18	17

7. Line dancing A line dance requires two steps forward and then one step back. If a wall is ten steps away, how many steps must the dancers take to reach the wall?

8. Book pages The total number of digits used to number all the pages of a book was 216. Find the number of pages in the book.

9. Quadratic equations In a quadratic equation of the form $ax^2 + bx + c = 0$, a, b, and c are positive real numbers that form a geometric sequence a, b, c, ... Describe the roots of the equation.

GUESS AND CHECK

One way to solve a problem is to guess the answer and then check to see whether it is correct. If it is not, you can keep guessing and checking until you get the right answer.

When a ball is dropped onto the ground, it rebounds to a certain percent of its original height. As the ball continues to bounce, the height of each rebound is the same percent of the height of the previous rebound.

For example, if a ball that rebounds to 70% of its original height is dropped from a height of 150 cm, the height of the first rebound is 0.7×150 or 105 cm. The height of the second rebound is 0.7×105 or 73.5 cm.

If a ball that rebounds to 75% of its original height is dropped from a height of 2 m, or 200 cm, the height, h centimetres, of the nth rebound is given by the following equation.

$$h = 200(0.75)^n$$

Which rebound has a height of about 20 cm?

Understand
the Problem

1. What information are you given?
2. What are you asked to find?
3. Do you need an exact or an approximate answer?

Think
of a Plan

Set up a table. Guess at the value of n. Use the guess to calculate the height of the nth rebound. If the calculated height is not about 20 cm, make another guess at the value of n and calculate the height again.

Carry Out
the Plan

Guess		Check
n	Calculation of h	Is h about 20 cm?
5	$200(0.75)^5 \doteq 47$	h too high
10	$200(0.75)^{10} \doteq 11$	h too low
9	$200(0.75)^9 \doteq 15$	h too low
8	$200(0.75)^8 \doteq 20$	20 checks!

The eighth rebound has a height of about 20 cm.

Look Back

Does the answer seem reasonable?

Is there a way to solve the problem graphically?

Apply, Solve, Communicate

Write your guess for each of questions 1–4. Compare your answer with a classmate's. Then, use your research skills to find the correct value.

1. Lake What is the area, in square kilometres, of the largest lake entirely in Canada?

2. Students How many students are enrolled in Grade 11 in Ontario?

3. Trail The Trans Canada Trail is the longest trail of its kind in the world.
a) How long is it?
b) What is the length of the trail in Ontario?

4. Airports a) How many airports are there in Canada?
b) How many have paved runways?

5. Number puzzle Copy the diagram shown below. Place five of the digits from 1 to 9 in the circles so that at least one of the two-digit numbers formed by each pair of numbers joined by a line segment is divisible by 13 or 7. The digits can be paired in either order.

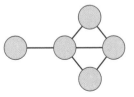

6. Cereal box The area of the front of a cereal box is 532 cm^2, the area of each side is 140 cm^2, and the area of the bottom is 95 cm^2. What is the volume of the box?

7. Finding digits What are the digits A, B, and C if the following equation is true?
$$(AA)^2 = BBCC$$

8. Number puzzle Copy the diagram. Place the numbers from 1 to 9 in the circles so that each line of three numbers adds to 18. The 6 and the 1 have been placed for you.

9. Word puzzle Each letter represents a different digit in this addition. Find the values of O, N, E, and T.

$$
\begin{array}{r}
ONE \\
ONE \\
ONE \\
+\ ONE \\
\hline
TEN
\end{array}
$$

10. True sentence Copy the figure. Replace each ■ with a spelled-out number to make the sentence true.

> In
> this
> triangle,
> there are
> ■ f's, ■ h's,
> and ■ t's.

11. Fire stations A county has 35 towns, as shown in the diagram. Each of the shortest line segments represents a road 10 km long. Regional planners are suggesting fire stations in some towns, so that no town is more than 10 km by road from a fire station. What is the minimum number of fire stations that must be built?

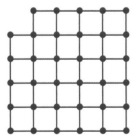

12. Magic square In a magic square, the sum of the numbers in each row, column, and diagonal is the same, in this case 15. Rearrange the numbers so that the sums for each row, column, and diagonal are all different, and none of the sums is 15.

6	1	8
7	5	3
2	9	4

13. Writing expressions By replacing each ■, use each digit from 0 to 9 once in three correct expressions of the form shown. Replace each ● with a symbol chosen from ×, ÷, +, or −.

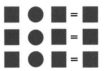

14. Whole numbers A, B, C, and D are whole numbers. The same number results when 4 is added to A, 4 is subtracted from B, C is divided by 4, and D is multiplied by 4. If A, B, C, and D add to 100, what are the values of A, B, C, and D?

15. Number sets The digits from 1 to 9 have been used to write three numbers, where the second number is twice the first number, and the third number is three times the first number.

> 1 9 2
> 3 8 4
> 5 7 6

a) Another set of three numbers that satisfies the same conditions can be found by rearranging the digits in the three given numbers. What is the set of numbers?
b) There are two other sets of three numbers that satisfy the same conditions. One set is a rearrangement of the digits of the other set. What are the two sets of numbers?

16. Formulating problems Write a problem that can be solved using the guess and check strategy. Have a classmate solve your problem.

USING THE STRATEGIES

1. Measurement The perimeters of a regular hexagon and an equilateral triangle are equal. What fraction of the area of the hexagon is the area of the triangle?

2. Keys In how many different ways can four different keys be arranged on a key ring?

3. Powers The powers of 3 are 3^0, 3^1, 3^2, 3^3, ..., or 1, 3, 9, 27, ... Using only addition or subtraction of powers of 3, the numbers 5, 26, and 35 can be expressed as follows.

$$5 = 3 + 1 + 1$$
$$26 = 27 - 1$$
$$35 = 27 + 9 - 1$$

Using only addition and subtraction, how many whole numbers from 1 to 50 cannot be expressed using three or fewer powers of 3?

4. Decimal What is the 53rd digit in the decimal form of $-\dfrac{1}{7}$?

5. Measurement
The points A, B, Q, D, and C lie on a circle. $\angle BAQ = 41°$ and $\angle DCQ = 37°$. What is the sum of the measures of $\angle P$ and $\angle AQC$?

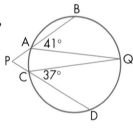

6. Digits How many three-digit whole numbers do not contain any of the digits 1, 2, 4, 5, 6, 8, and 9?

7. Tiling Suppose that a domino covers two squares on a checkerboard. If two squares in opposite corners of the checkerboard are removed, as shown, can the remaining portion of the board be completely covered with dominoes?

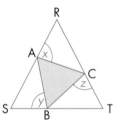

8. Lawn A circular lawn has a radius of 10 m. The lawn needs re-sodding. Sod can be purchased in strips that are 40 cm wide. What is the approximate length of sod needed?

9. Geometry In the diagram, $\triangle RST$ is isosceles, with $RS = RT$, and $\triangle ABC$ is equilateral. Express $\angle x$ in terms of $\angle y$ and $\angle z$.

10. Whole numbers In the sum, D, E, and F represent different whole numbers. Find the values of D, E, and F.

$$\begin{array}{r} D\ E\ F \\ D\ E\ F \\ +\ D\ E\ F \\ \hline E\ E\ E \end{array}$$

11. People walking About how many people can walk past a point on a 4-m wide road in one hour?

CUMULATIVE REVIEW: CHAPTERS 5 AND 6

Chapter 5

1. Change each radian measure to degree measure. Round to the nearest tenth of a degree, if necessary.

a) $\dfrac{\pi}{9}$ b) $\dfrac{3\pi}{8}$ c) 3.1

2. Find the exact radian measure, in terms of π, for each of the following angles.

a) 80°   b) 260°   c) 570°

3. Determine the exact value of $\sin \theta$, $\cos \theta$, and $\tan \theta$ for angle θ, if the point P$(-2, -3)$ is on the terminal arm and $0 \le \theta \le 2\pi$.

4. Find the exact value of each ratio.

a) $\cos 315°$ b) $\sin 240°$ c) $\tan 210°$

5. A periodic function f has a period of 15. If $f(5) = 13$ and $f(40) = -5$, determine

a) $f(110)$ b) $f(200)$ c) $f(-80)$

6. Sketch one cycle of the graph of each of the following. State the period and the amplitude in each case.

a) $y = \dfrac{1}{2}\cos 2x$ b) $y = 3\sin 2x$

7. Sketch one cycle of the graph of each of the following. State the period, the amplitude, the vertical translation, and the phase shift in each case.

a) $y = \sin\left(x - \dfrac{\pi}{3}\right) - 3$

b) $y = \dfrac{1}{2}\cos (3x - \pi) - 5$

8. Prove the identity.
$$(1 - \sin^2 x)(1 + \tan^2 x) = 1$$

9. Solve the trigonometric equation for $0° \le x \le 360°$. Round the solution to the nearest tenth of a degree, if necessary.
$$4\sin^2 x - 4\sin x - 3 = 0$$

Chapter 6

1. Given the formula for the nth term, state the first 5 terms of each sequence. Then, graph t_n or $f(n)$ versus n.

a) $t_n = 2(n - 1)$ b) $f(n) = n^2 + 5$
c) $t_n = 0.5(2)^{n-1}$

2. For the arithmetic sequence 9, 15, 21, ...,

a) find t_n b) find t_{25}

3. Find the number of terms in each sequence.

a) 251, 243, 235, ..., -205
b) $-4, -12, -36, ..., -8748$

4. Find the formula for the nth term and t_{12} for the geometric sequence. $-1, 2, -4, 8, ...$

5. Find S_{12} for the arithmetic series.
$-35 - 25 - 15 - ...$

6. Find the sum of the arithmetic series.
$21 + 23 + 25 + ... + 43$

7. Find S_{10} for the geometric series.
$-2 + 4 + -8 ...$

8. Find the sum of the geometric series.
$1280 - 640 + 320 - ... + 5$

9. **Measurement** The interior angles of a quadrilateral are in an arithmetic sequence. If the largest angle is 96°, find the other angles.

Complete questions 10 and 11 only if you studied section 6.4.

10. Write the first 5 terms determined by each recursion formula.

a) $t_1 = -6;\ t_n = t_{n-1} + 5$
b) $t_1 = 800;\ t_n = -0.25t_{n-1}$
c) $t_1 = -2;\ t_2 = -1;\ t_n = t_{n-1} \times t_{n-2}$

11. Write an explicit formula for the sequence determined by the recursion formula $t_1 = 10;\ t_n = t_{n-1} - 3$.

7 Compound Interest and Annuities

Specific Expectations	Functions	Functions & Relations
Derive the formulas for compound interest and present value, the amount of an ordinary annuity, and the present value of an ordinary annuity, using the formulas for the *n*th term of a geometric sequence and the sum of the first *n* terms of a geometric series.	7.2, 7.3, 7.4, 7.5, 7.6	7.2, 7.3, 7.4, 7.5, 7.6
Solve problems involving compound interest and present value.	7.2, 7.3, 7.4	7.2, 7.3, 7.4
Solve problems involving the amount and the present value of an ordinary annuity.	7.5, 7.6	7.5, 7.6
Demonstrate an understanding of the relationships between simple interest, arithmetic sequences, and linear growth.	7.1, 7.3	7.1, 7.3
Demonstrate an understanding of the relationships between compound interest, geometric sequences, and exponential growth.	7.2, 7.3	7.2, 7.3
Analyse the effects of changing the conditions in long-term savings plans.	7.2, 7.4, 7.5, 7.6	7.2, 7.4, 7.5, 7.6
Describe the manner in which interest is calculated on a mortgage and compare this with the method of interest compounded monthly and calculated monthly.	7.7, 7.8	7.7, 7.8
Generate amortization tables for mortgages, using spreadsheets or other appropriate software.	7.7, 7.8	7.7, 7.8
Analyse the effects of changing the conditions of a mortgage.	7.7, 7.8	7.7, 7.8
Communicate the solutions to problems and the findings of investigations with clarity and justification.	7.1, 7.2, 7.3, 7.4, 7.5, 7.6, 7.7, 7.8	7.1, 7.2, 7.3, 7.4, 7.5, 7.6, 7.7, 7.8

Making Financial Decisions

Financial decisions are encountered throughout your life. Information, experience, and understanding can enable you to make wise decisions for investing and borrowing. Buying a vehicle may require saving in advance, financing with a loan, or both. Real estate purchases usually involve paying part of the cost with a down payment and borrowing the remainder as a mortgage. If you start your own business, you might use money you have saved, and also arrange a loan. To save for retirement, you may invest money in a plan that provides an income at regular intervals.

In the Modelling Math questions on pages 510, 533, 542, and 569, you will solve the following problem and other problems involving making financial decisions.

Suppose that, on the day you were born, your parents invested $4000 in an account earning interest at a rate of 5.9% per annum, compounded semi-annually. How much would the investment be worth on your 19th birthday? If the rate were compounded monthly instead of semi-annually, what difference would it make to the amount on your 19th birthday?

Answer the following questions now.

Jeffrey wants to buy a second-hand car with money from his part-time job. He is deciding between two payment options: pay $150 per month for 7 years, or $200 per month for 4 years.

1. Which option costs less?

2. What reasons might lead Jeffery to choose the first option?

3. What reasons might lead him to choose the second option?

4. Which option do you think Jeffery should choose? Support your answer by referring to reasons from questions 2 and 3.

5. What additional information would help Jeffery make an informed decision?

Comparing Costs

1. Max can buy a stereo system, paying $599 now or paying $750 in a year.
a) What factors would you suggest Max consider when deciding which plan to accept? How might these vary from factors you would suggest to someone else?
b) What reasons would you give Max to convince him to consider these factors?
c) What assumptions are made about Max agreeing to either plan? What would you suggest doing about these assumptions?

2. Sadie wants to take soccer lessons for a year. The soccer club she chooses offers the following payment options.
Option A: Pay in a lump sum of $919, a week before classes start.
Option B: Pay $300 now, $350 in 4 months, and $350 again after another 4 months.
Option C: Pay $95 each month for the year.
a) What is the total amount Sadie will pay for the year for
i) option A?   **ii)** option B?   **iii)** option C?
b) Explain factors that could make it reasonable for Sadie to choose
i) the most expensive option
ii) the least expensive option

c) Research different plans for the cost of lessons that interest you. Decide which plan you would choose. Explain your process of arriving at your decision. Include a discussion of whether there is another choice you would consider.

3. Suppose you are arranging ice time for a recreational hockey team to practise. A local rink has three plans for ice use with these rates: $55/h for more than 2 h per week, $62/h for 1 h to 2 h per week, and $68/h for less than 1 h per week. Because of various commitments, the team members are not able to predict how many hours they will be able to practise during the season. Which plan would you book? Explain how you would justify your decision to the team.

Review of Prerequisite Skills

If you need help with any of the skills named in purple below, refer to Appendix A.

1. Write the next 3 terms.
a) $100, 100(0.05), 100(0.05)^2, 100(0.05)^3, \ldots$
b) $100, 106, 112, 118, \ldots$
c) $1 + 0.06, (1 + 0.06)^2, (1 + 0.06)^3, \ldots$

2. Solving linear equations Solve for r.
a) $120 = 100r$
b) $250 = 500r(2)$
c) $2500 = Pr(10)$
d) $I = 400r$
e) $300 = 2000rt$
f) $I = Prt$

3. Solving linear equations Solve for P.
a) $100 = P(0.02)$
b) $200 = P(0.05)(2)$
c) $400 = P(0.04)$
d) $I = Prt$

4. Solving linear equations Solve for t.
a) $80 = 1000(0.04)t$
b) $360 = 1200(0.06)t$
c) $216 = 450(0.08)t$
d) $5400 = 30\ 000(0.02)t$
e) $I = Prt$

5. Exponent rules Evaluate. Round to the nearest thousandth.
a) $(1.04)^4$
b) $(1.02)^{-18}$
c) $(1.055)^5$
d) $(1.098)^{-32}$
e) $(1.08)^{-7}$
f) $(1.065)^{11}$
g) $(1.015)^8$
h) $(1.045)^{20}$
i) $(1.225)^{-3}$
j) $(1.13)^{-6}$
k) $(1.01)^{-15}$
l) $(1.07)^{19}$

6. Evaluate.
a) $5000(1 + 0.035)^6$
b) $900\ 000(1 + 0.031)^{-30}$
c) $75\ 000(1 + 0.006)^{-8}$
d) $38\ 000(1 + 0.015)^4$
e) $142\ 000(1 + 0.0525)^{-7}$

7. Evaluate.
a) $\dfrac{300[(1.025)^{16} - 1]}{0.025}$

b) $\dfrac{1000[(1.007)^7 - 1]}{0.007}$

c) $\dfrac{48000[(1.009)^{12} - 1]}{0.009}$

8. Express each percent as a decimal.
a) 15%   b) 6.13%
c) 0.8%   d) 4.75%
e) 1.3%   f) 0.25%
g) 7%   h) 3.05%
i) 8.25%

9. Calculate.
a) 10% of 1000   b) 6% of 1250
c) 8.25% of 20 000   d) 7.6% of 12 390
e) 5.25% of $10 500   f) 12.5% of $2254
g) 4.5% of $2000   h) 0.9% of 27 355

10. Calculate the total cost of each, including 7% GST and 8% PST.
a) stereo system $1999
b) car $21 515
c) TV $1499
d) computer $1898

7.1 Investigation: Simple Interest, Arithmetic Sequences, and Linear Growth

When you invest money in a bank or other financial institution, you are paid interest for the use of your money. The financial institution uses your money to earn money. When you borrow money, you pay interest for the loan.

Interest can be calculated as **simple interest**, which means that only the money originally invested earns interest. If you invest $1000 at a simple interest rate of 5% per annum for several years, only the original investment of $1000 earns interest each year.

Simple Interest

The money you invest or borrow is called the **principal**. The **interest rate** is the percent of the principal that is earned, or paid, as interest. The sum of the principal and the interest is called the **amount**.

1. Suppose you invest $1000 that earns simple interest at a rate of 5% per annum.

a) How much interest would you earn in the first year?

b) What amount would you have after 1 year?

c) Since simple interest is being calculated, your interest from the first year is not reinvested. How much interest would you earn in the second year?

d) How much interest would you have altogether after 2 years?

e) What amount would you have after 2 years?

2. Complete a table like the following for 10 years for $1000 invested with simple interest at a rate of 5% per annum. Use your answers from question 1 in the second row.

Number of years	Principal ($)	Interest rate	Interest ($)	Amount ($)
1	1000	0.05	50	1050
2				
3				

3. Plot the points in the table from question 2, with the time, in years, along the horizontal axis and the amount, in dollars, along the vertical axis. Join the points.

4. Consider the values in the Amount column of the table.
a) What kind of sequence do you notice? Explain.
b) What is the first term in this sequence?
c) What is the difference between consecutive terms in this sequence? What is the difference called?
d) Use your answers from parts b) and c) to write a formula for the nth term for the sequence in the Amount column. Use the table to check your formula for t_n.

5. a) Substitute $n = 1, 2, 3, \ldots, 10$ into your formula for t_n from part d) of question 4 to find $t_1, t_2, t_3, \ldots, t_{10}$.
b) Graph the values from part a) with n along the horizontal axis and t_n along the vertical axis. Join the points.

6. Use the table from question 2.
a) Let P be the principal in dollars, r be the interest rate expressed as a decimal, and t be the time in years. Write an expression for finding the interest.
b) Let I be the interest in dollars. Use your expression from part a) to write a formula showing the relationship between I, P, r, and t. Check your formula by applying it to values in the table.
c) Let A be the amount in dollars. Write a formula showing the relationship between A, P, and I. Check your formula with the table.
d) Substitute for I in your formula from part c) to write a formula expressing A in terms of P, r, and t. Check this formula with the table.

7. a) Substitute $t = 1, 2, 3, \ldots, 10$ into your formula for A from part d) of question 6 to find the amount after each year up to 10 years.
b) Graph the values from part a) with t along the horizontal axis and A along the vertical axis. Join the points.

8. Consider your graphs from questions 3, 5, and 7.

a) What shape are your graphs?

b) What is the y-intercept of each graph?

c) What does each y-intercept represent for the investment of $1000?

d) What is the slope of each graph?

e) What does each slope represent for the investment of $1000?

f) Use the y-intercept and slope to write an equation in the form $y = mx + b$.

g) Graph the equation from part f).

9. If the points representing a relationship lie on a straight line, the relationship is **linear**.

a) Are the graphs linear or non-linear? Explain.

b) Why is it reasonable to join the points of each graph?

c) Use one graph to estimate the amount for a time from 1 to 10 years.

d) Use a formula to check your estimate.

e) Use your formula for t_n to find the amount after 20 years.

f) Use your formula for A to find the amount after 20 years. Compare this amount with the value for t_{20} in part e). Explain the result.

10. Suppose you invested $1000 earning simple interest at a rate of 6.75% per annum.

a) Predict the appearance of a graph representing this investment with the time in years along the horizontal axis and the amount in dollars along the vertical axis.

b) What amount would you have at the end of

i) 1 year?   **ii)** 2 years?   **iii)** 3 years?   **iv)** 4 years?   **v)** 5 years?

c) Name the type of sequence for the amounts in part b).

d) State the first term and the common difference for the sequence.

e) Write the formula for t_n of the sequence. Check your formula for $n = 1$, 3, and 5.

f) Graph the formula from part e) with n along the horizontal axis and t_n along the vertical axis. Join the points.

g) Write an equation in the form $y = mx + b$ to represent your graph.

h) Compare your graph with your prediction from part a).

11. Use your results to describe the relationship between simple interest, arithmetic sequences, and linear growth. Refer to your table, graphs, and formulas in your explanation.

7.2 Compound Interest

For **compound interest**, the interest is reinvested at regular intervals. The interest is added to the principal to earn interest for the next interval of time, or **compounding period**. If $4000 is invested at 6.25% per annum, compounded annually, the interest is added to the principal at the end of the year. The next year, interest is earned on the sum of the principal and the interest. Similarly, if $4000 is borrowed at 6.25% per annum, compounded annually, the interest is added to the principal at the end of the year.

Reasons for saving or borrowing are individual. Marc borrows $3000 to take a technology course. Daima compares plans for investing $10 000 in a Registered Retirement Savings Plan. Gabriella invests money so she can upgrade the computers in her business in 5 years.

INVESTIGATE & INQUIRE

Martin and Norma invest $2000 for their granddaughter Linda on her 12th birthday so she will have it on her 18th birthday for her education. The money is in an account earning interest at a rate of 8% per annum, compounded annually.

1. a) Use the formula for simple interest to find how much interest is earned in the first year.
b) What amount is in the account after 1 year?
c) Since compound interest is being calculated, the interest from the first year is reinvested with the principal for the next year. How much is the principal for the second year?
d) Use the formula for simple interest to find the interest on the principal for the second year, to the nearest cent.
e) What amount is in the account after 2 years?

2. Complete a table like the following for the investment until Linda's 18th birthday. Remember the money is withdrawn on her 18th birthday. Use your answers from question 1 in the second row.

Since the interest is reinvested, use the amount at the end of the year as the principal for the next year.

Birthday	Year number	Principal ($)	Interest rate	Interest ($)	Amount ($)
12	1	2000.00	0.08	160.00	2160.00
13	2	2160.00			
14	3				

3. a) What type of sequence is represented by the values in the Amount column of the table? Explain.
b) Write the formula for the nth term, t_n, for this type of sequence.

4. a) For the sequence representing the amount of Martin and Norma's investment, what is the value of
i) a?   **ii)** r?
b) Use the values of a and r to write a formula for the nth term, t_n, in this sequence.
c) To check the formula for the nth term, calculate
i) t_1 **ii)** t_2 **iii)** t_3 **iv)** t_4 **v)** t_7

5. a) Let A be the amount, P be the principal, i be the interest rate per compounding period, and n be the number of compounding periods. Substitute into the formula for the nth term from question 4b to write the formula for compound interest.
b) Check your formula by substituting values for P, i, and n, and calculating A for this investment.

6. a) If the principal invested is P and the annual interest rate is i, what is the amount after each number of years?
i) 1   **ii)** 2   **iii)** 3   **iv)** 4   **v)** 5
b) Write the amounts from part a) as a sequence.
c) What is the value of a for this sequence?
d) What is the value of r?

7. Use your results to describe the relationship between compound interest and the kind of sequence represented by Martin and Norma's investment.

The formula for a geometric sequence can be used to develop the formula for the amount accumulated, A, with compound interest,
$$A = P(1 + i)^n.$$
A is the amount at the end of the time for the investment or loan, P is the principal invested, i is the interest rate per compounding period, and n is the number of compounding periods.

EXAMPLE 1 Finding the Amount Compounded Annually

To take a technology course, Marc borrows $3000 at an interest rate of 4.75% per annum, compounded annually. He plans to pay back the loan in 5 years.
a) How much will Marc owe after 5 years?
b) How much interest will Marc pay for the loan?

SOLUTION 1 Paper-and-Pencil Method

a) Use the formula for compound interest.
Marc's loan is $3000, so $P = 3000$.
The interest rate is 4.75% per annum, compounded annually, so $i = 0.0475$.
The length of time for the loan is 5 years and the interest is compounded annually for 5 compounding periods, so $n = 5$.

Web Connection
www.school.mcgrawhill.ca/resources/
To investigate the power of compounding, visit the above web site. Go to **Math Resources**, then to *MATHEMATICS 11*, to find out where to go next. Write a report about simple and compound interest, outlining the similarities and differences.

$$A = P(1 + i)^n$$
Substitute known values: $= 3000(1 + 0.0475)^5$
Simplify: $= 3000(1.0475)^5$
  $\doteq 3783.48$

Marc will owe $3783.48 after 5 years.

b) The interest is the amount paid after 5 years less the money borrowed.
$3783.48 - 3000 = 783.48$
Marc will pay $783.48 interest for the loan.

SOLUTION 2 Graphing-Calculator Method

a) Change the mode settings to 2 decimal places. From the Finance menu, choose TVM Solver.
Enter the known values.
The investment is for 5 years, so N = 5.
The interest rate is 4.75% per annum, so I = 4.75.

Marc's loan is $3000, so PV = 3000.

The interest is compounded annually, so C/Y = 1.

The interest is calculated at the end of each compounding period, so select END.

Move the cursor to FV to find the future value of the loan, and press ALPHA SOLVE. Since Marc will pay out the amount, or future value, FV is negative.

Marc will owe $3783.48 after 5 years.

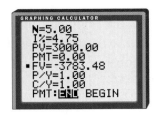

b) Marc will pay $783.48 interest for the loan, as shown in part b) of Solution 1.

EXAMPLE 2 Comparing the Effect of Different Compounding Periods

Daima is investing $10 000 in a Registered Retirement Savings Plan, or RRSP. She is considering a 9-year plan with an interest rate of 6% per annum, compounded semi-annually, or a 9-year plan with an interest rate of 5.95% per annum, compounded monthly. Which plan should Daima choose? Why?

Semi-annually means twice per year.

SOLUTION 1 Paper-and-Pencil Method

Use the formula for compound interest.

For 6% per annum, compounded semi-annually, Daima's investment is $10 000, so $P = 10\ 000$.

The interest is compounded semi-annually, so divide the interest rate per annum by 2 to find the interest rate per compounding period. $0.06 \div 2 = 0.03$, so $i = 0.03$.

The investment is for 9 years with interest compounded semi-annually, so multiply the number of years by 2 to find the number of compounding periods. $9 \times 2 = 18$, so $n = 18$.

Since i represents the interest rate per compounding period, the value of i is the interest rate per annum divided by the number of compounding periods in a year.

Since n represents the number of compounding periods, the value of n is the number of years times the number of compounding periods in a year.

$$A = P(1 + i)^n$$

Substitute known values: $= 10\ 000(1 + 0.03)^{18}$

Simplify: $= 10\ 000(1.03)^{18}$

        $\doteq 17\ 024.33$

At 6% per annum, compounded semi-annually, Daima's investment would be worth $17 024.33 after 9 years.

For 5.95% per annum, compounded monthly,
Daima's investment is $10 000, so $P = 10\ 000$.

The interest is compounded monthly, so divide the interest rate per annum by 12 to find the interest rate per compounding period.
$0.0595 \div 12 \doteq 0.004\ 958\ 333$, so $i \doteq 0.004\ 958\ 333$.

The investment is for 9 years with interest compounded monthly, so multiply the number of years by 12 to find the number of compounding periods.
$9 \times 12 = 108$, so $n = 108$.

$$A = P(1 + i)^n$$

Substitute known values: $\doteq 10\ 000(1 + 0.004\ 958\ 333)^{108}$
Simplify: $= 10\ 000(1.004\ 958\ 333)^{108}$
$\doteq 17\ 060.43$

At 5.95% per annum, compounded monthly, Daima's investment would be $17 060.43 after 9 years.

Daima's investment would be $17 024.33 at 6% per annum, compounded semi-annually, but $17 060.43 at 5.95% per annum, compounded monthly. She should choose the plan with 5.95% per annum, compounded monthly.

SOLUTION 2 Graphing-Calculator Method

Change the mode settings to 2 decimal places. From the Finance menu, choose the TVM Solver.

Enter the known values for 6% per annum, compounded semi-annually.
The investment is for 9 years, so N = 9.
The interest rate is 6% per annum, so I = 6.
Since the investment is paid out, PV is negative. Daima's investment is $10 000, so PV = −10 000.
The interest is compounded semi-annually, so C/Y = 2.
The interest is calculated at the end of each compounding period, so select END.

When C/Y is 2, the calculator automatically multiplies N by 2 and divides I by 2.

Move the cursor to FV to find the future value, and press ALPHA SOLVE.

At 6% per annum, compounded semi-annually, Daima's investment would be worth $17 024.33 after 9 years.

Enter the known values for 5.95% per annum, compounded monthly.

The investment is for 9 years, so N = 9.

The interest rate is 5.95% per annum, so I = 5.95.

Since the investment is paid out, PV is negative. Daima's investment is $10 000, so PV = −10 000.

The interest is compounded monthly, so C/Y = 12.

The interest is calculated at the end of each compounding period, so select END.

Move the cursor to FV to find the future value, and press ALPHA SOLVE.

At 5.95% per annum, compounded monthly, Daima's investment would be $17 060.43 after 9 years.

Since Daima's investment would be $17 024.33 at 6% per annum, compounded semi-annually, and $17 060.43 at 5.95% per annum, compounded monthly, she should choose the plan with 5.95% per annum, compounded monthly.

EXAMPLE 3 Finding the Interest Rate Using a Graphing Calculator

Gabriella hopes the $26 000 she is investing will be worth $40 000 in 5 years to upgrade the computers for her business. What rate of interest, to the nearest hundredth of a per cent, compounded quarterly, would Gabriella need to achieve this goal?

Quarterly means once every quarter of a year, 4 times a year, or every 3 months.

SOLUTION

Change the mode settings to 2 decimal places. From the Finance menu, choose TVM Solver.

Enter the known values.

The loan is for 5 years, so N = 5.

Since the investment is paid out, PV is negative. Gabriella's investment is $26 000, so PV = −26 000.

The amount wanted in 5 years is $40 000, so FV = 40 000.

The interest is compounded quarterly, so C/Y = 4.

The interest is calculated at the end of each compounding period, so select END.

Move the cursor to I to find the interest rate per annum, and press ALPHA SOLVE. Since the number of decimal places is set to 2, the graphing calculator rounds the interest rate to the nearest hundredth of a percent.

Gabriella will need an interest rate of 8.71% per annum, compounded quarterly, to achieve this goal.

You can check this interest rate by changing FV to 0, and then solving for FV.

Key Concepts

- Compound interest is a financial application of geometric sequences.
- The formula for the accumulated amount with compound interest is $A = P(1 + i)^n$, where A is the amount at the end of the time for the investment or loan, P is the principal invested, i is the interest rate per compounding period, and n is the number of compounding periods.
- In $A = P(1 + i)^n$, $i = r \div N$, where r is the interest rate per annum and N is the number of compounding periods per year, and $n = yN$, where y is the number of years.

Communicate Your Understanding

1. Explain the meanings of A, P, i, and n in the formula for compound interest.

2. Describe how you would find the amount and the interest for a $2000 investment after 5 years at an interest rate of 6% per annum, compounded semi-annually.

3. Suppose $60 000 is borrowed for 3 years at 8% per annum, compounded monthly. Describe how you would find the value of i and n to substitute into the formula for compound interest.

4. a) Describe how you would use a graphing calculator to find the interest rate, to the nearest hundredth of a percent, compounded quarterly, that would change the value of a loan from $12 000 to $15 000 in 6 years.

b) Describe how you would use a graphing calculator to find the length of time, to the nearest month, for $7500 invested at 7% per annum, compounded monthly, to be worth $10 000.

c) Explain the advantages of using a graphing calculator for parts a) and b).

Practise

A

1. The rate of interest for an investment is 6% per annum. What is the interest rate for each compounding period?
a) semi-annually   **b)** quarterly   **c)** monthly
d) daily in a year that is not a leap year

2. How many compounding periods are there for each loan?
a) compounding quarterly for 1 year
b) compounding annually for 1 year
c) compounding monthly for 3 years
d) compounding annually for 5 years
e) compounding semi-annually for 2 years
f) compounding quarterly for 6 years

3. What is the amount for each loan?
a) $500 at 5% per annum, compounded annually for 3 years
b) $45 500 at 10.5% per annum, compounded semi-annually for 5 years
c) $1000 at 4.75% per annum, compounded monthly for 4 years
d) $96 000 at 11% per annum, compounded quarterly for 2 years
e) $140 000 at 9.8% per annum, compounded annually for 7 years

4. Find the amount of the investment and the interest.
a) $2000 invested for 5 years at 12% per annum, compounded annually
b) $32 500 invested for 1 year at 8.25% per annum, compounded semi-annually
c) $10 000 invested for 2 years at 5.75% per annum, compounded quarterly
d) $8000 invested for 6 years at 10.5% per annum, compounded monthly

5. Find the amount of each investment.
a) $2200 for 5 years at 12% per annum, compounded monthly
b) $4400 for 7 years at 7.25% per annum, compounded annually
c) $12 600 for 4 years at 6.75% per annum, compounded quarterly
d) $500 000 for 10 years at 9.25% per annum, compounded semi-annually

Apply, Solve, Communicate

6. Investing Reza invested $1000 for a year at 6% per annum. What was the amount if the interest was compounded
a) semi-annually?
b) quarterly?
c) monthly?

7. Borrowing August borrowed $9500 for 3 years at 11.6% per annum, compounded quarterly.
a) How much did he owe at the end of 3 years?
b) How much interest did August pay for the loan?

B

8. Summer job Oscar invests $3200 he won for an essay contest. The investment pays 6.5% per annum, compounded monthly. How much will Oscar have after 18 months?

9. Stereo Zaineb is deciding whether to buy a stereo at $695 plus GST and PST now, or to invest the money to buy the stereo in a year. Her account pays 7.35% per annum, compounded monthly. If she can buy the stereo at the same price next year, how much would she save by investing the money?

10. GIC Anitha saved $8000 from her first job to buy a Guaranteed Investment Certificate, or GIC, at 5.75%, compounded annually. How much will the GIC be worth after 2 years?

11. Canada Savings Bond Marion is saving money for college. She has saved $1585 from her summer job. If she invests in a 2-year plan at her bank, she will earn 4.85% compounded semi-annually. If she buys Canada Savings Bonds, she will earn 6.3%, compounded annually, over 2 years.
a) Which is the better investment?
b) What is the difference in the interest?

12. RRSP Lila and Paul are investing $5000 in their Registered Retirement Savings Plans, RRSPs, at 6% per annum, compounded quarterly. How much will each investment be worth when they reach 60, if Lila makes the deposit on her 38th birthday and Paul makes the deposit on his 48th birthday?

13. Vacation Kate is saving for a vacation in the Bahamas. On January 1, she invested $750 at 8.2% per annum, compounded semi-annually. On July 1, she invested another $750 at the same rate. How much will she have from these investments on the next July 1st?

14. Travelling Noha is investing $2517 in an account compounded monthly. She wants to have $3000 in 3 years for a trip to Europe. What interest rate, to the nearest hundredth of a percent, compounded monthly, does she need?

15. Inquiry/Problem Solving Research current interest rates and interest rates 20 years ago.

a) Create a problem about a loan with these rates.

b) Trade problems with a classmate. Compare solutions.

16. Geometric sequence Use the formula for a geometric sequence to solve a question in this section. How does the solution with a geometric sequence compare with your original solution?

17. Comparing Simone is comparing an investment of $1550 for 2 years at 5%, compounded annually, 4.95%, compounded semi-annually, and 4.9%, compounded monthly.

a) Predict which rate would result in the greatest amount and which would result in the least amount.

b) Check your prediction and order the rates from least to greatest profit.

18. Application Joanne is investing $13 600 in an account that pays 8.2% per annum, compounded monthly. A friend has agreed to sell her a second-hand pick-up truck for her landscaping business for $15 900. She will need to pay the government GST on the price of the truck.

a) How long will it take, to the nearest month, to have enough in the account to buy the truck?

b) How many months sooner, to the nearest month, could she buy the truck if the interest rate went up to 9.3% per annum, compounded monthly?

19. Making financial decisions Suppose your parents invested $4000 on the day you were born, in an account earning interest at a rate of 5.9% per annum, compounded semi-annually. How much would the investment be worth on your 19th birthday? If the rate were compounded monthly instead of semi-annually, what difference would there be to the amount on your 19th birthday?

20. Communication a) At what interest rate, to the nearest hundredth of a percent, compounded monthly, would you double an investment of $50 000 in 10 years?

b) Explain your strategy for part a).

c) Use an example to illustrate your percent for part a).

C

21. Hammurabi's Code About 1800 B.C., Hammurabi, King of Babylonia, developed a system of laws with a code that permitted a maximum interest rate of 33% per annum, compounded annually, for loans of grain, and 20% per annum, compounded annually, for loans of silver.

a) To the nearest year, how long would it take for the value of a loan of grain to triple in value?

b) To the nearest year, how long would it take for the value of a loan of silver to triple in value?

22. True or false Classify each statement as true or false for investments earning compound interest. Justify your answer with an explanation or with examples.

a) As the compounding period increases, the amount increases.

b) A lower interest rate results in a higher interest.

c) A decrease in the length of time decreases the interest earned.

d) A lower interest rate can result in a higher amount if the number of compounding periods increases.

e) A lower interest rate always results in a higher amount if the number of compounding periods increases.

ACHIEVEMENT Check   Knowledge/Understanding   Thinking/Inquiry/Problem Solving   Communication   Application

Paul has deposited money in a savings account paying interest at 5% per annum, compounded semi-annually. If the money was invested at 5% per annum, compounded annually, would it amount to more or less in the same time period? If the money was invested at 5% per annum, compounded quarterly, would it amount to more or less in the same time period? Explain both in general and by example. Find an equivalent rate of interest to 5% per annum, compounded semi-annually if the new rate is compounded annually.

7.3 Investigation: Compound Interest, Geometric Sequences, and Exponential Growth

Banks and financial institutions offer a variety of accounts and investments. When you invest or borrow money, the interest rate can greatly affect the amount of interest you pay or receive. Researching and comparing interest rates is an important step in arranging a plan that suits you.

For **compound interest**, the money you invest or borrow is called the **principal**. The **interest rate** is the percent of the principal that is earned or paid as interest. The sum of the interest and the principal is the **amount**. The amount is the principal for the next compounding period.

CERTIFICATES OF DEPOSIT
WEEK OF 10 24
$ 1000 MIN. DEPOSIT

	RATE	ANNUAL YIELD
3 MONTHS	2.60%	2.63%
6 MONTHS	3.20%	3.25%
1 YEAR	3.94%	4.00%
18 MONTHS	4.67%	4.75%
2 YEARS	4.67%	4.75%
30 MONTH	4.67%	4.75%
3 YEARS	4.67%	4.75%
MONEY MARKET		2.50%
IRA		
	3 85	3 91
18 MOS	4 45	4 52
		4 52

Compound Interest

1. Complete a table like the following for $1000 invested with compound interest at a rate of 5% per annum, compounded annually for 10 years.

Number of years	Principal ($)	Interest rate	Interest ($)	Amount ($)
1	1000.00	0.05	50.00	1050.00
2	1050.00			1102.50
3				
4				

2. Plot the points in the table from question 1 with the time in years along the horizontal axis and the amount in dollars along the vertical axis. Join the points.

3. a) What kind of sequence is represented by the values in the Amount column? How do you know?
b) What is the value of a?
c) What is the value of r?
d) Use the values of a and r to write a formula for the nth term for the sequence in the Amount column. Use the table to check your formula for t_n.

4. a) Substitute $n = 1, 2, 3, \ldots, 10$ into your formula for t_n from part d) of question 3 to find $t_1, t_2, t_3, \ldots, t_{10}$.
b) Graph the values from part a) with n along the horizontal axis and t_n along the vertical axis. Join the points.

5. a) Use the formula for compound interest to find the amount after each year up to 10 years.
b) Graph the values from part a) with n along the horizontal axis and A along the vertical axis. Join the points.
c) Use your formula from part a) to write a formula with the variables x and y. Since n is represented along the horizontal axis, replace n with x. Since A is represented along the vertical axis, replace A with y.

6. Consider your graphs from questions 2, 4, and 5.
a) What shape are your graphs?
b) What is the y-intercept of each graph?
c) What does each y-intercept represent for the investment of $1000?
d) What do the points along the curve of each graph represent for the investment of $1000?

7. A function in which a variable is an exponent is called an **exponential function**. The graph of an exponential function models exponential growth.
a) Explain whether the formulas representing the investment are exponential functions.
b) Explain whether the graphs representing the investment are linear or non-linear.
c) Explain whether the graphs representing the investment model exponential growth.

8. a) Why is it reasonable to join the points of each graph?
b) Use one of your graphs to estimate the amount for a time from 1 to 10 years. Use your formula for t_n to check your estimate.
c) Use your formula for t_n to find the amount after 20 years.
d) Use your formula for A to find the amount after 20 years.
e) Compare the amounts from parts c) and d). Explain the result.

9. Suppose you invested $1000 earning compound interest at a rate of 6.75% per annum, compounded annually.
a) Predict the appearance of a graph representing this investment with the time in years along the horizontal axis and the amount in dollars along the vertical axis.

b) What amount would you have at the end of

i) 1 year?   **ii)** 2 years?   **iii)** 3 years?   **iv)** 4 years?   **v)** 5 years?

c) Name the sequence for the amounts in part b).

d) State the first term and the common ratio for the sequence.

e) Write the formula for t_n of the sequence. Check your formula for $n = 1$, 3, and 5.

f) Graph the formula from part e) with n along the horizontal axis and t_n along the vertical axis. Join the points.

g) Compare your graph with your prediction from part a).

10. Use your results to describe the relationship for compound interest, geometric sequences, and exponential growth. Refer to your table, graphs, and formulas in your explanation.

Comparing Compound Interest and Simple Interest

Two kinds of Canada Savings Bonds are regular Canada Savings Bonds and compound Canada Savings Bonds. Regular Canada Savings Bonds earn simple interest. Each year the interest is deposited into the owner's bank account or mailed to the owner. Compound Canada Savings Bonds earn compound interest, so the interest is reinvested and the whole amount of the bond is paid when it is cashed.

11. Copy and complete this table for a regular Canada Savings Bond at simple interest of 6% per annum, over 8 years.

Regular Savings Bond			
Year	Principal ($)	Interest ($)	Amount ($)
1	500.00	30.00	
2			560.00

12. Identify the type of sequence in the table. Write the formula for the nth term, t_n.

13. Use the formula to find t_1, t_5, and t_8.

14. Use the formula $A = P + Prt$ to find the amount after 5 years and 6 years.

15. Write the equation for the amount in the form $y = mx + b$.

16. Complete a table like the following for a compound Canada Savings Bond at an interest rate of 6% per annum, compounded annually, over 8 years.

Compound Savings Bond			
Year	Principal ($)	Interest ($)	Amount ($)
1	500.00	30.00	
2			561.80

17. Identify the sequence in the table. Write the formula for the nth term, t_n.

18. Use the formula to find t_1, t_5, and t_8.

19. Use the formula for compound interest to find the amount after 5 years and 6 years.

20. In the formula for t_n, replace n with x and t_n with y to write an equation with the variables x and y.

21. This graph compares amounts for the bonds in the above tables. Decide whether each graph represents linear or exponential growth. Explain your decision. Use the words simple interest, compound interest, arithmetic sequence, geometric sequence, linear growth, and exponential growth in your explanation.

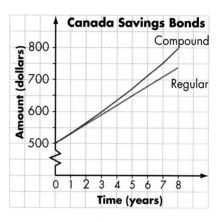

22. a) Describe the reasons someone might choose to buy regular Canada Savings Bonds.

b) Describe the reasons someone might choose to buy compound Canada Savings Bonds.

c) If you were investing $100 000, would you choose regular Canada Savings Bonds, compound Canada Savings Bonds, some of each, or neither? Explain how you would support this choice.

Web Connection
www.school.mcgrawhill.ca/resources/
To investigate the types of Canada Savings Bonds currently available and their interest rates, visit the above web site. Go to **Math Resources**, then to *MATHEMATICS 11*, to find out where to go next. Write a report about the current series of bonds available.

7.4 Present Value

Often, when people invest money, they have a goal for which they want a specific amount of money at a future date. Ramona hopes to buy a house in 3 years and estimates that a down payment of $70 000 should be sufficient. Ravi's grandparents want to have $150 000 for retirement in 9 years. The Elmview school orchestra plans a fundraising event to make an investment that will provide $8000 toward the cost of a trip to England in 2 years. Olivia compares investment options so she will have $26 000 in 4 years to start her own business.

The principal that is invested or borrowed is called the **present value** of the investment or loan. The present value that will result in a specific amount, with accumulated interest, can be calculated when the interest rate, the compounding period, and the time that interest is earned or paid are known.

INVESTIGATE & INQUIRE

Ramona hopes to buy a house in 3 years, and estimates that a down payment of $70 000 should be sufficient. She wants to know how much money to invest now, at 6.25% per annum, compounded annually, to obtain this down payment. The money Ramona invests now is the present value, or *PV*, of the investment.

The time line shows how the value of the investment increases each year.

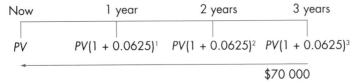

1. a) What is the interest rate per compounding period for Ramona's investment?
b) What expression on the time line represents the amount of the investment for each of the following?
i) now   **ii)** after 1 year   **iii)** after 2 years   **iv)** after 3 years

c) What is the amount of the investment after 3 years?

d) Write a formula with the amount from part c) and the equivalent expression from part b).

e) Calculate the present value of Ramona's investment, to the nearest cent.

2. a) Starting with the value of PV as the first term, list the values of Ramona's investment in order as a sequence.

b) What type of sequence do these values form? Explain.

c) For the sequence representing Ramona's investment, what is the value of

i) a?   **ii)** r?

d) Use the values of a and r to write the formula for the nth term, t_n, for the sequence.

e) Use t_n to calculate

i) t_1 **ii)** t_2 **iii)** t_3 **iv)** t_4

f) What do you notice about t_4?

3. To determine the formula for present value, let PV be the present value or first term, A be the amount at the end of an investment, i be the rate of interest per compounding period, and n be the number of compounding periods. Substitute into the formula for the nth term of the sequence for Ramona's investment. Use a negative exponent to express the formula as a product.

4. a) Write the formula for compound interest.

b) Since the present value, PV, of an investment is the principal, P, in the compound interest formula, substitute PV for P in the formula for compound interest.

c) Solve the formula for PV and use the exponent rules to write the resulting formula with a negative exponent.

5. Ramona wants to make another investment so that she will have $5000 in 5 years for renovations to the house she will buy. The interest rate for the investment is 7.5% per annum, compounded annually. Find the present value of the amount Ramona wants for renovations using

a) the formula for t_n

b) the formula for compound interest, remembering that the principal is the present value

c) the formula for PV

6. Use your results to describe the relationship between present value and the kind of sequence represented by Ramona's investment, and to describe the relationship between the formula for present value and the formula for compound interest.

The formula for a geometric sequence can be used to develop the formula for the present value, *PV*, of an investment or loan with compound interest.

$$PV = \frac{A}{(1 + i)^n} \text{ or } PV = A(1 + i)^{-n}$$

PV is the present value, *A* is the amount at the end of the investment, *i* is the rate of interest per compounding period, and *n* is the number of compounding periods.

EXAMPLE 1 Finding the Present Value With Interest Compounded Annually

Ravi's grandparents would like to have $150 000 when they retire in 9 years. How much should they invest now, at an interest rate of 5.75% per annum, compounded annually?

SOLUTION 1 Paper-and-Pencil Method

Use the formula for present value.

Ravi's grandparents want $150 000, so $A = 150\ 000$.
The interest rate is 5.75% per annum, compounded annually, so $i = 0.0575$.
The investment is for 9 years compounded annually, so $n = 9$.

$$PV = A(1 + i)^{-n}$$
Substitute known values: $= 150\ 000(1 + 0.0575)^{-9}$
Simplify: $= 150\ 000(1.0575)^{-9}$
$\doteq 90\ 691.77$

Ravi's grandparents should invest $90 691.77.

SOLUTION 2 Graphing-Calculator Method

Change the mode settings to 2 decimal places. From the Finance menu, choose TVM Solver.

Enter the known values.

The investment is for 9 years, so N = 9.

The interest rate is 5.75% per annum, so I = 5.75.

Ravi's grandparents want the amount, or future value, of $150 000, so FV = 150 000.

The interest is compounded annually, so C/Y = 1.

The payments are made at the end of each payment interval, so select END.

Move the cursor to PV to find the present value, and press ALPHA SOLVE. Since the investment is paid out, PV is negative.

Ravi's grandparents should invest $90 691.77.

EXAMPLE 2 Finding the Present Value With Interest Compounded Monthly

The Elmview school orchestra is planning a fundraising event to help finance a trip to England in 2 years. The orchestra plans to invest the money from this event for 24 months in an account with an interest rate of 4.5% per annum, compounded monthly. The orchestra hopes the money from this investment will provide $8000 toward the cost of the trip.

a) How much does the orchestra need to raise to achieve this goal?

b) How much interest should the orchestra earn to meet this goal?

SOLUTION 1 Paper-and-Pencil Method

a) Use the formula for present value.

The orchestra wants $8000, so $A = 8000$.

The interest rate is 4.5% per annum, compounded monthly, so divide the interest rate by 12.

$0.045 \div 12 = 0.003\ 75$, so $i = 0.003\ 75$.

The investment is for 2 years compounded monthly for $2 \times 12 = 24$ compounding periods, so $n = 24$.

$$PV = A(1 + i)^{-n}$$

Substitute known values: $\quad = 8000(1 + 0.00375)^{-24}$

Simplify: $\quad = 8000(1.00375)^{-24}$

$\quad \doteq 7312.68$

The orchestra needs to raise $7312.68 to achieve this goal.

b) The interest is the amount at the end of the investment less the amount invested.

$I = A - PV$

$\quad = 8000 - 7312.68$

$\quad = 687.32$

The orchestra should earn \$687.32 in interest to achieve this goal.

Solution 2 Graphing-Calculator Method

a) Change the mode settings to 2 decimal places. From the Finance menu, choose TVM Solver.

Enter the known values.

The interest rate is 4.5% per annum, so I = 4.5.

The orchestra wants the amount or future value of \$8000, so FV = 8000.

The interest is compounded monthly, so C/Y = 12.

The payments are made at the end of each payment interval, so select END.

Move the cursor to PV to find the present value, and press ALPHA SOLVE. Since the investment is paid out, PV is negative. The orchestra needs to raise \$7312.68 to achieve this goal.

b) The orchestra should earn \$687.32 in interest to achieve this goal, as shown in part b) of Solution 1.

Example 3 Comparing the Effects of Different Interest Rates

Olivia is making arrangements to start her own business in 4 years, and estimates she will need \$26 000. She compares investment plans: 5.1% per annum, compounded semi-annually, or 4.9% per annum, compounded quarterly. How much does Olivia need to invest for the better deal?

Solution 1 Paper-and-Pencil Method

Use the formula for present value.

For 5.1% per annum, compounded semi-annually:
Olivia wants \$26 000, so $A = 26\ 000$.
The interest rate is 5.1% per annum, compounded semi-annually.
$0.051 \div 2 = 0.0255$, so $i = 0.0255$.

The investment is for 4 years compounded semi-annually.
$4 \times 2 = 8$ compounding periods, so $n = 8$.

$$PV = A(1 + i)^{-n}$$

Substitute known values: $\quad = 26\ 000(1 + 0.0255)^{-8}$

Simplify: $\quad = 26\ 000(1.0255)^{-8}$

$\quad \doteq 21\ 256.32$

For 5.1% per annum, compounded semi-annually, Olivia must invest $21 256.32.

For 4.9% per annum, compounded quarterly:
Olivia wants $26 000, so $A = 26\ 000$.
The interest rate is 4.9% per annum, compounded quarterly.
$0.049 \div 4 = 0.012\ 25$, so $i = 0.012\ 25$.
The investment is for 4 years compounded quarterly.
$4 \times 4 = 16$ compounding periods, so $n = 16$.

$$PV = A(1 + i)^{-n}$$

Substitute known values: $\quad = 26\ 000(1 + 0.012\ 25)^{-16}$

Simplify: $\quad = 26\ 000(1.012\ 25)^{-16}$

$\quad \doteq 21\ 397.78$

For 4.9% per annum, compounded quarterly, Olivia must invest $21 397.78.

Olivia needs to invest $21 256.32 for the better deal.

SOLUTION 2 Graphing-Calculator Method

Change the mode settings to 2 decimal places. From the Finance menu, choose
TVM Solver.

Enter the known values for 5.1% per annum, compounded semi-annually.
The investment is for 4 years, so N = 4.
The interest rate is 5.1% per annum, so I = 5.1.
Olivia wants the amount or future value of $26 000, so FV = 26 000.
The interest is compounded semi-annually, so C/Y = 2.
The payments are made at the end of each payment interval,
so select END.

Move the cursor to PV to find the present value, and press
ALPHA SOLVE. Since the investment is paid out, PV is negative.

For 5.1% per annum, compounded semi-annually, Olivia must invest $21 256.32.

Enter the known values for 4.9% per annum, compounded quarterly.
The investment is for 4 years, so N = 4.
The interest rate is 4.9% per annum, so I = 4.9.
Olivia wants the amount or future value of $26 000, so FV = 26 000.
The interest is compounded quarterly, so C/Y = 4.
The payments are made at the end of each payment interval,
so select END.

Move the cursor to PV to find the present value, and press
ALPHA SOLVE. Since the present value is paid out, PV is negative.

For 4.9% per annum, compounded quarterly, Olivia must invest
$21 397.78.

Olivia needs to invest $21 256.32 for the better deal.

Key Concepts

- Present value is a financial application of geometric sequences.
- The formula for the present value, PV, of an investment or loan is

$$PV = \frac{A}{(1 + i)^n} \text{ or } PV = A(1 + i)^{-n},$$ where PV is the present value, A is the

amount at the end of the investment, i is the rate of interest per
compounding period, and n is the number of compounding periods.

Communicate Your Understanding

1. Would the present value of a $2000 investment for 2 years, at an interest
rate of 8%, compounded quarterly, be higher or lower than if the same
investment were compounded semi-annually? Give reasons for your answer.

2. Describe how you would find the present value to have $1000 in 3 years,
at an interest rate of 6% per annum, with the interest compounded

a) annually   b) semi-annually

c) quarterly   d) monthly

3. Explain why the formulas $PV = \dfrac{A}{(1 + i)^n}$ and $PV = A(1 + i)^{-n}$ result in

the same present value.

Practise

A

1. What is the rate of interest per compounding period for each investment?
a) 4.5% per annum, compounded semi-annually
b) 5.1% per annum, compounded quarterly
c) 8% per annum, compounded annually
d) 9% per annum, compounded monthly

2. An investment with a rate of interest of 6.25% per annum results in $12 000 in 7 years. What is the present value for each compounding period?
a) annually
b) semi-annually
c) quarterly
d) monthly

3. Consider the present values from question 2. State the relationship between the length of the compounding period and the present value of an investment.

4. What is the present value for each amount?
a) $9000 in 5 years, invested at 5.6% per annum, compounded semi-annually
b) $50 000 in 9 months, invested at 11% per annum, compounded quarterly
c) $100 000 in 3 years, invested at 3% per annum, compounded monthly
d) $78 840 in 9 years, invested at 4.8% per annum, compounded annually
e) $250 000 in a year, invested at 8.75% per annum, compounded quarterly

Apply, Solve, Communicate

5. To have $22 000 in 5 years, how much money must be invested today at 5.1% per annum, compounded semi-annually?

6. Buying a car How much money should Jessica put into an account paying 8% per annum, compounded semi-annually, to have $17 900 to buy a car in 2 years?

7. Education Sue wants to provide for her niece's education. How much should she invest on the day her niece is born to have $22 000 on her 18th birthday, if the money earns 7% per annum, compounded quarterly?

B

8. Down payment Samantha wants to have $40 000 available for a down payment on a house in 10 years. How much should she invest now at 6.25% per annum, compounded semi-annually?

9. Communication How much money should Gillian invest now to have $32 000 in 5 years if the money is invested at 8.25% per annum, compounded semi-annually?

10. Paying off a loan To pay his tuition, Nathan borrowed money at 3% per annum, compounded semi-annually. For this debt, he owes $5000 to be paid 2 years from now. He earned more at his summer job than he expected, so he wants to pay off the loan at its present value. How much would he pay?

11. A better deal Stephanie is choosing an investment plan that will pay $10 000 in 8 years. Does she need to invest more at 6.3% per annum, compounded quarterly, or at 6.3% per annum, compounded monthly? How much more does she need to invest?

12. Application An investment rate of 7.4% per annum, compounded annually, is advertised at a bank.
a) Predict an interest rate compounded quarterly that would result in a present value that is close to the present value for the advertised rate.
b) Use a graphing calculator to check your prediction with some examples.

13. Inquiry/Problem Solving Do you think banks prefer to advertise interest rates for loans that are compounded annually, semi-annually, monthly, or daily? Use the formula for present value, or examples of calculating the present value of investments, to support your answer.

14. a) Explain how the formula for present value is related to the formula for compound interest.
b) Describe how to use the formula for compound interest to check a solution for finding present value.
c) Follow your description from part b) to check a solution for a question in this section.
d) Explain why the formula for a geometric sequence can be used to solve a problem involving present value for compound interest.

C

15. Buying a van A mini-van sells for $32 000 plus GST and PST. A dealership predicts that in 3 years, the cost of the new model will increase by 15%. How much should you invest today at 7.25% per annum, compounded semi-annually, to buy the new model in 3 years?

16. Changing the rate Liam's goal is to save $20 000. What principal invested for 5 years at 6% per annum, compounded semi-annually, then for the next 3 years at 6.5% per annum, compounded quarterly, achieves this goal in 8 years?

17. Doubling Consider the formula for present value.
a) Predict the effect on the present value of an investment of doubling the amount wanted at the end of an investment.
b) Explain how the formula for present value justifies your prediction for part a).
c) Demonstrate your prediction with a few examples, using different compounding periods.
d) Does doubling the length of time for an investment affect the present value in the same way as doubling the amount wanted at the end of the investment? Justify your answer with an explanation or with a few examples.

18. Re-investing Marta is investing $6800 at an interest rate of 7% per annum, compounded quarterly, for 2 years. Then, she will invest the amount plus additional money at 6.5% per annum, compounded semi-annually, for 3 years. At the end of the second investment, she wants to have $15 000. How much extra must she invest for the second investment?

19. Formulating problems Research two interest rates that might be earned on an investment.
a) Use the rates to write a problem asking for the present value of $38 000.
b) Solve your problem.
c) Trade questions with a few classmates. Compare solutions.

PATTERN *Power*

The difference $10^2 - 10^1$ equals 90, when expressed in standard form.

1. Express each of the following differences in standard form.
a) $10^3 - 10^1$ **b)** $10^4 - 10^1$ **c)** $10^3 - 10^2$
d) $10^4 - 10^3$ **e)** $10^5 - 10^2$ **f)** $10^6 - 10^4$

2. For the difference $10^m - 10^n$ in standard form, where m and n are positive integers and $m > n$,
a) how many 9s are there?   **b)** how many 0s are there?

3. Use the pattern you found in question 2 to write each of the following differences in standard form.
a) $10^9 - 10^6$ **b)** $10^{11} - 10^{10}$ **c)** $10^{16} - 10^{12}$

4. Find and describe the pattern for the difference $10^a - 10^b$ in standard form, where a and b are negative integers and $a > b$.

5. Use the pattern you found in question 4 to write each of the following differences in standard form.
a) $10^{-5} - 10^{-6}$ **b)** $10^{-3} - 10^{-8}$ **c)** $10^{-1} - 10^{-7}$ **d)** $10^{-9} - 10^{-14}$

7.5 Amount of an Ordinary Annuity

Nigel is saving $700 each year for a trip. Rashid is saving $200 at the end of each month for university. Jeanine is depositing $875 at the end of each 3 months for 3 years. Marcel is saving for a home entertainment centre with equal payments at the end of every month for 18 months.

These investments by Nigel, Rashid, Jeanine, and Marcel are **annuities**. An **annuity** is a series of equal payments at regular intervals of time. For an **ordinary annuity**, each payment is made at the end of each **payment period**, or **payment interval**. A payment interval is the time between successive payments. The word annuity implies annual or yearly payments, however, payment intervals may be any length of time.

INVESTIGATE & INQUIRE

Last June 30, Nigel decided to save for a trip when he graduates. Starting next June 30, and for each of the following 3 years, he plans to deposit $700 into an account that pays 4.5% per annum, compounded annually. Complete the following to find the amount that Nigel will have when he makes the last deposit into this annuity.

The time line shows the value of each deposit at the time of Nigel's last deposit.

```
Now
June 30        June 30        June 30        June 30        June 30
  |              |              |              |              |
  └──────────────────────────────────────────────────────────→ 700(1 + 0.045)³
                 └───────────────────────────────────────────→ 700(1 + 0.045)²
                                └────────────────────────────→ 700(1 + 0.045)¹
                                               └─────────────→ 700
```

1. a) What is the interest rate per period?
b) How many years will the fourth deposit have been in the account? What will its value be at the time of the last deposit?

c) How many years will the third deposit have been in the account? What will its value be at the time of the last deposit?

d) How many years will the second deposit have been in the account? To the nearest tenth of a cent, what will its value be at the time of the last deposit?

e) How many years will the first deposit have been in the account? To the nearest tenth of a cent, what will its value be at the time of the last deposit?

f) When the last deposit is made, what is the sum of the deposits in Nigel's account, to the nearest cent?

2. a) What expression on the time line shows the value of each of the following deposits at the time of the last deposit?
i) the fourth deposit   ii) the third deposit
iii) the second deposit   iv) the first deposit

b) Start with the expression for the fourth deposit. Write the expressions from part a) in order as the terms of a series.

c) What type of series is the sum in part b)? Explain.

d) What is the first term of the series?

e) What is the common ratio of the series?

f) How many terms are in the series?

3. a) For the series representing Nigel's account, what is the value of
i) a?   ii) r?   iii) n?

b) Use the values of a, r, and n to write the formula for the sum of the series.

c) Use the formula to find the amount in Nigel's account at the time of the last deposit.

4. a) Nigel's investment is an ordinary annuity. To determine the formula for the amount of an ordinary annuity, let A be the total amount or sum of the series, R be the deposit, or payment, made at the end of each compounding period, i be the interest rate for each compounding period, and n be the number of compounding periods. Write the formula for the sum, A, of an ordinary annuity by substituting into the formula for the sum of the series for Nigel's investment. Simplify the denominator.

b) Use your formula from part a) to find the amount in Nigel's account at the time of the last deposit.

5. Use your results to describe the relationship between an ordinary annuity and the kind of series represented by the deposits into Nigel's account.

6. An ordinary annuity consists of a payment of $1000 made on July 20th in 6 successive years, with an interest rate of 7%, compounded annually. Use your formula from question 4 to find the amount of the ordinary annuity on the date of the last payment.

R is sometimes called the periodic rent.

The formula for the sum of a geometric series can be used to develop the formula for the amount, A, of an ordinary annuity.

$$A = \frac{R[(1+i)^n - 1]}{i}$$

A is the amount at the time of the last investment, R is the payment made at the end of each compounding period, n is the number of compounding periods, and i is the interest rate per compounding period.

EXAMPLE 1 Finding the Amount of an Annuity Compounded Annually

Starting in 4 months, Jeanine plans to deposit $875 on each July 31, October 31, January 31, and April 30, for 3 years, into an account. With an interest rate of 6%, compounded quarterly, how much will Jeanine have in her account when the last payment is made?

SOLUTION 1 Paper-and-Pencil Method

Use the formula for an ordinary annuity.

Each payment is $875, so $R = 875$.
There are 4 payments per year for 3 years, so $n = 12$.
The interest rate is 6% per annum, compounded quarterly
$0.06 \div 4 = 0.015$, so $i = 0.015$.

$$A = \frac{R[(1+i)^n - 1]}{i}$$

Substitute known values: $$= \frac{875[(1+0.015)^{12} - 1]}{0.015}$$

Simplify: $$= \frac{875[(1.015)^{12} - 1]}{0.015}$$

$$\doteq 11\ 411.06$$

Jeanine will have $11 411.06 in her account when the last payment is made.

MHR • Chapter 7

Solution 2 Graphing-Calculator Method

Change the mode settings to 2 decimal places. From the Finance menu, choose TVM Solver.

Enter the known values.
There are 4 payments a year for 3 years, so N = 12.
The interest rate is 6% per annum, so I = 6.
Since the deposits are paid out, PMT is negative. Each deposit is $875, so PMT = −875.
There are 4 payments per year, so P/Y = 4.
The interest is compounded quarterly, so C/Y = 4.
The payments are made at the end of each payment interval, so select END.

Move the cursor to FV to find the future value, and press ALPHA SOLVE.

Jeanine will have $11 411.06 in her account when the last payment is made.

Example 2 Finding the Monthly Payment of an Annuity

Marcel wants to buy a home entertainment centre that he sees priced at $3799 plus GST and PST. He plans to buy the centre in 18 months, and he assumes that the price will stay the same. He will make a payment into an account at the end of every month for 18 months. The interest rate is 9% per annum, compounded monthly.
a) How much will each of Marcel's payments be?
b) How much interest will Marcel have earned?

Solution 1 Paper-and-Pencil Method

a) Use the formula for an ordinary annuity.
With 7% GST and 8% PST, the cost of the home entertainment centre is 115% of $3799.
$1.15 \times 3799 = 4368.85$, so $A = 4368.85$.
There is a payment every month for 18 months, so $n = 18$.
The interest rate is 9% per annum, compounded monthly.
$0.09 \div 12$, so $i = 0.0075$.

$$A = \frac{R[(1 + i)^n - 1]}{i}$$

Substitute known values: $4368.85 = \dfrac{R[(1 + 0.0075)^{18} - 1]}{0.0075}$

Simplify: $4368.85 = \dfrac{R[(1.0075)^{18} - 1]}{0.0075}$

Solve for R: $R = \dfrac{4368.85(0.0075)}{1.0075^{18} - 1}$

Simplify: $\doteq 227.61$

Each of Marcel's payments will be $227.61.

b) Marcel will make 18 payments of $227.61.
$18 \times 227.61 = 4096.98$
Marcel's payments total $4096.98.
The interest is $4368.85 less $4096.98.
$4368.85 - 4096.98 = 271.87$
Marcel will have earned $271.87 in interest.

Solution 2 Graphing-Calculator Method

a) Change the mode settings to 2 decimal places. From the Finance menu, choose TVM Solver.
Enter the known values.
There is a payment every month for 18 months, so N = 18.
The interest rate is 9% per annum, so I = 9.
With 7% GST and 8% PST, the amount, or future value, is 115% of $3799, so FV = 1.15 × 3799.
There are 12 payments per year, so P/Y = 12.
The interest is compounded monthly, so C/Y = 12.
The payments are made at the end of each payment interval, so select END.
Move the cursor to PMT to find the payment, and press ALPHA SOLVE. Since the payment is paid out, PMT is negative.
Each of Marcel's payments will be $227.61.

Enter 1.15 × 3799 for FV and the calculator will determine the result when the cursor is moved to another variable.

b) Marcel will have earned $271.87 in interest, as shown in part b) of Solution 1.

Key Concepts

- An annuity is a sum of money paid as a series of regular payments. An ordinary annuity is an annuity for which each payment is made at the end of each compounding period.
- The amount of an annuity is a financial application of the sum of a geometric series.
- The formula for the amount, A, of an ordinary annuity is $A = \dfrac{R[(1 + i)^n - 1]}{i}$, where A is the amount at the time of the last payment, R is the payment made at the end of each compounding period, n is the number of compounding periods, and i is the interest rate per compounding period.

Communicate Your Understanding

1. At the end of each month for 2 years, $200 is deposited into an account with an interest rate of 6% per annum, compounded monthly. Describe how you would find the amount of money at the time of the last payment.

2. Starting in 6 months, $1000 is deposited twice a year into an account with an interest rate of 8% per annum, compounded semi-annually. Describe how you would find the amount in 4 years.

3. Explain how the relationship between an ordinary annuity and a geometric series is shown by their formulas.

4. Is the value for the time-value-of-money variable PMT on a graphing calculator positive or negative when you are calculating a deposit into an account? Explain.

Practise

A

1. Find the number of payments for each investment.
a) a deposit at the end of every year for 8 years
b) a deposit at the end of every month for 2 years
c) a deposit at the end of every 3 months for 15 months

2. For each interest rate, what value would you substitute for i in the annuity formula?
a) 4% per annum, compounded semi-annually
b) 3.25% per annum, compounded annually
c) 9% per annum, compounded quarterly
d) 6% per annum, compounded monthly

3. For each annuity, what value would you enter for P/Y using the TVM Solver on a graphing calculator?
a) a deposit at the end of each year for 3 years
b) a deposit at the end of each month for 5 years
c) a deposit at the end of each 3 months for 2 years

4. Find the amount of each investment.
a) $1500 at the end of each year for 6 years, at 7.1% per annum, compounded annually
b) $300 at the end of each 6 months, for 12 years at 4.95% per annum, compounded semi-annually

c) 36 monthly payments of $100 at the end of each month, for 3 years at 6% per annum, compounded monthly

5. Find the payment for each ordinary annuity.
a) 20 semi-annual payments giving an amount of $10 000 at 6% per annum, compounded semi-annually
b) an amount of $7000 with payments every 3 months for 5 years at 6.15% per annum, compounded quarterly
c) 36 monthly payments giving an amount of $4000 at 7% per annum, compounded monthly

Apply, Solve, Communicate

6. Saving Starting in 6 months, Lily will deposit $1000 into an account every March 1st and September 1st for 10 years. How much will she have at the time of the last payment if interest is 5.5% per annum, compounded semi-annually?

B

7. Bank account Marianna deposited $200 into her bank account at the end of each month for 8 months.
a) The account pays 2.9% per annum, compounded monthly. How much is in her account at the end of the 8 months?
b) If the amount deposited each month were doubled, how much would be in the account at the end of the 8 months?

8. Pensions David is planning to start saving for his pension by making the same deposit every 6 months starting 6 months after his 35th birthday. The plan he has chosen earns 9% per annum, compounded semi-annually. Use a graphing calculator to find how much each deposit must be to give him half a million dollars on his 60th birthday.

9. Savings account Tian opened a savings account on January 1st with a deposit of $150. The following July 1st, January 1st, and July 1st, he made 3 more deposits of $150 each. The account paid interest at 3.75% per annum, compounded semi-annually.
a) What amount did Tian have in his account at the end of the second year?
b) How much interest did Tian earn by the end of the second year?

10. Application Shannon plans to buy a new tractor in 3 years. Based on current prices, she predicts a new tractor, including taxes, will cost $90 000 in 3 years. How much should she invest at the end of each month at 9% per annum, compounded monthly, to have enough money to buy the tractor in 3 years?

11. Communication Describe a financial situation for paying an annuity as an investment. Research reasonable interest rates.
a) Write a problem with your information.
b) Solve the problem from part a).
c) Trade problems with a classmate. Compare solutions.

12. Making financial decisions At the end of grade 9, Rashid set up an annuity to save for university. At the end of each month, he invests $200 into an account bearing interest at 6.25% per annum, compounded monthly. How much money will he have at the end of grade 12?

C

13. Inquiry/Problem Solving Describe a strategy for deciding whether the result you calculate for an annuity is reasonable. Test your strategy.

14. Comparing solutions Pierre deposited $200 into his savings account at the end of each month for a year and a half. The account pays 3% per annum, compounded monthly.
a) Use the formula for the amount of an ordinary annuity to find how much Pierre has in his account at the end of 14 months.
b) Give the values of a, r, and n for solving the problem using the formula for the sum of a geometric series. Then, solve the problem with this formula.
c) Explain how your solutions in parts a) and b) are the same.

15. Saving for a car Nelida is purchasing a car for $30 000, including taxes. She hopes to replace it in 4 years with a similar car. She estimates that in 4 years, the price will have increased by 25%, and her present car will have lost 60% of its value. GST of 7% is charged on the difference between the trade-in value and the new car price. PST is charged on the price of the new car. She will start saving in 3 months, by making a payment every 3 months into an account paying 8% interest per annum, compounded quarterly.
a) How much should each payment be so that she can pay cash for the new car in 4 years?
b) Explain assumptions you made when finding the payment, and give your opinion about the importance of the assumptions.

7.6 Present Value of an Ordinary Annuity

A lottery advertising a $240 000 prize, in the form of $1000 every month for 20 years, might not have $240 000 when the winning ticket is drawn. Money could be invested in an ordinary annuity to pay $1000 every month for 20 years. The annuity would be earning interest that would be used for the prize of $1000 every month.

The amount of money invested now for an ordinary annuity is the **present value** of the annuity. The annuity provides equal payments at the end of each equal time interval. The present value can be calculated for a known interest rate, compounding period, number of payments, and amount of each payment.

INVESTIGATE & INQUIRE

Next year, Jane is going back to university for a Ph.D. in psychology. She wants to know how much money to deposit now into an account that pays 6% per annum, compounded annually, to provide a $5000 payment each year for 4 years, with the first payment due a year from now.

The time line shows the present value of each payment at the time of Jane's deposit.

	Now	1 year	2 years	3 years	4 years

$5000(1 + 0.06)^{-1}$

$5000(1 + 0.06)^{-2}$

$5000(1 + 0.06)^{-3}$

$5000(1 + 0.06)^{-4}$

1. a) What is the interest rate per period?

b) How many years will the first payment have been in the account? To the nearest tenth of a cent, how much must be deposited now for this payment?

c) How many years will the second payment have been in the account? To the nearest tenth of a cent, how much must be deposited now for this payment?

d) How many years will the third payment have been in the account? To the nearest tenth of a cent, how much must be deposited now for this payment?
e) How many years will the fourth payment have been in the account? To the nearest tenth of a cent, how much must be deposited now for this payment?
f) To the nearest cent, what is the sum of the amounts that must be deposited now for these payments?

2. a) Which expression on the time line shows the value of each of the following payments at the time of Jane's deposit?
i) the first payment   **ii)** the second payment
iii) the third payment   **iv)** the fourth payment
b) Start with the expression for the first payment. Write the expressions from part a) in order as the terms of a series.
c) What type of series is the sum in part b)? Explain.
d) What is the first term of the series?
e) What is the common ratio of the series?
f) How many terms are in the series?

3. a) For the series representing Jane's account, what is the value of
i) a?   **ii)** r?   **iii)** n?
b) Use the values of a, r, and n to write the formula for the sum of the series.
c) Use the formula to find the value of Jane's deposit.

4. a) Jane's investment is an ordinary annuity. To determine the formula for the present value of an ordinary annuity, let PV be the present value, R be the payment made at the end of each compounding period, i be the interest rate for each compounding period, and n be the number of compounding periods. Write the formula for the present value, PV, of an ordinary annuity by substituting into the formula for the sum of the series for Jane's investment. Use a negative exponent to express $\dfrac{1}{1+i}$ as $(1 + i)^{-1}$.
b) Use your formula from part a) to find the amount Jane must invest.

5. Use your results to describe the relationship between an ordinary annuity and the kind of series represented by the payments Jane plans to receive.

6. An ordinary annuity is invested at 4% per annum, compounded annually. The annuity is to pay $1000 a year for 3 years, starting in a year. Use your formula from question 4 to find the present value.

The formula for the sum of a geometric series can be used to develop the formula for the present value, PV, of an ordinary annuity.

$$PV = R\left[\frac{1 - (1 + i)^{-n}}{i}\right]$$

PV is the present value, R is the payment made at the end of each compounding period, n is the number of compounding periods, and i is the interest rate per compounding period.

EXAMPLE 1 Finding the Present Value Compounded Semi-Annually

Michael wants to make an investment so that he would receive $4000 every 6 months for 5 years, with the first payment due in 6 months. How much money should he invest now at 7% per annum, compounded semi-annually?

SOLUTION 1 Paper-and-Pencil Method

Use the formula for the present value of an ordinary annuity.

Each payment is $4000, so $R = 4000$.
There are 2 payments a year for 5 years, so $n = 10$.
The interest rate is 7% per annum, compounded semi-annually.
$0.07 \div 2 = 0.035$, so $i = 0.035$.

$$PV = R\left[\frac{1 - (1 + i)^{-n}}{i}\right]$$

Substitute known values:
$$= 4000\left[\frac{1 - (1 + 0.035)^{-10}}{0.035}\right]$$

Simplify:
$$= 4000\left[\frac{1 - (1.035)^{-10}}{0.035}\right]$$

$$\doteq 33\ 266.42$$

Michael should invest $33 266.42 now.

Solution 2 Graphing-Calculator Method

Change the mode settings to 2 decimal places. From the Finance menu, choose TVM Solver.

Enter the known values.
There are 2 payments a year for 5 years, so N = 10.
The interest rate is 7% per annum, so I = 7.
Michael will receive payments of $4000, so PMT = 4000.
There are 2 payments per year, so P/Y = 2.
The interest is compounded semi-annually, so C/Y = 2.
The payments are made at the end of each payment interval, so select END.

Move the cursor to PV to find the present value, and press ALPHA SOLVE. The investment is paid out, so PV is negative.

Michael should invest $33 266.42 now.

Example 2 Finding the Present Value Compounded Monthly

At the end of the season, the jeep Carmen is buying is offered with 0% financing for 48 months. The negotiated cost is $38 400, plus GST and PST. The total cost is divided into equal payments for 48 months, with the first payment on the date of purchase. Carmen will make the first payment, then invest an amount to provide the money each month for the remaining payments, which start in a month. How much must Carmen invest, at 6% per annum, compounded monthly, to have the amount each month for the payment?

Solution 1 Paper-and-Pencil Method

Use the formula for the present value of an ordinary annuity.

With 7% GST and 8% PST, the total cost is 115% of $38 400.
1.15 × 38 400 = 44 160
The total cost is $44 160.

Since there are 48 payments, divide the total cost by 48.
44 160 ÷ 48 = 920
Each payment is $920, so R = 920.

There are 12 payments a year for 4 years less the first payment already made, so $n = 47$.

The interest rate is 6% per annum, compounded monthly.

$0.06 \div 12 = 0.005$, so $i = 0.005$.

$$PV = R\left[\frac{1 - (1 + i)^{-n}}{i}\right]$$

Substitute known values: $\quad = 920\left[\frac{1 - (1 + 0.005)^{-47}}{0.005}\right]$

Simplify: $\quad = 920\left[\frac{1 - (1.005)^{-47}}{0.005}\right]$

$\quad \doteq 38\ 449.76$

Carmen must invest $38 449.76 at the time of the purchase.

Solution 2 Graphing-Calculator Method

Change the mode settings to 2 decimal places. From the Finance menu, choose TVM Solver.

Enter the known values.

There are 48 monthly payments, but the first payment is made at the time of the purchase, so N = 48 − 1.

The interest rate is 6% per annum, so I = 6.

With 7% GST and 8% PST, the total cost is 115% of $38 400. There are 48 payments, including the first at the time of the purchase. Carmen wants to receive the money for the payment each month, so

PMT = 1.15 × 38 4000 ÷ 48.

There are 12 payments per year, so P/Y = 12.

The interest is compounded monthly, so C/Y = 12.

The payments are made at the end of each payment interval, so select END.

Enter 1.15 × 38 4000 ÷ 48 for PMT and the calculator will determine the result when the cursor is moved to another variable.

Move the cursor to PV to find the present value, and press ALPHA SOLVE. Since Carmen is paying out the present value, PV is negative.

Carmen must invest $38 449.76 at the time of the purchase.

EXAMPLE 3 Finding the Value of the Payments

To provide an annual scholarship for 25 years, a donation of $50 000 is invested in an account for a scholarship that will start a year after the investment is made. If the money is invested at 5.5% per annum, compounded annually, how much is each scholarship?

SOLUTION 1 Paper-and-Pencil Method

Use the formula for the present value of an ordinary annuity.

The donation is $50 000, so $PV = 50\ 000$.
There is 1 payment per year for 25 years, so $n = 25$.
The interest rate is 5.5% per annum, compounded annually, so $i = 0.055$.

$$PV = R\left[\frac{1 - (1 + i)^{-n}}{i}\right]$$

Substitute known values: $50\ 000 = R\left[\dfrac{1 - (1 + 0.055)^{-25}}{0.055}\right]$

Simplify: $\qquad\qquad 50\ 000 = R\left[\dfrac{1 - (1.055)^{-25}}{0.055}\right]$

Solve for R: $\qquad\qquad R \doteq 3727.47$

Each scholarship is $3727.47.

SOLUTION 2 Graphing-Calculator Method

Change the mode settings to 2 decimal places. From the Finance menu, choose TVM Solver.

Enter the known values.
There is 1 payment per year for 25 years, so N = 25.
The interest rate is 5.5% per annum, so I = 5.5.
Since the donation is paid out, PV is negative. The donation is $50 000, so PV = −50 000.
The scholarship is annual, so P/Y = 1.
The interest is compounded annually, so C/Y = 1.
The payments are made at the end of each payment interval, so select END.

Move the cursor to PMT to find the payment, and press ALPHA SOLVE.
Each scholarship is $3727.47.

Key Concepts

- The present value of an annuity is the principal that must be invested now to provide a given series of equal payments at equal intervals of time.
- The present value of an annuity is a financial application of the sum of a geometric series.
- The formula for the present value, PV, of an ordinary annuity is $PV = R\left[\dfrac{1 - (1 + i)^{-n}}{i}\right]$, where PV is the present value, R is the payment made at the end of each compounding period, n is the number of compounding periods, and i is the interest rate per compounding period.

Communicate Your Understanding

1. Explain what n, R, and i represent in the formula for the present value of an ordinary annuity.

2. Describe how you would find how much money must be invested now at 5% per annum, compounded semi-annually, to provide 10 payments of $500 every 6 months, starting in half a year.

3. Describe how you would find 24 monthly payments resulting from an investment of $78 000 at 10% per annum, compounded monthly, with the first payment to be received a month from the time of the investment.

Practise

A

1. Jack deposited $11 718.83 into an account with an interest rate of 4.4% per annum, compounded monthly. He will receive $1000 each month, starting a month after the deposit.
a) What is the present value?
b) What is the compounding period?
c) What is the interest rate per annum?
d) What is the interest rate per compounding period?
e) What is the payment?

2. Find the present value of each investment, if the payments start in a year. The interest rate is 4.1% per annum, compounded annually.
a) 6 annual payments of $3000
b) 9 annual payments of $1500

3. For each investment plan, how much must be invested today at 6.25% per annum, compounded monthly, if the first payment is made a month from today?
a) 36 monthly payments of $250
b) 10 monthly payments of $900

4. For each investment, how much must be deposited now to receive 12 payments of $1000?

a) 6% per annum, compounded annually, with annual payments, starting in a year

b) 6% per annum, compounded semi-annually, with a payment every 6 months, starting in 6 months

c) 6% per annum, compounded quarterly, with a payment every 3 months, starting in 3 months

d) 6% per annum, compounded monthly, with monthly payments, starting in a month

5. Consider the present values from question 4. What happens to the present value of an ordinary annuity as the length of the compounding period decreases? Explain why this happens.

6. For each investment, $25 000 is deposited in an account. How much is each payment?

a) 8% per annum, compounded annually, with annual payments starting in a year

b) 8% per annum, compounded semi-annually, with payments every 6 months starting in 6 months

c) 8% per annum, compounded quarterly, with payment every 3 months starting in 3 months

d) 8% per annum, compounded monthly, with monthly payments starting in a month

7. Consider the payments for question 6. What happens to the payments as the length of the compounding periods decreases? Explain why this happens.

Apply, Solve, Communicate

8. Camp expenses Shelley is on a committee opening a camp next year. The committee estimates that the camp will need an annual supplement of $7000 at the beginning of each year for the first 5 years. The committee decides to invest money in a plan with an interest rate of 5.3% per annum, compounded annually, to cover the cost of supplements. How much should be invested now?

B

9. Lottery A lottery to raise funds for a hospital is advertising a $240 000 prize. The winner will receive $1000 every month for 20 years, starting a year from now.

a) If the interest rate is 8.9% per annum, compounded annually, how much must be invested now to have the money to pay this prize?

b) If the lottery were able to negotiate an interest rate of 9.3% per annum compounded annually, how much would be invested now?

10. Application When Jodi's grandmother retired, she decided to invest some money so she would receive $10 000 every 6 months for 10 years, starting in half a year. Her investment plan pays interest at 5.9% per annum, compounded semi-annually.
a) How much must she invest?
b) Draw a time line to illustrate the investment. Explain how the time line supports your answer.

11. Insurance Cora received an insurance settlement of $80 000, which she invested at 5.2% per annum, compounded monthly, to provide a payment each month for 10 years, starting next month.
a) How much will each payment be?
b) How much did Cora's insurance settlement give her altogether?

12. University A year before Morley started university, her parents invested $22 000 at 4.9% per annum, compounded annually. When she started university, she received the same amount of money every year for 4 years.
a) How much was each payment?
b) Use your research skills to find how close each payment is to the cost of tuition each year.

13. Triathlon Wray bought a bicycle for $2500, plus GST and PST, to compete in triathlons. He arranged to make a payment to the store at the end of every month for 2 years. The store is charging 11% interest per annum, compounded monthly.
a) How much is each payment?
b) How much interest is Wray paying?

14. Making financial decisions Josh has sold his business and plans to retire. He wants to deposit enough of the money from the sale to receive a payment of $4000 a month for 20 years, starting in a month. He is considering an investment with an interest rate of 5.4% per annum, compounded monthly.
a) How much does he need to deposit now for this investment?
b) What are some assumptions made in this plan?

C

15. Communication Describe the difference between your calculations for the ordinary annuities in this section and the ordinary annuities in the preceding section. Include at least one question from each section in your description.

16. Scholarship Marvin's graduating class raised $2198.74 to establish a fund for a scholarship of $200, starting the next year, for the student who contributed most to the school. The money is invested at 4.8% per annum, compounded annually.
a) For how many years will this scholarship be awarded?
b) Explain why this problem would be difficult to solve without a graphing calculator.

17. Inquiry/Problem Solving Natalie's family has decided that in 3 months, they will start depositing $160 every 3 months for 3 years. The account will earn interest at 5.7% per annum, compounded quarterly. Then, 3 months after the last deposit, they plan to withdraw money every 3 months for 24 equal payments for music lessons for Natalie. How much will each withdrawal be?

18. Estimation Choose one question in this section.
a) Estimate to decide whether your answer is reasonable.
b) Explain your estimation strategy to a few classmates. Then, write your opinion about whether it made sense to them.

19. True or false Classify each statement as true or false for the present value of an ordinary annuity. Justify your answer with an explanation or examples.
a) As the interest rate increases, but the compounding period does not change, the present value increases.
b) As the compounding period decreases, the amount of money that needs to be invested decreases.
c) As the payment received each month increases, the present value decreases.

20. Research the cost of arranging an annuity. How might this influence your decision to get an annuity?

ACHIEVEMENT Check   Knowledge/Understanding Thinking/Inquiry/Problem Solving Communication Application

Jonah signed a loan contract that requires a down payment of $1500 and payments of $200 a month for 10 years. The interest rate is 9% per annum, compounded monthly.
a) What is the cash value of the contract?
b) If Jonah missed the first 8 payments, what must he pay at the time the ninth payment is due to bring himself up to date?
c) If Jonah missed the first 8 payments, what must he pay at the time the ninth payment is due to pay off the contract completely?

7.7 Technology: Amortization Tables and Spreadsheets

Generally, people must borrow money when they purchase a car, house, or condominium, so they arrange a **loan** or **mortgage**. Loans and mortgages are agreements between a money lender and a borrower to finance a purchase. The agreement usually requires the borrower to repay the loan in equal payments at equal time intervals, with payments that include both principal and interest.

To **amortize** a mortgage means to repay the mortgage over a given period of time in equal payments at regular intervals. The period of time is known as the **amortization period**. The **term** of a mortgage is the length of time the mortgage agreement is in effect.

An **amortization table** can be created to show how much of the principal remains after each payment, and how much of each payment is interest or principal.

Suppose you borrow $10 000 at an interest rate of 6% per annum, compounded monthly, and agree to pay back the loan with equal payments at one month intervals over one year. You need to know the monthly payment and the principal that has been repaid at any time during the year.

You can use a graphing calculator to find the monthly payments for the loan and a spreadsheet to keep track of the amounts paid and owed. Change the mode settings to 2 decimal places. From the Finance menu, choose TVM Solver.

Enter the known values.
There are 12 payments a year, so N = 12.
The interest rate is 6% per annum, so I = 6.

The loan is for $10 000, so PV = 10 000.
A payment is made each month, so P/Y = 12.
The interest is compounded monthly, so C/Y = 12.
The payments are made at the end of each payment interval,
so select END.

Move the cursor to PMT to find the payment for the loan,
and press ALPHA SOLVE. Since the payments for the loan
are paid out, PMT is negative.

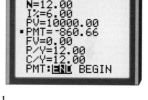

The payment for the loan is $860.66.

Use a spreadsheet to follow the progress of repaying the loan with an
amortization table. The following spreadsheet shows data for the $10 000
loan at the top left, and formulas for calculations with the loan. Then,
formulas are copied down into other rows.

	A	B	C	D	E	F
1	10000.00	Loan ($)				
2	0.06	Rate/Year				
3	=A2/12	Rate/Month				
4	860.66	Payment ($)				
5						
6					Principal	Remaining
7	Month	Principal ($)	Interest ($)	Payment ($)	Reduction ($)	Principal ($)
8	1	=A1	=B8*A$3	=A$4	=D8-C8	=B8-E8
9	=A8+1	=F8	=B9*A$3	=A$4	=D9-C9	=B9-E9
10	=A9+1	=F9	=B10*A$3	=A$4	=D10-C10	=B10-E10

For details of how
to create the
spreadsheets used
in this chapter, see
the Microsoft®
Excel and Corel®
Quattro® Pro
sections of
Appendix C.

1. a) Explain the value entered in each of these cells.
A1   A2   A4   A8
b) Explain the formula entered in each of these cells.
A3   A9   B8   B9   E8   F8
c) Make a generalization about what happens to the formulas in the cells
listed in part b) when they are copied into other rows.
d) Identify the cells with $ in the formulas.
e) Make a generalization about what happens to the formulas with $ when
they are copied into other rows.

2. The following spreadsheet shows the results of the calculations from the
formulas. The results of the calculations are seen on a spreadsheet. You can
see a formula for an individual cell by selecting the cell.

	A	B	C	D	E	F
1	10000.00	Loan ($)				
2	0.06	Rate/Year				
3	0.005	Rate/Month				
4	860.66	Payment ($)				
5						
6					Principal	Remaining
7	Month	Principal ($)	Interest ($)	Payment ($)	Reduction ($)	Principal ($)
8	1	10000.00	50.00	860.66	810.66	9189.34
9	2	9189.34	45.95	860.66	814.71	8374.63
10	3	8374.63	41.87	860.66	818.79	7555.84

a) For each of the following cells, explain how the formula in the first spreadsheet results in the value in the second spreadsheet.

A3   A9   B8   B9   E8   F8
C8   C9   C10   D8   D9   D10

b) Make a generalization about how the values in cells where formulas are copied are related.

c) Make a generalization about what happens to the values in cells where formulas with $ are copied.

3. a) Create the spreadsheet from question 1. Check your results with the spreadsheet from question 2.

b) Predict for how many rows you should copy the formula for this loan. Then, copy the formula down, and compare the results with your prediction.

c) Format the cells to express amounts of money to 2 decimal places.

4. a) Choose various months in the spreadsheet, and state how much of the principal has been repaid for each month. Include the row for the 12th month.

b) Explain whether the results in your spreadsheet are what you would expect.

A **mortgage** is a loan secured by real estate. Generally, a mortgage is arranged to finance the purchase of property. However, a mortgage can be set up for someone to borrow money for any reason, using property as security for the loan.

By Canadian law, the interest rate on mortgages is compounded semi-annually. Although other lengths of time can be arranged, most mortgages are paid monthly. This is contrary to the investments and loans in previous sections of this chapter, where the compounding period and the payment period are the same. Because of this difference, some calculations for mortgages are different.

Suppose you arrange a mortgage of $112 500 on a house, with an interest rate of 7.5% per annum, compounded semi-annually, and agree to make equal monthly payments. If you amortize the mortgage over 25 years, this means that the amount paid each month is the amount that would pay off the loan in 25 years if the mortgage continued. Generally, mortgages are for a term such as 3, 4, or 5 years. Planning a mortgage would require knowing the monthly payments, and you would probably want to know how much of the principal, the $112 500, had been repaid at various times during the mortgage.

Use a graphing calculator to find the monthly payments for the mortgage, and the monthly interest rate. The monthly interest rate, or rate per month, is calculated differently for a mortgage than for a loan because the compounding period and the payment period of a mortgage are not the same. A spreadsheet can show amounts paid and owed.

Choose the mode settings to 2 decimal places. From the Finance menu, choose TVM Solver.

Enter the known values.
There are 12 payments a year for 25 years, so N = 25 × 12.
The interest rate is 7.5% per annum, so I = 7.5.
The mortgage is $112 500, so PV = 112 500.
A payment is made each month, so P/Y = 12.
The interest is compounded semi-annually, so C/Y = 2.
The payments are made at the end of each payment interval, so select END.

For a mortgage, the value of P/Y is 12, but the value of C/Y is 2. When you store a value for P/Y, C/Y automatically changes to match P/Y, so you must go back to C/Y and enter 2.

Move the cursor to PMT to find the payment for the loan, and press ALPHA SOLVE. Since the payment is paid out, PMT is negative.

The monthly payment for this mortgage is $823.00.

GRAPHING CALCULATOR
```
N=300.00
I%=7.50
PV=112500.00
•PMT=-823.00
FV=0.00
P/Y=12.00
C/Y=2.00
PMT:END BEGIN
```

For finding the monthly interest rate, change the mode settings to 9 decimal places. From the Finance menu, choose TVM Solver. Enter values for $1 for 1 month of this mortgage.

Enter the known values.
The interest rate per month is being calculated, so N = 1.
The interest rate is 7.5% per annum, so I = 7.5.
Use the negative amount $1 to find the value per dollar paid out, so PV = −1.

The payments are monthly, so P/Y = 12.
The interest is compounded semi-annually, so C/Y = 2.
The payments are made at the end of each payment interval, so select END.

Move the cursor to FV to find the future value of $1, and press ALPHA SOLVE.

The future value of $1 is $1.006 154 524.

Since the future value is the amount with compound interest for the present value, or investment of $1, $A = P + i$, where i represents the monthly interest rate.

Amount equals principal plus interest. $\qquad A = P + i$

Substitute known values: $\qquad 1.006\ 154\ 524 = 1 + i$

Isolate i: $\qquad\qquad\qquad\qquad i = 0.006\ 154\ 524$

The interest rate per month is 0.006 154 524, or 0.615 452 4%.

Use a spreadsheet to follow the progress of repaying the mortgage with an amortization table. The following spreadsheet shows data for the $112 500 mortgage at the top left, and formulas for calculations for the mortgage. Then, formulas are copied down into other rows.

	A	B	C	D	E	F
1	112500.00	Mortgage ($)				
2	0.075	Rate/Year				
3	0.006154524	Rate/Month				
4	823.00	Payment ($)				
5						
6					Principal	Remaining
7	Month	Principal ($)	Interest ($)	Payment ($)	Reduction ($)	Principal ($)
8	1	=A1	=B8*A$3	=A$4	=D8-C8	=B8-E8
9	=A8+1	=F8	=B9*A$3	=A$4	=D9-C9	=B9-E9
10	=A9+1	=F9	=B10*A$3	=A$4	=D10-C10	=B10-E10

5. a) Explain the value entered in each of these cells.
A1   A2   A4   A8
b) For each of these cells, explain the formula and predict the value that it will calculate.
A9   B8   B9   E8   F8
c) Make a generalization about what happens to the formulas in cells listed in part b) where they are copied into other rows.
d) Make a prediction about what will happen for the values in cells where formulas listed in part b) are copied.

e) Make a generalization about what happens to the formulas with $ as they are copied into other rows.

f) Make a prediction about what will happen to the values in cells where formulas with $ are copied.

6. a) Create the spreadsheet shown on page 548.

b) Predict for how many rows you should copy the formulas for this mortgage. Then, copy the formulas down, and compare the results with your prediction.

c) Format the cells to express amounts of money to 2 decimal places.

7. a) Choose various months in the spreadsheet, and state how much of the principal has been repaid. Include the row for 25 years.

b) How close is the last value in the Remaining Principal column to $0.00? Because the monthly payment of $823.00 was obtained by rounding to the nearest cent, the values rounded make a difference over 25 years. Explain why.

c) What do you think would happen in the spreadsheet if all the decimal values for the monthly payment were used?

8. Explain whether the results in your spreadsheet are what you would expect.

EXAMPLE 1 Finding the Length of Time to Pay Off a Loan

Gurjeet is spending $1875.25 on furniture for her apartment. Her uncle offers to finance this purchase, charging an interest rate of 4% per annum, compounded monthly. Gurjeet agrees to pay $125 per month for the loan payment.

a) Use a spreadsheet to find how long it takes Gurjeet to pay off the loan.

b) What two methods can you use to find Gurjeet's final payment? How can you explain the difference in the results obtained by these two methods?

SOLUTION

a) Enter the data in a spreadsheet.

Gurjeet's loan is $1875.25, so enter 1875.25 in cell A1.

The interest rate per annum is 4%, so enter 0.04 in cell A2.

The interest rate per month is the annual interest rate divided by the number of months in a year.

$0.04 \div 12 \doteq 0.003\ 333\ 333$ The repeating decimal for the monthly interest rate is rounded to 9 decimal places to avoid an error in the spreadsheet because of rounding values used for calculations.

Gurjeet's monthly payment is $125, so enter 125.00 in cell A4.

Use formulas shown in the spreadsheets in the Investigate & Inquire.
Copy the formulas down until the rows show that the principal is repaid.

	A	B	C	D	E	F
1	1875.25	Loan ($)		Gurjeet's Loan		
2	0.04	Rate/Year				
3	0.003333333	Rate/Month				
4	125.00	Payment ($)				
5						
6					Principal	Remaining
7	Month	Principal ($)	Interest ($)	Payment ($)	Reduction ($)	Principal ($)
8	1	1875.25	6.25	125.00	118.75	1756.50
9	2	1756.50	5.86	125.00	119.14	1637.36
10	3	1637.36	5.46	125.00	119.54	1517.81
22	15	176.26	0.59	125.00	124.41	51.84
23	16	51.84	0.17	125.00	124.83	-72.98

The spreadsheet shows part of the amortization table for the mortgage.

The row for 16 months is the last row that shows principal owing.
The loan will be repaid in 16 months.

b) One method is to find the sum of the principal and the interest on this principal in the 16th month. So, the final payment is the sum of the values in cells B23 and C23.

$51.84 + 0.17 = 52.01$

Using the sum of the principal and the interest on this principal for the 16th month, Gurjeet's final payment is $52.01.

Another method is to find the difference between Gurjeet's usual payment of $125.00 and the negative remaining principal for the 16th month.

$125.00 - 72.98 = 52.02$

Using the difference between Gurjeet's usual payment and the negative remaining principal for the 16th month, Gurjeet's final payment is $52.02.
The difference in the results obtained by these two methods occurs because of rounding for calculations in the spreadsheet.

EXAMPLE 2 Comparing the Effects of Changing the Amortization Period

Robyn and Jonathon are working with a bank manager to arrange a mortgage of $200 000 to buy their new home. The bank will charge interest on their mortgage at 8.36% per annum, compounded semi-annually.

a) For a mortgage amortized over 25 years, find

i) the monthly payment
ii) the monthly interest rate
iii) the total interest
iv) the total amount of the payments

b) With the mortgage amortized over 25 years, how much do Robyn and Jonathon owe after 15 years?

c) For a mortgage amortized over 20 years, find the

i) monthly payments   **ii)** monthly interest rate

iii) total interest   **iv)** total amount of the payments

d) Compare Robyn and Jonathon's cost of amortizing their mortgage over 25 years with the cost of amortizing over 20 years.

SOLUTION

a) i) Use a graphing calculator to find the monthly payments. Change the mode settings to 2 decimal places. From the Finance menu, choose TVM Solver. Enter the known values.

There are 12 payments a year for 25 years, so N = 25 × 12.

The interest rate is 8.36% per annum, so I = 8.36.

The mortgage is $200 000, so PV = 200 000.

A payment is made each month, so P/Y = 12.

The interest is compounded semi-annually, so C/Y = 2.

The payments are made at the end of each payment interval, so select END.

Move the cursor to PMT to find the mortgage payment, and press ALPHA SOLVE. Since the mortgage payment is paid out, PMT is negative.

The monthly payments are $1572.63.

ii) Use a graphing calculator to find the interest rate per month. Change the mode settings to 9 decimal places. From the Finance menu, choose TVM Solver.

Enter the known values.

The interest rate per month is being calculated, so N = 1.

The interest rate is 8.36% per annum, so I = 8.36.

Use the negative amount $1 to find the value per dollar paid out, so PV = −1.

A payment is made each month, so P/Y = 12.

The interest is compounded semi-annually, so C/Y = 2.

The payments are made at the end of each payment interval, so select END.

Move the cursor to FV to find the future value of $1, and press ALPHA SOLVE.

The future value of $1 is $1.006 848 341.

Since the future value is the compound amount for the present value, or the investment for $1, $A = P + i$, where i represents the monthly interest rate.

$$A = P + i$$

Substitute known values: $\quad 1.006\ 848\ 341 = 1 + i$

Solve for i: $\qquad\qquad\qquad\quad i = 0.006\ 848\ 341$

The monthly interest rate is 0.006 848 341, or 0.684 834 1%.

iii) Enter the data in cells A1 to A4. Use formulas shown in the spreadsheet in the Investigate & Inquire. Copy the formulas down until the rows show the payments for 25 years.

	A	B	C	D	E	F
1	200000.00	Mortgage ($)	Robyn and Jonathon's 25-year mortgage			
2	0.0836	Rate/Year				
3	0.006848341	Rate/Month				
4	1572.63	Payment ($)				
5						
6					Principal	Remaining
7	Month	Principal ($)	Interest ($)	Payment ($)	Reduction ($)	Principal ($)
8	1	200000.00	1369.67	1572.63	202.96	199797.04
9	2	199797.04	1368.28	1572.63	204.35	199592.69
10	3	199592.69	1366.88	1572.63	205.75	199386.94
11	4	199386.94	1365.47	1572.63	207.16	199179.77
177	170	135717.33	929.44	1572.63	643.19	135074.14
178	171	135074.14	925.03	1572.63	647.60	134426.55
179	172	134426.55	920.60	1572.63	652.03	133774.51
180	173	133774.51	916.13	1572.63	656.50	133118.02
181	174	133118.02	911.64	1572.63	660.99	132457.03
182	175	132457.03	907.11	1572.63	665.52	131791.51
183	176	131791.51	902.55	1572.63	670.08	131121.43
303	296	7701.39	52.74	1572.63	1519.89	6181.50
304	297	6181.50	42.33	1572.63	1530.30	4651.21
305	298	4651.21	31.85	1572.63	1540.78	3110.43
306	299	3110.43	21.30	1572.63	1551.33	1559.10
307	300	1559.10	10.68	1572.63	1561.95	-2.85
308		Totals	271786.15	471789.00	200002.85	

In cell C308, enter the formula for the sum of the values in cells C8 to C307.

The sum in cell C308 shows that the total interest is $271 786.15.

iv) In cell D308, enter the formula for the sum of the values in cells D8 to D307. The sum in cell D308 shows that the total amount of the payments is $471 789.00.

The negative value in cell F307 indicates that Robyn and Jonathon overpaid by $2.85. If, however, the full decimal accuracy were entered the final remaining principal would be $0.00.

Variations because of rounding are also found by entering the formula for the sum of the values in cells E8 to E307 into cell E308. The sum for the principal reduction is $2.85 greater than the mortgage of $200 000.

b) Spreadsheet Method

Find the number of months in 15 years.

$15 \times 12 = 180$

The row in the amortization table for 180 months shows the remaining principal $128 394.92

After 15 years, Robyn and Jonathon owe $128 394.92.

Graphing-Calculator Method

Change the mode settings to 2 decimal places. From the Finance menu, choose TVM Solver.

Enter the known values.
After 15 years, 10 years of the 25-year amortization period remain with 12 payments per year, so N = 10 × 12.
The interest rate is 8.36% per annum, so I = 8.36.
Since the mortgage payment is paid out, PMT is negative. The monthly payments are $1572.63, so PMT = −1572.63.
A payment is made each month, so P/Y = 12.
The interest is compounded semi-annually, so C/Y = 2.
The payments are made at the end of each payment interval, so select END.

Move the cursor to PV to find the present value, and press ALPHA SOLVE.

After 15 years, Robyn and Jonathon owe $128 396.17.

The difference between the results for the spreadsheet method and the graphing-calculator method occurs because of rounding.

c) i) Use a graphing calculator to find the monthly payments. Change the mode settings to 2 decimal places. From the Finance menu, choose TVM Solver.

Enter the known values.
There are 12 payments a year for 20 years, so N = 20 × 12.
The interest rate is 8.36% per annum, so I = 8.36.
The mortgage is $200 000, so PV = 200 000.
A payment is made each month, so P/Y = 12.
The interest is compounded semi-annually, so C/Y = 2.
The payments are made at the end of each payment interval, so select END.

Move the cursor to PMT to find the mortgage payment, and press ALPHA SOLVE. Since the mortgage payment is paid out, PMT is negative.

The monthly payments are $1700.12.

ii) Since the monthly rate for amortizing the mortgage over 20 years is the same as for amortizing the mortgage over 25 years, use the monthly interest rate from part a) ii).

So, the monthly interest rate is 0.006 848 341, or 0.684 834 1%.

iii) Create a spreadsheet like this. Enter the formulas for sums of columns in row 248.

	A	B	C	D	E	F
1	200000.00	Mortgage ($)	Robyn and Jonathon's 20-year mortgage			
2	0.0836	Rate/Year				
3	0.006848341	Rate/Month				
4	1700.12	Payment ($)				
5						
6					Principal	Remaining
7	Month	Principal ($)	Interest ($)	Payment ($)	Reduction ($)	Principal ($)
8	1	200000.00	1369.67	1700.12	330.45	199669.55
9	2	199669.55	1367.41	1700.12	332.71	199336.83
10	3	199336.83	1365.13	1700.12	334.99	199001.84
11	4	199001.84	1362.82	1700.12	337.30	198664.55
244	237	6064.15	43.78	1700.12	1654.34	5029.80
245	238	5029.80	34.45	1700.12	1665.67	3364.13
246	239	3364.13	23.04	1700.12	1677.08	1687.05
247	240	1687.05	11.55	1700.12	1688.57	-1.52
248		Totals	208027.28	408028.80	200001.52	

The spreadsheet shows the first 3 months and the last 3 months of the amortization schedule for their mortgage.

The sum in cell C248 shows that the total interest is $208 027.28.

iv) The sum in cell D248 shows that the total amount of the payments is $408 028.80.

The negative value in cell F247 indicates that Robyn and Jonathon overpaid by $1.52. If the full decimal accuracy were entered the final remaining principal would be $0.00.

Variations because of rounding are also found by entering the formula for the sum of the values in cells E8 to E247 into cell E248. The sum for the principal reduction is $1.52 greater than the mortgage of $200 000.

d) For a mortgage amortized over 25 years, the sum of the payments is the value in cell D308 of the spreadsheet in part a). For a mortgage amortized over 20 years, the sum of the payments is the value in cell D248 of the spreadsheet in part c).

471 789.00 − 408 028.80 = 63 760.20

Robyn and Jonathan would save $63 760.20 by amortizing the mortgage over 20 years instead of 25 years.

Key Concepts

- A mortgage is a loan secured by real estate.
- An amortization table shows the progress of repaying a loan or a mortgage over a given period of time in equal payments at regular intervals. The payments generally include principal and interest.
- The term of a mortgage is the length of time the mortgage agreement is in effect.
- Interest on a mortgage is calculated differently than on a loan. In Canada, the interest on a mortgage is compounded semi-annually. For a loan, interest can be compounded monthly and calculated monthly.

Communicate Your Understanding

1. Explain the advantages of using spreadsheets to develop amortization schedules.

2. Describe how you would use a graphing calculator to find the monthly interest rate equivalent to the interest rate 6% per annum, compounded semi-annually.

3. Describe how you would set up an amortization table for a mortgage of $95 000, amortized over 25 years with a monthly interest rate of 0.776 438 3%.

4. Explain how you decide

a) how many decimal digits to format for each cell

b) how far to copy the formulas into the rows

Practise

A

1. Use a graphing calculator to find an equivalent monthly rate for each rate per annum, compounded semi-annually.

a) 6%   **b)** 10%   **c)** 5.5%   **d)** 20.46%

2. The equivalent monthly rates calculated in question 1 are for mortgages. Why are these calculations necessary?

3. Use a graphing calculator to find the monthly payment for each loan.

a) a car loan of $19 275 at 6% per annum, compounded monthly, with monthly payments for 5 years

b) a personal loan of $12 000 at 9% per annum, compounded annually, with yearly payments for 15 years

c) a mortgage of $275 000 with interest charged at 6.95% per annum, compounded semi-annually, with monthly payments for 15 years

4. Use a graphing calculator to determine the monthly interest rate for a loan at each rate per annum, compounded annually.

a) 10%   **b)** 7.5%   **c)** 5.25%

d) 3%   **e)** 18%

Apply, Solve, Communicate

5. Computer systems A company is selling computer systems, including a printer, scanner, and software for $3000. The cost is financed at a rate of 10% per annum, compounded monthly, paid off at $150 per month.
a) How long does it take to pay off the loan?
b) What amount is the final payment?

6. A store Kathleen is talking to a bank about a $134 000 mortgage for a store she is buying. The bank is offering an interest rate of 9.75%, with a 25-year amortization plan.
a) Use a graphing calculator to find Kathleen's monthly interest rate.
b) Use a spreadsheet to find Kathleen's mortgage payments.
c) How much would Kathleen owe after 4 years?

Since mortgages in Canada are given per annum, compounded semi-annually, this can be assumed in a mortgage rate.

B

7. A new home Andrea is ready to sign a mortgage worth $140 000 on her new home. Her bank manager says the interest rate will be 7%, on a mortgage repaid over 20 years.
a) Use a graphing calculator to find the monthly interest rate and payment.
b) Use a spreadsheet to create an amortization table.
c) If the bank increases her interest rate to 7.25%, but allows the same payment, how much would she owe on the mortgage after 20 years?
d) If the interest rate is changed to 7.25%, but the payment remains the same, how long would it take her to pay off the mortgage?

8. Town house The Khan family has a mortgage of $160 000 to finance their town house. The bank is charging interest at a rate of 6.25% amortized over 25 years.
a) Find the Khan family's monthly interest rate and payment.
b) What percent of each payment is used to pay interest
i) for the first payment?   **ii)** for the second payment?
iii) for the third payment?
c) What trend do you notice in the percent of the payments used to pay interest? Explain whether this trend makes sense.

9. Application The Vandenberghe family are refinancing their mortgage of $155 000 with a plan for repaying the mortgage over 25 years. The bank is charging interest at 9%. Use a spreadsheet to find the first time the value in the Principal Reduction column is greater than the value in the Interest column. What do these values mean about the mortgage?

10. Consider the spreadsheet at the right.

a) How much is owed on the loan after 4 months?

b) How long does it take to repay at least half the principal?

c) Create a spreadsheet so the amount borrowed will be repaid after

i) 6 months   **ii)** 2 years

	A	B	C	D	E	F
1	8000.00	Loan ($)				
2	0.08	Rate/Year				
3	0.006666667	Rate/Month				
4	695.91	Payment ($)				
5						
6					Principal	Remaining
7	Month	Principal ($)	Interest ($)	Payment ($)	Reduction ($)	Principal ($)
8	1	8000.00	53.33	695.91	642.58	7357.42
9	2	7357.42	49.05	695.91	646.86	6710.56
10	3	6710.56	44.74	695.91	651.17	6059.39
11	4	6059.39	40.40	695.91	655.51	5403.88
12	5	5403.88	36.03	695.91	659.88	4743.99
13	6	4743.99	31.63	695.91	664.28	4079.71
14	7	4079.71	27.20	695.91	668.71	3411.00
15	8	3411.00	22.74	695.91	673.17	2737.83
16	9	2737.83	18.25	695.91	677.66	2060.17
17	10	2060.17	13.73	695.91	682.18	1377.99
18	11	1377.99	9.19	695.91	686.72	691.27
19	12	691.27	4.61	695.91	691.30	-0.03

d) Explain your method for part c) to a few classmates. Then, describe their responses.

11. Comparing The Garcias have just taken out a mortgage for $125 000. Their bank charges interest at 8.5% for a mortgage amortized over 25 years. Their new neighbours, the Picards, are arranging a loan of $130 000, amortized over 20 years. The interest at their bank for this loan is 7% per annum, compounded annually.

a) Predict who has a greater monthly payment. Justify your prediction.

b) Use a spreadsheet to find who has a greater monthly payment and what the difference is between the payments.

c) Compare your spreadsheet calculations with your predictions.

12. Application When Pat graduated from university, he found a job in Thunder Bay. He negotiated a mortgage of $95 000 for a condominium. The bank's interest rate was 7.75% for a mortgage amortized over 25 years with a 5-year term. At the end of the term, he renegotiated his mortgage. Since the interest rate had decreased to 7.25%, he decided to reduce the remaining time from 20 years to 15 years.

a) Use a spreadsheet to find his new monthly payment.

b) Change your spreadsheet to show the amortization table if the time had remained at 20 years.

c) What is the difference between the payments on the spreadsheets for parts a) and b)?

d) What is the difference between the sum of the payments for parts a) and b)?

13. Farm After the Singhs made a down payment on a small farm, they took out a mortgage of $225 000, amortized over 25 years at 6.75%.
a) How much interest will be paid during the sixth year of their repayment schedule?
b) How much of the principal will remain after the sixth year?
c) Research mortgage rates for two different times, a decade apart. Repeat parts a) and b) for each of your researched rates.

14. Buying a jeep After Collin pays the down payment on a jeep, he has a debt of $15 000. The dealership offers a loan amortized over 5 years with an interest rate of 9% per annum, compounded monthly. If Collin can afford to pay only $300 per month, should he buy this car? Use spreadsheet calculations to support your answer.

C

15. Trial and error Design a spreadsheet to show the amortization of a loan of $20 000 with interest calculated at 8.5% per annum, compounded monthly. Monthly payments are made over 3 years. Instead of using a graphing calculator to find the monthly payment, use a trial-and-error method to find the exact payment for the loan, correct to the nearest hundredth of a cent.

16. Scatter plot The Soligos are planning a mortgage of $75 000 for their condominium. Their bank charges interest at 8.5% for a mortgage amortized over 15 years.
a) Use a spreadsheet to create an amortization table.
b) Use the values in the table to find the amount they owe after each year. Copy and complete the following table with these values.

Year	Amount owing
0	75 000.00
1	72 380.76
...	...
14	8 400.49
15	0.00

c) Enter the values for Year into L1 of a graphing calculator and the values for Amount owing into L2.
d) Create a scatter plot for the data.
e) Determine the equation of a curve of best fit for the data.
f) Use this equation to determine the amount owing after
i) 4 years   **ii)** 3 months
g) Determine how close the value after 4 years is to the value in the spreadsheet.

17. Communication a) What are the advantages of a 15-year mortgage? What are the disadvantages?
b) What are the advantages of a 25-year mortgage? What are the disadvantages?

7.8 Mortgages

A knowledge of mortgages enables you to make decisions about financial arrangements when buying property. The variety in amortization periods, terms, amounts of mortgages, and even interest rates, allows you to make or combine choices. These choices can have a significant impact on your payments.

INVESTIGATE & INQUIRE

This amortization table shows monthly mortgage payments for different interest rates. Each payment includes the interest on the principal still owed and some of the principal. A mortgage is arranged for a specific length of time, called the **term** of the mortgage.

Interest Rate (%)	Monthly Payment for each $1000 of a Mortgage			
	Amortization Period			
	10 years	15 years	20 years	25 years
5.0	10.581 493	7.881 238	6.571 250	5.816 050
5.5	10.821 941	8.137 981	6.843 913	6.103 915
6.0	11.065 099	8.398 828	7.121 884	6.398 066
6.5	11.310 931	8.663 695	7.405 004	6.698 238
7.0	11.559 399	8.932 494	7.693 106	7.004 158
7.5	11.810 465	9.205 137	7.986 021	7.315 549
8.0	12.064 090	9.481 529	8.283 575	7.632 135
8.5	12.320 234	9.761 579	8.585 592	7.953 635
9.0	12.578 856	10.045 189	8.891 895	8.279 774
9.5	12.839 914	10.332 261	9.202 305	8.610 276
10.0	13.103 367	10.622 699	9.516 644	8.944 872
10.5	13.369 173	10.916 402	9.834 734	9.283 297
11.0	13.637 287	11.213 269	10.156 396	9.625 292
11.5	13.907 667	11.513 201	10.481 456	9.970 606
12.0	14.180 269	11.816 096	10.809 741	10.318 996

1. Use the amortization table.

a) What happens to the monthly payment as the interest rate increases? Why does this make sense?

b) What happens to the monthly payment as the number of years for the amortization period decreases? Why does this make sense?

c) How much is the monthly payment for an interest rate of 7%, amortized over 25 years for each amount?

By Canadian law, the interest rate on mortgages is compounded semi-annually.

i) $1000   **ii)** $2000   **iii)** $3000   **iv)** $4000   **v)** $5000
vi) $100 000   **vii)** $200 000   **viii)** $300 000   **ix)** $400 000   **x)** $500 000

d) How much is the monthly payment for an interest rate of 11.5%, amortized over 10 years for each amount?

i) $1000   **ii)** $10 000   **iii)** $100 000   **iv)** $1 000 000

2. Suppose you have a mortgage of $100 000 and the current interest rate is 9% for a 3-year term.

a) What would your monthly mortgage payment be if your mortgage is amortized over 25 years?

b) What would you have paid by the end of 1 year?

c) What would you have paid by the end of the 3-year term?

3. Suppose you have a mortgage of $100 000 and the current interest rate is 9% for a 3-year term.

a) What would your mortgage payment be per month if your mortgage is amortized over 20 years?

b) What would you have paid by the end of 1 year?

c) What would you have paid by the end of the 3-year term?

4. Suppose you have a mortgage of $100 000 and the current interest rate is 9% for a 3-year term.

a) What would your mortgage payment be per month if your mortgage is amortized over 15 years?

b) What would you have paid by the end of 1 year?

c) What would you have paid by the end of the 3-year term?

5. Suppose you have a mortgage of $100 000 and the current interest rate is 9% for a 3-year term.

a) What would your mortgage payment be per month if your mortgage is amortized over 10 years?

b) What would you have paid by the end of 1 year?

c) What would you have paid by the end of the 3-year term?

6. a) Make a generalization about the effect of the amortization period on the amount of a mortgage.

b) Explain why someone might want a shorter amortization period.

c) Explain why someone might want a longer amortization period.

7. What factors might affect a decision whether to increase a down payment to decrease the mortgage payments?

8. a) Why do you think the amounts in the table are not rounded to the nearest cent?

b) Use an example to show what would happen if the amounts were rounded to the nearest cent.

EXAMPLE 1 Finding Mortgage Payments

Victor is buying a house for $196 500. He makes a down payment of 25% of the price and negotiates a mortgage at 7.5%, amortized over 25 years, for the balance of the price.

a) How much is Victor's mortgage?
b) How much are Victor's monthly payments?

SOLUTION 1 Paper-and-Pencil and Table Method

a) The down payment is 25% of the price, or 25% of $196 500.
25% of 196 500 = 49 125
The mortgage is the price of $196 500 less the down payment of $49 125.
196 500 − 49 125 = 147 375
Victor's mortgage is $147 375.

b) Use the amortization table in the Investigate & Inquire. Since the table gives monthly mortgage payments for each $1000, divide the mortgage of $147 375 by 1000 to find how many thousands are in $147 375.
147 375 ÷ 1000 = 147.375

At 7.5%, amortized over 25 years, the monthly payment for each $1000 is $7.315 549, so multiply 147.375 times $7.315 549.
147.375 × 7.315 549 ≐ 1078.13

Victor's monthly mortgage payments are $1078.13.

> Round the mortgage payment to the nearest cent. Additional decimal places are kept for the calculations so that you are not calculating with values rounded more than the payment.

SOLUTION 2 Graphing-Calculator Method

a) Change the mode settings to 2 decimal places. From the Finance menu, choose TVM Solver.

Enter the known values.
There are 12 payments a year for 25 years, so N = 25 × 12.
The interest rate is 7.5% per annum, so I = 7.5.
The mortgage is the house price of $196 500 less the down payment of 25%, so PV = 196 500 − 0.25 × 196 500.
A payment is made each month, so P/Y = 12.
The interest is compounded semi-annually, so C/Y = 2.
The payments are made at the end of each payment interval, so select END.

Remember the interest rate for mortgages in Canada is compounded semi-annually.

Move the cursor to PMT to find the mortgage payment, and press ALPHA SOLVE. Since the mortgage payment is paid out, PMT is negative.

Victor's mortgage is $147 375.

b) Victor's monthly payments are $1078.13, as shown in Solution 1.

EXAMPLE 2 Finding the Cost of a Mortgage

The Pin family is negotiating a mortgage on a condominium for $85 000, amortized over 20 years, at an interest rate of 8.9%.
a) What is the Pins' monthly mortgage payment?
b) Mortgages are renegotiated after the end of the term, which is often 5 years. However, if the Pin family did keep the mortgage for 20 years and the interest rate remained unchanged, how much would it cost?

SOLUTION 1 Graphing-Calculator Method

a) Change the mode settings to 2 decimal places. From the Finance menu, choose TVM Solver.

Enter the known values.
There are 12 payments a year for 20 years, so N = 20 × 12.
The interest rate is 8.9% per annum, so I = 8.9.
The mortgage is $85 000, so PV = 85 000.
A payment is made each month, so P/Y = 12.
The interest is compounded semi-annually, so C/Y = 2.

The payments are made at the end of each payment interval, so select END.

Move the cursor to PMT to find the mortgage payment, and press ALPHA SOLVE. Since the mortgage payment is paid out, PMT is negative.

The Pins' monthly payment is $750.58.

b) There are 12 monthly payments for 20 years, each $750.58.
12 × 20 × 750.58 = 180 139.20
The mortgage would cost $180 139.20.

Solution 2 **Spreadsheet Method**

a) The Pins' monthly payment of $750.58 is found with a graphing calculator, as in Solution 1.

b) Use a graphing calculator to find the interest rate per month. Change the mode settings to 9 decimal places. From the Finance menu, choose TVM Solver.
Enter values for $1 for 1 month of this mortgage.

Enter the known values.
The interest rate per month is being calculated, so N = 1.
The interest rate is 8.9% per annum, so I = 8.9.
Use the negative amount $1 to find the value per dollar paid out, so PV = −1.
The payments are monthly, so P/Y = 12.
The interest is compounded semi-annually, so C/Y = 2.
The payments are made at the end of each payment interval, so select END.

Move the cursor to FV to find the future value of $1, and press ALPHA SOLVE.

The future value of $1 is $1.007 282 775.

Since the future value is the amount compounded for the present value, or the investment for $1, $A = P + i$, where i represents the monthly interest rate.

$$A = P + i$$

Substitute known values: $\quad 1.007\ 282\ 775 = 1 + i$

Isolate i: $\qquad\qquad\qquad\qquad i = 0.007\ 282\ 775$

So, the interest rate per month is 0.007 282 775, or 0.728 277 5%.

Create a spreadsheet with formulas like those shown below. Format the cells to express amounts of money to 2 decimal places.

The Pin family's mortgage is $85 000, so enter 85 000 in cell A1.
The interest rate per annum is 8.9%, so enter 0.089 in cell A2.
The calculated interest rate per month as a decimal is 0.007 282 775, so enter 0.007 282 775 in cell A3.
The monthly payment is $750.58, so enter 750.58 in cell A4.

	A	B	C	D	E	F
1	85000.00	Mortgage ($)				
2	0.089	Rate/Year				
3	0.007282775	Rate/Month				
4	750.58	Payment ($)				
5						
6					Principal	Remaining
7	Month	Principal ($)	Interest ($)	Payment ($)	Reduction ($)	Principal ($)
8	1	=A1	=B8*A$3	=A$4	=D8-C8	=B8-E8
9	=A8+1	=F8	=B9*A$3	=A$4	=D9-C9	=B9-E9
10	=A9+1	=F9	=B10*A$3	=A$4	=D10-C10	=B10-E10
244	=A243+1	=F243	=B244*A$3	=A$4	=D244-C244	=B244-E244
245	=A244+1	=F244	=B245*A$3	=A$4	=D245-C245	=B245-E245
246	=A245+1	=F245	=B246*A$3	=A$4	=D246-C246	=B246-E246
247	=A246+1	=F246	=B247*A$3	=A$4	=D247-C247	=B247-E247
248		Total		=SUM(D8:D247)		

Copy the formulas down until the rows show payments for 20 years.

	A	B	C	D	E	F
1	85000.00	Mortgage ($)				
2	0.089	Rate/Year				
3	0.007282775	Rate/Month				
4	750.58	Payment ($)				
5						
6					Principal	Remaining
7	Month	Principal ($)	Interest ($)	Payment ($)	Reduction ($)	Principal ($)
8	1	85000.00	619.04	750.58	131.54	84868.46
9	2	84868.46	618.08	750.58	132.50	84735.95
10	3	84735.95	617.11	750.58	133.47	84602.49
244	237	2945.62	21.45	750.58	729.13	2216.49
245	238	2216.49	16.14	750.58	734.44	1482.06
246	239	1482.06	10.79	750.58	739.79	742.27
247	240	742.27	5.41	750.58	745.17	-2.90
248		Total		180139.20		

The sum in cell D248 is 180 139.20.

If the Pin family kept the mortgage for 20 years, it would cost $180 139.20.

The solution with a graphing calculator shows that the mortgage would cost $180 139.20, instead of $180 138.12. Such differences occur with rounding.

EXAMPLE 3 Comparing the Effects of the Frequency of Payments

Noela is getting a $100 000 mortgage, with an interest rate of 9.2%, amortized over 15 years, on the cottage she is buying. She is able to choose whether to make the payments monthly or biweekly. If Noela pays biweekly, she plans to pay half the amount of the monthly payment. Use a graphing calculator to decide how this choice of the frequency of payments would affect the length of time taken to pay off the mortgage.

SOLUTION

To find the length of time for biweekly payments, first calculate the monthly payment.

Biweekly means every 2 weeks.

Change the mode settings to 2 decimal places. From the Finance menu, choose TVM Solver.

Enter the known values for monthly payments.
There are 12 payments a year for 15 years, so N = 15 × 12.
The interest rate is 9.2% per annum, so I = 9.2.
The mortgage is $100 000, so PV = 100 000.
A payment is made each month, so P/Y = 12.
The interest is compounded semi-annually, so C/Y = 2.
The payments are made at the end of each payment interval, so select END.

Move the cursor to PMT to find the mortgage payment, and press ALPHA SOLVE.
Since the mortgage payment is paid out, PMT is negative.

Each monthly payment would be $1015.96.

Enter the values for biweekly payments.
The interest rate is 9.2% per annum, so I = 9.2.
The mortgage is $100 000, so PV = 100 000.
Since the payment is paid out, PMT is negative. Noela plans that each biweekly payment would be half the amount of the monthly payment, so PMT = −1015.96 ÷ 2.
There are 52 weeks in a year and payments are made every 2 weeks, so P/Y = 52 ÷ 2.
The interest is compounded semi-annually, so C/Y = 2.
The payments are made at the end of each payment interval, so select END.

When you store a value to P/Y, C/Y automatically changes to match, so go back to C/Y and enter 2.

Move the cursor to N to find the number of payments, and press ALPHA SOLVE.

For biweekly payments, the number of payments would be 331.37.

Since there are 26 biweekly payments in a year, divide the number of payments by 26 to find the number of years.
331.37 ÷ 26 = 12.745

With these biweekly payments, it would take 12.745 years to pay off the mortgage.

To find the number of months in 0.745 years, multiply 0.745 times 12.
0.745 × 12 = 8.94
For these biweekly payments it would take 12 years and 8.94 months, or 12 years and 9 months with the last payment less than $507.98. With monthly payments, it would take 15 years to pay off the mortgage.

Key Concepts

- The total amount paid for a mortgage depends on the amount of the mortgage, the interest rate, and the frequency of payments. Changing the term of a mortgage affects the total amount of interest and the total amount of the payments.
- Monthly mortgage payments can be calculated using an amortization table, a graphing calculator, or a spreadsheet program.
- Mortgage payments are usually made monthly, but other lengths of time such as biweekly are possible.

Communicate Your Understanding

1. List five terms related to mortgages, and explain each.
2. Describe how you would use the amortization table in the Investigate & Inquire to find the monthly mortgage payment for a mortgage of $60 000, at an interest rate of 6.5%, amortized over 25 years.
3. Describe how you would use a graphing calculator to find the monthly mortgage payment for a mortgage of $91 000, at an interest rate of 9.9%, amortized over 18 years.
4. Explain why the compounding period is not given in the examples.

Practise

A

1. What is the amount of each down payment?
a) 75% down on a price of $214 000
b) 29% down on a price of $92 800
c) 80% down on a price of $579 900
d) 50% down on a price of $74 440

2. What is the mortgage for each property?
a) a house costing $161 500 with 79% down
b) a condominium costing $73 000 with $40 000 down
c) a farm costing $850 000 with 90% down
d) a cabin costing $19 000 with $12 000 down

3. For each of the following mortgages, use the amortization table in the Investigate & Inquire to find the monthly payment for each $1000.
a) 10% interest on a mortgage, amortized over 25 years
b) 5.5% interest on a mortgage, amortized over 10 years
c) 10.50% interest on a mortgage, amortized over 15 years

4. Why is the amount not needed for the mortgages in question 3?

5. For each mortgage, use the amortization table in the Investigate & Inquire to find the monthly payment.
a) 9% interest on a $52 000 mortgage, amortized over 20 years
b) 6.5% interest on a $154 800 mortgage, amortized over 25 years
c) 11.5% interest on a $87 200 mortgage, amortized over 10 years
d) 8% interest on a $600 000 mortgage, amortized over 15 years

6. Use a graphing calculator to find the monthly payment for each mortgage.
a) 7.13% interest on a $52 000 mortgage, amortized over 25 years
b) 14.42% interest on a $154 800 mortgage, amortized over 15 years
c) 8.7% interest on a $87 200 mortgage, amortized over 23 years
d) 6.49% interest on a $250 000 mortgage, amortized over 16 years

7. For each mortgage in question 6, what would be the total cost, assuming that the mortgage was kept for the full amortization period?

Apply, Solve, Communicate

8. Country lot Jesse is arranging a mortgage for $78 000 to buy a lot in the country. The rate is 8.5%, amortized over 25 years.
a) What are Jesse's monthly mortgage payments?
b) If Jesse kept the mortgage for 25 years, how much would it cost?

9. Offices A group of investors bought two units in an industrial plaza to rent to companies. The two units sold for $780 000. The investors made a down payment and arranged a mortgage for the remaining 70% of the cost. The mortgage is at 9.5%, paid monthly for a 9-year term and amortized over 20 years.

a) What is the monthly payment?

b) How much will the investors owe after 9 years?

10. Condominium a) Josie would like to buy a condo priced at $150 000. She has $80 000 for a down payment. Find her monthly mortgage payments for a 12% mortgage, amortized over

i) 10 years   **ii)** 15 years   **iii)** 20 years   **iv)** 25 years

b) What happens to the mortgage payment as the amortization period increases? Why do you think this happens?

11. Assuming a mortgage a) Three years ago, the Edward family arranged a mortgage of $81 000 on their cottage. The mortgage is amortized over 25 years, at a rate of 6% for a 5-year term. What is the monthly payment?

b) The Samuel family is buying the cottage, and assuming the Edwards' mortgage. How much is still owed on this mortgage?

> Assuming a mortgage means taking over the previous owner's mortgage without changing the arrangements.

12. Application Naomi and Kevin are buying a house for $345 500 with a down payment of $260 000. The owners agreed to take back the mortgage, which means Naomi and Kevin will arrange the mortgage with them, and make the payments to them. The owners are offering an interest rate of 5%, amortized over 15 years with either monthly payments or biweekly payments. If the biweekly offer is accepted, each payment will be half what the monthly payment would be. Which plan would allow the mortgage to be paid off faster? How much faster would it be?

13. Assumptions A $100 000 mortgage loan is taken at 6.5% per annum, amortized over 20 years.

a) What is the monthly payment of the loan?

b) Assuming the interest rate remains the same over the period of the loan, how much is the total amount paid for the loan? How likely is this assumption? Explain reasons for your answer.

14. Inquiry/Problem Solving a) The Camerons are buying a farm for $700 000. After a down payment of $425 000, they are considering a mortgage for the remainder at an interest rate of 6.5%, amortized over 25 years. What would their mortgage payments be?

b) If the mortgage rate changes to 6% before they make their arrangements, what would their mortgage payments be?

c) If the mortgage rate changes to 7% before they make their arrangements, what would their mortgage payments be?

d) Make a generalization about how mortgage rates affect mortgage payments. Why does this make sense?

15. Communication a) What are the advantages of using the amortization table to find the monthly payment for a mortgage?

b) What are the advantages of using a graphing calculator to find the monthly payment for a mortgage?

16. Making financial decisions In 1982, Barbara bought a house worth $158 000 with a down payment of $94 000 and a mortgage on the balance at 19.25% per annum, amortized over 15 years. Since the mortgage rate was unusually high, Barbara arranged a 1-year term.

a) What were her monthly mortgage payments?

b) How much would she have paid for the house altogether if the mortgage had continued for 15 years?

c) A year later, when the mortgage rate was 13%, Barbara negotiated a new mortgage, amortized over 14 years, for the remainder she owed. How much were her mortgage payments then?

d) What factors might influence people as they negotiate the term of a mortgage?

C

17. Paying off a mortgage Mark and Sandy bought a house for $192 000 with a down payment of $120 000. Their mortgage rate is 9%, amortized over 25 years, for a 5-year term.

a) What is their monthly mortgage payment?

b) After the 5-year term, they renegotiate for a 3-year term at 6.5%. What is their new monthly payment?

c) If they continue to make the same payments as they did for the 9% rate, how long would it take to paid off the mortgage?

18. Estimating a) Use the amortization table in the Investigate & Inquire to estimate the monthly payment for each $1000 of a mortgage, amortized over 25 years, with an interest rate of

i) 6.3%   ii) 19.5%

b) Explain why your estimation strategies in part a) are reasonable.

c) Use your estimates from part a) to calculate the monthly payments on a mortgage of $200 000, amortized over 25 years with an interest rate of

i) 6.3%   ii) 19.5%

d) Use technology to find the monthly mortgage payments at these rates.

i) 6.3%   ii) 19.5%

19. Increase in interest rates a) Predict, or state from your understanding of mortgages, how an increase in interest rates affects the length of time required to pay off the mortgage.

b) Explain how you arrived at your prediction.

c) Devise a way to verify your prediction. Apply this method.

d) Explain how your answer from part c) compares with your prediction from part a).

20. Research a) Research financial information about mortgages with regard to the following questions.

• How are mortgage rates influenced by the economy?

• How do mortgage rates influence house sales?

• What techniques are suggested for determining the amount you can afford to carry in a mortgage?

• What additional costs are involved in buying real estate and setting up a mortgage?

• What different types of mortgages are available?

b) Use your research to describe a situation in which someone would need to make a financial decision about a mortgage. Explain what you would recommend for the situation, and justify your explanation with facts and with logical reasoning.

ACHIEVEMENT Check Knowledge/Understanding Thinking/Inquiry/Problem Solving Communication Application

In July 1979, mortgage rates in Canada averaged around 11%. Two years later, they reached the unprecedented level of 22%. For a $100 000 mortgage to be repaid over 25 years, find the required monthly payment at the two different interest rates. What total amount would be paid in each case? Determine the total amount that would be paid using the rates and terms offered for mortgages currently.

Investigate & Apply

The Cost of Car Ownership

A car is the first major purchase that many people make.

1. Ghita plans to buy a 5-year old car for $12 500. She has $2500 saved for a down payment.
a) The car dealership offers her the option of paying off the balance in 24 monthly instalments of $495. How much extra will she pay for the car if she chooses this option?
b) Ghita's bank will give her a car loan with interest at 8.5% per annum, compounded quarterly, for a term of two years. The amount will be divided by 24 to obtain the monthly instalments. What will her monthly payments be if she takes the bank loan?
c) Which source of financing costs Ghita less, the car dealership or the bank? Do some research to find out whether this is generally the case.

2. Many factors affect how much it costs to operate a car. These include
• distance driven per month
• type of driving, city or highway, for which the car is used
• frequency of repairs needed
What is the difference between fixed car expenses and variable car expenses? Give examples of each.

3. Depreciation is a frequently-overlooked car expense. One method of computing the annual depreciation is the double-declining balance method. For this method, divide 100% by the estimated life of the car, and then double the result to obtain the rate of depreciation. Use this method to find the value of the car at the end of the third year that Ghita owns it if she can reasonably expect to keep it for 5 years.

4. a) Estimate the total operating costs in the first year. State your assumptions.
b) Do you think that the operating costs would increase or decrease in the second year, and by how much? Explain.

5. Prepare a report that categorizes and displays all the costs involved in owning and operating the car for a two-year period. Include your assumptions.

REVIEW OF **KEY CONCEPTS**

■ **7.2** Compound Interest

Refer to the Key Concepts on page 507.

1. Find the amount of each investment.
a) $400 at 6% per annum, compounded monthly, for 5 years
b) $1500 at 4.25% per annum, compounded semi-annually, for 4 years
c) $3000 at 5.5% per annum, compounded quarterly, for 6 years
d) $2000 at 6.5% per annum, compounded annually, for 20 years

2. College To save for college, Charlene invested her summer earnings of $4200 in an account with 6.2% interest per annum, compounded semi-annually. She plans to leave the money in the account for 4 years.
a) What amount of money will she have at the end of the 4 years?
b) How much interest will Charlene have earned?
c) Charlene had the option of investing the money at 7% per annum, compounded annually for the same term. Do you think she made the right choice? Why or why not? Support your answer with calculations.

3. Comparing plans Eric is considering two investments plans: Plan A with 5.95% interest per annum, compounded quarterly, and Plan B with 5% interest per annum, compounded monthly. He is investing $2300 for 3 years.
a) If he invests $1000 in Plan A and $1300 in Plan B, how much will he have after 3 years?
b) If he invests $1300 in Plan A and $1000 in Plan B, how much will he have after 3 years?
c) Which investment option do you think is better? Why?

4. Different rates Cecelia has $1000 to invest for the next 5 years. She can invest it at 4.65% per annum, compounded semi-annually, or at 5.15% per annum, compounded annually, or at 4.25% per annum, compounded quarterly. If you were her financial advisor, which option would you recommend? How would you convince her to follow your advice?

7.4 Present Value

Refer to the Key Concepts on page 522.

5. Find the present value of each amount.
a) $3000 in 5 years at 4.9% per annum, compounded quarterly
b) $2500 in 2 years at 5.5% per annum, compounded annually
c) $2500 in 5 years at 3.9% per annum, compounded semi-annually
d) $9000 in 8 years at 5.25% per annum, compounded quarterly

6. Furniture Amy wants to have $6500 to buy furniture when she moves into an apartment in 3 years. How much should she invest today at 6.3% per annum, compounded monthly?

7. Tuition Stephan will need $5500 for college tuition in 4 years. He has saved $3200 from his summer job. What interest rate compounded semi-annually does he need when he invests the $3200?

8. Buying a car Mackenzie's uncle just bought a new car and has agreed to let her buy it from him in 6 years for $9800. How much should she invest now in an account with interest at 4.95% per annum, compounded quarterly, so she can buy the car?

7.5 Amount of an Ordinary Annuity

Refer to the Key Concepts on page 531.

9. Find the amount for each ordinary annuity.
a) a payment of $1500 at the end of every 3 months at 8% per annum, compounded quarterly
b) a payment of $4000 at the end of every year for 12 years with interest at 6.5% per annum, compounded annually
c) a payment of $700 at the end of every month, into an account that pays 10% per annum, compounded monthly
d) a payment of $2800 at the end of every 6 months for 5 years with an interest rate of 9% per annum, compounded semi-annually

10. Travelling Gail's grandfather saved for a trip by depositing $400 at the end of each month for 18 months. The account earns 4.8% per annum, compounded monthly. How much will be in the account when the last deposit is made?

11. RRSP Barry plans to invest $7000 in his RRSP on each March 1 for 8 years, starting next year. The interest rate is 8.3% per annum, compounded annually. How much will this investment be worth when the last payment is made?

12. Savings Nancy deposits $300 into her savings account at the end of every 3 months for 2 years. If the account pays 3.5% per annum, compounded quarterly, how much money will Nancy have in her account at the end of 2 years?

7.6 Present Value of an Ordinary Annuity

Refer to the Key Concepts on page 540.

13. Find the present value for each of the following.
a) a payment of $12 500 every year for 5 years, starting in a year, with interest at 6% per annum, compounded annually
b) a payment of $600 every month for 24 months, starting in a month, with interest at 10% per annum, compounded monthly
c) a payment of $3000 every 6 months for 3 years, starting in half a year, with interest at 7.5% per annum, compounded semi-annually

14. How much is each payment for an investment of $32 000?
a) an interest rate of 12% per annum, compounded semi-annually, with payments every 6 months, starting in a half year
b) an interest rate of 8.2% per annum, compounded annually, with payments every year, starting in a year
c) an interest rate of 9.5% per annum, compounded monthly, with payment every month starting in a month
d) an interest rate of 4.3% per annum, compounded quarterly, with payments every 3 months starting in 3 months

15. Living expenses Marcia wants to receive $10 000 every 6 months, for 3 years for living expenses when she goes back to school, starting 6 months from now. How much money must she invest now at 4.25% per annum, compounded semi-annually?

16. Loan Sandy is putting aside enough money to pay back a loan at a rate of $3000 a year for 5 years, starting in a year. To cover this loan, how much must Sandy put into an account that pays an interest rate of 7.2% per annum, compounded annually?

17. Prize Denis won a prize of $25 000. He has decided to invest in an account paying 6.25% per annum, compounded monthly. How much can he withdraw from the account each month over the next 3 years, starting in a month?

7.7 Technology: Amortization Tables and Spreadsheets
Refer to the Key Concepts on page 555.

18. Use a graphing calculator to find the monthly interest rate equivalent to each of the following semi-annual rates.

a) 6%   **b)** 10%   **c)** 5.5%   **d)** 20.46%

19. Comparing principals The Wrights arranged a $70 000 mortgage at an interest rate of 8.5%, amortized over 25 years. Use a spreadsheet to find how many months it will take for the principal to be less than half the original principal.

> Recall that since mortgages in Canada are given per annum, compounded semi-annually, this can be assumed in a mortgage rate.

20. Lower rates Coral and Sam are looking at a house that would require them to take out a $102 000 mortgage. They would choose to amortize it over 25 years. The current mortgage rate is 7.6%. Use a spreadsheet to find out how much they would pay over a 5-year term.

21. Cost of a mortgage Dale has negotiated a $67 000 mortgage on his home at 9.25% interest, amortized over 20 years.
a) Use a spreadsheet to find how much interest he will pay on the mortgage by the end of the fourth year.
b) How much of the principal will he pay by the end of the fourth year?

7.8 Mortgages
Refer to the Key Concepts on page 566.

22. Determine the monthly mortgage payment.
a) $100 000 at 11.5%, amortized over 25 years
b) $54 000 at 6%, amortized over 15 years
c) $93 000 at 12%, amortized over 10 years
d) $26 000 at 5.5%, amortized over 20 years

23. Condominium Carla bought a condominium and arranged a mortgage of $155 000 at 12%, amortized over 15 years.
a) What is her monthly mortgage payment?
b) What will the mortgage cost if it continues for 15 years?
c) What would the monthly mortgage payments be if she was able to increase the down payment by $40 000?

24. Cash To get some cash to expand his business, Todd took a $120 000 mortgage on his house at 6.25%, amortized over 25 years.
a) What is the monthly payment of the loan?
b) Assuming the interest rate remains constant, how much will Todd pay for the loan?

25. Selling Glen and Christine set up a mortgage on their house for $79 000 at 8.5%, amortized over 20 years, with a 5-year term.
a) How much were their mortgage payments?
b) When the term ended, they decided to pay off the mortgage and sell the house. What would it cost to pay off the mortgage?

26. A decision Would you choose a $50 000 mortgage at 5%, amortized over 20 years or over 25 years? Explain your choice. Include calculations in your explanation.

CHAPTER TEST

Achievement Chart

Category	Knowledge/Understanding	Thinking/Inquiry/Problem Solving	Communication	Application
Questions	All	4, 8, 9	2, 8, 9	4, 7, 8, 9

1. **GIC** Jean Paul is saving for a car. He puts $4500 in a Guaranteed Investment Certificate paying 5.25% per annum, compounded quarterly. How much money will he have available to buy a car 3 years from now?

2. **Long-term savings** Melanie is deciding which of two long-term savings plans to choose for investing $9000: 6.2% per annum, compounded semi-annually, or 5.75% per annum, compounded quarterly. If she plans to leave the money in the plan for 8 years, which is the better option? Justify your answer.

3. **University** Faris needs $5000 for university in 3 years. His parents plan to invest some money in an account paying interest at a rate of 7.1% per annum, compounded quarterly. How much should they invest now to have $5000 in 3 years?

4. **Loan** Sophia is borrowing some money to pay the set-up costs for her web site. She predicts she can afford to pay $17 000 for the loan in 2 years. The plan she is arranging offers a loan at 9.8% per annum, compounded monthly. At this rate, how much can she borrow?

5. **Retirement** Sam's aunt is investing $1000 for retirement every 6 months, starting 6 months from now. The account pays 7% per annum interest, compounded semi-annually. How much will she have saved in 10 years?

6. **Lottery** Brooke won $100 000 in a lottery. The prize will be paid in yearly installments of $10 000 each year for 10 years. What is the present value of her winnings, if current interest rates are 6.4% per annum, compounded annually?

7. Artist fund A company is investing money to start a fund to send young artists to special classes. They want to invest an amount that would continue for 10 years to give $3000 a month for the art students to share, starting in a month. If the money is invested at 8.2% per annum, compounded monthly, how much must they invest?

8. House Ina is negotiating a mortgage of $145 000 for her new house.
a) What are Ina's monthly payments for a mortgage at 5% per annum, compounded semi-annually, amortized over
i) 25 years?   **ii)** 20 years?   **iii)** 15 years?
b) How much would Ina pay for the mortgage amortized over 15 years if it were kept for 15 years?

ACHIEVEMENT Check   Knowledge/Understanding   Thinking/Inquiry/Problem Solving   Communication   Application

9. Instead of paying $4000 at the end of 5 years and $3000 at the end of 10 years, Erica agrees to make equal monthly payments for 10 years. Find the monthly payment if the interest charged is 8% per annum, compounded monthly. Explain your method.

CHALLENGE PROBLEMS

1. Buy or Lease A company is considering the possibility of acquiring new computer equipment for $400 000 cash. The salvage value is estimated to be $50 000 at the end of the 6-year life of the equipment. Maintenance costs will be $4000 per month, payable at the end of each month. The company could lease the equipment for $12 000 per month payable at the end of each month. Under the 6-year lease agreement the lessor would pay the maintenance costs. If the company can earn 18% per annum, compounded monthly, on its capital, advise the company whether to buy or to lease.

2. Mining property A company is considering whether or not to develop a mining property. It is estimated that an immediate expenditure of $7 000 000 will be needed to bring the property into production. After that, the net cash inflow will be $1 700 000 at the end of each year for the next 10 years. An additional expenditure of $3 200 000 at the end of 11 years will have to be made to restore the property to an attractive condition. On projects of this type, the company would expect to earn at least 20% per annum, compounded semi-annually on its capital. Advise whether the company should proceed.

3. Mortgages The Andersons have a $200 000 mortgage with monthly payments over 20 years at 10%. Because they both get paid weekly, they decide to switch to weekly payments (at the same interest rate). Compare their weekly payments to their monthly payments.

4. Trust fund Shani is investing $500 000 in a trust fund from which her three children are to receive equal amounts when they each reach 21 years of age. The children are now 19, 15, and 13 years old. If this trust fund earns interest at 7% per annum, compounded semi-annually, how much will each child receive?

PROBLEM SOLVING STRATEGY

USE LOGIC

The ability to think logically is an important skill that you will use in any profession you choose and in everyday life. This skill can be improved with practice.

Most counterfeit coin problems include a balance scale with two pans. In the present problem, there is a scale with a single pan. There are three large bags of gold coins, with an unknown number of coins in each bag. Two of the bags contain only real coins that have masses of 60 g each. The other bag contains only counterfeit coins that have masses of 61 g each. You can remove coins from the bags, but since the mass of a real coin and the mass of a counterfeit coin are almost the same, you cannot tell a real coin from a counterfeit coin without using the scale. What is the minimum number of times you can use the scale in order to be certain which bag contains the counterfeit coins?

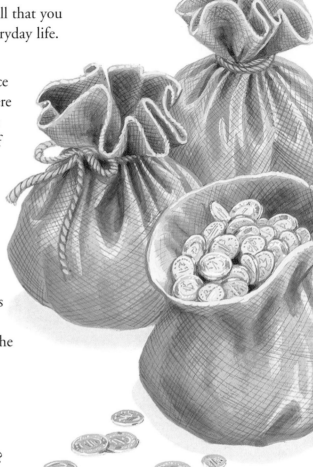

Understand the Problem

1. What information are you given?
2. What are you asked to find?
3. Do you need an exact or an approximate answer?

Think of a Plan

Think of different ways of removing coins from the bags and finding the masses of the coins.
Label the bags A, B, and C.
You could remove and find the mass of one coin from each bag. This method requires using the scale three times, unless you are lucky enough to find that the first coin has a mass of 61 g or the second coin has a mass of 61 g. Another method is to take one coin from each of two bags and find the total mass of the two coins by using the scale only once. If you are lucky, and the total mass is 120 g, you know that the two coins came from the bags of real coins. However, if the total mass is 121 g, you do not know which coin is counterfeit.

One way to use the scale only once, and to be certain which bag contains the counterfeit coins, is to remove 1 coin from bag A, 2 coins from bag B, and 3 coins from bag C. Then, find the total mass of the 6 coins.

If the total mass is 361 g, bag A has the counterfeit coins, since
$61 + 120 + 180 = 361$.

If the total mass is 362 g, bag B has the counterfeit coins, since
$60 + 122 + 180 = 362$.

If the total mass is 363 g, bag C has the counterfeit coins, since
$60 + 120 + 183 = 363$.

So, the minimum number of times you can use the scale in order to be certain which bag contains the counterfeit coins is one.

Look Back

Is there another method that will give the same answer?

Use Logic

1. Organize the information.
2. Draw conclusions from the information.
3. Check that your answer is reasonable.

Apply, Solve, Communicate

1. Cards There are four cards on a table, as shown in the diagram. Each card is coloured red or blue on one side, and has a circle or square on the other side.

To determine whether every red card has a square on its other side, what is the minimum number of cards you must turn over, and which cards are they?

2. Number grid Copy the grid into your notebook. Arrange the numbers 1, 2, 3, and 4 on the 4 by 4 grid so that
• each row and each column includes the numbers 1, 2, 3, and 4
• no identical numbers are next to each other in a row, a column, or a diagonal from corner to corner
Two numbers have been placed for you. Find three solutions.

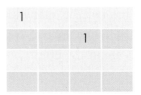

3. Months You are given the following information.

January = 2	June = 1
February = 3	July = 0
March = 1	August = 2
April = 2	September = 5
May = 1	October = 4

Find the values of November and December.

4. Hockey games The chart gives the standings in a four-team hockey league after each team played every other team.

Team	Won	Lost	Tied	Goals For	Goals Against
Lions	3	0	0	7	0
Tigers	1	1	1	1	1
Bears	0	1	2	2	7
Rams	0	2	1	2	4

Determine the score of each game.

5. Car speed A circular track is 1 km long. A car makes one lap of the track at 30 km/h. Is there a speed that the car can travel on the second lap to average 60 km/h for the two laps? Explain.

6. Probability There are four balls in a bag. One is yellow, one is green, and two are white. Someone takes two balls from the bag, looks at them, and states that one of them is white. What is the probability that the second ball taken from the bag is also white?

7. Puppies Max and Sheba belong to a litter of puppies. Max has as many sisters as he has brothers. Sheba has twice as many brothers as she has sisters. How many females and how many males are in the litter?

8. Coordinate geometry The coordinates of the endpoints of one diagonal of a square are (8, 11) and (4, 5). Determine the coordinates of the endpoints of the other diagonal without graphing.

9. Whole numbers What is the largest list of whole numbers less than 100 such that no number in the list is the sum of two other numbers in the list?

10. Cube faces Three faces of a cube are shown. Each face of the cube has been divided into four squares. Each square is coloured red, blue, or yellow. No two squares of the same colour touch along an edge anywhere on the cube. How many squares of each colour are on the cube?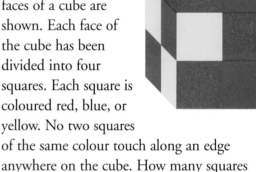

11. Whole numbers The sum of 17 consecutive whole numbers is 306. Find the greatest of the 17 whole numbers.

12. Odd digits a) Using only the odd digits, how many different three-digit numbers can be made?
b) Determine the sum of these three-digit numbers without adding them.

13. Formulate problems Write a problem that can be solved using logic. Have a classmate solve your problem.

USING THE STRATEGIES

1. Real number What is the smallest value the expression $x + 10x^2$ can have, if x is a real number?

2. Pattern Determine the pattern and find the next number in this sequence.
1, 256, 2187, 4096, 3125, 1296, ...

3. Three Find two numbers whose quotient and difference both equal 3.

4. Product Find a positive number such that the product of $\frac{1}{5}$ of the number and $\frac{1}{9}$ of the number equals the number.

5. Water containers You have a 24-L container that is full of water. You also have three empty containers that can hold 5 L, 11 L, and 13 L. How can you use the containers to divide the water into three equal parts?

6. Number puzzle If you reduce a certain number by 8 and multiply the result by 8, you get the same answer as if you reduce the number by 9 and multiply the result by 9.
a) What is the number?
b) Use algebra to explain.

7. Fenced field A rectangular field is half as wide as it is long and is completely enclosed by x metres of fencing. Express the area of the field in terms of x.

8. Solutions Find the number of solutions to the equation $2x + 3y = 715$, if x and y must be positive integers.

9. Hockey The chart gives the standings in a four-team hockey league after each team played every other team once. Determine the score of each game.

Team	Won	Lost	Tied	Goals For	Goals Against
Spartans	2	0	1	5	2
Penguins	2	1	0	6	3
Ravens	1	2	0	1	3
Eagles	0	2	1	0	4

10. Power of 2 Evaluate $79\ 999\ 999\ 999^2$.

11. Two numbers The sum of two numbers is 4 and their product is 6. What is the sum of the squares of the reciprocals of these numbers?

12. System of equations Solve the following system of equations.
$$a + b + c = 9$$
$$ab + bc + ac = 26$$
$$abc = 24$$

13. Same digits A four-digit number, whose digits are the same, is divided by the sum of the digits.

For example, $\dfrac{4444}{4+4+4+4} = \dfrac{4444}{16}$
$$= 277.75$$

Explain why the result is 277.75 for any four-digit number whose digits are the same.

14. Perfect Find n such that $\dfrac{n}{2}$ is a perfect square and $\dfrac{n}{3}$ is a perfect cube.

C H A P T E R

8 Loci and Conics

Specific Expectations	Functions	Functions & Relations
Construct a geometric model to represent a described locus of points; determine the properties of the geometric model; and use the properties to interpret the locus.		8.1, 8.2
Explain the process used in constructing a geometric model of a described locus.		8.1, 8.2
Determine an equation to represent a described locus.		8.2
Construct geometric models to represent the locus definitions of the conics.		8.3, 8.4, 8.5, 8.6, 8.7
Determine equations for conics from their locus definitions, by hand for simple particular cases.		8.4, 8.5, 8.6, 8.7
Identify the standard forms for the equations of parabolas, circles, ellipses, and hyperbolas having centres at $(0, 0)$ and at (h, k).		8.4, 8.5, 8.6, 8.7
Identify the type of conic, given its equation in the form $ax^2 + by^2 + 2gx + 2fy + c = 0$.		8.8
Determine the key features of a conic whose equation is given in the form $ax^2 + by^2 + 2gx + 2fy + c = 0$, by hand in simple cases.		8.8
Sketch the graph of a conic whose equation is given in the form $ax^2 + by^2 + 2gx + 2fy + c = 0$.		8.8
Illustrate the conics as intersections of planes with cones, using concrete materials or technology.		8.3
Describe the importance, within applications, of the focus of a parabola, an ellipse, or a hyperbola.		8.5, 8.6, 8.7
Pose and solve problems drawn from a variety of applications involving conics, and communicate the solutions with clarity and justification.		8.4, 8.5, 8.6, 8.7
Solve problems involving the intersections of lines and conics.		8.9

Motion in Space

In the Modelling Math questions on pages 616, 634, 640, and 663, you will solve the following problem and other problems that involve motion in space.

Halley's Comet orbits the sun about every 76 years. The comet travels in an elliptical path, with the sun at one of the foci. At the closest point, or perihelion, the distance of the comet from the sun is 8.8×10^7 km. At the furthest point, or aphelion, the distance of the comet from the sun is 5.3×10^9 km. Write an equation of the ellipse that models the motion of Halley's Comet. Assume that the sun is on the x-axis.

Use your research skills to answer the following questions now.

1. In what year did Halley's Comet most recently reach its perihelion?

2. Halley's Comet is known as a short-period comet, because it completes an orbit of the sun in a period of less than 200 years. Identify the comet with the shortest known period. State the period of this comet, and its distances from the sun at the perihelion and aphelion of its orbit.

3. Edmund Halley (1656–1742) was not the first astronomer to observe the comet now known as Halley's Comet, so why is it named after him?

Communications Satellites

In 1972, Canada became the first country to use satellites for communications within its own borders. The *Anik A1* satellite carried radio and television programs to all parts of the country. The satellite was in a geostationary orbit, which means that it moved in time with the Earth's rotation and stayed a constant distance above a fixed point on the Earth's surface. Therefore, the satellite's orbit was circular.

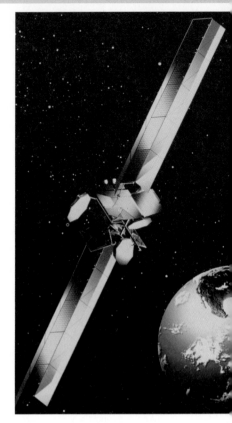

Anik F1, launched in 2000, carries the bulk of Canada's television signals. This satellite is in a geostationary orbit 35 880 km above a point on the Earth's equator, at a longitude of 107.3°W. Canada has five teleports for uplinking signals from TV and radio stations to geostationary satellites. The teleports are in Montréal, Toronto, Calgary, Edmonton, and Vancouver.

1. The radius of the Earth is about 6370 km. To the nearest kilometre, how far does *Anik F1* travel in one day?

2. At what velocity is *Anik F1* revolving about the Earth's axis, to the nearest metre per second?

3. At what velocity is a point on the Earth's equator revolving about the Earth's axis, to the nearest metre per second?

4. How many times faster, to the nearest tenth, is *Anik F1* revolving than a point on the Earth's equator?

5. The velocity, v metres per second, of a satellite about the Earth's axis is related to the distance from the centre of the Earth, r metres, and the acceleration of gravity, g metres per second per second, by the following equation.
$$v = \sqrt{gr}$$

a) Solve the equation for g.
b) Calculate g, at the location of the satellite, to the nearest hundredth of a metre per second per second.
c) The value of g at the Earth's surface is about 9.8 m/s^2. How many times as great is this value as the value of g at the location of the satellite, to the nearest whole number?

6. Which Canadian teleport is closest to a longitude of 107.3°W?

Review of Prerequisite Skills

If you need help with any of the skills named in purple below, refer to Appendix A.

1. Simplifying expressions Expand and simplify.
a) $5x(x-3) - 2x^2(x-1)$
b) $(2x-3)^2 - (4x+5)^2$
c) $2(3x+2)^2 - (x-1)^2$
d) $3(3x-1)^2 + 5(1-4x)^2$

2. Length of a line segment Find the length of the line segment joining each pair of points. Round answers to the nearest tenth.
a) $(2, 5)$ and $(7, 2)$ b) $(-2, 3)$ and $(5, 1)$
c) $(5, -2)$ and $(0, -3)$
d) $(-1, -3)$ and $(-6, -9)$

3. Midpoint formula Find the coordinates of the midpoint of the line segment with the given endpoints.
a) $(-2, 5)$ and $(4, -3)$
b) $(3, -2)$ and $(7, 10)$
c) $(-3, -5)$ and $(2, -4)$

4. Graphing equations Graph each equation.
a) $y = x - 4$ b) $y = 2x - 1$
c) $y = -x + 2$ d) $2x - y = -1$

5. Solve for y.
a) $3x + y = 4$ b) $x - 4y = 2$
c) $y^2 = 25$ d) $x^2 + y^2 = 25$

6. Solving quadratic equations by factoring Solve by factoring.
a) $x^2 - x - 6 = 0$ b) $2x^2 + 3x - 2 = 0$
c) $4x^2 - 13x + 3 = 0$ d) $6x^2 - 5x + 1 = 0$
e) $6x^2 - 13x - 5 = 0$ f) $5x^2 - 36x + 7 = 0$

7. The quadratic formula Solve using the quadratic formula. Round answers to the nearest tenth, if necessary.
a) $x^2 + 3x - 10 = 0$ b) $2x^2 + 5x = 3$

c) $3x^2 + 2 = -7x$ d) $2x^2 + x - 4 = 0$
e) $4x^2 + x - 2 = 0$ f) $12x^2 = 5 - 16x$

8. Determine the constant to be added to each polynomial to make it a perfect square trinomial.
a) $x^2 + 12x$ b) $x^2 - 8x$ c) $x^2 + 3x$
d) $x^2 - 6x$ e) $x^2 - 5x$ f) $x^2 + x$

9. Rewriting in the Form $y = a(x - h)^2 + k$, $a = 1$ Rewrite each of the following the form $y = a(x - h)^2 + k$. State the maximum or minimum value of y, and the the value of x when it occurs.
a) $y = x^2 + 4x - 5$ b) $y = x^2 - 6x - 10$
c) $y = -x^2 - x + 30$ d) $y = x^2 - 11x + 2$
e) $y = -x^2 - 8x$ f) $y = x^2 + 5x$

10. Rewriting in the Form $y = a(x - h)^2 + k$, $a \neq 1$ Rewrite each of the following the form $y = a(x - h)^2 + k$. State the maximum or minimum value of y, and the the value of x when it occurs.
a) $y = 2x^2 + 8x - 16$ b) $y = -3x^2 + 6x + 6$
c) $y = 3x^2 + 6x - 8$ d) $y = 4x^2 - 12x$
e) $y = 0.1x^2 + 2x + 1$ f) $y = -0.2x^2 - 6x$

11. Solving linear systems Solve and check.
a) $3x + 2y = 7$ b) $4x - y = 7$
   $2x + y = 4$ $3x + 2y = -3$
c) $3x + 2y = 11$ d) $3x + 4y = -18$
   $x - 3y = 11$ $5x - 2y = -4$

12. Write an equation in standard form for the line through the given points.
a) $(5, 6)$ and $(4, 8)$ b) $(-3, 8)$ and $(0, -1)$
c) $(-2, -5)$ and $(-4, -6)$
d) $(-3, -4)$ and $(2, -2)$

8.1 Technology: Constructing Loci Using *The Geometer's Sketchpad®*

A **locus** is a set of points defined by a given rule or condition. *Locus* comes from Latin and means place or location.

An example of a locus is the set of points that are all the same distance from a fixed point. This locus is a circle, where each point on the circumference is the same distance from the fixed point. The fixed point is the centre, and the distance is the radius.

In this section, *The Geometer's Sketchpad®* will be used to construct several loci.

Points Equidistant From Two Given Points

You are given a line segment AB. Follow the steps below to find the locus of points that are equidistant from points A and B. Using *The Geometer's Sketchpad®*, circles can be used to find the points for this locus.

1. Construct a line segment AB in the lower part of the screen.

2. Construct a line segment CD near the top of the screen.

3. With line segment CD selected, choose **Point On Object** from the **Construct** menu. Label this point E.

4. Select points C and E. From the **Construct** menu, choose **Segment**.

5. Select point A and line segment CE. From the **Construct** menu, choose **Circle By Center and Radius**.

6. Select point B and line segment CE. From the **Construct** menu, choose **Circle By Center and Radius**.

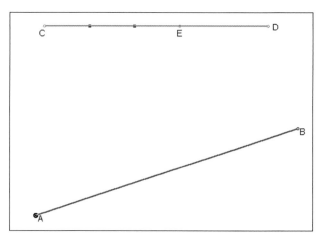

7. Select both circles. From the **Construct** menu, choose **Point At Intersection**. Label these points F and G.

8. Select points F and A. From the **Construct** menu, choose **Segment**. Measure the length of FA. Construct FB and measure its length. Are these lengths equal? If so, then point F satisfies the definition of the locus. Are the lengths of GA and GB also equal to these lengths? If so, then point G also satisfies the condition of the locus.

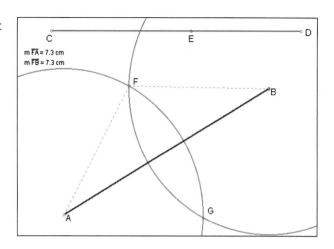

m FA = 7.3 cm
m FB = 7.3 cm

9. Select points F and G. From the **Display** menu, choose **Trace Points**.

10. Select the two circles. From the **Display** menu, choose **Hide Circles**.

11. Drag point E back and forth along line segment CD. The path of the points F and G represents the locus of points that are equidistant from point A and point B.

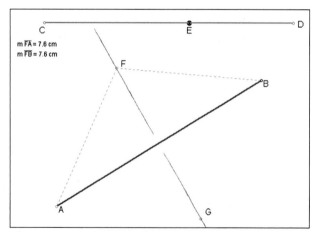

m FA = 7.6 cm
m FB = 7.6 cm

12. Select point F and point G. From the **Construct** menu, choose **Segment**.

13. Select line segments FG and AB. From the **Construct** menu, choose **Point At Intersection**. Label this point H.

14. Measure ∠AHF, ∠AHG, ∠BHF, and ∠BHG. Drag point A, point B, or point E to a new location. Do the measures of these angles change?

15. Measure line segments AH and BH. How do the lengths compare? Will they always compare in this way?

16. Describe the locus of points obtained from this construction. To test the construction and your description, drag point A or point B to a new location and create the new locus.

17. Extension a) Select Point E and line segment CD. From the **Edit** menu, choose **Action button** and sub-option **Animation**. Change the first and third drop boxes, as shown in the window.

Click on the **Animate button**. A new button, Animate , will be created on the screen. Double click this button. Point C will be dragged along the line segment and the locus will be left on the screen.

To test the construction, drag either of points A or B to a new location and double click on the Animate button.

b) Select points F and G. From the **Display** menu choose **Trace Points**. This will remove the checkmark beside **Trace Points,** so that this feature is deactivated. Select point F and point E in that order. From the **Construct** menu, choose **Locus**. This will cause point E to be moved along line segment CD. The various locations of point F will not only be displayed on the screen, but will remain there. Repeat this process but select point G and point E. Drag one of the endpoints of AB to a new location, and describe what happens.

Points Equidistant From the Sides of an Angle

You are given ∠BAC, formed by drawing two rays AB and AC. Point D is on AB and point E is on AC, so that AD = AE. Use the following steps to find the locus of points that are equidistant from points D and E.

18. Click the **Line Tool** on the toolbox and hold the left mouse button down. A window will open, giving a choice of a line segment, a ray, or a line. Choose the ray, and construct ray AB in the same way as a line segment is constructed. Construct a second ray AC.

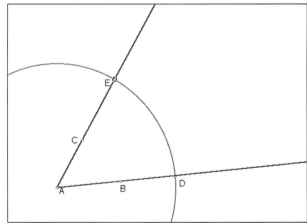

19. Select ray AB. From the **Construct** menu, choose **Point On Object**. Label this point D.

20. Select points A and D, in that order. From the **Construct** menu, choose **Circle by Center and Radius.**

21. Select the circle and ray AC. From the **Construct** menu, choose **Point At Intersection**. Label the new point E. Now, points D and E are the same distance from the vertex, A.

22. Select the circle. Choose **Hide Circle** from the **Display** menu.

23. Create a line segment FG at the bottom of the screen. With line segment FG selected, choose **Point on Object** from the **Construct** menu. Label this new point H.

24. Select points F and H. From the **Construct** menu, choose **Segment.**

25. Select point D and line segment FH. From the **Construct** menu, choose **Circle By Center And Radius.**

26. Select Point E and line segment FH. From the **Construct** menu, choose **Circle By Center And Radius.**

27. Select the two circles. From the **Construct** menu, choose **Point At Intersection**. Label these points I and J.

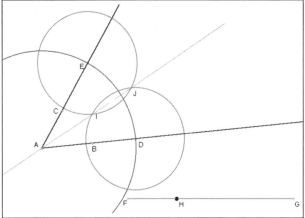

28. With points I and J selected, choose **Trace Points** from the **Display** menu.

29. Drag point H along line segment FG. The locus of points forms a line passing through points A, I, and J.

30. Select points J, A, and D in that order. From the **Measure** menu, choose **Angle**. Select points J, A, and E in that order. From the **Measure** menu, choose **Angle**. How do the measures of the angles compare? Will they always compare in this way?

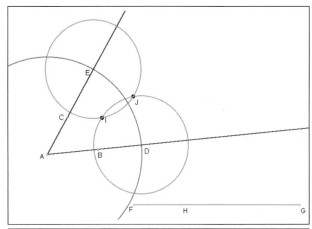

31. Describe the locus of points created by this construction. To test the construction and your description, drag any of points A, B, or D to a new location, and drag point H along FG.

32. Extension a) Select point H and line segment FG. From the **Edit** menu, choose **Action Button**. Choose the **Animation** sub-option. Change the first drop box to "once" and the third drop box to "slowly." Click on Animate. A new button, ⟨⏵ Animate⟩, appears on the screen. Double click this button to get the locus of points.

b) Click points J and I. From the **Display** menu, choose **Trace Points**. This will deactivate the trace feature for these two points. Select point J and point H in that order. From the **Construct** menu, choose **Locus**. Select point I and point H. From the **Construct** menu, choose **Locus**. The locus of points traced out by points I and J, as point H is dragged along FG, is shown and remains on the screen. Increase and decrease the measure of the original angle by dragging point C or point B. Does the construction work for an obtuse angle?

Points Equidistant From Two Parallel Lines

To construct the locus of points that are equidistant from two parallel lines, we will first construct a point on each line so that the line segment joining the points is perpendicular to the parallel lines.

33. Construct a line through points A and B. Make sure the **Line Tool** is selected in the toolbox at the left side of the screen. Construct a point C above the line.

34. Select point C and the line through A and B. From the **Construct** menu, choose **Parallel Line**.

35. With the new line selected, choose **Point On Object** from the **Construct** menu. Label this point D.

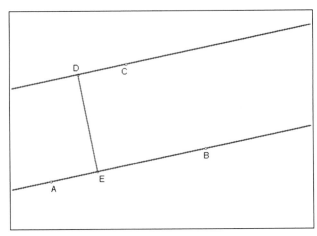

36. Select point D and the line through A and B. From the **Construct** menu, choose **Perpendicular Line**.

37. Select this new line and the line through A and B. From the **Construct** menu, choose **Point At Intersection**. Label this point E.

38. Select the line through D and E. From the **Display** menu, choose **Hide Line**.

39. Select points D and E. Change the **Line Tool** to line segment. From the **Construct** menu, choose **Segment**.

40. Construct a line segment FG at the bottom of the screen. With line segment FG selected, choose **Point On Object** from the **Construct** menu. Label this point H.

41. Select points F and H. From the **Construct** menu, choose **Segment.**

42. Select point D and line segment FH. From the **Construct** menu, choose **Circle By Center and Radius.**

43. Select point E and line segment FH. From the **Construct** menu, choose **Circle By Center and Radius.**

44. Select both circles. From the **Construct** menu, choose **Point At Intersection**. Label the points I and J. With the two points selected, choose **Trace Points** from the **Display** menu.

45. Drag point H along line segment FG.

46. Describe the locus of points. To test the construction and your description, drag any of points A, B, or C to a new location, and drag point H again.

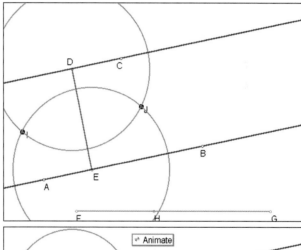

47. Extension a) Select point H and line segment FG. From the **Edit** menu, choose **Action Button**. Choose the **Animation** sub-option. Change the first drop box to "once" and the third drop box to "slowly." Click on Animate. A new button, $\boxed{\leftrightarrows \text{Animate}}$, appears on the screen. Double click this button to get the locus of points.

b) Click points J and I. From the **Display** menu, choose **Trace Points** to remove the trace feature. Select point J and point H in that order. From the **Construct** menu, choose **Locus**. Select point I and point H. From the **Construct** menu, choose **Locus**. The locus of points traced out by points I and J, as point H is dragged along FG, is

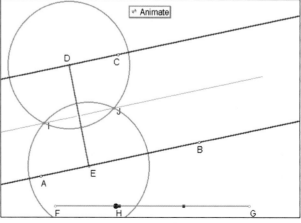

shown and remains on the screen. To test the construction, drag A or B to a new location to change the slope of the original parallel lines. Dragging point C changes the original distance between the parallel lines.

8.2 Equations of Loci

A locus is a set of points defined by a rule or condition. For example, if a dog is attached by a 10-m leash to a post in the middle of a large yard, then the locus of the farthest points that the dog can reach is a circle with a radius of 10 m.

In analytic geometry, equations can be used to describe loci.

INVESTIGATE & INQUIRE

1. a) Draw a diagram that shows six points that are exactly 2 units from the *x*-axis.
b) How would you change your diagram to show all the points that are exactly 2 units from the *x*-axis?
c) Write an equation, or equations, to describe the points in part b), that is, the locus of points that are exactly 2 units from the *x*-axis.

2. a) How are the lines $y = x + 5$ and $y = x - 7$ related?
b) Draw a diagram to show the two lines and the locus of points that are equidistant from both of them.
c) Write an equation to describe this locus.

3. a) Use a diagram to represent the locus of points that are equidistant from both axes.
b) Write an equation, or equations, to describe this locus.

To determine an equation to represent a described locus, use the following steps.
Step 1 Construct a diagram showing the given information.
Step 2 Locate several points that satisfy the rule or condition.
Step 3 Draw a curve or line using the located points.
Step 4 Write an equation for the locus.

EXAMPLE 1 **Constructing a Locus**

For $\angle ABC = 60°$, construct a geometric model to represent the locus of points in the interior of the angle that are equidistant from the sides.

SOLUTION

Step 1 Construct $\angle ABC = 60°$.

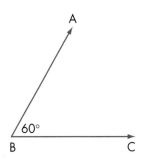

Step 2 Use two clear plastic rulers to locate several points whose perpendicular distances from each side of the angle are equal.

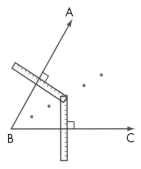

Step 3 Draw a ray through the points.

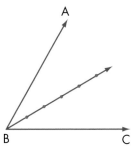

Note that, in Example 1, the locus of points in the interior of the angle that are equidistant from the sides is the angle bisector.

EXAMPLE 2 Determining an Equation for a Locus

a) Determine an equation to represent the locus of points equidistant from the points A(4, 3) and B(−2, 1).

b) Determine the relationship between the line segment AB and the locus of points equidistant from the points A(4, 3) and B(−2, 1).

SOLUTION

a) *Step 1* Plot points A(4, 3) and B(−2, 1) on a coordinate grid.

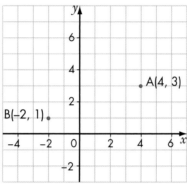

Step 2 Locate several points that are equidistant from A and B.

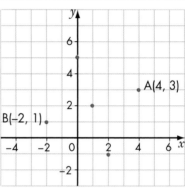

Step 3 Draw a line through the points.

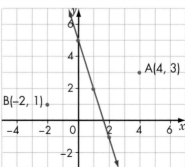

Step 4 To determine an equation for the locus of points, use the formula for the length of a line segment, $l = \sqrt{(x_2 - x_1)^2 + (y_2 - y_1)^2}$.

Let $P(x, y)$ be any point equidistant from $A(4, 3)$ and $B(-2, 1)$.

So, $PA = PB$.

$$PA = \sqrt{(x_2 - x_1)^2 + (y_2 - y_1)^2}$$
$$ = \sqrt{(x - 4)^2 + (y - 3)^2}$$
$$PB = \sqrt{(x_2 - x_1)^2 + (y_2 - y_1)^2}$$
$$ = \sqrt{(x - (-2))^2 + (y - 1)^2}$$
$$ = \sqrt{(x + 2)^2 + (y - 1)^2}$$

Since $PA = PB$,

$$\sqrt{(x - 4)^2 + (y - 3)^2} = \sqrt{(x + 2)^2 + (y - 1)^2}$$

Square both sides: $\quad (x - 4)^2 + (y - 3)^2 = (x + 2)^2 + (y - 1)^2$

Expand: $\quad x^2 - 8x + 16 + y^2 - 6y + 9 = x^2 + 4x + 4 + y^2 - 2y + 1$

Simplify: $\quad -12x - 4y + 20 = 0$

Divide both sides by -4: $\quad 3x + y - 5 = 0$

An equation of the locus is $3x + y - 5 = 0$.

b) The slope of the line segment joining $A(4, 3)$ and $B(-2, 1)$ is

$$m_{AB} = \frac{y_2 - y_1}{x_2 - x_1}$$
$$\phantom{m_{AB}} = \frac{1 - 3}{-2 - 4}$$
$$\phantom{m_{AB}} = \frac{-2}{-6}$$
$$\phantom{m_{AB}} = \frac{1}{3}$$

The slope of the locus can be found by writing the equation from part a) in the form $y = mx + b$, where m is the slope.

$$3x + y - 5 = 0$$
$$y = -3x + 5$$

So, the slope of the locus is -3.

Since the slopes $\dfrac{1}{3}$ and -3 are negative reciprocals, the locus and the line segment AB are perpendicular.

The midpoint of the line segment joining $A(4, 3)$ and $B(-2, 1)$ can be found using the midpoint formula, $\left(\dfrac{x_1 + x_2}{2}, \dfrac{y_1 + y_2}{2} \right)$.

$$\left(\frac{x_1 + x_2}{2}, \frac{y_1 + y_2}{2}\right) = \left(\frac{4 + (-2)}{2}, \frac{3 + 1}{2}\right)$$
$$= \left(\frac{2}{2}, \frac{4}{2}\right)$$
$$= (1, 2)$$

The point (1, 2) satisfies the equation of the locus, $3x + y - 5 = 0$, since $3(1) + 2 - 5 = 0$. So, the locus intersects the segment AB at its midpoint.

The locus of points equidistant from the points A(4, 3) and B(−2, 1) is the right bisector of the line segment AB.

Key Concepts

• A locus is a set of points defined by a rule or condition.
• To determine an equation to represent a described locus, use the following steps.
  Step 1 Construct a diagram showing the given information.
  Step 2 Locate several points that satisfy the rule or condition.
  Step 3 Draw a curve or line using the located points.
  Step 4 Write the equation.

Communicate Your Understanding

1. Write an equation to represent the locus of points that are 7 units from the point (0, 0) on a coordinate grid.
2. Does a linear equation represent the locus of points that are equidistant from the vertices of a rectangle? Explain.

Practise

A

1. Sketch and describe the locus of points in the plane that are
a) in the interior of a right angle and equidistant from the sides
b) equidistant from two parallel lines 6 cm apart
c) n units away from a given line l

d) equidistant from two concentric circles, one with radius 6 cm and the other with radius 4 cm
e) equidistant from the vertices of a square

2. Determine an equation that represents the locus of points equidistant from $x = -2$ and $x = 4$.

3. a) How are the lines $y = 2x + 1$ and $y = 2x - 3$ related?
b) Determine an equation for the locus of points equidistant from the two lines.

4. Determine an equation for the locus of points equidistant from each pair of points.
a) $(4, -3)$ and $(2, -5)$ **b)** $(2, 4)$ and $(5, -2)$
c) $(-3, 5)$ and $(2, -1)$

5. a) On a coordinate grid, construct two circles, one with radius 3 cm, the other with radius 7 cm, and both with centre $(0, 0)$.
b) Determine an equation to represent the locus of points equidistant from the two circles.

Apply, Solve, Communicate

6. Ambulance station Two hospitals are located at $(4, -1)$ and $(3, 7)$. An ambulance station is to be built such that it is equidistant from the two hospitals. Determine an equation for the locus of points equidistant from the two hospitals.

7. Home purchase Andrea is purchasing a new home. She would like to be close to downtown and to her office, and the same distance from them. City Hall is downtown and is located at $(-1, -3)$. Andrea's office is located at $(4, -5)$. Find the equation of the locus of points equidistant from downtown and Andrea's office.

B

8. Highway driving Sketch the locus of points traced by each of the following on a car being driven down a highway.
a) the centre of a wheel
b) a point on the outer edge of a tire

9. a) Verify that the points $(-3, 4)$ and $(4, 3)$ lie on the circle $x^2 + y^2 = 25$.
b) Determine an equation for the locus of points equidistant from the points $(-3, 4)$ and $(4, 3)$.
c) Determine the relationship between the centre of the circle and the locus of points equidistant from the points $(-3, 4)$ and $(4, 3)$.

10. Inquiry/Problem Solving Sketch the set of ordered pairs. Then, write an equation of a locus that all the points in each set might satisfy.
a) $\{(-5, 0), (5, 0), (0, -5), (0, 5)\}$
b) $\{(0, 0), (-1, 1), (1, 1), (-2, 4), (2, 4), (-3, 9), (3, 9)\}$
c) $\{(0, 0), (1, 1), (4, 2), (9, 3), (16, 4), (25, 5)\}$

11. Determine an equation, or equations, to represent the locus of points equidistant from each pair of lines.

a) $y = x$ and $y = -x$

b) $y = 2x + 2$ and $y = -2x + 2$

c) $y = 2x$ and $y = 0.5x$

12. Determine an equation for the locus of points equidistant from each pair of graphs.

a) $y = -\sqrt{x}$ and $y = \sqrt{x}$

b) $y = \sqrt{x} + 4$ and $y = -\sqrt{x} - 6$

13. Determine an equation for the locus of points equidistant from the points on the graph of $y = 2(x - 3)^2 + 2$.

14. Flower bed The outside edge of a fountain is the locus of points 2 m from the centre. The outside edge of a flower bed is the locus of points 3 m from the centre of the fountain. There is no gap between the flower bed and the fountain. Sketch the flower bed and calculate its area.

15. Describe and sketch the locus of points in the plane that are 13 units from the origin and 12 units from the y-axis.

16. Communication Sketch each of the following loci in the plane. Explain how your sketch represents each locus.

a) the points 2 cm or less from a given line

b) the points at least 3 cm from a given point

c) the points from 2 cm to 5 cm from a given point

17. Application Loci can be described in three dimensions. Describe each of the following loci in two dimensions and in three dimensions.

a) the locus of points 3 cm from a given point

b) the locus of points 2 cm from a given line

c) the locus of points equidistant from two given points

d) the locus of points equidistant from two parallel lines

8.3 Technology: Loci and Conics

The diagram shows a double cone. The two cones have one point in common.

The intersection of a double cone and a plane is called a **conic section** or a **conic**.

The circle, ellipse, parabola, and hyperbola are the cross sections that can be formed when a double cone is sliced by a plane. The different cross sections are formed by changing the angle and location of the plane.

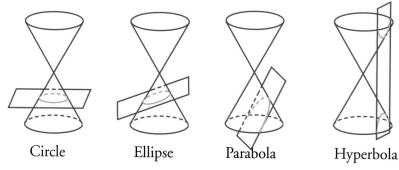

Circle   Ellipse   Parabola   Hyperbola

In this section, the locus definitions of an ellipse, a hyperbola, and a parabola will be used to construct geometric models of these conics using *The Geometer's Sketchpad®.*

The Ellipse

An **ellipse** is the set or locus of points P in the plane such that the sum of the distances from P to two fixed points, F_1 and F_2, is a constant.

$$F_1P + F_2P = k$$

One method of constructing this locus using *The Geometer's Sketchpad®* is described in the following steps.

1. Construct a line segment in the lower portion of the screen. Label the endpoints F1 and F2.

2. Construct line segment CD in the upper portion of the screen. CD must be longer than F1F2.

3. With line segment CD selected, choose **Point on Object** from the **Construct** menu. Label this point E.

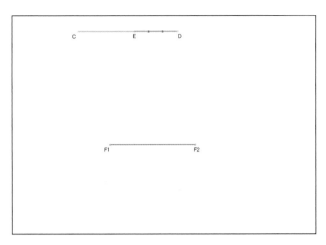

4. Select points C and E. From the **Construct** menu, choose **Segment**.

5. Select points D and E. From the **Construct** menu, choose **Segment**.

6. Select point F1 and segment CE. From the **Construct** menu, choose **Circle By Center And Radius**.

7. Select point F2 and segment DE. From the **Construct** menu, choose **Circle By Center And Radius**. If necessary, drag point E so that the two circles intersect.

8. Select the two circles. From the **Construct** menu, choose **Point At Intersection**. Label the points P1 and P2.

9. Select F1 and P1. From the **Construct** menu, choose **Segment**.

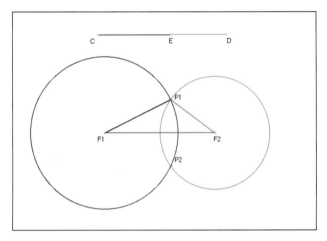

10. Select F2 and P1. From the **Construct** menu, choose **Segment**.

11. Select line segment CD. From the **Display** menu, choose **Hide Segment**.

12. Select line segment CE, line segment F1P1, and the circle with centre F1. Right click and choose a colour to be applied to these three objects.

13. Select line segment ED, line segment F2P1, and the circle with centre F2. Right click and choose a different colour to be applied to these three objects.

14. Select points P1 and P2. From the **Display** menu, choose **Trace Points**. Drag point E along line segment CD to generate the locus of points. Describe the resulting shape.

15. Explain why the sum of F1P1 + F2P1 = CD. *Hint:* Check the colour used for F1P1 and CE, and then the colour used for F2P1 and DE.

16. Test the construction by changing the length of the line segment joining F1 and F2. Describe how you would change the length of this line segment in order to get ellipses that are wider and narrower.

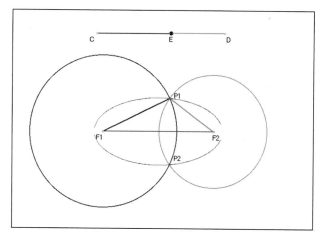

17. Extension a) Select point E and line segment CD. You may need to choose **Show All Hidden** from the **Display** menu. From the **Edit** menu, choose **Action Button**. Choose the **Animation** sub-option. Change the first drop box to "once" and the third to "slowly." Click on Animate to close the window. Double click on ⇌ Animate to show an ellipse. Experiment with the placement of points F1 and F2 and the length of line segment CD to obtain wider and narrower ellipses.

b) From the **Graph** menu, choose **Show Grid**. Place F1 and F2 on the *x*-axis equal distances from the origin. When the construction is completed, what is the distance between the points of intersection of the locus and the *x*-axis? How is this distance related to the length of line segment CD?

c) Select point P1 and point P2. From the **Edit** menu, choose **Trace Points**. This will remove the trace feature. Delete the Animate button. Select point P1 and point E in that order. From the **Construct** menu, choose **Locus**. Repeat for P2 and point E. The result is a locus of points for each of P1 and P2. What happens to this locus of points if you drag point F1?

The Hyperbola

The **absolute value** of a real number is its distance from zero on a real number line. Since distance is always positive, the absolute value of a number is always positive.

For a real number represented by x, the absolute value is written $|x|$, which means the positive value of x.

The absolute value of –2 is 2, or $|-2| = 2$.
Similarly $|4 - 9| = |-5|$
$$= 5$$
And $|9 - 4| = |5|$
$$= 5$$

A **hyperbola** is the set or locus of points P in the plane such that the absolute value of the difference of the distances from P to two fixed points, F_1 and F_2, is a constant.

$$|F_1P - F_2P| = k$$

One method of constructing this locus using *The Geometer's Sketchpad®* is described in the following steps.

18. Change the **Line Tool** to line and construct a line through points A and B near the top of the screen. Place points A and B as close to the centre of the top of the screen as possible.

19. From the **Construct** menu, choose **Point On Object**. Label this point C. If necessary, move C so that it is not between A and B.

20. Change the **Line Tool** to line segment. Near the bottom of your screen, construct a line segment. Name the endpoints F1 and F2.

21. Measure the lengths of AB and F1F2. If necessary drag A or B so that AB is the shorter segment.

22. Construct line segments AC and BC. The difference between the lengths of these two line segments will always equal the length of line segment AB and will serve as the constant for the locus.

23. Select point F1 and line segment AC. From the **Construct** menu, choose **Circle By Center And Radius**.

24. Select point F2 and line segment BC. From the **Construct** menu, choose **Circle By Center And Radius**.

25. Select the two circles. From the **Construct** menu, choose **Point At Intersection**. Label these two points P1 and P2.

26. Select points P1 and F1. From the **Construct** menu, choose **Segment**.

27. Select points P1 and F2. From the **Construct** menu, choose **Segment**.

28. Select and measure the line segments AC, BC, P1F1, and P1F2. From the **Measure** menu, choose **Calculate**. Use the calculator to find $|m(AC) - m(BC)|$ and $|m(P1F1) - m(P1F2)|$.

29. Drag point C along the line through A and B. Notice which segments and calculated values are equal.

30. Select P1 and P2. From the **Display** menu, choose **Trace Points**.

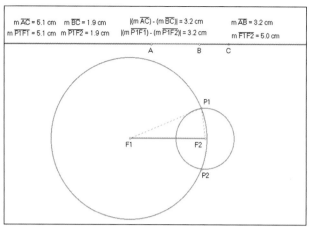

31. Select the two circles. From the **Display** menu, choose **Hide Circles**.

32. Drag point C along the line through A and B. For the best effect, start at one side of the screen and drag point C to the other side slowly.

33. Describe the locus of points generated by this construction.

34. Extension a) Change the length of line segment F1F2. How can you make the curves flatten? Can you get a similar effect by changing the length of line segment AB?

b) Select point C and the line through A and B. From the **Edit** menu, choose **Action Button**. Choose the **Animation** sub-option. Change the first drop box to "once" and the third drop box to "slowly." Click on Animate. Double click on ⇌ Animate.

c) Select Point P1 and point P2. From the **Edit** menu, choose **Trace Points**. This will remove the trace feature. Delete the Animate button. Select point P1 and point C in that order. From the **Construct** menu, choose **Locus**. Repeat for P2 and point C. The result is a locus of points for each of P1 and P2. What happens to this locus of points if you drag point F1? What happens when the length of F1F2 is shorter than the length of AB?

The Parabola

A **parabola** is the set or locus of points P in the plane such that the distance from P to a fixed point F equals the distance from P to a fixed line l.

$$PF = PD$$

The fixed point F is called the **focus**. The fixed line l is called the **directrix**.

One method of constructing this locus using *The Geometer's Sketchpad*® is described in the following steps.

35. Construct a line AB near the bottom of the screen. Label this line d for directrix.

36. Construct a point not on line d. Label it F, for focus.

37. Select point A and line d. From the **Construct** menu, choose **Perpendicular Line.**

38. With this perpendicular line selected, choose **Point On Object** from the **Construct** menu. Label this point D.

39. Select points A and D. Change the line tool to a segment tool. From the **Construct** menu, choose **Segment.**

40. Select point F and segment AD. From the **Construct** menu, choose **Circle By Center And Radius.** This creates a set of points a fixed distance from the focus.

41. We now create a set of points that are the same distance from line d. Select points A and D in that order. From the **Transform** menu, choose **Mark Vector "A→D".**

42. Select line d. From the **Transform** menu, choose **Translate....** Choose the option **By Marked Vector.** Press OK.

43. Select this new line through D and the circle. From the **Construct** menu, choose **Point At Intersection.** Label these points P and Q.

44. With points P and Q selected, choose **Trace Points** from the **Display** menu.

45. Select the circle and the line through D and P. From the **Display** menu, choose **Hide Objects.**

46. Drag point D back and forth along line AD. The path of the points P and Q represents the locus of points that are equidistant from point F and line d.

47. Describe the locus of points you constructed.

48. To test this construction, place point F at different positions. Describe the locus when the focus, point F, is closer to the directrix, line AB. Describe the locus when point F is farther away.

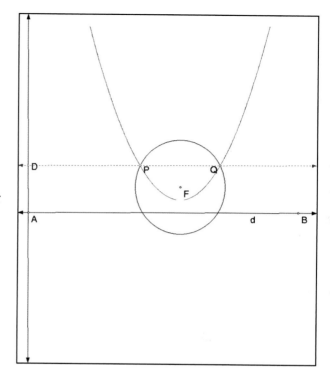

49. Extension a) To start with a grid, choose **Show Grid** from the **Graph** menu. Construct the directrix as a line parallel to and below the *x*-axis. Place point F on the *y*-axis the same number of units above the origin as the directrix is below the *x*-axis. When the construction is complete, where is the vertex of the parabola?

b) Select point D and line AD. From the **Edit** menu, choose **Action Button**. Choose the **Animation** sub-option. Change the first drop box to "once" and the third drop box to "slowly." Click on Animate. Double click ⇌ Animate to get the locus of points.

c) Select points P and Q. From the **Edit** menu, choose **Trace Points**. This will remove the tracing of points P and Q. Click on points P and D. From the **Construct** menu, choose **Locus**. A locus of points will be created. Then select points Q and D. Again choose **Locus** from the **Construct** menu to complete the shape on the screen. Drag point F to new locations and watch the locus of points change. Describe the effect that the position of F has on the locus.

8.4 The Circle

The epicentre of an earthquake is where the earthquake originates on the Earth's surface. Seismologists find an epicentre by taking seismic readings from three recording stations in different locations. Each reading indicates the distance of the epicentre from the station where the reading is recorded. However, a single reading does not indicate the direction of the epicentre relative to that station. In this section, circles will be used to determine the location of an epicentre.

INVESTIGATE & INQUIRE

Recall that the equation of a circle having centre O(0, 0) and radius r is $x^2 + y^2 = r^2$. An equation of a circle with centre O(0, 0) and radius 6 is $x^2 + y^2 = 36$.

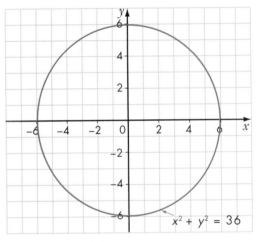

$$x^2 + y^2 = 36$$

The graph shows a circle with centre C(2, 3) and radius 6. Point P(x, y) is on the circle. CP is a radius of the circle.

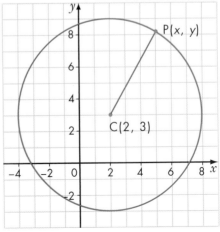

1. The formula for the length of a line segment is $l = \sqrt{(x_2 - x_1)^2 + (y_2 - y_1)^2}$, which can be rewritten as $\sqrt{(x_2 - x_1)^2 + (y_2 - y_1)^2} = l$. Using the second version of the formula, substitute the coordinates of P for x_2 and y_2, and the coordinates of C for x_1 and y_1. Substitute the value of the radius for l.

2. Square both sides of the equation. Do not simplify.

3. Describe how the equation of the circle $x^2 + y^2 = 36$ compares with the equation of the circle you found in question 2.

4. Describe how the graph of the circle with the equation $x^2 + y^2 = 36$ compares with the graph of the circle with the equation you found in question 2.

5. What transformation can be applied to $x^2 + y^2 = 36$ to give the equation of the circle with centre C(2, 3) and radius 6?

6. Write an equation of the circle with centre C(4, 5) and radius 6.

7. Write an equation of the circle with centre C(-2, -6) and radius 2.

8. Write an equation of the circle with centre C(h, k) and radius r.

9. When you use a compass to draw a circle, how are you using the locus definition of a circle to construct a geometric model of a circle?

A **circle** can be defined geometrically as the set or locus of points in the plane that are equidistant from a fixed point. The fixed point is called the **centre**. The distance from the centre to any point on the circle is called the **radius**.

To derive the equation of a circle with centre O(0, 0) and radius r, use the formula for the length of a line segment.
Let P(x, y) be any point on the circle.
Then, OP = r.
The formula for the length of a line segment is $l = \sqrt{(x_2 - x_1)^2 + (y_2 - y_1)^2}$.
Substitute r for l, (x, y) for (x_2, y_2), and (0, 0) for (x_1, y_1).

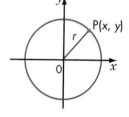

$$l = \sqrt{(x_2 - x_1)^2 + (y_2 - y_1)^2}$$

Substitute: $\quad r = \sqrt{(x - 0)^2 + (y - 0)^2}$

Simplify: $\quad r = \sqrt{x^2 + y^2}$

Square both sides: $\quad r^2 = x^2 + y^2$

or $x^2 + y^2 = r^2$

The standard form of the equation of a circle with centre (0, 0) and radius r is $x^2 + y^2 = r^2$.

EXAMPLE 1 Writing an Equation of a Circle, Given its Centre and Radius

Write an equation for the circle with centre (0, 0) and radius 5.
Sketch the circle.

SOLUTION

An equation for the circle, with centre (0, 0) and radius r, is
$$x^2 + y^2 = r^2$$
Substitute 5 for r: $x^2 + y^2 = 5^2$
$$x^2 + y^2 = 25$$
An equation for the circle with centre (0, 0) and radius 5
is $x^2 + y^2 = 25$.

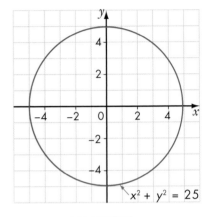

The formula for the length of a line segment can also be used to derive
the equation of a circle with centre (h, k) and radius r.

Let $P(x, y)$ be any point on a circle with centre $C(h, k)$.
Then, $CP = r$.
The formula for the length of a line segment is $l = \sqrt{(x_2 - x_1)^2 + (y_2 - y_1)^2}$.
Substitute r for l, (x, y) for (x_2, y_2), and (h, k) for (x_1, y_1).

$$l = \sqrt{(x_2 - x_1)^2 + (y_2 - y_1)^2}$$
Substitute: $\quad r = \sqrt{(x - h)^2 + (y - k)^2}$
Square both sides: $\quad r^2 = (x - h)^2 + (y - k)^2$
$$\text{or } (x - h)^2 + (y - k)^2 = r^2$$

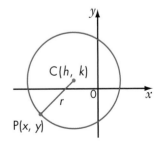

The standard form of the equation of a circle with centre (h, k) and
radius r is $(x - h)^2 + (y - k)^2 = r^2$.

Comparing $(x - h)^2 + (y - k)^2 = r^2$ with $x^2 + y^2 = r^2$,
• if h is positive, the circle is translated h units to the right
• if h is negative, the circle is translated h units to the left
• if k is positive, the circle is translated k units upward
• if k is negative, the circle is translated k units downward

EXAMPLE 2 Writing an Equation of a Circle, Given its Centre and Radius

Write an equation in standard form for the circle with centre (3, −1) and radius 4. Sketch a graph of the circle, and state the domain and range.

SOLUTION

Substitute known values into the standard form of the equation of a circle.

In this case, $h = 3$, $k = -1$, and $r = 4$.

$$(x - h)^2 + (y - k)^2 = r^2$$
$$(x - 3)^2 + (y - (-1))^2 = 4^2$$
$$(x - 3)^2 + (y + 1)^2 = 16$$

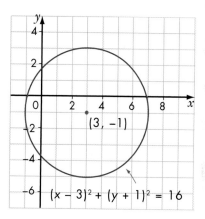

So, an equation of a circle with centre (3, −1) and radius 4 is $(x - 3)^2 + (y + 1)^2 = 16$.

The domain is $-1 \le x \le 7$.

The range is $-5 \le y \le 3$.

EXAMPLE 3 Writing an Equation of a Circle, Given its Centre and a Point on the Circle

Write an equation, in standard form, for the circle with centre C(−2, 3) and passing through point P(4, 1).

SOLUTION

Draw a diagram.

Use the Pythagorean theorem to find r^2.

$$r^2 = (4 - (-2))^2 + (1 - 3)^2$$
$$= 6^2 + (-2)^2$$
$$= 40$$

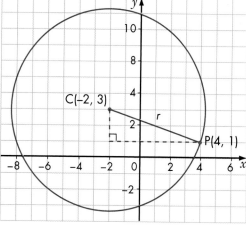

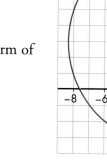

Substitute known values into the general form of the equation of a circle.

In this case, $h = -2$, $k = 3$, and $r^2 = 40$.

$$(x - h)^2 + (y - k)^2 = r^2$$
$$(x - (-2))^2 + (y - 3)^2 = 40$$
$$(x + 2)^2 + (y - 3)^2 = 40$$

An equation of the circle with centre C(−2, 3) and passing through point P(4, 1) is $(x + 2)^2 + (y - 3)^2 = 40$.

EXAMPLE 4 Basketball

The centre circle on a basketball court has a radius of 1.8 m. The length of
the court is 26 m and the width is 14 m. A coordinate grid is placed on the
court so that the court is in the first quadrant, the origin is at one corner of
the court, and two sides of the court lie along the *x*- and *y*-axes. The side of
each grid square represents 2 m. Write the equation of the centre circle if
a) a longer side of the court is on the *x*-axis
b) a shorter side of the court is on the *x*-axis

SOLUTION

a) Draw a diagram.
The centre of the circle is at the centre of the
court, (13, 7).
The equation of the circle is
$$(x - h)^2 + (y - k)^2 = r^2$$
so, $(x - 13)^2 + (y - 7)^2 = 1.8^2$
$$(x - 13)^2 + (y - 7)^2 = 3.24$$
The equation of the centre circle is
$(x - 13)^2 + (y - 7)^2 = 3.24$.

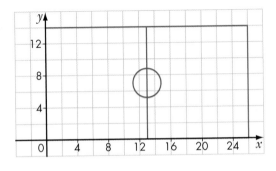

b) Draw a diagram.
The centre of the circle is at (7, 13).
The equation of the circle is
$$(x - h)^2 + (y - k)^2 = r^2$$
$$(x - 7)^2 + (y - 13)^2 = 3.24$$
The equation of the centre circle is
$(x - 7)^2 + (y - 13)^2 = 3.24$.

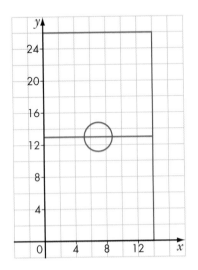

Note that circles can be graphed using a graphing calculator.
To graph $(x - 2)^2 + (y + 3)^2 = 16$, first solve the equation for y.

$$(x - 2)^2 + (y + 3)^2 = 16$$
$$(y + 3)^2 = 16 - (x - 2)^2$$
$$y + 3 = \pm\sqrt{16 - (x - 2)^2}$$
$$y = -3 \pm \sqrt{16 - (x - 2)^2}$$

Then, enter both of the resulting equations in the Y= editor.

$$Y_1 = -3 + \sqrt{16 - (x - 2)^2} \qquad Y_2 = -3 - \sqrt{16 - (x - 2)^2}$$

To see a circular shape, adjust the viewing window using the Zsquare instruction.

 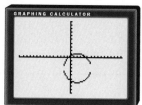

Note that the conics in this chapter can also be graphed using a graphing software program, such as Zap-a-Graph. For details of how to do this, refer to the Zap-a-Graph section of Appendix C.

Another way to graph a circle is to use the Circle instruction, which requires the coordinates of the centre and the radius. The graph shown is for the circle with centre $(2, -3)$ and radius 4. The Zsquare instruction was used to adjust the viewing window.

Key Concepts

• A circle is the set or locus of points in the plane equidistant from a fixed point. The fixed point is called the centre. The distance from the centre to any point on the circle is called the radius.
• The standard form of the equation of a circle with centre $(0, 0)$ and radius r is $x^2 + y^2 = r^2$.
• The standard form of the equation of a circle with centre (h, k) and radius r is $(x - h)^2 + (y - k)^2 = r^2$.

Communicate Your Understanding

1. For the circle $x^2 + y^2 = 25$, state whether each of the following statements is true or false. Explain your reasoning.

a) The circle is not a function.

b) The circle has infinite lines of symmetry.

c) The point $(6, -1)$ lies on the circle.

2. Describe the similarities and differences in the graphs of the circles $(x - 3)^2 + (y + 2)^2 = 25$ and $(x + 3)^2 + (y - 2)^2 = 25$.

3. Which of the following equations model the graph shown? Justify your choices.

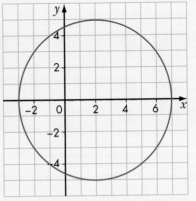

a) $x^2 + (y - 2)^2 = 25$ **b)** $(x + 2)^2 + y^2 = 25$

c) $(x - 2)^2 + y^2 = 25$ **d)** $(2 - x)^2 + y^2 = 25$

4. Describe how you would write an equation for each of the following circles.

a) centre $(2, 4)$, radius 3

b) centre $(-1, -2)$ and passing through the point $(3, 1)$

Practise

A

1. Write an equation in standard form for the circle with centre $(0, 0)$ and the given radius.

a) radius 3 **b)** radius 7.3

c) radius $\sqrt{2}$ **d)** radius $3\sqrt{5}$

2. Write an equation in standard form for each of the following circles, given the centre and the radius.

a) centre $(2, 5)$, radius 3

b) centre $(-1, 3)$, radius 4

c) centre $(3, -2)$, radius 5

d) centre $(0, 2)$, radius 8

e) centre $(-3, -4)$, radius $\sqrt{7}$

f) centre $(-5, 0)$, radius $2\sqrt{5}$

3. Sketch each circle and state the domain and range.

a) $x^2 + y^2 = 100$

b) $x^2 + y^2 = 80$

c) $(x - 3)^2 + (y - 5)^2 = 64$

d) $(x + 4)^2 + y^2 = 49$

e) $x^2 + (y + 3)^2 = 121$

f) $(x + 2)^2 + (y - 7)^2 = 50$

4. For each of the following circles, write an equation in standard form.

a)

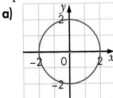

b)

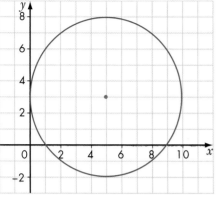

c)

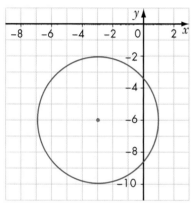

d)

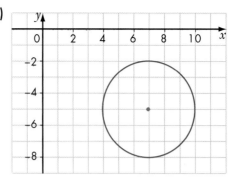

5. Write an equation in standard form for the circle with the given centre and passing through the given point.
a) centre (2, 3) and passing through (5, 7)
b) centre (−2, 4) and passing through (3, 8)
c) centre (3, −2) and passing through (−2, −3)
d) centre (−4, −1) and passing through (3, 5)

6. Write an equation in standard form for the circle with the given centre and intercept.
a) centre (4, −5), x-intercept 3
b) centre (2, 4), y-intercept −1

Apply, Solve, Communicate

7. Application The centre circle on a soccer pitch has a radius of about 9 m. The length and the width of the pitch can vary. For one pitch, the length is 100 m and the width is 60 m. A coordinate grid is placed on the pitch, with the side length of each grid square representing 1 m. Write the equation of the centre circle if

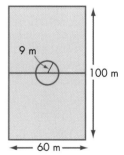

a) the origin is at the centre of the circle
b) the origin is at one corner of the pitch, with a longer side along the positive x-axis and a shorter side along the positive y-axis
c) the origin is at one corner of the pitch, with a shorter side along the negative x-axis and a longer side along the negative y-axis

B

8. a) Graph the family of circles $(x − 1)^2 + (y + 3)^2 = r^2$ for $r = 1, 2, 3$.
b) How are the graphs alike? How are they different?

9. Motion in space A satellite is in a geostationary orbit 35 880 km above a point on the equator. Write an equation that represents the path of the satellite. Assume that the origin is at the centre of the Earth and that the radius of the Earth is 6370 km.

10. Given the endpoints of a diameter, write an equation for the circle.
a) (−3, 5) and (1, 3)
b) (2, −1) and (−2, 3)
c) (−2, −3) and (−5, 1)
d) (2, 5) and (−3, −4)

11. Water ripples When a stone is dropped into a lake, circular waves or ripples are created. The entry point of the stone marks the centre of the circles. Suppose the radius increases at a rate of 20 cm/s. Write an equation for the outermost circle 2.5 s after the stone hits the water.

12. Play area a) Louis' dog, Grover, is tied to a pole in the backyard. The length of the leash is 7 m. Write an equation to represent the dog's maximum circular play area, if the pole is located at the origin of the coordinate grid system.
b) If Grover's water bowl is located at the point (3, −2), does he have access to water?

13. Search and rescue A private plane was forced to make an emergency landing. Two hours later, a search plane located the downed plane at the coordinates (3, −1), but the pilot was missing. The pilot of the search plane defined the search area by assuming that the missing pilot left immediately after landing and walked in a straight line at an average speed of 3 km/h. Write an equation for the circle that models the search area.

14. A point $(k, 2)$ lies on the circle $(x - 2)^2 + (y - 3)^2 = 12$. What are the possible values of k?

15. Technology Graph each circle using a graphing calculator.
a) $(x - 4)^2 + (y - 2)^2 = 36$
b) $x^2 + (y - 2)^2 = 9$
c) $(x + 3)^2 + (y - 1)^2 = 49$

16. Coins The Royal Canadian Mint used a centimetre grid to design a frame to hold a set of coins. Write an equation in standard form for the outline of each coin.

Coin	Diameter (mm)	Centre
Toonie	28.00	(0, 5)
Loonie	26.50	(3.5, 3.5)
50¢	27.13	(−3.5, 3.5)
25¢	23.88	(1.5, −4.5)
10¢	18.03	(3.5, −3)
5¢	21.20	(−1.5, −4.5)
1¢	19.05	(−3.5, −3)

8.5 The Ellipse

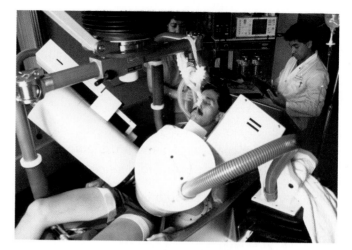

Kidney stones are crystal-like objects that can form in the kidneys. Traditionally, people have undergone surgery to remove them. In a process called lithotripsy, kidney stones can now be removed without surgery. To remove the stones, doctors can use a lithotripter, which means "stone crusher" in Greek.

A lithotripter is elliptically-shaped, and its design makes use of the properties of the ellipse to provide a safer method for removing kidney stones. This method will be explained in Example 5.

Web Connection
www.school.mcgrawhill.ca/resources/
To learn more about lithotripters, visit the above web site. Go to **Math Resources**, then to *MATHEMATICS 11*, to find out where to go next. Write a brief report about how lithotripters work.

An **ellipse** is the set or locus of points P in the plane such that the sum of the distances from P to two fixed points F_1 and F_2 is a constant.
$$F_1P + F_2P = k$$

The two fixed points, F_1 and F_2, are called the **foci** (plural of focus) of the ellipse. The line segments F_1P and F_2P are called **focal radii** of the ellipse.

INVESTIGATE & INQUIRE

You will need two clear plastic rulers, a sheet of paper, and a pencil for this investigation.

Step 1 Draw a 10-cm line segment near the centre of the piece of paper. Label the two endpoints F_1 and F_2.

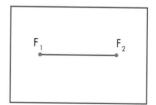

Step 2 Choose a length, k centimetres, which is greater than the length F_1F_2. You may want to make k less than 20 cm.

Step 3 Choose a pair of lengths, a centimetres and b centimetres, where $k = a + b$.

Step 4 Use both rulers to mark two points that are *a* centimetres from F_1 and *b* centimetres from F_2.

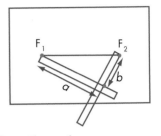

Step 5 Repeat steps 3 and 4 using different values for *a* and *b* until you have marked enough points to define a complete curve.

Step 6 Draw a smooth curve through the points. The curve is an example of an ellipse.

1. How many axes of symmetry does the ellipse have?

2. In relation to F_1 and F_2, where is the point of intersection of the axes of symmetry?

3. a) Why must *k* be greater than the length of the line segment F_1F_2?
b) What diagram would result if *k* equalled F_1F_2?

4. In step 4, you located two points for the chosen values of *a* and *b*. Are there any values of *a* and *b* for which only one point can be marked? If so, what are the values of *a* and *b*? Where is the location of the point on the ellipse?

The diagram at the right shows another method for drawing an ellipse. Pushpins at F_1 and F_2 are used to fasten a loop of string, which is over twice as long as F_1F_2. If you hold a pencil tight against the string, and move the pencil along the string, you will trace an ellipse.

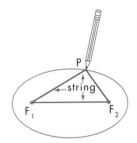

Since the length of string is a constant, the sum of the distances F_1P and F_2P is a constant for all positions of P.

EXAMPLE 1 Finding the Equation of an Ellipse From its Locus Definition

Use the locus definition of the ellipse to find an equation of an ellipse with foci $F_1(-4, 0)$ and $F_2(4, 0)$, and with the constant sum of the focal radii equal to 10.

SOLUTION

Let $P(x, y)$ be any point on the ellipse.

The locus definition of the ellipse can be stated algebraically as $F_1P + F_2P = 10$.

Use the formula for the length of a line segment, $l = \sqrt{(x_2 - x_1)^2 + (y_2 - y_1)^2}$, to rewrite F_1P and F_2P.

$$l = \sqrt{(x_2 - x_1) + (y_2 - y_1)^2}$$
$$F_1P = \sqrt{(x - (-4))^2 + (y - 0)^2}$$
$$= \sqrt{(x + 4)^2 + y^2}$$
$$F_2P = \sqrt{(x - 4)^2 + (y - 0)^2}$$
$$= \sqrt{(x - 4)^2 + y^2}$$

Substitute : $\qquad \sqrt{(x + 4)^2 + y^2} + \sqrt{(x - 4)^2 + y^2} = 10$

Isolate a radical: $\qquad \sqrt{(x + 4)^2 + y^2} = 10 - \sqrt{(x - 4)^2 + y^2}$

Square both sides: $\qquad (x + 4)^2 + y^2 = 100 - 20\sqrt{(x - 4)^2 + y^2} + (x - 4)^2 + y^2$

Simplify: $\qquad x^2 + 8x + 16 + y^2 = 100 - 20\sqrt{(x - 4)^2 + y^2} + x^2 - 8x + 16 + y^2$

Isolate the radical: $\qquad 16x - 100 = -20\sqrt{(x - 4)^2 + y^2}$

Divide both sides by 4: $\qquad 4x - 25 = -5\sqrt{(x - 4)^2 + y^2}$

Square both sides: $\qquad 16x^2 - 200x + 625 = 25((x - 4)^2 + y^2)$

Simplify: $\qquad 16x^2 - 200x + 625 = 25(x - 4)^2 + 25y^2$

$\qquad 16x^2 - 200x + 625 = 25(x^2 - 8x + 16) + 25y^2$

$\qquad 16x^2 - 200x + 625 = 25x^2 - 200x + 400 + 25y^2$

$\qquad 225 = 9x^2 + 25y^2$

Divide each side by 225: $\qquad 1 = \dfrac{x^2}{25} + \dfrac{y^2}{9}$

The equation of the ellipse is $\dfrac{x^2}{25} + \dfrac{y^2}{9} = 1$.

An ellipse has two axes of symmetry. The longer line segment is called the **major axis**, and the shorter line segment is called the **minor axis**. The endpoints on the major axis are the **vertices** of the ellipse. The endpoints of the minor axes are the **co-vertices** of the ellipse.

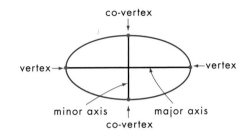

The ellipse in Example 1 can be modelled graphically, as shown.

Note that the coordinates of the vertices are (5, 0) and (−5, 0).

The length of the major axis is 10.

The coordinates of the co-vertices are (0, 3) and (0, −3).

The length of the minor axis is 6.

The equation can be written as $\dfrac{x^2}{5^2} + \dfrac{y^2}{3^2} = 1$.

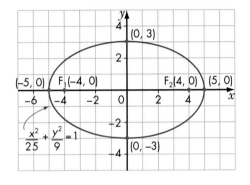

In this form of the equation, notice that 5 is half the length of the major axis, or half the constant sum of the focal radii, and 3 is half the length of the minor axis.

In the diagram, half the length of the major axis is denoted by a. Half the length of the minor axis is denoted by b. Half the distance between the two foci, which are on the major axis, is denoted by c.

If P is a point on the ellipse when $y = 0$, the focal radii $F_1P = a − c$ and $F_2P = a + c$.

$$F_1P + F_2P = a − c + a + c$$
$$= 2a$$

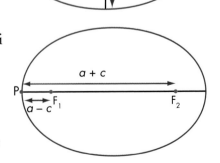

So, the constant sum of the distances from a point on the ellipse to the two foci is $2a$, which is the length of the major axis.

Let P be a point on the ellipse when $x = 0$.
Since $F_1P + F_2P = 2a$, and $F_1P = F_2P$, then
$F_1P = a$ and $F_2P = a$.

Using the Pythagorean theorem, $a^2 = b^2 + c^2$,
with $a^2 > b^2$.
Notice that, in Example 1, $a = 5$, $b = 3$, and $c = 4$,
and that $5^2 = 3^2 + 4^2$.

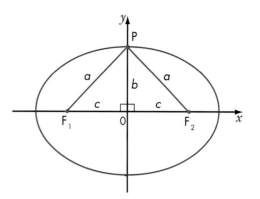

Therefore, the major axis has a length of $2a$ units, the minor axis has
a length of $2b$ units, and the distance between the two foci is $2c$ units.
The following diagrams show how the key points of ellipses centred at
the origin are labelled.

Horizontal major axis   **Vertical major axis**

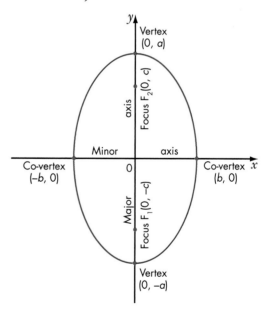

These results can be summarized as follows.

Ellipse with centre at the origin and major axis along the x-axis.

The standard form of the equation of an ellipse centred at the origin, with the major axis along the *x*-axis is

$$\frac{x^2}{a^2} + \frac{y^2}{b^2} = 1, \ a > b > 0$$

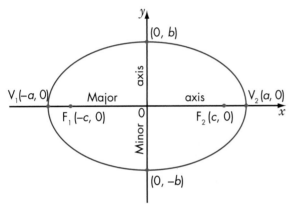

Ellipse with centre at the origin and major axis along the y-axis.

The standard form of the equation of an ellipse centred at the origin, with the major axis along the *y*-axis is

$$\frac{x^2}{b^2} + \frac{y^2}{a^2} = 1, \ a > b > 0$$

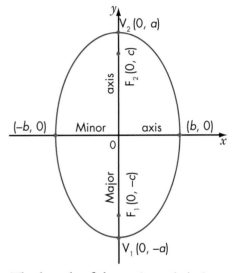

The length of the major axis is $2a$.
The length of the minor axis is $2b$.
The vertices are $V_1(-a, 0)$ and $V_2(a, 0)$.
The co-vertices are $(0, -b)$ and $(0, b)$.
The foci are $F_1(-c, 0)$ and $F_2(c, 0)$.
$a^2 = b^2 + c^2$

The length of the major axis is $2a$.
The length of the minor axis is $2b$.
The vertices are $V_1(0, -a)$ and $V_2(0, a)$.
The co-vertices are $(-b, 0)$ and $(b, 0)$.
The foci are $F_1(0, -c)$ and $F_2(0, c)$.
$a^2 = b^2 + c^2$

EXAMPLE 2 Sketching the Graph of an Ellipse With Centre (0, 0)

Sketch the graph of the ellipse $4x^2 + y^2 = 36$. Label the foci.

SOLUTION

Rewrite $4x^2 + y^2 = 36$ in standard form.

$$4x^2 + y^2 = 36$$

Divide both sides by 36: $\dfrac{x^2}{9} + \dfrac{y^2}{36} = 1$

Since the denominator of x^2 is less than the denominator of y^2, the

equation is in the form $\dfrac{x^2}{b^2} + \dfrac{y^2}{a^2} = 1$.

The ellipse is centred at the origin and the major axis is on the y-axis.
Since $a^2 = 36$, $a = 6$, and the vertices are $V_1(0, -6)$ and $V_2(0, 6)$.
Since $b^2 = 9$, $b = 3$, and the co-vertices are $(-3, 0)$ and $(3, 0)$.

$$a^2 = b^2 + c^2$$
$$6^2 = 3^2 + c^2$$
$$36 - 9 = c^2$$
$$27 = c^2$$
$$3\sqrt{3} = c$$

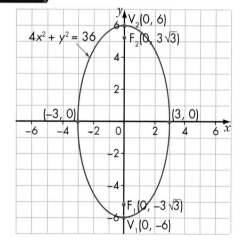

The foci are $F_1(0, -3\sqrt{3})$ and $F_2(0, 3\sqrt{3})$, or
approximately $(0, -5.2)$ and $(0, 5.2)$.

Plot the vertices and co-vertices.
Draw a smooth curve through the points.
Label the foci and the graph.

An ellipse may not be centred at the origin. As in the case of the circle, an
ellipse can have a centre (h, k). The translation rules that apply to the circle
also apply to the ellipse.

The standard form of the equation of an ellipse with centre (h, k) and the major axis parallel to the x-axis is

$$\frac{(x-h)^2}{a^2} + \frac{(y-k)^2}{b^2} = 1, \ a > b > 0$$

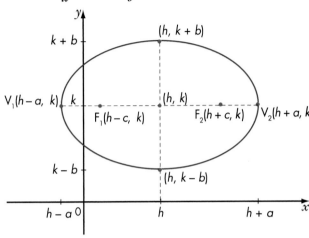

The length of the major axis is $2a$.
The length of the minor axis is $2b$.
The vertices are $V_1(h-a, k)$ and $V_2(h+a, k)$.
The co-vertices are $(h, k-b)$ and $(h, k+b)$.
The foci are $F_1(h-c, k)$ and $F_2(h+c, k)$.
$a^2 = b^2 + c^2$

The standard form of the equation of an ellipse with centre (h, k) and the major axis parallel to the y-axis is

$$\frac{(x-h)^2}{b^2} + \frac{(y-k)^2}{a^2} = 1, \ a > b > 0$$

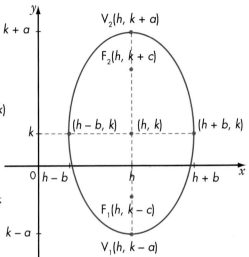

The length of the major axis is $2a$.
The length of the minor axis is $2b$.
The vertices are $V_1(h, k-a)$ and $V_2(h, k+a)$.
The co-vertices are $(h-b, k)$ and $(h+b, k)$.
The foci are $F_1(h, k-c)$ and $F_2(h, k+c)$.
$a^2 = b^2 + c^2$

EXAMPLE 3 Sketching the Graph of an Ellipse With Centre (h, k)

Sketch the graph of the ellipse $\dfrac{(x+1)^2}{25} + \dfrac{(y-3)^2}{16} = 1$. Label the foci.

SOLUTION

Since the denominator of $(x+1)^2$ is greater than the denominator of $(y-3)^2$, the equation $\dfrac{(x+1)^2}{25} + \dfrac{(y-3)^2}{16} = 1$ is in the form

$\dfrac{(x-h)^2}{a^2} + \dfrac{(y-k)^2}{b^2} = 1.$

The ellipse is centred at (h, k), or $(-1, 3)$, and the major axis is parallel to the x-axis.

$a^2 = 25$, so $a = 5$
$b^2 = 16$, so $b = 4$
The major axis, which is parallel to the x-axis, has a length of $2a$, or 10.
The minor axis, which is parallel to the y-axis, has a length of $2b$, or 8.

The vertices are $V_1(h - a, k)$ and $V_2(h + a, k)$.
Substitute the values of h, k, and a.
The vertices are $V_1(-1 - 5, 3)$ and $V_2(-1 + 5, 3)$, or $V_1(-6, 3)$ and $V_2(4, 3)$.

The co-vertices are $(h, k - b)$ and $(h, k + b)$.
Substitute the values of h, k, and b.
The co-vertices are $(-1, 3 - 4)$ and $(-1, 3 + 4)$, or $(-1, -1)$ and $(-1, 7)$.

The foci are $F_1(h - c, k)$ and $F_2(h + c, k)$.
To find c, we use $a^2 = b^2 + c^2$, with $a = 5$ and $b = 4$.
$$a^2 = b^2 + c^2$$
$$5^2 = 3^2 + c^2$$
$$25 = 16 + c^2$$
$$25 - 16 = c^2$$
$$9 = c^2$$
$$3 = c$$
The foci are $F_1(-1 - 3, 3)$ and $F_2(-1 + 3, 3)$, or $F_1(-4, 3)$ and $F_2(2, 3)$.

Plot the vertices and co-vertices.
Draw a smooth curve through the points.
Label the foci and the graph.

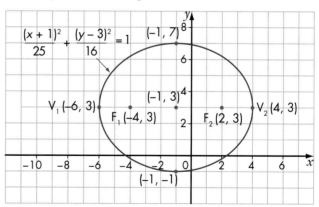

EXAMPLE 4 Writing an Equation of an Ellipse

Write an equation in standard form for each ellipse
a) The coordinates of the centre are $(2, -1)$. The major axis has a length of
16 units and is parallel to the x-axis. The minor axis has a length of 4 units.
b) The coordinates of the centre are $(-2, 5)$. The ellipse passes through the
points $(-5, 5)$, $(1, 5)$, $(-2, -2)$, and $(-2, 12)$.

SOLUTION

a) The centre is $(2, -1)$, so $h = 2$, and $k = -1$. The major axis is parallel to the x-axis.

The length of the major axis is 16, so $a = 8$. The length of the minor axis is 4, so $b = 2$.

Substitute known values into the general formula for an ellipse whose major axis is parallel to the x-axis.

$$\frac{(x-h)^2}{a^2} + \frac{(y-k)^2}{b^2} = 1$$

$$\frac{(x-2)^2}{8^2} + \frac{(y-(-1))^2}{2^2} = 1$$

$$\frac{(x-2)^2}{64} + \frac{(y+1)^2}{4} = 1$$

The equation of the ellipse is $\dfrac{(x-2)^2}{64} + \dfrac{(y+1)^2}{4} = 1$.

b) Plot the points $(-5, 5)$, $(1, 5)$, $(-2, -2)$, and $(-2, 12)$.

Draw a smooth curve through the points.
Label the centre $(-2, 5)$.
The centre is $(-2, 5)$ so $h = -2$ and $k = 5$.
From the sketch of the ellipse, the major axis is parallel to the y-axis, and is 14 units in length, so $a = 7$. The minor axis is 6 units in length, so $b = 3$.

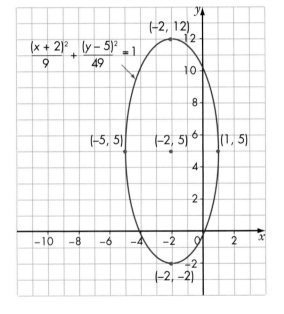

Substitute known values into the general formula for an ellipse whose major axis is parallel to the y-axis.

$$\frac{(x-h)^2}{b^2} + \frac{(y-k)^2}{a^2} = 1$$

$$\frac{(x-(-2))^2}{3^2} + \frac{(y-5)^2}{7^2} = 1$$

$$\frac{(x+2)^2}{9} + \frac{(y-5)^2}{49} = 1$$

The equation of the ellipse is

$$\frac{(x+2)^2}{9} + \frac{(y-5)^2}{49} = 1.$$

Note that ellipses can be graphed using a graphing calculator. As with circles, the equations of ellipses must first be solved for y.

$$\frac{(x+1)^2}{9} + \frac{(y-2)^2}{4} = 1$$

$$36 \times \frac{(x+1)^2}{9} + 36 \times \frac{(y-2)^2}{4} = 36 \times 1$$

$$4(x+1)^2 + 9(y-2)^2 = 36$$

$$9(y-2)^2 = 36 - 4(x+1)^2$$

$$(y-2)^2 = \frac{1}{9}(36 - 4(x+1)^2)$$

$$y - 2 = \pm\sqrt{\frac{1}{9}(36 - 4(x+1)^2)}$$

$$y = 2 \pm \frac{1}{3}\sqrt{36 - 4(x+1)^2}$$

Then, enter both of the resulting equations in the Y= editor.

$$Y_1 = 2 + \frac{1}{3}\sqrt{36 - 4(x+1)^2} \quad \text{and} \quad Y_2 = 2 - \frac{1}{3}\sqrt{36 - 4(x+1)^2}$$

Adjust the window variables and use the Zsquare instruction.

The window variables include Xmin = –5.3, Xmax = 5.3, Ymin = –2, and Ymax = 5.

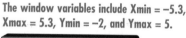

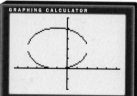

EXAMPLE 5 **Lithotripsy**

Lithotripsy is being used to provide an alternative method for removing kidney stones. A kidney stone is carefully positioned at one focus point of the elliptically shaped lithotripter. Shock waves are sent from the other focus point. The reflective properties of the ellipse cause the shock waves to intensify, destroying the kidney stone located at the focus. Suppose that the length of the major axis of a lithotripter is 60 cm, and half of the length of the minor axis is 25 cm.

a) Write an equation of the semi-ellipse. Assume that the centre is at the origin and that the major axis is along the x-axis.

b) How far must the kidney stone be from the source of the shock waves, to the nearest tenth of a centimetre?

SOLUTION

a) Draw a diagram.
The ellipse is centred at the origin with major axis along the x-axis, so the equation is of the form

$$\frac{x^2}{a^2} + \frac{y^2}{b^2} = 1$$

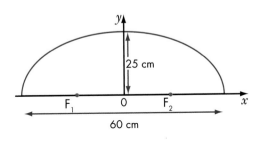

The length of the major axis is 60, so $a = 30$.
Half the length of the minor axis is 25, so $b = 25$.
An equation of the semi-ellipse is

$$\frac{x^2}{30^2} + \frac{y^2}{25^2} = 1,\ y \geq 0$$

or $\dfrac{x^2}{900} + \dfrac{y^2}{625} = 1,\ y \geq 0$

b) Find the coordinates of the foci.
The foci are $F_1(-c, 0)$ and $F_2(c, 0)$.

$$a^2 = b^2 + c^2$$
$$30^2 = 25^2 + c^2$$
$$900 - 625 = c^2$$
$$275 = c^2$$
$$\sqrt{275} = c$$
$$5\sqrt{11} = c$$

The foci are $F_1(-5\sqrt{11}, 0)$ and $F_2(5\sqrt{11}, 0)$.
The distance between the two foci is
$10\sqrt{11} \doteq 33.2$

The kidney stone must be 33.2 cm from the source of the shock waves, to the nearest tenth of a centimetre.

Key Concepts

• An ellipse is the set or locus of points P in the plane such that the sum of the distances from P to two fixed points F_1 and F_2 is a constant.

• The standard form of the equation of an ellipse, centred at the origin, with $a > b > 0$, is either $\frac{x^2}{a^2} + \frac{y^2}{b^2} = 1$ (major axis on the x-axis) or $\frac{x^2}{b^2} + \frac{y^2}{a^2} = 1$ (major axis on the y-axis).

• The standard form of the equation of the ellipse, with centre (h, k) and with $a > b > 0$, is either $\frac{(x-h)^2}{a^2} + \frac{(y-k)^2}{b^2} = 1$ (major axis parallel to the x-axis) or $\frac{(x-h)^2}{b^2} + \frac{(y-k)^2}{a^2} = 1$ (major axis parallel to the y-axis).

Communicate Your Understanding

1. If the ellipses $\frac{(x+3)^2}{4} + \frac{(y-2)^2}{9} = 1$ and $\frac{(x+3)^2}{9} + \frac{(y-2)^2}{4} = 1$ were graphed, what features would be the same? What features would be different? Explain.

2. State whether each of the following statements is always true, sometimes true, or never true for an ellipse. Explain your reasoning.

a) The length of the major axis is greater than the length of the minor axis.

b) The ellipse is a function.

c) The ellipse has infinitely many axes of symmetry.

d) For the ellipse, $\frac{x^2}{a^2} + \frac{y^2}{b^2} = 1$, $a < 0$ and $b < 0$.

3. Describe how you would use the locus definition of the ellipse to find an equation of an ellipse with centre $(0, 0)$, foci $F_1(0, -4)$ and $F_2(0, 4)$, and with the sum of the focal radii equal to 10.

4. Describe how you would sketch the graph of $\frac{(x+2)^2}{9} + \frac{(y+4)^2}{16} = 1$.

Practise

A

1. Use the locus definition of the ellipse to write an equation in standard form for each ellipse.
a) foci $(-3, 0)$ and $(3, 0)$, with sum of focal radii 10
b) foci $(0, -3)$ and $(0, 3)$, with sum of focal radii 10
c) foci $(-8, 0)$ and $(8, 0)$, with sum of focal radii 20
d) foci $(0, 8)$ and $(0, -8)$, with sum of focal radii 20

2. Sketch the graph of each ellipse. Label the coordinates of the centre, the vertices, the co-vertices, and the foci.
a) $\dfrac{x^2}{25} + \dfrac{y^2}{16} = 1$ **b)** $\dfrac{x^2}{4} + \dfrac{y^2}{36} = 1$
c) $x^2 + 16y^2 = 64$ **d)** $4x^2 + y^2 = 36$
e) $25x^2 + 9y^2 = 225$

3. For each of the following ellipses,
i) find the coordinates of the centre
ii) find the lengths of the major and minor axes
iii) find the coordinates of the vertices and co-vertices
iv) find the coordinates of the foci
v) find the domain and range
vi) write an equation in standard form

a)
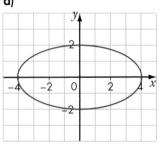

b)
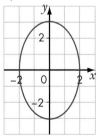

4. Write an equation in standard form for each ellipse with centre $(0, 0)$.
a) The major axis is on x-axis, the length of the major axis is 14, and the length of the minor axis is 6.
b) The length of the minor axis is 6, and the coordinates of one vertex are $(-5, 0)$.
c) The length of the major axis is 12, and the coordinates one co-vertex are $(5, 0)$.
d) The coordinates of one vertex are $(-8, 0)$, and the coordinates of one focus are $(\sqrt{55}, 0)$.
e) The coordinates of one focus are $(0, 2\sqrt{10})$, and the length of the minor axis is 6.

5. Sketch the graph of each ellipse by finding the coordinates of the centre, the lengths of the major and minor axes, and the coordinates of the foci, the vertices, and the co-vertices.
a) $\dfrac{(x+2)^2}{25} + \dfrac{(y-3)^2}{9} = 1$
b) $\dfrac{(x-3)^2}{49} + \dfrac{(y+1)^2}{81} = 1$
c) $(x+1)^2 + 9(y-3)^2 = 36$
d) $16(x-3)^2 + (y+2)^2 = 16$

6. For each of the following ellipses,
i) find the coordinates of the centre
ii) find the lengths of the major and minor axes
iii) find the coordinates of the vertices and co-vertices
iv) find the coordinates of the foci
v) find the domain and range
vi) write an equation in standard form

a)

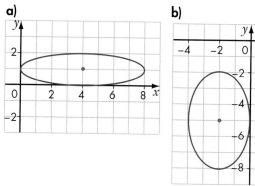

b)

c)

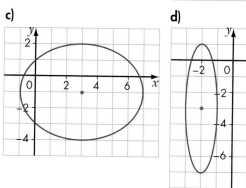

d)

7. Write an equation in standard form for each ellipse.

a) centre $(2, -3)$, major axis of length 12, minor axis of length 4

b) centre $(3, -2)$ and passing through $(-4, -2)$, $(10, -2)$, $(3, 1)$, and $(3, -5)$

c) centre $(-1, -2)$ and passing through $(-5, -2)$, $(3, -2)$, $(-1, 4)$, and $(-1, -8)$

d) foci at $(0, 0)$ and $(0, 8)$, and sum of focal radii 10

e) foci at $(-1, -1)$ and $(9, -1)$, and sum of focal radii 26

Apply, Solve, Communicate

B

8. Whisper Chamber Statuary Hall, located in the United States Capitol, has elliptical walls. Because of the reflective property of the ellipse, the hall is known as the Whisper Chamber.

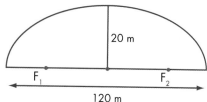

a) President John Quincy Adams' desk was located at one of the focus points, and he was able to listen in on many private conversations. Where would the conversations have to take place for Adams to hear them? Explain.

b) Write an equation of the ellipse that models the shape of Statuary Hall. Assume that the length of the major axis is 120 m and the length of the semi-minor axis (half the minor axis) is 20 m.

9. Kepler's First Law Johannes Kepler was a physicist who devised the three laws of planetary motion. Kepler's First Law states that all planets orbit the sun in elliptical paths, with the centre of the sun at one focus. The distance from the sun to a planet continually changes. The Earth is closest to the sun in January. The closest point, or perihelion, is 1.47×10^8 km from the sun. The Earth is farthest from the sun in July. The farthest point, or aphelion, is 1.52×10^8 km from the sun. Write an equation of the ellipse that models the Earth's orbit about the sun. Assume that the centre of the ellipse is at the origin and that the major axis is along the x-axis.

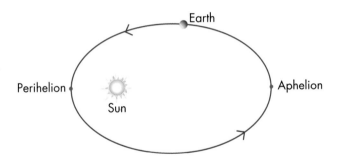

10. Motion in space Halley's Comet orbits the sun about every 76 years. The comet travels in an elliptical path, with the sun at one of the foci. At the closest point, or perihelion, the distance of the comet to the sun is 8.8×10^7 km. At the furthest point, or aphelion, the distance of the comet from the sun is 5.3×10^9 km. Write an equation of the ellipse that models the path of Halley's Comet. Assume that the sun is on the x-axis.

11. Application The Earth's moon orbits the Earth in an elliptical path. The perigee, the point where the moon is closest to the Earth, is approximately 363 000 km from the Earth. The apogee, the point where the moon is furthest from the Earth, is approximately 405 000 km from the Earth. The Earth is located at one focus. Write an equation of the ellipse that models the moon's orbit about the Earth. Assume that the Earth is on the y-axis.

12. Show that the equation of the of the circle $x^2 + y^2 = 49$ can be written in the standard form of an equation of an ellipse.

13. *Sputnik I* The first artificial Earth-orbiting satellite was *Sputnik I*, launched into an elliptical orbit by the USSR in 1957. If this orbit is modelled with the centre of the ellipse at the origin and the major axis along the x-axis, then the length of the major axis is 1180 km, and the length of the minor axis is 935 km. The Earth is at one focus.
a) Write an equation of the ellipse that models the orbit of the satellite.
b) What is the closest distance of the satellite to the Earth?
c) What is the furthest distance of the satellite from the Earth?

14. Coin set The twelve "Hopes and Aspirations" Canadian millennium quarters can be purchased in an elliptically shaped souvenir set.
a) If the ellipse has vertices (0, 7) and (30.94, 7), and co-vertices (15.47, 0) and (15.47, 14), write an equation of the ellipse.
b) What is the length of the major axis?

c) What is the length of the minor axis?

d) Sketch a scale diagram of the souvenir set.

15. Spotlight When a spotlight shines on a stage, the spotlight illuminates an area in the shape of an ellipse. Assume that one focus of the ellipse is (−1, 5), and the sum of the focal radii is 6. Write the equation of the ellipse if the major axis is parallel to the x-axis.

16. Jupiter Like all planets, Jupiter has an elliptical orbit, with the centre of the sun located at a focus. The diagram gives the approximate minimum and maximum distances from Jupiter to the sun, in millions of kilometres. Write an equation of the ellipse that models Jupiter's orbit around the sun. Assume that the centre of the sun is on the x-axis.

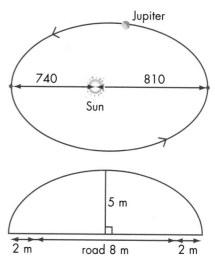

17. Covered entrance A semi-elliptical covering is to be built over an 8-m-wide road and the 2-m-wide sidewalks on either side of it that lead to an arts centre. If there is a maximum clearance of 5 m over the road, what will be the minimum clearance over the road, to the nearest hundredth of a metre?

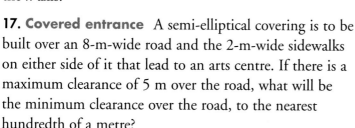

18. Technology Use a graphing calculator to graph each ellipse.

a) $\dfrac{(x-2)^2}{16} + \dfrac{(y+1)^2}{4} = 1$ b) $\dfrac{(x+3)^2}{4} + \dfrac{(y+1)^2}{25} = 1$ c) $\dfrac{x^2}{36} + \dfrac{(y-3)^2}{16} = 1$

19. Communication a) Use a graphing calculator to graph the family of ellipses $\dfrac{x^2}{25} + \dfrac{y^2}{b^2} = 1$ for $b = 1, 2,$ and 3.

b) Graph the family for $b = \dfrac{1}{2}, \dfrac{1}{3}, \dfrac{1}{4}$.

c) How are the graphs alike? How are they different?

d) What happens to the ellipses as b gets closer to 0?

20. Distorted circle An ellipse can be thought of as a distorted circle.

a) Sketch the graphs of the circle $x^2 + y^2 = 1$ and the ellipse $\dfrac{x^2}{25} + \dfrac{y^2}{9} = 1$ on the same grid.

b) By what factor has the circle expanded horizontally to form the ellipse in part a)?

c) By what factor has the circle expanded vertically to form the ellipse in part a)?

d) Sketch the graphs of the circle $x^2 + y^2 = 1$ and the ellipse $\dfrac{x^2}{4} + \dfrac{y^2}{36} = 1$ on the same grid.

e) By what factor has the circle expanded horizontally to form the ellipse in part d)?

f) By what factor has the circle expanded vertically to form the ellipse in part d)?

g) How can you recognize the horizontal and vertical stretch factors from the equation of the ellipse?

C

21. Eccentricity Ellipses can be long and narrow, or nearly circular. The amount of elongation, or "flatness," of an ellipse is measured by a number called the eccentricity. To calculate the eccentricity, e, use the formula $e = \dfrac{c}{a}$.

a) Calculate the eccentricity of each of the following ellipses. Round answers to the nearest hundredth.

i) $\dfrac{x^2}{9} + \dfrac{y^2}{4} = 1$ ii) $\dfrac{x^2}{16} + \dfrac{y^2}{25} = 1$ iii) $\dfrac{x^2}{36} + y^2 = 1$ iv) a circle

b) **Inquiry/Problem Solving** Find the greatest and least possible eccentricities for an ellipse. Explain your reasoning.

22. Standard form Consider an ellipse with its major axis along the x-axis, foci at $(-c, 0)$ and $(c, 0)$, x-intercepts at $(-a, 0)$ and $(a, 0)$, and y-intercepts at $(0, -b)$ and $(0, b)$.

a) Use the distance formula to show that for a point (x, y) on the ellipse,
$\sqrt{(x-c)^2 + y^2} + \sqrt{(x+c)^2 + y^2} = 2a$.

b) Isolate one radical term in the equation and derive $\dfrac{x^2}{a^2} + \dfrac{y^2}{a^2 - c^2} = 1$.

c) Derive $\dfrac{x^2}{a^2} + \dfrac{y^2}{b^2} = 1$, the standard form for the ellipse with its major axis along the x-axis.

ACHIEVEMENT Check Knowledge/Understanding Thinking/Inquiry/Problem Solving Communication Application

The roof of an ice arena is in the form of a semi-ellipse. It is 100 m across and 21 m high. What is the length of a stabilizing beam 5 m below the roof?

8.6 The Hyperbola

Some ships navigate using a radio navigation system called LORAN, which is an acronym for LOng RAnge Navigation. A ship receives radio signals from pairs of transmitting stations that send signals at the same time. The LORAN equipment detects the difference in the arrival times of the signals and uses the locus definition of the hyperbola to determine the ship's location.

To determine the equations of hyperbolas, the absolute values of numbers are used. The **absolute value** of a real number is its distance from zero on a real number line. For a real number represented by x, the absolute value is written $|x|$, which means the positive value, or magnitude, of x.

The diagram shows that the absolute value of -3 is 3 and the absolute value of 3 is 3.

$|-3| = 3 \qquad |3| = 3$

-4 -3 -2 -1 0 1 2 3 4

3 units   3 units

Web Connection

www.school.mcgrawhill.ca/resources/

To learn more about LORAN, visit the above web site. Go to **Math Resources**, then to MATHEMATICS 11, to find out where to go next. Write a brief report about the origins of LORAN.

A **hyperbola** is the set or locus of points P in the plane such that the absolute value of the difference of the distances from P to two fixed points F_1 and F_2 is a constant.

$$\left| F_1P - F_2P \right| = k$$

The two fixed points, F_1 and F_2, are called the **foci** of the hyperbola. The line segments F_1P and F_2P are called the **focal radii** of the hyperbola.

You will need two clear plastic rulers, a sheet of paper, and a pencil.

Step 1 Draw a 10-cm line segment near the centre of the piece of paper. Label the two endpoints F_1 and F_2.

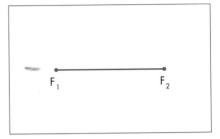

Step 2 Choose a length, k centimetres, which is less than the length F_1F_2. For this investigation, use $k = 4$ cm.

Step 3 Choose a pair of lengths, a centimetres and b centimetres, such that the absolute value of their difference equals 4, that is, $|a - b| = 4$. For example, choose $a = 9$ cm and $b = 5$ cm, or $a = 5$ cm and $b = 9$ cm, since $|9 - 5| = 4$ and $|5 - 9| = 4$.

Step 4 Use both rulers to mark two points that are 9 cm from F_1 and 5 cm from F_2. Then, use both rulers to mark two points that are 5 cm from F_1 and 9 cm from F_2.

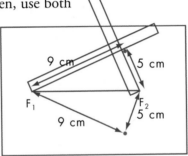

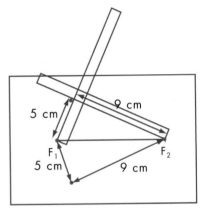

Step 5 Repeat steps 3 and 4 using different values of a and b, such that $|a - b| = 4$, until you have marked enough points to define two curves.

Step 6 Draw a smooth curve through each set of points. The two curves form a hyperbola.

1. How many axes of symmetry does the hyperbola have?

2. In relation to F_1 and F_2, where is the point of intersection of the axes of symmetry?

3. In Step 4, you marked four points for the chosen values of a and b. Are there any values of a and b for which only two points can be marked? If so, what are the values of a and b? Describe the locations of the points on the hyperbola.

4. The vertices of this hyperbola are located on F_1F_2. How is the distance between the vertices of the two curves related to the value of k? Explain.

EXAMPLE 1 Finding the Equation of a Hyperbola From its Locus Definition

Use the locus definition of the hyperbola to find an equation of the hyperbola with foci $F_1(-5, 0)$ and $F_2(5, 0)$, and with the constant difference between the focal radii equal to 8.

SOLUTION

Let $P(x, y)$ be any point on the hyperbola.
The locus definition of the hyperbola can be stated algebraically as
$|F_1P - F_2P| = 8$.
Use the formula for the length of a line segment, $l = \sqrt{(x_2 - x_1)^2 + (y_2 - y_1)^2}$, to rewrite F_1P and F_2P.

$$l = \sqrt{(x_2 - x_1)^2 + (y_2 - y_1)^2}$$
$$F_1P = \sqrt{(x - (-5))^2 + (y - 0)^2}$$
$$= \sqrt{(x + 5)^2 + y^2}$$
$$F_2P = \sqrt{(x - 5)^2 + (y - 0)^2}$$
$$= \sqrt{(x - 5)^2 + y^2}$$

Without loss in generality, we assume that $F_1P > F_2P$ and substitute.

Substitute: $\quad \sqrt{(x + 5)^2 + y^2} - \sqrt{(x - 5)^2 + y^2} = 8$

Isolate a radical: $\qquad \sqrt{(x + 5)^2 + y^2} = 8 + \sqrt{(x - 5)^2 + y^2}$

Square both sides: $\qquad (x + 5)^2 + y^2 = 64 + 16\sqrt{(x - 5)^2 + y^2} + (x - 5)^2 + y^2$

$\qquad\qquad x^2 + 10x + 25 + y^2 = 64 + 16\sqrt{(x - 5)^2 + y^2} + x^2 - 10x + 25 + y^2$

Isolate the radical: $\qquad 20x - 64 = 16\sqrt{(x - 5)^2 + y^2}$

Divide both sides by 4: $\qquad 5x - 16 = 4\sqrt{(x - 5)^2 + y^2}$

Square both sides: $\qquad 25x^2 - 160x + 256 = 16((x - 5)^2 + y^2)$

Simplify: $\qquad 25x^2 - 160x + 256 = 16(x^2 - 10x + 25 + y^2)$

$\qquad\qquad 25x^2 - 160x + 256 = 16x^2 - 160x + 400 + 16y^2$

$\qquad\qquad 9x^2 - 16y^2 = 144$

Divide both sides by 144: $\qquad \dfrac{x^2}{16} - \dfrac{y^2}{9} = 1$

An equation of the hyperbola is $\dfrac{x^2}{16} - \dfrac{y^2}{9} = 1$.

The hyperbola in Example 1 can be modelled graphically, as shown. The hyperbola has two axes of symmetry. The points $(-4, 0)$ and $(4, 0)$ lie on one axis of symmetry. The line segment joining these two points is called the **transverse axis**. In this case, the length of the transverse axis is 8. The endpoints of the transverse axis are the **vertices** of the hyperbola.

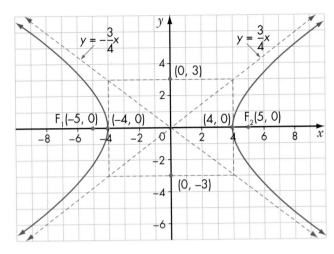

The points $(0, -3)$ and $(0, 3)$ lie on the other axis of symmetry. The line segment joining these two points is called the **conjugate axis**. In this case, the length of the conjugate axis is 6. The endpoints of the conjugate axis are called the **co-vertices** of the hyperbola.

The point of intersection of the transverse axis and the conjugate axis is called the **centre** of the hyperbola. In this case, the centre is the origin, $(0, 0)$.

The lines $x = -4$ and $x = 4$ form a rectangle with the lines $y = -3$ and $y = 3$. The graph of the hyperbola lies between the diagonals of this rectangle. The diagonals of this rectangle are called the asymptotes of the hyperbola. These are the lines that the hyperbola approaches for large values of x and y.

The equation of the hyperbola from Example 1 can be written as

$$\frac{x^2}{4^2} - \frac{y^2}{3^2} = 1$$

In this form of the equation, notice that 4 is half the length of the transverse axis, or half the difference between the focal radii, and 3 is half the length of the conjugate axis. Notice also that the equations of the asymptotes are

$y = \frac{3}{4}x$ and $y = -\frac{3}{4}x$.

The coordinates of the foci are $(-5, 0)$ and $(5, 0)$.
Notice that $4^2 + 3^2 = 5^2$.

The standard form of the equation of a hyperbola centred at the origin, with the transverse axis along the x-axis and the conjugate axis along the y-axis, is

$$\frac{x^2}{a^2} - \frac{y^2}{b^2} = 1$$

• The length of transverse axis along the x-axis is $2a$.
• The length of conjugate axis along the y-axis is $2b$.
• The vertices are $V_1(-a, 0)$ and $V_2(a, 0)$.
• The co-vertices are $(0, -b)$ and $(0, b)$.
• The foci are $F_1(-c, 0)$ and $F_2(c, 0)$, which are on the transverse axis.
• The equations of the asymptotes are
$$y = \frac{b}{a}x \text{ and } y = -\frac{b}{a}x.$$
• $a^2 + b^2 = c^2$

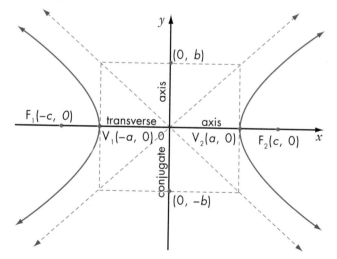

The standard form of the equation of a hyperbola centred at the origin, with the transverse axis along the y-axis and the conjugate axis along the x-axis, is

$$\frac{y^2}{a^2} - \frac{x^2}{b^2} = 1$$

• The length of transverse axis along the y-axis is $2a$.
• The length of conjugate axis along the x-axis is $2b$.
• The vertices are $V_1(0, -a)$ and $V_2(0, a)$.
• The co-vertices are $(-b, 0)$ and $(b, 0)$.
• The foci are $F_1(0, -c)$ and $F_2(0, c)$, which are on the transverse axis.
• The equations of the asymptotes are
$$y = \frac{a}{b}x \text{ and } y = -\frac{a}{b}x.$$
• $a^2 + b^2 = c^2$

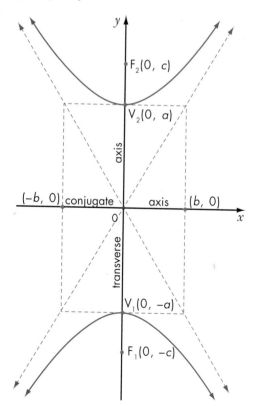

EXAMPLE 2 Sketching the Graph of a Hyperbola With Centre (0, 0)

Sketch the graph of the hyperbola $\dfrac{y^2}{36} - \dfrac{x^2}{4} = 1$. Label the foci and the asymptotes.

SOLUTION

Since the equation is in the form $\dfrac{y^2}{a^2} - \dfrac{x^2}{b^2} = 1$, the centre is (0, 0) and the

transverse axis is along the y-axis.

Since $a^2 = 36$, $a = 6$, and the vertices are $V_1(0, -6)$ and $V_2(0, 6)$.
Since $b^2 = 4$, $b = 2$, and the co-vertices are $(-2, 0)$ and $(2, 0)$.
The length of the transverse axis is $2a = 2(6)$
$$= 12$$
The length of the conjugate axis is $2b = 2(2)$
$$= 4$$

The foci are $F_1(0, -c)$ and $F_2(0, c)$.
To find c, use $a^2 + b^2 = c^2$, where $a = 6$ and $b = 2$.
$$c^2 = a^2 + b^2$$
$$= 6^2 + 2^2$$
$$= 36 + 4$$
$$= 40$$
$$c = \sqrt{40}$$
$$= 2\sqrt{10}$$

The coordinates of the foci are $(0, -2\sqrt{10})$ and $(0, 2\sqrt{10})$, or
approximately $(0, -6.32)$ and $(0, 6.32)$.

The equations of the asymptotes are $y = \dfrac{a}{b}x$ and $y = -\dfrac{a}{b}x$.

Substituting for a and b gives
$$y = \dfrac{6}{2}x, \text{ or } y = 3x, \text{ and } y = -\dfrac{6x}{2}, \text{ or } y = -3x.$$

To sketch the graph of the hyperbola, plot the vertices and the
co-vertices. Use the lines $y = 6$, $y = -6$, $x = 2$, and $x = -2$ to
construct a rectangle. Sketch the asymptotes by extending the
diagonals of the rectangle.

Since the transverse axis is along the y-axis, the hyperbola must
open up and down. To sketch a branch of the hyperbola, start at
a vertex and approach the asymptotes. Sketch the other branch
in the same way. Label the foci and the asymptotes.

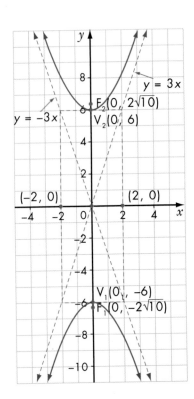

A hyperbola may not be centred at the origin. As in the case of the circle and the ellipse, a hyperbola can have a centre (h, k). The translation rules that apply to the circle and the ellipse also apply to the hyperbola.

The standard form of the equation of a hyperbola centred at (h, k), with the transverse axis parallel to the x-axis and the conjugate axis parallel to the y-axis, is

$$\frac{(x-h)^2}{a^2} - \frac{(y-k)^2}{b^2} = 1$$

• The length of the transverse axis parallel to the x-axis is $2a$.
• The length of the conjugate axis parallel to the y-axis is $2b$.
• The vertices are $V_1(h - a, k)$ and $V_2(h + a, k)$.
• The co-vertices are $(h, k - b)$ and $(h, k + b)$.
• The foci are $F_1(h - c, k)$ and $F_2(h + c, k)$, which are on the transverse axis.
• $a^2 + b^2 = c^2$

The standard form of the equation of a hyperbola centred at (h, k), with the transverse axis parallel to the y-axis, and conjugate axis parallel to the x-axis, is

$$\frac{(y-k)^2}{a^2} - \frac{(x-h)^2}{b^2} = 1$$

• The length of the transverse axis parallel to the y-axis is $2a$.
• The length of the conjugate axis parallel to the x-axis is $2b$.
• The vertices are $V_1(h, k - a)$ and $V_2(h, k + a)$.
• The co-vertices are $(h - b, k)$ and $(h + b, k)$.
• The foci are $F_1(h, k - c)$ and $F_2(h, k + c)$, which are on the transverse axis.
• $a^2 + b^2 = c^2$

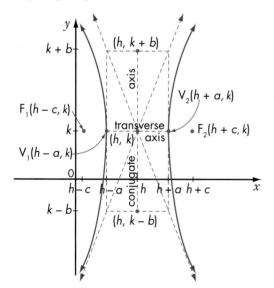

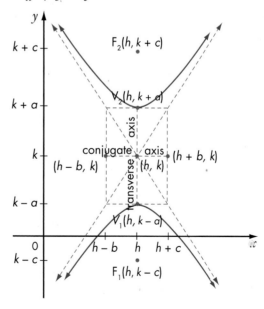

EXAMPLE 3 Sketching the Graph of a Hyperbola With Centre (h, k)

Sketch the graph of the hyperbola $\dfrac{(x-2)^2}{36} - \dfrac{(y+1)^2}{16} = 1$. Label the foci.

SOLUTION

The centre is $C(h, k) = (2, -1)$.

Since the equation is in the form $\dfrac{(x-h)^2}{a^2} - \dfrac{(y-k)^2}{b^2} = 1$, the transverse axis
is parallel to the x-axis.

$a^2 = 36$, so $a = 6$
$b^2 = 16$, so $b = 4$

The length of the transverse axis is $2a$, or 12.
The length of the conjugate axis is $2b$, or 8.

The vertices are $V_1(h - a, k)$ and $V_2(h + a, k)$.
Substitute the values of h, k, and a.
The vertices are $V_1(2 - 6, -1)$ and $V_2(2 + 6, -1)$, or $V_1(-4, -1)$ and
$V_2(8, -1)$.

The co-vertices are $(h, k - b)$ and $(h, k + b)$.
Substitute the values of h, k, and b.
The co-vertices are $(2, -1 - 4)$ and $(2, -1 + 4)$, or $(2, -5)$ and $(2, 3)$.

The foci are $F_1(h - c, k)$ and $F_2(h + c, k)$.
To find c, use $a^2 + b^2 = c^2$, with $a = 6$ and $b = 4$.

$c^2 = 6^2 + 4^2$
$\quad = 36 + 16$
$\quad = 52$
$c = \sqrt{52}$
$\quad = 2\sqrt{13}$

```
GRAPHING CALCULATOR
2-2√(13)
         -5.211102551
2+2√(13)
         9.211102551
■
```

The coordinates of the foci are $(2 - 2\sqrt{13}, -1)$ and $(2 + 2\sqrt{13}, -1)$, or
approximately $(-5.21, -1)$ and $(9.21, -1)$.

To sketch the graph of the hyperbola, plot the vertices and the co-vertices.
Then, construct the rectangle that goes through all four of these points.
Sketch the asymptotes by extending the diagonals of the rectangle.

Since the transverse axis is parallel to the x-axis, the hyperbola must open left and right. To sketch a branch of the hyperbola, start at a vertex and approach the asymptotes. Sketch the other branch in the same way. Label the foci.

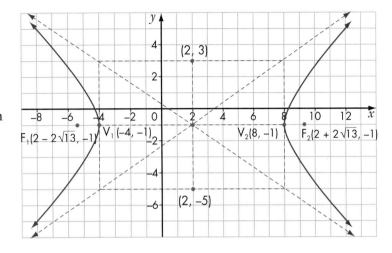

EXAMPLE 4 Writing an Equation of a Hyperbola

Write an equation for the hyperbola with vertices $(2, -2)$ and $(2, 8)$ and co-vertices $(0, 3)$ and $(4, 3)$. Sketch the hyperbola.

SOLUTION

The vertices $(2, -2)$ and $(2, 8)$ lie on the transverse axis, so the transverse axis is parallel to the y-axis.

The centre of the hyperbola is located midway between the vertices, and is also located midway between the co-vertices.
Using the midpoint formula with the two vertices $(2, -2)$ and $(2, 8)$ gives

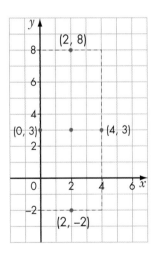

$$\left(\frac{x_1 + x_2}{2}, \frac{y_1 + y_2}{2}\right) = \left(\frac{2 + 2}{2}, \frac{-2 + 8}{2}\right)$$
$$= (2, 3)$$

Since the transverse axis is parallel to the y-axis, and the centre is not at the origin, the standard form of the equation of the ellipse is
$$\frac{(y - k)^2}{a^2} - \frac{(x - h)^2}{b^2} = 1.$$

Since the centre is $(2, 3)$, $h = 2$ and $k = 3$.

The length of the transverse axis, which is parallel to the y-axis, is 10.
Since $2a = 10$, $a = 5$.
The length of the conjugate axis, which is parallel to the x-axis, is 4.
Since $2b = 4$, $b = 2$.

The equation of the hyperbola is $\dfrac{(y-3)^2}{25} - \dfrac{(x-2)^2}{4} = 1$.

The foci are $F_1(h, k - c)$ and $F_2(h, k + c)$.
To find c, use $a^2 + b^2 = c^2$, with $a = 5$ and $b = 2$.

$$c^2 = a^2 + b^2$$
$$= 5^2 + 2^2$$
$$= 25 + 4$$
$$= 29$$
$$c = \sqrt{29}$$

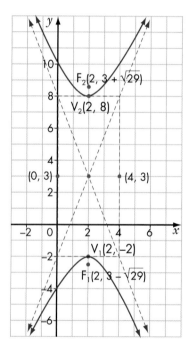

GRAPHING CALCULATOR
3-√(29)
        -2.385164807
3+√(29)
        8.385164807

The foci are located at $(h, k - c)$ and $(h, k + c)$. Thus, the coordinates of the foci are $(2, 3 - \sqrt{29})$ and $(2, 3 + \sqrt{29})$, or approximately $(2, -2.39)$ and $(2, 8.39)$.

To sketch the graph of the hyperbola, plot the vertices and the co-vertices. Then, construct the rectangle that goes through all four of these points. Sketch the asymptotes by extending the diagonals of the rectangle.

Since the transverse axis is parallel to the y-axis, the hyperbola must open up and down. To sketch a branch of the hyperbola, start at a vertex and approach the asymptotes. Sketch the other branch in the same way. Label the foci.

Note that hyperbolas can be graphed using a graphing calculator. As with circles and ellipses, the equations of hyperbolas must first be solved for y.

For example, solving $\dfrac{y^2}{4} - \dfrac{x^2}{9} = 1$, results in $y = \pm\dfrac{2}{3}\sqrt{9 + x^2}$.

Enter both of the resulting equations in the Y= editor.

$Y_1 = \dfrac{2}{3}\sqrt{9 + x^2}$ and $Y_2 = -\dfrac{2}{3}\sqrt{9 + x^2}$

Adjust the window variables, if necessary, and use the Zsquare instruction.

 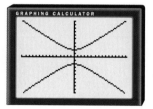

Key Concepts

• A hyperbola is the set, or locus, of points P in the plane such that the absolute value of the difference of the distances from P to two fixed points F_1 and F_2 is a constant.

$$|F_1P - F_2P| = k$$

• The standard form of the equation of a hyperbola, with centre at the origin, is either $\dfrac{x^2}{a^2} - \dfrac{y^2}{b^2} = 1$ (transverse axis along the x-axis) or $\dfrac{y^2}{a^2} - \dfrac{x^2}{b^2} = 1$ (transverse axis along the y-axis).

• The standard form of the equation of a hyperbola, centred at (h, k), is either $\dfrac{(x-h)^2}{a^2} - \dfrac{(y-k)^2}{b^2} = 1$ (transverse axis parallel to the x-axis)

or $\dfrac{(y-k)^2}{a^2} - \dfrac{(x-h)^2}{b^2} = 1$ (transverse axis parallel to the y-axis).

Communicate Your Understanding

1. Describe how you would use the locus definition of the hyperbola to find an equation of the hyperbola with foci $F_1(0, -5)$ and $F_2(0, 5)$, and with the constant difference between the focal radii equal to 8.

2. Describe how you would sketch the graph of $\dfrac{(x-2)^2}{25} - \dfrac{(y+4)^2}{9} = 1$.

3. Describe how you would write an equation in standard form for the hyperbola with vertices $(3, -4)$ and $(3, 6)$ and co-vertices $(1, 1)$ and $(5, 1)$.

4. How would you know if

a) an equation in standard form models an ellipse or a hyperbola?

b) a graph models an ellipse or a hyperbola?

5. Is a hyperbola a function? Explain.

Practise

A

1. Use the locus definition of the hyperbola to find an equation for each of the following hyperbolas, centred at the origin.
a) foci at $(-5, 0)$ and $(5, 0)$, with the difference between the focal radii 6
b) foci at $(0, -5)$ and $(0, 5)$, with the difference between the focal radii 6

2. For each hyperbola, determine
i) the coordinates of the centre
ii) the coordinates of the vertices and co-vertices
iii) the lengths of the transverse and conjugate axes
iv) an equation in standard form
v) the coordinates of the foci
vi) the domain and range

a)

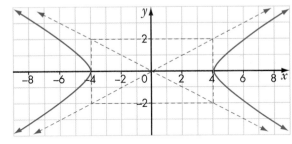

b)

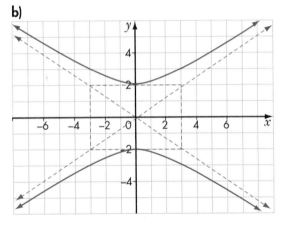

c)

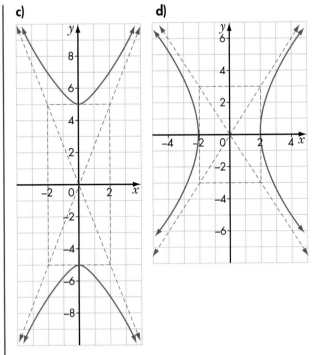

d)

3. Sketch the graph of each hyperbola. Label the coordinates of the centre, the vertices, the co-vertices, and the foci, and the equations of the asymptotes.

a) $\dfrac{x^2}{25} - \dfrac{y^2}{36} = 1$ **b)** $\dfrac{y^2}{16} - \dfrac{x^2}{49} = 1$

c) $25x^2 - 4y^2 = 100$ **d)** $4y^2 - 9x^2 = 36$

4. Determine an equation in standard form for each of the following hyperbolas.
a) vertices $(0, -4)$ and $(0, 4)$, co-vertices $(-5, 0)$ and $(5, 0)$
b) vertices $(-3, 0)$ and $(3, 0)$, foci $(-5, 0)$ and $(5, 0)$
c) foci $(-5, 0)$ and $(5, 0)$, with constant difference between focal radii 4
d) foci $(0, -3)$ and $(0, 3)$, with constant difference between focal radii 5

5. For each hyperbola, determine
i) the coordinates of the centre
ii) the coordinates of the vertices and co-vertices
iii) the lengths of the transverse and conjugate axes
iv) an equation in standard form
v) the coordinates of the foci

a)

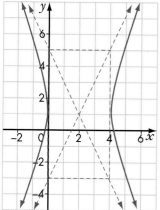

b)

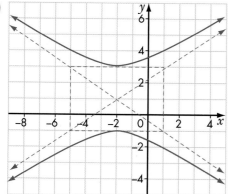

c)

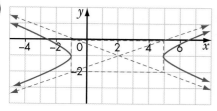

d)

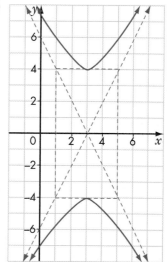

6. Sketch the graph of each hyperbola. Label the coordinates of the centre, the vertices, the co-vertices, and the foci.

a) $\dfrac{(x-3)^2}{16} - \dfrac{(y+1)^2}{9} = 1$

b) $\dfrac{(y+2)^2}{81} - \dfrac{(x-1)^2}{25} = 1$

c) $16x^2 - 9(y-2)^2 = 144$

d) $4(y-2)^2 - 9(x-5)^2 = 36$

7. Determine an equation in standard form for each of the following hyperbolas.
a) vertices $(-2, -5)$ and $(6, -5)$, foci $(-4, -5)$ and $(8, -5)$
b) centre $(0, 3)$, one vertex $(0, 5)$, and one focus $(0, 6)$
c) centre $(-3, 1)$, one focus $(-5, 1)$, and length of conjugate axis 4
d) centre $(-2, 2)$, one focus $(-2, 5)$, and length of conjugate axis 6

Apply, Solve, Communicate

B

8. Marine biology A ship is monitoring the movement of a pod of whales with its radar. The radar screen can be modelled as a coordinate grid with the ship at the centre (0, 0). The pod appears to be moving along a curve such that the absolute value of the difference of its distances from (2, 7) and (2, −3) is always 6. Write an equation in standard form to describe the path of the pod.

9. Motion in space Some comets travel on hyperbolic paths and never return. Suppose the hyperbolic path of a comet is modelled on a grid with a scale of 1 unit = 3 000 000 km. The vertex of the path is at (2, 0), and the focus is at (6, 0). Write an equation in standard form to model the path of the comet.

10. Conjugate hyperbolas a) Two hyperbolas are centred at (2, 1). One has a transverse axis parallel to the x-axis, and the other has a transverse axis parallel to the y-axis. They share the same pair of asymptotes. The equation of one hyperbola is $\dfrac{(y-1)^2}{25} - \dfrac{(x-2)^2}{9} = 1$. Find the lengths of the conjugate and transverse axes, and the equation in standard form, of the other hyperbola.

b) The two hyperbolas in part a) are known as conjugate hyperbolas. Write equations in standard form for another pair of conjugate hyperbolas with their common centre not at the origin. Explain your reasoning.

11. Application Two park rangers were stationed at separate locations on the same side of a lake. The rangers were 6 km apart. Both rangers saw a lightning bolt strike the ground on the other side of the lake and heard the clap of thunder.

a) If one ranger heard the clap of thunder 4 s before the other, write an equation that describes all the possible locations of the thunder clap. Place the two rangers on the y-axis, with the midpoint between the rangers at the origin. The speed of sound is approximately 0.35 km/s.

b) Sketch the possible locations where the lightning bolt hit the ground. Include the ranger stations in the sketch.

12. Technology Use a graphing calculator to graph each of the following.

a) $\dfrac{(x+2)^2}{9} - \dfrac{(y-1)^2}{4} = 1$ b) $\dfrac{(x+3)^2}{4} - \dfrac{(y+1)^2}{16} = 1$

13. Navigation The LORAN system uses the locus definition of the hyperbola. The system measures the difference in the arrival times at a ship or aircraft of radio signals from two stations. The difference in the arrival times is converted to a difference in the distances of the ship or aircraft from the two stations. Two pairs of stations sending radio signals define two hyperbolas. The exact location of the ship or aircraft is a point of intersection of the two hyperbolas.

a) On a grid with a scale in kilometres, plot and label the foci $F_1(-6, 0)$ and $F_2(6, 0)$ of a hyperbola. The foci represent two stations, F_1 and F_2, that send radio signals to a ship positioned at point P. Find an equation of the hyperbola, if the difference in the arrival times of the radio signals at P is converted to the difference in distances $|F_1P - F_2P| = 8$.

b) On the same grid as in part a), plot and label the foci $F_3(0, -5)$ and $F_4(0, 5)$ of another hyperbola. The foci represent another two stations, F_3 and F_4, that send radio signals to the same ship at point P. Find an equation of this hyperbola, if $|F_3P - F_4P| = 4$.

c) Sketch the two hyperbolas and estimate the coordinates of the points of intersection. How many points of intersection are there?

d) Use a graphing calculator to find the coordinates of the points of intersection of the two hyperbolas, to the nearest tenth. If the second quadrant of the grid is the only quadrant that contains a body of water, what are the coordinates of the ship?

14. Communication a) Use a graphing calculator to graph the family of hyperbolas $\dfrac{x^2}{25} - \dfrac{y^2}{b^2} = 1$ for $b = 1, 2,$ and 3.

b) Now, graph for $b = \dfrac{1}{2}, \dfrac{1}{3}, \dfrac{1}{4}$.

c) How are the graphs alike? How are they different?

d) What happens to the hyperbola as b gets closer to 0?

C

15. Inquiry/Problem Solving The eccentricity, e, of a hyperbola is defined as $e = \dfrac{c}{a}$. Since $c > a$, it follows that $e > 1$.

a) Describe the general shape of a hyperbola whose eccentricity is close to 1.

b) Describe the shape if e is very large.

16. Equilateral hyperbolas The following are examples of equilateral hyperbolas.

i) $x^2 - y^2 = 4$ ii) $x^2 - y^2 = 9$ iii) $y^2 - x^2 = 16$ iv) $y^2 - x^2 = 25$

a) What do the equations have in common?

b) Graph each hyperbola. State the equations of its asymptotes.

c) What do the graphs have in common that is different from the other hyperbolas you have graphed?

17. Standard form Consider a hyperbola with its transverse axis parallel to the x-axis, foci at $(-c, 0)$ and $(c, 0)$, and vertices at $(-a, 0)$ and $(a, 0)$.

a) Use the formula for the length of a line segment, $l = \sqrt{(x_2 - x_1)^2 + (y_2 - y_1)^2}$, to show that, for a point $P(x, y)$ on the hyperbola,

$\sqrt{(x - c)^2 + y^2} - \sqrt{(x + c)^2 + y^2} = 2a$ and $\sqrt{(x + c)^2 + y^2} - \sqrt{(x - c)^2 + y^2} = 2a$.

b) From the equations in part a), derive $\dfrac{x^2}{c^2 - b^2} - \dfrac{y^2}{b^2} = 1$.

c) Derive $\dfrac{x^2}{a^2} - \dfrac{y^2}{b^2} = 1$, the standard form for the hyperbola centred at $(0, 0)$ with transverse axis parallel to the x-axis.

18. Inverse variation When one variable increases and another variable decreases, such that their product is a constant, the relationship can be represented by $xy = k$, $k \neq 0$. The relationship can be stated verbally as y varies inversely as x. The graph of the inverse variation is one branch of a hyperbola, for $x, y > 0$. Complete the following using a graphing calculator.

a) Graph $xy = 2$, $xy = 5$, and $xy = 10$. What are the similarities and differences for $xy = k$?

b) Graph $xy = -2$, $xy = -5$, and $xy = -10$. Compare these graphs with the graphs in part a).

c) Explain why the graphs in parts a) and b) do not go through the origin.

d) Can you graph $xy = 0$? Explain.

ACHIEVEMENT Check   Knowledge/Understanding   Thinking/Inquiry/Problem Solving   Communication   Application

Find an equation for the locus of points such that the product of the slopes of the lines from a point on the locus to the points $(6, 0)$ and $(-6, 0)$ is 4.

8.7 The Parabola

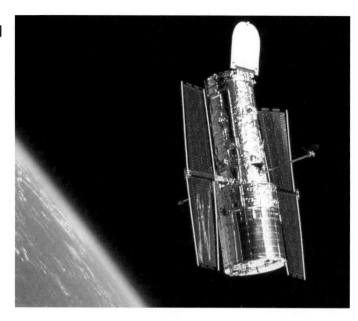

The Hubble Space Telescope orbits the Earth at an altitude of approximately 600 km. The telescope takes about ninety minutes to complete one orbit.

Since it orbits above the Earth's atmosphere, the telescope can perform its scientific work without the negative effects of the atmosphere. The primary reflector of the Hubble Space Telescope is parabolic.

A **parabola** is the set or locus of points P in the plane such that the distance from P to a fixed point F equals the distance from P to a fixed line l.

$$PF = PD$$

The fixed point F is called the **focus**. The fixed line l is called the **directrix**.

The vertex of a parabola is located midway between the focus and the directrix. The parabola never crosses the directrix. We will use this information to help sketch parabolas.

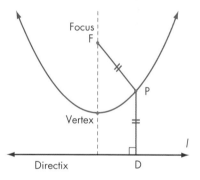

INVESTIGATE & INQUIRE

You will need two clear plastic rulers, a sheet of paper, and a pencil.

Step 1 Draw a 15-cm line segment near the bottom of the piece of paper. Label the line l. Mark a point about 4 cm above the middle of the segment. Label this point F.

Step 2 Choose a length, *k* cm, that is greater than or equal to half the distance from the point F to the line *l*. You may want to make *k* less than 8 cm.

Step 3 To locate a point P that is *k* cm from line *l* and *k* cm from point F, place one ruler so that its 0 mark is on line *l* and the ruler is perpendicular to line *l*. Place the other ruler so that its 0 mark is on point F. Adjust the positions of the rulers to locate P that is *k* cm from *l* and *k* cm from F.

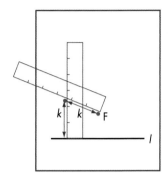

Step 4 Mark a second point that is also *k* cm from line *l* and *k* cm from point F.

Step 5 Repeat steps 3 and 4 using different values for *k* until you have marked enough points to define a complete curve.

Step 6 Draw a smooth curve through the points. The curve is an example of a parabola.

1. How many axes of symmetry does the parabola have?

2. Steps 3 and 4 instructed you to mark two points for the chosen distance, *k*. Are there any values of *k* for which only one point can be marked? If so, describe the location of the point on the parabola.

3. What is the relationship between the vertex of the parabola, line *l*, and point F?

4. In Step 2, why must *k* be greater than or equal to half the distance from line *l* to the point F?

EXAMPLE 1 Finding the Equation of a Parabola From its Locus Definition

Use the locus definition of the parabola to find an equation of the parabola with focus F(0, 3) and directrix $y = -3$.

SOLUTION

Draw a diagram.
The vertex is located midway between the focus and directrix, so V(0, 0) is the vertex.
Since the parabola never crosses the directrix, the parabola must open up.

Let P(x, y) be any point on the parabola.
The focus is F(0, 3).
Let D(x, -3) be any point on the directrix.

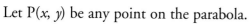

The locus definition of the parabola can be stated algebraically as PF = PD.
Use the formula for the length of a line segment, $l = \sqrt{(x_2 - x_1)^2 + (y_2 - y_1)^2}$, to rewrite PF and PD.

$$PF = \sqrt{(x - 0)^2 + (y - 3)^2}$$
$$= \sqrt{x^2 + (y - 3)^2}$$
$$PD = \sqrt{(x - x)^2 + (y - (-3))^2}$$
$$= \sqrt{(y + 3)^2}$$

Substitute: $\sqrt{x^2 + (y - 3)^2} = \sqrt{(y + 3)^2}$

Square both sides: $x^2 + (y - 3)^2 = (y + 3)^2$

Expand: $x^2 + y^2 - 6y + 9 = y^2 + 6y + 9$

Solve for y: $x^2 = 12y$

$$\frac{1}{12}x^2 = y$$

An equation of the parabola is $y = \dfrac{1}{12}\, x^2$.

In Example 1, note that, when the focus is (0, 3) and the directrix is $y = -3$, the equation of the parabola is $y = \dfrac{1}{12}\, x^2$, and that $12 = 4 \times 3$.

In general, if the focus is F(0, p) and the directrix is $y = -p$, then the equation of the parabola is $y = \dfrac{1}{4p}\, x^2$.

The standard form of the equation of a parabola with its vertex at the origin and a horizontal directrix is

$$y = \frac{1}{4p} x^2$$

The vertex is V(0, 0).
If $p > 0$, the parabola opens up.
If $p < 0$, the parabola opens down.
The focus is F(0, p).
The equation of the directrix is $y = -p$.
The axis of symmetry is the y-axis.

The standard form of the equation of a parabola with its vertex at the origin and a vertical directrix is

$$x = \frac{1}{4p} y^2$$

The vertex is V(0, 0).
If $p > 0$, the parabola opens right.
If $p < 0$, the parabola opens left.
The focus is F(p, 0).
The equation of the directrix is $x = -p$.
The axis of symmetry is the x-axis.

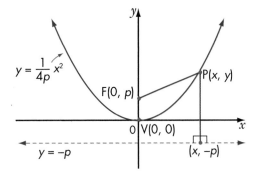

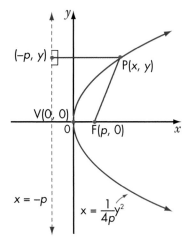

EXAMPLE 2 Writing an Equation of a Parabola With Vertex (0, 0)

Write an equation in standard form for the parabola with focus (4, 0) and directrix $x = -4$.

SOLUTION

The vertex is located midway between the focus and the directrix, so the vertex is V(0, 0). Since the parabola never crosses the directrix, the parabola opens right.

The standard form of the equation of a parabola opening right with vertex at the origin is $x = \dfrac{1}{4p} y^2$.

The directrix is $x = -p$, so $p = 4$.

Since $p = 4$, the equation is $x = \dfrac{1}{16} y^2$.

The equation can be modelled graphically.

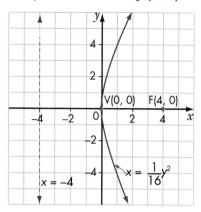

EXAMPLE 3 Determining the Characteristics of a Parabola From its Equation

For the parabola $y = -\dfrac{1}{8}x^2$, determine the direction of the opening, the coordinates of the vertex and the focus, and the equation of the directrix.

SOLUTION

The equation is in the form $y = \dfrac{1}{4p}x^2$, so the graph opens up or down.

Find the value of p.

Rewrite $y = -\dfrac{1}{8}x^2$ as $y = \dfrac{1}{4(-2)}x^2$.

So, $p = -2$.
Since $p < 0$, the graph opens down.
The vertex is $V(0, 0)$.
The focus is $F(0, p)$, or $F(0, -2)$.
The equation of the directrix is $y = -p$, or $y = 2$.

The equation can be modelled graphically.

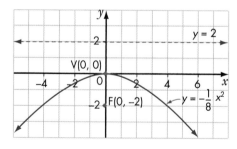

Recall that we can translate a parabola $y = ax^2$, with vertex at the origin, to $y = a(x - h)^2 + k$ by translating h units to the left or right and k units up or down. This translation results in a parabola with vertex (h, k). The equation of the resulting parabola can be expressed in standard form.

The standard form of the equation of a parabola with vertex V(h, k) and a horizontal directrix is

$$y - k = \frac{1}{4p}(x - h)^2$$

The vertex is V(h, k).
If $p > 0$, the parabola opens up.
If $p < 0$, the parabola opens down.
The focus is F(h, $k + p$).
The equation of the directrix is $y = k - p$.
The equation of the axis of symmetry is $x = h$.

The standard form of the equation of a parabola with vertex V(h, k) and a vertical directrix is

$$x - h = \frac{1}{4p}(y - k)^2$$

The vertex is V(h, k).
If $p > 0$, the parabola opens right.
If $p < 0$, the parabola opens left.
The focus is F($h + p$, k).
The equation of the directrix is $x = h - p$
The equation of the axis of symmetry is $y = k$.

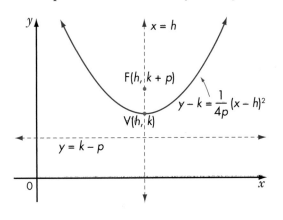

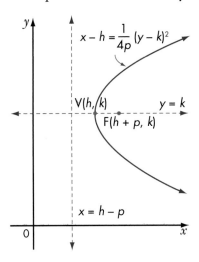

EXAMPLE 4 Writing an Equation of a Parabola With Vertex (h, k)

Write an equation in standard form for a parabola with focus F(-2, 6) and directrix $x = 4$.

SOLUTION

Since the parabola never crosses the directrix, the parabola opens left.

The standard form of the equation is $x - h = \dfrac{1}{4p}(y - k)^2$.

The vertex is located midway between the focus and the directrix.
The coordinates of the vertex are V(1, 6).
Since the vertex is V(1, 6), $h = 1$ and $k = 6$.

Use the focus to find the value of p.

The focus $F(h + p, k)$ is $(-2, 6)$.

So, $(1 + p, 6) = (-2, 6)$

$$1 + p = -2$$
$$p = -3$$

Now substitute known values into the standard form of the equation.

$$x - h = \frac{1}{4p}(y - k)^2$$

$$x - 1 = \frac{1}{4(-3)}(y - 6)^2$$

$$x - 1 = -\frac{1}{12}(y - 6)^2$$

The equation can be modelled graphically.

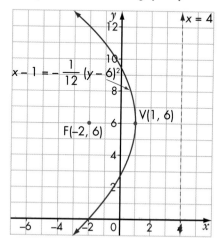

A equation of the parabola in standard form is $x - 1 = -\dfrac{1}{12}(y - 6)^2$.

Recall that parabolas can be graphed using a graphing calculator. If the equation of a parabola that opens left or right is given in standard form, first solve the equation for y. For example, solving $x = -\dfrac{1}{4}y^2$ for y results in $y = \pm\sqrt{-4x}$.

Enter both of the resulting equations in the Y= editor.

$Y_1 = \sqrt{-4x}$ $Y_2 = -\sqrt{-4x}$.

Adjust the window variables if necessary, and use the Zsquare instruction.

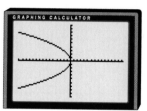

Key Concepts

- A parabola is the set or locus of points P in the plane such that the distance from P to a fixed point F equals the distance from P to a fixed line l.

$$PF = PD$$

The fixed point F is called the focus. The fixed line l is called the directrix.

- The vertex of a parabola is located midway between the focus and the directrix.

- The standard form of a parabola with vertex at the origin is $y = \frac{1}{4p}x^2$ (opens up if $p > 0$, or down if $p < 0$), or $x = \frac{1}{4p}y^2$ (opens right if $p > 0$, or left if $p < 0$).

- The standard form of a parabola with vertex (h, k) is $y - k = \frac{1}{4p}(x - h)^2$ (opens up if $p > 0$, or down if $p < 0$), or $x - h = \frac{1}{4p}(y - k)^2$ (opens right if $p > 0$, or left if $p < 0$).

Communicate Your Understanding

1. In your own words, define the following terms.

a) vertex **b)** directrix **c)** focus

2. Describe how you would use the locus definition to find the equation of a parabola with focus F(2, 0) and directrix $x = -2$.

3. Describe the relationship between the axis of symmetry and the directrix of any parabola.

4. Describe the similarities and differences between the parabolas $y - 3 = \frac{1}{8}(x + 2)^2$ and $x - 3 = \frac{1}{8}(y + 2)^2$.

5. Describe how you would determine an equation in standard form for a parabola with focus F(1, 3) and directrix $y = 1$.

Practise

A

1. Use the locus definition of the parabola to write an equation for each of the following parabolas.

a) focus $(0, -2)$, directrix $y = 2$

b) focus $(-1, -3)$, directrix $x = -4$

2. Determine the coordinates of the vertex for each of the following parabolas.

a) focus $(6, 3)$, directrix $x = 2$

b) focus $(3, 0)$, directrix $y = 3$

c) focus $(-4, 2)$, directrix $x = -1$

d) focus $(-3, -4)$, directrix $y = -2$

3. Write an equation in standard form for the parabola with the given focus and directrix. Sketch the parabola.

a) focus $(0, 6)$, directrix $y = -6$

b) focus $(0, -4)$, directrix $y = 4$

c) focus $(-8, 0)$, directrix $x = 8$

d) focus $(0, -2)$, directrix $y = 2$

e) focus $(1, 0)$, directrix $x = -1$

f) focus $(0, 3)$, directrix $y = -3$

g) focus $(-5, 0)$, directrix $x = 5$

4. Write an equation in standard form for the parabola with the given focus and directrix. Sketch the parabola.

a) focus $(6, -2)$, directrix $x = 0$

b) focus $(0, 4)$, directrix $y = 5$

c) focus $(2, 2)$, directrix $x = -5$

d) focus $(-1, -4)$, directrix $y = 2$

e) focus $(-3, 5)$, directrix $x = 1$

5. For each of the following parabolas, determine the direction of the opening, the coordinates of the vertex and the focus, and the equation of the directrix. Sketch the graph, and determine the domain and range.

a) $y = \dfrac{1}{16}x^2$

b) $y = -\dfrac{1}{8}x^2$

c) $x = \dfrac{1}{8}y^2$

d) $x = -\dfrac{1}{16}y^2$

e) $y - 3 = \dfrac{1}{4}(x + 2)^2$

f) $x + 2 = \dfrac{1}{10}(y - 5)^2$

g) $x = \dfrac{1}{5}(y + 1)^2$

h) $y + 3 = -\dfrac{1}{20}(x - 2)^2$

i) $x - 2 = -\dfrac{1}{12}(y + 6)^2$

Apply, Solve, Communicate

6. Headlights Automobile headlights contain a parabolic reflector. A bulb with two filaments is used to produce low and high beams. The filament at the focus of the parabola produces the high beam. Light from the filament at the focus is reflected from the parabolic reflector to produce parallel light rays, projecting the light a greater distance. Suppose the filament for the high beams is 5 cm from the vertex of the reflector. Write an equation in standard form that models the parabola. Assume that the vertex is at the origin and that the filament is 5 units to the right of the origin on the *x*-axis.

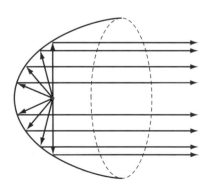

B

7. Parabolic reflector TV technicians use parabolic reflectors to pick up the sounds from the playing field at sporting events. The reflector focuses the incoming sound waves on a microphone, which is located at the focus of the reflector. Suppose the microphone is located 15 cm from the vertex of the reflector.

a) Write an equation in standard form for the parabolic reflector. Assume that the vertex is at the origin and that the microphone is to the left of the vertex on the *x*-axis.

b) Find the width of the reflector, to the nearest centimetre, at a horizontal distance of 30 cm from the vertex.

8. Application The stream of water from some water fountains follows a path in the form of a parabolic arch. For one fountain, the maximum height of the water is 8 cm, and the horizontal distance of the water flow is 12 cm.

a) Find an equation in standard form that models the continuous flow of water. Assume that the water spout is located at the origin.

b) At a horizontal distance of 10 cm from the origin, what is the height of the water, to the nearest tenth of a centimetre?

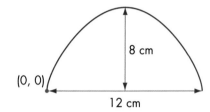

9. Skateboard ramp For a skateboarding competition, the organizers would like to use a parabolic ramp with a depth of 5 m, and a width of 15 m. Assume that the starting point of a skateboarder at the top of the ramp is (0, 0), as shown.

a) Find an equation in standard form that models the parabolic ramp.

b) Find the depth of the ramp, to the nearest tenth of a metre, at a horizontal distance of 10 m from the origin.

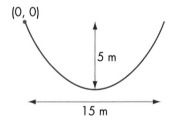

10. Parabolic antenna A parabolic antenna is 320 cm wide at a distance of 50 cm from its vertex. Determine the distance of the focus from the vertex.

11. Motion in space A spacecraft is in a circular orbit 150 km above Earth. When it reaches the velocity needed to escape the Earth's gravity, the spacecraft will follow a parabolic path with the focus at the centre of the Earth, as shown. Suppose the spacecraft reaches its escape velocity above the North Pole. Assume the radius of the Earth is 6400 km. Write an equation in standard form for the parabolic path of the spacecraft.

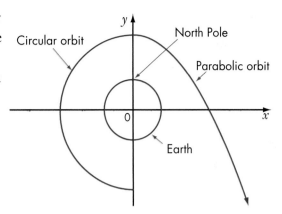

12. Hubble Space Telescope The primary reflector of the Hubble Space Telescope is parabolic and has a diameter of 4.27 m and a depth of 0.75 m.
a) If a camera is recording pictures at the focus of the reflector, how far is the camera from the vertex, to the nearest hundredth of a metre?
b) Write an equation in standard form that models the parabolic reflector. Sketch the location of the reflector on the coordinate axes for this equation.

13. Technology Use a graphing calculator to graph each parabola.

a) $y + 3 = \dfrac{1}{4}(x - 2)^2$

b) $x + 1 = \dfrac{1}{12}(y - 5)^2$

c) $x - 3 = \dfrac{1}{8}(y + 2)^2$

14. Communication a) Use a graphing calculator to graph the family of parabolas of the form $y = \dfrac{1}{4p}x^2$ for $p = 1, 2, 3$.

b) Now graph for $p = \dfrac{1}{2}, \dfrac{1}{3}, \dfrac{1}{4}$.

c) How are the graphs alike? How are they different?
d) What happens to the parabola as p gets closer to 0?
e) What happens to the foci as p gets closer to 0?

C

15. Inquiry/Problem Solving Use the locus definition of the parabola to derive the equation in standard form for a parabola with vertex $(0, 0)$, focus $(0, p)$ and directrix $y = -p$.

16. Standard form Use the locus definition of the parabola to derive the equation in standard form for a parabola with vertex $(0, 0)$, focus $(p, 0)$ and directrix $x = -p$.

A parabolic bridge is 40 m across and 25 m high. What is the length of a stabilizing beam across the bridge at a height of 16 m?

CAREER CONNECTION *Communications*

The need for people to communicate with each other has been an important aspect of human history. Modern communication between people can take many forms, including travelling to see each other, mailing a letter, or making a phone call.

For a country as large as Canada to compete economically, a highly developed communications industry is essential. Sending information by electronic means, including radio, television, and the Internet, is an increasingly important aspect of the communications industry.

1. TV satellite dish A satellite dish picks up TV signals from a satellite. The signals travel in parallel paths. When the signals reach the dish, they are reflected to the focus, where the detector is located. Suppose that the focus is located 20 cm from the vertex.
a) Find an equation in standard form that models the shape of the satellite dish. Sketch the location of the dish on the coordinate axes for this equation.
b) Find the width of the dish 20 cm from the vertex.

2. Research Use your research skills to investigate each of the following.
a) a career that interests you in the communications industry, including the education and training required and the type of work involved
b) the work of Marshall McLuhan (1911–1980), a Canadian who was world famous for his work on communications and the media

8.8 Conics With Equations in the Form $ax^2 + by^2 + 2gx + 2fy + c = 0$

The CF-18 Hornet is a supersonic jet flown in Canada. It has a maximum speed of Mach 1.8. The speed of sound is Mach 1. When a plane like the Hornet breaks the sound barrier, it produces a shock wave in the shape of a cone. For a plane flying parallel to the ground, the shock wave, or sonic boom, intersects the ground in a branch of a hyperbola.

The standard form of the equation of a conic provides a convenient way to identify the conic and sketch the graph. The equation of a conic can also be written in the form $ax^2 + by^2 + 2gx + 2fy + c = 0$.

For example, an equation in standard form for a hyperbola is

$$\frac{(x+2)^2}{9} - \frac{(y-1)^2}{4} = 1$$

Multiply both sides by 36: $\qquad 4(x+2)^2 - 9(y-1)^2 = 36$

Expand and simplify: $\qquad 4(x^2 + 4x + 4) - 9(y^2 - 2y + 1) = 36$

$$4x^2 + 16x + 16 - 9y^2 + 18y - 9 = 36$$

$$4x^2 - 9y^2 + 16x + 18y - 29 = 0$$

In general, $ax^2 + by^2 + 2gx + 2fy + c = 0$ is a quadratic equation when a and b are not both equal to zero.

INVESTIGATE & INQUIRE

1. Copy and complete the table by expanding and simplifying each equation. Write each result in the form $ax^2 + by^2 + 2gx + 2fy + c = 0$.

	Standard Form	$ax^2 + by^2 + 2gx + 2fy + c = 0$
a)	$\dfrac{(x-3)^2}{4} + \dfrac{(y+2)^2}{25} = 1$	
b)	$(x-3)^2 + (y-1)^2 = 9$	
c)	$y + 3 = 2(x-1)^2$	
d)	$\dfrac{(x-1)^2}{9} - \dfrac{(y+1)^2}{16} = 1$	

2. Identify each equation in standard form in question 1 as the equation of a circle, ellipse, hyperbola, or parabola.

3. a) For the circle written in the form $ax^2 + by^2 + 2gx + 2fy + c = 0$, what do you notice about the values of a and b?
b) For a circle, if $f = 0$ and $g = 0$, what are the coordinates of the centre?

4. a) For the ellipse written in the form $ax^2 + by^2 + 2gx + 2fy + c = 0$, what do you notice about the signs of a and b?
b) For an ellipse, if $f = 0$ and $g = 0$, what are the coordinates of the centre?

5. a) For the parabola written in the form $ax^2 + by^2 + 2gx + 2fy + c = 0$, what do you notice about the values of a and b?
b) What is always true for the value of a or b for a parabola?

6. a) For the hyperbola written in the form $ax^2 + by^2 + 2gx + 2fy + c = 0$, what do you notice about the signs of a and b?
b) For a hyperbola, if $f = 0$ and $g = 0$, what are the coordinates of centre?

7. If you are given the equation of a conic in the form $ax^2 + by^2 + 2gx + 2fy + c = 0$, how can you identify the type of conic, without rewriting the equation in standard form?

8. Without rewriting in standard form, identify each of the following conics.
a) $3x^2 + 24x - y + 50 = 0$
b) $16x^2 - y^2 - 32x - 10y - 25 = 0$
c) $4x^2 + y^2 + 8x - 4y + 4 = 0$
d) $x^2 + y^2 - 6x - 2y + 1 = 0$

The type of conic represented by an equation in the form $ax^2 + by^2 + 2gx + 2fy + c = 0$ can be identified using the signs and values of a and b.

- For a circle, $a = b$.
- For an ellipse, a and b have the same sign, and $a \neq b$.
- For a parabola, either $a = 0$ or $b = 0$.
- For a hyperbola, a and b have opposite signs.

EXAMPLE 1 Sketching the Graph of a Conic

a) Identify the type of conic whose equation is $4x^2 + 9y^2 - 16x + 18y - 11 = 0$.
b) Write the equation in standard form.
c) Determine the key features and sketch the graph.

SOLUTION

a) Since $a = 4$ and $b = 9$, a and b have the same sign, with $a \neq b$. The conic is an ellipse.

b) To write the equation in standard form, complete the square for both variables.

$$4x^2 + 9y^2 - 16x + 18y - 11 = 0$$

Add 11 to both sides:
$$4x^2 + 9y^2 - 16x + 18y = 11$$

Group the x and y terms:
$$4x^2 - 16x + 9y^2 + 18y = 11$$

Remove common factors:
$$4(x^2 - 4x) + 9(y^2 + 2y) = 11$$

Complete the square:
$$4(x^2 - 4x + 4 - 4) + 9(y^2 + 2y + 1 - 1) = 11$$

$$4(x - 2)^2 - 16 + 9(y + 1)^2 - 9 = 11$$

$$4(x - 2)^2 + 9(y + 1)^2 = 36$$

Divide both sides by 36:
$$\frac{(x - 2)^2}{9} + \frac{(y + 1)^2}{4} = 1$$

The equation in standard form is $\dfrac{(x - 2)^2}{9} + \dfrac{(y + 1)^2}{4} = 1$.

c) The equation is in the form $\dfrac{(x - h)^2}{a^2} + \dfrac{(y - k)^2}{b^2} = 1$.

The ellipse is centred at (h, k), or $(2, -1)$, and the major axis is parallel to the x-axis.

$a^2 = 9$, so $a = 3$

$b^2 = 4$, so $b = 2$

The major axis, which is parallel to the x-axis, has a length of $2a$, or 6.
The minor axis, which is parallel to the y-axis, has a length of $2b$, or 4.

The vertices are $V_1(h - a, k)$ and $V_2(h + a, k)$.
Substitute the values of h, k, and a.
The vertices are $V_1(2 - 3, -1)$ and $V_2(2 + 3, -1)$, or $V_1(-1, -1)$ and $V_2(5, -1)$.

The co-vertices are $(h, k - b)$ and $(h, k + b)$.
Substitute the values of h, k, and b.
The co-vertices are $(2, -1 - 2)$ and $(2, -1 + 2)$, or $(2, -3)$ and $(2, 1)$.

The foci are $F_1(h - c, k)$ and $F_2(h + c, k)$.

To find c, we use $a^2 = b^2 + c^2$, with $a = 3$ and $b = 2$.

$$a^2 = b^2 + c^2$$
$$3^2 = 2^2 + c^2$$
$$9 = 4 + c^2$$
$$5 = c^2$$
$$\sqrt{5} = c$$

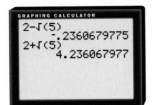

The coordinates of the foci are $(2 - \sqrt{5}, -1)$ and $(2 + \sqrt{5}, -1)$, or approximately $(-0.24, -1)$ and $(4.24, -1)$.

Plot the vertices and co-vertices.
Draw a smooth curve through the points.
Label the foci and the graph.

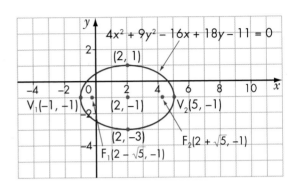

Example 2 **Sketching the Graph of a Conic**

a) Identify the type of conic whose equation is $y^2 + 8x + 2y - 15 = 0$.
b) Write the equation in standard form.
c) Determine the key features and sketch the graph.

Solution

a) Since $a = 0$ and $b \neq 0$, the conic is a parabola.

b) To write the equation in standard form, complete the square for the y variable.

$$y^2 + 8x + 2y - 15 = 0$$

Add 15 to both sides: $\qquad y^2 + 8x + 2y = 15$

Group the y terms: $\qquad y^2 + 2y + 8x = 15$

Complete the square: $\qquad y^2 + 2y + 1 - 1 + 8x = 15$
$$(y + 1)^2 - 1 + 8x = 15$$
$$(y + 1)^2 + 8x = 16$$

Rearrange: $\qquad\qquad 8x - 16 = -(y + 1)^2$

Remove a common factor: $\qquad 8(x - 2) = -(y + 1)^2$

Divide both sides by 8: $\qquad x - 2 = -\dfrac{1}{8}(y + 1)^2$

The equation in standard form is $x - 2 = -\dfrac{1}{8}(y + 1)^2$.

c) The equation is in the form $x - h = \dfrac{1}{4p}(y - k)^2$.

The vertex is V(h, k).
$h = 2$, $k = -1$, so the vertex is V(2, −1).

Find the value of p.

From the equation, $\dfrac{1}{4p} = -\dfrac{1}{8}$

$$4p = 4(-2)$$
$$p = -2$$

$p < 0$, so the parabola opens left.
The focus F($h + p$, k) is $(2 + (-2), -1)$ or $(0, -1)$.

The directrix is $x = h - p$

$$x = 2 - (-2)$$
$$x = 4$$

The axis of symmetry is $y = k$ or $y = -1$.
Sketch and label the graph.

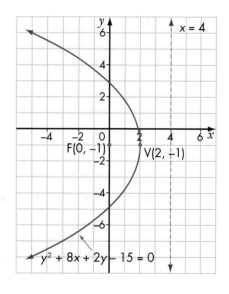

EXAMPLE 3 Shock Wave

A shock wave from an aircraft that breaks the sound barrier intersects
the ground in a curve with the equation $x^2 - 4y^2 + 4x + 24y - 36 = 0$.
a) Identify the type of conic.
b) Write the equation in standard form.
c) Determine the key features and sketch the graph.

SOLUTION

a) Since a and b have opposite signs, the conic is a hyperbola.

b) To write the equation in standard form, complete the square for both variables.

$$x^2 - 4y^2 + 4x + 24y - 36 = 0$$

Add 36 to both sides: $x^2 - 4y^2 + 4x + 24y = 36$

Group the x and y terms: $x^2 + 4x - 4y^2 + 24y = 36$

Remove a common factor: $x^2 + 4x - 4(y^2 - 6y) = 36$

Complete the square: $x^2 + 4x + 4 - 4 - 4(y^2 - 6y + 9 - 9) = 36$

$$(x + 2)^2 - 4 - 4(y - 3)^2 + 36 = 36$$

$$(x + 2)^2 - 4(y - 3)^2 = 4$$

Divide both sides by 4: $\dfrac{(x + 2)^2}{4} - (y - 3)^2 = 1$

The equation in standard form is $\dfrac{(x + 2)^2}{4} - (y - 3)^2 = 1$.

c) The equation is in the form $\dfrac{(x - h)^2}{a^2} - \dfrac{(y - k)^2}{b^2} = 1$.

The centre is C(h, k) = (−2, 3).
The transverse axis is parallel to the x-axis.
$a^2 = 4$, so $a = 2$
$b^2 = 1$, so $b = 1$

The vertices are $V_1(h - a, k)$ and $V_2(h + a, k)$.
Substitute the values for h, k, and a.
$V_1(-2 - 2, 3)$ and $V_2(-2 + 2, 3)$ or $V_1(-4, 3)$ and $V_2(0, 3)$.

The co-vertices are $(h, k - b)$ and $(h, k + b)$.
Substitute the values for h, k, and b.
The coordinates of the co-vertices are $(-2, 3 - 1)$ and $(-2, 3 + 1)$ or $(-2, 2)$ and $(-2, 4)$.

The length of the transverse axis is
$$2a = 2(2)$$
$$= 4$$
The length of the conjugate axis is
$$2b = 2(1)$$
$$= 2$$

The coordinates of the foci are $F_1(h - c, k)$ and $F_2(h + c, k)$.
To find c, use $c^2 = a^2 + b^2$, with $a = 2$ and $b = 1$.

$$c^2 = 2^2 + 1^2$$
$$= 5$$
$$c = \sqrt{5}$$

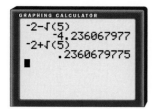

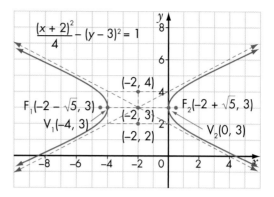

The coordinates of the foci are $(-2 - \sqrt{5}, 3)$
and $(-2 + \sqrt{5}, 3)$, or approximately $(-4.24, 3)$
and $(0.24, 3)$.
Sketch and label the graph.

Key Concepts

• The equation of a conic can be written in the form
$ax^2 + by^2 + 2gx + 2fy + c = 0$, where a and b are not both equal to zero.
• The type of conic represented by an equation in the form
$ax^2 + by^2 + 2gx + 2fy + c = 0$ can be identified using the signs and values
of a and b.
  * For a circle, $a = b$.
  * For an ellipse, a and b have the same sign, and $a \neq b$.
  * For a parabola, $a = 0$ or $b = 0$.
  * For a hyperbola, a and b have opposite signs.
• To graph a conic section whose equation is in the form
$ax^2 + by^2 + 2gx + 2fy + c = 0$, first use the method of completing the square to
write the equation in standard form.

Communicate Your Understanding

1. Identify each of the following conics.
a) $x^2 + 4y^2 - 16 = 0$ **b)** $2x^2 - 2x - 6y - 3 = 0$
c) $x^2 + y^2 - 4x + 8y - 44 = 0$ **d)** $x^2 - 9y^2 - 14x + 36y + 4 = 0$
2. Describe how you would write an equation in standard form for the conic
defined by $4x^2 + 25y^2 - 16x + 50y - 9 = 0$.

Practise

A

1. Identify the type of conic by inspection.
a) $x^2 - 2y^2 - 6x + 4y - 2 = 0$
b) $2x^2 + y^2 - 6x - 4y - 3 = 0$
c) $x^2 + y^2 - 5x + 4y + 3 = 0$
d) $3y^2 + 6x - 6y - 9 = 0$
e) $2x^2 - 3y^2 - 6x - 1 = 0$
f) $3x^2 - 4y^2 + 3x + 6y - 1 = 0$
g) $2x^2 - 6x + 9y = 0$

2. For each of the following equations,
i) identify the type of conic
ii) write the equation in standard form
iii) determine the key features and sketch the graph
a) $x^2 + y^2 - 2x - 6y - 15 = 0$
b) $4x^2 + y^2 + 24x - 4y - 24 = 0$
c) $x^2 + 6x - 8y + 25 = 0$
d) $y^2 - 4y - 8x + 12 = 0$
e) $x^2 + 16y^2 + 8x - 96y + 144 = 0$
f) $x^2 + y^2 + 4x - 6y - 23 = 0$
g) $2x^2 - 2y^2 + 4x - 4y + 1 = 0$
h) $y^2 - 4y + 4x + 8 = 0$
i) $x^2 - 2y^2 - 6x - 4y - 2 = 0$

3. For each conic, write an equation in standard form and in the form
$ax^2 + by^2 + 2gx + 2fy + c = 0$.

a)
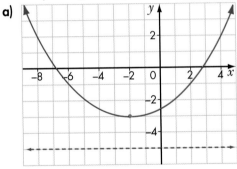

b)
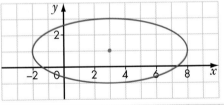

c)
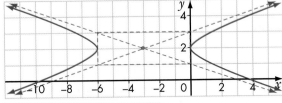

d)
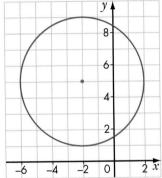

e)
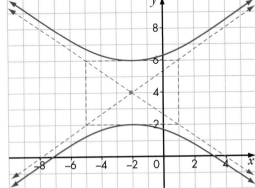

f)
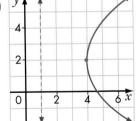

Apply, Solve, Communicate

B

4. Inquiry/Problem Solving a) For the equation $x^2 + by^2 - 4 = 0$, determine the value(s) of b that result in an equation of
i) a circle   **ii)** a parabola
iii) an ellipse   **iv)** a hyperbola
b) Give an example to illustrate each answer in part a).

5. a) For the equation $ax^2 - y^2 + 9 = 0$, determine the value(s) of a that will result in an equation of
i) a circle   **ii)** a parabola
iii) an ellipse   **iv)** a hyperbola
b) Give an example to illustrate each answer in part a).

6. Inquiry/Problem Solving The three squares in the diagram are centred at the same point. The red border has the same area as the smallest square.
a) How are p and q related?
b) Graph the relation.

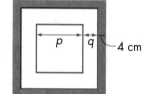

7. Application When a plane breaks the sound barrier, a shock wave in the shape of a cone is produced. If the plane is flying parallel to the ground, the shock wave intersects the ground in a branch of a hyperbola.
a) Use your knowledge of the intersection of a plane and a cone to explain why the intersection is a branch of a hyperbola.
b) If the intersection of a shock wave with the ground can be modelled by the equation $x^2 + 25y^2 - 8x + 100y + 91 = 0$, describe how the plane is flying.
c) Is it possible for the intersection of a shock wave with the ground to be modelled by the equation $4x^2 + 4y^2 + 36y + 5 = 0$? Explain.

8. Degenerate conics The rules for identifying a type of conic from its equation do not always apply. For example, in $x^2 + y^2 + 2x - 4y + 5 = 0$, $a = b$, so the equation appears to model a circle.

Completing the square gives

$$x^2 + y^2 + 2x - 4y + 5 = 0$$
$$x^2 + 2x + y^2 - 4y + 5 = 0$$
$$x^2 + 2x + 1 - 1 + y^2 - 4y + 4 - 4 + 5 = 0$$
$$(x + 1)^2 + (y - 2)^2 = 0$$

The ordered pair $(-1, 2)$ satisfies the equation. The solution is $(-1, 2)$. Therefore, the graph of $x^2 + y^2 + 2x - 4y + 5 = 0$ is a point, not a circle. Because the equation appears to model a circle, the graph is referred to as a degenerate circle.

Changing the equation to $x^2 + y^2 + 2x - 4y + 6 = 0$ and completing the square gives $(x + 1)^2 + (y - 2)^2 = -1$.

This equation has no solution, since the sum of two squares cannot be negative. So, the equation $x^2 + y^2 + 2x - 4y + 6 = 0$ is degenerate.

i) Identify the type of conic that each of the following equations appears to model.

ii) Verify that each equation is degenerate.

iii) Graph the equation, if possible.

a) $4x^2 + y^2 - 8x + 2y + 6 = 0$

b) $x^2 + y^2 - 2x - 6y + 10 = 0$

c) $3x^2 + 4y^2 - 6x - 24y + 39 = 0$

d) $9x^2 - y^2 + 18x + 6y = 0$

LOGIC *Power*

What are the four moves that X should not play, if X wants to stop O from winning?

8.9 Intersection of Lines and Conics

The centre circle of a hockey rink has a radius of 4.5 m. A diameter of the centre circle lies on the centre red line.

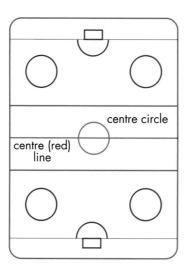

INVESTIGATE & INQUIRE

1. A coordinate grid is superimposed on a hockey rink so that the origin is at the centre of the centre circle and the *x*-axis lies along the red line. The side length of each grid square represents 1 m.

a) Write the equation of the centre circle.

b) Graph the centre circle on grid paper or use a graphing calculator.

2. The puck is passed across the centre line along each of the following lines. Graph each line and find the number of points in which it intersects the centre circle. Find the coordinates of any points of intersection, rounding to the nearest tenth, if necessary.

a) $y = x$ **b)** $y = 8 - x$ **c)** $y = 4.5$

3. a) What is the maximum number of points in which a line and a circle can intersect?

b) What is the minimum number of points in which a line and a circle can intersect?

4. Without graphing each of the following lines, state the number of points in which it intersects the centre circle.

a) $y = 6$ **b)** $y = -4.5$ **c)** $y = -1$ **d)** $y = 3x$

Recall that a system of equations consists of two or more equations that are considered together. If the graphs of two equations in a system are a straight line and a conic, the system will have no solution, one solution, or two solutions.

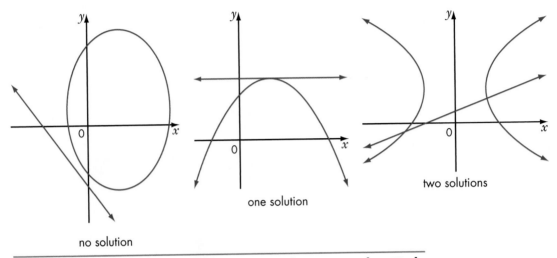

no solution

one solution

two solutions

Example 1 Finding Points of Intersection of a Line and a Circle

Find the coordinates of the points of intersection of the line $y = x - 1$ and the circle $x^2 + y^2 = 25$.

Solution 1 Paper-and-Pencil Method

Solve the system of equations by substitution.

$y = x - 1$ (1)
$x^2 + y^2 = 25$ (2)

Substitute $x - 1$ for y in (2): $x^2 + (x - 1)^2 = 25$
Expand: $x^2 + x^2 - 2x + 1 = 25$
Simplify: $2x^2 - 2x - 24 = 0$
Divide both sides by 2: $x^2 - x - 12 = 0$
Factor: $(x - 4)(x + 3) = 0$
Use the zero product property: $x - 4 = 0$ or $x + 3 = 0$
                                    $x = 4$ or $x = -3$

Substitute these values of x into (1).

For $x = 4$,   For $x = -3$,
$y = x - 1$ $y = x - 1$
   $= 4 - 1$ $= -3 - 1$
   $= 3$ $= -4$

Check in (1).   Check in (2).

For (4, 3),   For (4, 3),

L.S. $= y$ R.S. $= x - 1$ L.S. $= x^2 + y^2$ R.S. $= 25$

    $= 3$ $= 4 - 1$ $= 4^2 + 3^2$

                 $= 3$ $= 25$

For (−3, −4)   For (−3, −4),

L.S. $= y$ R.S. $= x - 1$ L.S. $= x^2 + y^2$ R.S. $= 25$

    $= -4$ $= -3 - 1$ $= (-3)^2 + (-4)^2$

                 $= -4$ $= 25$

The coordinates of the points of intersection are (4, 3) and (−3, −4).

Solution 2   Graphing-Calculator Method

Solve the equation of the circle for y.

$$x^2 + y^2 = 25$$
$$y^2 = 25 - x^2$$
$$y = \pm\sqrt{25 - x^2}$$

So, in the same viewing window, graph $y = x - 1$, $y = \sqrt{25 - x^2}$, and $y = -\sqrt{25 - x^2}$.

To make the circle look like a circle, use the Zsquare instruction.

To find the points of intersection, use the intersect operation.

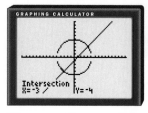 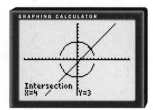

The coordinates of the points of intersection are (−3, −4) and (4, 3). These coordinates can be checked by substitution into the original equations.

Example 2   Finding Points of Intersection of a Line and an Ellipse

Find the coordinates of the points of intersection of the line $2x + y = 1$ and the ellipse $4x^2 + y^2 = 25$.

SOLUTION

Solve the system of equations by substitution.

$2x + y = 1$ (1)
$4x^2 + y^2 = 25$ (2)

Solve for y in (1):
$$2x + y = 1$$
$$y = 1 - 2x$$

Substitute $1 - 2x$ for y in (2): $4x^2 + (1 - 2x)^2 = 25$

Expand: $4x^2 + 1 - 4x + 4x^2 = 25$

Simplify: $8x^2 - 4x - 24 = 0$

Divide both sides by 4: $2x^2 - x - 6 = 0$

Factor: $(2x + 3)(x - 2) = 0$

Use the zero product property: $2x + 3 = 0$ or $x - 2 = 0$

$$x = -\frac{3}{2} \quad \text{or} \quad x = 2$$

Substitute for x in (1) to find y.

For $x = -\dfrac{3}{2}$,   For $x = 2$,

$$2\left(-\frac{3}{2}\right) + y = 1 \qquad 2(2) + y = 1$$
$$-3 + y = 1 \qquad\qquad 4 + y = 1$$
$$y = 4 \qquad\qquad\qquad y = 1 - 4$$
$$= -3$$

The system can be modelled graphically.

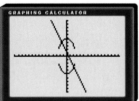

Check in (1).

For $\left(-\dfrac{3}{2}, 4\right)$,   For $(2, -3)$,

L.S. $= 2x + y$ R.S. $= 1$ L.S. $= 2x + y$ R.S. $= 1$

$= 2\left(-\dfrac{3}{2}\right) + 4$ $= 2(2) - 3$

$= 1$ $= 1$

Check in (2).

For $\left(-\dfrac{3}{2}, 4\right)$,   For $(2, -3)$,

L.S. $= 4x^2 + y^2$ R.S. $= 25$ L.S. $= 4x^2 + y^2$ R.S. $= 25$

$= 4\left(-\dfrac{3}{2}\right)^2 + 4^2$ $= 4(2)^2 + (-3)^2$

$= 9 + 16$ $= 16 + 9$

$= 25$ $= 25$

The coordinates of the points of intersection are $\left(-\dfrac{3}{2}, 4\right)$ and $(2, -3)$.

EXAMPLE 3 Finding Points of Intersection of a Line and a Parabola

Find the coordinates of the points of intersection of the parabola
$y - 4 = -(x + 1)^2$ and
a) the line $y = -4x + 4$
b) the line $y = 3x + 13$

SOLUTION

a) Solve the system of equations by substitution.

$y = -4x + 4$ (1)

$y - 4 = -(x + 1)^2$ (2)

Substitute $-4x + 4$ for y in (2): $-4x + 4 - 4 = -(x + 1)^2$

Expand: $-4x + 4 - 4 = -(x^2 + 2x + 1)$

Simplify: $-4x = -x^2 - 2x - 1$

                               $x^2 - 2x + 1 = 0$

Factor: $(x - 1)(x - 1) = 0$

Use the zero product property: $x - 1 = 0$ or $x - 1 = 0$

                           $x = 1$ or $x = 1$

Substitute for x in (1) to find y.

$y = -4x + 4$

$ = -4(1) + 4$

$ = 0$

Check in (1).

L.S. $= y$ R.S. $= -4x + 4$

$ = 0$ $= -4(1) + 4$

                      $= 0$

Check in (2).

L.S. $= y - 4$ R.S. $= -(x + 1)^2$

$ = 0 - 4$ $= -(1 + 1)^2$

$ = -4$ $= -4$

The coordinates of the point of intersection are (1, 0).

The system can be modelled graphically.

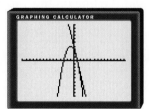

b) Solve the system of equations.

$$y = 3x + 13 \qquad (1)$$
$$y - 4 = -(x + 1)^2 \qquad (2)$$

Substitute $3x + 13$ for y in (2): $\quad 3x + 13 - 4 = -(x + 1)^2$

Expand: $\qquad\qquad\qquad\qquad 3x + 13 - 4 = -(x^2 + 2x + 1)$

Simplify: $\qquad\qquad\qquad\qquad\quad 3x + 9 = -x^2 - 2x - 1$

$$x^2 + 5x + 10 = 0$$

For $x^2 + 5x + 10 = 0$, $a = 1$, $b = 5$, and $c = 10$.
Substitute these values into the quadratic formula.

$$x = \frac{-b \pm \sqrt{b^2 - 4ac}}{2a}$$

$$= \frac{-5 \pm \sqrt{5^2 - 4(1)(10)}}{2(1)}$$

$$= \frac{-5 \pm \sqrt{25 - 40}}{2}$$

$$= \frac{-5 \pm \sqrt{-15}}{2}$$

The system can be modelled graphically.

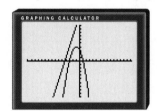

Since no real number is the square root of a negative number, there are no real solutions.
The parabola and the line do not intersect.

Example 4 Finding Points of Intersection of a Line and a Hyperbola

Find the coordinates of the points of intersection of the line $y = x + 3$ and the hyperbola $y^2 - 4x^2 = 4$. Round answers to the nearest tenth.

Solution

Solve the system of equations by substitution.

$$y = x + 3 \qquad (1)$$
$$y^2 - 4x^2 = 4 \qquad (2)$$

Substitute $x + 3$ for y in (2): $\quad (x + 3)^2 - 4x^2 = 4$

Expand: $\qquad\qquad\qquad\qquad x^2 + 6x + 9 - 4x^2 = 4$

Simplify: $\qquad\qquad\qquad\qquad\quad -3x^2 + 6x + 5 = 0$

Multiply by -1: $\qquad\qquad\qquad 3x^2 - 6x - 5 = 0$

Solve using the quadratic formula.

$$x = \frac{-b \pm \sqrt{b^2 - 4ac}}{2a}$$

$$= \frac{6 \pm \sqrt{(-6)^2 - 4(3)(-5)}}{2(3)}$$

$$= \frac{6 \pm \sqrt{36 + 60}}{6}$$

$$= \frac{6 \pm \sqrt{96}}{6}$$

$$\doteq 2.6 \text{ or } -0.6$$

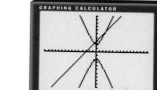

The system can be modelled graphically.

Substitute for x in (1) to find y.

For $x = 2.6$, $y = 2.6 + 3$ For $x = -0.6, y = -0.6 + 3$

$\qquad\qquad = 5.6$ $\qquad\qquad\qquad\qquad = 2.4$

The coordinates of the points of intersection are (2.6, 5.6) and (−0.6, 2.4), to the nearest tenth.

Example 5 Flight Path

A pilot is flying a small aircraft at 200 km/h directly toward the centre of an intense weather system. She decides to change direction to avoid the worst of the weather. Taking the aircraft as the origin of a coordinate grid, with the side length of each grid square representing 1 km, the weather system can be modelled by the relation $(x - 20)^2 + (y - 10)^2 = 49$. The new direction of the aircraft is along the line $y = 0.2x$ in the first quadrant. Will the aircraft completely avoid the weather system? If not, for what amount of time will the aircraft be within the weather system, to the nearest tenth of a minute?

Solution 1 Paper-and-Pencil Method

Solve the system of equations.

$(x - 20)^2 + (y - 10)^2 = 49$ $\qquad$ (1)

$y = 0.2x$ $\qquad$ (2)

Substitute $0.2x$ for y in (1).

$$(x - 20)^2 + (0.2x - 10)^2 = 49$$

$$x^2 - 40x + 400 + 0.04x^2 - 4x + 100 = 49$$

$$1.04x^2 - 44x + 451 = 0$$

Solve using the quadratic formula.

$$x = \frac{-b \pm \sqrt{b^2 - 4ac}}{2a}$$

$$= \frac{44 \pm \sqrt{(-44)^2 - 4(1.04)(451)}}{2(1.04)}$$

So, $x \doteq 24.87$ or $x \doteq 17.43$.

Substitute in (2).

For $x = 24.87$,   For $x = 17.43$,

$y = 0.2(24.87)$ $y = 0.2(17.43)$

$\doteq 4.97$ $\doteq 3.49$

The coordinates of the points of intersection are approximately (17.43, 3.49) and (24.87, 4.97).

So, the aircraft will not completely avoid the weather system.

Use the formula for the length of a line segment to find the distance between the two points of intersection.

$$d = \sqrt{(x_2 - x_1)^2 + (y_2 - y_1)^2}$$

$$= \sqrt{(24.87 - 17.43)^2 + (4.97 - 3.49)^2}$$

$$\doteq 7.59$$

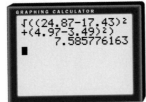

The aircraft will be within the weather system for a distance of about 7.59 km.

Use time $= \dfrac{\text{distance}}{\text{speed}}$ to find the amount of time within the weather system.

$$\text{time} = \frac{7.59}{200}$$

$$= 0.037\ 95$$

Convert this number of hours to minutes.

$0.037\ 95 \times 60 = 2.277$

So, the aircraft will be within the weather system for 2.3 min, to the nearest tenth of a minute.

SOLUTION 2 Graphing-Calculator Method

Solve $(x - 20)^2 + (y - 10)^2 = 49$ for y.

$$(x - 20)^2 + (y - 10)^2 = 49$$
$$(y - 10)^2 = 49 - (x - 20)^2$$
$$y - 10 = \pm \sqrt{49 - (x - 20)^2}$$
$$y = 10 \pm \sqrt{49 - (x - 20)^2}$$

In the same viewing window, graph $y = 0.2x$, $y = 10 + \sqrt{49 - (x - 20)^2}$, and $y = 10 - \sqrt{49 - (x - 20)^2}$.

Adjust the window variables and use the Zsquare instruction. To find the points of intersection, use the Intersect operation.

The window variables include Xmin $\doteq$ 4.8, Xmax $\doteq$ 35.2, Ymin = 0, Ymax = 20.

The coordinates of the points of intersection are approximately (17.43, 3.49) and (24.87, 4.97). So, the aircraft will not completely avoid the weather system.

The time that the aircraft will be within the weather system can be found using the formula for the length of a line segment and the formula $\text{time} = \dfrac{\text{distance}}{\text{speed}}$, as shown in Solution 1.

The aircraft will be within the weather system for 2.3 min, to the nearest tenth of a minute.

Key Concepts

• If the graphs of two equations in a system are a straight line and a conic, the system will have no solution, one solution, or two solutions.
• To find the points of intersection of a linear equation and a quadratic equation algebraically,
a) solve the linear equation for one variable
b) substitute the expression for that variable in the quadratic equation
c) solve the quadratic equation to find any real value(s) of the remaining variable
d) substitute any real value(s) of the variable found from part c) into the linear equation to find the value(s) of the other variable
• To solve a system of a straight line and a conic graphically, graph both equations and determine the coordinates of the point or points of intersection, if they exist.

Communicate Your Understanding

1. Describe how you would find the points of intersection of $x - y = 2$ and $x^2 + y^2 = 16$ algebraically.
2. Describe how you would solve the system $2x^2 + y^2 = 36$ and $2x - y = 1$ graphically.
3. If the graphs of a linear equation and a quadratic equation do not intersect, describe the algebraic solution to the system.

Practise

A

1. Solve each system of equations.

a) $y - 9 = 0$
$y - x^2 = 0$

b) $y + 25 = 0$
$x^2 + y = 0$

c) $y = 2x$
$y - x^2 = 0$

d) $y - x = 0$
$x^2 + y^2 - 32 = 0$

e) $y = 4x$
$x^2 - y + 4 = 0$

f) $y + 3x = 0$
$x^2 + y^2 = 10$

e) $4x^2 + y^2 = 64$
$\dfrac{1}{2}x + y = 10$

f) $y = 4x^2$
$12x - y - 9 = 0$

g) $y = x - 5$
$x^2 + y^2 = 4$

h) $9x^2 + 4y^2 = 36$
$2x + y = -4$

i) $x + y = 4$
$y = \dfrac{1}{2}x^2$

2. Solve each system of equations.

a) $y = x^2 + 3$
$3x + y = 1$

b) $x^2 + y^2 = 36$
$2x + y = -12$

c) $x^2 - y^2 = 9$
$x - y = 6$

d) $x^2 + 4y^2 = 16$
$5x - y = 2$

3. Solve each system of equations. Round solutions to the nearest hundredth, if necessary.

a) $x^2 + y^2 = 25$
$2x + y = -5$

b) $y = x^2 + 5$
$x + y = 2$

c) $x^2 + 4y^2 = 100$
$x - y = 5$

d) $x^2 - y^2 = 64$
$2x + y = 14$

e) $x^2 + y^2 = 16$
$x - 2y = 7$

f) $2y - x^2 = 0$
$2x + y = -2$

g) $x^2 + y^2 = 17$
$x + 2y = 2$

h) $25x^2 - 36y^2 = 900$
$x + 2y = -6$

i) $9x^2 + y^2 = 81$
$x + 3y = 3$

j) $x^2 + y^2 = 13$
$2x - y = 7$

k) $x^2 - 64y^2 = 1$
$x + 8y = 0$

l) $4x^2 - y^2 - 9 = 0$
$2x - y = 0$

4. Determine the length of the chord AB in each circle, to the nearest hundredth.

a)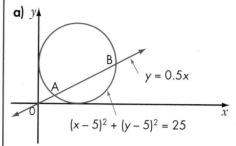
$(x - 5)^2 + (y - 5)^2 = 25$
$y = 0.5x$

b)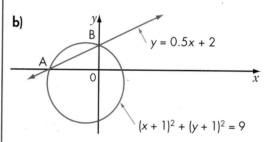
$y = 0.5x + 2$
$(x + 1)^2 + (y + 1)^2 = 9$

Apply, Solve, Communicate

5. Air-traffic control A radar screen has a range of 50 km. The screen can be modelled on a coordinate grid as a circle centred at the origin with radius 50.

a) Write the equation that describes the edge of the radar screen's range.

b) A small aircraft flies over the area covered by the radar at 180 km/h, on a path given by $y = -0.5x - 10$. Determine the length of time, to the nearest minute, that the plane is within radar range.

B

6. Communication In a city park, a sprinkler waters an area with a circumference that can be modelled by the equation $x^2 + y^2 = 20$. The sprinkler is located at the origin. A path through the park can be modelled by the equation $2x - y = 12$. Will the sprinkler spray people walking along the path? Explain.

7. Application The radar detector of a lighthouse has a range that can be modelled by the equation $x^2 + y^2 = 5$, if the lighthouse is located at the origin of a grid. A boat is travelling on a path defined by the equation $x - y = 2$. Will the boat be detected on the radar screen at the lighthouse? Explain.

8. UV index The UV index from 08:00 to 18:00 on one sunny July day in Central Ontario could be modelled by the quadratic function $y = -0.15(x - 13)^2 + 7.6$, where y is the UV index and x is the time of day, in hours on the 24-hour clock.

Web Connection
www.school.mcgrawhill.ca/resources/
To learn more about the UV index, visit the above web site. Go to **Math Resources**, then to *MATHEMATICS 11*, to find out where to go next. Write a brief report about how the UV index is determined.

a) Write a system of equations to determine the time period when the UV index was 7 or more.

b) Solve the system of equations from part a) and state the time period.

9. Radar The radar on a marine police launch has a range of 28 km. While the launch is at anchor in a bay, the radar shows a boat travelling on a linear path given by $y = 0.7x + 20$. If the boat is visible on the radar screen for 2 h, at what speed is it travelling, to the nearest kilometre per hour?

10. Power consumption The power consumption of a college cafeteria varies through the day. The power consumption can be approximated by the quadratic function $y = -0.25(x - 12)^2 + 50$, where y is the power consumption, in kilowatts (kW), and x is the time of day, in hours on the 24-hour clock.

a) Write a system of equations that could be used to determine the times of day when the power consumption is at least 40 kW.

b) Solve the system of equations from part a) to find the time period, to the nearest hour.

11. Inquiry/Problem Solving The line $x + 3y - 5 = 0$ intersects the circle $(x - 5)^2 + (y - 5)^2 = 25$ in two points, A and B.

a) Find the coordinates of the endpoints of the chord AB.

b) Verify that the right bisector of the chord AB passes through the centre of the circle.

12. In the figure, the circle has its centre at the origin and a radius of 5 units. Determine the exact length of the chord MN.

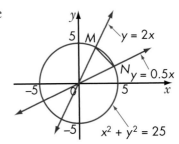

13. Theatre A spotlight illuminates an elliptical area on a theatre stage. The shape of the ellipse can be modelled by the equation $2x^2 + y^2 = 9$ on a grid with 1 unit equal to 1 m. An actor is crossing the stage along a path with the equation $y = 2x - 1$. Find the length of the actor's path illuminated by the spotlight, to the nearest tenth of a metre.

C

14. Hyperbolic mirror A hyperbolic mirror has the shape of one branch of a hyperbola. Light rays directed at the focus of a hyperbolic mirror are reflected toward the other focus of the hyperbola, as shown in the diagram.

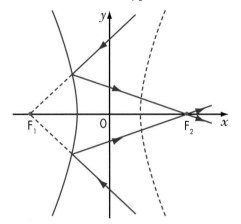

A mirror can be modelled by the left branch of the hyperbola with the equation $\dfrac{x^2}{9} - \dfrac{y^2}{16} = 1$. A light source is located at $(0, -8)$. Find the coordinates of the point on the mirror at which light from the source is reflected to the point $(5, 0)$. Round the coordinates to the nearest tenth.

15. Solve each system of equations.
a) $x + y^2 = 2$
   $2y - 2\sqrt{2} = x(\sqrt{2} + 2)$
b) $x^2 + y^2 = 1$
   $y = 3x + 1$
   $x^2 + (y + 1)^2 = 4$

Investigate & Apply

Confocal Conics

Confocal conics are conic sections that share a common focus. For example, the elliptical orbits of the planets are confocal. They all have the sun as one of their foci.

1. Write an equation for a circle, an ellipse, a hyperbola, and a parabola that have a focus at $(0, 2)$. Note that the focus of a circle is the centre.

2. a) The diagram shows the graphs of a circle, an ellipse, a parabola, and a hyperbola that have a focus at $(0, 1)$ and a vertex at the origin. Only the upper branch of the hyperbola is shown.

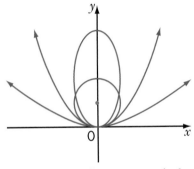

Write an equation for a circle, an ellipse, a parabola, and a hyperbola that satisfy the conditions.
b) Why do only one circle and one parabola satisfy the conditions?
c) How many equations are possible for ellipses that satisfy the conditions? Use examples to justify your answer.
d) How many equations are possible for hyperbolas whose upper branches satisfy the conditions? Use examples to justify your answer.

REVIEW OF **KEY CONCEPTS**

8.1-8.2 Equations of Loci

Refer to the Key Concepts on page 598.

1. Sketch the locus of points in the plane that are 2 cm from a circle of radius 5 cm.

2. a) How are the lines $y = -2x - 3$ and $y = -2x - 9$ related?
b) Graph the two lines.
c) Graph the locus of points that are equidistant from the two lines.
d) Write an equation to represent this locus.

3. Find an equation of the locus of points equidistant from each pair of points.
a) (2, 1) and (−3, −5)
b) (−3, 4) and (6, 1)

4. Determine an equation of the locus of points equidistant from the graphs of $y = \sqrt{x + 2}$ and $y = -\sqrt{x + 2}$.

8.4 The Circle

Refer to the Key Concepts on page 613.

5. Sketch and label each circle. Then, state the domain and range.
a) $x^2 + y^2 = 81$
b) $x^2 + y^2 = 40$
c) $(x - 3)^2 + (y + 5)^2 = 25$
d) $(x + 1)^2 + (y - 4)^2 = 60$

6. Write an equation in standard form for each circle.
a) centre (4, −2) and radius 3
b) centre (2, −4) and passing through the point (3, 0)
c) endpoints of a diameter (1, −4) and (7, 2)

7. Earthquake The epicentre of an earthquake is found to be 50 km from a seismic recording station. The direction of the epicentre cannot be determined from a reading at one station. Write an equation to model all possible locations of the earthquake epicentre if the recording station is located at (−2, 3).

8.5 The Ellipse

Refer to the Key Concepts on page 631.

8. Use the locus definition of the ellipse to write an equation in standard form for each ellipse.
a) foci $(-2, 0)$ and $(2, 0)$, sum of focal radii 6
b) foci $(0, -3)$ and $(0, 3)$, sum of focal radii 8

9. Sketch the graph of each ellipse. Label the coordinates of the centre, the vertices, the co-vertices, and the foci. State the domain and range.

a) $\dfrac{x^2}{16} + \dfrac{y^2}{4} = 1$

b) $\dfrac{x^2}{9} + \dfrac{y^2}{25} = 1$

c) $\dfrac{(x-3)^2}{36} + \dfrac{(y+1)^2}{9} = 1$

d) $\dfrac{(x+1)^2}{16} + \dfrac{(y-2)^2}{49} = 1$

10. Write an equation in standard form for each ellipse.
a) centre $(0, 0)$, major axis along x-axis, length of major axis 10, length of minor axis 5
b) centre $(0, 0)$, one vertex is $(0, -7)$, one focus is $(0, 5)$
c) foci at $(-2, -2)$ and $(-2, 7)$, with sum of focal radii 20
d) centre $(2, -1)$ and passing through $(-3, -1)$, $(7, -1)$, $(2, 1)$ and $(2, -4)$

11. Mercury Planets orbit the sun in elliptical paths with the sun at one focus. The least distance from the sun to Mercury is 4.60 million kilometres, and the greatest distance from the sun to Mercury is 6.98 million kilometres. Write an equation of the ellipse that models Mercury's orbit about the sun. Assume that the sun is on the x-axis.

8.6 The Hyperbola

Refer to the Key Concepts on page 647.

12. Use the locus definition of the hyperbola to find an equation of each hyperbola.
a) foci $(-5, 0)$ and $(5, 0)$, and $|F_1P - F_2P| = 4$
b) foci $(0, -4)$ and $(0, 4)$, and $|F_1P - F_2P| = 2$

13. Sketch the graph of each hyperbola. Label the coordinates of the centre, the vertices, the co-vertices, and the foci. State the domain and range.

a) $\dfrac{x^2}{25} - \dfrac{y^2}{4} = 1$

b) $\dfrac{x^2}{4} - \dfrac{y^2}{36} = 1$

c) $\dfrac{(x+2)^2}{25} - \dfrac{(y-1)^2}{16} = 1$

d) $\dfrac{(x-1)^2}{49} - \dfrac{(y-3)^2}{36} = 1$

14. Determine an equation in standard form for each hyperbola.
a) vertices $(-4, 0)$, $(4, 0)$, co-vertices $(0, -3)$, $(0, 3)$
b) vertices $(-5, -2)$, $(-5, 6)$, foci $(-5, -4)$, $(-5, 8)$
c) centre $(-2, 1)$, one focus $(-4, 1)$, length of conjugate axis 4
d) foci $(0, -6)$ and $(0, 6)$, with constant difference between focal radii 4

15. Roof arches Hyperbolic arches anchored to the ground support the roof of a sports complex. These arches span a distance of 60 m and have a maximum height of 20 m.
a) Find a possible equation for a hyperbola to model one of these arches.
b) What is the height of the arch at a horizontal distance of 25 m from the maximum?

8.7 The Parabola
Refer to the Key Concepts on page 660.

16. Write an equation in the standard form for the parabola with the given focus and directrix.
a) focus $(0, 3)$, directrix $y = -3$
b) focus $(-3, 2)$, directrix $x = 1$
c) focus $(5, -1)$, directrix $y = 1$
d) focus $(1, 2)$, directrix $y = 4$
e) focus $(3, -2)$, directrix $x = -1$

17. For each of the following parabolas, determine the coordinates of the vertex and the focus, and the equation of the directrix. Sketch the graph and determine the domain and range.

a) $y = \dfrac{1}{8}x^2$

b) $x = -\dfrac{1}{12}y^2$

c) $x - 3 = \dfrac{1}{2}(y+1)^2$

d) $y + 1 = \dfrac{1}{4}(x-5)^2$

18. Use the locus definition of the parabola to write an equation for each parabola.

a) focus $(4, -2)$, directrix $x = 2$ **b)** focus $(-2, -3)$, directrix $y = 1$
c) focus $(4, 1)$, directrix $x = 1$

19. Satellite dish The focus of a parabolic satellite dish is 25 cm from the vertex.
a) Determine an equation in standard form to represent the dish, if the parabola opens up, the vertex is at the origin, and the focus is on the y-axis.
b) Determine the width of the dish 10 cm from the vertex. Round the width to the nearest tenth of a centimetre.

20. Football A football player kicks the ball during a field goal attempt. The ball reaches a maximum height of 10 m and travels a horizontal distance of 50 m. Write an equation in standard form to model the parabolic path of the football.

8.8 Conics With Equations in the Form $ax^2 + by^2 + 2gx + 2fy + c = 0$

Refer to the Key Concepts on page 671.

21. For each of the following equations,
i) identify the conic
ii) write the equation in standard form
iii) determine the key features and sketch the graph
a) $x^2 + y^2 + 6x - 4y + 12 = 0$ **b)** $4x^2 + 25y^2 + 8x - 100y + 4 = 0$
c) $2x^2 - y - 12x + 28 = 0$ **d)** $9x^2 + 4y^2 - 72x + 8y + 112 = 0$
e) $x^2 + y^2 - 6x - 10y + 23 = 0$ **f)** $9x^2 - y^2 - 36x - 8y - 16 = 0$
g) $y^2 - x + 4y + 7 = 0$ **h)** $x^2 - 4y^2 + 6x + 16y - 23 = 0$

22. For each conic, write an equation in standard form and in the form $ax^2 + by^2 + 2gx + 2fy + c = 0$.

a)

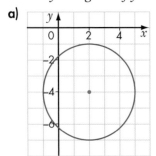

b)

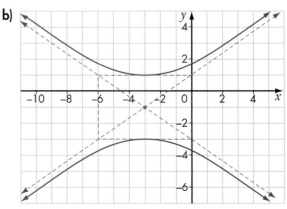

c)

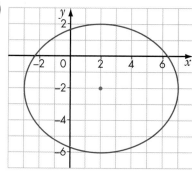

d)

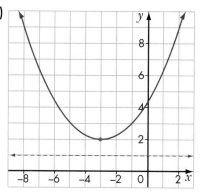

8.9 Intersections of Lines and Conics

Refer to the Key Concepts on page 684.

23. Solve each system of equations. Round answers to the nearest tenth, if necessary.

a) $x^2 + y^2 = 25$
$2x + y = -5$

b) $y = x^2 - 3$
$2x + y = -3$

c) $x^2 + 9y^2 = 81$
$x - 2y = 6$

d) $x^2 - y^2 = 16$
$2x + y = 6$

e) $x^2 + y^2 = 64$
$x - y = -2$

f) $4x^2 + y^2 = 100$
$2x - y = 5$

g) $y = 4x^2$
$4x - y = 1$

h) $x^2 + y^2 = 25$
$\frac{1}{2}x - y = 7$

i) $9x^2 - y^2 = 36$
$3x + y = -12$

24. Air-traffic control The radar signals from an air traffic control centre at a small airport can be modelled by the equation $x^2 + y^2 = 30$ on a grid with 1 unit equal to 1 km. A plane is flying along a path represented by the equation $y = \frac{1}{2}x + 3$. Find the length of the plane's path that the control centre can monitor, to the nearest tenth of a kilometre.

CHAPTER TEST

Achievement Chart

Category	Knowledge/Understanding	Thinking/Inquiry/Problem Solving	Communication	Application
Questions	All	10, 13, 15	5, 6, 8, 9, 15	9, 11–15

1. Determine the equation of the locus of points equidistant from the graphs of $y = 8$ and $y = -4$.

2. Determine equations for the locus of points equidistant from the graphs of $y = x + 1$ and $y = -x + 1$.

3. Determine an equation for the locus of points that are 5 units from $(-1, -2)$.

4. Use the locus definition of the parabola to write an equation of the parabola with focus $(-2, 5)$ and directrix $y = 1$.

5. Sketch the circle $(x + 7)^2 + (y - 2)^2 = 64$. State the domain and range.

6. Sketch the ellipse $\dfrac{(x - 2)^2}{36} + \dfrac{(y + 3)^2}{9} = 1$. Label the coordinates of the centre, the foci, the vertices, and the co-vertices. State the domain and range.

7. Use the locus definition of the ellipse to write an equation of the ellipse centred at the origin, with foci $(0, -12)$ and $(0, 12)$, and with the sum of the focal radii 26.

8. Sketch the hyperbola $\dfrac{(y + 4)^2}{16} - \dfrac{(x - 2)^2}{9} = 1$. Label the coordinates of the centre, the vertices, the co-vertices, and the foci. State the domain and range.

9. Sketch the parabola $x - 2 = \dfrac{1}{2}(y - 3)^2$. Label the coordinates of the vertex and the focus, and the equation of the directrix. Determine the domain and range.

10. Write an equation in standard form for each conic.
a) the circle with a diameter that has endpoints $(-1, 4)$ and $(3, 6)$
b) the hyperbola with foci $(-5, 0)$ and $(5, 0)$, with $|F_1P - F_2P| = 3$
c) the ellipse with centre $(-1, 2)$, major axis parallel to the y-axis, length of the major axis 10, length of the minor axis 8

11. For each of the following equations,

i) identify the conic

ii) write the equation in standard form

iii) determine the key features and sketch the graph

a) $9x^2 + y^2 + 36x - 9 = 0$

b) $3x^2 + 12x - 4y - 12 = 0$

c) $x^2 + y^2 - 2x + 4y - 5 = 0$

d) $9x^2 - 4y^2 - 8y + 32 = 0$

12. Solve each system of equations. Round solutions to the nearest tenth, if necessary.

a) $x^2 + y^2 = 20$
$x - 2y = 8$

b) $x - 2 = (y - 3)^2$
$y + x = 5$

c) $x^2 - y^2 = 1$
$y = 2x + 3$

13. Solar energy Solar energy is collected using parabolic mirrors, which concentrate the sun's rays at the focus. The focus of one type of parabolic mirror is located 5 m above the vertex. Assume that the mirror opens up and that its vertex is at the origin.

a) Write an equation in standard form for the parabolic mirror.

b) Find the diameter of the mirror 10 m from the vertex. Round the diameter to the nearest tenth of a metre.

14. Pond A pond in a botanical garden is built in the shape of an ellipse, with a major axis of length 6 m and a minor axis of length 4 m. Assume that the centre of the pond is at the origin of a coordinate grid on which 1 unit represents 1 m, and that the major axis is along the x-axis.

a) Write an equation in standard form to represent the ellipse.

b) A bridge across one end of the pond can be represented by the equation $y = x + 2$. Find the length of the bridge, to the nearest tenth of a metre.

ACHIEVEMENT Check   Knowledge/Understanding   Thinking/Inquiry/Problem Solving   Communication   Application

15. A balloon arch is being built for a celebration. The arch is to span 4 m and have a height of 2.2 m. Find a possible equation for the arch if it is in the shape of

a) a parabola

b) a semi-ellipse

c) a hyperbola

CHALLENGE PROBLEMS

1. Locus Q(−1, 0), R(3, 0), and P are the vertices of △PQR. P traces the locus of points so that the area of △PQR is equal to the square of the distance from P to the *y*-axis. Find the equation of the locus.

2. Circle A circle is centered on the *y*-axis. If the circle also passes through (−3, −1) and (4, 6), will it also intersect the line $3x − 4y − 13 = 0$?

3. Circle and hyperbola A circle is centred at one focus of the hyperbola $x^2 − y^2 = k^2$. If the *y*-axis intersects the circle in exactly one point, find the length of the line segment between the circle's *x*-intercepts in terms of *k*.

4. Design An ellipse is drawn with centre (0, 0) and major axis along the *y*-axis. A second ellipse is drawn with centre (0, 0) and major axis equal to the minor axis of the first ellipse. A third ellipse is drawn with centre (0, 0) and major axis equal to the minor axis of the second ellipse. A fourth ellipse is drawn with centre (0, 0) and major axis equal to the minor axis of the third ellipse. The first ellipse passes through (0, 5). The second ellipse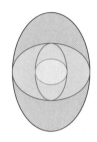
passes through $\left(2, \dfrac{3\sqrt{3}}{2}\right)$. The third ellipse passes through (0, 3), (0, −3),

and $\left(\dfrac{4\sqrt{2}}{3}, 1\right)$. The fourth ellipse passes through $(\sqrt{2}, 1)$. Find

an equation in standard form for each ellipse.

5. Area Find the area of the rectangle with the greatest area that can be

inscribed in the ellipse with the equation $\dfrac{x^2}{25} + \dfrac{y^2}{9} = 1$.

6. Proof Prove that the area of a square inscribed in the ellipse
$\dfrac{x^2}{a^2} + \dfrac{y^2}{b^2} = 1$ is $\dfrac{4a^2 b^2}{a^2 + b^2}$.

PROBLEM SOLVING STRATEGY

USE A DATA BANK

You must locate information to solve some problems. There are many sources of information, including the Internet, the media, print data banks, and experts.

The brightness of planets and stars as they appear from the Earth is described on a scale of apparent magnitudes. A lower apparent magnitude signifies a brighter object. A difference of 5 magnitudes corresponds to a factor of 100 times as bright. So, a star of magnitude 3 is 100 times as bright as a star of magnitude 8, and a star of magnitude 0 is 100 times as bright as a star of magnitude 5.

A difference of 1 magnitude corresponds to a factor of $\sqrt[5]{100}$ times as bright. A difference of 2 magnitudes corresponds to $\sqrt[5]{100} \times \sqrt[5]{100}$ or $\left(\sqrt[5]{100}\right)^2$ times as bright. A positive difference of d magnitudes corresponds to $\left(\sqrt[5]{100}\right)^d$ times as bright.

Aside from our sun, the brightest star in the sky is Sirius. How many times as bright as the North Star, Polaris, is Sirius, to the nearest whole number?

Understand
the Problem

1. What information are you given?
2. What are you asked to find?
3. Do you need an exact or an approximate answer?

Think
of a Plan

Locate values for the apparent magnitudes of Sirius and Polaris. Find the positive difference, d, between these magnitudes. Substitute the difference into the formula $\left(\sqrt[5]{100}\right)^d$ to find how many times as bright Sirius is as Polaris.

The apparent magnitude of Sirius is −1.5, and the apparent magnitude of Polaris is 2.

The positive difference, $d = 2 - (-1.5)$
$$= 3.5$$

Substitute for d in the formula.

$$\left(\sqrt[5]{100}\right)^d = \left(\sqrt[5]{100}\right)^{3.5}$$
$$= \left(100^{\frac{1}{5}}\right)^{3.5}$$
$$= 100^{0.7}$$
$$\doteq 25$$

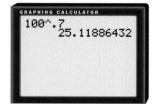

GRAPHING CALCULATOR
```
100^.7
        25.11886432
```

So, Sirius is 25 times as bright as Polaris, to the nearest whole number.

Check by estimation.
A difference of 1 magnitude means a factor of $\sqrt[5]{100}$, which is about 2.5.
A difference of 3 magnitudes is a factor of about $2.5 \times 2.5 \times 2.5$, or about 15.
A difference of 4 magnitudes is a factor of about $2.5 \times 2.5 \times 2.5 \times 2.5$, or about 40.
So, a difference of 3.5 magnitudes is a factor between about 15 and 40.

Look Back Does the answer seem reasonable?
Is there a way to solve the problem graphically?

Use a Data Bank
1. Locate the information you need.
2. Solve the problem.
3. Check that the answer is reasonable.

Apply, Solve, Communicate

Locate the necessary information and solve the problems.

1. Moons Most planets in our solar system have moons. Which planets have moons that are larger than the planet Mercury?

2. Land area Suppose the land area of each province were divided equally among all the people living in that province. In which province would a person receive
a) the most land?   **b)** the least land?

3. **Species at risk** Species at risk are grouped into five categories: extinct, extirpated, endangered, threatened, or vulnerable.
a) What does each category mean?
b) What things are included when the word *species* is used?
c) What is the total number of species now at risk in Canada?
d) What is the average annual rate of increase in the species at risk in Canada?
e) What is the current estimate of the rate of extinction of species worldwide

4. **Astronomy** Some objects in our solar system can be brighter than the brightest stars in the night sky. The moon and the planet Venus can be the brightest objects. How many times as bright as the planet Venus is a full moon, when Venus is at its brightest? Round your answer to the nearest whole number.

5. **Driving distance** a) What is the shortest driving distance from Sudbury to Brownsville, Texas?
b) How long would the trip take if you could average 100 km/h and you drove, at most, 10 h a day?

6. **National debt** If Canada's current national debt were divided equally among all Canadians, what would be each person's share?

7. **Coastal communities** What percent of Canadians live in coastal communities?

8. **Identifying people** Verification systems based on hand geometry have been developed to identify people. These systems are suited for applications that do not require extreme security.
a) How do these systems use hand measurements to identify people?
b) Where could these systems be used?

9. **Renewable energy** a) What percent of Canadian energy comes from renewable sources, such as solar and wind power?
b) What is the projected percent increase in renewable energy use over the next 25 years?
c) What country produces the most wind energy? What is Canada's wind energy production as a percent of that country's wind energy production?

10. **Canada** a) What percent of the world's population lives in Canada?
b) What percent of the world's energy consumption is used by Canadians?

11. **Oil supply** a) Graph the actual or projected annual world oil production for every tenth year from 1950 to 2050.
b) On the same set of axes or in the same viewing window, graph the actual or projected annual world oil consumption for every tenth year from 1950 to 2050.
c) Use the data to decide if oil consumption is projected to exceed oil production before the year 2051.

12. **Formulating problems** Write a problem using information from a data bank. Have a classmate locate the information and solve your problem.

USING THE STRATEGIES

1. Numbers If $A \times B = 24$, $C \times D = 32$, $B \times D = 48$, and $B \times C = 24$, what is the value of $A \times B \times C \times D$?

2. Measurement The diameter of the circle is the same as the radius of the semicircle. What fraction of the area of the semicircle is shaded?

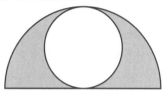

3. Number property The number 153 has an interesting property.
$$1^3 + 5^3 + 3^3 = 1 + 125 + 27$$
$$= 153$$
There are two other three-digit numbers that begin with 3 and also have this property. There is also one three-digit number, beginning with 4, that has the same property. Find these numbers.

4. Real numbers If a, b, and c are three different real numbers whose sum is zero and whose product is two, what is the value of $a^3 + b^3 + c^3$?

5. Positive integers The positive integers are arranged in the pattern shown. If this pattern is continued, what number will fall in the nineteenth column of the sixty-fifth row?

```
 1
 2    3
 4    5    6
 7    8    9   10
11   12   13   14   15
```

6. Equation Place the fewest possible mathematical symbols between digits to make the following equation true.
$$1\ 2\ 3\ 4\ 5\ 6\ 7\ 8\ 9 = 100$$

7. Rolls of steel Three rolls of steel are held together with a band. The diameter of each roll of steel is 1 m. What is the length of the band?

8. Time If it is 9 a.m. now, what time will it be 99 999 999 999 h from now?

9. Circles Find the equation of the largest circle that can be inscribed in a quadrant of a circle with centre at the origin and radius r.

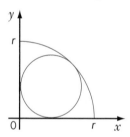

10. Locus In a 2 by 2 square, find the locus of points for which the sum of the squares of the distances from the four vertices is 16.

11. Mount Everest Estimate the number of dump trucks full of rocks needed to move Mount Everest.

CUMULATIVE REVIEW: CHAPTERS 7 AND 8

Chapter 7

1. Leila is borrowing $65 000 for 4 years. She is deciding between a loan at 6.95% per annum, compounded monthly, and a loan at 7% per annum, compounded annually.
a) Predict which loan is the better deal.
b) How much interest is paid on the loan that is the better deal?

2. To have the tuition of $7250 when she returns to school in a year and a half, Fatima will make a payment into an investment account every 3 months. She will start the payments 3 months from now, and continue until she returns to school. The account earns interest at a rate of 5.2% per annum, compounded quarterly.
a) How much must Fatima's payments be?
b) By the day of the last payment, how much interest will she have earned?

3. Leo's business plans are based on his receiving investment income of $25 000 every 6 months, starting 6 months from now. How much must he pay today into an investment earning interest at a rate of 8.5% per annum, compounded semi-annually.

4. Randall is buying a house for $242 000. His down payment is 55% of the price. The mortgage rate for a 5-year term is 8.2% per annum, compounded semi-annually, amortized over 25 years, and paid monthly.
a) For how much is the mortgage?
b) How much are the monthly payments?
c) Use a spreadsheet to find how much he will owe at the end of the term.

Chapter 8

1. a) Graph the locus of points that are equidistant from the lines $y = 3x + 7$ and $y = 3x - 5$.
b) Write an equation to describe this locus.

2. Sketch the circle $(x - 3)^2 + (y + 5)^2 = 81$. State the domain and range.

3. Use the locus definition of the ellipse to find an equation of the ellipse with foci $(-12, 0)$ and $(12, 0)$, and with the sum of the focal radii 26.

4. Sketch each conic. Label the coordinates of the centre, the foci, the vertices, and the co-vertices. State the domain and range.
a) $\dfrac{(x-5)^2}{144} + \dfrac{(y+11)^2}{121} = 1$
b) $\dfrac{(x+1)^2}{4} - \dfrac{(y+3)^2}{9} = 1$

5. Sketch the parabola $x + 2 = \dfrac{1}{2}(y - 4)^2$. Label the coordinates of the vertex and the focus, and the equation of the directrix. Determine the domain and range.

6. For each of the following equations,
i) identify the conic
ii) write the equation in standard form
iii) determine the key features and sketch the graph
a) $4x^2 + y^2 + 24x - 4y + 36 = 0$
b) $x^2 - 8x + 8y + 8 = 0$
c) $x^2 + y^2 + 4x - 6y - 3 = 0$
d) $2x^2 - y^2 - 4x - 6y - 3 = 0$

7. Solve each system of equations. Round solutions to the nearest tenth, if necessary.
a) $x^2 + y^2 = 16$
  $x + y = 1$
b) $x + 4 = (y - 1)^2$
  $y + x + 1 = 0$

APPENDIX A REVIEW OF PREREQUISITE SKILLS

Common factors

To factor $3xy + 9xz - 6x$, remove the greatest common factor, which is $3x$.
$3xy + 9xz - 6x = 3x(y + 3z - 2)$

1. Factor.

a) $4x + 6xy$

b) $3xy + 12xy^2 + 6x^3y^2$

c) $5m^2 - 30m$

d) $2xy - 6x^2y + 8xy^3$

e) $6c^3 - 4c^2d^2 + 2c^2d$

f) $2y^5 - 4y^3 + 8y^2$

g) $5ax + 10ay - 5az$

h) $3pqr + 4pqs - 5pqt$

i) $27xy - 18yz + 9y^2$

Evaluating radicals

Since $8 \times 8 = 64$, $\sqrt{64} = 8$.
Since $0.6 \times 0.6 = 0.36$, $\sqrt{0.36} = 0.6$.

1. Evaluate.

a) $\sqrt{16}$

b) $\sqrt{81}$

c) $\sqrt{0.25}$

d) $\sqrt{1.44}$

e) $\sqrt{0.04}$

f) $\sqrt{6.25}$

g) $\sqrt{196}$

h) $\sqrt{0.49}$

2. Evaluate.

a) $\sqrt{8^2 - 2(7)(2)}$

b) $\sqrt{5^2 - 2(-4)(7)}$

c) $\sqrt{4^2 + 4(-3)(-7)}$

Evaluating expressions

To evaluate the expression $3x^2 + 2y$ for $x = 2$ and $y = -1$, substitute 2 for x and -1 for y in the expression. Then, simplify using the order of operations.

$$3x^2 + 2y = 3(2)^2 + 2(-1)$$
$$= 3(4) + 2(-1)$$
$$= 12 - 2$$
$$= 10$$

1. Evaluate for $x = -2$, $y = 3$, and $z = 2$.

a) $2x + 3$

b) $3x + 2y - 2z$

c) $2(x + z)$

d) $2x^2 + y^2 - z^2$

e) $2yz - 3xy + 4$

f) $x^2 + 3y^2$

g) $(xy)^2$

h) $5(x + y - z)$

i) $2x(y + z)$

2. Evaluate for $x = 3$, $y = 4$, and $z = -2$.

a) $x - y + z$

b) $3y + 2z$

c) $2x - 3y - 4z$

d) $2xyz - 3$

e) $xy - yz + xz$

f) $x^2 + y^2 - z^2$

g) $4(3z + y)$

h) $3z(4x - 2y)$

i) $(x - y)(y + z)$

Exponent rules

To multiply powers with the same base, add the exponents.
To divide powers with the same base, subtract the exponents.
To raise a power to a power, multiply the exponents.

1. Simplify, using the exponent rules. Express each answer in exponential form.

a) $3^2 \times 3^4$
b) $2^3 \times 2^2$
c) $5^2 \times 5^4 \times 5$

d) $4^5 \div 4^2$
e) $3^6 \div 3$
f) $(4^2)^3$

g) $(2^5)^2$
h) $x^3 \times x^5$
i) $z^9 \div z^3$

j) $(y^2)^7$
k) $2x^2 \times 4x^4$
l) $(-3y)(-5y^5)$

m) $(2x^3)^4$
n) $(-3z^2)^3$
o) $-12m^6 \div (-4m^3)$

Factoring $ax^2 + bx + c$, $a = 1$

To factor $x^2 - 8x + 12$, where $a = 1$, $b = -8$, and $c = 12$, use a table to find two integers whose product is 12 and whose sum is -8. The only two integers with a product of 12 and a sum of -8 are -6 and -2.
So, $x^2 - 8x + 12 = (x - 6)(x - 2)$.

Product of 12		Sum
12	1	13
-12	-1	-13
6	2	8
* -6	-2	-8
4	3	7
-4	-3	-7

1. Factor.

a) $x^2 - x - 20$
b) $y^2 + 3y - 10$
c) $n^2 - 5n - 36$

d) $m^2 + 9m + 18$
e) $x^2 - 11x + 30$
f) $c^2 - 2c - 24$

g) $16 + 6y - y^2$
h) $x^2 + 12xy + 32y^2$
i) $c^2 - 3cd - 28d^2$

Factoring $ax^2 + bx + c$, $a \neq 1$

To factor $3x^2 + 13x + 10$, where $a = 3$, $b = 13$, and $c = 10$, break up the middle term. Find two integers whose product is $a \times c$, or 30, and whose sum is b, or 13. The only two integers with a product of 30 and a sum of 13 are 10 and 3.

Product of 30		Sum
30	1	31
-30	-1	-31
15	2	17
-15	-2	-17
* 10	3	13
-10	-3	-13
6	5	11
-6	-5	-11

1. Factor.

a) $3x^2 - 2x - 8$
b) $2c^2 + 7c - 4$

c) $4m^2 - 11m + 6$
d) $5y^2 + 8y + 3$

e) $3n^2 + n - 2$
f) $6x^2 - 17x - 3$

g) $3x^2 - 5xy - 12y^2$
h) $5x^2 - 14x + 8$

i) $4x^2 + 23x + 15$
j) $2p^2 + pq - q^2$

Trigonometric ratios

To find the length of AC in $\triangle ABC$, use the Pythagorean Theorem.

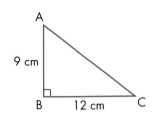

$$(AC)^2 = 9^2 + 12^2$$
$$= 81 + 144$$
$$= 225$$
$$AC = \sqrt{225}$$
$$= 15$$

$$\sin A = \frac{\text{opposite}}{\text{hypotenuse}} \qquad \cos A = \frac{\text{adjacent}}{\text{hypotenuse}} \qquad \tan A = \frac{\text{opposite}}{\text{adjacent}}$$
$$= \frac{12}{15} \qquad\qquad\quad = \frac{9}{15} \qquad\qquad\quad = \frac{12}{9}$$
$$= \frac{4}{5} \qquad\qquad\quad = \frac{3}{5} \qquad\qquad\quad = \frac{4}{3}$$

$$\sin C = \frac{\text{opposite}}{\text{hypotenuse}} \qquad \cos C = \frac{\text{adjacent}}{\text{hypotenuse}} \qquad \tan C = \frac{\text{opposite}}{\text{adjacent}}$$
$$= \frac{9}{15} \qquad\qquad\quad = \frac{12}{15} \qquad\qquad\quad = \frac{9}{12}$$
$$= \frac{3}{5} \qquad\qquad\quad = \frac{4}{5} \qquad\qquad\quad = \frac{3}{4}$$

1. For each of the following triangles, find the indicated trigonometric ratios.

a)

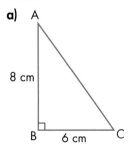

sin A	sin C
cos A	cos C
tan A	tan C

b)

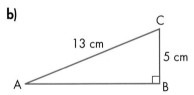

sin A	sin C
cos A	cos C
tan A	tan C

Exponent rules

To multiply powers with the same base, add the exponents.
To divide powers with the same base, subtract the exponents.
To raise a power to a power, multiply the exponents.

1. Simplify, using the exponent rules. Express each answer in exponential form.

a) $3^2 \times 3^4$ b) $2^3 \times 2^2$ c) $5^2 \times 5^4 \times 5$

d) $4^5 \div 4^2$ e) $3^6 \div 3$ f) $(4^2)^3$

g) $(2^5)^2$ h) $x^3 \times x^5$ i) $z^9 \div z^3$

j) $(y^2)^7$ k) $2x^2 \times 4x^4$ l) $(-3y)(-5y^5)$

m) $(2x^3)^4$ n) $(-3z^2)^3$ o) $-12m^6 \div (-4m^3)$

Factoring $ax^2 + bx + c$, $a = 1$

To factor $x^2 - 8x + 12$, where $a = 1$, $b = -8$, and $c = 12$, use a
table to find two integers whose product is 12 and whose sum
is -8. The only two integers with a product of 12 and a sum of
-8 are -6 and -2.
So, $x^2 - 8x + 12 = (x - 6)(x - 2)$.

Product of 12		Sum
12	1	13
-12	-1	-13
6	2	8
* -6	-2	-8
4	3	7
-4	-3	-7

1. Factor.

a) $x^2 - x - 20$ b) $y^2 + 3y - 10$ c) $n^2 - 5n - 36$

d) $m^2 + 9m + 18$ e) $x^2 - 11x + 30$ f) $c^2 - 2c - 24$

g) $16 + 6y - y^2$ h) $x^2 + 12xy + 32y^2$ i) $c^2 - 3cd - 28d^2$

Factoring $ax^2 + bx + c$, $a \neq 1$

To factor $3x^2 + 13x + 10$, where $a = 3$, $b = 13$, and $c = 10$,
break up the middle term. Find two integers whose product is
$a \times c$, or 30, and whose sum is b, or 13. The only two integers
with a product of 30 and a sum of 13 are 10 and 3.

Product of 30		Sum
30	1	31
-30	-1	-31
15	2	17
-15	-2	-17
* 10	3	13
-10	-3	-13
6	5	11
-6	-5	-11

1. Factor.

a) $3x^2 - 2x - 8$ b) $2c^2 + 7c - 4$

c) $4m^2 - 11m + 6$ d) $5y^2 + 8y + 3$

e) $3n^2 + n - 2$ f) $6x^2 - 17x - 3$

g) $3x^2 - 5xy - 12y^2$ h) $5x^2 - 14x + 8$

i) $4x^2 + 23x + 15$ j) $2p^2 + pq - q^2$

Finding an angle in a right triangle

To calculate the measure of $\angle Z$, to the nearest degree, use the tangent ratio.

$\tan Z = \dfrac{4}{9}$

$\angle Z \doteq 24°$

$\angle Z$ is 24°, to the nearest degree.

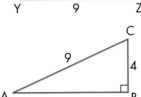

1. Find the measure of $\angle A$, to the nearest degree.

a)

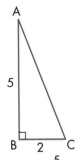

b)

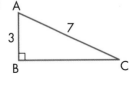

c)

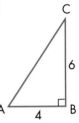

d)

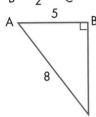

e)

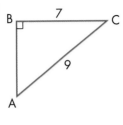

f)

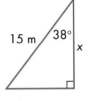

Finding a side length in a right triangle

To calculate the length f, to the nearest tenth of a metre, use the sine ratio.

$\sin 44° = \dfrac{f}{18}$

$18\sin 44° = f$

$12.5 \doteq f$

The length of f is 12.5 m, to the nearest tenth of a metre.

1. Find x, to the nearest tenth of a metre.

a)

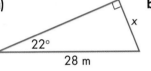
22°
28 m

b)

x
46°
32 m

c)

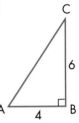
15 m 38°
x

d)

9 m
65°
x

e)

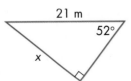
21 m
52°
x

f)

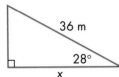
36 m
28°
x

Graphing equations

To graph the line $y = 2x + 3$ using a table of values, choose suitable values
for x, say $\{-2, -1, 0, 1, 2\}$. Complete a table of values by finding the value
of y for each value of x.

Plot the points on a grid and draw a line through the points.

$y = 2x + 3$

x	y
-2	-1
-1	1
0	3
1	5
2	7

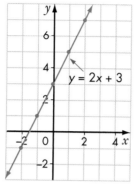

$y = 2x + 3$

1. Graph each equation using a table of values.

a) $x + y = 6$ **b)** $x - y = 3$ **c)** $y = x + 4$

d) $y = x - 2$ **e)** $y = 3x - 2$ **f)** $y = 4x + 3$

Graphing quadratic functions

To sketch the graph of $y = 2(x + 3)^2 - 8$, stretch the graph of
$y = x^2$ vertically by a factor of 2. Then, translate 3 units to the
left and 8 units downward.

The coordinates of the vertex are $(-3, -8)$.
The equation of the axis of symmetry is $x = -3$.
The domain is the set of real numbers.
The range is $y \geq -8$.
The minimum value of the function is -8 when $x = -3$.
The graph crosses the y-axis at $(0, 10)$, so the y-intercept is 10.
The graph appears to cross the x-axis at $(-5, 0)$ and $(-1, 0)$, so
the x-intercepts are -5 and -1.

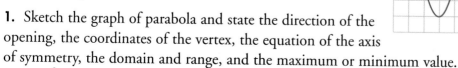

$y = x^2$

$y = 2(x + 3)^2 - 8$

1. Sketch the graph of parabola and state the direction of the
opening, the coordinates of the vertex, the equation of the axis
of symmetry, the domain and range, and the maximum or minimum value.

a) $y = x^2 + 2$ **b)** $y = -x^2 - 1$ **c)** $y = 2x^2$

d) $y = -4x^2 + 1$ **e)** $y = -2(x - 1)^2 + 4$ **f)** $y = (x + 5)^2 - 3$

g) $y = 0.5(x + 2)^2 + 3$ **h)** $y = -(x + 1)^2 - 5$

Inequalities

To graph the integers $x > -3$ on a number line, graph the integers greater than -3.

To graph the integers $x \leq 2$ on a number line, graph the integers less than or equal to 2.

Note that the red arrow means that the integers continue without end.

1. Graph the following integers on a number line.

a) $x \geq -4$ **b)** $x < 6$ **c)** $x \leq -2$ **d)** $x > 3$

Length of a line segment

To find the exact length of the line segment joining $(3, 5)$ and $(1, -4)$, and the approximate length, to the nearest tenth, use the length formula.

$$
\begin{aligned}
l &= \sqrt{(x_2 - x_1)^2 + (y_2 - y_1)^2} \\
&= \sqrt{(1 - 3)^2 + (-4 - 5)^2} \\
&= \sqrt{(-2)^2 + (-9)^2} \\
&= \sqrt{4 + 81} \\
&= \sqrt{85} \\
&\doteq 9.2
\end{aligned}
$$

The exact length is $\sqrt{85}$ and the approximate length is 9.2, to the nearest tenth.

1. Find the exact and the approximate length of the line segment joining each pair of points.

a) A(6, 8) and B(2, 3)   **b)** W(3, 4) and X(−3, 2)

c) E(−1, 7) and F(−4, 3)   **d)** R(−8, 4) and T(5, −2)

e) U(1.5, −0.2) and V(−0.6, 2.4)

Midpoint formula

To determine the coordinates of the midpoint of the line segment with endpoints A(−3, 8) and B(5, 2), use the formula for the midpoint.

$$
\left(\frac{x_1 + x_2}{2}, \frac{y_1 + y_2}{2} \right) = \left(\frac{-3 + 5}{2}, \frac{8 + 2}{2} \right)
$$
$$
= (1, 5)
$$

The coordinates of the midpoint are (1, 5).

1. Find the coordinates of the midpoint of each line segment.

a) P(1, −8) and Q(−4, 4)　　　　　**b)** S(5, −1) and T(3, 3)

c) M(1, −4) and N(4, 0)　　　　　**d)** G(−3, −1) and H(−5, 2)

e) V(2.7, 3.1) and W(3.3, −4.3)　　**f)** $A\left(2\frac{1}{2}, 1\frac{1}{2}\right)$ and $B\left(-1\frac{1}{2}, 2\frac{1}{2}\right)$

Reflections

To draw the image of △DEF with vertices D(2, 5), E(4, 2), and F(6, 6) after a reflection in the *x*-axis, draw △DEF. Locate D′ so that the perpendicular distance from D′ to the *x*-axis equals the perpendicular distance from D to the *x*-axis. The coordinates of D′ are (2, −5). Locate E′ and F′ in the same way.

D(2, 5) → D′(2, −5)

E(4, 2) → E′(4, −2)

F(6, 6) → F′(6, −6)

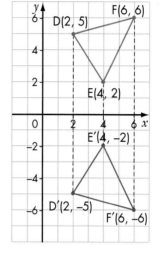

Join points D′, E′, and F′. △D′E′F′ is the image of △DEF after a reflection in the *x*-axis.

1. Copy each figure onto a grid and draw its reflection image in the *x*-axis.

a)

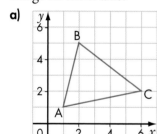

b)

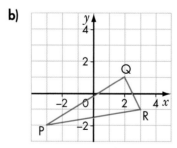

2. Copy each figure onto a grid and draw its reflection image in the *y*-axis.

a)

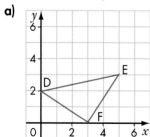

b)

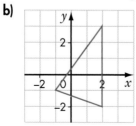

Rewriting in the form $y = a(x - h)^2 + k$, $a = 1$

To rewrite the equation $y = x^2 + 10x + 24$ in the form $y = a(x - h)^2 + k$, first determine what must be added to $x^2 + 10x$ to make it a perfect square trinomial. The square of half the coefficient of x is 25. Since 25 must be added to the original function, 25 must also be subtracted to keep the value of the function the same.

$$y = x^2 + 10x + 24$$

Add and subtract the square of half the coefficient of x: $\qquad = x^2 + 10x + 25 - 25 + 24$

Group the perfect square trinomial: $\qquad = (x^2 + 10x + 25) - 25 + 24$

Write the perfect square trinomial as the square of a binomial: $= (x + 5)^2 - 1$

The equation in the form $y = a(x - h)^2 + k$ is $y = (x + 5)^2 - 1$.
The minimum value of the function is -1 when $x = -5$.

1. Write each function in the form $y = a(x - h)^2 + k$. Give the maximum or minimum value of the function and the value of x when it occurs.

a) $y = x^2 + 6x + 4$ b) $y = x^2 + 4x + 9$ c) $y = x^2 - 12x - 7$

d) $y = x^2 - 10x + 5$ e) $y = x^2 + 8x - 2$ f) $y = x^2 - 2x - 11$

Rewriting in the form $y = a(x - h)^2 + k$, $a \neq 1$

To express $y = 4x^2 - 8x + 7$ in the form $y = a(x - h)^2 + k$, factor the coefficient of x^2 from the first two terms. Then, complete the square as you would for $a = 1$.

$$y = 4x^2 - 8x + 7$$

Group the terms containing x: $\qquad = [4x^2 - 8x] + 7$

Factor the coefficient of x^2 from the first two terms: $\qquad = 4[x^2 - 2x] + 7$

Complete the square inside the brackets: $\qquad = 4[x^2 - 2x + 1 - 1] + 7$

Write the perfect square trinomial as the square of a binomial: $= 4[(x - 1)^2 - 1] + 7$

Expand to remove the square brackets: $\qquad = 4(x - 1)^2 - 4 + 7$

Simplify: $\qquad = 4(x - 1)^2 + 3$

So, the equation in the form $y = a(x - h)^2 + k$ is $y = 4(x - 1)^2 + 3$.
The minimum value of the function is 3, when $x = 1$.

1. Write each function in the form $y = a(x - h)^2 + k$. Give the maximum or minimum value of the function and the value of x when it occurs.

a) $y = 3x^2 + 6x - 6$ b) $y = -2x^2 - 8x$ c) $y = 2x^2 - 12x + 3$

d) $y = -4x^2 + 16x - 9$ e) $y = 3x^2 - 24x$ f) $y = 4x^2 + 24x - 5$

Simplifying expressions

To simplify $2(x^2 - 2x + 1) - 3(2x^2 - x - 4) - (3x^2 + 5x + 2)$, remove brackets and collect like terms.

$$2(x^2 - 2x + 1) - 3(2x^2 - x - 4) - (3x^2 + 5x + 2)$$
$$= 2(x^2 - 2x + 1) - 3(2x^2 - x - 4) - 1(3x^2 + 5x + 2)$$
$$= 2x^2 - 4x + 2 - 6x^2 + 3x + 12 - 3x^2 - 5x - 2$$
$$= -7x^2 - 6x + 12$$

1. Simplify.

a) $3(x^2 - 2x) + 4(3x^2 - x)$ **b)** $2(4a^2 + 3a) - (6a^2 - a)$ **c)** $-5(2c^2 - 3c) + 4c^2 - 2c$

d) $3a^2 - a + 2 - 3(a^2 + a - 2)$ **e)** $2(x^2 - x + 3) + 6x^2 - 4$ **f)** $3x^2 + 4(x^2 + 2x - 3) + 4x$

2. Simplify.

a) $3(x^2 + 4x - 2) + 2(4x^2 - x - 3) - (5x^2 + 4x - 3)$

b) $-2(3x^2 - 5x + 4) - 3(2x^2 + 4x - 1) + 2(x^2 - 2x + 4)$

c) $-4(x^2 - 3x - 1) - 5(x^2 + 3x + 2) - (-4x^2 + 3x - 2)$

d) $6x^2 - (3x^2 - 2x + 1) - 4(2x^2 + 3x - 2) + 2(5x^2 + 3x - 6)$

Solving linear equations

To solve the equation $3(x - 4) - 8 = 12 - 5x$, follow the order of operations.
First expand to remove the brackets.

$$3(x - 4) - 8 = 12 - 5x$$
$$3x - 12 - 8 = 12 - 5x$$
$$3x - 20 = 12 - 5x$$
$$3x - 20 + 20 = 12 - 5x + 20$$
$$3x = 32 - 5x$$
$$3x + \boxed{} = 32 - 5x + \boxed{}$$
$$8x = 32$$
$$\frac{8x}{8} = \frac{32}{8}$$
$$x = 4$$

Check by substituting $x = 4$ into the original equation.

L.S. $= 3(x - 4) - 8$	R.S. $= 12 - 5x$
$= 3(4 - 4) - 8$	$= 12 - 5(4)$
$= 3(0) - 8$	$= 12 - 20$
$= 0 - 8$	$= -8$
$= -8$	

Since L.S. = R.S., the solution is $x = 4$.

To solve $\dfrac{x-1}{3} + \dfrac{x+2}{6} = 7$, multiply all terms by the lowest common denominator, 6.

Write the equation:
$$\frac{x-1}{3} + \frac{x+2}{6} = 7$$

Multiply both sides by 6:
$$6 \times \frac{(x-1)}{3} + 6 \times \frac{(x+2)}{6} = 6 \times 7$$

Simplify:
$$2(x-1) + (x+2) = 42$$

Expand:
$$2x - 2 + x + 2 = 42$$

Simplify:
$$3x = 42$$

Divide both sides by 3:
$$\frac{3x}{3} = \frac{42}{3}$$
$$x = 14$$

Check by substituting $x = 14$ back into the original equation.

$$\text{L.S.} = \frac{x-1}{3} + \frac{x+2}{6} \qquad \text{R.S.} = 7$$
$$= \frac{14-1}{3} + \frac{14+2}{6}$$
$$= \frac{13}{3} + \frac{16}{6}$$
$$= \frac{26}{6} + \frac{16}{6}$$
$$= \frac{42}{6}$$
$$= 7$$

Since L.S. = R.S., the solution is $x = 14$.

1. Solve and check.
   a) $3(2x + 3) = -3$
   b) $3(x + 1) + 10 = 8 - 2x$
   c) $2(x + 1) = (3x - 2) + 1$
   d) $5(x + 4) - (x + 2) = 8x + 2$
   e) $7(x - 1) - 2(x - 6) = 2(x - 5) + 6$

2. Solve and check.
   a) $\dfrac{x+1}{2} - \dfrac{1-x}{5} = 1$
   b) $\dfrac{y+3}{4} - \dfrac{y+1}{2} = -4$
   c) $\dfrac{3z}{4} - \dfrac{2z}{3} - \dfrac{5}{6} = z$
   d) $\dfrac{2x+6}{3} = \dfrac{5+3x}{2}$

Solving linear systems

To solve by elimination, multiply one or both equations by numbers to obtain two equations in which the coefficients of one variable are the same or opposites.

$3x + 2y = 6$ (1)
$2x - 3y = 17$ (2)

Multiply (1) by 2: $6x + 4y = 12$
Multiply (2) by 3: $\underline{6x - 9y = 51}$
Subtract: $13y = -39$
                   $y = -3$
Substitute -3 for y in (1). $3x + 2(-3) = 6$
                        $3x - 6 = 6$
                          $3x = 12$
                           $x = 4$

The solution is $(4, -3)$.
Check by substituting 4 for x and -3 for y in both of the original equations.

1. Solve each system of equations by elimination. Check each solution.

a) $4x + 3y = -5$
    $3x + 8y = 2$

b) $3a - b = 17$
    $2a + 3b = -7$

c) $\dfrac{x}{4} + \dfrac{y}{3} = 1$
    $\dfrac{x}{2} - \dfrac{y}{3} = 9$

d) $\dfrac{2x}{5} + \dfrac{2y}{3} = -4$
    $\dfrac{3x}{5} - \dfrac{5y}{3} = 2$

e) $0.5x - 0.2y = 2.4$
    $0.6x + 0.3y = 1.8$

f) $0.4x + 0.5y = 1.1$
    $0.8x - 0.2y = -1.4$

Solving proportions

Proportions can be solved using the cross-product rule, which states that

if $\dfrac{a}{b} = \dfrac{c}{d}$, then $a \times d = b \times c$.

To solve $0.8:1.2 = 3.2:x$, first write the proportion in fraction form,
$\dfrac{0.8}{1.2} = \dfrac{3.2}{x}$.
Then, use the cross-product rule.
$$\dfrac{0.8}{1.2} = \dfrac{3.2}{x}$$
$0.8x = 1.2 \times 3.2$
$0.8x = 3.84$
   $x = 4.8$

1. Solve for x. Express each answer as a decimal. Round to the nearest hundredth, if necessary.

a) $\dfrac{x}{2.4} = 3.6$

b) $\dfrac{4.1}{x} = 2.5$

c) $1.8 = \dfrac{x}{4.5}$

d) $\dfrac{x}{1.2} = \dfrac{3.4}{2}$

e) $\dfrac{0.6}{x} = \dfrac{5.2}{1.8}$

f) $\dfrac{3.5}{1.4} = \dfrac{2.5}{x}$

g) $6.2 : x = 1.2 : 2.4$

h) $2.4 : 0.2 = 3.3 : x$

Solving quadratic equations by factoring

To solve $x^2 - 5x = 6$ by factoring, first write the equation in the form $ax^2 + bx + c = 0$.

$$x^2 - 5x - 6 = 0$$

Factor the left side: $\qquad\qquad (x - 6)(x + 1) = 0$

Use the zero product property: $\quad x - 6 = 0 \ \text{ or } \ x + 1 = 0$

$$x = 6 \quad \text{ or } \quad x = -1$$

The roots are 6 and -1.

To solve $2x^2 + 9x = -10$ by factoring, first write the equation in the form $ax^2 + bx + c = 0$.

$2x^2 + 9x + 10 = 0$. Find two integers whose product is $a \times c$, or 20, and whose sum is b, or 9. The only two integers with a product of 20 and a sum of 9 are 4 and 5.

$$2x^2 + 9x + 10 = 0$$

Break up the middle term: $\qquad\qquad\quad 2x^2 + 4x + 5x + 10 = 0$

Group terms: $\qquad\qquad\qquad\qquad (2x^2 + 4x) + (5x + 10) = 0$

Remove common factors: $\qquad\qquad 2x(x + 2) + 5(x + 2) = 0$

Remove a common binomial factor: $\qquad (2x + 5)(x + 2) = 0$

Use the zero product property: $\quad 2x + 5 = 0 \ \text{ or } \ x + 2 = 0$

$$2x = -5 \quad \text{ or } \quad x = -2$$

$$x = -\frac{5}{2}$$

The roots are $-\dfrac{5}{2}$ and -2.

1. Solve by factoring.

a) $x^2 + 7x - 30 = 0$

b) $y^2 - 3y + 2 = 0$

c) $b^2 + 9b + 20 = 0$

d) $a^2 + 8a + 15 = 0$

e) $4x^2 - 20x = -25$

f) $3x^2 + x = 2$

g) $25y^2 - 9 = 0$

h) $3x^2 - 10x - 8 = 0$

i) $9x^2 - 4x = 0$

Solving quadratic equations by graphing

To solve $x^2 + 3x - 4 = 0$ by graphing, graph the related quadratic function
$y = x^2 + 3x - 4$ using paper and pencil, a graphing calculator, or graphing
software.

Find the values of the x-intercepts.

x	y
2	6
1	0
0	-4
-1	-6
-2	-6
-3	-4
-4	0
-5	6

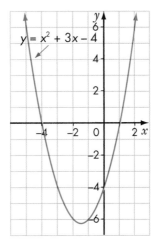

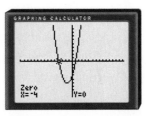

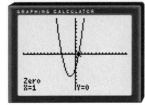

The graph intersects the x-axis at $(1, 0)$
and $(-4, 0)$.
The roots of the equation $x^2 + 3x - 4 = 0$ are 1 and -4.

1. Solve by graphing.

a) $x^2 - 4x - 5 = 0$ **b)** $x^2 + 5x + 4 = 0$ **c)** $x^2 - 16 = 0$

d) $-x^2 - 7x + 8 = 0$ **e)** $x^2 - 6x + 5 = 0$ **f)** $x^2 + 8x = -12$

Stretches and shrinks

To graph $y = x^2$, $y = \frac{1}{2}x^2$, and $y = 2x^2$, set up tables of values, or use a
graphing calculator or graphing software.

$y = x^2$

x	y
3	9
2	4
1	1
0	0
-1	1
-2	4
-3	9

$y = \frac{1}{2}x^2$

x	y
3	4.5
2	2
1	0.5
0	0
-1	0.5
-2	2
-3	4.5

$y = 2x^2$

x	y
3	18
2	8
1	2
0	0
-1	2
-2	8
-3	18

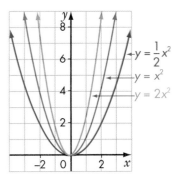

The y-coordinates of $y = 2x^2$ are two times the corresponding y-coordinates
of $y = x^2$.

The y-coordinates of $y = \frac{1}{2}x^2$ are half the corresponding y-coordinates of $y = x^2$.

1. Sketch the graph of each pair of parabolas. Describe how the graphs in each pair are related.

a) $y = x^2$ and $y = 3x^2$

b) $y = x^2$ and $y = \frac{1}{4}x^2$

2. Describe what happens to the point $(3, 9)$ on the graph of $y = x^2$ when the following transformation is applied to the parabola.

a) vertical stretch of scale factor 4

b) vertical shrink of scale factor $\frac{1}{3}$

The quadratic formula

To solve $5x^2 + 4x - 1 = 0$, substitute $a = 5$, $b = 4$, and $c = -1$ into the quadratic formula.

$$x = \frac{-b \pm \sqrt{b^2 - 4ac}}{2a}$$

$$= \frac{-4 \pm \sqrt{4^2 - 4(5)(-1)}}{2(5)}$$

$$= \frac{-4 \pm \sqrt{16 + 20}}{10}$$

$$= \frac{-4 \pm \sqrt{36}}{10}$$

$$= \frac{-4 \pm 6}{10}$$

So, $x = \dfrac{-4 + 6}{10}$ or $x = \dfrac{-4 - 6}{10}$

$\qquad = \dfrac{2}{10} \qquad\qquad = -\dfrac{10}{10}$

$\qquad = \dfrac{1}{5} \qquad\qquad = -1$

The roots are $\dfrac{1}{5}$ and -1.

To solve $x^2 - 3x - 1 = 0$, substitute $a = 1$, $b = -3$, and $c = -1$ into the quadratic formula.

$$x = \frac{-b \pm \sqrt{b^2 - 4ac}}{2a}$$

$$= \frac{3 \pm \sqrt{(-3)^2 - 4(1)(-1)}}{2(1)}$$

$$= \frac{3 \pm \sqrt{13}}{2}$$

The exact roots are $\dfrac{3 + \sqrt{13}}{2}$ and $\dfrac{3 - \sqrt{13}}{2}$.

1. Solve using the quadratic formula.

a) $2x^2 - 3x + 1 = 0$ **b)** $10x^2 - 21x + 9 = 0$ **c)** $7x^2 - 3x = 0$

d) $2x^2 - x - 3 = 0$ **e)** $3x^2 - 5x + 2 = 0$

2. Solve using the quadratic formula. Express answers as exact roots.

a) $5x^2 + 2x - 2 = 0$ **b)** $3x^2 - 2x - 2 = 0$ **c)** $2x^2 - 8x + 7 = 0$

d) $4x^2 + 4x - 14 = 0$ **e)** $10x^2 - 4x - 4 = 0$

Translations

To translate $\triangle ABC$ 3 units to the right and 4 units up, translate each vertex of the triangle 3 units to the right and 4 units up. Join the new points to form the translation image, $\triangle A'B'C'$ are congruent, since corresponding side lengths and angles are equal.

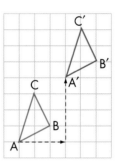

1. Draw each triangle on grid paper. Then, draw the translation image for the given translation.

a)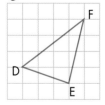

2 left, 4 down

b)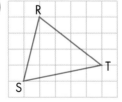

4 right, 3 up

c)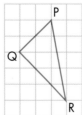

2 left, 5 down

Trigonometric ratios

To find the length of AC in $\triangle ABC$, use the Pythagorean Theorem.

$$(AC)^2 = 9^2 + 12^2$$
$$= 81 + 144$$
$$= 225$$
$$AC = \sqrt{225}$$
$$= 15$$

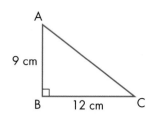

$\sin A = \dfrac{\text{opposite}}{\text{hypotenuse}}$ $\cos A = \dfrac{\text{adjacent}}{\text{hypotenuse}}$ $\tan A = \dfrac{\text{opposite}}{\text{adjacent}}$

$= \dfrac{12}{15}$ $= \dfrac{9}{15}$ $= \dfrac{12}{9}$

$= \dfrac{4}{5}$ $= \dfrac{3}{5}$ $= \dfrac{4}{3}$

$\sin C = \dfrac{\text{opposite}}{\text{hypotenuse}}$ $\cos C = \dfrac{\text{adjacent}}{\text{hypotenuse}}$ $\tan C = \dfrac{\text{opposite}}{\text{adjacent}}$

$= \dfrac{9}{15}$ $= \dfrac{12}{15}$ $= \dfrac{9}{12}$

$= \dfrac{3}{5}$ $= \dfrac{4}{5}$ $= \dfrac{3}{4}$

1. For each of the following triangles, find the indicated trigonometric ratios.

a)

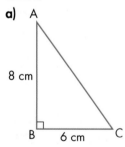

sin A   sin C
cos A   cos C
tan A   tan C

b)

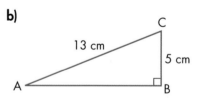

sin A   sin C
cos A   cos C
tan A   tan C

Note: Unless otherwise stated, all keystrokes are for the TI-83 Plus or TI-83 graphing calculator. Where stated, keystrokes are provided for the TI-92 Plus or TI-92 graphing calculator.

Function or Instruction and Description	Keystroke(s), Menu, or Screen
Circle instruction The Circle instruction allows you to draw a circle with centre (X, Y) and a given radius.	**EXAMPLE:** For the circle $(x - 1)^2 + (y + 2)^2 = 25$, the centre is at $(1, -2)$ and the radius is 5. Clear any previous drawings by pressing [2nd] [PRGM] 1 [ENTER]. To draw the circle, first use the ZSquare instruction to adjust the window variables to make the circle circular. Press [2nd] [PRGM] to display the Draw menu. You will see: 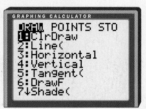 Select 9:Circle to choose the Circle instruction. Press 1 [,] −2 [,] 5 [ )] to enter the values for the x-coordinate of the centre, the y-coordinate of the centre, and the radius. You will see:  Press [ENTER].

Function or Instruction and Description	Keystroke(s), Menu, or Screen

You will see:

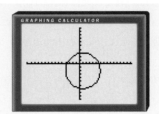

cSolve function (TI-92PLUS or TI-92)

The cSolve function, on the TI-92 Plus or TI-92, solves a polynomial equation for complex solutions.

To select the cSolve function from the Algebra menu, press F2 A to display the Complex submenu.

You will see:

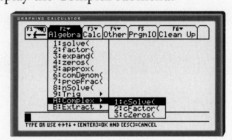

Press 1 to select the cSolve function.

Example:

To solve $x^2 + 2x + 2 = 0$ for x,

press F2 X ^ 2 + 2 × X + 2 = 0 , X) ENTER .

You will see:

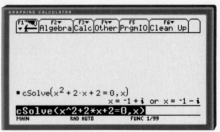

Function or Instruction and Description	Keystroke(s), Menu, or Screen

common denominator function

(TI-92 PLUS or TI-92)

The common denominator function allows you to express a sum or difference of two or more rational expressions with different denominators as a single rational expression.

To select the common denominator function from the Algebra menu, press (F2) 6.

EXAMPLE:

To add $\dfrac{3x+2}{4} + \dfrac{5x-2}{3}$,

press (F2) 6 (( ) 3 X (+) 2 ( )) (÷) 4 (+)
(( ) 5 X (−) 2 ( )) (÷) 3 ( )) (ENTER).
You will see:

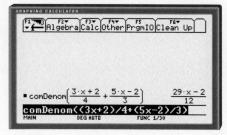

DrawInv instruction

The DrawInv instruction draws the inverse of a function.

To select the DrawInv instruction from the DRAW menu, press (2nd) (PRGM) 8.

EXAMPLE:

To draw the function $y = x^2 + 3$ and its inverse, input the function as Y1 in the Y= editor.

Clear any previous drawings by pressing (2nd) (PRGM) 1 (ENTER).

Press (2nd) (PRGM) 8 to select the DrawInv instruction.

Press (VARS) (▶) 1 1 to enter Y1 as the function to draw.

Function or Instruction and Description	Keystroke(s), Menu, or Screen

You will see:

Press ⬚ENTER⬚.

You will see:

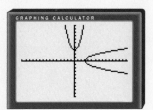

Expand function (TI-92 PLUS or TI-92)

The expand function, on the TI-92 Plus or TI-92, expands and simplifies a product of polynomials.

To select the expand function from the Algebra menu, press ⬚F2⬚ 3.

EXAMPLE:

To expand $(4x + 3y)(x - y)$,

press ⬚F2⬚ 3 ⬚(⬚ 4 ⬚×⬚ X ⬚+⬚ 3 ⬚×⬚ Y ⬚)⬚ ⬚(⬚ X ⬚−⬚ Y ⬚)⬚ ⬚)⬚ ⬚ENTER⬚.

You will see:

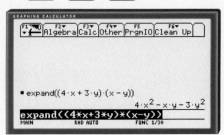

Function or Instruction and Description	Keystroke(s), Menu, or Screen

Factor function (TI-92 PLUS or TI-92)

The factor function, on the TI-92 Plus or TI-92, factors a polynomial.

To select the factor function from the Algebra menu, press F2 2.

EXAMPLE:

To factor $3x^2 - 5x + 2$,

press F2 2 3 × X ^ 2 − 5 × X + 2) ENTER.

You will see:

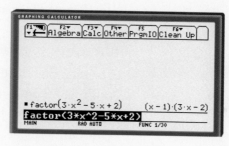

FINANCE menu

To access the FINANCE menu on the TI-83 Plus, press APPS 1.

To access the FINANCE menu on the TI-83, press 2nd x^{-1}.

You will see:

Function or Instruction and Description	Keystroke(s), Menu, or Screen

Format settings

Format settings define the appearance of a graph on the display. Format settings apply to all graphing modes.

To display the format settings, press (2nd) (ZOOM).
The default settings are shown here.

To change a format setting, press (▼) (,) (▶) (,)
(▲) (,) and (◀) as necessary, to move the cursor to the setting you want to select, and press e to select it.

For example, the default setting GridOff means that no grid is displayed on a graph. If you wish to display a grid, press (▼) (,) (▼) (,) (▶), and (ENTER) to select GridOn.

▶Frac function

The ▶Frac function displays an answer as a fraction.

To display an answer as a fraction, press (MATH) to display the MATH menu. Press 1 to select ▶Frac to convert an answer to a fraction.

EXAMPLE:

To find the exact point of intersection of $y = 5x + 3$ and $y = -2x - 5$, enter the functions into the Y= editor and graph them. Use the intersect operation to find the point of intersection, $x = -1.142857$ and $y = -2.714286$.

To find the exact value of x, press (2nd) (MODE) (X,T,θ,*n*) (MATH)
1 (ENTER).

To find the exact value of y, press (ALPHA) 1 (MATH) 1 (ENTER).

You will see:

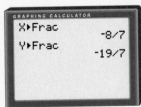

Function or Instruction and Description	Keystroke(s), Menu, or Screen

graph styles

Varying graph styles allow you to visually differentiate functions that are graphed together.

To set the graph style for a function, press $\boxed{\text{Y=}}$ to display the Y= editor.

Use the $\boxed{\blacktriangleleft}$ key to move the cursor to the left of the function to be graphed. Press the $\boxed{\text{ENTER}}$ key repeatedly to choose an appropriate graph style.

Graph styles available may vary depending on which graphing mode you are in.

In Func (function) graphing mode, the following styles are available.

$\boxed{\diagdown}$	Line	A solid line connects points. This is the default style in Connected mode.
$\boxed{\diagdown}$	Thick	A thick solid line connects points.
$\boxed{\blacktriangledown}$	Above	Shading covers the area above the graph.
$\boxed{\blacktriangle}$	Below	Shading covers the area below the graph.
$\boxed{\text{-0}}$	Path	A circular cursor traces the leading edge of the graph and draws a path.
$\boxed{0}$	Animate	A circular cursor traces the leading edge of the graph without drawing a path.
$\boxed{\therefore}$	Dot	A small dot represents each point. This is the default style in dot mode.

EXAMPLE:

To graph $y = x^2$ and $y = x^2 - 3$ on the same set of axes, enter the functions in the Y= editor as $Y1 = x^2$ and $Y2 = x^2 - 3$. The default graph style is line. Leave $Y1$ as line style. Move the cursor to the left of $Y2$ using the arrow keys. Press $\boxed{\text{ENTER}}$ repeatedly (6 times) to change the graph style to dot style.

Change the window variables to values such as Xmin = -8, Xmax = 8, Ymin = -5, Ymax = 10. Press the $\boxed{\text{GRAPH}}$ key to display the graphs.

You will see:

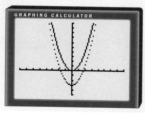

Function or Instruction and Description	Keystroke(s), Menu, or Screen
Intersect operation The intersect operation finds the coordinates of a point at which two or more functions intersect.	To find an intersection, display the graphs on the screen. The point of intersection must appear on the display to use the intersect operation. **EXAMPLE:** To find the point of intersection of $y = 3x - 2$ and $y = -x + 6$, use the Y= editor to input both functions and graph them. Press (2nd) (TRACE), to display the CALCULATE menu. To select the intersect operation, press 5.

You will see:

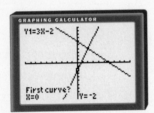

Press (▼) or (▲), if necessary, to move the cursor to the first function, and then press (ENTER).

You will see:

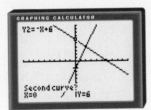

Press (▼) or (▲), if necessary, to move the cursor to the second function, and then press (ENTER).

You will see:

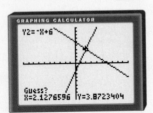

Function or Instruction and Description	Keystroke(s), Menu, or Screen

Press (▶) or (◀) to move the cursor to the point that is your guess for the point of intersection, and then press (ENTER).

You will see:

Notice that the cursor appears on the solution, and the coordinates of the solution are displayed.

LIST MATH menu

The LIST MATH menu is used to perform operations on one or more lists of data.

To display the LIST MATH menu, press (2nd) (STAT) (▶) (▶)

You will see:

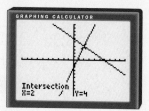

LIST OPS menu

The LIST OPS menu is used to perform operations on one or more lists of data.

To display the LIST OPS menu, press (2nd) (STAT) (▶)

You will see:

Function or Instruction and Description	Keystroke(s), Menu, or Screen

Maximum operation

The maximum operation finds the maximum of a function within a specified interval.

EXAMPLE:

To find the maximum of $y = -2x^2 + 6x + 2$, input the function in the Y= editor.

Press (2nd) (TRACE) to display the CALCULATE menu. Press 4 to select the maximum operation.

You will see:

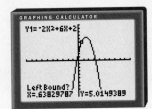

Press (▼) or (▲), if necessary, to move the cursor onto the function.

Press (▶) or (◀) to move the cursor to the left of the maximum (or enter a value). Select the x-value for the left bound of the interval by pressing (ENTER).

Press (▶) or (◀) to move the cursor to the right of the maximum (or enter a value). Select the x-value for the right bound of the interval by pressing (ENTER).

Press (▶) or (◀) (or enter a value) to select an x-value for a guess at the maximum, and then press (ENTER).

You will see:

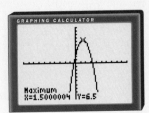

The vertex of $y = -2x^2 + 6x + 2$ is (1.5, 6.5).

The maximum value of $y = -2x^2 + 6x + 2$ is 6.5 when $x = 1.5$.

Function or Instruction and Description	Keystroke(s), Menu, or Screen

Minimum operation

The minimum operation finds the minimum of a function within a specified interval.

EXAMPLE:

To find the minimum of $y = 2x^2 + 5x - 4$, input the function in the Y= editor.

Press (2nd) (TRACE) to display the CALCULATE menu. Press 3 to select the minimum operation.

You will see:

Press (▼) or (▲), if necessary, to move the cursor onto the function.

Press (▶) or (◀) to move the cursor to the left of the minimum (or enter a value). Select the x-value for the left bound of the interval by pressing (ENTER).

Press (▶) or (◀) to move the cursor to the right of the minimum (or enter a value). Select the x-value for the right bound of the interval by pressing (ENTER).

Press (▶) or (◀) (or enter a value) to select an x-value for a guess at the minimum, and then press (ENTER).

You will see:

The vertex of $y = 2x^2 + 5x - 4$ is $(-1.25, -7.125)$.

The minimum value of $y = 2x^2 + 5x - 4$ is -7.125 when $x = -1.25$.

Function or Instruction and Description	Keystroke(s), Menu, or Screen

Mode settings

Mode settings control the way the calculator displays and interprets numbers and graphs.

To display the mode settings, press (MODE).

The default settings are shown here.

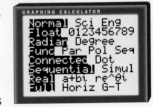

To change a mode setting, press (▼), (▶), (▲), and (◀) as necessary, to move the cursor to the setting you want to select, and press (ENTER) to select it.

For example, the default setting Radian means that angles are measured in radians. If you wish angles to be measured in degrees, press (▼), (▼), (▶), and (ENTER) to select Degree.

sequence function

The sequence function can be used to generate elements of a sequence.

To select the sequence function from the LIST OPS menu, press (2nd) (STAT) (▶) 5.

EXAMPLE:

To generate the first 5 terms of the sequence $t_n = 3n - 2$, change the mode settings to Seq(Sequence) mode by pressing (MODE) (▼) (▼) (▼) (▶) (▶) (▶).
Select the sequence function and input the expression to be used to generate the sequence. The variable, n, can be input by pressing the (X,T,θ,n) key.

Press (,) and input the variable to evaluate the expression for, in this case, n. Press (,) and input the term number to start at, in this case, 1. Press (,) and input the term number to end at, in this case, 5. Press (,) and input the increment value for the term numbers, in this case, 1. If the increment in n is not given, the default value is 1. Press ( )) and then (ENTER).

You will see:

```
GRAPHING CALCULATOR
seq(3n-2,n,1,5,1
)
        {1 4 7 10 13}
```

Function or Instruction and Description	Keystroke(s), Menu, or Screen

Sequence Y= editor

The sequence Y= editor can be used to enter and display sequences for $u(n)$, $v(n)$, and $w(n)$.

Change the mode settings to Seq(Sequence) mode by pressing (MODE) (▼) (▼) (▼) (▶) (▶) (▶).

To display the sequence Y= editor, press (Y=).

You will see:

The value for nMin defines the minimum n value to evaluate. You can edit this value, if desired.

You can change the graph style by moving the cursor to the left of $u(n)$, $v(n)$, or $w(n)$ using the (◀) key. Press (ENTER) repeatedly to alternate between dot, line, and thick styles. The default style in Seq(Sequence) mode is dot style.

EXAMPLE:

To input $f(n) = 4n + 3$ using the sequence Y= editor, make sure the calculator is in Seq graphing mode.

Press the (Y=) key.

To enter the function in $u(n)$, press 4 (X,T,θ,n) (+) 3 (ENTER).

You will see:

Function or Instruction and Description	Keystroke(s), Menu, or Screen

SinReg (sinusoidal regression) instruction

The sinusoidal regression instruction is used to fit the equation of a trigonometric function to a given set of data. The values for a, b, c, and d are displayed for a function of the form $y = a \sin (bx + c) + d$.

To use the sinusoidal regression instruction, enter the data points as two lists using the STAT EDIT menu. Choose suitable window variables using the window editor, or adjust the window automatically using the ZoomStat instruction. Graph L2 versus L1 on a scatter plot using the STAT PLOTS menu.

To select the sinusoidal regression instruction, press STAT ▶ to access the STAT CALC menu. Select C:SinReg.

You will see:

Specify the Xlist name, such as L1, by pressing 2nd 1.
Specify the Ylist name, such as L2, by pressing , 2nd 2.

Press , VARS ▶ 1 to list possible Y= variables. To store the regression equation in Y1, select 1:Y1 and press ENTER.

You will see:

Press ENTER.

You will see, for example:

The regression equation is stored in the Y= editor. If you wish to view the curve of best fit on your scatter plot, press GRAPH.

Function or Instruction and Description	Keystroke(s), Menu, or Screen

Solve function (TI-92 PLUS or TI-92)

The solve function, on the TI-92 Plus or TI-92, solves an algebraic equation for an indicated variable.

To select the solve function from the Algebra menu, press (F2) 1.

EXAMPLE:

To solve $3x - 4y = -12$ for y,

press (F2) 1 3 (×) X (−) 4 (×) (Y=) ((-)) 12 (,) Y ( )) (ENTER).

You will see:

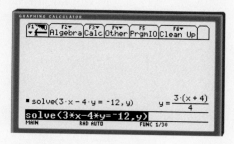

Standard viewing window

The standard viewing window is the portion of the coordinate plane often used for graphs shown on the calculator screen.

Press (MODE) (▼) (▼) (▼) (ENTER) to ensure that the calculator is in Func graphing mode.

To display a graph using the standard viewing window, press (ZOOM) 6.

To display the current window variables, press (WINDOW).

You will see:

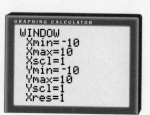

These are the window variables for the standard viewing window.

Function or Instruction and Description	Keystroke(s), Menu, or Screen

STAT EDIT menu

The STAT EDIT menu is used when you wish to store, edit, and view lists of data in the stat list editor.

To display the STAT EDIT menu, press (STAT).

You will see:

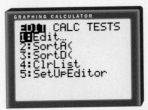

To display the stat list editor, press (STAT) 1:Edit.

You will see:

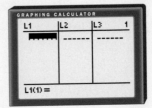

Lists of data can be stored in lists named L_1 through L_6.

To clear data from a specific list, for example L_1, press (STAT) 4:ClrList (2nd) 1 (ENTER).

To clear data from all lists, press (2nd) (+) 4:ClrAllLists.

EXAMPLE:

Enter the table shown in lists L_1 and L_2.

Press (2nd) (+) 4:ClrAllLists (ENTER) to clear all data from lists L_1 to L_6.

L_1	L_2
2	0
3	12
4	24
5	36

Press (STAT) 1:Edit to display the stat list editor.

To enter the data in L_1, press 2 (ENTER) 3 (ENTER) 4 (ENTER) 5 (ENTER).

To enter the data in L_2, press (▶) 0 (ENTER) 12 (ENTER) 24 (ENTER) 36 (ENTER).

You will see:

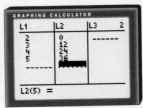

Function or Instruction and Description	Keystroke(s), Menu, or Screen

STAT PLOTS menu

The STAT PLOTS menu allows you to plot data in a scatterplot, xyLine, histogram, modified box plot, regular box plot, or normal probability plot.

To display the STAT PLOTS menu, press $\boxed{\text{2nd}}$ $\boxed{\text{Y=}}$.

EXAMPLE:

Plot the data in a scatter plot.

L1	L2
2	0
3	12
4	24
5	36

Enter the data in lists L1 and L2 using the STAT EDIT menu.

Press $\boxed{\text{2nd}}$ $\boxed{\text{Y=}}$ to display the STAT PLOTS menu.

You will see:

Press $\boxed{\text{ENTER}}$ to select Plot1 or use the $\boxed{\blacktriangledown}$ key to select Plot2 or Plot3 and press $\boxed{\text{ENTER}}$. To turn on a plot, press $\boxed{\text{ENTER}}$.

You will see:

To select a scatter plot, press $\boxed{\blacktriangledown}$ $\boxed{\text{ENTER}}$.

If the Xlist is not already L1, press $\boxed{\blacktriangledown}$ $\boxed{\text{2nd}}$ 1 $\boxed{\text{ENTER}}$.

If the Ylist is not already L2, press $\boxed{\blacktriangledown}$ $\boxed{\text{2nd}}$ 2 $\boxed{\text{ENTER}}$.

Choose the type of mark for the data points, by pressing $\boxed{\blacktriangleright}$ or $\boxed{\blacktriangleleft}$ to highlight the desired mark, and then press $\boxed{\text{ENTER}}$.

Function or Instruction and Description	Keystroke(s), Menu, or Screen

To display the plot, press (ZOOM) 9 to select the ZoomStat instruction.

You will see:

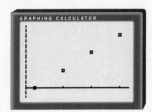

sum function

The sum function can be used to find the sum of the elements of a series.

Change the mode settings to Seq(Sequence) mode by pressing (MODE) (▼) (▼) (▼) (▶) (▶) (▶).

To select the sum function from the LIST MATH menu, press (2nd) (STAT) (▶) (▶) 5.

EXAMPLE:

To find the sum of the first 20 terms of the series $5 + 8 + 11 + \dots$, use the sequence function to generate the first 20 terms of the series defined by $f(n) = 3n + 2$ and press (STO▸) (2nd) 1 (ENTER) to store the terms in L1.

You will see:

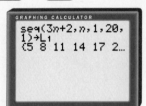

To find the sum of the series, press (2nd) (STAT) (▶) (▶) 5 (2nd) 1 ( ) (ENTER).

You will see:

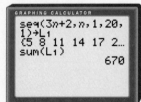

Function or Instruction and Description	Keystroke(s), Menu, or Screen

TABLE SETUP screen

A table of values can be calculated and displayed for any function.

EXAMPLE:

To display a table of values for the function $y = 2x^2 - 8$, enter the function into the Y= editor.

If you wish the table of values to be generated automatically, press (2nd) (WINDOW) to display the TABLE SETUP screen. To define the initial value for the independent variable, x, set TblStart to the initial value you want for your table of values, for example, −3.

Set $\triangle$Tbl to the value of the desired increment for the independent variable, for example, 1.

If you wish the values for both the independent variable, x, and the dependent variable, y, to be displayed automatically, select Indpnt: Auto and Depend: Auto.

For this example, you will see:

Use the (TABLE) key to display the table of values, by pressing (2nd) (GRAPH).

For this example, you will see:

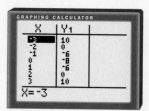

Function or Instruction and Description	Keystroke(s), Menu, or Screen

TEST menu

The TEST menu allows you to use relational operators such as =, ≠, >, ≥, <, and ≤.

To display the TEST menu, press [2nd] [MATH].

You will see:

test submenu

(TI-92 PLUS or TI-92)

The test submenu allows you to use relational operators such as =, ≠, >, ≥, <, and ≤.

To display the test submenu, press [2nd] 5 to display the MATH menu. Press 8 or use the arrow keys to highlight 8:Test and press [ENTER].

You will see:

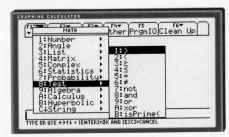

You can now select the desired operator by using the arrow keys and pressing [ENTER] or by pressing the number or letter to the left of the operator.

To cancel the submenu without making a selection, press [◄] or [ESC].

Trace instruction

The TRACE instruction allows you to move the cursor along the graph of a function. The coordinates of points on the graph are displayed in the viewing window.

For a graph drawn using an equation in the Y= editor, press the [TRACE] key to display the cursor on the graph. Press [◄] or [►] to move the cursor along the graph. The x- and y-coordinates of points on the graph will be displayed at the bottom of the screen.

For a scatter plot drawn using the STAT PLOTS menu, press the [TRACE] key and [◄] or [►] to move the cursor from one point to the next on the scatter plot. The coordinates of each point will be displayed.

Function or Instruction and Description	Keystroke(s), Menu, or Screen

TVM Solver

The TVM Solver displays the time-value-of-money (TVM) variables. By entering four of the values, the TVM Solver will solve for the fifth variable.

The five TVM variables are:

N - Total number of payment periods

I% - Annual interest rate

PV - Present value

PMT - Payment amount

FV - Future value or amount

P/Y is the number of payment periods per year in a transaction.

C/Y is the number of compounding periods per year in the same transaction.

PMT:END BEGIN indicates whether the payment is made at the end or at the beginning of a payment period.

Change the mode settings to 2 decimal places.

To access the TVM Solver on the TI-83 Plus, press the (APPS) key and choose 1:Finance to display the FINANCE menu . Choose 1:TVM Solver.

To access the TVM Solver on the TI-83, press (2nd) (x⁻¹) to display the FINANCE menu. Choose 1:TVM Solver.

You will see a screen like the following:

Enter the known values for four TVM variables.

You must enter a value or a 0 for the cursor to move to the next variable.

Enter a value for P/Y and C/Y (if C/Y is not the same as P/Y).

Select END or BEGIN.

Place the cursor on the TVM variable you wish to solve for.

Press (ALPHA) [SOLVE]. [SOLVE] is located above the (ENTER) key.

Function or Instruction and Description	Keystroke(s), Menu, or Screen

You will see, for example:

A square (■) in the left column indicates the solution variable.

Note: When entering payment values, cash received is input as a positive number. Cash paid out is input as a negative number.

Value operation

The value operation evaluates a function for a specified value of x.

EXAMPLE:

To evaluate the function $y = x^2 + 3x - 4$ for $x = -3$, input the function in the Y= editor.

Press (2nd) (TRACE) to display the CALCULATE menu.

Press (ENTER) to select the value operation.

You will see:

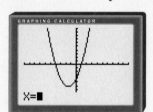

Enter the value -3 for x.

Press (ENTER).

You will see:

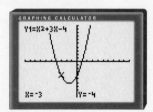

Note: The value for which you evaluate a function must lie between Xmin and Xmax of the viewing window used for the graph.

Function or Instruction and Description	Keystroke(s), Menu, or Screen

Window variables

The window variables define the current viewing window.

To display the current window variable values, press (WINDOW).
To change a window variable value, press (▼) or (▲) to move the cursor to the window variable you want to change. Enter the new value and then press (ENTER).

Y= editor

The Y= editor is used to define or edit a function.

Press (MODE) (▼) (▼) (▼) (ENTER) to ensure that the calculator is in Func graphing mode.
To display the Y= editor, press (Y=).
To move the cursor to the next function, press (ENTER) or (▼).
To move the cursor from one function to another, press (▼) or (▲).
To erase a function, highlight the function and press (CLEAR).
The independent variable is X. To input X, press (X,T,θ,n) or (ALPHA) (STO▸).

When you input the first character of a function, the = is highlighted. This indicates that the function is selected. To deselect a function, move the cursor to the = symbol of the function and press (ENTER).

EXAMPLE:

To input $y = 3x - 2$ using the Y= editor, press the (Y=) key.

You will see:

Press 3 (X,T,θ,n) (—) 2 (ENTER).

You will see:

Function or Instruction and Description	Keystroke(s), Menu, or Screen

Zero operation

The zero operation finds the zeros or *x*-intercepts of functions. If a function has two or more *x*-intercepts, they must be found separately by repeated use of the zero operation.

EXAMPLE:

To find the *x*-intercepts of $y = x^2 + 2x - 8$, input the function in the Y= editor.

Press [2nd] [TRACE] to display the CALCULATE menu.
Press [▼] [ENTER] to select the zero operation.

You will see:

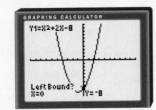

Press [▼] or [▲] , if necessary, to move the cursor onto the function.

To find the left *x*-intercept, press [▶] or [◀] to move the cursor to the left of the left *x*-intercept (or enter a value). Select the *x*-value by pressing [ENTER].

Press [▶] or [◀] to move the cursor to a location between the left *x*-intercept and the right *x*-intercept (or enter a value). Select the *x*-value by pressing [ENTER].

Press [▶] or [◀] (or enter a value) to select an *x*-value for a guess at the left *x*-intercept, and then press [ENTER].

You will see:

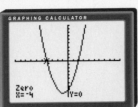

To find the right *x*-intercept, press [▶] or [◀] to move the cursor to a location between the left *x*-intercept and the right *x*-intercept (or enter a value). Select the *x*-value by pressing [ENTER].

Function or Instruction and Description	Keystroke(s), Menu, or Screen

Press (▶) or (◀) to move the cursor to the right of the right *x*-intercept (or enter a value). Select the *x*-value by pressing (ENTER).

Press (▶) or (◀) (or enter a value) to select an *x*-value for a guess at the right *x*-intercept, and then press (ENTER).

You will see:

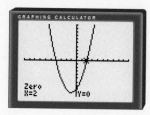

Zoom In instruction

The Zoom In instruction allows you to magnify the part of the graph that surrounds the cursor location.

Press (ZOOM) to display the ZOOM menu.

Select 2:Zoom In. The zoom cursor (+) is displayed on your graph.

Use the arrow keys to move the cursor to the point that is to be the centre of the new viewing window. Press (ENTER).

The viewing window is adjusted and updated and the selected function is replotted, centred on the cursor location.

You can zoom in on the graph again by pressing (ENTER) to zoom in at the same point or by moving the cursor to the point you want as the centre of the new viewing window and then pressing (ENTER).

ZOOM menu

The ZOOM menu contains instructions that allow you to adjust the viewing window quickly.

To display the ZOOM menu, press (ZOOM).

You will see:

Function or Instruction and Description	Keystroke(s), Menu, or Screen

ZoomStat instruction

The ZoomStat instruction redefines the window variables to display all statistical data points on the screen.

Press (ZOOM) to display the ZOOM menu.

Press 9 to display all your data using the ZoomStat instruction.

ZSquare instruction

The ZSquare instruction allow you to adjust the viewing window, so that graphs are plotted using equal-sized-scales on both the x- and y-axes.

Press (ZOOM) to access the ZOOM menu.

Press 5 to display a graph using the ZSquare instruction.

EXAMPLE:

To graph $y = x^2 - 4$ using the ZSquare instruction, input the function using the Y= editor.

Press (ZOOM) to access the ZOOM menu.

You will see:

Press 5 to display the graph using the ZSquare instruction.

You will see:

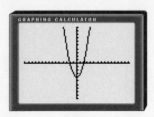

APPENDIX C COMPUTER SOFTWARE

THE GEOMETER'S SKETCHPAD®

Preferences

Before you begin using *The Geometer's Sketchpad*®, you may need to change some of the default settings in the program. Click on the Display menu and choose Preferences.

A window like the one shown will open up. Be sure that the Distance units are set to "cm" and not "inches." You should also change the Precision options to "tenths" instead of "units". If you are working with angle measures, make sure that the Angle Unit is "degrees," not "radians." Finally, click on the Autoshow Labels for Points button in the upper left corner of the window.

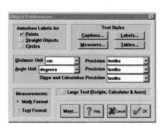

Tools and Menus

On the left side of the screen there is a set of features called a Toolbox. To select a tool from the Toolbox, click on the tool. The active tool will be highlighted.

The **Selection Tool** is used to select objects on the screen. To select an object, click on the Selection Tool and then on the object, which will be highlighted. To highlight more than one object, press and hold the SHIFT key and click on the objects.

The **Point Tool** is used to create new points on the screen. Clicking anywhere on the screen creates a point.

The **Circle Tool** is used to create a circle on the screen. To use this tool, click on a point and hold the left mouse button. This creates the centre point of a circle. As you drag the mouse, the circle will be created. Release the mouse button to fix the size of the circle. A second method of creating a circle is given later in this appendix.

The **Line Tool** is used to create a line segment. Click on the first endpoint and drag out to the second endpoint. Later in this appendix, you will see that the Line Tool can also be used to create a ray or a line, and that a line segment can also be created using another method.

The **Text Tool** is used to show, hide, or change labels assigned to objects on the screen.

Across the top of the screen is a set of menus. A brief description of some of these menu items follows.

The File Menu

Click on **New Sketch** when you are ready to start a new drawing.

The **Open** command allows you to open a file saved on disk.

Save As allows you to save a drawing to disk.

The **Print** command sends a copy of your drawing to the printer.

Exit is used to quit *The Geometer's Sketchpad*®.

The Display Menu

At the start of every drawing, you can use the **Display** menu to set the **Line Style** to Thick and the "**Color**" to your choice.

The Preferences command at the bottom of the menu was referred to earlier. It is used to set up your preferences for *The Geometer's Sketchpad*®.

The **Display** menu can be used not only to set preferences, but also for tracing objects. A trace can be applied to either points or lines. In this graphic, the blue line through point E is highlighted. If point C is dragged along line segment AB, the path of the blue line will be left on the screen.

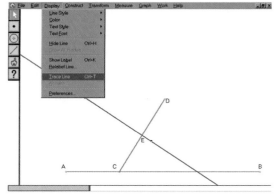

The Edit Menu

The **Edit** menu shows commands that allow you to cut and paste as well as create action buttons on the screen.

In this graphic, point C and line segment AB are highlighted. Creating an **animation** button will cause point C to move along line segment AB.

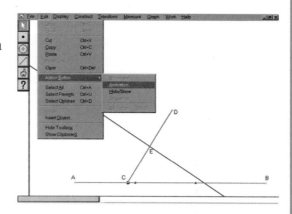

The Construct Menu

The **Construct** menu shows commands that apply to the object highlighted on the screen. In the graphic, a line segment is highlighted, so the menu gives you the choice of creating a point somewhere on the segment, or creating the midpoint of the segment, and so on. The commands will be explained as they are needed later in this appendix.

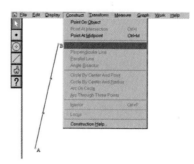

The Measure Menu

The **Measure** menu shows commands that apply to the object highlighted on the screen. The graphic shows the commands that apply to the same line segment as above. Other commands become available for other objects and will be explained as needed.

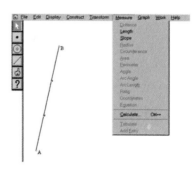

Constructions and Measurements

Constructing and Measuring a Line Segment

1. Click on the **Point Tool** on the left side of the screen. With this tool activated, clicking anywhere on the screen creates a point.

2. Create two points on the screen.

3. Highlight both points. To do this, click on one point, and then press SHIFT and click on the other point.

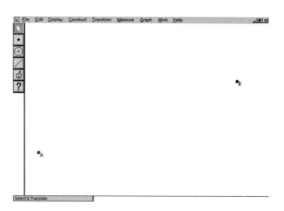

4. From the **Construct** menu, choose the **Segment** command. You will create a line segment joining the points.

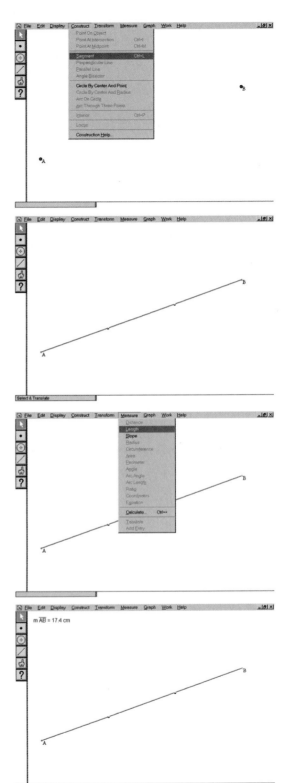

5. Using the **Selection Tool**, click on the line segment. You will see two small squares on the line segment to indicate that it is highlighted.

6. From the **Measure** menu, choose the **Length** command. The length of the segment will be displayed in the upper left corner of the screen.

Creating a New Line Segment From an Existing One

In the construction of conics, you often need to have two line segments whose total length is a constant. To do this, you divide an existing line segment into two smaller line segments.

1. Construct a line segment AB.

2. Highlight line segment AB. From the **Construct** menu, choose the **Point On Object** command.

3. Highlight endpoint A and the new point C.

4. From the **Construct** menu, choose the **Segment** option.

5. Repeat steps 3 and 4 for points B and C.

6. Drag point C along line segment AB. Notice that the length of AC plus the length of BC always equals the length of AB.

Constructing a Line

1. Click on the **Line Tool** on the left side of the screen and hold down the left mouse button. A window will pop up to the right. The first option is a line segment, the second is a ray, and the third is line. Choose the line option.

2. Click on the Point Tool and create two points anywhere on the screen.

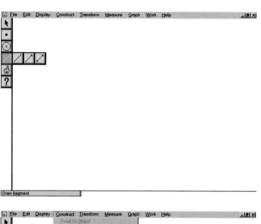

3. Click on the **Construct** menu and choose the **Line** command. A line through the two points will appear on the screen.

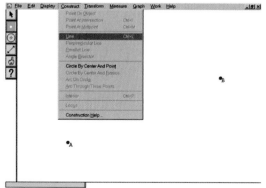

Constructing a Midpoint of a Line Segment

1. Construct a line segment, AB.

2. From the **Construct** menu, choose the **Point At Midpoint** command. You will create a new point, C, in the centre of the line segment.

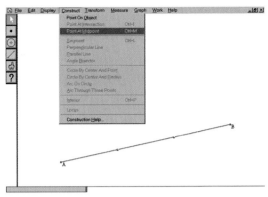

3. Construct two new segments from A to C and from B to C.

4. Measure the lengths of AB, AC, and BC. What do you notice about the lengths of AC and BC? How do the lengths of these segments compare to the length of AB?

5. Are these properties preserved when you drag point A to a new location?

Constructing a Right Bisector of a Line Segment

The right bisector of a line segment is a line that intersects a line segment at its midpoint at a 90° angle.

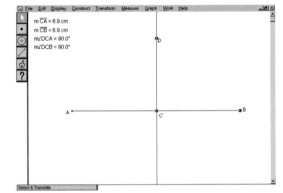

1. Construct a line segment, AB.

2. From the **Construct** menu, choose the **Point at Midpoint** command to construct the midpoint, C.

3. Click on the midpoint and the line segment so that both are highlighted.

4. Click on the **Construct** menu and choose the **Perpendicular Line** command.

5. With this line highlighted, choose the **Point on Object** command from the **Construct menu** to create point D.

6. Measure the line segments AC and BC, and ∠DCA and ∠DCB.

Constructing a Circle With a Given Centre and Radius

For the construction of conics, diagrams such as
the one shown are often used. In this graphic,
point D and line segment AC are highlighted.
When the **Circle By Center And Radius**
command is executed from the Construct
menu, a circle is drawn with centre D and
radius equal to the length of line segment AC.

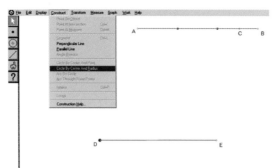

1. Construct the circle.

2. Drag point C along line segment AB.

3. Describe what happens to the radius of the circle?

ZAP-A-GRAPH

Defining a Function

Zap-a-Graph function definition is
accomplished through menus.

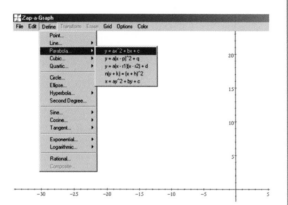

- Click the **Define** menu.

- Scroll down to the type of function that you
  wish to work with. In this example, the
  Parabola option is chosen.

- Once a form of an equation has been chosen,
  a dialog box opens asking you to enter the
  coefficients for the function.

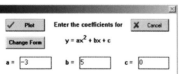

- Click the **Plot** button, and the graph is
  displayed with the equation shown at the bottom of the screen.

Setting the Scale

The default scale values are 5 on the x-axis and y-axis. To change these values:

- From the **Grid** menu choose **Scale**.
- In the dialog box, enter the scale for x-axis and for the y-axis.
- Click the **OK** button.

The screen shows the menu option as well as the dialog box that allows you to set the scale.

Finding a Maximum or Minimum

Zap-a-Graph performs analysis on functions. For a quadratic function, the program gives the vertex of a parabola as well as the x- and y-intercepts.

- With the graph on the screen, from the **Options** menu choose **Analyze**.
- The program displays the intercepts and the vertex in a window.

Transforming Functions

Zap-a-Graph devotes one menu to transformations.

To translate a function:

- Define a function, for example, $y = -3x^2 + 5x$.
- From the **Transform** menu, choose **Translate**.
- A dialog box is presented that allows you to define the horizontal translation and the vertical translation.

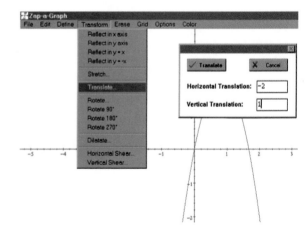

To reflect a function in the *x*-axis:

- Define the function $y = -3x^2 + 5x$.
- From the **Transform** menu, choose **Reflect in x-axis**.

The screen shows the menu option and the effect of the reflection.

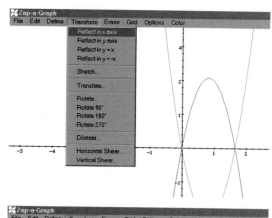

To stretch the function vertically:

- From the **Transform** menu choose **Stretch**.

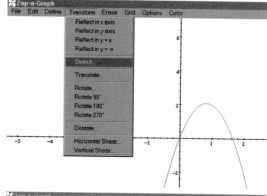

- Enter the Vertical Stretch Factor.
- Click the **Stretch** button.

The screen shows the stretch dialog box and the resulting graph.

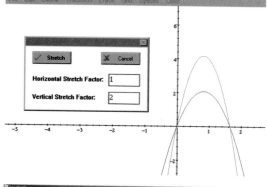

To stretch the function horizontally:

- From the **Transform** menu choose **Stretch**.
- Enter the Horizontal Stretch Factor.
- Click the **Stretch** button.

The screen shows the stretch dialog box and the resulting graph.

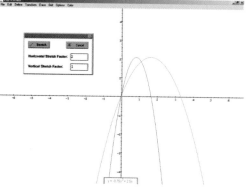

Combining Transformations

Transformations can be combined by performing individual transformations in the correct order. For example, start with the quadratic function $y = x^2$ and develop the graph of $y = 3(x + 2)^2 - 3$. The original function is shown in the screen in blue. It undergoes three transformations: a vertical stretch of factor 3 (red), a horizontal translation 2 units left (purple), followed by followed by a vertical translation 3 units down (green).

- From the **Define** menu, choose **Parabola** and choose the form $y = ax^2 + bx + c$.

- Leave $a = 1$, $b = 0$, and $c = 0$.

- Click the **Plot** button.

- From the **Transform** menu, choose **Stretch**.

- Enter 3 for the Vertical Stretch Factor.

- Click the **Stretch** button.

- From the **Transform** menu, choose **Translate**.

- Enter −2 for the Horizontal Translation.

- Click the **Translate** button.

- From the **Transform** menu, choose **Translate**.

- Enter −3 for the Vertical Translation.

- Click the **Translate** button.

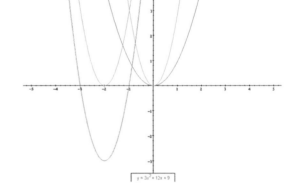

Note: Zap-a-Graph can graph $y = 3(x + 2)^2 - 3$ in another way. Use the equation form $y = a(x - p)^2 + q$ and set $a = 3$, $p = -2$, and $q = -3$.

Inverse of a Function

You can graph the inverse of a function by reflecting in the line $y = x$.

To graph the inverse of a quadratic function:

- Define and plot a quadratic function, for example, $y = -3x^2 + 5x$.

- From the **Transform** menu, choose **Reflect in $y = x$**.

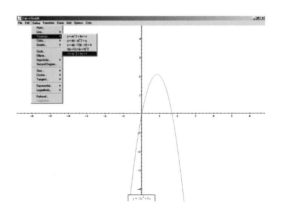

- Click the **Reflect** button.

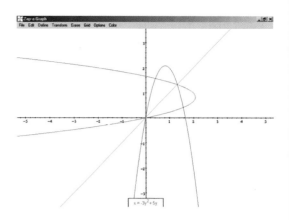

Graphing an Ellipse or a Hyperbola

The **Define** menu in Zap-a-Graph allows you to graph conics.

To graph an ellipse:

- From the **Define** menu choose **Ellipse**.

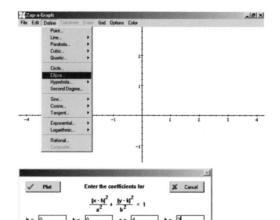

- Set $a = 4$ and $b = 2$. Enter values for h and k when the centre of the ellipse is not at the origin.

- Click the **Plot** button.
- Use different values for a and b to see the effect on the graph. You may have to change the scale on your screen.
- Use different values for h and k to move your graph to different positions on the screen.

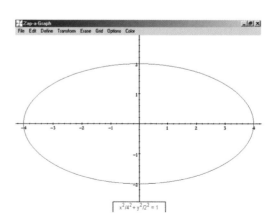

To graph a hyperbola:

- From the **Define** menu choose **Hyperbola** and choose the appropriate form. The screen shows the form with the transverse axis on the *x*-axis.

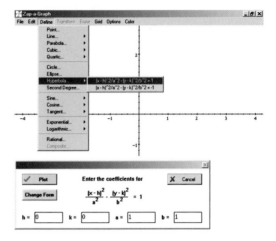

- Enter values for *h* and *k* only when the centre of the hyperbola is not at the origin.

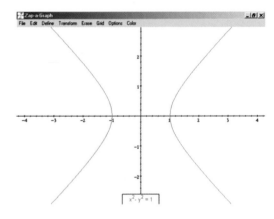

- Click the **Plot** button.

- Use different values for *a* and *b* to see the effect on the graph. You may have to change the scale on the screen.
- Use different values for *h* and *k* to move the graph to different positions on the screen.
- From the **Define** menu, choose the **Hyperbola** option and select the option that places the transverse axis on the *y*-axis. Experiment with different values of *h*, *k*, *a* and *b* to obtain graphs that are wider or narrower, or are in different positions on the screen.

Graphing Trigonometric Functions

Trigonometric functions can be graphed and transformed in the same way as polynomial functions. The scale may also be changed to accommodate fractions of π.

To set the scale in fractions of π:

- From the **Grid** menu, select **Scale**.
- For the *x*-scale, enter the letter p so that the symbol π appears. Enter $\pi/2$.
- For the *y*-scale, enter 1.
- Click the **OK** button.

To define a sine function

- From the **Define** menu, select **Sine**.
- Click the **Plot** button.

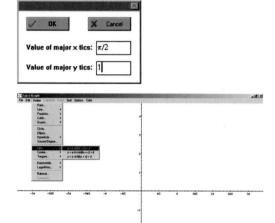

Transforming Trigonometric Functions

To translate a sine function horizontally:

- From the **Transform** menu, select **Translate**.
- Enter the value $\pi/2$ for the Horizontal Translation.
- Click the **Translate** button.

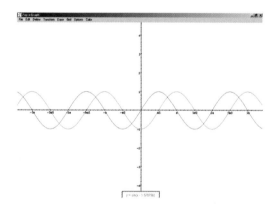

To stretch a sine function vertically:

- From the **Transform** menu choose **Stretch**.
- Enter the value 3 for the Vertical Stretch Factor.
- Click the **Stretch** button.

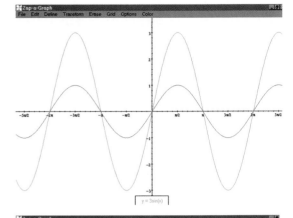

To stretch a sine function horizontally:

- From the **Transform** menu choose **Stretch**.
- Enter the value 1/2 for the Horizontal Stretch Factor.
- Click the **Stretch** button.

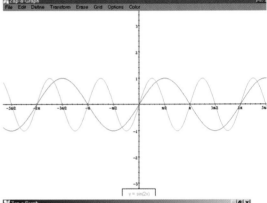

To translate a sine function vertically:

- From the **Transform** menu choose **Translate**.
- Enter the value 2 for the Vertical Translation.
- Click the **Translate** button.

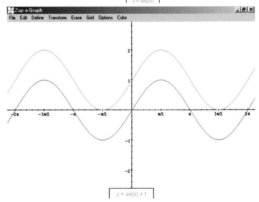

Combining Transformations of Trigonometric Functions

Start with the function $f(x) = \sin x$ and develop the graph of $y = 2\sin\left(x - \dfrac{\pi}{2}\right) + 1$. The graph of the original function is shown in blue. It undergoes three transformations: a vertical stretch of factor 2 (red), a horizontal translation of $\dfrac{\pi}{2}$ units right (purple), and a vertical translation of 1 unit up (green).

- From the **Define** menu choose **Sine**.
- Leave $a = 1$, $b = 1$, $c = 0$, and $d = 0$.
- From the **Transform** menu choose **Stretch**.
- Enter 2 for the Vertical Stretch Factor.
- Click the **Stretch** button.
- From the **Transform** menu choose **Translate**.
- Enter $\pi/2$ for the Horizontal Translation.
- Click the **Translate** button.
- From the **Transform** menu choose **Translate**.
- Enter 1 for the Vertical Translation.
- Click the **Translate** button.

As each transformation is performed, the equation of the function is updated at the bottom of the screen.

Note: Zap-a-Graph can graph $y = 2\sin\left(x - \dfrac{\pi}{2}\right) + 1$ directly. Set $a = 2$, $b = 1$, $c = -\pi/2$, and $d = 1$.

MICROSOFT® EXCEL SPREADSHEETS

These methods apply for Microsoft® Excel 97. Methods may vary slightly for other versions.

A spreadsheet screen includes a menu, toolbars, a name box, a formula bar, rows and columns of cells, and scroll bars.

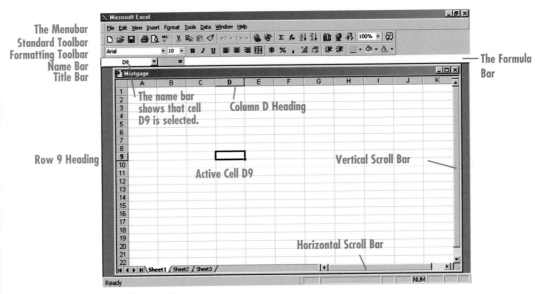

The Menubar
Standard Toolbar
Formatting Toolbar
Name Bar
Title Bar

Row 9 Heading

The name bar shows that cell D9 is selected.

Column D Heading

Active Cell D9

Vertical Scroll Bar

Horizontal Scroll Bar

The Formula Bar

To move from cell to cell:

• Click with the cursor.

To display a menu:

• Click on the menu heading in the **Menu toolbar**.

To perform standard tasks such as save, print, and check spelling:

• Click buttons on the **Standard toolbar**.

To perform formatting tasks, such as select the font, change the type size, colour type, and fill a cell with colour:

• Click buttons on the **Formatting toolbar**.

To display or remove the Formula Bar or Status Bar:

• Go to the **View** menu.

• If it is not checked, choose a bar to display it.

To display or remove a toolbar:

• From the **View** menu, choose **Toolbars**.

• If it is not checked, choose a toolbar to display it.

Getting Help

To get help:

- From the **Help** menu, make a choice.

- Follow the steps.

Opening, Saving, and Closing a Spreadsheet

By saving a document frequently, you can avoid losing your work.

To open a new screen:

- From the **File** menu, choose **New**.

To save a spreadsheet:

- From the **File** menu, choose **Save**.

- Make choices for naming and locating the spreadsheet. Click **Save**.

To close a spreadsheet:

- From the **File** menu, choose **Close**.

Selecting Cells

You can select single cells, groups of cells, rows, or columns.

To select a cell:

- Click the cell. The name of the cell appears in the **Name Bar**.

To select adjacent cells:

- Click the first cell and drag to the last cell.

- Or click the first cell. Hold down the **Shift** key, move the cursor to the last cell, and click the last cell.

To select cells that are not adjacent:

- Hold down the **Control** key and click the cells.

To select a row or a column:

- Click the row or column heading.

Using Formulas

You can enter formulas in a spreadsheet to perform calculations.

To enter a formula with operations:

- Select the cell for the result.
- In the **Formula Bar**, type =. Then, type the formula with the operation symbols. Include brackets for the order of operations.
  For example, the formula **=6^3** calculates 6^3; the formula, **=6^3+4-3** calculates $6^3 + 4 - 3$

+ plus sign for addition,
− minus sign for subtraction
− negation sign for negative values
* asterisk for multiplication
/ forward slash for division
^ caret for exponents
% percent sign

| A1 | ▼ | = | =6^3+(4-3) |

Investment

	A	B	C	D
1	217			
2				
3				

- Click ✓, or press **Enter**, or click a different cell in the spreadsheet.
- To cancel entering the formula, click ✗ in the **Formula Bar**.

To enter a formula that copies the value from one cell into another cell:

- Select the cell for the result.
- **Enter** the formula. For example, selecting cell D3 and entering the formula **=A2** copies the value from cell A2 into cell D3.

| D3 | ▼ | = | =A2 |

Finances

	A	B	C	D	
1	850				
2	740				
3	2100			740	
4					

- Click ✓, or press **Enter**, or click another cell. If you change the value in cell A2, the value in cell D3 is changed to match.

To perform an operation that includes a cell reference:

- Select the cell for the result.
- Enter the formula with the cell reference. For example, selecting cell C4 and entering the formula **=B4/2** divides the value in cell B4 by 2 and shows the result in cell C4.

| C4 | ▼ | = | =B4/2 |

Financial Planning

	A	B	C	D
1				
2				
3				
4		5000	2500	
5				

- Click ✓, or press **Enter**, or click another cell. If you change the value in cell B4, the result in cell C4 is changed to the new value in cell B4 divided by 2.

To change the formula for a cell:

- Select the cell. Change the type that appears in the **Formula Bar**.
- Click ✓ in the **Formula Bar**, or press **Enter**, or click another cell.

To delete the formula for a cell:

- Select the cell. From the **Edit** menu, choose **Clear**.
- Or select the cell and press **Delete**.

Entering Functions

You can use functions such as the **Sum Function** to perform calculations.

To use the sum function **=SUM(** to add the values in a range of cells:

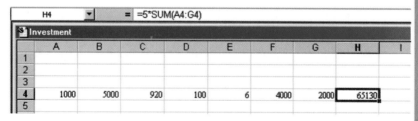

- Select the cell for the sum.

- Enter the sum function with the name of the first cell in the range and the name of the last cell in the range separated by a colon. For example, the function **=SUM(B4:B9)** adds the values in cells B4 to B9. This is an alternative to entering the formula **=B4+B5+B6+B7+B8+B9**.

To use the **Sum Function** button to add the values in a range of cells:

- Select the cell for the sum.
- Click the ∑ button on the **Standard Toolbar**.
- If **Formula Bar** shows the function **=SUM(**, enter the names of the first and last cells separated by a colon.
- If the formula contains different values that do not define the range you want, change the formula. For example, the function **=SUM(A4:G4)** adds the values in cells A4 to G4.

To perform an operation that includes a function:

- Select the cell for the result.

- Enter the formula with the sum function. For example, **=5*SUM(A4:G4)** multiplies the sum of the values in cells A4 to G4 by 5.

Copying Cell Content

To copy content into cells to the right or down:

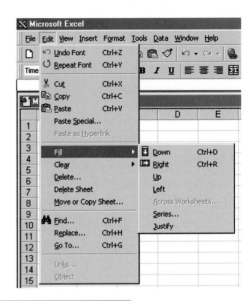

- Select the cell with the content to be copied and drag the mouse from the first to the last cell where you want to copy the content. From the **Edit** menu, choose **Fill**.

- Then, choose **Down** to copy the formula in the cells below, or **Right** to copy it in the cells to the right. The formula will change to correspond to the cell where it is moved. For example, if the formula **=D2–C2** is copied from cell E2 to cell E3, it will become **=D3–C3**.

To move content into another cell:

- Select the cell with the content. From the **Edit** menu, choose **Cut**.

- Select the target cell.

- From the **Edit** menu, choose **Paste**.

Many operations in a spreadsheet can be accomplished in different ways.

To copy content into another cell:

- Select the cell with the content. From the **Edit** menu, choose **Copy**.

- Select the target cell.

- From the **Edit** menu, choose **Paste**.

To enter a formula that does not change as it is moved or copied:

- Type the formula with $ between the letter for the column and the number for the row. For example, if the formula **=A$4** is copied from cell D8 to cell D9, it remains **=A$4**. If the formula **=B8*A$4** is copied from cell C8 to cell C9, it becomes **=B9*A$4**.

To check the formula in a cell:

- Select the cell. The formula will appear in the **Formula Bar**.

E9	▼		=	=B9*A$4	

Investing

	A	B	C	D	E	F
1						
2						
3						
4	3500					
5						
6						
7						
8		2		3500	7000	
9		3		3500	10500	
10						

Formatting Type and Numbers

Numerous options for formatting type and numbers are available using the **Formatting Toolbar** or the **Format** menu.

To align numbers:

- Select the cells.

- Click the button on the **Formatting Toolbar** to position the type aligned to the left, centred, or aligned to the right.

	A	B	C
1			
2			
3			
4			

To change the width of a cell:

- Drag the boundary on the right side of the column heading.

To insert cells:

- Select a cell.

- From the **Insert** menu, choose **Cells**. Click the appropriate box.

- Click **OK**.

To delete cells:

- Select a cell.

- From the **Edit** menu, choose **Delete**. Click the appropriate box.

- Click **OK**.

To format numbers:

- Select the cells.
- From the **Format** menu, choose **Cells**.
- Click the **Number** tab.
- For the **Category**, click **Number**.
- Choose the number of decimal places. Make sure the box for a comma separator is not checked.
- Click **OK**.

To customize formatting for numbers:

- Select the cells.
- From the **Format** menu, choose **Cells**.
- Click the **Number** tab.
- For the **Category**, click **Custom**.
- Choose the formatting. Replace the comma with a space.
- Click **OK**.

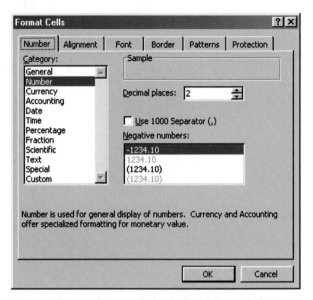

A space will appear between the hundreds and thousands digits, but not farther to the left. For example, a space will not appear between the hundred thousands and millions digits.

Printing and Copying

To prepare for printing or copying a spreadsheet:

- From the **File** menu, choose **Page Setup**.
- Click the **Page** tab. Click **Portrait** or **Landscape**.
- Click **OK**.

To print part of a spreadsheet:

- Select the part of the spreadsheet you want to print.
- From the **File** menu, choose **Print**.
- Click **OK**.

To copy a spreadsheet to another document:

- From the **Edit** menu, choose **Copy**.
- Go to the document where you want to copy the spreadsheet and paste it.

To copy part of a spreadsheet to another document:

- First select the part of the spreadsheet you want to copy.

COREL® QUATTRO® PRO SPREADSHEETS

These methods apply for Corel® Quattro® Pro 8. Methods may vary slightly for other versions.

A spreadsheet screen includes a menu, toolbars, a name box, a formula bar, rows and columns of cells, and scroll bars.

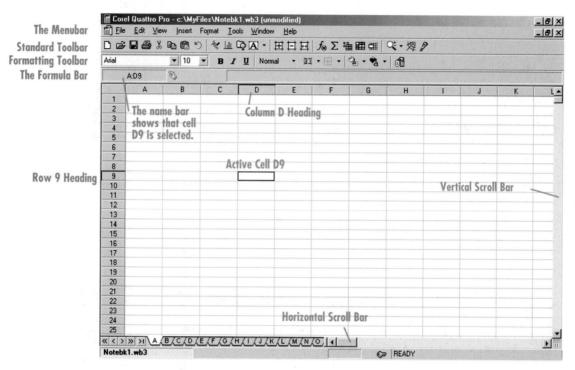

To move from cell to cell:

- Click with the cursor.

To display a menu:

- Click on the menu heading in the **Menu Toolbar**.

To perform standard tasks such as save, print, and check spelling:

- Click buttons on the **Standard Toolbar**.

To perform formatting tasks, such as select the font, change the type size, colour type, and fill a cell with colour:

- Click buttons on the **Formatting Toolbar**.

To display or remove a toolbar:

- From the **View** menu, choose **Toolbars**.
- If it is not checked, choose a toolbar to display it.

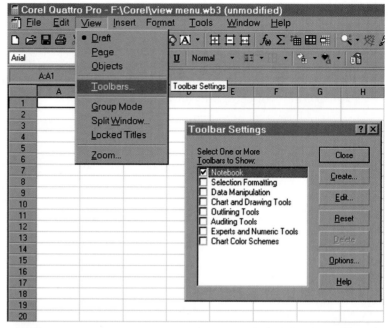

Getting Help

To get help:

- From the **Help** menu, make a choice.
- Follow the steps.

Opening, Saving, and Closing a Spreadsheet

By saving a document frequently, you can avoid losing your work.

To open a new screen:

- From the **File** menu, choose **New**.

To save a spreadsheet:

- From the **File** menu, choose **Save**.
- Make choices for naming and locating the spreadsheet. Click **Save**.

To close a spreadsheet:

- From the **File** menu, choose **Close**.

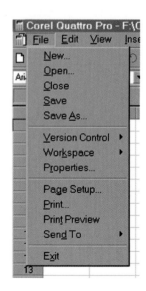

Selecting Cells

You can select single cells, groups of cells, rows, or columns.

To select a cell:

- Click the cell. The name of the cell appears in the **Name Bar**.

To select adjacent cells:

- Click the first cell and drag to the last cell.
- Or click the first cell. Hold down the **Shift** key, move the cursor to the last cell, and click the last cell.

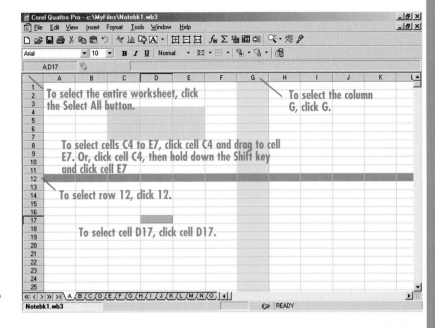

To select cells that are not adjacent:

- Hold down the **Control** key and click the cells.

To select a row or a column:

- Click the row or column heading.

Entering Formulas

You can enter formulas in a spreadsheet to perform calculations.

To enter a formula with operations:

- Select the cell for the result.
- Type + followed by the formula with the operation symbols. As you type, the expression will appear in the cell and in the **Formula Bar**. Include brackets for the order of operations. For example, the formula **+6^3** calculates 6^3, the formula **+6^3+4−3** calculates $6^3 + 4 − 3$.
- Press **Enter**, or click a different cell in the spreadsheet.
- To cancel entering the formula, press **Delete**.

+ plus sign for addition,
− minus sign for subtraction
− negation sign for negative values
* asterisk for multiplication
/ forward slash for division
^ caret for exponents
% percent sign for percent

To enter a formula that copies the value from one cell into another cell:

	A	B	C	D	E
A:D3			+A2		
1	850				
2	740				
3	2100			740	
4					

- Select the cell for the result.
- Enter the formula. For example, selecting cell D3 and entering the formula **+A2** copies the value from cell A2 into cell D3.
- Press **Enter**, or click another cell. If you change the value in cell A2, the value in cell D3 is changed to match.

To perform an operation that includes a cell reference:
- Select the cell for the result.
- Enter the formula with the cell reference. For example, selecting cell C4 and entering the formula **+B4/2** divides the value in cell B4 by 2 and shows the result in C4.
- Press **Enter**, or click another cell. If you change the value in cell B4, the result in cell C4 is changed to the new value in cell B4 divided by 2.

To change the formula for a cell:
- Select the cell. Change the type that appears in the **Formula Bar**.
- Press **Enter**, or click another cell.

To delete the formula for a cell:
- Select the cell. From the **Edit** menu, choose **Clear**.
- Or select the cell and press **Delete**.

Entering Functions

You can use functions such as the **Sum Function** to perform calculations.

	A	B	C	D	E
A:B10		@SUM(B4..B9)			
1					
2					
3					
4		5			
5		2			
6		8			
7		9			
8		3			
9		4			
10		31			
11					

To use the sum function @SUM(to add the values in a range of cells:
- Select the cell for the sum.
- Enter the sum function with the name of the first cell in the range and the name of the last cell in the range separated by two periods For example, the function **@SUM(B4..B9)** adds the values in cells B4 to B9. This is an alternative to entering the formula +B4+B5+B6+B7+B8+B9.

To use the **Sum Function** button to add the values in a range of cells:

- Select the cell for the sum.
- Click the Σ button on the **Standard Toolbar.**
- If **Formula Bar** shows the function @SUM(, enter the names of the first and last cells separated by two periods.

Many operations in a spreadsheet can be accomplished in different ways.

A:H4			+5*@SUM(A4..G4)					
	A	B	C	D	E	F	G	H	I
1									
2									
3									
4	1000	5000	920	100	6	4000	2000	65130	
5									

- If the formula contains different values that do not define the range you want, change the formula. For example, the function @SUM(A4..G4) adds the values in cells A4 to G4.

To perform an operation that includes a function:

- Select the cell for the result.
- Enter the formula with the sum function. For example, **+5*@SUM(A4..G4)** multiplies the sum of the values in cells A4 to G4 by 5.

Copying Cell Content

Corel® Quattro® Pro uses either a copy and paste feature or the **Copy Cells** option in the **Edit** menu.

To copy content into other cells:

- Select the cell(s) with the content to be copied.
- From the **Edit** menu choose **Copy.**
- Click the first cell of the range that you want to copy into and drag the mouse to the last cell.
- From the **Edit** menu, choose **Paste.**

To use the **Copy Cells** feature:

- Select the cell with the content to be copied.
- From the **Edit** menu, choose **Copy Cells.**
- A **Copy Cells** window opens. Click the arrow to the right of the **To:** window.
- Click the first target cell and drag the mouse to the last cell in the range.
- Press **Enter** to set the range. Click **OK.**

To move content into another cell:

- Select the cell with the content. From the **Edit** menu, choose **Cut**.
- Select the target cell. From the **Edit** menu, choose **Paste**.

To copy content into another cell:

- Select the cell with the content. From the **Edit** menu, choose **Copy**.
- Select the target cell. From the **Edit** menu, choose **Paste**.

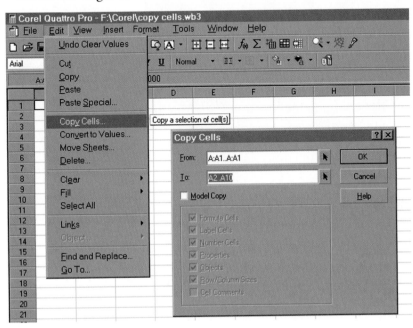

The formula will change to correspond to the cell where it is moved. For example, if the formula **+D2–C2** is copied from cell E2 to cell E3, it will become **+D3–C3**.

To enter a formula that does not change as it is moved or copied:

- Type the formula with $ between the letter for the column and the number for the row. For example, if the formula **+A$4** is copied from cell D8 to cell D9, it remains **+A$4**. If the formula **+B8*A$4** is copied from cell C8 to cell C9, it becomes **+B9*A$4**.

A:E9			+B9*A$4		
A	**B**	**C**	**D**	**E**	**F**
3500					
	2		3500	7000	
	3		3500	10500	

To check the formula in a cell:

- Select the cell. The formula will appear in the **Formula Bar**.

Formatting Type and Numbers

Numerous options for formatting type and numbers are available using the **Formatting Toolbar** or the **Format** menu.

To align numbers:

- Select the cells.
- Click the button on the **Formatting Toolbar** to position the type aligned to the left, centred, or aligned to the right.

To change the width of a cell:

- Drag the boundary on the right side of the column heading.

	A	B	C	D	E	F	G	
1								
2								
3								
4								
5								
6								
7								

To insert cells:

- Select a cell.
- From the **Insert** menu, choose **Cells**.
- Click **OK**.

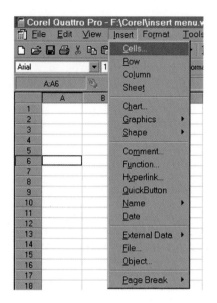

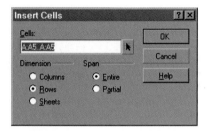

To delete cells:

- Select a cell.
- From the **Edit** menu, choose **Delete**. Click the appropriate box.
- Click **OK**.

To format numbers:

- Select the cells.
- From the **Format** menu, choose **Selection**.
- Click the **Numeric Format** tab.
- Click the **Fixed** button.
- Choose the number of decimal places.
- Click **OK**.

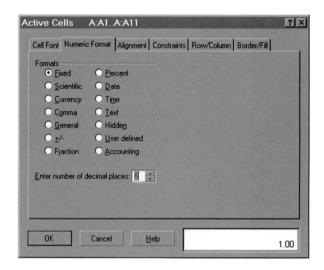

Printing and Copying

To prepare for printing or copying a spreadsheet:

- From the **File** menu, choose **Page Setup**.
- Click the **Page** tab. Click **Portrait** or **Landscape**.
- Click **OK**.

To print part of a spreadsheet:

- Select the part of the spreadsheet you want to print.
- From the **File** menu, choose **Print**.
- Click **OK**.

To copy a spreadsheet to another document:

- From the **Edit** menu, choose **Copy**.
- Go to the document where you want to copy the spreadsheet and paste it.

To copy part of a spreadsheet to another document:

- First select the part of the spreadsheet you want to copy.

ANSWERS

Chapter 1

Getting Started, p. 2

1. a) $250 \le f \le 21\ 000$ **b)** $760 \le f \le 1520$ **2. a)** yes
b) no **3. a)** no **b)** yes **4.** grasshopper **5. a)** $1000 \le f \le 20\ 000$
b) $\frac{1}{11}$ **6. a)** the soprano **b)** probably not

Review of Prerequisite Skills, p. 3

1. a) $24t - 30$ **b)** $-4w - 13$ **c)** $-4m + 32$ **d)** $10y - 26$
e) $6x^2 + 4x + 26$ **f)** $-x - 18y$ **g)** $-7x^2 + 4xy + 11y^2$ **2. a)** -3
b) 3 **c)** 5 **d)** 2 **e)** 5 **f)** -4 **g)** 3 **3. a)** $-\frac{3}{2}$ **b)** 19 **c)** $\frac{9}{4}$ **d)** -8 **e)** 5
f) 7 **g)** 10.5 **h)** 9.25 **i)** -27 **j)** 7 **4. a)** $7t^2(1 - 2t)$ **b)** $12x^5(3x^2 + 2)$
c) $2x(2y - z + 5)$ **d)** $4x(2x^2 - 4x + 1)$ **e)** $3xy(3x + 2 - y)$
f) $5a(2ab + b - 3)$ **5. a)** $(x + 3)(x + 4)$ **b)** $(y - 4)(y + 2)$
c) $(d + 5)(d - 2)$ **d)** $(x - 3)(x - 5)$ **e)** $(w - 9)(w + 9)$ **f)** $t(t - 4)$
g) $(y - 5)^2$ **h)** $(x - 8)(x + 5)$ **6. a)** $(x + 3)(2x + 1)$
b) $(x - 1)(2x - 1)$ **c)** $(t - 5)(3t + 4)$ **d)** $(y - 1)(2y - 5)$
e) $(2x + 1)(3x - 1)$ **f)** $(2x + 3)^2$ **g)** $(3a - 4)(3a + 4)$
h) $(2s - 3)(3s + 1)$ **i)** $(u + 2)(2u + 3)$ **j)** $(3x - 1)^2$
k) $(x + 4)(3x - 5)$ **l)** $2v(2v + 5)$ **7. a)** $2, -1$ **b)** $3, -3$ **c)** $7, 0$ **d)** 2
e) $-2, -4$ **f)** no real roots **g)** $2, \frac{1}{2}$ **h)** $-\frac{1}{2}, -3$

Section 1.1, pp. 9–10

1. a) 2^7 **b)** 2^4 **c)** 2^{12} **d)** 2^8 **e)** 2^{3+m} **f)** 2^{7-y} **g)** 2^{x-4} **h)** 2^{xy} **i)** 2^1 **j)** 2^3
k) 2^{-3} **l)** 2^{-4} **2. a)** $\frac{1}{9}$ **b)** 1 **c)** $\frac{1}{16}$ **d)** $\frac{1}{4}$ **e)** -1 **f)** 25 **g)** -4 **h)** $-\frac{1}{64}$
3. a) a^7 **b)** m^8 **c)** b^{12} **d)** a^5b^2 **e)** x^8y^5 **f)** $\frac{1}{x^2}$ **g)** $\frac{1}{m^9}$ **h)** $\frac{1}{y}$ **i)** a^5 **j)** $\frac{1}{ab^2}$
4. a) x^3 **b)** m^6 **c)** t^6 **d)** $\frac{1}{y^2}$ **e)** m^4 **f)** t^5 **5. a)** x^6 **b)** a^8b^{12} **c)** $\frac{1}{x^2}$ **d)** 1
e) $\frac{a^2}{b^4}$ **f)** $\frac{1}{x^6y^9}$ **6. a)** $\frac{x^3}{8}$ **b)** $\frac{a^4}{b^4}$ **c)** $\frac{x^{10}}{y^{15}}$ **d)** $\frac{3}{x}$ **e)** $\frac{a^4}{b^6}$ **7. a)** $15m^6$
b) $-20a^4b^6$ **c)** $30a^2b^5$ **d)** $24m^4n^7$ **e)** 42 **f)** $-\frac{6}{y}$ **g)** $\frac{6}{a^4b^2}$ **h)** $5x^3$
i) $5ab^2$ **j)** $2m^4$ **k)** $\frac{2a^3}{b}$ **l)** $\frac{4a^2}{b^5}$ **m)** $7x^8$ **n)** $\frac{9a^7}{b^4}$ **o)** $-\frac{3n^8}{m^3}$ **p)** $\frac{4y^2}{x^8}$
8. a) $4m^6$ **b)** $-64x^6$ **c)** $9m^6n^4$ **d)** $\frac{c^6}{25d^6}$ **e)** $\frac{a^2b^6}{8}$ **f)** $\frac{y^8}{81x^{12}}$ **g)** $\frac{16x^2}{9y^2}$
h) $-\frac{8a^6}{27y^9}$ **i)** $\frac{81a^4}{b^{16}}$ **j)** $\frac{n^6}{4m^4}$ **k)** $27b^6$ **l)** $\frac{4x^{10}}{y^{12}}$ **9. a)** 3 **b)** 1 **c)** $\frac{3}{8}$
d) 12 **e)** 3 **10.** 12 000 years ago **11. a)** 1024 s **b)** $\frac{1}{32}$ s
12. a) $2^{-2} \times (2^2 + 2^2) - 2^0 = 2^0$ **b)** $(3^{-4} - 3^{-2}) \div (3^0 - 3^2) = 3^{-4}$
13. 20^{100} is greater, since $400^{40} = (20^2)^{40} = 20^{80}$. **14. a)** $\frac{37}{35}$

b) $\frac{12}{65}$ **c)** -1 **d)** 0 **15. a)** none **b)** all **16. a)** 2 **b)** 4 **c)** ± 3 **d)** 5
17. all values except $x = 0$

Section 1.2, pp. 16–18

1. a) $\sqrt[3]{2}$ **b)** $\sqrt{37}$ **c)** $\sqrt{x}$ **d)** $(\sqrt[5]{a})^3$ **e)** $(\sqrt[3]{6})^4$ **f)** $(\sqrt[4]{6})^3$ **g)** $\frac{1}{\sqrt{7}}$ **h)** $\frac{1}{\sqrt[5]{9}}$
i) $\frac{1}{(\sqrt[7]{x})^3}$ **j)** $\frac{1}{(\sqrt[5]{b})^6}$ **k)** $\sqrt{3x}$ **l)** $3\sqrt{x}$ **2. a)** $7^{\frac{1}{2}}$ **b)** $34^{\frac{1}{2}}$ **c)** $(-11)^{\frac{1}{3}}$
d) $a^{\frac{2}{5}}$ **e)** $6^{\frac{4}{3}}$ **f)** $b^{\frac{4}{3}}$ **g)** $x^{-\frac{1}{2}}$ **h)** $a^{-\frac{1}{3}}$ **i)** $x^{-\frac{4}{5}}$ **j)** $2^{\frac{1}{3}}b$ **k)** $3^{\frac{1}{2}}x^2$ **l)** $5^{\frac{5}{4}}t^{\frac{3}{4}}$
3. a) 2 **b)** 5 **c)** $\frac{1}{2}$ **d)** -2 **e)** 5 **f)** $-\frac{1}{3}$ **g)** $\frac{1}{2}$ **h)** 0.2 **i)** 3 **j)** 0.1 **k)** $\frac{2}{3}$
l) $\frac{3}{2}$ **4. a)** 4 **b)** 8 **c)** 243 **d)** 27 **e)** $\frac{1}{8}$ **f)** 4 **g)** $-\frac{1}{32}$ **h)** $\frac{1}{9}$ **i)** 1 **j)** 1
k) $\frac{1000}{27}$ **l)** $\frac{4}{9}$ **5. a)** not possible **b)** 1000 **c)** $\frac{9}{4}$ **d)** 3 **e)** -3 **f)** 4
g) -32 **h)** 4 **i)** not possible **j)** 5 **k)** $\frac{11}{6}$ **l)** 27 **m)** not possible
n) $\frac{50}{9}$ **o)** 5 **p)** 3 **q)** 5 **r)** 2 **s)** 3 **t)** 12 **6. a)** x **b)** x **c)** $3^{\frac{1}{4}}x^{\frac{3}{2}}$
d) $2^{\frac{7}{2}}x^6$ **e)** $3x^2$ **f)** x^2y **g)** a^4b^3 **h)** $-3x^{\frac{1}{3}}$ **i)** $3a^2b$ **j)** $\frac{9x^4}{y^6}$ **k)** $x^{\frac{11}{6}}$
l) $x^{\frac{17}{12}}$ **m)** $x^{\frac{19}{15}}$ **n)** $a^{\frac{4}{3}}b^3$ **o)** $a^{\frac{3}{8}}b^{\frac{5}{8}}$ **7.** Estimates may vary.
a) 2.05 **b)** 21.67 **c)** 0.19 **d)** 1.71 **e)** 0.31 **f)** 2.68
8. 2.7 m/s **9. a)** Expand and simplify the equation
$r^2 + d^2 = (r + h)^2$ to obtain $d = (2rh + h^2)^{\frac{1}{2}}$. **b)** 357 km,
1609 km **10.** 65 km **11. a)** 27.6 m **12.** 277.2, 293.6, 311.1,
329.6, 349.2, 370.0, 392.0, 415.3, 440.0, 466.1, 493.8,
523.2 **13. a)** 5 **b)** 3 **c)** 2, 4, 6, ... **d)** 4 **e)** 2 **f)** 1, 3, 5, ...
14. a) 4, 5 **c)** $2^{\frac{1}{2}}$, $5^{\frac{1}{2}}$ **d)** $2^{\frac{1}{2}}$, $8^{\frac{1}{2}}$; $5^{\frac{1}{2}}$, $10^{\frac{1}{2}}$
e) 11 different-sized squares can be made. $2^{\frac{1}{2}}$, $8^{\frac{1}{2}}$; $5^{\frac{1}{2}}$,
$10^{\frac{1}{2}}$, $17^{\frac{1}{2}}$; $13^{\frac{1}{2}}$

Section 1.3, pp. 23–25

1. a) 4 **b)** 3 **c)** 7 **d)** 3 **e)** 4 **f)** 3 **g)** 3 **h)** 5 **i)** 2 **j)** 4 **k)** 4 **l)** 3 **m)** 4
n) x any even integer **o)** m any odd integer **2. a)** 4 **b)** -1 **c)** -6
d) 1 **e)** 1 **f)** -2 **g)** 1 **h)** 1 **i)** all values of x **3. a)** 3 **b)** 4 **c)** 2 **d)** 4
e) 2 **f)** -2 **4. a)** 0 **b)** -3 **c)** -3 **d)** 2 **e)** 1 **f)** -2 **5. a)** 1 **b)** 2 **c)** 6
d) -3 **e)** 5 **f)** 1 **6. a)** 5 **b)** -2 **c)** 2 **d)** -1 **e)** -2 **f)** 2 **7. a)** $\frac{3}{2}$
b) $\frac{2}{3}$ **c)** $\frac{1}{3}$ **d)** $-\frac{1}{2}$ **e)** $-\frac{1}{3}$ **f)** $-\frac{1}{4}$ **g)** 16 **h)** $-\frac{5}{2}$ **i)** $\frac{1}{4}$ **8. a)** $\frac{1}{2}$ **b)** $\frac{3}{4}$
c) $\frac{1}{2}$ **d)** $-\frac{1}{2}$ **e)** $\frac{2}{3}$ **f)** $-\frac{1}{2}$ **9. a)** 1 **b)** 9 **c)** $\frac{15}{2}$ **d)** -3 **e)** -6 **f)** -4
10. a) 0 **b)** 4 **c)** $\frac{1}{2}$ **d)** $-\frac{3}{2}$ **e)** $\frac{1}{4}$ **f)** -1 **11. a)** 4 **b)** 1 **c)** 5 **d)** -1 **e)** 3
f) -3 **g)** -2 **h)** 4 **i)** 2 **12.** The equation is true for all values of

x, since each term equals 2^{6x+6}. **13.** 6 years **14. a)** 56 years
b) 84 years **c)** 140 years **15. a)** $\frac{1}{8}$ **b)** 26 days **16. a)** 2 m
b) 11% **17. a)** 5 h **b)** 20.4 years **c)** 30 s **18.** 59.6 h **19. a)** -1
b) 1 **c)** 1 **20. a)** 2, -3 **b)** 1, 2 **c)** 4, -3 **21.** 16 days **22. a)** -2
b) 7 **c)** 4 **23.** $x = -17$, $y = 2$

Career Connection: Microbiology, p. 26

1. a) The number of bacteria, N, equals the initial number of
bacteria, N_0, times the doubling factor, 2^{t+7}. **b)** 21 h, 42 h,
56 h **c)** 6 h **d)** $N = 15\ 000(2)^{t+6}$

Technology Extension: Solving Exponential Equations With a Graphing Calculator, p. 27

1. The zeros of the left-hand side of the rewritten equation
occur where the previous two equations intersect. **3. a)** The
lower bound for any interval in which a zero is to be found
is reported as a zero of the function; the graph of the
function coincides with the graph of $y = 0$. **5. a)** There are
no solutions. **b)** Every value of x is a solution.

Section 1.4, pp. 29–34

1. a) $7x^2 + 2x + 1$ **b)** $5t^2 + t - 1$ **c)** $13m^2 + 8mn + 3n^2$
d) $y^2 - 4xy + x^2$ **e)** $9xy + 3x + 4$ **f)** $7x - 2xy - 2y$
2. a) $2x^2 - 12x + 5$ **b)** $-s^2 + 9s - 16$ **c)** $-x^2 - 7xy - 7y^2$
d) $-7r^2 + 5rs - 10s^2$ **e)** $2x + 5y - 4z$ **f)** $3m + 4n - 4$
3. $-x^2 - 7x + 10$ **4.** $5x + 6y - 13$ **5.** $3t^2 - 7t + 13$
6. $-2m^2 + 3m$ **7.** $x - 4y + z$ **8. a)** $6x + 8$ **b)** $-10 + 15x$
c) $8y^2 - 12y$ **d)** $-9m - 6n$ **e)** $8st - 10t^2$ **f)** $8b^2 + 4b - 4$
g) $-2q^2 + 10b + 8$ **h)** $6p^3 - 3p^2 + 12p$ **i)** $-4g - 12g^2 + 12g^3$
9. a) $-x + 7$ **b)** $3y^2 - 32y + 35$ **c)** $13x - 15y + 7$
d) $-20a - 20b + 32c$ **e)** $2x - 29$ **f)** $t - 21$ **g)** $1 - 2s - s^2$
h) $10x^2 + 14x$ **i)** $7a^2 - 12a$ **j)** $-4m^2 + m + 3$ **k)** $-4x^2 + 11x$
l) $20r^2 + 13r$ **10. a)** $12x - 69$ **b)** $18x - 6$ **c)** $22t + 6$
d) $-16y - 38$ **e)** $3x^2 - 8x$ **f)** $-3y^3 + 10y^2 - y - 2$
11. a) $x^2 - x - 42$ **b)** $t^2 + 3t - 40$ **c)** $y^2 - 12y + 27$
d) $12y^2 + 17y - 7$ **e)** $8x^2 + 34x + 21$ **f)** $15 - 14m - 8m^2$
g) $-10x^2 + 76x + 32$ **h)** $12x^2 - 60x + 75$ **i)** $36 - 25x^2$
12. a) $56x^2 - 33xy - 14y^2$ **b)** $6s^2 - 7st - 3t^2$
c) $12x^2 - 55xy + 50y^2$ **d)** $18w^2 + 21wx - 99x^2$
e) $15x^4 - 2x^3 - 8x^2$ **f)** $-3m^4 - 4m^3 + 4m^2$
g) $9x^2 - 24xy + 16y^2$ **h)** $-50x^2 + 72y^2$ **i)** $5 - 5x^2y^2$
13. a) $2x^2 + 2x + 5$ **b)** $-t^2 - 13t - 16$ **c)** $12x^2 + 53x - 54$
d) $-y^2 - 43y + 52$ **e)** $-m^2 - 44m - 51$ **f)** $8x^2 + 65x - 18$
g) $12y^2 - 8y - 102$ **h)** $10t^2 + 40t + 9$ **i)** $x^2 - 14xy + y^2 + 5$
j) $3r^2 + 17rt - 24t^2$ **14. a)** The area of a rectangle with
dimensions $2x + 1$ by $x + 2y + 3$ is found by adding the
areas of the constituent rectangles with dimensions given by
the constituent terms of the polynomials.
b) $2x^2 + 4xy + 7x + 2y + 3$ **15. a)** $x^3 + 5x^2 + 10x + 12$
b) $y^3 - 3y^2 - 3y + 10$ **c)** $6m^3 + 13m^2 - 6m - 8$
d) $2t^3 - 9t^2 - 19t - 7$ **e)** $x^4 + x^3 - 7x^2 - 7x + 4$

f) $y^4 - 4y^3 + 7y^2 - 7y + 2$ **g)** $3a^4 - 7a^3 - 9a^2 + 18a - 10$
h) $3x^6 - 14x^3 - 49$ **i)** $x^4 - 8x^3 + 18x^2 - 8x + 1$
j) $4n^4 - 4n^3 - 3n^2 + 2n + 1$
k) $4a^2 + b^2 + 9c^2 - 4ab - 6bc + 12ca$
l) $2x^4 - 5x^3 + 12x^2 - 11x + 3$ **m)** $-2x^3 - 8x^2 + 11x - 16$
n) $3x^2 + 7y^2 - 13z^2 - 14xy + 2yz - 10zx - x + y + 2z$
o) $6x^3 - 4x^2 - 13x + 5$ **16. a)** $x^3 - 7x - 6$ **b)** $x^3 - 7x - 6$
c) $x^3 - 7x - 6$ **d)** no **17. a)** $8x^3 - 30x^2 + 13x + 15$
b) $2x^3 - 3x^2y - 11xy^2 + 6y^3$
c) $a^2 + b^2 + c^2 + d^2 + 2ab + 2bc + 2cd + 2da + 2ac + 2bd$
18. a) $10x^2 + 18x - 18$ **b)** $2x^3 + 5x^2 - 21x - 36$ **c)** 598 cm²;
748 cm³ **19.** $2xy + x - y - 2$ **20.** No; for example, the
product $(x + y)(x - y) = x^2 - y^2$ is not a trinomial.

21. a) $x^2 - \dfrac{1}{x^2}$ **b)** $y^2 - \dfrac{6}{y^2} + 1$

Section 1.5, pp. 40–43

1. a) $t^2 + 2t - 5$, $t \neq 0$ **b)** $\dfrac{2a+3}{4a}$, $a \neq 0$ **c)** $y(2y^2 + y - 3)$, $y \neq 0$

d) $\dfrac{7n^3 - 2n^2 + 3n + 4}{n}$, $n \neq 0$ **e)** $\dfrac{m - 2n}{n}$, m, $n \neq 0$

f) $\dfrac{y^2}{3x}$; x, $y \neq 0$ **g)** $\dfrac{4}{bc}$, a, b, $c \neq 0$ **h)** $-\dfrac{x}{5y}$, x, y, $z \neq 0$

i) $\dfrac{3(m-4)}{m}$, $m \neq 0$ **2. a)** $\dfrac{x}{x+4}$, $x \neq -4$ **b)** $\dfrac{2t(t+5)}{t-5}$,

$t \neq 0$, 5 **c)** $\dfrac{1}{2x}$, $x \neq 0$, 3 **d)** $\dfrac{m+2}{m+4}$, $m \neq 1$, -4 **e)** $\dfrac{x}{x+4}$, $x \neq -4$

f) $\dfrac{y}{y+2}$, $y \neq 0$, -2 **g)** $\dfrac{2}{x-3}$, $x \neq 0$, 3 **h)** $\dfrac{1}{4x^2 - 3}$, $x \neq 0$, $\pm\dfrac{\sqrt{3}}{2}$

i) $\dfrac{1}{2x - 4y}$, x, $y \neq 0$, $x \neq 2y$ **3. a)** 6, $t \neq 6$ **b)** $\dfrac{m+6}{2m-6}$, $m \neq 3$

c) $\dfrac{5}{3}$, $x \neq 2$ **d)** $\dfrac{a+2}{a-3}$, $a \neq 0$, 3 **e)** $\dfrac{4}{3}$, $x \neq 0$, $-\dfrac{1}{2}$ **f)** $\dfrac{x-1}{x+1}$,

$x \neq 0$, -1 **g)** $\dfrac{4}{5}$, $x \neq -y$ **h)** $\dfrac{2ab + 4b}{3a - 3}$, $a \neq 0$, 1 **i)** $\dfrac{5x}{2y}$, $y \neq 0$, -2

4. a) $\dfrac{1}{m-3}$, $m \neq 2$, 3 **b)** $y + 5$, $y \neq -5$ **c)** $\dfrac{2}{x-9}$, $x \neq 9$, -3

d) $\dfrac{r-2}{5}$, $r \neq -2$ **e)** $\dfrac{a}{a+1}$, $a \neq -1$ **f)** $\dfrac{x+3}{2xy}$, $x \neq 0$, 3, $y \neq 0$

g) $\dfrac{2}{2w+1}$, $w \neq -1$, $-\dfrac{1}{2}$ **h)** $\dfrac{t-2}{2t}$, $t \neq 0$, $\dfrac{2}{3}$ **i)** $\dfrac{2z}{3z-4}$, $z \neq \pm\dfrac{4}{3}$

j) $\dfrac{5x - 2y}{3x}$, $x \neq 0$, $-y$ **5. a)** -1, $y \neq 2$ **b)** -1, $x \neq 3$ **c)** $-\dfrac{1}{4}$, $t \neq \dfrac{1}{2}$

d) $-\dfrac{2}{3}$, $w \neq \dfrac{3}{5}$ **e)** -1, $x \neq \pm 1$ **f)** $-\dfrac{1}{2}$, $y \neq \pm\dfrac{1}{2}$ **6. a)** $\dfrac{x+2}{x+3}$,

$x \neq -2$, -3 **b)** $\dfrac{a+3}{a-5}$, $a \neq 4$, 5 **c)** $\dfrac{m-2}{m+5}$, $m \neq 3$, -5 **d)** $\dfrac{y-3}{y+5}$,

$y \neq \pm 5$ **e)** $\dfrac{x-4}{x-6}$, $x \neq 6$ **f)** $\dfrac{n+1}{n+3}$, $n \neq 2$, -3 **g)** $\dfrac{p+4}{p-4}$, $p \neq \pm 4$

h) $\frac{2t+1}{t-2}$, $t \neq 1, 2$ **i)** $\frac{3v+1}{2v+1}$, $v \neq -\frac{1}{2}, -\frac{3}{2}$ **j)** $\frac{3x-2}{4x+3}$, $x \neq -\frac{3}{4}, \frac{3}{2}$ **k)** $\frac{z-2}{3z-1}$, $z \neq \frac{1}{3}$ **l)** $\frac{m-n}{2m-3n}$, $m \neq -\frac{1}{2}n, \frac{3}{2}n$

7. a) $\frac{x^2+3x+2}{x+1}$ **b)** $x+2$, $x \neq -1$ **c)** 3:2 **8. a)** -1, $x \neq 1$

b) not possible **c)** not possible **d)** 1, $t \neq \frac{7}{3}$ **e)** $\frac{t-s}{s+t}$, $s \neq -t$

f) $\frac{x}{2}$ **9. a)** $x = y$ **b)** $x = -\frac{y}{3}$ **c)** $x = 0$ **d)** $x = 2$ **e)** $x = \pm 1$

f) $x = \pm\frac{3y}{2}$ **10. a)** No; the values differ when $x = 3$, for example. **b)** No; the second expression is not defined when $x = 0$, whereas the first expression is. **c)** Yes; the expressions have the same value for all x. **d)** No; the values differ when $x = 2$, for example. **e)** No; the values differ when $x = 2$, for example. **f)** No; the expressions differ by a factor of -1.

11. $\frac{x^3}{6x^2} = \frac{x}{6}$, $x > 0$ **12.** $\frac{\frac{4}{3}\pi r^3}{4\pi r^2} = \frac{r}{3}$, $r > 0$ **13. a)** $n+1$

b) $(n+1)(n+3)$ **c)** $n+3$ **d)** 13 **e)** 440 **f)** 35 **14.** $\frac{5x}{5x+4}$, $x > 0$

15. $\frac{(x+4)(x-1)}{5x}$, $x > 1$ **16.** Answers will vary. **a)** $\frac{1}{x-1}$

b) $\frac{y+2}{y^2+3y}$ **c)** $\frac{a}{8a^2+2a-3}$ **d)** $\frac{x}{x^3+x^2-3x-3}$ **17. a)** The graphs of the two functions appear to coincide. **b)** The tables of values are the same except at $x = 0$, where one of the functions is undefined. **18. a)** $\frac{rh}{3(r+s)}$ **b)** $r = 6$, $h = 8$, $s = 10$

Section 1.6, pp. 50–52

1. a) $\frac{8y}{3}$, $y \neq 0$ **b)** $-\frac{x}{4}$, $x \neq 0$ **c)** $\frac{1}{9n^3}$, $n \neq 0$ **d)** $-\frac{8m}{3}$

2. a) $\frac{x}{4}$, $x \neq 0$ **b)** $-\frac{y}{2}$, $y \neq 0$ **c)** $-\frac{9m^2}{4}$, $m \neq 0$ **d)** $\frac{20t^2}{9}$, $t \neq 0$

e) $-\frac{32}{9x^3}$, $x \neq 0$ **f)** $-\frac{2}{3r}$, $r \neq 0$ **3. a)** $\frac{4x^2y}{3}$, $x, y \neq 0$

b) $\frac{16m}{5n}$, $m, n \neq 0$ **c)** $\frac{9}{xt^2}$, $x, y, t \neq 0$ **d)** $\frac{1}{2a^3b^3}$, $a, b \neq 0$

e) $\frac{9}{2mt}$, $m, t \neq 0$ **f)** $-\frac{45a}{32c^2}$, $a, b, c \neq 0$ **4. a)** $\frac{2xy^2}{3ab}$, $a, b, x, y \neq 0$

b) $\frac{4nx}{3}$, $m, n, x, y \neq 0$ **c)** $\frac{3x}{4}$, $x, y \neq 0$ **d)** $16a^2$, $a, b \neq 0$

e) $-9x^2y^3$, $x, y \neq 0$ **f)** $-\frac{2ab}{9c}$, $a, b, c \neq 0$ **5. a)** $\frac{1}{2}$, $x \neq 4$

b) $\frac{2m+4}{y+1}$, $y \neq -1$ **c)** $\frac{y-2}{2}$, $y \neq -1$ **d)** 2, $x \neq -1, 2$

e) $-\frac{a}{6b}$, $a \neq -b$, $a, b \neq 0$ **f)** $\frac{9m^2}{5}$, $m \neq 0, -4$ **6. a)** $\frac{8}{5}$, $x \neq \pm 1$

b) $\frac{15}{4}$, $m \neq 0, -3$ **c)** $-\frac{1}{a}$, $a \neq 0, -2$ **d)** $\frac{3(x+2)}{4}$, $x \neq 2, -3$

e) $\frac{2}{y(y-3)}$, $y \neq 0, \pm 3$ **f)** $\frac{2(m+5)}{m-4}$, $m \neq 5, \pm 4$ **g)** $\frac{1}{3x}$;

$x \neq 0, \frac{3}{2}$, $y \neq 0$ **h)** $x+2$, $x \neq -2, -\frac{1}{2}$ **7. a)** $\frac{x+2}{x-1}$,

$x \neq -6, -3, 1, 5$ **b)** $\frac{a+4}{a+2}$, $a \neq -2, \pm 3$ **c)** $\frac{m+1}{m}$,

$m \neq -5, 0, 3, 4$ **d)** $\frac{4a-5}{2a+3}$, $a \neq \frac{1}{3}, \pm\frac{3}{2}$ **e)** 1, $x \neq -\frac{1}{2}, \frac{5}{2}, 3$

f) $\frac{(4w+1)(4w+5)}{(4w+7)(4w+3)}$, $w \neq -\frac{7}{4}, -\frac{5}{4}, -\frac{3}{4}, \frac{2}{3}, \frac{3}{2}$

8. a) $\frac{x+2y}{x-2y}$, $x \neq -4y, \pm 2y, 3y, 5y$ **b)** $\frac{x}{x+6y}$, $x \neq -6y, \pm 3y, 7y$

c) $\frac{a+7b}{a-9b}$, $a \neq -8b, -6b, 2b, 9b$ **d)** $\frac{3s+5t}{5s+t}$,

$s \neq -\frac{5t}{3}, -\frac{t}{5}, \frac{6t}{5}, \frac{5t}{4}$ **9. a)** $\frac{10x^2}{27}$ **b)** $\frac{22x^2}{9}$ **c)** $\frac{33}{5}$

d) No; the answer in c) is independent of x. **10.** $x^2 - 5x + 6$

11. $3y + 2$ **12. a)** $\frac{(6x-9)(2x+4)}{2}$ **b)** $\frac{(2x-3)(3x+6)}{2}$ **c)** 2

13. $\frac{x-2}{3(x-3)}$ **14.** Neither b nor d can be 0 or the fractions will be undefined, and c cannot be 0 since division by 0 is undefined. **15.** $(x+3)(2x-1)$ **16. a)** The product simplifies to $\frac{1}{2y+3}$, and so the value of the product approaches 0 as y increases. **b)** The quotient simplifies to $\frac{3}{2}$, and so the value of the quotient remains at $\frac{3}{2}$ as y increases.

17. $\frac{3x+y}{x-y}, \frac{x+2y}{x+y}; \frac{3x+y}{x+y}, \frac{x+2y}{x-y}$ **18.** $\frac{4}{2x-1}, \frac{(x-1)^2}{x+3}$; $\frac{4(x-1)}{2x-1}, \frac{x-1}{x+3}; \frac{4(x-1)^2}{2x-1}, \frac{1}{x+3}; \frac{4(x-1)^2}{x+3}, \frac{1}{2x-1}$

Section 1.7, pp. 58–61

1. a) $\frac{1}{y}$, $y \neq 0$ **b)** $\frac{8}{x^2}$, $x \neq 0$ **c)** $\frac{9}{x+3}$, $x \neq -3$ **d)** $\frac{x-y}{x-2}$, $x \neq 2$

2. a) $\frac{2x+11}{2}$ **b)** $\frac{5y-7}{3}$ **c)** $\frac{-a-3}{a}$, $a \neq 0$ **d)** $\frac{x-2y}{3x}$, $x \neq 0$

e) $\frac{3x^2+4}{x+1}$, $x \neq -1$ **f)** $\frac{t-5}{7}$ **g)** $\frac{4z+3}{2z-1}$, $z \neq \frac{1}{2}$ **h)** $\frac{5x-1}{x^2-1}$, $x \neq \pm 1$

i) $\frac{7x+3}{x^2+5x+6}$, $x \neq -2, -3$ **j)** $\frac{-7y-2}{2x^2+3x+1}$, $x \neq -1, -\frac{1}{2}$

3. a) 60 **b)** 36 **c)** 120 **d)** 60 **4. a)** $\frac{4x}{3}$ **b)** $\frac{11a}{12}$ **c)** $\frac{2x-5y+7}{10}$

d) $-\frac{11m}{24}$ **5. a)** $\frac{20m+29}{14}$ **b)** $\frac{16x-1}{12}$ **c)** $\frac{-4y-1}{12}$

d) $\dfrac{-16x+11y}{10}$ e) $\dfrac{3t+1}{2}$ f) $\dfrac{-12a+19b}{18}$ g) $\dfrac{10x+39}{30}$

6. a) $\dfrac{-1}{x-2}$, $x \neq 2$ b) $\dfrac{2}{x-1}$, $x \neq 1$ c) $\dfrac{-5}{2a-3}$, $a \neq \dfrac{3}{2}$

d) $\dfrac{-y+1}{4y-3}$, $y \neq \dfrac{3}{4}$ e) $\dfrac{8x-4}{x^2-9}$, $x \neq \pm 3$ f) $\dfrac{3x^2}{4x^2-9}$, $x \neq \pm\dfrac{3}{2}$

7. a) $\dfrac{1191}{s}$ b) $\dfrac{685}{s}$ c) $\dfrac{1876}{s}$ d) 2.68 h 8. a) $x+8$ b) $\dfrac{11x}{5}$

c) 15 cm by 23 cm; 23 cm by 33 cm 9. $x(x+2)$

10. a) $\dfrac{6x^2+5x+1}{2x+1}$ b) $\dfrac{4x^2-4x-3}{2x+1}$ c) $\dfrac{-2x^2-9x-4}{2x+1}$

d) $\dfrac{2x^2+9x+4}{2x+1}$ e) They differ by a factor of -1.

11. a) $\dfrac{\pi d^2}{4}$ b) $\dfrac{\pi(d+1)^2}{4}$ c) $\dfrac{\pi(2d+1)}{4}$ d) 16.5 cm²

12. a) $\dfrac{(2x+1)(x-3)}{16}$ b) $\dfrac{(x-1)(x-3)}{4}$ c) $\dfrac{3(x-3)(2x-1)}{16}$

d) $\dfrac{4x-1}{4}$ e) $\dfrac{3(x-3)(2x-1)}{16}$ f) They are equal.

13. a) $\dfrac{n(n+1)}{2}$ b) 15, 21, 28, 36, 45 c) A perfect square

results. d) $\dfrac{(n+1)(n+2)}{2}$ e) $(n+1)^2$ f) $(n+1)^2$ is a perfect square.

Section 1.8, pp. 67–69

1. a) $\dfrac{24xy}{12x^2y^2}$, $x, y \neq 0$ b) $\dfrac{12x^3y}{12x^2y^2}$, $x, y \neq 0$ c) $\dfrac{20x}{12x^2y^2}$, $x, y \neq 0$

d) $\dfrac{-2y^3}{12x^2y^2}$, $x, y \neq 0$ 2. a) $20a^2b^3$ b) $6m^2n^2$ c) $12x^3y^2$ d) $60s^2t^2$

3. a) $\dfrac{23}{10x}$, $x \neq 0$ b) $\dfrac{1}{y}$, $y \neq 0$ c) $\dfrac{2x^2+x-4}{2x^3}$, $x \neq 0$

d) $\dfrac{15n^2+8mn^2-10}{10m^2n^3}$, $m, n \neq 0$ e) $\dfrac{x^2+5x-2}{x}$, $x \neq 0$

f) $\dfrac{3m-2mn-n+4}{mn}$, $m, n \neq 0$ g) $\dfrac{-30x^2-10x+1}{15x^2}$, $x \neq 0$

h) $\dfrac{-3x^2-2y^2+5xy-4x-y}{xy}$, $x, y \neq 0$ 4. a) $6(m+2)$

b) $15(y-1)(y+2)$ c) $12(m-2)(m-3)$ d) $20(2x-3)$

5. a) $\dfrac{21}{4(x+3)}$, $x \neq -3$ b) $\dfrac{-5}{3(y-5)}$, $y \neq 5$ c) $\dfrac{t}{3(t-4)}$, $t \neq 4$

d) $\dfrac{8}{3(m+1)}$, $m \neq -1$ e) $\dfrac{1}{12(y-2)}$, $y \neq 2$ f) $\dfrac{11}{6(2a+1)}$, $a \neq -\dfrac{1}{2}$

6. a) $\dfrac{5x+7}{(x+1)(x+2)}$, $x \neq -1, -2$ b) $\dfrac{m^2-3m+15}{(m-3)(m+2)}$, $m \neq -2, 3$

c) $\dfrac{8x-3}{x(x-1)}$, $x \neq 0, 1$ d) $\dfrac{11t-1}{5(t-1)}$, $t \neq 1$ e) $\dfrac{10x-x^2}{(x-2)(x+2)}$,

$x \neq \pm 2$ f) $\dfrac{15-n}{(3n-1)(2n+3)}$, $n \neq -\dfrac{3}{2}, \dfrac{1}{3}$ g) $\dfrac{5x-7}{4(x-1)(x-2)}$,

$x \neq 1, 2$ h) $\dfrac{2t^2-9t-5}{6(t+5)(t-4)}$, $t \neq -5, 4$ i) $\dfrac{-s^2+16s-10}{5(s-6)(s-1)}$, $s \neq 1, 6$

j) $\dfrac{11m^2-31m}{12(m-5)(m-2)}$, $m \neq 2, 5$ 7. a) $(x+2)^2$

b) $(y-2)(y+2)(y+4)$ c) $(t+3)(t-4)(t+1)$

d) $2(x-2)(x+1)(x-4)$ e) $(m+3)^2(m-5)$

8. a) $\dfrac{2x+7}{(x+3)(x+2)}$, $x \neq -3, -2$ b) $\dfrac{-3y+16}{(y-4)(y+4)}$, $y \neq \pm 4$

c) $\dfrac{3x^2+5x}{(x-5)(x+1)}$, $x \neq -1, 5$ d) $\dfrac{9a-2a^2}{(a-3)(a-4)}$, $a \neq 3, 4$

e) $\dfrac{2x+6}{(2x+1)(x+1)}$, $x \neq -1, -\dfrac{1}{2}$ f) $\dfrac{18n-9}{(2n-1)(3n-1)}$,

$n \neq \dfrac{1}{3}, \dfrac{1}{2}$ 9. a) $\dfrac{3}{(m+1)(m+4)}$, $m \neq -4, -3, -1$

b) $\dfrac{-2x-8}{(x+2)^2(x-2)}$, $x \neq \pm 2$ c) $\dfrac{a^2-6a-10}{(a-5)(a+5)(a-4)}$, $a \neq \pm 5, 4$

d) $\dfrac{6m^2-26m}{(m-3)(m-6)(m-5)}$, $m \neq 3, 5, 6$

e) $\dfrac{7x-3}{(3x+1)(x+1)(x-1)}$, $x \neq -\dfrac{1}{3}, \pm 1$

f) $\dfrac{2y^2-15y}{(2y-3)^2(2y+3)}$, $y \neq \pm\dfrac{3}{2}$ 10. a) $\dfrac{t^2-3t-2}{(t-1)(t-4)}$, $t \neq 1, 4$

b) $\dfrac{y^2+4y+1}{(y-1)(y+2)}$, $y \neq -2, 1$ c) $\dfrac{-2x^2-2x-3}{(x+3)(x+1)}$, $x \neq -3, -1$

d) $\dfrac{4n+7}{3(n+4)(n-4)}$, $n \neq \pm 4$ e) $\dfrac{-4}{(m+3)(m-4)(m-1)}$,

$m \neq -3, 1, 4$ f) $\dfrac{5a+1}{(a-1)(a+1)^2}$, $a \neq \pm 1$

g) $\dfrac{5w^2-11w+5}{(w+1)(w+4)(w-2)}$, $w \neq -4, -1, 2$

h) $\dfrac{10x^2+3x}{(2x+1)(x+1)(3x+1)}$, $x \neq -1, -\dfrac{1}{2}, -\dfrac{1}{3}$

i) $\dfrac{-20z-26}{(2z-5)(2z+5)(2z+1)}$, $z \neq \pm\dfrac{5}{2}, -\dfrac{1}{2}$

11. a)

Expressions	Product	LCM	GCF	LCM × GCF
$3x, 5x$	$15x^2$	$15x$	x	$15x^2$
$12, 8$	96	24	4	96
$15y^2, 9y$	$135y^3$	$45y^2$	$3y$	$135y^3$
$a+1, a-1$	a^2-1	a^2-1 for even a, $\dfrac{a^2-1}{2}$ for odd a	1 for even a, 2 for odd a	a^2-1
$2t-2, 3t-3$	$6(t-1)^2$	$6(t-1)$	$t-1$	$6(t-1)^2$

b) They are equal. c) Any factors common to both terms appear once each in the LCM and the GCF. All other factors appear once in the LCM. Thus the product LCM × GCF contains all factors of both expressions and so is equal to the product of both expressions. **12. a)** $\dfrac{45}{s}$ **b)** $\dfrac{45}{2s}$ **c)** $\dfrac{135}{2s}$

d) 6.75 h **13.** Answers will vary. **14. a)** $\dfrac{6m+3}{m+1}$, $m \neq -1, -2$

b) $\dfrac{-2x}{x+5}$, $x \neq -5$, 1 **c)** $\dfrac{2x-2}{x-4}$, $x \neq 3, 4, -1$ **d)** $\dfrac{2y-7}{y-4}$, $y \neq 2, 4$

e) $\dfrac{2z^2+3z-9}{2(z+1)}$, $z \neq \pm1, -\dfrac{3}{2}$ **f)** $\dfrac{x^2-x+18}{(x-2)(x+3)}$, $x \neq 1, 2, -3$

15. Answers may vary. **a)** $\dfrac{2}{x+2} + \dfrac{3}{x+1}$ **b)** $\dfrac{1}{2x-3} + \dfrac{1}{3x-2}$

c) $\dfrac{x}{x-3} + \dfrac{1}{x-1}$ **d)** $\dfrac{4x+3}{2(2x+3)} + \dfrac{3}{2(2x-3)}$

Technology Extension: Radical Expressions and the Graphing Calculator, pp. 70–71

1. a) $\dfrac{1}{2x^3y^2}$ **b)** $\dfrac{-3a^5}{c^2}$ **c)** $\dfrac{-5p^5}{2qrs^4}$ **2. a)** $\dfrac{2x^2-3x+4}{x}$

b) $\dfrac{t}{4t^2+2t-1}$ **c)** $\dfrac{m+5}{3-m}$ **d)** $\dfrac{4x^2+3x-2}{5x^2+x+2}$ **3. a)** $\dfrac{x+5}{x-5}$

b) $\dfrac{x+y}{x-y}$ **c)** $\dfrac{3n-1}{2n+5}$ **d)** $\dfrac{5m-6}{4m-5}$ **e)** $\dfrac{x^2+2}{x^2-2}$ **f)** $\dfrac{a(a+b)}{3a+2b}$

4. a) $\dfrac{-2n}{5m}$ **b)** $\dfrac{2b}{5x}$ **c)** $\dfrac{1}{6a}$ **d)** $\dfrac{x+1}{x-2}$ **e)** $\dfrac{3x-2}{4x+3}$ **f)** $\dfrac{(4p-1)(2p-1)}{(3p+1)(3p+2)}$

5. a) $\dfrac{s}{6t}$ **b)** $\dfrac{-9my^2z}{2n}$ **c)** $\dfrac{4}{9}$ **d)** $\dfrac{(y-6)(y+5)}{(y-2)(y+4)}$ **e)** $\dfrac{3m-1}{2(3m-2)}$

f) $\dfrac{(4x+1)(3x+1)}{(3x+2)(5x+3)}$ **6. a)** $\dfrac{4q}{3} + \dfrac{1}{12}$ **b)** $\dfrac{16q+1}{12}$

7. a) $\dfrac{24x-13}{10}$ **b)** $\dfrac{2y^2+5y+3}{y^3}$ **c)** $\dfrac{25}{12(n-3)}$ **d)** $\dfrac{t}{2(t-1)}$

e) $\dfrac{5n^2+7n-2}{(n+2)^2(n-2)}$ **f)** $\dfrac{3c^2-2c+3}{2(2c+1)(c-1)(c+1)}$ **8. a)** $\dfrac{q}{3} - \dfrac{5}{12}$

b) $\dfrac{4q-5}{12}$ **9. a)** $\dfrac{z+13}{30}$ **b)** $\dfrac{9x-2}{6x^2}$ **c)** $\dfrac{4}{3(2r+3)}$

d) $\dfrac{x^2-17x}{(3x+4)(2x-1)}$ **e)** $\dfrac{-y^2}{(y-2)(y-1)(y-3)}$

f) $\dfrac{t^2-21t+21}{(3t+1)^2(2t-7)}$ **10. a)** $\dfrac{(x+3)(x-1)}{12}$ **b)** $\dfrac{7x+5}{6}$

c) $\dfrac{2(7x+5)}{(x+3)(x-1)}$ **d)** No; one of the side lengths of the "rectangle" would be 0, and the area would be 0.

Section 1.9, pp. 78–81

1. a) $y < 2$ **b)** $w > -1$ **c)** $x \geq 3$ **d)** $z \leq -3$ **e)** $x > -2$ **f)** $t > -4$

g) $m \leq 3$ **h)** $n \geq 0$ **2. a)** $x > \dfrac{1}{2}$ **b)** $x < -\dfrac{2}{3}$ **c)** $y \leq -1$ **d)** $z \geq 5$

e) $x < 2$ **f)** $x > 1$ **g)** $x \leq 0$ **h)** $x \geq -\dfrac{1}{4}$ **3. a)** $x \leq 3$ **b)** $x > 2$

c) $x < -2$ **d)** $x > 1$ **e)** $y \geq -4$ **f)** $z \leq \dfrac{5}{4}$ **g)** $x > 4$ **h)** $x < \dfrac{5}{3}$

4. a) $x < 1$ **b)** $x > 1$ **c)** $y \leq -4$ **d)** $c \geq 2$ **e)** $x \leq 1$ **f)** $x > \dfrac{1}{3}$

g) $x \geq \dfrac{3}{2}$ **h)** $t < -\dfrac{3}{5}$ **5. a)** $y < -3$ **b)** $w > 2$ **c)** $x \geq \dfrac{3}{2}$ **d)** $z \leq -8$

e) $x > 3$ **f)** $x < -6$ **g)** $q \geq 2$ **h)** $n \geq 3$ **6. a)** $a > -2$ **b)** $x \geq 1$

c) $y < -1$ **d)** $n \geq -2$ **e)** $x > -\dfrac{1}{2}$ **f)** $x < 0$ **7. a)** $x < 1$ **b)** $x \leq \dfrac{3}{4}$

c) $z > 6$ **d)** $x \geq 1$ **8.** 8 **9.** $16 < x < 34$

10. a) $C = 1.55n + 12.25$ **b)** $12.25 + 1.55n \leq 20$; 5

11. $22.5 \leq x < 45$ **12. a)** At the end of a week in which he works t hours, Mario has $15t - 75$ dollars.

b) $15t - 75 \geq 450$; 35 **13.** 63 **14.** The population of Aylmer will exceed the population of Paris at the beginning of the year 2021. **15. a)** $x > 7$ **b)** $x < 9$ **c)** $x > \dfrac{7}{3}$, because the area has to be greater than 0. **16.** $\dfrac{5}{2} \leq x \leq 3$ **17.** $x \neq 1$ **18. a)** \$63

b) The net profit is expected to be between \$90 000 and \$121 500. **19.** Answers may vary. **20.** Between 15:00 and 17:45, Hakim was farther from Hamilton than Jason was. **21. a)** $2 = -2$ **b)** No real values of x satisfy the equation. **c)** $2 > -2$ **d)** All real values of x satisfy the inequality.

Technology Extension: Solving Inequalities With a Graphing Calculator, p. 82

1. $x > 3$ **8. a)** $x \geq 2$ **b)** $x < 3$ **c)** $x > 1$ **d)** $x \leq -3$ **e)** $x \geq 1$

f) $x \leq 4$ **g)** $x \geq -2$ **h)** $x < -3$ **i)** $x \leq 3$ **j)** $x > 2$ **k)** $x < -1$ **l)** $x \geq 3$

m) $x \leq -\dfrac{1}{2}$ **n)** $x > \dfrac{1}{3}$

Review of Key Concepts, pp. 85–89

1. a) $\dfrac{1}{25}$ **b)** 1 **c)** $\dfrac{1}{27}$ **d)** $\dfrac{1}{81}$ **e)** $\dfrac{1}{25}$ **f)** -3 **g)** 16 **h)** 4 **i)** $\dfrac{2}{3}$ **2. a)** m^7

b) $\dfrac{1}{y^5}$ **c)** t^3 **d)** $\dfrac{1}{m^5}$ **e)** x^8y^{12} **f)** 1 **g)** $\dfrac{x^4}{y^6}$ **h)** $\dfrac{m^{12}}{n^8}$ **i)** $\dfrac{x^6}{y^4}$ **3. a)** $10x^5y^7$

b) $9ab$ **c)** $12m^4$ **d)** $-5x$ **e)** $4a^{10}b^6$ **f)** $\dfrac{-m^9n^3}{27}$ **g)** $\dfrac{27m^6}{8n^9}$ **h)** $\dfrac{9x^6}{4y^8}$

i) $-2x^3y^3$ **j)** $\dfrac{6b^2}{5a}$ **k)** $\dfrac{-5t^6}{2s^7}$ **l)** $\dfrac{a^8b^4}{9}$ **4. a)** $\sqrt{6}$ **b)** $\dfrac{1}{\sqrt{5}}$ **c)** $(\sqrt[5]{7})^3$

d) $\dfrac{1}{(\sqrt[3]{10})^4}$ **5. a)** $(-8)^{\frac{1}{3}}$ **b)** $m^{\frac{5}{3}}$ **c)** $x^{\frac{2}{3}}$ **d)** $2^{\frac{1}{5}}a^{\frac{2}{5}}$ **6. a)** 5 **b)** $\dfrac{1}{3}$ **c)** $\dfrac{1}{7}$

d) 1 **e)** 0.3 **f)** $-\dfrac{1}{2}$ **g)** 5 **h)** 9 **i)** $-\dfrac{1}{8}$ **j)** $\dfrac{243}{32}$ **k)** $\dfrac{1}{243}$ **l)** $\dfrac{25}{9}$ **m)** 16

n) -2 **o)** 2 **7. a)** $y^{\frac{2}{3}}$ **b)** $3m^2$ **c)** $-2x^{\frac{1}{3}}$ **d)** x^2 **e)** $-4x$ **f)** $-4x^{\frac{1}{3}}$

8. $\dfrac{4}{25}$ cubic units **9. a)** 6 **b)** 3 **c)** 4 **d)** 3 **e)** -2 **f)** $\dfrac{1}{4}$ **g)** 0 **h)** 3

10. a) 6 **b)** 6 **c)** -7 **d)** $\dfrac{1}{3}$ **e)** $\dfrac{5}{2}$ **f)** $-\dfrac{7}{3}$ **11. a)** 5 **b)** 4 **c)** 2 **12.** 30 h

13. a) $13x^2 - x - 5$ **b)** $6x^2 - 3xy - 4y^2$ **14. a)** $-2y^2 + y - 4$

b) $-2m^2 - 2mn + n^2$ **15. a)** $7x - 1$ **b)** $11s - 23t + 5$

c) $-x^2 - 2x$ **d)** $y^2 + 14y$ **16. a)** $-18y + 34$

b) $-2x^3 + 3x^2 + 24x - 3$ **17. a)** $y^2 - 17y + 72$

b) $-6x^2 - 10x + 56$ **c)** $27x^2 - 18x + 3$ **d)** $8x^2 - 14xy - 15y^2$

18. a) $2m^2 - 6m - 7$ **b)** $15x^2 + 64$ **c)** $60y^2 - 10y + 27$

d) $-2x^2 + 22xy - 14y^2$ **19. a)** $x^3 - 6x^2 + 11x - 6$

b) $6t^3 + t^2 - 3t - 1$ **c)** $x^4 + x^3 - 5x - 3$

d) $6z^4 + 2z^3 - 11z^2 + 8z - 3$

20. $(2x+3)(x+1) - (2x+1)(x-1) = 6x+4$ **21. a)** $\frac{x}{x+3}$,
$x \neq -3$ **b)** $\frac{4y-5x}{2}$, $y \neq 0$ **c)** $\frac{5}{7}$, $x \neq y$ **d)** -2, $x \neq \frac{5}{3}$ **e)** $\frac{1}{w-4}$,
$w \neq 0, 4$ **f)** $\frac{3}{4}$, $m \neq 0, 1$ **g)** $\frac{1}{t-1}$, $t \neq 1, 2$ **h)** $2a+3$, $a \neq 5$
i) $\frac{y+3}{y+4}$, $y \neq 3, -4$ **j)** $\frac{2n-3}{4n+1}$, $n \neq -\frac{1}{3}, -\frac{1}{4}$
22. a) $\frac{2x^2+4x+2}{x+1} = 2(x+1)$ **b)** $2(x+1):(x+1) = 2:1$

23. a) $\frac{4x}{3}$, $x, y \neq 0$ **b)** $-2a^2b$, $a, b \neq 0$ **c)** $\frac{5ax}{8b}$, $a, b, x, y \neq 0$
d) $\frac{b}{6x^2}$, $b, x, y \neq 0$ **e)** $\frac{5}{4}$, $x \neq \pm 1$ **f)** $\frac{14(m+2)}{3(m+3)}$, $m \neq -3, n \neq 1$
g) $\frac{(t+2)(t-3)}{3}$, $t \neq \pm 2, 3$ **h)** $\frac{(x+2)(2x+1)}{(x-1)(x-2)}$, $x \neq 1, 2, 3, \frac{1}{2}$
i) $\frac{(2y-1)(4y-1)(y+1)}{(3y-2)(4y+1)(y-1)}$, $y \neq \pm 1, \frac{1}{3}, \frac{2}{3}, \pm\frac{1}{4}$

24. a) $\frac{2t^2-3t+1}{2t-1}$ **b)** $\frac{3t^2-2t-1}{3t+1}$
c) $\frac{2t^2-3t+1}{2t-1} \times \frac{3t^2-2t-1}{3t+1} = (t-1)^2$, $t \neq -\frac{1}{3}, \frac{1}{2}$
d) It is a square; the length and the width both equal $t-1$.

25. a) $\frac{-2}{x}$, $x \neq 0$ **b)** $\frac{5m-4}{m-2}$, $m \neq 2$ **c)** $\frac{z-2}{z^2}$, $z \neq 0$ **d)** $\frac{t}{12}$
e) $\frac{26x-1}{20}$ **f)** $\frac{-5a}{12}$ **g)** $\frac{2}{2y-5}$, $y \neq \frac{5}{2}$ **h)** $\frac{2x^2+3}{x^2-4}$, $x \neq \pm 2$
i) $\frac{1}{x+2}$, $x \neq -1, -2$ **26.** $\frac{3x+1}{4}$ **27. a)** $\frac{y+4}{y^2}$, $y \neq 0$
b) $\frac{4y^2-5xy+2x^2}{x^2y^2}$, $x, y \neq 0$ **c)** $\frac{3a+4}{6(a-1)}$, $a \neq 1$
d) $\frac{-2(x+5)}{(x+3)(x+1)}$, $x \neq -1, -3$
e) $\frac{t-3}{(t-1)(t+1)(t+2)}$, $t \neq \pm 1, -2$ **f)** $\frac{3(x-1)}{(3x+1)(x-2)}$,
$x \neq -1, -\frac{1}{3}, 2$ **28. a)** $y < 6$ **b)** $w > 2$ **c)** $x \geq -1$ **d)** $z \leq 2$
e) $k > -2$ **f)** $t > -8$ **g)** $m \leq 4$ **h)** $n \geq -4$ **29. a)** $x > -4$ **b)** $y < 0$
c) $m \leq -1$ **d)** $z > 11$ **e)** $b > 8$ **f)** $q > 1$ **g)** $b \leq 2$ **h)** $n \leq 4$
30. a) $m \leq 3$ **b)** $w < 2$ **c)** $x < 17$ **d)** $z \leq 1$ **e)** $y \leq -2$ **f)** $n < 3$
31. a) $x > 8$ **b)** $w \leq -15$ **c)** $m < 2$ **d)** $p \leq -2$ **e)** $x > 0$ **f)** $w \leq 5$
g) $y > -2$ **h)** $k \leq 4$ **32. a)** $x \geq 7$ **b)** $x < 9$ **33.** 256 or more

Chapter Test, pp. 90–91

1. a) $\frac{1}{25}$ **b)** 100 **c)** $\frac{3}{10}$ **2. a)** s^6t^9 **b)** $-6a^3b$ **c)** $9a^4b^{10}$ **d)** $-5n^5$
e) $\frac{-s^2t^7}{16}$ **3. a)** $-\frac{1}{1000}$ **b)** $\frac{1}{27}$ **c)** $\frac{9}{4}$ **4.** 7 square units **5. a)** 4

b) 9 **c)** -2 **d)** $\frac{5}{2}$ **e)** -2 **f)** 0 **6. a)** -3 **b)** -4 **c)** -2 **d)** 1 **e)** 4
7. a) $9x^2 - 3x - 18$ **b)** $-4y^2 - 12y + 2$ **8. a)** $-5t^2 - 31t$ **b)** $-19w$
c) $x^2 + 6x - 55$ **d)** $6x^2 - 21xy + 9y^2$ **e)** $-8s^2 - 24st - 18t^2$
f) $-4x^2 - 19x + 16$ **g)** $9x^2 + 2xy - 26y^2$ **9. a)** $\frac{3}{5}$, $x \neq y$
b) $\frac{2}{3}$, $y \neq 0, -2$ **c)** $\frac{t+4}{t+3}$, $t \neq -3, 4$ **d)** $\frac{2m+3}{3m+5}$, $m \neq 1, -\frac{5}{3}$
10. a) $\frac{x-1}{x+2}$, $x \neq -4, -3, \pm 2$ **b)** $\frac{(a-1)(a-3)}{(a+1)(a-2)}$,
$a \neq -1, -\frac{1}{2}, \frac{2}{3}, 2, 3$ **c)** $\frac{-8n-1}{12}$ **d)** $\frac{5}{2x-3}$, $x \neq \frac{3}{2}$
e) $\frac{-x-20}{(x+1)(x+4)(x-4)}$, $x \neq \pm 4, -1$ **11. a)** $z \geq -8$ **b)** $x > -1$
c) $z \leq -3$ **d)** $y \leq 12$ **e)** $h < 8$ **f)** $y < 5$ **12. a)** $\frac{1}{4}$ **b)** 15.9 years
13. a) $\frac{13x+1}{3}$ **b)** 2, 5, 8 **14. a)** $\frac{2ab}{a+b}$, $\frac{3abc}{ab+bc+ca}$,
$\frac{4abcd}{abc+bcd+cda+dab}$ **b)** $\frac{1260}{337}$

Challenge Problems, p. 92

1. e) $\sqrt[8]{x^7}$ **2. d)** 11 **3. c)** $27a^2$ **4. d)** $x \leq -1$ or $x > 2$ **5. b)** 2
6. 6 **7.** 2 **8.** (2, 5), (3, 5) **9.** $xyz = 7^4$ or 2401

Problem Solving Strategy: Model and Communicate Solutions, p. 94

1. 22 000 **2. a)** yes **b)** $400 **3.** 384 **4.** 400 cm²

Problem Solving: Using the Strategies, p. 95

1. Monday, Tuesday, Wednesday **2.** 400 **3.** 9, 81
4. 512 cm² **5.** 9 **8.** D = 1, F = 9 **10.** Answers may vary.
11. F = 14, E = 10 **12.** 9 km

Chapter 2

Getting Started: Store Profits, p. 98

1. $225 **2.** 25 **3.** 11 **4.** 39 **5.** 20, 21, 22, 23, 24, 25, 26, 27, 28, 29, and 30 **6. a)** no **b)** The company should hire 29 staff members. **c)** $315 172

Review of Prerequisite Skills, p. 99

1. a) 11 **b)** 15 **c)** 0.3 **d)** 1.3 **e)** 0.04 **f)** 4 **g)** 13 **h)** 20 **i)** 7 **j)** 10
2. a) vertex: $(0, -8)$, axis of symmetry: $x = 0$, minimum value: -8, y-intercept: -8, x-intercepts: $2, -2$ **b)** vertex: $(0, 6)$, axis of symmetry: $x = 0$, maximum value: 6, y-intercept: 6, x-intercepts: $\sqrt{2}, -\sqrt{2}$ **c)** vertex: (2, 3), axis of symmetry: $x = 2$, minimum value: 3, y-intercept: 7 **d)** vertex: $(-1, 8)$, axis of symmetry: $x = -1$, maximum value: 8, y-intercept: 6, x-intercepts: $1, -3$ **3. a)** $-2, 3$ **b)** $-4, -1$ **c)** $-2, 2$

d) 0, 6 **4. a)** 3, −4 **b)** 5 **c)** −3, −5 **d)** 8, −4 **e)** 3, $-\frac{1}{2}$ **f)** $\frac{1}{3}$

g) $\frac{3}{2}$, $-\frac{4}{3}$ **h)** 4, $-\frac{3}{2}$ **i)** $-\frac{3}{5}$, −1 **j)** 4, $-\frac{1}{4}$ **k)** $\frac{3}{2}$, 0 **l)** $\frac{5}{3}$, $-\frac{5}{3}$

5. a) −2, −4 **b)** 5, −3 **c)** 1, $-\frac{3}{4}$ **d)** $\frac{3}{2}$, −1 **e)** $\frac{2}{3}$, $-\frac{3}{2}$ **f)** $\frac{4}{3}$, $-\frac{5}{2}$

6. a) $\frac{3+\sqrt{5}}{2}$, $\frac{3-\sqrt{5}}{2}$; 2.62, 0.38 **b)** $\frac{-3+\sqrt{21}}{2}$, $\frac{-3-\sqrt{21}}{2}$;

0.79, −3.79 **c)** $\frac{1+\sqrt{41}}{4}$, $\frac{1-\sqrt{41}}{4}$; 1.85, −1.35

d) $\frac{1+\sqrt{13}}{6}$, $\frac{1-\sqrt{13}}{6}$; 0.77, −0.43 **e)** $\frac{-7+\sqrt{33}}{8}$, $\frac{-7-\sqrt{33}}{8}$;

−0.16, −1.59 **f)** $\frac{1+\sqrt{5}}{2}$, $\frac{1-\sqrt{5}}{2}$; 1.62, −0.62

g) $\frac{-5+\sqrt{17}}{4}$, $\frac{-5-\sqrt{17}}{4}$; −0.22, −2.28 **h)** $\frac{3+\sqrt{69}}{10}$, $\frac{3-\sqrt{69}}{10}$;

1.13, −0.53 **7. a)** 25 **b)** 36 **c)** 1 **d)** 16 **e)** 49 **f)** 4 **g)** 225 **h)** 81

8. a) $y=(x+1)^2-6$; minimum of −6 at $x=-1$
b) $y=(x-2)^2+2$; minimum of 2 at $x=2$ **c)** $y=(x+3)^2-7$;
minimum of −7 at $x=-3$ **d)** $y=-(x-4)^2+10$; maximum
of 10 at $x=4$ **e)** $y=-(x+3)^2+12$; maximum of 12 at
$x=-3$ **f)** $y=-(x-1)^2-4$; maximum of −4 at $x=1$
g) $y=(x+5)^2-25$; minimum of −25 at $x=-5$
h) $y=-(x-2)^2+5$; maximum of 5 at $x=2$
9. a) $y=2(x-2)^2-5$; minimum of −5 at $x=2$
b) $y=3(x+1)^2-10$; minimum of −10 at $x=-1$
c) $y=-2(x+3)^2+9$; maximum of 9 at $x=-3$
d) $y=-4(x-1)^2+2$; maximum of 2 at $x=1$
e) $y=2(x-5)^2-39$; minimum of −39 at $x=5$
f) $y=-3(x-3)^2+32$; maximum of 32 at $x=3$
g) $y=6(x-1)^2-6$; minimum of −6 at $x=1$
h) $y=-5(x+2)^2+22$; maximum of 22 at $x=-2$

Section 2.1, pp. 106–109

1. a) $2\sqrt{3}$ **b)** $2\sqrt{5}$ **c)** $3\sqrt{5}$ **d)** $5\sqrt{2}$ **e)** $2\sqrt{6}$ **f)** $3\sqrt{7}$ **g)** $10\sqrt{2}$ **h)** $4\sqrt{2}$
i) $2\sqrt{11}$ **j)** $2\sqrt{15}$ **k)** $3\sqrt{2}$ **l)** $3\sqrt{6}$ **m)** $8\sqrt{2}$ **n)** $3\sqrt{10}$ **o)** $5\sqrt{5}$ **2. a)** $\sqrt{2}$
b) $\sqrt{5}$ **c)** $2\sqrt{5}$ **d)** $2\sqrt{2}$ **e)** $\sqrt{11}$ **f)** $\frac{\sqrt{7}}{2}$ **g)** $\frac{2\sqrt{5}}{3}$ **h)** 6 **i)** $9\sqrt{3}$ **j)** 15 **k)** 2

l) $\frac{2}{3}$ **3. a)** $2\sqrt{5}$ **b)** $3\sqrt{2}$ **c)** $5\sqrt{3}$ **d)** $\sqrt{77}$ **e)** $4\sqrt{21}$ **f)** 54 **g)** $12\sqrt{3}$
h) $30\sqrt{2}$ **i)** $36\sqrt{5}$ **j)** $56\sqrt{2}$ **k)** 6 **l)** 42 **4. a)** $2+3\sqrt{5}$ **b)** $3-\sqrt{6}$
c) $3+\sqrt{2}$ **d)** $4-\sqrt{3}$ **e)** $-2-\sqrt{2}$ **f)** $-3+\sqrt{3}$ **5. a)** $3i$ **b)** $5i$ **c)** $9i$
d) $i\sqrt{5}$ **e)** $i\sqrt{13}$ **f)** $i\sqrt{23}$ **g)** $2i\sqrt{3}$ **h)** $2i\sqrt{10}$ **i)** $3i\sqrt{6}$ **j)** $-2i\sqrt{5}$ **k)** $-2i\sqrt{5}$
l) $-2i\sqrt{15}$ **6. a)** −25 **b)** −6 **c)** −4 **d)** −12 **e)** 10 **f)** 18 **7. a)** $-i$
b) 1 **c)** id **d)** −20 **e)** −5 **f)** ig **g)** −12 **h)** 64 **i)** 18 **j)** −2 **k)** 5 **l)** 6
m) −12 **n)** −50 **o)** 40 **8. a)** $4+2i\sqrt{5}$ **b)** $7-3i\sqrt{2}$ **c)** $10+5i\sqrt{3}$
d) $11-3i\sqrt{7}$ **e)** $-2-3i\sqrt{10}$ **f)** $-6-2i\sqrt{13}$ **9. a)** $3+4i\sqrt{5}$
b) $1-2i\sqrt{6}$ **c)** $5-2id$ **d)** $4+i\sqrt{3}$ **e)** $-2+i\sqrt{2}$ **f)** $-3-i\sqrt{2}$
10. a) 5 **b)** 5 **c)** 5 **d)** −5 **e)** −5 **f)** −5 **11. a)** irrational, real,
complex **b)** pure imaginary, imaginary, complex **c)** irrational,

real, complex **d)** imaginary, complex **12.** $\frac{\sqrt{15}}{2}$ **13.** $15\sqrt{3}$ cm

14. a) 100 **b)** 144 **15. a)** $-1, -i, 1, i, -1, -i, 1, i, -1, -i, 1$
b) The values $-1, -i, 1, i$ repeat. **c)** Write n as $4k+j$, where
k and j are integers and $0 \le j \le 3$. Then, if $j=0$, $i^n=1$; if
$j=1$, $i^n=i$; if $j=2$, $i^n=-1$; and if $j=3$, $i^n=-i$. **d)** $1, -1, i,$
$-i$ **16. a)** $\sqrt{5}$; $2\sqrt{5}$; $3\sqrt{5}$ **b)** The length of the diagonal is the
product of $\sqrt{5}$ and the width. The length of the diagonal is
one half the product of $\sqrt{5}$ and the length. **c)** $75\sqrt{5}$ cm
d) The length of the diagonal is the product of $\sqrt{5}$ and the
square root of half the area. **e)** $110\sqrt{5}$ cm **17.** Yes; $x=-i\sqrt{3}$ is
a solution, since $(-i\sqrt{3})^2+3=-3+3=0$. **18. a)** $2\sqrt[3]{2}$
b) $2\sqrt[3]{4}$ **c)** $3\sqrt[3]{2}$ **d)** $3\sqrt[3]{3}$ **19. a)** $\sqrt{7}$ **b)** $\sqrt{2}$ **c)** $3\sqrt{2}$ **d)** $\sqrt{6}$ **20. a)** If
both $a<0$ and $b<0$, then $\sqrt{ab}$ is positive, whereas $\sqrt{a}\sqrt{b}$ is
negative. Thus, $\sqrt{ab} \ne \sqrt{a}\sqrt{b}$. **b)** Division by 0 is undefined.

Section 2.2, pp. 115–119

1. a) 9 **b)** 100 **c)** $\frac{9}{4}$ **d)** $\frac{25}{4}$ **e)** $\frac{1}{4}$ **f)** $\frac{1}{4}$ **g)** 0.16 **h)** 0.000 625

i) 1.44 **j)** 46.9225 **k)** $\frac{1}{9}$ **l)** $\frac{1}{144}$ **2. a)** minimum of −43 at
$x=-6$ **b)** maximum of 10 at $x=3$ **c)** minimum of −87 at
$x=10$ **d)** maximum of 44 at $x=-7$ **e)** maximum of 35 at
$x=-5$ **f)** minimum of −18 at $x=-3$ **g)** minimum of −1 at
$x=2$ **h)** maximum of 3 at $x=-1$ **i)** minimum of −59 at
$x=3$ **j)** minimum of −1.7 at $x=-2$ **k)** minimum of −0.1 at
$x=-4$ **l)** maximum of 3.6 at $x=-3$ **3. a)** minimum of $-\frac{5}{4}$

at $x=-\frac{3}{2}$ **b)** minimum of $-\frac{9}{4}$ at $x=\frac{1}{2}$ **c)** minimum of $-\frac{1}{3}$ at

$x=-\frac{1}{3}$ **d)** maximum of −8 at $x=\frac{1}{2}$ **e)** maximum of 7 at

$x=3$ **f)** maximum of $-\frac{7}{8}$ at $x=\frac{3}{4}$ **g)** maximum of $\frac{25}{4}$ at

$x=-\frac{5}{2}$ **h)** minimum of 0.97 at $x=0.1$ **i)** maximum of −1.92

at $x=-0.2$ **j)** minimum of 1.5 at $x=-1$ **k)** maximum of $\frac{4}{3}$

at $x=\frac{2}{3}$ **l)** minimum of −0.18 at $x=0.6$ **m)** maximum of
−3.19 at $x=0.9$ **n)** maximum of −1.6875 at $x=1.25$
4. a) −36 **b)** −6, 6 **5. a)** −20.25 **b)** −4.5, 4.5 **6. a)** 132.25
b) 11.5, 11.5 **7. a)** $y=x^2-8x+35$ **b)** 19; 4
8. a) $y=375-10x-x^2$ **b)** 400; −5 **9. a)** $17.50 **b)** $9187.50
10. a) 84.5 **b)** 6.5, 6.5 **11. a)** minimum of −9 at $x=0$
b) maximum of 25 at $x=0$ **12.** 21.125 cm^2
13. a) 9.375 square units **b)** 1.25 **14. a)** 62.1 m **b)** 3.5 s
15. a) 3.0625 square units **b)** 0.875 **16. a)** 3.23 m **b)** 17.5 m
c) 2 m **17.** 0.05 mm^2/h; 2.5 h **18. a)** 32.52 m **b)** 55 m
19. a) 306.25 m^2 **b)** 17.5 m by 17.5 m **c)** 612.5 m^2 **d)** 17.5 m
by 35 m **20. a)** Mercury: $h=-2t^2+20t+2$;
Neptune: $h=-6t^2+20t+2$; Venus: $h=-4.5t^2+20t+2$
b) Mercury: 52 m, 5 s; Neptune: 18.7 m, 1.7 s;

Venus: 24.2 m, 2.2 s **21.** k must be an even integer.
22. k must be divisible by 4.

Section 2.3, pp. 128–133

1. a) 25 **b)** 49 **c)** 12.25 **d)** 6.25 **e)** $\frac{4}{9}$ **f)** $\frac{1}{9}$ **g)** 0.49 **h)** 0.0009

2. a) 0, -6 **b)** 11, 9 **c)** 3, -1 **d)** 9, -1 **e)** $\frac{3}{2}$, $-\frac{5}{2}$ **f)** 0, $\frac{2}{3}$ **g)** 0, $-\frac{3}{2}$
h) 1.6, -0.6 **i)** -0.3, -0.5 **3. a)** $-3 + \sqrt{5}$, $-3 - \sqrt{5}$ **b)** $2 + \sqrt{15}$,
$2 - \sqrt{15}$ **c)** $-4 + \sqrt{23}$, $-4 - \sqrt{23}$ **d)** $5 + 2\sqrt{7}$, $5 - 2\sqrt{7}$
e) $\frac{7 + \sqrt{13}}{2}$, $\frac{7 - \sqrt{13}}{2}$ **f)** $\frac{5 + \sqrt{17}}{2}$, $\frac{5 - \sqrt{17}}{2}$ **g)** $\frac{-1 + \sqrt{13}}{2}$,
$\frac{-1 - \sqrt{13}}{2}$ **h)** $10 + 4\sqrt{6}$, $10 - 4\sqrt{6}$ **4. a)** $\frac{-4 + \sqrt{6}}{2}$, $\frac{-4 - \sqrt{6}}{2}$
b) $\frac{3 + \sqrt{3}}{3}$, $\frac{3 - \sqrt{3}}{3}$ **c)** $\frac{-3 + \sqrt{57}}{12}$, $\frac{-3 - \sqrt{57}}{12}$ **d)** 2, $-\frac{1}{3}$
e) $\frac{-1 + \sqrt{31}}{5}$, $\frac{-1 - \sqrt{31}}{5}$ **f)** $\frac{-1 + \sqrt{6}}{5}$, $\frac{-1 - \sqrt{6}}{5}$ **g)** $-1 + 3\sqrt{3}$,
$-1 - 3\sqrt{3}$ **h)** $\frac{1 + \sqrt{10}}{3}$, $\frac{1 - \sqrt{10}}{3}$ **5. a)** 0.41, -2.41 **b)** 3.73,
0.27 **c)** 2.19, -3.19 **d)** 3.58, 0.42 **e)** -0.72, -2.78
f) 4.10, -1.10 **g)** -0.13, -3.87 **h)** 3.81, -1.31

6. a) $\frac{-1 + i\sqrt{23}}{2}$, $\frac{-1 - i\sqrt{23}}{2}$ **b)** $1 + i\sqrt{7}$, $1 - i\sqrt{7}$ **c)** $-3 + 2i\sqrt{2}$,
$-3 - 2i\sqrt{2}$ **d)** $\frac{3 + i\sqrt{39}}{4}$, $\frac{3 - i\sqrt{39}}{4}$ **e)** $\frac{5 + i\sqrt{7}}{4}$, $\frac{5 - i\sqrt{7}}{4}$
f) $\frac{-4 + 2i\sqrt{2}}{3}$, $\frac{-4 - 2i\sqrt{2}}{3}$ **g)** $-1 + i$, $-1 - i$ **h)** $\frac{3 + i\sqrt{11}}{2}$,
$\frac{3 - i\sqrt{11}}{2}$ **7. a)** 4, -7 **b)** 1, $-\frac{3}{2}$ **c)** $-\frac{1}{3}$, $-\frac{3}{4}$ **d)** $\frac{5}{2}$, $\frac{5}{2}$
8. a) 5, -8 **b)** 4, -3 **c)** 6, 6 **d)** 5, -6 **e)** 4, -1 **f)** 5, 0 **g)** 7, -2
h) 4, -4 **i)** -5, -5 **j)** 8, -2 **9. a)** 3, $-\frac{1}{4}$ **b)** 4, $\frac{1}{4}$ **c)** $-\frac{4}{3}$, $-\frac{4}{3}$
d) 3, $-\frac{5}{3}$ **e)** $\frac{5}{2}$, $-\frac{5}{2}$ **f)** $\frac{1}{2}$, -5 **g)** $\frac{1}{4}$, $-\frac{1}{2}$ **h)** $\frac{1}{2}$, $-\frac{2}{3}$ **i)** $-\frac{1}{2}$, $-\frac{2}{3}$
j) $\frac{6}{5}$, -10 **k)** $\frac{4}{3}$, $-\frac{4}{3}$ **l)** -3, -3 **10. a)** $\frac{5}{3}$, -3 **b)** 0, $-\frac{7}{3}$ **c)** $\frac{3}{2}$, $\frac{3}{2}$
d) 3, $-\frac{1}{4}$ **e)** $\frac{2}{3}$, -5 **f)** $\frac{5}{2}$, 0 **g)** 1, $-\frac{5}{6}$ **h)** $\frac{1}{3}$, $-\frac{3}{2}$ **i)** 4, $-\frac{3}{4}$ **j)** 1, $\frac{8}{9}$
k) -2, -2 **l)** 1, $-\frac{1}{3}$ **11. a)** $\frac{5}{2}$, $\frac{1}{2}$ **b)** $\frac{7}{3}$, -4 **c)** 3, $-\frac{5}{2}$ **d)** 1, $\frac{3}{2}$
e) 4, $-\frac{5}{3}$ **f)** 3, $-\frac{1}{4}$ **g)** 1, $-\frac{4}{5}$ **h)** $\frac{5}{3}$, $-\frac{3}{2}$ **12. a)** $\frac{-3 + \sqrt{6}}{3}$,
$\frac{-3 - \sqrt{6}}{3}$; -0.18, -1.82 **b)** $\frac{-3 + \sqrt{3}}{2}$, $\frac{-3 - \sqrt{3}}{2}$; -0.63, -2.37
c) $\frac{3 + \sqrt{2}}{2}$, $\frac{3 - \sqrt{2}}{2}$; 2.21, 0.79 **d)** $-3 + \sqrt{5}$, $-3 - \sqrt{5}$; -0.76,
-5.24 **e)** $\frac{3 + \sqrt{7}}{2}$, $\frac{3 - \sqrt{7}}{2}$; 2.82, 0.18 **f)** $\frac{\sqrt{22}}{2}$, $-\frac{\sqrt{22}}{2}$; 2.35,

-2.35 **g)** $\frac{3 + \sqrt{21}}{6}$, $\frac{3 - \sqrt{21}}{6}$; 1.26, -0.26 **h)** $\frac{-3 + \sqrt{13}}{2}$,
$\frac{-3 - \sqrt{13}}{2}$; 0.30, -3.30 **i)** $\frac{-3 + \sqrt{15}}{3}$, $\frac{-3 - \sqrt{15}}{3}$; 0.29, -2.29
j) $\frac{5 + 3\sqrt{5}}{2}$, $\frac{5 - 3\sqrt{5}}{2}$; 5.85, -0.85 **k)** $\frac{-3 + \sqrt{17}}{4}$, $\frac{-3 - \sqrt{17}}{4}$;
0.28, -1.78 **l)** $\frac{1 + \sqrt{57}}{14}$, $\frac{1 - \sqrt{57}}{14}$; 0.61, -0.47 **m)** $\frac{-1 + \sqrt{73}}{12}$,
$\frac{-1 - \sqrt{73}}{12}$; 0.63, -0.80 **n)** $\frac{1 + \sqrt{11}}{2}$, $\frac{1 - \sqrt{11}}{2}$; 2.16, -1.16

13. a) $-1 + i$, $-1 - i$ **b)** $2 + 2i$, $2 - 2i$ **c)** $\frac{-5 + i\sqrt{7}}{2}$, $\frac{-5 - i\sqrt{7}}{2}$
d) $\frac{3 + i\sqrt{3}}{2}$, $\frac{3 - i\sqrt{3}}{2}$ **e)** $\frac{1 + 3i\sqrt{3}}{2}$, $\frac{1 - 3i\sqrt{3}}{2}$ **f)** $\frac{3 + 3i\sqrt{3}}{2}$,
$\frac{3 - 3i\sqrt{3}}{2}$ **g)** $\frac{-3 + i\sqrt{15}}{4}$, $\frac{-3 - i\sqrt{15}}{4}$ **h)** $\frac{2 + i\sqrt{2}}{3}$, $\frac{2 - i\sqrt{2}}{3}$
i) $\frac{-5 + i\sqrt{15}}{10}$, $\frac{-5 - i\sqrt{15}}{10}$ **j)** $\frac{2 + i}{5}$, $\frac{2 - i}{5}$ **14. a)** $3i$, $-3i$ **b)** $4i$,
$-4i$ **c)** $5i$, $-5i$ **d)** $10i$, $-10i$ **e)** $2i\sqrt{5}$, $-2i\sqrt{5}$ **f)** $2i\sqrt{3}$, $-2i\sqrt{3}$
g) $2i\sqrt{2}$, $-2i\sqrt{2}$ **h)** $i\sqrt{6}$, $-i\sqrt{6}$ **15.** 40 m by 110 m **16.** 0.5 m
17. 4.3 cm **18.** 3.6 m by 5.6 m **19.** 14 or -15 **20.** 1.45
21. a) $7 + 2\sqrt{3}$, $7 - 2\sqrt{3}$ **b)** 3.536, 10.464 **22.** $1 + 3\sqrt{3}$,
$1 - 3\sqrt{3}$ **23.** 8, 24 **24.** 5, 8 **25.** Yes, it is possible, but not if
the building is to be rectangular. **26. a)** yes; 10 m by 10 m
or 5 m by 20 m **b)** no **27. a)** no **b)** yes; 11 cm by 11 cm
c) yes; 9 cm by 13 cm **28. a)** $\frac{-7 + 3\sqrt{5}}{2}$, $\frac{-7 - 3\sqrt{5}}{2}$
b) $\frac{9 + \sqrt{73}}{2}$, $\frac{9 - \sqrt{73}}{2}$ **c)** $2 + \sqrt{6}$, $2 - \sqrt{6}$ **d)** $-3 + 2\sqrt{2}$,
$-3 - 2\sqrt{2}$ **29. a)** 4.1 s **b)** 2 s **30.** 9 cm, 12 cm, 15 cm
31. a) 15, 16 **b)** $\frac{31 + \sqrt{41}}{2}$, $\frac{31 - \sqrt{41}}{2}$ **c)** $\frac{31 + i\sqrt{39}}{2}$,
$\frac{31 - i\sqrt{39}}{2}$ **d)** $\frac{55}{2}$, $\frac{7}{2}$ **32.** On the TI-83 Plus, an error
message appears, saying there is no sign change. On the
TI-92, an error message appears, saying there are no
solutions. **33.** 7, -7 **34. a)** yes; after 1 s and after
9 s **b)** yes; after 5 s **c)** no **35.** no **36. a)** $\frac{3}{5}$, -7 **b)** 4, -4 **c)** 2
d) 6, -3 **37. a)** $-2 + \sqrt{7}$, $-2 - \sqrt{7}$; 0.65, -4.65
b) $\frac{-11 + \sqrt{61}}{2}$, $\frac{-11 - \sqrt{61}}{2}$; -1.59, -9.41 **c)** $\frac{-7 + \sqrt{73}}{4}$,
$\frac{-7 - \sqrt{73}}{4}$; 0.39, -3.89 **d)** $\frac{11 + \sqrt{55}}{3}$, $\frac{11 - \sqrt{55}}{3}$; 6.14, 1.19
38. a) $\frac{-17 + i\sqrt{15}}{4}$, $\frac{-17 - i\sqrt{15}}{4}$ **b)** $\frac{9 + 5i\sqrt{7}}{8}$, $\frac{9 - 5i\sqrt{7}}{8}$
c) $\frac{-2 + i\sqrt{3}}{2}$, $\frac{-2 - i\sqrt{3}}{2}$ **d)** $\frac{-1 + \sqrt{19}}{3}$, $\frac{-1 - \sqrt{19}}{3}$ **39.** 0.89 m

40. a) $-1 + \sqrt{k+1},\ -1 - \sqrt{k+1}$ **b)** $\dfrac{1 + \sqrt{1 + k^2}}{k}$,

$\dfrac{1 - \sqrt{1 + k^2}}{k}$ for $k \neq 0$; 0 for $k = 0$ **c)** $\dfrac{k + \sqrt{k^2 + 4}}{2}$,

$\dfrac{k - \sqrt{k^2 + 4}}{2}$ **41. a)** $x^2 - 5 = 0$ **b)** $x^2 + 4 = 0$ **42.** 5, 9

43. $\dfrac{-b + \sqrt{b^2 - 4c}}{2},\ \dfrac{-b - \sqrt{b^2 - 4c}}{2}$ **44. a)** $\sqrt{t}, -\sqrt{t}$

b) $\sqrt{\dfrac{b}{a}}, -\sqrt{\dfrac{b}{a}}$ **c)** $0, -\dfrac{t}{r}$ **d)** $\dfrac{m + \sqrt{m^2 + 4t}}{2}, \dfrac{m - \sqrt{m^2 + 4t}}{2}$

45. a) $b^2 - 4ac = 0$ **b)** $b^2 - 4ac > 0$ **c)** $b^2 - 4ac < 0$

Career Connection: Publishing, p. 133

1. a) The company must sell 6000 books at $26 each, or 6500 books at $24 each. **b)** No; the maximum revenue is $156 250.

Technology Extension: Solving Quadratic Equations, p. 134

1. a) $\dfrac{-3 + \sqrt{5}}{2}, \dfrac{-3 - \sqrt{5}}{2}$ **b)** "false" (no solutions) **c)** $\dfrac{-1 + \sqrt{17}}{4}$,

$\dfrac{-1 - \sqrt{17}}{4}$ **d)** "false" (no solutions) **2. a)** $\dfrac{-3 + \sqrt{5}}{2}, \dfrac{-3 - \sqrt{5}}{2}$

b) $1 + i\sqrt{3}, 1 - i\sqrt{3}$ **c)** $\dfrac{-1 + \sqrt{17}}{4}, \dfrac{-1 - \sqrt{17}}{4}$ **d)** $\dfrac{-1 + i\sqrt{5}}{3}$,

$\dfrac{-1 - i\sqrt{5}}{3}$ **3.** The cSolve function finds both real and imaginary roots, whereas the solve function finds only real

roots. **4. a)** $\dfrac{-5 + \sqrt{13}}{2}, \dfrac{-5 - \sqrt{13}}{2}$ **b)** $2 + \sqrt{6}, 2 - \sqrt{6}$

c) $\dfrac{3 + i\sqrt{15}}{2}, \dfrac{3 - i\sqrt{15}}{2}$ **d)** $\dfrac{-3 + i\sqrt{19}}{2}, \dfrac{-3 - i\sqrt{19}}{2}$ **e)** $-4 + \sqrt{19}$,

$-4 - \sqrt{19}$ **f)** $\dfrac{5 + i\sqrt{7}}{2}, \dfrac{5 - i\sqrt{7}}{2}$ **g)** $\dfrac{-3 + \sqrt{17}}{2}, \dfrac{-3 - \sqrt{17}}{2}$

h) $4 + i, 4 - i$ **5. a)** $\dfrac{-1 + \sqrt{13}}{3}, \dfrac{-1 - \sqrt{13}}{3}$ **b)** $\dfrac{-2 + i}{2}, \dfrac{-2 - i}{2}$

c) $\dfrac{5 + 3\sqrt{5}}{5}, \dfrac{5 - 3\sqrt{5}}{5}$ **d)** $\dfrac{3 + \sqrt{33}}{12}, \dfrac{3 - \sqrt{33}}{12}$ **e)** $\dfrac{4 + \sqrt{46}}{10}$,

$\dfrac{4 - \sqrt{46}}{10}$ **f)** $\dfrac{3 + 3i\sqrt{7}}{4}, \dfrac{3 - 3i\sqrt{7}}{4}$ **g)** $-1 + \sqrt{6}, -1 - \sqrt{6}$

h) $\dfrac{1 + i\sqrt{119}}{12}, \dfrac{1 - i\sqrt{119}}{12}$ **6. a)** yes; 6.3 m by 23.7 m **b)** yes;

10 m by 15 m **c)** no **7. a)** no **b)** yes; after 2 s

Section 2.4, pp. 139–142

1. a) $11\sqrt{5}$ **b)** $5\sqrt{3}$ **c)** $9\sqrt{2}$ **d)** $6\sqrt{7}$ **e)** $-\sqrt{10}$ **f)** 0 **g)** $4\sqrt{5}$

2. a) $8\sqrt{3} + 2\sqrt{6}$ **b)** $4\sqrt{5} + 4\sqrt{7}$ **c)** $7\sqrt{2} - \sqrt{10}$ **d)** $8\sqrt{6} - 5\sqrt{13}$

e) $6\sqrt{11} + 3\sqrt{14}$ **f)** $13 + 9\sqrt{7}$ **g)** $-1 - 2\sqrt{11}$ **3. a)** $5\sqrt{3}$ **b)** $5\sqrt{5}$

c) $\sqrt{2}$ **d)** $11\sqrt{2}$ **e)** $12\sqrt{3}$ **f)** $5\sqrt{6} + 2\sqrt{2}$ **g)** $5\sqrt{7} + 7\sqrt{3}$ **4. a)** $12\sqrt{7}$

b) $7\sqrt{2}$ **c)** $31\sqrt{3}$ **d)** $12\sqrt{2}$ **e)** $\sqrt{5}$ **f)** $9\sqrt{5}$ **g)** $24\sqrt{3} - 20\sqrt{2}$

5. a) $2\sqrt{5} + 4\sqrt{2}$ **b)** $3\sqrt{2} - \sqrt{3}$ **c)** $2\sqrt{3} + 6$ **d)** $12\sqrt{3} - 2\sqrt{6}$

e) $\sqrt{6} + 4\sqrt{2}$ **f)** $12\sqrt{3} + 6\sqrt{5}$ **g)** $23 + 4\sqrt{30}$ **h)** $16 + \sqrt{3}$

i) $26 + 14\sqrt{14}$ **j)** $28 + 6\sqrt{3}$ **k)** $13 - 4\sqrt{10}$ **l)** 1 **m)** 4 **n)** -17

6. a) $\dfrac{\sqrt{3}}{3}$ **b)** $\dfrac{2\sqrt{5}}{5}$ **c)** $\dfrac{2\sqrt{7}}{7}$ **d)** $\dfrac{\sqrt{2}}{2}$ **e)** $\dfrac{5\sqrt{15}}{6}$ **f)** $\dfrac{2}{3}$ **g)** 2 **h)** $\sqrt{15}$ **i)** $\sqrt{2}$

j) $\dfrac{3\sqrt{15}}{20}$ **k)** $\dfrac{7\sqrt{33}}{6}$ **l)** $\dfrac{\sqrt{10}}{5}$ **7. a)** $\dfrac{2 - \sqrt{2}}{2}$ **b)** $\dfrac{3 + 3\sqrt{5}}{4}$

c) $-\dfrac{3\sqrt{2} + 2\sqrt{3}}{3}$ **d)** $\dfrac{2\sqrt{6} - 2\sqrt{3}}{3}$ **e)** $\sqrt{5} + \sqrt{2}$ **f)** $3 - \sqrt{6}$

g) $\dfrac{24 - 2\sqrt{6}}{23}$ **h)** $3 - 2\sqrt{2}$ **i)** $-\dfrac{5\sqrt{2} + 2\sqrt{3} + 2\sqrt{5} + \sqrt{30}}{4}$

j) $\dfrac{52 + 7\sqrt{35}}{43}$ **8.** $10\sqrt{5}$ **9.** $(\sqrt{3} + 1)^2, \sqrt{3}(\sqrt{3} + 1)$,

$(\sqrt{3} + 1)(\sqrt{3} - 1), (1 - \sqrt{3})^2$ **10. a)** $6\sqrt{8} + \sqrt{8} - 5\sqrt{8}$ **b)** It is

twice as large as the others. **11.** 1 **12. a)** $8\sqrt{6} - 6$

b) $8\sqrt{2} + 2\sqrt{3}$ **13.** $13 - 4\sqrt{10}$ **14.** $38\sqrt{15} - 38\sqrt{2}$

15. $2\sqrt{7} + 2\sqrt{5}$ **16. a)** $x^2 - 6x + 7 = 0$ **b)** $x^2 + 2x - 11 = 0$

c) $4x^2 - 8x - 9 = 0$ **17. a)** $5\sqrt[3]{2}$ **b)** $5\sqrt[3]{3}$ **c)** $19\sqrt[3]{4}$ **d)** $13\sqrt[3]{2}$

e) $-\sqrt[3]{2}$ **f)** $\sqrt[3]{4}$ **g)** $3\sqrt[3]{5}$ **h)** $4\sqrt[3]{6}$ **18.** $97 + 56\sqrt{3}$ **19. a)** $8 + 4\sqrt{13}$

b) $3\sqrt{2} + 2\sqrt{5}$ **20.** The statement is sometimes true. It is true if and only if one or both of a and b are equal to zero.

Technology Extension: Radical Expressions and Graphing Calculators, p. 143

1. a) $5\sqrt{17}$ **b)** $7\sqrt{6}$ **c)** $13\sqrt{3}$ **d)** $4\sqrt{5}$ **e)** $14\sqrt{15}$ **f)** $120\sqrt{3}$ **2. a)** $\dfrac{\sqrt{3}}{3}$

b) $\dfrac{5\sqrt{7}}{28}$ **c)** $-\dfrac{\sqrt{2}}{2}$ **d)** $4\sqrt{2}$ **e)** $\dfrac{5\sqrt{6}}{18}$ **f)** $\dfrac{2\sqrt{6}}{3}$ **3. a)** $5\sqrt{5}$ **b)** $24\sqrt{2}$ **c)** $\sqrt{7}$

d) $-\sqrt{6}$ **e)** $21\sqrt{3}$ **f)** $9\sqrt{10}$ **g)** $2\sqrt{5}$ **h)** $-5\sqrt{11}$ **4. a)** $5\sqrt{2} + 5\sqrt{3}$

b) $6\sqrt{3} - 3\sqrt{2}$ **c)** $18 + 60\sqrt{2}$ **d)** $84\sqrt{3} - 56\sqrt{2}$ **e)** -7

f) $226 + 40\sqrt{22}$ **g)** $304 - 60\sqrt{15}$ **h)** $26\sqrt{6} - 54$ **5. a)** $2\sqrt{3} - 2\sqrt{2}$

b) $1 + \sqrt{2}$ **c)** $\dfrac{6\sqrt{2} + 3\sqrt{6}}{2}$ **d)** $\dfrac{8 + \sqrt{10}}{9}$ **e)** $-3 - 2\sqrt{2}$ **f)** $\dfrac{-6 - \sqrt{14}}{2}$

6. $\dfrac{48 - 12\sqrt{2}}{7}$

Section 2.5, pp. 150–152

1. a) $7 - 2i$ **b)** $3 - 11i$ **c)** $2 - 5i$ **d)** $1 + 6i$ **e)** $-2 + 13i$

f) $-11 + 5i$ **g)** $-3 - i$ **h)** $24 - 3i$ **i)** $14 - 17i$ **j)** $7 - 12i$

2. a) $8 - 6i$ **b)** $-6 + 3i$ **c)** $-20 - 12i$ **d)** $8 - 2i$ **e)** $14 + 2i$

f) $29 - 3i$ **g)** $-7 - 19i$ **h)** $26i$ **i)** $-3 + 4i$ **j)** $-7 - 24i$ **k)** $-2i$ **l)** 4

3. a) $-2i$ **b)** $-\dfrac{4i}{3}$ **c)** $-\dfrac{7i}{4}$ **d)** $\dfrac{6i}{5}$ **e)** $\dfrac{5i}{2}$ **f)** $\dfrac{3i}{7}$ **4. a)** $1 - 3i$

b) $-2 - 2i$ **c)** $\dfrac{2 - 5i}{2}$ **d)** $\dfrac{4 + 3i}{3}$ **e)** $\dfrac{-3 + 4i}{2}$ **5. a)** $3 - 2i$

b) $7 + 3i$ **c)** $5 + 4i$ **d)** $6 - 7i$ **6. a)** $\dfrac{6+3i}{5}$ **b)** $1 - 2i$ **c)** $\dfrac{-4+6i}{13}$

d) $\dfrac{3+4i}{25}$ **e)** $\dfrac{11+7i}{10}$ **f)** $\dfrac{2-4i}{5}$ **g)** $\dfrac{-5+12i}{13}$ **h)** $\dfrac{1+7i}{4}$

7. a) $-1 + i, -1 - i$ **b)** $2 + 2i, 2 - 2i$ **c)** $3 + i, 3 - i$
d) $-2 + i\sqrt{2}, -2 - i\sqrt{2}$ **e)** $1 + i\sqrt{5}, 1 - i\sqrt{5}$ **f)** $4 + i\sqrt{3}, 4 - i\sqrt{3}$

8. $-2i, -4, 16, 256$ **9. a)** $(35 + 4j)$ V **b)** $\dfrac{275 - 165j}{17}$ Ω

c) $\dfrac{33 + 11j}{2}$ A **10. a)** $-1 + i, -1 - i, 1 - i, 1 + i$

b) $-1 + i, -1 - i, 1 - i, 1 + i$; The pattern is cyclic.
11. a) $9 + 6i$ **b)** 0 **c)** $9 - 6i$ **12. a)** $7 + 2i$ **b)** $-2 - 36i$
c) $-8 + 9i$ **d)** $-17 - 18i$ **e)** 10 **f)** $20 + 40i$ **13.** $(a + bi)(a - bi)$

14. $\dfrac{a - bi}{a^2 + b^2}$ **15. a)** $x^2 - 2x + 2 = 0$ **b)** $4x^2 - 12x + 13 = 0$

16. Since the equation has imaginary roots, $b^2 - 4ac < 0$, and so the term $\sqrt{b^2 - 4ac}$ in the quadratic formula will be purely imaginary. The roots $\dfrac{-b + \sqrt{b^2 - 4ac}}{2a}$ and $\dfrac{-b - \sqrt{b^2 - 4ac}}{2a}$ are then complex conjugates. **17. a)** $x = 5, y = -4$ **b)** $x = 3,$ $y = -6$ **c)** $x = 6.5, y = 3.5$ **d)** $x = 2, y = -1$ **18. a)** $a = 0$ **b)** b could be any real number. **19.** reflection in the real axis
20. a) $2, -2$ **b)** $i, -i$ **c)** $1, -1, 2i, -2i$ **d)** $\sqrt{2}, -\sqrt{2}, \sqrt{3}, -\sqrt{3}$

e) $\sqrt{3}, -\sqrt{3}, i\sqrt{2}, -i\sqrt{2}$ **f)** $\dfrac{\sqrt{6}}{3}, -\dfrac{\sqrt{6}}{3}, 1, -1$ **g)** $i, -i, \dfrac{i\sqrt{6}}{2}, -\dfrac{i\sqrt{6}}{2}$

h) $\dfrac{\sqrt{6}}{2}, -\dfrac{\sqrt{6}}{2}, i\sqrt{2}, -i\sqrt{2}$ **i)** $\dfrac{\sqrt{2}}{2}, -\dfrac{\sqrt{2}}{2}, \dfrac{i\sqrt{2}}{2}, -\dfrac{i\sqrt{2}}{2}$

j) $0, 0, \dfrac{2}{3}, -\dfrac{2}{3}$ **21.** No; imaginary roots occur in complex conjugate pairs. **22. a)** $1, -1; i, -i$ **b)** $1, -1, i, -i$

c) $\dfrac{1+i}{\sqrt{2}}, -\dfrac{1+i}{\sqrt{2}}, \dfrac{-1+i}{\sqrt{2}}, \dfrac{1-i}{\sqrt{2}}$

Review of Key Concepts, pp. 154–157

1. a) $3\sqrt{2}$ **b)** $4\sqrt{2}$ **c)** $10\sqrt{5}$ **2. a)** $\sqrt{5}$ **b)** $\sqrt{7}$ **c)** $6\sqrt{6}$ **d)** 10
3. a) $2\sqrt{15}$ **b)** $30\sqrt{2}$ **4. a)** $2 + 3\sqrt{5}$ **b)** $1 + \sqrt{2}$ **c)** $2 - \sqrt{5}$ **5. a)** $7i$
b) $3i\sqrt{2}$ **c)** $4i\sqrt{5}$ **6. a)** -30 **b)** 36 **c)** -16 **d)** -3 **7. a)** $5 - 6i$
b) $3 + 2i\sqrt{5}$ **c)** $2 + i\sqrt{3}$ **8.** $30\sqrt{3}$ **9.** $20\sqrt{3}$ **10. a)** 25 **b)** 64
c) 6.25 **d)** 0.09 **11. a)** minimum of -12 at $x = -3$
b) minimum of -15 at $x = 6$ **c)** maximum of 18 at $x = -4$
d) maximum of 24 at $x = 5$ **e)** minimum of 10.5 at $x = -1.5$
f) maximum of 3 at $x = -2$ **g)** minimum of -3.25 at $x = 2.5$
h) maximum of -3.75 at $x = 0.5$ **i)** minimum of -1.25 at
$x = -0.25$ **j)** maximum of 1.125 at $x = -1.5$ **12. a)** -72.25
b) $-8.5, 8.5$ **13.** 55.125 cm² **14.** 56.25 m² **15. a)** 2.5 m
b) 0 m **c)** 2.5 m **16. a)** $16; (x + 4)^2$ **b)** $81; (y - 9)^2$ **c)** $\dfrac{1}{4}$;
$\left(m + \dfrac{1}{2}\right)^2$ **d)** $12.25; (r - 3.5)^2$ **e)** $\dfrac{9}{16}; \left(t + \dfrac{3}{4}\right)^2$ **f)** 0.0004;

$(w - 0.02)^2$ **17. a)** $-2 + \sqrt{3}, -2 - \sqrt{3}; -0.27, -3.73$
b) $3 + \sqrt{13}, 3 - \sqrt{13}; 6.61, -0.61$ **c)** $4 + \sqrt{5}, 4 - \sqrt{5}; 6.24,$
1.76 **d)** $5 + \sqrt{11}, 5 - \sqrt{11}; 8.32, 1.68$ **e)** $\dfrac{3+\sqrt{29}}{2}, \dfrac{3-\sqrt{29}}{2}$;
$4.19, -1.19$ **f)** $\dfrac{-7+3\sqrt{13}}{2}, \dfrac{-7-3\sqrt{13}}{2}; 1.91, -8.91$

g) $\dfrac{-5+\sqrt{21}}{2}, \dfrac{-5-\sqrt{21}}{2}; -0.21, -4.79$ **h)** $\dfrac{3+\sqrt{3}}{2}, \dfrac{3-\sqrt{3}}{2}$;
$2.37, 0.63$ **i)** $\dfrac{-5+\sqrt{13}}{6}, \dfrac{-5-\sqrt{13}}{6}; -0.23, -1.43$

j) $\dfrac{-1+\sqrt{31}}{6}, \dfrac{-1-\sqrt{31}}{6}; 0.76, -1.09$ **k)** $1, -\dfrac{1}{6}$ **l)** $2 + \sqrt{2},$

$2 - \sqrt{2}; 3.41, 0.59$ **18. a)** $-2 + i, -2 - i$ **b)** $\dfrac{1+i\sqrt{3}}{2}$,

$\dfrac{1-i\sqrt{3}}{2}$ **c)** $\dfrac{-3+i\sqrt{11}}{2}, \dfrac{-3-i\sqrt{11}}{2}$ **d)** $\dfrac{-1+i\sqrt{5}}{2}, \dfrac{-1-i\sqrt{5}}{2}$

e) $\dfrac{-3+i\sqrt{7}}{8}, \dfrac{-3-i\sqrt{7}}{8}$ **f)** $-2 + i, -2 - i$ **19. a)** $-4, -9$

b) $8, -7$ **c)** $6, -6$ **d)** $3, 2$ **e)** $-12, -12$ **f)** $\dfrac{3}{2}, -\dfrac{3}{2}$ **g)** $-\dfrac{1}{2}, -2$

h) $-\dfrac{1}{3}, -3$ **i)** $2, -\dfrac{1}{3}$ **j)** $2, \dfrac{1}{3}$ **k)** $\dfrac{3}{2}, \dfrac{4}{3}$ **l)** $\dfrac{5}{3}, -\dfrac{4}{3}$

20. a) $10, -4$ **b)** $6, -4$ **c)** $\dfrac{-9+3\sqrt{5}}{2}, \dfrac{-9-3\sqrt{5}}{2}; -1.15,$

-7.85 **d)** $5 + \sqrt{34}, 5 - \sqrt{34}; 10.83, -0.83$ **e)** $-3 + \sqrt{3},$

$-3 - \sqrt{3}; -1.27, -4.73$ **f)** $3, \dfrac{1}{2}$ **g)** $\dfrac{4}{5}, -1$

h) $\dfrac{1+\sqrt{7}}{3}, \dfrac{1-\sqrt{7}}{3}; 1.22, -0.55$ **i)** $\dfrac{1+\sqrt{61}}{4}, \dfrac{1-\sqrt{61}}{4}; 2.20,$

-1.70 **j)** $\dfrac{-4+\sqrt{10}}{2}, \dfrac{-4-\sqrt{10}}{2}; -0.42, -3.58$ **k)** $\dfrac{3+3\sqrt{33}}{16},$

$\dfrac{3-3\sqrt{33}}{16}; 1.26, -0.89$ **l)** $\dfrac{-5+\sqrt{73}}{6}, \dfrac{-5-\sqrt{73}}{6}; 0.59, -2.26$

21. a) $\dfrac{5+i\sqrt{11}}{2}, \dfrac{5-i\sqrt{11}}{2}$ **b)** $\dfrac{-1+i\sqrt{2}}{2}, \dfrac{-1-i\sqrt{2}}{2}$

c) $\dfrac{1+i\sqrt{11}}{3}, \dfrac{1-i\sqrt{11}}{3}$ **d)** $\dfrac{-9+i\sqrt{7}}{4}, \dfrac{-9-i\sqrt{7}}{4}$ **e)** $\dfrac{1+i\sqrt{79}}{4},$

$\dfrac{1-i\sqrt{79}}{4}$ **f)** $\dfrac{-1+2i\sqrt{6}}{5}, \dfrac{-1-2i\sqrt{6}}{5}$ **22. a)** $4i, -4i$ **b)** $3i, -3i$

c) $\dfrac{2i}{3}, -\dfrac{2i}{3}$ **23.** 2.5 m **24.** yes **25. a)** yes; 5, 12 **b)** yes; 4, 13

c) no **26. a)** $5\sqrt{2}$ **b)** $4\sqrt{3} + 3\sqrt{6}$ **c)** $7\sqrt{5}$ **d)** $-\sqrt{3}$ **e)** $2\sqrt{2}$ **f)** $\sqrt{5}$
g) $\sqrt{10} - \sqrt{2}$ **h)** $-9\sqrt{7}$ **27. a)** $\sqrt{6} + 5\sqrt{3}$ **b)** $2\sqrt{5} - 2\sqrt{3}$
c) $-7 - 11\sqrt{10}$ **d)** $17 + 4\sqrt{15}$ **e)** 4 **f)** 4 **28. a)** $\dfrac{\sqrt{5}}{5}$ **b)** $\dfrac{\sqrt{6}}{3}$ **c)** $\dfrac{2\sqrt{2}}{3}$

d) $\dfrac{\sqrt{30}}{4}$ **29. a)** $1 + \sqrt{3}$ **b)** $\dfrac{4\sqrt{5} - 4\sqrt{2}}{3}$ **c)** $\dfrac{-2\sqrt{6} - 10\sqrt{3}}{23}$

d) $\dfrac{48-7\sqrt{21}}{51}$ **30.** $(4-\sqrt{5})^2 = 21-8\sqrt{5}$ **31. a)** $12-3i$

b) $-2-6i$ **c)** $-25+19i$ **d)** $-5-12i$ **32. a)** $-\dfrac{3i}{2}$ **b)** $2i$ **c)** $3-4i$

d) $\dfrac{-2-5i}{3}$ **33. a)** $\dfrac{6-2i}{5}$ **b)** $\dfrac{-3+i}{5}$ **c)** $\dfrac{-7+6i}{17}$ **d)** $\dfrac{-1-13i}{10}$

34. a) $-1+i\sqrt{6}, -1-i\sqrt{6}$ **b)** $2+i\sqrt{7}, 2-i\sqrt{7}$ **c)** $-1+i\sqrt{2}$, $-1-i\sqrt{2}$ **35.** $-2i, -4-2i, 12+14i$

Chapter Test, pp. 158–159

1. a) $5\sqrt{2}$ **b)** $2\sqrt{11}$ **c)** $4\sqrt{5}$ **d)** $\sqrt{35}$ **e)** $6\sqrt{2}$ **f)** $2\sqrt{3}$ **g)** $30\sqrt{5}$
h) $4-\sqrt{10}$ **2. a)** $6i$ **b)** $-4i\sqrt{3}$ **c)** -45 **d)** $6-6i$ **e)** $5+3i\sqrt{2}$
f) $4-i\sqrt{3}$ **3. a)** maximum of 9 at $x=-1$ **b)** minimum of $-\dfrac{41}{4}$
at $x=\dfrac{7}{2}$ **c)** maximum of $\dfrac{65}{8}$ at $x=\dfrac{5}{4}$ **4. a)** $\dfrac{7+\sqrt{33}}{4}$,
$\dfrac{7-\sqrt{33}}{4}$; 3.19, 0.31 **b)** $\dfrac{2+\sqrt{19}}{3}, \dfrac{2-\sqrt{19}}{3}$; 2.12, −0.79

5. a) $\dfrac{1+3i}{5}, \dfrac{1-3i}{5}$ **b)** $\dfrac{-2+i\sqrt{6}}{2}, \dfrac{-2-i\sqrt{6}}{2}$ **6. a)** $4, -\dfrac{1}{2}$

b) $\dfrac{2}{3}, -3$ **7. a)** $\dfrac{-3+\sqrt{29}}{2}, \dfrac{-3-\sqrt{29}}{2}$ **b)** $\dfrac{4+\sqrt{11}}{5}, \dfrac{4-\sqrt{11}}{5}$

c) $\dfrac{5+3i\sqrt{3}}{2}, \dfrac{5-3i\sqrt{3}}{2}$ **d)** $\dfrac{-3+5i\sqrt{3}}{6}, \dfrac{-3-5i\sqrt{3}}{6}$ **8.** 1.8 cm

9. factoring, completing the square, using the quadratic
formula **10. a)** 0, 4 **b)** $k<0$ or $k>4$ **c)** $0<k<4$ **11. a)** $3\sqrt{3}$
b) $7\sqrt{7}$ **12. a)** $14\sqrt{3}$ **b)** $-7+5\sqrt{3}$ **c)** $22-12\sqrt{2}$
13. a) $\dfrac{2\sqrt{7}}{7}$ **b)** $\dfrac{12+3\sqrt{3}}{-13}$ **c)** $\dfrac{10\sqrt{6}-5\sqrt{3}}{21}$ **14.** 7 square units

15. a) $13-8i$ **b)** $-2+8i$ **c)** $27+24i$ **d)** $-\dfrac{5i}{3}$ **e)** $\dfrac{-4-5i}{2}$

f) $\dfrac{1-18i}{25}$

Challenge Problems, p. 160

1. $y=-\dfrac{1}{2}x^2+3x+8$ **2.** $-k, -\dfrac{1}{k}$ **3.** 64 **4.** $k\geq -3+2\sqrt{2}$ or
$k\leq -3-2\sqrt{2}$ **5.** 16, −16, 8, −8 **6.** (1, 9), (2, 8), (3, 7),
(4, 6), (5, 5), (6, 4), (7, 3), (8, 2), (9, 1) **7.** AB = 5, XY = 2

Problem Solving Strategy: Look for a Pattern, pp. 162–163

1. a) 2002, 3003, 4004, 5005, 6006 **b)** When the second
factor in the product is $7n$, the product equals $1000n + n$.
c) Since $98 = 7(14)$, the product equals $1000(14) + 14$, or
14 014. **2.** 49 **3. a)** $y=4x+6$; 42, 23, 214 **b)** $y=x^2+2$;
66, 11, 227 **c)** $y=2x-1$; 9, 19, 64 **d)** $y=\dfrac{x+3}{2}$; 7, 37, 51

4. a) 35 **b)** n^2+2n **c)** 960, 2600 **d)** 21st **5.** 8 **6.** 17, 10 **8.** 5
7. a) equal **b)** The sum is divisible by 9. **c)** Answers may vary;
111 211 111, 12 222 122 221, 211 211 121 121 112

9. 65 536 **10.** Q **11.** 7

Problem Solving: Using the Strategies, p. 164

1. 252 km from Ottawa at 15:36 **2.** 1.8 **3.** Label three sides
with 0 and label three sides with 6. **4.** $\dfrac{qst}{prx}$ **5.** 13 cm,

14 cm, 15 cm **6.** Cut four rods into lengths of 5, 5, and
3. Cut five rods into lengths of 4, 4, and 5. Cut three rods
into lengths 3, 3, 3, and 4. **7.** 10 240 **8.** R = −2, S = 4,
T = −6 or R = 6, S = 4, T = 2 **9. a)** 37

b) $\dfrac{100d+10d+d}{d+d+d} = \dfrac{111d}{3d} = 37$ **c)** yes; yes; yes

10. 24 km **11.** 7π cm^2

Cumulative Review: Chapters 1 and 2, p. 165

Chapter 1

1. a) $-\dfrac{1}{64}$ **b)** 4 **c)** $\dfrac{32}{3}$ **d)** $\dfrac{1}{3}$ **e)** $-\dfrac{1}{8}$ **f)** $\dfrac{25}{4}$ **2. a)** $8x^4$ **b)** $-3a^5$

c) $-\dfrac{8z^6}{y^3}$ **d)** $\dfrac{x^5}{3y^4}$ **3. a)** $-\dfrac{1}{2}$ **b)** 1 **4. a)** $2x^2+6x+1$

b) $5a^2-2ab-b^2$ **5. a)** $-8y^3+12y^2+3y$
b) $6z^3-7z^2-15z-4$ **c)** $-4a^2+11ab+11b^2$

6. a) $\dfrac{2}{t+4}, t\neq -2, -4$ **b)** $\dfrac{2x-1}{3x+1}, x\neq 1, -\dfrac{1}{3}$

c) $\dfrac{3(t+3)}{3(t-3)}, t\neq 3, -3$ **d)** $\dfrac{(2x+1)(x-1)}{(x-3)(2x-1)}$,

$x\neq 3, 1, \dfrac{1}{2}, -\dfrac{3}{2}, -2$ **e)** $\dfrac{3(y+3)}{(1+3y)(1-3y)}, y\neq \dfrac{1}{3}, -\dfrac{1}{3}$

f) $\dfrac{10}{(m-1)(m+3)(2m+1)}, m\neq 1, -\dfrac{1}{2}, -3$

7. a) $k\leq 5$ **b)** $m>5$ **c)** $q\geq 7$

Chapter 2

1. a) $12\sqrt{7}$ **b)** $2\sqrt{2}$ **c)** $\dfrac{3+\sqrt{5}}{2}$ **d)** $3i\sqrt{6}$ **e)** $-2i\sqrt{6}$ **f)** 48 **g)** 10

h) $2-3i\sqrt{3}$ **i)** $2+\sqrt{5}$ **2. a)** minimum of −7.25 at $x=2.5$

b) maximum of $\dfrac{10}{3}$ at $x=\dfrac{2}{3}$ **3. a)** $-1, -\dfrac{4}{3}$ **b)** $\dfrac{4}{3}, -\dfrac{1}{2}$

4. a) $\dfrac{-3+i\sqrt{7}}{2}, \dfrac{-3-i\sqrt{7}}{2}$ **b)** $\dfrac{3+\sqrt{33}}{4}, \dfrac{3-\sqrt{33}}{4}$ **c)** $4+2\sqrt{3}$,

$4-2\sqrt{3}$ **5. a)** $2+\sqrt{15}, 2-\sqrt{15}$ **b)** $1, -\dfrac{7}{3}$ **6.** 1.5 m

7. a) $-5\sqrt{3}-7\sqrt{5}$ **b)** $3\sqrt{2}+3\sqrt{7}$ **c)** $24-12\sqrt{6}-3\sqrt{10}+2\sqrt{15}$

d) $\dfrac{\sqrt{10}}{3}$ **e)** $\dfrac{12-5\sqrt{6}}{6}$ **f)** $-1-3i$ **g)** $27+5i$ **h)** $9+40i$

i) $\dfrac{-15+6i}{29}$ **j)** $2-i$

Chapter 3

Getting Started: Human Physiology, p. 168

1. b) the first quadrant, since $t \geq 0$ **2. b)** The y-coordinates of $y = 30x$ are 6 times those of $y = 5x$. **3. b)** time, in terms of the number of litres of blood pumped **4. b)** equal

5. a) $y = 12x$ **6.** $y = \dfrac{x}{12}$; the number of breaths divided by 12 produces the number of minutes.

Review of Prerequisite Skills, p. 169

1. a) $(-3, 2)$ **b)** $(-3, -8)$ **c)** $(-6, -2)$ **d)** $(2, -2)$ **e)** $(1, -4)$ **f)** $(-12, 6)$ **g)** $(-13, -9)$ **h)** $(8, 10)$ **2. a)** 12 units to the left **b)** 8 units upward **c)** 8 units to the left and 1 unit downward **d)** 6 units to the left and 5 units downward **e)** 11 units to the right and 21 units downward **f)** 7 units to the right and 19 units downward **3. a)** A′$(-3, 0)$, B′$(-6, -4)$, C′$(0, -6)$ **b)** A′$(4, 5)$, B′$(1, 1)$, C′$(7, -1)$ **c)** A′$(6, -1)$, B′$(3, -5)$, C′$(9, -7)$ **4. a)** A′$(-2, -4)$, B′$(-4, -1)$, C′$(5, 2)$ **b)** A′$(2, 4)$, B′$(4, 1)$, C′$(-5, -2)$ **c)** A′$(2, -4)$, B′$(4, -1)$, C′$(-5, 2)$ **5. a)** vertical stretch by a factor of 3 **b)** vertical compression by a factor of $\dfrac{1}{3}$ **6.** reflection in the x-axis **7. a)** 3 units upward **b)** 7 units downward **c)** 4 units to the right **d)** 6 units to the left **e)** 3 units to the left and 8 units downward **f)** 7 units to the right and 2 units upward **8. a)** opens up; vertex: $(0, -4)$; axis of symmetry: $x = 0$; domain: all real numbers; range: $y \geq -4$; minimum value: -4 **b)** opens down; vertex: $(0, 5)$; axis of symmetry: $x = 0$; domain: all real numbers; range: $y \leq 5$; maximum value: 5 **c)** opens up; vertex: $(2, 3)$; axis of symmetry: $x = 2$; domain: all real numbers; range $y \geq 3$; minimum value: 3 **d)** opens down; vertex: $(-3, -5)$; axis of symmetry: $x = -3$; domain: all real numbers; range: $y \leq -5$; maximum value: -5 **9. c)** equal **f)** congruent parabolas at differing locations **g)** For the function $y = x$, the graph is the collection of all ordered pairs of the form (x, x); the translation of the graph upward by 1 unit produces the collection $(x, x + 1)$, which is equivalent to the collection $(x - 1, x)$, produced by the translation of 1 unit to the left.

Section 3.1, pp. 178–181

1. a) a function **b)** not a function **c)** a function **d)** not a function **2. a)** domain: $\{0, 1, 2, 3\}$, range: $\{5, 6, 7, 8\}$ **b)** domain: $\{1, 2\}$, range: $\{3, 4, 5\}$ **c)** domain: $\{-2, -1, 0, 1, 2\}$, range: $\{-1, 0, 2\}$ **d)** domain: $\{-2, 0, 3, 4, 7\}$, range: $\{1\}$ **e)** domain: $\{-2, -1, 0, 1, 2\}$, range: $\{-5, -4, -1, 3, 5\}$ **f)** domain: $\{0\}$, range: $\{-1, 0, 2, 4\}$ **3. a)** domain: the set of real numbers, range: the set of real numbers **b)** domain: $-4 \leq x \leq 3$, range: $\{2\}$ **c)** domain: $\{-3\}$, range: the set of real numbers **d)** domain: the set of real

numbers, range: $y \geq 3$ **e)** domain: $x \geq 1$, range: the set of real numbers **f)** domain: $-4 \leq x \leq 5$, range: $-4 \leq y \leq 1$ **g)** domain: $-4 \leq x \leq 4$, range: $-4 \leq y \leq 4$ **h)** domain: $-2 \leq x \leq 2$, range: $-3 \leq y \leq 3$ **4. b)** yes **5. a)** a function **b)** a function **c)** not a function **6. a)** 3 **b)** 0 **c)** -4 **d)** -5 **e)** -7 **7. a)** 10 **b)** 4 **c)** 1 **d)** -5 **e)** 5.5 **8. a)** -1 **b)** 34 **c)** -1 **d)** 4.25 **e)** -1.75 **9. a)** 5 **b)** 176 **c)** 33 **d)** 5 **e)** 41 **10. a)** 5 **b)** 1 **c)** -59 **d)** -31 **e)** -4 **11.** $(-1, -8)$, $(0, -1)$, $(2, 13)$, $(5, 34)$ **12. a)** 5 **b)** -2 **c)** 13 **d)** -5 **e)** 2.5 **13.** a point **14. a)** dependent variable: C, independent variable: p **b)** Yes; there is only one cost for each number of pens purchased **15. a)** a function; dependent variable: time, independent variable: speed **b)** not a function **c)** a function; dependent variable: revenue, independent variable: number of tickets sold **16.** 0 **17.** No; the vertical line test fails. **18. a)** $\{-5, -1, 0, 5\}$ **b)** $\{7, 47, 67\}$ **19. a)** salary of $400/week plus 5% commission; salary of $400/week plus 6% commission **b)** dependent variable: E, independent variable: s **c)** domain: $s \geq 0$, range: $E \geq 400$ **d)** $500, $670 **20.** domain: $r \geq 0$, range: $A \geq 0$ **21. a)** domain: $-1 \leq x \leq 1$, range: $-1 \leq y \leq 1$ **b)** no **22. a)** $S(p) = 0.75p$ **b)** $56 **23. a)** $S(x) = 6x^2$ **b)** 37.5 cm^2 **24. a)** $A(x) = x(100 - 2x)$ **b)** 1250 m^2, 25 m by 50 m **25.** Yes; for every value of x there is only one value of y. **26. a)** 1, -5 **b)** -2 **c)** 0, -4 **d)** -1, -3 **28. a)** Yes; jogging for a given period of time results in exactly one value for the distance covered. **b)** Yes; at any given time, she can be only one distance from the starting point. **29.** It is the slope of the line segment that joins the two points $(1, 1)$ and $(4, 19)$ on the graph. **30.** -5 **31.** $A = \dfrac{C^2}{4\pi}$ **32. a)** $8a + 3$ **b)** $-1 - 3n$ **c)** $m^2 - 2m + 2$ **d)** $8k^2 + 8k - 1$ **e)** $9t^2 + 6t - 4$ **f)** $12w^2 - 32w + 25$

Section 3.2, pp. 182–183

1. a) 0, 1, 2, 3, 4 **b)** The square root is defined only for positive values of x, and is defined as the positive square root. **c)** $x \geq 0$ **d)** No; the function is always increasing. **e)** $y \geq 0$ **2. a)** 9, 4, 1, 0, 1, 4, 9 **b)** the set of all real numbers **c)** $y \geq 0$ **3.** The graph of $f(x) = \sqrt{x}$ is the reflection of the right half of the graph of $f(x) = x^2$ in the line $y = x$. **4. b)** It is the reflection of $y = \sqrt{x}$ in the x-axis, and $y = \sqrt{x}$ appears only in the first quadrant. **c)** $x \geq 0$ **d)** $y \leq 0$ **e)** The graphs of $f(x) = \sqrt{x}$ and $f(x) = -\sqrt{x}$, taken together, are the reflection of the graph of $f(x) = x^2$ in the line $y = x$. **5. a)** 4, 3, 2, 1, $\dfrac{1}{2}$, $\dfrac{1}{3}$, $\dfrac{1}{4}$; $-4, -3, -2, -1, -\dfrac{1}{2}, -\dfrac{1}{3}, -\dfrac{1}{4}$ **b)** For $x > 0$, $y = \dfrac{1}{x}$ is positive and so has its graph in the first quadrant. For $x < 0$, $y = \dfrac{1}{x}$ is negative and so has its graph in the third quadrant. **6.** Table 1: 1, 0.1, 0.01, 0.001; 1, 10, 100, 1000; Table 2: $-1, -0.1, -0.01, -0.001$; $-1, -10, -100, -1000$

7. a) decreases **b)** No; $\dfrac{1}{x}$ does not equal 0 for any value of x.

c) increases **d)** No; $\dfrac{1}{x}$ gets greater and greater as x gets closer and closer to 0 through positive values. **8. a)** increases

b) No; $-\dfrac{1}{x}$ does not equal 0 for any value of x. **c)** decreases

d) No; $-\dfrac{1}{x}$ gets greater and greater in negative value as x gets closer and closer to 0 through negative values. **9. a)** the set of real numbers, excluding 0 **b)** the set of real numbers, excluding 0 **10. b)** all real numbers **c)** all real numbers

11. The graph of $f(x) = \dfrac{1}{x}$ is the reciprocal of the graph of $f(x) = x$.

Section 3.3, pp. 189–193

1. a) translate the graph of $y = f(x)$ upward 5 units
b) translate the graph of $y = f(x)$ downward 6 units
c) translate the graph of $y = f(x)$ to the right 4 units
d) translate the graph of $y = f(x)$ to the left 8 units
e) translate the graph of $y = f(x)$ upward 3 units **f)** translate the graph of $y = f(x)$ downward 7 units **g)** translate the graph of $y = f(x)$ to the left 3 units and downward 5 units
h) translate the graph of $y = f(x)$ to the right 6 units and upward 2 units **i)** translate the graph of $y = f(x)$ to the right 5 units and downward 7 units **j)** translate the graph of $y = f(x)$ to the left 2 units and upward 9 units **2. a)** $h = 0$, $k = 6$ **b)** $h = 0$, $k = -8$ **c)** $h = 3$, $k = 0$ **d)** $h = -5$, $k = 0$
e) $h = -2$, $k = -4$ **f)** $h = 7$, $k = 7$ **3. a)** domain: all real numbers, range: all real numbers **b)** domain: all real numbers, range: all real numbers **c)** domain: all real numbers, range: $y \geq -3$ **d)** domain: all real numbers, range: $y \geq 0$
e) domain: all real numbers, range: $y \geq -1$ **f)** domain: $x \geq -1$, range: $y \geq 0$ **g)** domain: $x \geq 0$, range: $y \geq -5$ **h)** domain: $x \geq 3$, range: $y \geq 6$ **5. a)** $y = \sqrt{x - 5}$ **b)** $y = \sqrt{x + 4}$
c) $y = \sqrt{x + 2} + 4$ **d)** $y = (x + 4)^2 - 4$ **9.** $f(\text{P-C1}) = -5t^2 + 210$, $f(\text{TBW}) = -5t^2 + 148$, $f(\text{CG}) = -5t^2 + 126$ **c)** translate the graph of $f(\text{P-C1})$ downward 62 units **d)** translate the graph of $f(\text{CG})$ upward 22 units **e)** translate the graph of $f(\text{P-C1})$ downward 84 units **10. a)** $C = 45 + 35t$ **b)** $C = 40 + 35t$
c) The graph for Elena is a vertical translation of the graph for Mario 5 units upward. **11.** Equal; the graph of $f(x) = 2x + 3$ is a straight line with slope 2. If a unit step is taken from the line to the right, two unit steps must be taken upward to reach the line again. **12. a)** vertical translation upward 1000 units **b)** $S(y) = 26\,000 + 2250y$
c) domain: $0 \leq y \leq 45$, range: $26\,000 \leq S \leq 127\,250$
13. a) A closed dot represents a function value; an open dot represents the absence of a function value. **b)** domain: all real numbers, range: all integers **14. b)** $5, $6, $6, $8
15. a) horizontal translation to the left 7 units **b)** vertical translation downward 12 units **16.** vertical translation downward 2 units **17.** The translations produce a collection

of points of the form $(x, x + 3)$ or of the form $(x - 3, x)$, and either collection represents the function $f(x) = x + 3$.

18. a) $C(x) = \dfrac{20}{50 + x} \times 100\%$ **c)** $33\dfrac{1}{3}\%$

d) $C(x) = \dfrac{20}{40 + x} \times 100\%$ **f)** horizontal translation to the left 10 units

Career Connection: Veterinary Medicine, p. 193

1. b) The graph giving the number of human years equivalent to the age of a dog is a vertical translation upwards 5 units of the graph giving the number of human years equivalent to the age of a cat. **c)** They differ by 5. **d)** Yes; the slopes of the graphs are equal. **e)** No; a dog ages more rapidly, since after 3 years the number of human years equivalent to its age is 32, whereas the number of human years equivalent to a 3-year-old cat is 27.

Section 3.4, pp. 203–206

2. a) $y = -\sqrt{x + 3}$ **b)** $y = -x^2 + 3$ **c)** $y = -\sqrt{x} - 3$ **d)** $y = -2x + 2$
3. a) $y = -x - 3$ **b)** $y = \sqrt{2 - x}$ **c)** $y = \sqrt{-x} - 2$ **d)** $y = (x + 4)^2$
5. a) $-f(x) = 4 - 2x$, $f(-x) = -2x - 4$ **c)** $-f(x)$: $(2, 0)$, $f(-x)$: $(0, -4)$ **d)** For each function, both the domain and range are the set of all real numbers. **6. a)** $-f(x) = 3x - 2$,
$f(-x) = 3x + 2$ **c)** $-f(x)$: $\left(\dfrac{2}{3}, 0\right)$, $f(-x)$: $(0, 2)$
7. a) $-f(x) = 4x - x^2$, $f(-x) = x^2 + 4x$ **c)** $-f(x)$: $(0, 0)$, $(4, 0)$; $f(-x)$: $(0, 0)$ **8. a)** $-f(x) = 9 - x^2$, $f(-x) = x^2 - 9$ **c)** $-f(x)$: $(3, 0)$, $(-3, 0)$; $f(-x)$: $(0, -9)$ **9. a)** $-f(x) = (x + 4)(2 - x)$, $f(-x) = (x - 4)(x + 2)$ **c)** $f(x)$: domain: all real numbers, range: $y \geq -9$; $-f(x)$: domain: all real numbers, range: $y \leq 9$; $f(-x)$: domain: all real numbers, range: $y \geq -9$
10. a) $-f(x) = -\sqrt{x + 4}$, $f(-x) = \sqrt{4 - x}$ **c)** $f(x)$: domain: $x \geq -4$, range: $y \geq 0$; $-f(x)$: domain: $x \geq -4$, range: $y \leq 0$; $f(-x)$: domain: $x \leq 4$, range: $y \geq 0$ **11. a)** $-f(x) = -\sqrt{x} - 4$, $f(-x) = \sqrt{-x} + 4$ **c)** $f(x)$: domain: $x \geq 0$, range: $y \geq 4$; $-f(x)$: domain: $x \geq 0$, range: $y \leq -4$; $f(-x)$: domain: $x \leq 0$, range: $y \geq 4$ **12. a)** AD, since the equation has a positive slope
b) $y = -0.9x + 146$; this is $f(-x)$, because one is the reflection of the other in the y-axis. **13.** They are mirror images in the x-axis. If one is $f(x)$, the other is $-f(x)$.
14. a) $(0, -6)$, $(-2, 0)$, $(3, 0)$ **15. a)** $y = 0.7x + 1.9$ **b)** 1.9 m
c) 5.4 m **d)** left half: domain: $-2.7 \leq x \leq 0$, range: $0 \leq y \leq 1.9$; right half: domain: $0 \leq x \leq 2.7$, range: $0 \leq y \leq 1.9$ **17.** $(0, 0)$; each coordinate must satisfy the equation $a = -a$. **18. a)** $-f(x) = -\sqrt{x - 3}$, $f(-x) = \sqrt{-x - 3}$, $-f(-x) = -\sqrt{-x - 3}$ **c)** $f(x)$: domain: $x \geq 3$, range: $y \geq 0$; $-f(x)$: domain: $x \geq 3$, range: $y \leq 0$; $f(-x)$: domain: $x \leq -3$, range: $y \geq 0$; $-f(-x)$: domain: $x \leq -3$, range: $y \leq 0$
19. a) $-f(x) = -(x + 1)^2$, $f(-x) = (1 - x)^2$, $-f(-x) = -(1 - x)^2$ **c)** $f(x)$: domain: all real numbers, range: $y \geq 0$; $-f(x)$: domain: all real numbers, range: $y \leq 0$; $f(-x)$: domain: all real numbers, range: $y \geq 0$; $-f(-x)$: domain: all

real numbers, range: $y \le 0$ **20. a)** $n = p^2 - 8$ **b)** $n = -p^2 + 8$
c) mirror images in the p-axis **d)** The domain of $n_1(p)$ is the
set of all positive integers. The range is $p^2 - 8$, where p is a
positive integer. The domain of $n_2(p)$ is the set of all positive
integers. The range is $-p^2 + 8$, where p is a positive integer.
21. a) $-f(x) = -\sqrt{25 - x^2}$, $f(-x) = \sqrt{25 - x^2}$ **c)** $-f(x)$: $(-5, 0)$,
$(5, 0)$, $f(-x)$: $(0, 5)$ **d)** $f(x)$: domain: $-5 \le x \le 5$, range:
$0 \le y \le 5$; $-f(x)$: domain: $-5 \le x \le 5$, range: $-5 \le y \le 0$;
$f(-x)$: domain: $-5 \le x \le 5$, range: $0 \le y \le 5$; $-f(-x)$:
domain: $-5 \le x \le 5$, range: $-5 \le y \le 0$ **22.** One; there is
only one y-intercept. **23. a)** opposites **b)** opposites

Section 3.5, pp. 215–220

1. a) $f^{-1} = \{(2, 0), (3, 1), (4, 2), (5, 3)\}$ **b)** $g^{-1} = \{(-3, -1),$
$(-2, 1), (4, 3), (0, 5), (1, 6)\}$ **2. a)** $f^{-1} = \{(3, -2), (2, -1),$
$(0, 0), (-2, 4)\}$; a function **b)** $g^{-1} = \{(-2, 4), (1, 2), (3, 1),$
$(-2, 0), (-3, -3)\}$; not a function **4. a)** $x = \dfrac{f(x) - 2}{3}$

b) $x = \dfrac{12 - 3f(x)}{2}$ **c)** $x = \dfrac{3 - f(x)}{4}$ **d)** $x = 4f(x) - 3$

e) $x = 2f(x) + 10$ **f)** $x = \pm\sqrt{y - 3}$ **5. a)** $f^{-1}(x) = x + 1$

b) $f^{-1}(x) = 2x$ **c)** $f^{-1}(x) = x - 3$ **d)** $f^{-1}(x) = \dfrac{3}{4}x$

e) $f^{-1}(x) = \dfrac{x - 1}{2}$ **f)** $f^{-1}(x) = 3x - 2$ **g)** $g^{-1}(x) = \dfrac{2x + 8}{5}$

h) $h^{-1}(x) = 5x - 5$ **6. a)** $f^{-1}(x) = x - 2$ **b)** $f^{-1}(x) = \dfrac{x}{4}$

c) $f^{-1}(x) = \dfrac{x + 2}{3}$ **d)** $f^{-1}(x) = x$ **e)** $f^{-1}(x) = 3 - x$

f) $f^{-1}(x) = 3x + 2$ **7. a)** $f^{-1}(x) = \dfrac{x + 5}{2}$, function

b) $f^{-1}(x) = 4x - 3$, function **c)** $f^{-1}(x) = 4x - 12$, function
d) $f^{-1}(x) = 5 - x$, function **8. a)** yes **b)** yes **c)** yes **d)** no **e)** no
f) no **g)** yes **9. a)** reflections in the line $y = x$ **b)** reflections in
the line $y = x$ **10. i) a)** $f^{-1}(x) = \pm\sqrt{x + 3}$ **c)** $f(x)$: domain: all
real numbers, range: $y \ge -3$; $f^{-1}(x)$: domain: $x \ge -3$, range:
all real numbers **ii) a)** $f^{-1}(x) = \pm\sqrt{x - 1}$ **c)** $f(x)$: domain: all
real numbers, range: $y \ge 1$; $f^{-1}(x)$: domain: $x \ge 1$, range: all
real numbers **iii) a)** $f^{-1}(x) = \pm\sqrt{-x}$ **c)** $f(x)$: domain: all real
numbers, range: $y \le 0$; $f^{-1}(x)$: domain $x \le 0$, range: all real
numbers **iv) a)** $f^{-1}(x) = \pm\sqrt{-x - 1}$ **c)** $f(x)$: domain: all real
numbers, range $y \le -1$; $f^{-1}(x)$: domain: $x \le -1$, range: all
real numbers **v) a)** $f^{-1}(x) = \pm\sqrt{x + 2}$ **c)** $f(x)$: domain: all real
numbers, range: $y \ge 0$; $f^{-1}(x)$: domain: $x \ge 0$, range: all real
numbers **vi) a)** $f^{-1}(x) = \pm\sqrt{x - 1}$ **c)** $f(x)$: domain: all real
numbers, range: $y \ge 0$; $f^{-1}(x) =$ domain: $x \ge 0$, range: all real
numbers **11. a)** $y = \dfrac{x - 3}{2}$ **b)** $y = \pm\sqrt{x - 4}$ **12. a)** no

b) no **13. a)** $f^{-1}(x) = \dfrac{x + 3}{2}$, domain: all real numbers,

range: all real numbers **b)** $f^{-1}(x) = \dfrac{2 - x}{4}$, domain: all real

numbers, range: all real numbers **c)** $f^{-1}(x) = \dfrac{x + 6}{3}$, domain:

all real numbers, range: all real numbers **d)** $f^{-1}(x) = 2x + 6$,
domain: all real numbers, range: all real numbers
e) $f^{-1}(x) = \pm\sqrt{x}$ **f)** $f^{-1}(x) = \pm\sqrt{x - 2}$ **g)** $f^{-1}(x) = \pm\sqrt{x + 4}$

h) $f^{-1}(x) = \pm\dfrac{\sqrt{2x + 2}}{2}$ **i)** $f^{-1}(x) = \pm\sqrt{x + 3}$ **j)** $f^{-1}(x) = \pm\sqrt{x} - 2$

14. i) a) $f^{-1}(x) = \sqrt{x}$ **c)** $f(x)$: domain: $x \ge 0$, range: $y \ge 0$;
$f^{-1}(x)$: domain: $x \ge 0$, range: $y \ge 0$ **ii) a)** $f^{-1}(x) = \sqrt{x + 2}$
c) $f(x)$: domain: $x \ge 0$, range: $y \ge -2$; $f^{-1}(x)$: domain:
$x \ge -2$, range: $y \ge 0$ **iii) a)** $f^{-1}(x) = -\sqrt{x - 4}$ **c)** $f(x)$: domain:
$x \le 0$, range: $y \ge 4$; $f^{-1}(x)$: domain: $x \ge 4$, range: $y \le 0$
iv) a) $f^{-1}(x) = \sqrt{3 - x}$ **c)** $f(x)$: domain: $x \ge 0$, range: $y \le 3$;
$f^{-1}(x)$: domain: $x \le 3$, range: $y \ge 0$ **v) a)** $f^{-1}(x) = \sqrt{x + 4}$
c) $f(x)$: domain: $x \ge 4$, range: $y \ge 0$; $f^{-1}(x)$: domain: $x \ge 0$,
range: $y \ge 4$ **vi) a)** $f^{-1}(x) = -\sqrt{x} - 3$ **c)** $f(x)$: domain: $x \le -3$,
range: $y \ge 0$; $f^{-1}(x)$: domain: $x \ge 0$, range: $y \le -3$
15. a) $f^{-1}(x) = x^2 + 2$, $x \ge 0$ **b)** $f^{-1}(x) = 3 - x^2$, $x \ge 0$
c) $f^{-1}(x) = \pm\sqrt{x^2 - 9}$, $x \ge 3$ **16. i) a)** $f^{-1}(x) = \pm\sqrt{x - 3}$

c) domain: $x \ge 0$ **ii) a)** $f^{-1}(x) = \pm\dfrac{\sqrt{2x}}{2}$ **c)** domain: $x \ge 0$

iii) a) $f^{-1}(x) = \pm\sqrt{x + 1}$ **c)** domain: $x \ge 0$ **iv) a)** $f^{-1}(x) = \pm\sqrt{-x}$
c) domain: $x \ge 0$ **v) a)** $f^{-1}(x) = \pm\sqrt{1 - x}$ **c)** $x \ge 0$
vi) a) $f^{-1}(x) = \pm\sqrt{x + 2}$ **c)** $x \ge 2$ **vii) a)** $f^{-1}(x) = \pm\sqrt{x + 4}$ **c)** $x \ge 4$
viii) a) $f^{-1}(x) = \pm\sqrt{-x} - 5$ **c)** $x \ge -5$ **17. a)** $f^{-1}(x) = \dfrac{1}{x}$ **b)** Yes;
for every value of x there is only one corresponding value
of y. **18. a)** $f^{-1}(x) = x^2$, $x \ge 0$ **b)** yes **19. a)** $c(d) = 50 + 0.15d$

b) $d(c) = \dfrac{c - 50}{0.15}$ **c)** The inverse represents the distance that

can be driven for a given rental cost. **d)** $c \ge 50$

20. a) $f(x) = 2\pi x$, $x \ge 0$ **b)** $f^{-1}(x) = \dfrac{x}{2\pi}$ **c)** yes

d) The inverse represents the length of the radius for a circle
of a given circumference. **21. a)** $f(x) = 4\pi x^2$, $x \ge 0$

b) $f^{-1}(x) = \dfrac{\sqrt{\pi x}}{2\pi}$ **c)** domain: $x \ge 0$, range: $y \ge 0$ **d)** yes **e)** The

inverse represents the length of the radius for a sphere of a
given surface area. **22. a)** $f(x) = 0.7x$ **b)** $f^{-1}(x) = \dfrac{x}{0.7}$ **c)** The

inverse represents the original selling price as a function of
the sale price. **23. a)** $u(c) = 0.7c$ **b)** $c(u) = 1.43u$ **c)** $214.50

24. a) $T(d) = 35d + 20$ **b)** $d(T) = \dfrac{T - 20}{35}$ **c)** 2 km

25. a) $E(s) = 400 + 0.05s$ **b)** $s(E) = 20E - 8000$ **c)** The inverse
represents the amount of sales as a function of earnings.

d) $3500 **26. a)** 128.6° **b)** $n(i) = \dfrac{360}{180 - i}$ **c)** decagon

27. b) $t(h) = \dfrac{\sqrt{400 - 5h}}{5}$ **c)** yes, since the domain of h is

restricted to $t \geq 0$ **d)** The inverse represents the time for an object to fall from a height of 80 m to h metres above the ground. **e)** 3 s **f)** 4 s **28. a)** $-f(x) = -2x + 4$, $f(-x) = -2x - 4$, $f^{-1}(x) = \dfrac{x+4}{2}$ **c)** $-f(x)$: (2, 0), $f(-x)$: (0, −4), $f^{-1}(x)$: (4, 4) **29. a)** $-f(x) = 3x - 2$, $f(-x) = 3x + 2$, $f^{-1}(x) = \dfrac{2-x}{3}$ **c)** $-f(x)$: $\left(\dfrac{2}{3}, 0\right)$, $f(-x)$: (0, 2), $f^{-1}(x)$: $\left(\dfrac{1}{2}, \dfrac{1}{2}\right)$ **30. a)** $-f(x) = -\sqrt{x+3}$, $f(-x) = \sqrt{3-x}$, $f^{-1}(x) = x^2 - 3$, $x \geq 0$ **31.** $y = 2$ **33.** Answers may vary. $f(x) = x$, $f(x) = \dfrac{1}{x}$, $f(x) = 1 - x$, $f(x) = \dfrac{3-x}{2x+1}$

34. a) Yes; the function f could be $f(x) = x^2 - 6x + 11$, for example. Then, $f(2) = 3$, and since $f(4) = 3$, (3, 4) is included in f^{-1}. **b)** No; if the inverse includes (4, 2), then f must include (2, 4). This contradicts the fact that f is a function, since f already includes (2, 3). **35.** 4 square units **36. a)** Yes; to each value of x there corresponds only one value of y, namely k. **b)** No; the inverse is a vertical line, $x = k$, which is not a function. **37.** the original function **38.** the reciprocal **39. a)** infinitely many ways **b)** infinitely many ways

Section 3.6, pp. 229–232

1. a) domain: $-4 \leq x \leq 4$, range: $0 \leq y \leq 8$ **b)** domain: $-4 \leq x \leq 4$, range: $0 \leq y \leq 2$ **c)** domain: $-2 \leq x \leq 2$, range: $0 \leq y \leq 4$ **d)** domain: $-8 \leq x \leq 8$, range: $0 \leq y \leq 4$ **e)** domain: $-8 \leq x \leq 8$, range: $0 \leq y \leq 8$ **f)** domain: $-8 \leq x \leq 8$, range: $0 \leq y \leq 2$ **g)** domain: $-2 \leq x \leq 2$, range: $0 \leq y \leq 8$ **h)** domain: $-2 \leq x \leq 2$, range: $0 \leq y \leq 2$ **2. a)** domain: $-3 \leq x \leq 3$, range: $0 \leq y \leq 18$ **b)** domain: $-3 \leq x \leq 3$, range: $0 \leq y \leq 3$ **c)** domain: $-1 \leq x \leq 1$, range: $0 \leq y \leq 9$ **d)** domain: $-9 \leq x \leq 9$, range: $0 \leq y \leq 9$ **e)** domain: $-6 \leq x \leq 6$, range: $0 \leq y \leq 9$ **f)** domain: $-1 \leq x \leq 1$, range: $0 \leq y \leq 3$

3. a) $g(x) = 2f(x)$, $h(x) = 3f(x)$ **b)** $g(x) = f(2x)$, $h(x) = f\left(\dfrac{1}{2}x\right)$

4. i) b) vertical expansion by a factor of 2; vertical compression by a factor of $\dfrac{1}{2}$ **c)** (0, 0) **ii) b)** vertical expansion by a factor of 3; vertical compression by a factor of $\dfrac{1}{2}$ **c)** (0, 0) **iii) b)** vertical expansion by a factor of 3; vertical expansion by a factor of 1.5 **c)** (0, 0) **iv) b)** vertical expansion by a factor of 4 or horizontal compression by a factor of 2; vertical compression by a factor of $\dfrac{1}{4}$ or horizontal expansion by a factor of $\dfrac{1}{4}$ **c)** (0, 0) **5. a)** vertical expansion by a factor of 3 **b)** vertical compression by a factor of $\dfrac{1}{2}$ **c)** vertical expansion by a factor of 2 **d)** vertical compression

by a factor of $\dfrac{1}{3}$ **e)** horizontal compression by a factor of $\dfrac{1}{2}$ **f)** horizontal expansion by a factor of 2 **g)** horizontal compression by a factor of $\dfrac{1}{4}$ **6. a)** vertical expansion by a factor of 3 and horizontal compression by a factor of $\dfrac{1}{2}$ **b)** vertical compression by a factor of $\dfrac{1}{2}$ and horizontal expansion by a factor of 3 **c)** vertical expansion by a factor of 4 and horizontal expansion by a factor of 2 **d)** vertical compression by a factor of $\dfrac{1}{3}$ and horizontal compression by a factor of $\dfrac{1}{3}$ **e)** vertical expansion by a factor of 2 and horizontal compression by a factor of $\dfrac{1}{4}$ **f)** vertical expansion by a factor of 5 and horizontal expansion by a factor of 2 **7. a)** $f(x) = \dfrac{1}{2}x^2$ **b)** $f(x) = 2x + 2$ **c)** $f(x) = 2\sqrt{x}$ **8. a)** −4, 2 **b)** −8, 4 **c)** −2, 1 **9. a)** 38.4 m, 57.6 m, 256 m **b)** domain: $0 \leq s \leq 120$, ranges: $0 \leq d \leq 86.4$, $0 \leq d \leq 129.6$, $0 \leq d \leq 576$ **d)** The three functions can be obtained from $y = x^2$ by vertical compressions or horizontal expansions. **10. a)** $a = 4$, $k = 1$ **b)** $a = 1$, $k = 3$ **c)** $a = \dfrac{1}{2}$, $k = \dfrac{1}{3}$ **d)** $a = 2$, $k = 4$ **11. a)** domain: $-4 \leq x \leq 4$, range: $0 \leq y \leq 12$ **b)** domain: $-4 \leq x \leq 4$, range: $0 \leq y \leq 2$ **c)** domain: $-2 \leq x \leq 2$, range: $0 \leq y \leq 4$ **d)** domain: $-8 \leq x \leq 8$, range: $0 \leq y \leq 4$ **e)** domain: $-4 \leq x \leq 4$, range: $0 \leq y \leq 8$ **f)** domain: $-1 \leq x \leq 1$, range: $0 \leq y \leq 4$ **12. a)** 0, −1, 2 **b)** 0, −3, 6 **c)** 0, −6, 12 **d)** 0, $-\dfrac{3}{2}$, 3 **13.** The first is a vertical compression of the second by a factor of $\dfrac{1}{3}$. **14. b)** horizontal compression by a factor of 0.4 **c)** The point is invariant under the transformation because it is the height from which the object is dropped. **d)** Earth: domain: $0 \leq t \leq 2$, range: $0 \leq h \leq 20$, Moon: domain: $0 \leq t \leq 5$, range: $0 \leq h \leq 20$ **16. a)** horizontal compression by a factor of $\dfrac{1}{4}$ **b)** vertical expansion by a factor of 2 **c)** equal **d)** The transformations are equivalent for this function.

Section 3.7, pp. 240–243

1. a) vertical expansion by a factor of 2, translation upward 3 units **b)** vertical compression by a factor of $\dfrac{1}{2}$, translation downward 2 units **c)** translation left 4 units and upward 1 unit **d)** vertical expansion by a factor of 3, translation right 5 units **e)** horizontal expansion by a factor of 2, translation downward 6 units **f)** horizontal compression by a factor of $\dfrac{1}{2}$, translation upward 1 unit **2. a)** vertical expansion by a factor of 2, reflection in the x-axis **b)** vertical compression by a factor of $\dfrac{1}{3}$, reflection in the x-axis **c)** horizontal

compression by a factor of $\frac{1}{4}$, reflection in the y-axis
d) horizontal expansion by a factor of 2, reflection in the y-axis **3. a)** horizontal compression by a factor of $\frac{1}{2}$, reflection in the x-axis **b)** vertical expansion by a factor of 3, horizontal compression by a factor of $\frac{1}{2}$, reflection in the y-axis **c)** vertical compression by a factor of $\frac{1}{2}$, horizontal expansion by a factor of 3, reflection in the x-axis **d)** vertical expansion by a factor of 4, translation right 6 units and upward 2 units **e)** vertical expansion by a factor of 2, reflection in the x-axis, translation downward 3 units **f)** reflection in the x-axis, translation right 3 units and upward 1 unit **g)** vertical expansion by a factor of 3, horizontal compression by a factor of $\frac{1}{2}$, translation downward 6 units **h)** vertical compression by a factor of $\frac{1}{2}$, horizontal expansion by a factor of 2, translation downward 4 units **4. a)** domain: $-2 \le x \le 6$, range: $2 \le y \le 6$; no invariant points **b)** domain: $-8 \le x \le 0$, range: $-4 \le y \le 0$; no invariant points **c)** domain: $-6 \le x \le 2$, range: $-3 \le y \le -1$; no invariant points **d)** domain: $-3 \le x \le 1$, range: $3 \le y \le 7$; no invariant points **e)** domain: $-6 \le x \le 2$, range: $-8 \le y \le 0$; invariant points: $(-6, 0)$, $(2, 0)$ **f)** domain: $-2 \le x \le 6$, range: $-2 \le y \le 2$; no invariant points **g)** domain: $-4 \le x \le 12$, range: $0 \le y \le 4$; invariant point: $(0, 4)$ **h)** domain: $-1 \le x \le 3$, range: $-2 \le y \le 0$; no invariant points **5. a)** domain: $1 \le x \le 7$, range: $0 \le y \le 8$ **b)** domain: $-2 \le x \le 4$, range: $-4 \le y \le 4$ **c)** domain: $-1 \le x \le 2$, range: $-6 \le y \le 2$ **d)** domain: $-2 \le x \le 4$, range: $-11 \le y \le 13$ **e)** domain: $-4 \le x \le 2$, range: $-1 \le y \le 7$ **f)** domain: $-4 \le x \le 2$, range: $-6 \le y \le 10$ **7. a)** horizontal compression by a factor of $\frac{1}{2}$, translation right 4 units **b)** reflection in the y-axis, translation left 1 unit and downward 1 unit
c) horizontal compression by a factor of $\frac{1}{3}$, translation left 4 units and upward 5 units **d)** vertical expansion by a factor of 2, horizontal compression by a factor of $\frac{1}{4}$, reflection in the x-axis, translation right 2 units **e)** reflection in the y-axis, translation right 2 units **f)** horizontal compression by a factor of $\frac{1}{2}$, translation left 4 units and downward 4 units
g) reflection in the y-axis, translation right 4 units and upward 5 units **h)** horizontal compression by a factor of $\frac{1}{3}$, translation right 2 units and upward 8 units **8. a)** domain: all real numbers, range: $y \ge 1$ **b)** domain: all real numbers, range: $y \ge -4$ **c)** domain: all real numbers, range: $y \ge 3$ **d)** domain: all real numbers, range: $y \le -3$ **e)** domain: all real numbers, range: $y \ge 2$ **f)** domain: all real numbers, range: $y \le -2$ **9. a)** domain: $x \ge 5$, range: $y \ge -4$ **b)** domain:

$x \ge -3$, range: $y \ge 2$ **c)** domain: $x \ge 1$, range: $y \ge -2$
d) domain: $x \ge 3$, range: $y \ge 1$ **e)** domain: $x \le 0$, range: $y \le 5$
f) domain: $x \le 4$, range: $y \le -3$ **10.** $y = -(x - 2)^2 - 3$
11. $k(x) = 3(x - 4)^2 - 2$ **12.** $g(x) = \sqrt{\dfrac{6 - x}{2}}$
13. b) equal sides: $2\sqrt{34}$, other side: 12, height: 10
c) vertical expansion by a factor of $\frac{5}{3}$, translation upward 10 units; vertical expansion by a factor of $\frac{5}{3}$, reflection in the x-axis, translation upward 10 units
14. a) vertical expansion by a factor of 5, reflection in the x-axis, translation right 4 units and upward 80 units **b)** The vertex $(0, 0)$ of $y = x^2$ is translated to $(4, 80)$. Since the new parabola opens downward, the maximum height is 80 m after 4 s. **15. a)** -2, 6 **b)** 1, -3 **c)** 2, -6 **d)** -4, 12 **e)** -3, 5 **f)** 0, -8 **16. a)** domain: $-4 \le x \le 4$, range: $-2 \le y \le 0$
b) domain: $-2 \le x \le 2$, range: $0 \le y \le 4$ **c)** domain: $-4 \le x \le 4$, range: $5 \le y \le 7$ **d)** domain: $-4 \le x \le 4$, range: $-5 \le y \le -3$ **e)** domain: $-4 \le x \le 4$, range: $-5 \le y \le -1$
f) domain: $-10 \le x \le -2$, range: $-3 \le y \le 5$ **17. a)** $a = 3$, $k = -1$, $h = 0$, $q = 0$ **b)** $a = \frac{1}{3}$, $k = \frac{1}{2}$, $h = 6$, $q = -1$ **c)** $a = -2$, $k = -2$, $h = -7$, $q = 4$ **18. a)** The point $(5, 0)$ is in common, and means that an object falls 125 m on Earth in the same time that an object falls 20 m on the moon. **c)** horizontal compression by a factor of 4, translation upward 300 units; or vertical expansion by a factor of 16 **d)** The point $(5, 0)$ is in common, and means that an object falls 320 m on Jupiter in the same time that an object falls 20 m on the moon.
e) moon: domain: $0 \le t \le 5$, range: $0 \le h \le 20$; Jupiter: domain: $0 \le t \le 5$, range: $0 \le h \le 320$ **19. a)** vertical expansion by a factor of 2, translation left 1 unit
b) horizontal compression by a factor of $\frac{1}{4}$, translation left 1 unit **c)** equal **20.** reflection in the line $y = x$, translation left 2 units and downward 3 units **21. a)** equal **b)** No; the result is $y = 2x^2 + 12x + 13$.

Review of Key Concepts, pp. 246–253

1. a) a function **b)** not a function **c)** not a function **d)** a function **e)** a function **f)** not a function **2. a)** domain: all real numbers, range: $y \ge -4$ **b)** domain: $-3 \le x \le 5$, range: $-2 \le y \le 4$ **c)** domain: $-4 \le x \le 5$, range: $-2 \le y \le 7$
d) domain: all real numbers, range: $y \le 3$ **3. a)** -5 **b)** -17 **c)** 7 **d)** 10 **e)** 19 **f)** 5.5 **g)** 7.3 **h)** -293 **i)** $-14\ 993$ **4. a)** -5 **b)** -8 **c)** -2 **d)** 20 **e)** -7 **f)** -5 **g)** -8.12 **h)** 7252 **i)** 19 892 **5. a)** $\{-17, -9, 3, 11\}$ **b)** $\{-34, -6, -9, -41\}$ **6. a)** $15a + 2$ **b)** $32 - 7n$ **c)** $12k^2 + 40k + 28$ **7. a)** dependent variable: C, independent variable: b **b)** Yes; only one cost is associated with a given number of bottles. **8. a)** $SA = 4\pi r^2$ **b)** domain: $r \ge 0$, range: $SA \ge 0$ **9. a)** $x \ge 0$ **b)** $y \ge 0$ **10.** The graph of $y = \sqrt{x}$ is the reflection of the right half of $y = x^2$ in the line

$y = x$. **11. a)** $x \neq 0$ **b)** $y \neq 0$ **12.** The graph of $y = \dfrac{1}{x}$ is the reciprocal of the graph of $y = x$, except at the origin where $\dfrac{1}{x}$ is undefined. **13. a)** translation downward 3 units **b)** translation to the left 6 units **c)** translation to the right 4 units and downward 5 units **d)** translation downward 5 units **14. a)** $y = (x + 3)^2 - 2$ **b)** $y = \sqrt{x - 1} + 2$ **17. a)** domain: $x \geq 0$, range: $y \geq -4$ **b)** domain: $x \geq -5$, range: $y \geq 0$ **c)** domain: $x \geq 2$, range: $y \geq 3$ **d)** domain: $x \geq -4$, range: $y \geq -3$ **18. a)** domain: all real numbers, range: $y \geq 0$ **b)** domain: all real numbers, range: $y \geq 0$ **c)** domain: all real numbers, range: $y \geq -6$ **d)** domain: all real numbers, range: $y \geq 4$ **19. a)** a fixed cost, independent of the number of days **b)** domain: $0 \leq d \leq 62$, range: $100 \leq C \leq 7540$ **c)** vertical translation upward 100 units **d)** $C(d) = 120d + 60$ **21. a)** f: domain: $x \geq -2$, range: $y \geq 0$; h: domain: $x \leq 2$, range: $y \geq 0$ **b)** f: domain: all real numbers, range: all real numbers; k: domain: all real numbers, range: all real numbers **c)** f: domain: all real numbers, range: $y \geq 6$; g: domain: all real numbers, range: $y \geq 6$ **22. a)** $-f(x) = 2x - 3$, $f(-x) = 2x + 3$ **c)** $-f(x)$: $\left(\dfrac{3}{2}, 0\right)$, $f(-x)$: $(0, 3)$ **23. a)** $-f(x) = -x^2 - 6x$, $f(-x) = x^2 - 6x$ **c)** $-f(x)$: $(0, 0)$, $(-6, 0)$, $f(-x)$: $(0, 0)$ **24. a)** $-f(x) = 4 - x^2$, $f(-x) = x^2 - 4$ **c)** $-f(x)$: $(-2, 0)$, $(2, 0)$, $f(-x)$: $(0, -4)$ **25. a)** $-f(x) = 4 - \sqrt{x}$, $f(-x) = \sqrt{-x} - 4$ **c)** $-f(x)$: $(16, 0)$, $f(-x)$: $(0, -4)$ **26. a)** $-f(x) = -\sqrt{x} - 5$, $f(-x) = \sqrt{-x} - 5$ **c)** $-f(x)$: $(5, 0)$, $f(-x)$: none **27. a)** $y = 1.6x + 4$ **b)** 4 m **c)** 5 m **d)** left half: domain: $-2.5 \leq x \leq 0$, range: $0 \leq y \leq 4$; right half: domain: $0 \leq x \leq 2.5$, range: $0 \leq y \leq 4$ **28. b)** Yes; to each x-coordinate, there corresponds only one y-coordinate. **29. a)** $f^{-1}(x) = x - 7$; f: domain: all real numbers, range: all real numbers; f^{-1}: domain: all real numbers, range: all real numbers **b)** $f^{-1}(x) = 3x + 4$; f: domain: all real numbers, range: all real numbers; f^{-1}: domain: all real numbers, range: all real numbers **c)** $f^{-1}(x) = \dfrac{x + 1}{3}$; f: domain: all real numbers range: all real numbers; f^{-1}: domain: all real numbers, range: all real numbers **d)** $f^{-1}(x) = \pm\sqrt{x + 5}$; f: domain: all real numbers, range: $y \geq -5$; f^{-1}: domain: $x \geq -5$, range: all real numbers **e)** $f^{-1}(x) = \pm\sqrt{x - 2}$; f: domain: all real numbers, range: $y \geq 0$; f^{-1}: domain: $x \geq 0$, range: all real numbers **f)** $f^{-1}(x) = x^2 + 3$; f: domain: $x \geq 3$, range: $y \geq 0$; f^{-1}: domain: $x \geq 0$, range: $y \geq 3$ **31. a)** yes **b)** no **32. a)** $f^{-1}(x) = \pm\sqrt{3 - x}$ **c)** $x \geq 0$ **e)** f: domain: $x \geq 0$, range: $y \leq 3$; f^{-1}: domain: $x \leq 3$, range: $y \geq 0$ **33. a)** $(-1, -1)$ **b)** $(2, 2)$ **34. a)** $s(p) = 0.6p$ **b)** $p(s) = \dfrac{s}{0.6}$ **c)** The inverse finds the original price as a function of the sale price. **35. a)** domain: $-4 \leq x \leq 4$, range: $-6 \leq y \leq 6$ **b)** domain: $-8 \leq x \leq 8$, range: $-2 \leq y \leq 2$ **c)** domain: $-8 \leq x \leq 8$, range: $-4 \leq y \leq 4$ **36. i) b)** vertical expansion by

a factor of 4; vertical compression by a factor of 0.5 **c)** $(0, 0)$ **ii) b)** vertical expansion by a factor of 2; vertical compression by a factor of 0.5 **c)** $(0, 0)$ **iii) b)** vertical expansion by a factor of 2; vertical compression by a factor of $\dfrac{1}{4}$ **c)** $(-3, 0)$

37. a) 6, -4 **b)** 3, -2 **c)** 12, -8 **38. a)** domain: $-6 \leq x \leq 6$, range: $0 \leq y \leq 12$ **b)** domain: $-6 \leq x \leq 6$, range: $0 \leq y \leq 3$ **c)** domain: $-2 \leq x \leq 2$, range: $0 \leq y \leq 6$ **39. a)** vertical expansion by a factor of 4 **b)** horizontal compression by a factor of $\dfrac{1}{3}$ **c)** vertical compression by a factor of $\dfrac{1}{2}$, horizontal compression by a factor of $\dfrac{1}{2}$ **d)** vertical expansion by a factor of 3, horizontal expansion by a factor of 2 **41. a)** vertical expansion by a factor of 3, translation left 2 units and downward 4 units **b)** vertical expansion by a factor of 2, reflection in the x-axis, translation upward 5 units **c)** vertical compression by a factor of 0.5, horizontal compression by a factor of $\dfrac{1}{4}$, translation upward 2 units **d)** horizontal compression by a factor of $\dfrac{1}{2}$, translation left 1 unit **e)** horizontal expansion by a factor of 2, reflection in the x-axis, translation right 3 units and upward 1 unit **f)** vertical expansion by a factor of 3, reflection in the y-axis, translation right 4 units and downward 7 units **42. a)** domain: all real numbers, range: $y \geq -4$ **b)** domain: all real numbers, range: $y \leq 6$ **c)** domain: all real numbers, range: $y \geq -3$ **d)** domain: all real numbers, range: $y \leq 3$ **43. a)** domain: $x \geq 3$, range: $y \geq 4$ **b)** domain: $x \geq -4$, range: $y \leq -5$ **c)** domain: $x \leq 2$, range: $y \leq -2$ **d)** domain: $x \leq 6$, range: $y \geq 2$ **44.** $g(x) = 4\sqrt{-x - 5} + 4$ **46. a)** $N = 1.06S + 2000$ **b)** Yes; if the productivity raise is given first, then this amount would be included when calculating the 6% increase.

Chapter Test, pp. 254–256

1. a) no **b)** yes **c)** yes **2. a)** domain: $-3 \leq x \leq 2$, range: $-4 \leq y \leq 5$ **b)** domain: $x \geq 0$, range: $y \geq -4$ **c)** domain: $0 \leq x \leq 4$, range: $-2 \leq y \leq 2$ **3. a)** -47 **b)** -5 **c)** -47 **d)** 3 **e)** -95 **f)** 2.5 **g)** $3 - 18a^2$ **h)** $1 - 8a^2 + 8a$ **4. a)** Domain: $x \geq 0$, range: $y \geq 0$; the function increases continuously from the origin to the right. **b)** Domain: $x \neq 0$, range: $y \neq 0$; each branch in the first and third quadrants is decreasing. **6. a)** translation upward 4 units **b)** translation to the right 2 units and downward 3 units **c)** reflection in the x-axis, translation to the left 5 units and downward 1 unit **d)** vertical compression by a factor of $\dfrac{1}{3}$, horizontal compression by a factor of $\dfrac{1}{3}$, reflection in the y-axis, translation upward 5 units **e)** vertical expansion by a factor of 2, horizontal compression by a factor of $\dfrac{1}{2}$, reflection in the x-axis, translation left 3 units and upward 6 units

f) vertical expansion by a factor of 2, reflection in the y-axis, translation right 2 units and downward 4 units

7. a) $y = \sqrt{x+2} - 3$; domain: $x \geq -2$, range: $y \geq -3$
b) $y = \dfrac{1}{2}x$; domain: all real numbers, range: all real numbers **c)** $y = -(x+2)^2 + 4$; domain: all real numbers, range: $y \leq 4$ **8. a)** $f^{-1}(x) = \dfrac{x+5}{3}$; a function
b) $f^{-1}(x) = \pm\sqrt{x+7}$; not a function **10.** $y = 2(x+3)^2 + 4$; domain: all real numbers, range: $y \geq 4$ **11.** $y = -\sqrt{2x+8}$; domain: $x \geq -4$, range: $y \leq 0$ **12. a)** $-f(x) = 10x - x^2$, $f(-x) = x^2 + 10x$ **c)** $-f(x)$: $(0, 0)$, $(10, 0)$, $f(-x)$: $(0, 0)$
d) $f(x)$: domain: all real numbers, range: $y \geq -25$; $-f(x)$: domain: all real numbers, range: $y \leq 25$; $f(-x)$: domain: all real numbers, range: $y \geq -25$ **13. a)** $-f(x) = 3 - \sqrt{x}$, $f(-x) = \sqrt{-x} - 3$ **c)** $-f(x)$: $(9, 0)$, $f(-x)$: $(0, -3)$ **d)** $f(x)$: domain: $x \geq 0$, range: $y \geq -3$; $-f(x)$: domain: $x \geq 0$, range: $y \leq 3$; $f(-x)$: domain: $x \leq 0$, range: $y \geq -3$ **14. a)** 40
b) $h = \dfrac{25g^2}{81}$; The inverse finds the number of hours required to manufacture a given number of doors. **c)** 193 h
15. a) \$1000 increase: vertical translation upward 1000 units; 2.75% raise: vertical expansion by a factor of 1.0275 **b)** \$1000 increase: $S(t) = 26\,000 + 3000t - 150t^2$, $S(t) = 500\sqrt{t - 10} + 41\,000$; 2.75% raise: $25\,687.5 + 3082.5t - 154.125t^2$, $S(t) = 513.75\sqrt{t - 10} + 41\,100$ **c)** $0 \leq t \leq 10$ and $10 < t \leq 45$
d) The \$1000 increase is better since it gives you a salary of \$37 250, while the 2.75% raise gives you a salary of only \$37 246.88. **e)** The 2.75% raise is better since it gives you a salary of \$42 641.25, while the \$1000 increase gives you a salary of only \$42 500.

Challenge Problems, p. 257

1. a) neither **b)** even **c)** neither **d)** odd **e)** neither
2. $3x - 2y + 6 = 0$ **3.** When n is odd, f is symmetric about the origin, and when n is even, f is symmetric about the y-axis.
4. $f^{-1}(x) = \dfrac{b - dx}{cx - a}$ **5.** 5 **6.** $b = -7$, $c = -12$, $d = 25$

Problem Solving Strategy: Solve Rich Estimation Problems, p. 260

1–18. Answers will vary.

Problem Solving: Using the Strategies, p. 261

1. A: 60 kg, E: 61 kg, B: 62 kg, D: 63 kg, C: 64 kg **2.** 25°
3. X = 2, Y = 7 **4.** 1¢, 2¢, 4¢, 7¢ **5.** A: 11 kg, B: 13 kg, C: 14 kg, D: 9 kg, E: 7 kg **6.** 11 **7.** 12 km **8.** 19, 22, 7; 4, 16, 28; 25, 10, 13 **9.** 9 **10. a)** 48 or 60

Chapter 4

Getting Started, p. 264

1. 8.5×10^{-5} degrees **2.** 1.0×10^{14} km **3.** 10.7 light-years
4. 1.6×10^{-4} degrees

Review of Prerequisite Skills, p. 265

1. a) 10 cm; $\sin A = \dfrac{4}{5}$, $\cos A = \dfrac{3}{5}$, $\tan A = \dfrac{4}{3}$; $\sin C = \dfrac{3}{5}$, $\cos C = \dfrac{4}{5}$, $\tan C = \dfrac{3}{4}$ **b)** 5 m; $\sin X = \dfrac{5}{13}$, $\cos X = \dfrac{12}{13}$, $\tan X = \dfrac{5}{12}$; $\sin Z = \dfrac{12}{13}$, $\cos Z = \dfrac{5}{13}$, $\tan Z = \dfrac{12}{5}$ **c)** 29 m; $\sin L = \dfrac{21}{29}$, $\cos L = \dfrac{20}{29}$, $\tan L = \dfrac{21}{20}$; $\sin M = \dfrac{20}{29}$, $\cos M = \dfrac{21}{29}$, $\tan M = \dfrac{20}{21}$ **d)** 24 cm; $\sin P = \dfrac{7}{25}$, $\cos P = \dfrac{24}{25}$, $\tan P = \dfrac{7}{24}$; $\sin R = \dfrac{24}{25}$, $\cos R = \dfrac{7}{25}$, $\tan R = \dfrac{24}{7}$
2. a) 0.454 **b)** 0.559 **c)** 4.705 **d)** 0.993 **e)** 0.839 **f)** 0.883
3. a) 37° **b)** 72° **c)** 24° **d)** 66° **e)** 78° **f)** 18° **4. a)** 7.8 cm
b) 7.5 cm **c)** 10.8 m **d)** 4.1 m **e)** 12.9 cm **f)** 94.6 m **5. a)** 39°
b) 50° **c)** 37° **d)** 57° **e)** 39° **f)** 44° **6. a)** 37.17 **b)** 161.81
c) 78.37 **d)** 17.59 **e)** 2.16 **f)** 0.99

Section 4.1, pp. 272–275

1. a) $\angle A = 57°$, $a = 47$ cm, $b = 30$ cm **b)** $\angle F = 49°$, $d = 80$ m, $e = 52$ m **c)** $\angle T = 42°$, $\angle U = 48°$, $u = 11$ m
d) $\angle P = 32°$, $\angle R = 58°$, $q = 15$ cm **2. a)** $\angle X = 26.5°$, $x = 4.3$ cm, $y = 8.6$ cm **b)** $\angle M = 32.6°$, $m = 13.0$ m, $n = 24.1$ m **c)** $\angle G = 31.5°$, $\angle I = 58.5°$, $i = 20.9$ m
d) $\angle K = 43.6°$, $\angle L = 46.4°$, $j = 99.8$ cm **3. a)** $\angle Y = 63.8°$, $\angle Z = 26.2°$, $y = 8.5$ cm **b)** $\angle L = 53°$, $k = 7.4$ cm, $l = 9.8$ cm **c)** $\angle C = 34.9°$, $a = 5.9$ m, $c = 3.3$ m
d) $\angle D = 50.7°$, $\angle F = 39.3°$, $e = 23.5$ cm **4. a)** 49.4°
b) 12.4° **c)** 44.7° **d)** 71.6° **5.** 7.3 m **6.** 27.4 cm **7.** 13.5 m
8. 88.6 cm **9.** 3.7 cm **10.** 41 m **11.** 242 m **12.** 15 950 km
13. 50.6 m **14.** 49.7 m² **15. a)** 37 670 km **16.** 491 m
17. The radius, r, of any parallel of latitude is equal to the radius of the equator, r_e, times the cosine of the latitude angle, θ: $r = r_e \cos \theta$. The length of the parallel of latitude is $2\pi r$. So $2\pi r = 2\pi r_e \times \cos \theta$, and $2\pi r_e$ is the length of the equator. **18.** 26 299 cm³ **19.** 58° **20. a)** If the base of $\triangle ABC$ is a, then the height is $c \sin B$, and the area is equal to $0.5bh = 0.5ac \sin B$. **b)** If the base of $\triangle ABC$ is a, then the height is $b \sin C$, and the area is equal to $0.5bh = 0.5ab \sin C$. **c)** $A = 0.5bc \sin A$ **d)** The area is half the product of two side lengths and the sine of the contained angle. **e)** the length of two sides and the measure of the contained angle **f)** 28.9 m² **g)** 168.3 cm² **h)** yes

Section 4.2, pp. 281–282

1. a) $\sin A = \dfrac{4}{5}$, $\cos A = \dfrac{3}{5}$ **b)** $\sin A = \dfrac{12}{13}$, $\cos A = -\dfrac{5}{13}$
2. a) -0.8090 **b)** 0.9659 **c)** 0.2250 **d)** -0.0349 **e)** -0.7034
f) 0.8545 **g)** 0.0209 **h)** -0.3923 **i)** 0.4003 **3. a)** $30°$ or $150°$
b) $60°$ **c)** $120°$ **d)** $14.9°$ or $165.1°$ **e)** $62.9°$ **f)** $124.1°$ **g)** $35.0°$
or $145.0°$ **h)** $95.0°$ **i)** $90°$ **j)** $180°$ **4.** $90°$ **5.** $45°$ and $135°$
6. a) $\dfrac{24}{25}$ **b)** $\dfrac{7}{24}$, $-\dfrac{7}{24}$ **7.** The sine of the angle starts at 0,
increases to 1 when the angle is 90°, and decreases to 0
when the book is closed. The cosine of the angle starts at −1,
increases to 0 when the angle is 90°, and increases to 1 when
the book is closed. **8. a)** 0 cm, 7 cm, 10 cm, 7 cm, 0 cm
b) The pendulum starts in its central position at 0 s, swings
out to its amplitude in 0.5 s, and back to its central position
in 1 s. **9. a)** Answers may vary; (30°, 60°), (60°, 30°),
(20°, 70°). **b)** They are complementary. **c)** (150°, 60°),
(135°, 45°), (120°, 30°) **d)** The difference is 90°.
10. a) $36.9°$ **b)** $22.6°$ **c)** $126.9°$ **d)** $163.7°$ **11.** They are equal.
12. Construct right triangle ABC, where $\angle A = 90°$, $\angle B = \theta$,
$\angle C = 90° - \theta$. Then: **a)** $\cos \theta = \cos B = \dfrac{c}{a}$ and
$\sin (90° - \theta) = \sin C = \dfrac{c}{a}$. Thus, $\sin (90° - \theta) = \cos \theta$.
b) $\sin \theta = \sin B = \dfrac{b}{a}$ and $\cos (90° - \theta) = \cos C = \dfrac{b}{a}$. Thus,
$\cos (90° - \theta) = \sin \theta$.

Section 4.3, pp. 290–294

1. a) 13.3 **b)** 38.6 **c)** 2.4 **d)** 11.6 **e)** 73.1 **f)** 16.5 **2. a)** $79.1°$
b) $51.9°$ **c)** $67.6°$ **d)** $25.3°$ **e)** $110.2°$ **f)** $36.0°$ **3. a)** $\angle B = 56°$,
$b = 4.7$ m, $c = 3.6$ m **b)** $\angle P = 52.0°$, $\angle Q = 99.5°$,
$q = 13.0$ cm **c)** $\angle L = 64.3°$, $\angle N = 53.7°$, $m = 16.6$ m
d) $\angle U = 33.9°$, $u = 40.6$ km, $w = 60.4$ km **e)** $\angle Y = 46.6°$,
$\angle Z = 41.1°$, $x = 4.3$ cm **f)** $\angle F = 115.9°$, $\angle G = 37.4°$,
$\angle H = 26.7°$ **4. a)** 12.4 m **b)** 31.0 mm **c)** 13.7 m **d)** 4.8 cm
e) 6.9 m **f)** 8.2 cm **5. a)** $104.3°$ **b)** $74.2°$ **c)** $20.6°$ **d)** $7.8°$
6. $\angle A = 40.7°$, $\angle B = 103.2°$, $\angle C = 36.1°$, $b = 18.7$ cm,
$c = 11.3$ cm **7. a)** 20.3 cm **b)** 13.7 cm^2 **8.** 383 km **9.** 73 m
10. 98 km **11.** 31.3 m^2 **12. a)** 9.0 b) 9.0 **c)** The cosine law
gives $x^2 = 5.2^2 + 7.3^2 - 2(5.2)(7.3) \cos 90°$, but since
$\cos 90° = 0$, this becomes $x^2 = 5.2^2 + 7.3^2$, which is
equivalent to the Pythagorean theorem. **13.** Each ratio is
equal to b. **14.** 13 m **16.** 5.8 m^2 **17.** 991.4 m^2 **18.** 193 cm^3
19. 480 m^3 **20.** $\angle P = 59.5°$, $\angle Q = 22.4°$, $\angle R = 98.1°$

Career Connection: Surveying, p. 295

1. 820 m

Section 4.4, pp. 308–311

1. a) $x = 46.7°$, $y = 133.3°$ **b)** $x = 130.1°$, $y = 49.9°$
c) $x = 161.7°$, $y = 18.3°$ **d)** $x = 46.0°$, $y = 134.0°$ **e)** $x = 56.9°$,
$y = 123.1°$ **f)** $x = 120.0°$, $y = 60.0°$ **2. a)** one; $\angle B = 33.9°$,

$\angle C = 104.1°$ **b)** two; $\angle A = 6.0°$, $\angle C = 147.0°$ or
$\angle A = 120.0°$, $\angle C = 33.0°$ **c)** one; $\angle Q = 90°$, $\angle R = 60°$
d) two; $\angle K = 102.8°$, $\angle L = 39.9°$ or $\angle K = 2.6°$,
$\angle L = 140.1°$ **e)** No triangles are possible. **f)** two;
$\angle A = 65.1°$, $\angle C = 66.9°$ or $\angle A = 18.9°$, $\angle C = 113.1°$
g) one; $\angle Y = 34.3°$, $\angle Z = 25.7°$ **h)** No triangles are possible.
3. a) $\angle B = 34.4°$, $\angle C = 100.6°$, $c = 41.7$ cm **b)** $\angle X = 14.5°$,
$\angle Z = 132.8°$, $z = 73.3$ cm **c)** $\angle P = 91.6°$, $\angle Q = 48.1°$,
$p = 54.4$ cm or $\angle P = 7.8°$, $\angle Q = 131.9°$, $p = 7.4$ cm
d) $\angle F = 33.7°$, $\angle H = 41.3°$, $h = 4.2$ cm **e)** No triangles are
possible. **f)** $\angle D = 32.3°$, $\angle F = 76.5°$, $d = 16.7$ cm or
$\angle D = 5.3°$, $\angle F = 103.5°$, $d = 2.9$ cm **g)** No triangles are
possible. **h)** $\angle M = 84.5°$, $\angle N = 52.7°$, $m = 23.1$ cm or
$\angle M = 9.9°$, $\angle N = 127.3°$, $m = 4.0$ cm
4. a) $\angle BCD = 54.3°$, $\angle BDA = 125.7°$ **b)** 18.7 cm **5.** 7.2 m
6. 5.3 cm **7. a)** 31.1 m **b)** 14.0 m **8.** 162 m **9.** 15.2 m
10. a) 24.78 **b)** 14.64 **11.** 4 km or 9 km **12.** 1.5 m
13. 3.6 h **14.** 0.004 s **15.** 20.4 m **16. a)** $(0, 12)$ **b)** $(0, 5.76)$
or $(0, 18.24)$ **c)** none **d)** $(0, 26.14)$ **17.** $\triangle ABC$: $AB = 4\sqrt{2}$,
$AC = 2\sqrt{5}$, $BC = 6$, $\angle ABC = 45°$, $\angle ACB = 63.43°$,
$\angle CAB = 71.57°$ $\triangle ABD$: $AB = 4\sqrt{2}$, $AD = 2\sqrt{5}$, $BD = 2$,
$ABD = 45°$, $\angle ADB = 116.57°$, $\angle BAD = 18.43°$
19. $286°$, $14°$

Review of Key Concepts, pp. 313–315

1. a) $\angle A = 42.9°$, $a = 9.0$ cm, $b = 13.2$ cm **b)** $\angle E = 36.6°$,
$\angle F = 53.4°$, $f = 4.2$ cm **c)** $\angle F = 65°$, $e = 2.2$ cm, $d = 5.3$ cm
d) $\angle L = 35.4°$, $\angle M = 54.6°$, $k = 15.2$ cm **2.** $60.9°$ **3.** 8.7 m
4. 146 m **5.** $35.3°$ **6.** $\sin \theta = \dfrac{21}{29}$, $\cos \theta = \dfrac{20}{29}$ **7.** $\sin \theta = \dfrac{3}{5}$,
$\cos \theta = -\dfrac{4}{5}$ **8. a)** 0.9994 **b)** -0.1736 **c)** 0.7738 **d)** -0.9598
9. a) $38.0°$ or $142.0°$ **b)** $74.0°$ **c)** $82.0°$ or $98.0°$ **d)** $154.0°$
10. a) $60°$ **b)** $60°$ **11. a)** $\angle A = 54.1°$, $a = 37.4$ cm,
$c = 44.2$ cm **b)** $\angle R = 60.9°$, $\angle S = 52.9°$, $\angle T = 66.2°$
c) $\angle E = 46.1°$, $\angle G = 66.1°$, $g = 12.4$ m **d)** $\angle X = 77.3°$,
$\angle Y = 44.4°$, $w = 27.4$ cm **e)** $\angle D = 30.9°$, $e = 47.0$ cm,
$f = 35.5$ cm **f)** $\angle A = 21.4°$, $\angle C = 25.2°$, $b = 19.5$ m
g) $\angle Y = 11.8°$, $\angle Z = 39.4°$, $z = 25.5$ m **h)** $\angle S = 37.6°$,
$\angle T = 112.2°$, $\angle U = 30.2°$ **i)** $\angle P = 118.4°$, $\angle R = 30.2°$,
$q = 9.0$ m **12. a)** 30 cm^2 **b)** 113 m^2 **13. a)** 3.1 m **b)** 9.6 m
14. a) $47.8°$ **b)** $104.3°$ **15.** 139 m **16.** 31.2 km on a course
of $194°$ **17.** $\$4273$ **18. a)** two; $\angle H = 58.8°$, $\angle I = 101.2°$,
$i = 5.7$ cm or $\angle H = 121.2°$, $\angle I = 38.8°$, $i = 3.7$ cm **b)** No
triangles are possible. **c)** one; $\angle C = 20.3°$, $\angle A = 55.2°$,
$a = 3.3$ m **d)** two; $\angle K = 111.0°$, $\angle M = 42.9°$, $k = 8.9$ m or
$\angle K = 16.8°$, $\angle M = 137.1°$, $k = 2.8$ m **19.** $\angle QRS = 62.7°$,
$\angle QSP = 117.3°$, $RS = 3.6$ cm

Chapter Test, pp. 316–317

1. a) $\angle B = 50.2°$, $\angle C = 39.8°$, $b = 1.9$ m **b)** $\angle D = 58.2°$,
$e = 59.1$ cm, $f = 31.1$ cm **2.** 9.7 cm **3.** $\sin \theta = \dfrac{3}{5}$,

$\cos \theta = -\dfrac{4}{5}$ **4. a)** 0.9910 **b)** 0.9403 **c)** 0.5075 **d)** −0.9348

5. a) 41.9° or 138.1° **b)** 123.5° **6.** No; the cosine of an acute angle is positive. **7.** Yes; the angle, to the nearest tenth of a degree, could be 157.8°. **8. a)** $\angle$P = 68.4°, m = 14.3 cm, n = 10.6 cm **b)** $\angle$B = 63.0°, $\angle$C = 83.9°, d = 2.6 m **c)** $\angle$R = 48.7°, q = 17.6 m, r = 13.3 m **d)** $\angle$K = 35.0°, $\angle$M = 23.8°, l = 18.0 cm **e)** $\angle$A = 118.7°, $\angle$B = 35.1°, $\angle$C = 26.2° **f)** $\angle$H = 44.2°, $\angle$I = 37.0°, i = 26.0 cm **9.** 38.4 m **10. a)** No triangles are possible. **b)** two; $\angle$T = 56.2°, $\angle$U = 94.8°, u = 7.2 cm or $\angle$T = 123.8°, $\angle$U = 27.2°, u = 3.3 cm **c)** one; $\angle$Y = 18.5°, $\angle$Z = 65.2°, z = 2.3 m **d)** two; $\angle$F = 72.1°, $\angle$H = 66.2°, f = 10.3 cm or $\angle$F = 24.5°, $\angle$H = 113.8°, f = 4.5 cm **11.** 38 m or 52 m **12.** 938 km

Challenge Problems, p. 318

1. e) $\dfrac{10}{\sqrt{3}}$ **2. b)** $\dfrac{1}{\sin \theta}$ **3.** $100\sqrt{3}$ or $50\sqrt{3}$ **4.** 9 **5.** 120° **6.** $3r^2$ cm² **7.** 8

Problem Solving Strategy: Use a Diagram, pp. 320–321

1. 67 m **2.** 65 cm **3.** 20 s **4.** Timid swimmer and Bold swimmer cross in boat. Timid returns in boat; Timid and Bold cross in boat. Bold swimmer returns in boat. Bold swimmer and Timid cross in boat. Bold swimmer returns in boat. Bold swimmer and Timid cross in boat. Bold swimmer returns in boat. Bold and Bold swimmer cross in boat. **5.** Train A is heading to the right and train B is heading to the left. Train A pulls into the siding with 20 cars, leaving 20 cars on the track. Train B pushes these cars back until it clears the way for train A to back out of the siding. Train B unhitches 20 of its cars. Train A picks them up and pulls them down the track. Train B backs its remaining 20 cars into the siding and train A backs down the track to pick up its 20 cars. Train A now pulls 60 cars—its 40 plus 20 from train B in the middle—to the right. Train B exits the siding, then backs up the track to get 20 cars from train A, pulling them forward and backing them into the siding. Train B backs up to collect its remaining 20 cars, pulls forward, and hitches up the 20 cars from Train A in the siding, pulling them onto the track. Train A backs up to hitch up its 20 cars. Both trains proceed to their destinations. **6.** 16.4 m **7.** $3\sqrt{2}$ cm **8.** (3, 10), (9, 6) **9.** 24 **10.** 5.8 cm or 10.0 cm **13.** 1300 m

Problem Solving: Using the Strategies, p. 322

1. Use the initial four tires for 12 000 km each, and label the remaining five tires with the letters A, B, C, D, and E. Rotate the tires as follows, using each group for 3000 km: (A, B, C, D), (B, C, D, E), (C, D, E, A), (D, E, A, B),

(E, A, B, C). **2.** 7 **3.** 12 cm **4.** three other pairs; 17 and 71, 37 and 73, and 79 and 97 **5.** Answers may vary. **6.** 6 km/h **7.** 9 cm² **8.** D **9.** 1982 **10.** 12 **11.** 25 **12.** 13 **13.** 844 m² **14.** Answers may vary.

Cumulative Review: Chapters 3 and 4, p. 323

Chapter 3

1. a) 1 **b)** 4 **c)** 19 **d)** −1 **3. a)** $f^{-1}(x) = \dfrac{x-5}{2}$; a function

b) $f^{-1}(x) = \pm\sqrt{x+4}$; not a function **5. a)** domain: $x \geq 0$, range: $y \geq 0$; domain: $x \geq 2$, range: $y \leq 4$ **b)** domain: all real numbers, range: $y \geq 0$; domain: all real numbers, range: $y \geq -5$ **6. a)** vertical expansion by a factor of 5, reflection in the x-axis, vertical translation 3 units upward **b)** horizontal expansion by a factor of 2, vertical expansion by a factor of 4, vertical translation 2 units upward **c)** vertical expansion by a factor of 2, horizontal translation 3 units left, vertical translation 4 units downward **d)** horizontal compression by a factor of $\dfrac{1}{7}$, horizontal translation 5 units right, vertical translation 1 unit upward **e)** horizontal compression by a factor of $\dfrac{1}{8}$, reflection in the x-axis, horizontal translation 7 units left **f)** horizontal compression by a factor of $\dfrac{1}{2}$, vertical compression by a factor of $\dfrac{1}{3}$, horizontal translation 3 units right, vertical translation 9 units downward **7. a)** $d(t) = 800t$ **b)** $t(d) = \dfrac{d}{800}$ **c)** The inverse represents the time required to travel a given distance.

Chapter 4

1. 57.9° **2. a)** 140.6° **b)** 16.6° or 163.4° **3.** $\sin \theta = \dfrac{21}{29}$, $\cos \theta = -\dfrac{20}{29}$ **4. a)** $\angle$J = 34°, j = 12.5 cm, k = 18.3 cm **b)** $\angle$B = 78.6°, $\angle$C = 57.9°, a = 8.8 cm **c)** no triangles **5.** 10.6 m **6.** 39 km

Chapter 5

Getting Started, p. 326

2. 15 h **3.** 8 **4. a)** 67% **b)** 34% **5.** vertical translation downward **6.** reflection in the line $y = 12$

Review of Prerequisite Skills, p. 327

1. a) sin E = 0.949, cos E = 0.316, tan E = 3, sin F = 0.316, cos F = 0.949, tan F = 0.333 **b)** sin A = 0.831, cos A = 0.556, tan A = 1.497, sin C = 0.556, cos C = 0.831, tan C = 0.668 **2. a)** $\sin \theta = 0.651$, $\cos \theta = 0.759$ **b)** $\sin \theta = 0.625$, $\cos \theta = -0.781$ **c)** $\sin \theta = 0.124$, $\cos \theta = -0.992$ **d)** $\sin \theta = 1$, $\cos \theta = 0$ **e)** $\sin \theta = 0$, $\cos \theta = -1$

7 units **c)** translation upward 9 units and to the left 5 units
d) translation to the right 6 units and downward 3 units
6. a) vertical expansion by a factor of 2 and translation
upward 5 units **b)** horizontal compression by a factor of $\frac{1}{2}$,
translation left 1 unit and downward 2 units **c)** vertical
compression by a factor of $\frac{1}{3}$, horizontal expansion by a
factor of 2, translation to the right 4 units **d)** vertical
expansion by a factor of 3, reflection in the x-axis,
translation to the right 1 unit and upward 5 units
8. a) $-6, 7$ **b)** $-5, -8$ **c)** $10, 8$ **d)** $-9, 9$ **e)** $3, \frac{5}{3}$ **f)** $-\frac{2}{3}, -\frac{4}{5}$
g) $\frac{1}{4}, -\frac{5}{2}$ **h)** $0, \frac{2}{3}$ **i)** $\frac{5}{2}$ **j)** $-\frac{3}{4}, \frac{3}{4}$

Section 5.1, pp. 334–339

1. a) $60°$ **b)** $45°$ **c)** $360°$ **d)** $90°$ **e)** $135°$ **f)** $270°$ **g)** $720°$ **h)** $150°$
i) $10°$ **j)** $660°$ **k)** $210°$ **l)** $900°$ **2. a)** $\frac{2\pi}{9}$ **b)** $\frac{5\pi}{12}$ **c)** $\frac{\pi}{18}$ **d)** $\frac{2\pi}{3}$
e) $\frac{5\pi}{4}$ **f)** $\frac{7\pi}{4}$ **g)** $\frac{11\pi}{6}$ **h)** $\frac{4\pi}{3}$ **i)** 3π **j)** 6π **3. a)** $143.2°$ **b)** $100.3°$
c) $20.1°$ **d)** $71.6°$ **e)** $128.6°$ **f)** $49.1°$ **g)** $235.4°$ **h)** $28.6°$
i) $179.9°$ **j)** $69.3°$ **4. a)** 1.05 **b)** 2.62 **c)** 1.40 **d)** 2.53 **e)** 4.01
f) 5.67 **g)** 0.98 **h)** 2.24 **i)** 4.46 **j)** 5.42 **5. a)** 11.9 cm
b) 26.8 cm **c)** 26.2 m **6. a)** 1.38 rad, $78.8°$ **b)** 1.89 rad,
$108.2°$ **c)** 2.29 rad, $131.0°$ **7. a)** 11.4 cm **b)** 12.1 cm
c) 16.0 m **8. a)** 49.2 m **b)** 7.0 m **c)** 2.56 rad **9.** $\frac{5}{100}$ rad or
0.05 rad **10.** $\frac{\pi}{3}$ rad or 1.05 rad **11.** 100π rad/s or
314.16 rad/s **12.** 5π rad or 15.71 rad **13. a)** $\frac{7\pi}{8}$ **b)** $\frac{3\pi}{8}$
14. $\frac{\pi}{5}$ rad or 0.63 rad, $\frac{3\pi}{10}$ rad or 0.94 rad **15.** 57.3
16. a) 732π **b)** 10 h **c)** 28 h **17.** $\frac{35\pi}{9}$ rad or 12.2 rad
18. a) $6600°$ **b)** $\frac{110\pi}{3}$ rad or 115 rad **19. a)** $6°$ **b)** $\frac{\pi}{30}$ rad or
0.10 rad **20. a)** 2 r/day **b)** 11π rad or 34.56 rad
21. a) $\frac{50\pi}{3}$ rad or 52.36 rad **b)** $1200°$ **c)** $36\ 000°$, $120\ 000°$,
$150\ 000°$ **d)** 160π rad or 503 rad **22. a)** $\frac{\pi}{15}$ rad/s
b) 1206 m **23.** 311 km **24. a)** 76.4 r/min **b)** 1600 cm
25. a) 3142 s **b)** 14.4 km/s **26.** $18\ 400$ km
27. a) 1 **b)** $\frac{1}{2}$ **c)** $\frac{1}{2}$ **d)** $-\frac{1}{2}$ **28. a)** 12.3 r/s **b)** 2315 rad
29. a) equal **b)** greater for the person on the equator
30. a) $1, 0$ **b)** $90°$, $\frac{\pi}{2}$ rad **c)** 1 rad $= \frac{200}{\pi}$ grad,
1 rad $= 63.66$ grad

Career Connection: Crafts, p. 340

1. a) 18.85 rad/s **b)** 30.2 cm

Section 5.2, pp. 348–350

1. a) $\sin\theta = \frac{15}{17}$, $\cos\theta = \frac{8}{17}$, $\tan\theta = \frac{15}{8}$ **b)** $\sin\theta = \frac{5}{\sqrt{34}}$,
$\cos\theta = -\frac{3}{\sqrt{34}}$, $\tan\theta = -\frac{5}{3}$ **c)** $\sin\theta = -\frac{3}{5}$, $\cos\theta = -\frac{4}{5}$,
$\tan\theta = \frac{3}{4}$ **d)** $\sin\theta = -\frac{5}{13}$, $\cos\theta = \frac{12}{13}$, $\tan\theta = -\frac{5}{12}$
e) $\sin\theta = -\frac{7}{\sqrt{53}}$, $\cos\theta = -\frac{2}{\sqrt{53}}$, $\tan\theta = \frac{7}{2}$ **f)** $\sin\theta = -\frac{2}{\sqrt{13}}$,
$\cos\theta = \frac{3}{\sqrt{13}}$, $\tan\theta = -\frac{2}{3}$ **2. a)** $\sin\theta = \frac{5}{\sqrt{61}}$, $\cos\theta = \frac{6}{\sqrt{61}}$,
$\tan\theta = \frac{5}{6}$ **b)** $\sin\theta = \frac{8}{\sqrt{65}}$, $\cos\theta = -\frac{1}{\sqrt{65}}$, $\tan\theta = -8$
c) $\sin\theta = -\frac{5}{\sqrt{29}}$, $\cos\theta = -\frac{2}{\sqrt{29}}$, $\tan\theta = \frac{5}{2}$ **d)** $\sin\theta = -\frac{1}{\sqrt{37}}$,
$\cos\theta = \frac{6}{\sqrt{37}}$, $\tan\theta = -\frac{1}{6}$ **e)** $\sin\theta = -\frac{2}{\sqrt{5}}$, $\cos\theta = \frac{1}{\sqrt{5}}$,
$\tan\theta = -2$ **f)** $\sin\theta = -\frac{3}{\sqrt{10}}$, $\cos\theta = -\frac{1}{\sqrt{10}}$, $\tan\theta = 3$
g) $\sin\theta = \frac{1}{\sqrt{2}}$, $\cos\theta = \frac{1}{\sqrt{2}}$, $\tan\theta = 1$ **h)** $\sin\theta = \frac{3}{\sqrt{10}}$,
$\cos\theta = -\frac{1}{\sqrt{10}}$, $\tan\theta = -3$ **3. a)** $\frac{1}{2}$ **b)** -1 **c)** $-\frac{1}{2}$ **d)** $-\frac{1}{\sqrt{3}}$
e) $-\frac{1}{\sqrt{2}}$ **f)** $\frac{1}{\sqrt{2}}$ **g)** $\frac{\sqrt{3}}{2}$ **h)** $-\frac{\sqrt{3}}{2}$ **4. a)** $-\frac{1}{\sqrt{2}}$ **b)** $-\frac{1}{\sqrt{3}}$ **c)** $\frac{\sqrt{3}}{2}$ **d)** $\frac{1}{\sqrt{2}}$
e) $\sqrt{3}$ **f)** $-\frac{\sqrt{3}}{2}$ **g)** $\frac{1}{2}$ **h)** $-\frac{1}{\sqrt{2}}$ **5. a)** $6(\sqrt{2}-1)$ m **b)** 2.5 m
6. a) $\cos\theta = -\frac{3}{5}$, $\tan\theta = -\frac{4}{3}$ **b)** $\sin\theta = -\frac{\sqrt{5}}{3}$, $\tan\theta = \frac{\sqrt{5}}{2}$
c) $\sin\theta = -\frac{5}{\sqrt{29}}$, $\cos\theta = \frac{2}{\sqrt{29}}$ **d)** $\cos\theta = -\frac{2\sqrt{10}}{7}$,
$\tan\theta = \frac{3}{2\sqrt{10}}$ **7. a)** $\cos\theta = \frac{2\sqrt{2}}{3}$, $\tan\theta = \frac{1}{2\sqrt{2}}$ or
$\cos\theta = -\frac{2\sqrt{2}}{3}$, $\tan\theta = -\frac{1}{2\sqrt{2}}$ **b)** $\sin\theta = \frac{4}{5}$, $\tan\theta = \frac{4}{3}$ or
$\sin\theta = -\frac{4}{5}$, $\tan\theta = -\frac{4}{3}$ **c)** $\sin\theta = \frac{1}{\sqrt{17}}$, $\cos\theta = \frac{4}{\sqrt{17}}$ or
$\sin\theta = -\frac{1}{\sqrt{17}}$, $\cos\theta = -\frac{4}{\sqrt{17}}$ **d)** $\sin\theta = \frac{\sqrt{3}}{2}$, $\tan\theta = -\sqrt{3}$
or $\sin\theta = -\frac{\sqrt{3}}{2}$, $\tan\theta = \sqrt{3}$ **e)** $\sin\theta = -\frac{8}{\sqrt{89}}$, $\cos\theta = \frac{5}{\sqrt{89}}$
or $\sin\theta = \frac{8}{\sqrt{89}}$, $\cos\theta = -\frac{5}{\sqrt{89}}$ **f)** $\cos\theta = \frac{\sqrt{11}}{6}$,
$\tan\theta = -\frac{5}{\sqrt{11}}$ or $\cos\theta = -\frac{\sqrt{11}}{6}$, $\tan\theta = \frac{5}{\sqrt{11}}$
8. a) $6(\sqrt{3}-\sqrt{2})$ m, 1.9 m **b)** $6(\sqrt{2}-1)$ m, 2.5 m **9.** 40 m
10. The pair anchored at a $45°$ angle, by 4.2 m. **11. a)** $45°$,
$135°$ **b)** $60°$, $300°$ **c)** $45°$, $225°$ **d)** $60°$, $120°$ **e)** $135°$, $225°$

f) 120°, 300° g) 150°, 210° h) 210°, 330° i) 135°, 315°
j) 150°, 330° k) 120°, 240° **12.** 0, 1, 0; 0, −1, 0; −1, 0, not
defined; 0, 1, 0 **13. a)** $\dfrac{1}{\sqrt{2}}$ or $-\dfrac{1}{\sqrt{2}}$ **b)** 1 or −1 **14. a)** $\dfrac{\sqrt{3}}{2}$ or
$-\dfrac{\sqrt{3}}{2}$ **b)** $\dfrac{1}{2}$ or $-\dfrac{1}{2}$ **15.** 1.39 **16.** 0.43 **17. a)** $\dfrac{1}{\sqrt{2}}$ or $-\dfrac{1}{\sqrt{2}}$
b) $-\dfrac{1}{\sqrt{2}}$ or $\dfrac{1}{\sqrt{2}}$ **c)** −1 **18. a)** $\dfrac{\sqrt{6}}{8}$ **b)** $\dfrac{\sqrt{3}}{2}$ **c)** 1 **d)** $\dfrac{\sqrt{3}}{2}$ **e)** 10

Extension: Angles of Rotation, p. 354

1. a) $\dfrac{1}{2}, \dfrac{\sqrt{3}}{2}, \dfrac{1}{\sqrt{3}}$ **b)** $\dfrac{1}{\sqrt{2}}, \dfrac{1}{\sqrt{2}}, 1$ **c)** $-\dfrac{1}{2}, \dfrac{\sqrt{3}}{2}, -\dfrac{1}{\sqrt{3}}$ **d)** 1, 0, not
defined **e)** $\dfrac{\sqrt{3}}{2}, \dfrac{1}{2}, \sqrt{3}$ **f)** $-\dfrac{1}{\sqrt{2}}, -\dfrac{1}{\sqrt{2}}, 1$ **g)** 0, −1, 0 **h)** $-\dfrac{\sqrt{3}}{2}$,
$-\dfrac{1}{2}, \sqrt{3}$ **2. a)** $-\dfrac{1}{\sqrt{2}}$ **b)** $-\sqrt{3}$ **c)** $-\dfrac{\sqrt{3}}{2}$ **d)** 1 **e)** $\dfrac{1}{\sqrt{3}}$ **f)** $\dfrac{1}{2}$
3. a) $-\dfrac{\sqrt{3}}{2}, \dfrac{1}{2}, -\sqrt{3}$ **b)** $\dfrac{1}{\sqrt{2}}, -\dfrac{1}{\sqrt{2}}, -1$ **c)** −1, 0, not defined
d) $-\dfrac{1}{2}, -\dfrac{\sqrt{3}}{2}, \dfrac{1}{\sqrt{3}}$ **e)** 1, 0, not defined **f)** $\dfrac{1}{2}, \dfrac{\sqrt{3}}{2}, \dfrac{1}{\sqrt{3}}$
g) $-\dfrac{1}{\sqrt{2}}, -\dfrac{1}{\sqrt{2}}, 1$ **h)** 0, −1, 0 **4. a)** 0 **b)** 0 **c)** 1 **d)** $-\dfrac{\sqrt{3}}{2}$
e) $-\dfrac{1}{\sqrt{2}}$ **f)** $\dfrac{1}{\sqrt{2}}$

Section 5.3, pp. 359–362

1. a) periodic **b)** periodic **c)** not periodic **d)** not periodic
2. a) period: 5, amplitude: 3 **b)** period: 2, amplitude: 1
c) period: 2, amplitude: 1.5 **d)** period: 8, amplitude: 3
e) period: 3, amplitude: 4.5 **f)** period: 5, amplitude: 1
3. Answers will vary. **4. a)** −2 **b)** −2 **c)** 9 **d)** 9 **5. a)** maximum:
4, minimum: 0, amplitude: 2, period: 8, domain:
−4 ≤ x ≤ 12, range: 0 ≤ y ≤ 4 **b)** maximum: 1, minimum:
−2, amplitude: 1.5, period: 3, domain: all real numbers,
range: −2 ≤ y ≤ 1 **6. b)** period: 2 min, amplitude: 25 m
c) domain: 0 ≤ t ≤ 6, range: 0 ≤ d ≤ 50 **d)** period: smaller
$\left(\dfrac{20}{11}\right)$, amplitude: equal **7. a)** yes **b)** no **c)** yes **8. b)** period:
30 s, amplitude: 16 m **c)** domain: 0 ≤ t ≤ 120, range:
2 ≤ d ≤ 34 **9. b)** period: 365 days, amplitude: 18.5
10. period: 12 h 25 min, amplitude: 1.2 m **13.** The length
of the interval that gives the domain is a whole number
multiple of the period. **14.** The amplitude is half the length
of the interval that gives the range. **15. a)** 24 h **b)** half the
length of the equator, or about 20 100 km

Section 5.4, pp. 363–366

1. 1, $\dfrac{\sqrt{3}}{2}, \dfrac{1}{2}$, 0, $-\dfrac{1}{2}, -\dfrac{\sqrt{3}}{2}$, −1, $-\dfrac{\sqrt{3}}{2}, -\dfrac{1}{2}$, 0; 1, 0.9, 0.5, 0,
−0.5, −0.9, −1, −0.9, −0.5, 0 **3. a)** 1 **b)** $\dfrac{\pi}{2}$ rad or 90°

4. a) −1 **b)** $\dfrac{3\pi}{2}$ rad or 270° **5.** 1 **6. a)** periodic **8. a)** periodic
10. a) yes **b)** 2π **11.** Use the vertical line test over one period.
12. a) all real numbers **b)** −1 ≤ y ≤ 1 **13.** 1, $\dfrac{\sqrt{3}}{2}, \dfrac{1}{2}$, 0, $-\dfrac{1}{2}$,
$-\dfrac{\sqrt{3}}{2}$, −1, $-\dfrac{\sqrt{3}}{2}, -\dfrac{1}{2}$, 0, $\dfrac{1}{2}, \dfrac{\sqrt{3}}{2}$, 1; 1, 0.9, 0.5, 0, −0.5, −0.9,
−1, −0.9, −0.5, 0, 0.5, 0.9, 1 **15.** 1 **16. a)** periodic
18. a) periodic **20. a)** yes **b)** 2π **21.** Use the vertical line test
over one period. **22. a)** all real numbers **b)** −1 ≤ y ≤ 1
23. Both graphs are periodic; one is a translation of the
other. **24.** 0, 1, 1.7, 2.7, 5.7, not defined, −5.7, −2.7, −1.7,
−1, −0.6, 0, 1, 1.7, 2.7, 5.7, not defined, −5.7, −2.7, −1.7,
−1, −0.6, 0 **26.** tan x increases without bound.
27. a) undefined **b)** The line x = 90° is a vertical asymptote.
28. The value of tan x increases from negative infinity to
positive infinity. **29. a)** undefined **b)** The line x = 270° is a
vertical asymptote. **30. a)** periodic **32. a)** periodic
34. a) yes **b)** π **35.** no; no **36.** Use the vertical line test over
one period. **37. a)** all real numbers, except odd integer
multiples of $\dfrac{\pi}{2}$ **b)** all real numbers

Section 5.5, pp. 375–377

1. a) domain: 0 ≤ x ≤ 2π, range: −3 ≤ y ≤ 3 **b)** domain:
0 ≤ x ≤ 2π, range: −5 ≤ y ≤ 5 **c)** domain: 0 ≤ x ≤ 2π, range:
−1.5 ≤ y ≤ 1.5 **d)** domain: 0 ≤ x ≤ 2π, $-\dfrac{2}{3} \le y \le \dfrac{2}{3}$ **2. a)** 60°,
$\dfrac{\pi}{3}$ rad **b)** 90°, $\dfrac{\pi}{2}$ rad **c)** 540°, 3π rad **d)** 540°, 3π rad **e)** 2160°,
12π rad **f)** 45°, $\dfrac{\pi}{4}$ rad **3. a)** domain: 0 ≤ x ≤ π, range:
−1 ≤ y ≤ 1 **b)** domain: $0 \le x \le \dfrac{2\pi}{3}$, −1 ≤ y ≤ 1 **c)** domain:
$0 \le x \le \dfrac{\pi}{3}$, −1 ≤ y ≤ 1 **d)** domain: 0 ≤ x ≤ 8π, range:
−1 ≤ y ≤ 1 **e)** domain: $0 \le x \le \dfrac{8\pi}{3}$, range: −1 ≤ y ≤ 1
f) domain: 0 ≤ x ≤ 6π, range: −1 ≤ y ≤ 1 **4. a)** y = 6sin (2x)
b) $y = 1.5\sin\left(\dfrac{3}{2}x\right)$ **c)** $y = 0.8\sin\left(\dfrac{2}{3}x\right)$ **d)** $y = 4\sin\left(\dfrac{1}{3}x\right)$
5. a) y = 3cos (2x) **b)** $y = 0.5\cos\left(\dfrac{1}{2}x\right)$ **c)** $y = 4\cos\left(\dfrac{1}{2}x\right)$
d) $y = 2.5\cos\left(\dfrac{2}{5}x\right)$ **7. a)** y = 4sin x **b)** $y = 2\sin\left(\dfrac{1}{2}x\right)$
c) $y = 5\sin\left(\dfrac{1}{2}x\right)$ **d)** y = 3sin (3x) **8. a)** $y = 3\cos\left(\dfrac{1}{4}x\right)$
b) $y = 3\cos\left(\dfrac{3}{5}x\right)$ **c)** $y = 4\cos\left(\dfrac{1}{2}x\right)$ **d)** y = 8cos (2x)
10. a) (nπ, 0), with n any integer **b)** (0, 0) **c)** (0, 0) **d)** (0, 0)
11. a) $\left(n\pi + \dfrac{\pi}{2}, 0\right)$, with n any integer **b)** (0, 1) **c)** none
d) none **12. a)** domain: all real numbers, range:
−3.5 ≤ y ≤ 3.5, amplitude: 3.5, period: 2π **b)** domain: all real
numbers, range: −1 ≤ y ≤ 1, amplitude: 1, period: $\dfrac{4\pi}{5}$

c) domain: all real numbers, range: $-2 \leq y \leq 2$, amplitude: 2, period: 12π **d)** domain: all real numbers, range: $-\frac{1}{4} \leq y \leq \frac{1}{4}$, amplitude: $\frac{1}{4}$, period: 4π **13. a)** yellow: $y = \sin\left(\frac{\pi}{290}x\right)$, green: $y = \sin\left(\frac{\pi}{265}x\right)$, violet: $y = \sin\left(\frac{\pi}{205}x\right)$ **b)** yellow: $y = \sin(0.011x)$, green: $y = \sin(0.012x)$, violet: $y = \sin(0.015x)$ **14. a)** $45° < x < 225°$ **b)** $0° \leq x < 45°$, $225° < x \leq 360°$ **c)** $45°, 225°$ **15. a)** amplitude: 10, period: $\frac{1}{440}$ **b)** vertical stretch by a factor of 10, horizontal compression by a factor of $\frac{1}{880\pi}$ **16. a)** $\frac{1}{2\pi}$ Hz **b)** $\frac{1}{2\pi}$ Hz **c)** $\frac{1}{\pi}$ Hz **d)** $\frac{1}{4\pi}$ Hz **17. b)** $h(t) = 0.25\cos\left(\frac{\pi}{2}t\right)$ **c)** horizontal compression by a factor of $\frac{1}{60}$; $h(t) = 0.25\cos(30\pi t)$

18. a) $y = \sin\left(\frac{2\pi}{23}t\right)$, $y = \sin\left(\frac{\pi}{14}t\right)$, $y = \sin\left(\frac{2\pi}{33}t\right)$ **c)** 58 years **19. a)** amplitude: 170, period: $\frac{1}{60}$

b) 60 Hz

Section 5.6, pp. 387–391

1. a) vertical translation: 3 units upward; phase shift: none **b)** vertical translation: 1 unit downward; phase shift: none **c)** vertical translation: none; phase shift: 45° to the right **d)** vertical translation: none; phase shift: $\frac{3\pi}{4}$ to the right **e)** vertical translation: 1 unit upward; phase shift: 60° to the right **f)** vertical translation: 4 units upward; phase shift: $\frac{\pi}{3}$ to the left **g)** vertical translation: 0.5 units downward; phase shift: $\frac{3\pi}{8}$ to the left **h)** vertical translation: 4.5 units downward; phase shift: 15° to the right **2. a)** vertical translation: 6 units upward; phase shift: none **b)** vertical translation: 3 units downward; phase shift: none **c)** vertical translation: none; phase shift: $\frac{\pi}{2}$ to the left **d)** vertical translation: none; phase shift: 72° to the left **e)** vertical translation: 2 units downward; phase shift: 30° to the right **f)** vertical translation: 1.5 units upward; phase shift: $\frac{\pi}{6}$ to the left **g)** vertical translation: 25 units upward; phase shift: 110° to the left **h)** vertical translation: 3.8 units downward, phase shift: $\frac{5\pi}{12}$ to the right **3. a)** amplitude: 3, period: 2π, domain: $0 \leq x \leq 2\pi$, range: $-1 \leq y \leq 5$ **b)** amplitude: 2, period: 2π, domain: $0 \leq x \leq 2\pi$, range: $-4 \leq y \leq 0$ **c)** amplitude: 1.5, period: 2π, domain: $0 \leq x \leq 2\pi$, range: $-2.5 \leq y \leq 0.5$ **d)** amplitude: $\frac{1}{2}$, period: 2π, domain: $0 \leq x \leq 2\pi$, range: $\frac{1}{2} \leq y \leq \frac{3}{2}$ **4. a)** amplitude: 2, period: 2π, domain: $0 \leq x \leq 2\pi$, range: $-2 \leq y \leq 2$, phase shift: π to the

right **b)** amplitude: 1, period: 2π, domain: $0 \leq x \leq 2\pi$, range: $-1 \leq y \leq 1$, phase shift: $\frac{\pi}{2}$ to the right **c)** amplitude: $\frac{1}{2}$, period: 2π, domain: $0 \leq x \leq 2\pi$, range: $-\frac{1}{2} \leq y \leq \frac{1}{2}$, phase shift: $\frac{\pi}{2}$ to the left **d)** amplitude: 3, period: 2π, domain: $0 \leq x \leq 2\pi$, range: $-3 \leq y \leq 3$, phase shift: $\frac{\pi}{4}$ to the left **e)** amplitude: 1, period: 2π, domain: $0 \leq x \leq 2\pi$, range: $-1 \leq y \leq 1$, phase shift: $\frac{\pi}{2}$ to the left **5. a)** amplitude: 2, period: 2π, vertical translation: 3 units downward, phase shift: none **b)** amplitude: 0.5, period: π, vertical translation: 1 unit downward, phase shift: none **c)** amplitude: 6, period: $\frac{2\pi}{3}$, vertical translation: none, phase shift: $\frac{\pi}{9}$ to the right **d)** amplitude: 5, period: π, vertical translation: 1 unit upward, phase shift: $\frac{\pi}{6}$ to the right **6. a)** amplitude: 1, period: 2π, vertical translation: 3 units upward, phase shift: none **b)** amplitude: 1, period: $\frac{2\pi}{3}$, vertical translation: none, phase shift: $\frac{\pi}{2}$ to the right **c)** amplitude: 3, period: $\frac{\pi}{2}$, vertical translation: 5 units upward, phase shift: $\frac{\pi}{4}$ to the right **d)** amplitude: 0.8, period: 3π, vertical translation: 7 units downward, phase shift: $\frac{\pi}{3}$ to the right **7. a)** amplitude: 1, period: π, domain: $0 \leq x \leq \pi$, range: $-1 \leq y \leq 1$, phase shift: $\frac{\pi}{4}$ to the left **b)** amplitude: 2, period: π, domain: $0 \leq x \leq \pi$, range: $-1 \leq y \leq 3$, phase shift: $\frac{\pi}{4}$ to the right **c)** amplitude: 3, period: 4π, domain: $0 \leq x \leq 4\pi$, range: $-5 \leq y \leq 1$, phase shift: π to the right **d)** amplitude: 4, period: 6π, domain: $0 \leq x \leq 6\pi$, range: $-8 \leq y \leq 0$, phase shift: 2π to the left **e)** amplitude: 3, period: π, domain: $0 \leq x \leq \pi$, range: $-1 \leq y \leq 5$, phase shift: $\frac{\pi}{4}$ to the right **8. a)** amplitude: 1, period: π, domain: $0 \leq x \leq \pi$, range: $-1 \leq y \leq 1$, phase shift: $\frac{\pi}{4}$ to the right **b)** amplitude: 1, period: 4π, domain: $0 \leq x \leq 4\pi$, range: $-3 \leq y \leq -1$, phase shift: 2π to the right **c)** amplitude: 2, period: $\frac{2\pi}{3}$, domain: $0 \leq x \leq \frac{2\pi}{3}$, range: $0 \leq y \leq 4$, phase shift: $\frac{\pi}{3}$ to the right **d)** amplitude: 3, period: π, domain: $0 \leq x \leq \pi$, range: $-4 \leq y \leq 2$, phase shift: 2π to the right **9. a)** $y = 8\sin x - 6$ **b)** $y = 7\cos 2x + 2$ **c)** $y = \sin \frac{1}{2}(x - \pi) + 3$ **d)** $y = 10\cos 4\left(x + \frac{\pi}{2}\right)$ **10. a)** range: $0 \leq y \leq 4$ **b)** range: $-3 \leq y \leq -1$ **c)** range: $-3 \leq y \leq 3$ **d)** range: $-5 \leq y \leq 3$ **e)** range: $-1 \leq y \leq 3$ **f)** range: $-0.5 \leq y \leq 7$ **g)** range: $-2 \leq y \leq 2$ **11. a)** $a = 2$, $k = 3$, $d = \frac{\pi}{3}$, $c = -1$ **b)** $a = 5$, $k = \frac{1}{2}$, $d = \pi$,

$c = 2$ **12. a)** $y = 5\sin\frac{\pi}{6}(t-3) + 16$ **c) i)** 08:00, 20:00

ii) 14:00, 02:00 **iii)** 17:00, 23:00, 05:00, 11:00 **d) i)** 16 m

ii) 20.3 m **e) i)** 16:13, 23:47, 04:13, 11:47 **ii)** 18:46, 21:14, 06:46, 09:14 **iii)** between 17:47 and 22:13, and between 05:47 and 10:13, each day **13. b)** 4.8 cm **c)** $\frac{\pi}{6}$ or 0.52 s

14. a) $y = 1.5\cos\frac{\pi}{6}t$ **c)** 3.75 m, 4.5 m, 3.75 m, 3 m

d) $y = 1.5\cos\frac{\pi}{6}(t-6)$ **15. b)** $y = 15\sin\frac{5\pi}{2}t$

16. b) $y = 7\sin\frac{\pi}{8}(t-4) + 8.5$ **d)** $y = 7\sin\frac{\pi}{10}(t-5) + 8.5$

17. a) $y = 10\sin\frac{\pi}{6}t$ **b)** $y = 10\sin\frac{\pi}{6}(t+2)$

c) $y = 10\sin\frac{\pi}{6}(t+3)$ **18. a)** No; the amplitude is determined by the value of a. **b)** No; the period is determined by the value of k. **c)** Yes; the maximum value is $c + |a|$ and the minimum value is $c - |a|$. **d)** No; the phase shift is determined by the value of d. **19. b)** 12.2 h **c)** 8.4 h **21. a)** $0 < |k| < 1$ **b)** $0 < |a| < 1$ **c)** $c = a\sin(kd)$ **d)** $|a| < |c|$ **22. a)** equal **23. a)** equal

Section 5.7, pp. 398–401

1. Answers may vary. **a)** $\sin x$ **b)** $1 - \cos^2 x$ **c)** $1 - \sin^2 x$

d) $\frac{\sin^2 x}{\cos^2 x}$ **e)** $\frac{\sin^2 x}{\cos x}$ **f)** $\cos^2 x$ **g)** $\sin^2 x$ **h)** $\sin^2 x$ **i)** 1 **5. a)** Each

formula gives $\frac{2g}{\sqrt{3\omega^2}}$. **8. a)** $\sin\left(-\frac{\pi}{6}\right) = -\frac{1}{2}$,

$\sqrt{\sin^2\left(-\frac{\pi}{6}\right)} = \frac{1}{2}$; LHS $\neq$ RHS **b)** $\cos\frac{2\pi}{3} = -\frac{1}{2}$,

$\sqrt{\cos^2\frac{2\pi}{3}} = \frac{1}{2}$; LHS $\neq$ RHS **11. a)** an identity **b)** not an

identity **c)** not an identity **14.** No; the left-hand side is never negative.

Section 5.8, pp. 408–410

1. a) 0, π, 2π **b)** $\frac{2\pi}{3}$, $\frac{4\pi}{3}$ **c)** $\frac{\pi}{4}$, $\frac{5\pi}{4}$ **d)** $\frac{5\pi}{4}$, $\frac{7\pi}{4}$ **e)** $\frac{\pi}{6}$, $\frac{11\pi}{6}$

f) $\frac{4\pi}{3}$, $\frac{5\pi}{3}$ **2. a)** 270° **b)** 45°, 315° **c)** 60°, 120° **d)** 135°, 225°

e) 210°, 330° **f)** 135°, 315° **3. a)** 60°, 300° **b)** 30°, 150°

c) 210°, 330° **d)** 90°, 270° **e)** 45°, 135°, 225°, 315° **f)** 210°,

270°, 330° **g)** 30°, 90°, 150° **4. a)** $\frac{3\pi}{2}$; 4.71 **b)** $\frac{\pi}{6}$, $\frac{5\pi}{6}$, $\frac{3\pi}{2}$;

0.52, 2.62, 4.71 **c)** 0, $\frac{2\pi}{3}$, π, $\frac{4\pi}{3}$, 2π; 0, 2.09, 3.14, 4.19,

6.28 **d)** no solutions **e)** 0, π, $\frac{3\pi}{2}$, 2π; 0, 3.14, 4.71, 6.28

f) $\frac{\pi}{6}$, $\frac{\pi}{2}$, $\frac{5\pi}{6}$, $\frac{3\pi}{2}$; 0.52, 1.57, 2.62, 4.71 **5. a)** 41.4°, 318.6°

b) 19.5°, 160.5° **c)** 60°, 109.5°, 250.5°, 300° **d)** 19.5°, 160.5°

e) no solutions **f)** 41.8°, 138.2°, 210°, 330° **6. a)** 0,

1.23, π, 5.05, 2π **b)** 1.23, $\frac{2\pi}{3}$, $\frac{4\pi}{3}$, 5.05 **c)** 1.82, 4.46

d) $\frac{2\pi}{3}$, 2.42, 3.86, $\frac{4\pi}{3}$ **7. a)** 28.9° **b)** $0° < r < 48.8°$

8. a) $A = \frac{1}{2}bh = \frac{1}{2}b\left(\frac{b\tan x}{2}\right) = \frac{b^2\tan x}{4}$ **b)** 68.2° **9. c)** 15°

or 75° **10. a)** 0, 1.33, π, 4.47, 2π **b)** 1.11, 1.25, 4.25, 4.39

c) 1.11, 4.25 **d)** 0.46, 0.59, 3.61, 3.73 **11. a)** 30°, 45°, 150°,

225° **b)** 45°, 180°, 225° **12. b)** $30° \leq x \leq 150°$

c) $0° \leq x \leq 30°$, $150° \leq x \leq 360°$ **13. b)** $k = 2$ **14. a)** 45°,

225° **b)** 90°, 270° **c)** 15°, 75°, 195°, 255° **d)** 30°, 150°, 210°,

330° **e)** 22.5°, 157.5°, 202.5°, 337.5° **f)** 30°, 60°, 210°, 240°

g) 75°, 105°, 255°, 285° **h)** 0°, 90°, 180°, 270°, 360° **i)** 60°,

300° **15. a)** 0°, 120°, 240°, 360° **b)** 90°, 270° **c)** 0°, 45°,

135°, 180°, 225°, 315°, 360° **16. b)** 90°, 270°

Review of Key Concepts, pp. 412–417

1. a) 120° **b)** 450° **c)** 108° **d)** 40° **e)** 480° **f)** 720° **2. a)** $\frac{5\pi}{18}$

b) $\frac{3\pi}{2}$ **c)** $\frac{3\pi}{6}$ **d)** $\frac{7\pi}{6}$ **e)** $\frac{5\pi}{4}$ **f)** 4π **3. a)** 200.5° **b)** 25.7° **c)** 43.0°

d) 81.8° **e)** 83.1° **f)** 96.9° **4. a)** 0.52 **b)** 2.09 **c)** 1.22 **d)** 0.81

e) 4.04 **f)** 5.50 **5. a)** 8.4 cm **b)** 30.2 cm **6. a)** 1.33 rad, 76.4°

b) 2.29 rad, 131.0° **7. a)** 7.7 cm **b)** 5.6 m **8.** 80π rad/s,

251.33 rad/s **9. a)** $\frac{3\pi}{10}$ **b)** 36°, 54° **10. a)** $\sin\theta = \frac{5}{\sqrt{41}}$,

$\cos\theta = \frac{4}{\sqrt{41}}$, $\tan\theta = \frac{5}{4}$ **b)** $\sin\theta = \frac{7}{\sqrt{53}}$, $\cos\theta = -\frac{2}{\sqrt{53}}$,

$\tan\theta = -\frac{7}{2}$ **c)** $\sin\theta = -\frac{2}{\sqrt{5}}$, $\cos\theta = -\frac{1}{\sqrt{5}}$, $\tan\theta = 2$

d) $\sin\theta = -\frac{4}{\sqrt{65}}$, $\cos\theta = \frac{7}{\sqrt{65}}$, $\tan\theta = -\frac{4}{7}$ **11. a)** $\frac{\sqrt{3}}{2}$

b) 1 **c)** $-\frac{1}{2}$ **d)** $-\frac{\sqrt{3}}{2}$ **12. a)** $\frac{1}{2}$ **b)** $-\frac{1}{2}$ **c)** -1 **d)** $-\frac{\sqrt{3}}{2}$

13. a) $\cos\theta = -\frac{\sqrt{21}}{5}$, $\tan\theta = -\frac{2}{\sqrt{21}}$ **b)** $\sin\theta = -\frac{\sqrt{33}}{7}$,

$\tan\theta = \frac{\sqrt{33}}{4}$ **c)** $\sin\theta = -\frac{5}{\sqrt{61}}$, $\cos\theta = \frac{6}{\sqrt{61}}$ **14. a)** 30°, 150°

b) 45°, 315° **c)** 120°, 300° **d)** 30°, 330° **15. a)** The pattern of

y-values in one section of the graph repeats at regular

intervals. **b)** period: $\frac{5\pi}{2}$, amplitude: 2 **c) i)** 2 **ii)** −2 **iii)** 2

16. Answers will vary. **17. b)** period: 2π, amplitude: 1, range:

$-1 \leq y \leq 1$, for both $y = \sin x$ and $y = \cos x$. **18. b)** period:

180°, domain: $-180° \leq x \leq 450°$, $x \neq -90°$, 90°, 270°, 450°,

range: all real numbers **19. a)** domain: $0 \leq x \leq 2\pi$,

range: $-4 \leq y \leq 4$ **b)** domain: $0 \leq x \leq 2\pi$, range: $-\frac{1}{2} \leq y \leq \frac{1}{2}$

c) domain: $0 \leq x \leq \frac{2\pi}{3}$, range: $-1 \leq y \leq 1$ **d)** domain:

$0 \leq x \leq \pi$, range: $-1 \leq y \leq 1$ **e)** domain: $0 \leq x \leq 6\pi$, range:

$-1 \leq y \leq 1$ **f)** domain: $0 \leq x \leq \frac{8\pi}{3}$, range: $-1 \leq y \leq 1$

20. a) $180°$, π rad **b)** $240°$, $\dfrac{4\pi}{3}$ rad **c)** $720°$, 4π rad

d) $1440°$, 8π rad **21. a)** $y = 4\sin\left(\dfrac{2}{3}x\right)$ **b)** $y = 2\sin(2x)$

23. a) range: $-5 \le y \le 5$, amplitude: 5, period: $\dfrac{\pi}{2}$ **b)** range:

$-0.5 \le y \le 0.5$, amplitude: 0.5, period: $\dfrac{2\pi}{3}$ **c)** range:

$-3 \le y \le 3$, amplitude: 3, period: 6π **d)** range: $-2 \le y \le 2$,

amplitude: 2, period: 3π **24. a)** $y = 12\sin(\pi t)$

25. a) domain: $0 \le x \le 2\pi$, range: $-7 \le y \le 1$, amplitude: 4,

period: 2π, phase shift: none **b)** domain: $0 \le x \le 2\pi$, range:

$-1 \le y \le 5$, amplitude: 3, period: 2π, phase shift: none

c) domain: $0 \le x \le 2\pi$, range: $-2 \le y \le 2$, amplitude: 2,

period: 2π, phase shift: $\dfrac{\pi}{2}$ to the right **d)** domain:

$0 \le x \le 2\pi$, range: $-\dfrac{1}{2} \le y \le \dfrac{1}{2}$, amplitude: $\dfrac{1}{2}$, period: 2π,

phase shift: $\dfrac{\pi}{2}$ to the left **26. a)** domain: $0 \le x \le \dfrac{4\pi}{3}$,

range: $-2 \le y \le 2$, amplitude: 2, period: $\dfrac{4\pi}{3}$, phase shift:

none **b)** domain: $0 \le x \le 8\pi$, range: $-3 \le y \le 3$, amplitude:

3, period: 8π, phase shift: none **27. a)** domain: $0 \le x \le \pi$,

range: $-2 \le y \le 2$, amplitude: 2, period: π, phase shift: $\dfrac{\pi}{4}$ to

the right **b)** domain: $0 \le x \le \pi$, range: $-4 \le y \le 2$,

amplitude: 3, period: π, phase shift: $\dfrac{\pi}{4}$ to the left **c)** domain:

$0 \le x \le 6\pi$, range: $2 \le y \le 4$, amplitude: 1, period: 6π,

phase shift: 2π to the right **d)** domain: $0 \le x \le 4\pi$, range:

$-1 \le y \le 3$, amplitude: 2, period: 4π, phase shift: π to the

right **e)** domain: $0 \le x \le 4\pi$, range: $-\dfrac{5}{2} \le y \le -\dfrac{3}{2}$,

amplitude: $\dfrac{1}{2}$, period: 4π, phase shift: 2π to the right

f) domain: $0 \le x \le \dfrac{2\pi}{3}$, range: $0 \le y \le 4$, amplitude: 2,

period: $\dfrac{2\pi}{3}$, phase shift: $\dfrac{\pi}{3}$ to the right

28. a) $y = 3\sin(x - \pi) - 4$ **b)** $y = 5\cos 2\left(x + \dfrac{\pi}{2}\right) + 1$

29. a) range: $-5 \le y \le 1$ **b)** range: $0 \le y \le 4$

c) range: $-\dfrac{1}{2} \le y \le \dfrac{1}{2}$ **d)** range: $-4 \le y \le 6$ **e)** range:

$-3 \le y \le 1$ **f)** range: $-4 \le y \le 3.5$

30. a) $y = 9.5\sin\dfrac{\pi}{5}(t - 2.5) + 10.7$

31. a) $d(t) = 6\sin\dfrac{\pi}{6}(t + 3) + 14$ **35. a)** $\dfrac{\pi}{2}, \dfrac{3\pi}{2}$ **b)** $\dfrac{\pi}{6}, \dfrac{5\pi}{6}$

c) $\dfrac{3\pi}{4}, \dfrac{7\pi}{4}$ **d)** $\dfrac{\pi}{4}, \dfrac{3\pi}{4}$ **e)** $\dfrac{\pi}{6}, \dfrac{11\pi}{6}$ **f)** $\dfrac{4\pi}{3}, \dfrac{5\pi}{3}$ **g)** $\dfrac{3\pi}{4}, \dfrac{5\pi}{4}$

h) $0, 2\pi$ **i)** $\dfrac{\pi}{3}, \dfrac{4\pi}{3}$ **36. a)** $0, \pi, 2\pi$ **b)** $\dfrac{2\pi}{3}, \pi, \dfrac{4\pi}{3}$ **c)** $\dfrac{\pi}{2}$ **d)** $0, \pi, 2\pi$

e) $\dfrac{\pi}{6}, \dfrac{\pi}{2}, \dfrac{5\pi}{6}$ **f)** $\dfrac{\pi}{3}, \pi, \dfrac{5\pi}{3}$ **g)** $0, 2\pi$ **h)** $\dfrac{7\pi}{6}, \dfrac{11\pi}{6}$ **i)** no solutions

37. a) $36.9°, 323.1°$; 0.64 rad, 5.64 rad **b)** $48.6°, 131.4°$;

0.85 rad, 2.29 rad **c)** $109.5°, 250.5°$; 1.91 rad, 4.37 rad

d) $30°, 41.8°, 138.2°, 150°, \dfrac{\pi}{6}$ rad, 0.73 rad, 2.41 rad,

$\dfrac{5\pi}{6}$ rad

Chapter Test, pp. 418–419

1. a) $300°$ **b)** $210°$ **c)** $97.4°$ **2. a)** $\dfrac{\pi}{3}$ or 1.05 **b)** 3.58 **c)** 5.46

3. a) $\sin\theta = \dfrac{3}{\sqrt{13}}$, $\cos\theta = -\dfrac{2}{\sqrt{13}}$, $\tan\theta = -\dfrac{3}{2}$

b) $\sin\theta = -\dfrac{3}{\sqrt{34}}$, $\cos\theta = \dfrac{5}{\sqrt{34}}$, $\tan\theta = -\dfrac{3}{5}$ **4. a)** 1 **b)** $-\dfrac{\sqrt{3}}{2}$

c) $-\dfrac{1}{2}$ **d)** $-\dfrac{1}{2}$ **5. a)** $225°, 315°$ **b)** $30°, 330°$ **c)** $135°, 315°$

6. b) period: π, domain: $-2\pi \le x \le \pi$, $x \ne -\dfrac{3\pi}{2}, -\dfrac{\pi}{2}, \dfrac{\pi}{2}$,

range: all real numbers **7. a)** domain: $0 \le x \le \dfrac{2\pi}{3}$, range:

$-4 \le y \le 4$, amplitude: 4, period: $\dfrac{2\pi}{3}$, phase shift: none

b) domain: $0 \le x \le 4\pi$, range: $-3 \le y \le 3$, amplitude: 3,

period: 4π, phase shift: none **c)** domain: $0 \le x \le \pi$, range:

$-2 \le y \le 2$, amplitude: 2, period: π, phase shift: $\dfrac{\pi}{4}$ to the

right **d)** domain: $0 \le x \le 4\pi$, range: $-2 \le y \le 6$, amplitude: 4,

period: 4π, phase shift: π to the left **e)** domain: $0 \le x \le \dfrac{2\pi}{3}$,

range: $-5 \le y \le 1$, amplitude: 3, period: $\dfrac{2\pi}{3}$, phase shift: $\dfrac{\pi}{3}$

to the right **8. a)** range: $-7 \le y \le 1$, amplitude: 4, period:

π, phase shift: none **b)** range: $-2 \le y \le 2$, amplitude: 2,

period: 4π, phase shift: π to the right **c)** range: $-4 \le y \le 2$,

amplitude: 3, period: π, phase shift: $\dfrac{\pi}{2}$ to the left

9. a) $d(t) = 2\sin\dfrac{\pi}{6}(t - 3)$ **10. b)** 1 **d)** $\dfrac{1}{2}$ **e)** 20 for each

12. a) $\dfrac{\pi}{6}, \dfrac{11\pi}{6}$ **b)** $\dfrac{\pi}{2}, \dfrac{3\pi}{2}$ **c)** $\dfrac{3\pi}{2}$ **d)** $\dfrac{\pi}{6}, \dfrac{5\pi}{6}, \dfrac{7\pi}{6}, \dfrac{11\pi}{6}$

13. a) 3200 **b)** February **c)** May and October

Challenge Problems, p. 420

1. more than 16 times **2.** 3π **3.** $-\sin^2\theta$

4. $\dfrac{3}{10}$ **5.** $-\dfrac{12}{\sqrt{481}}$ **6.** 3 **7.** $\dfrac{1}{3}$

Problem Solving Strategy: Solve a Simpler Problem, p. 422

1. -100 **2.** 55 **3.** $11\ 111\ 111\ 100\ 000\ 000\ 000$

4. 728 units **5.** 46 **6.** 37 **8.** 14 **9. a)** $\dfrac{n(n+1)}{2}$ **b)** 741

c) 1562 **10.** $1\ 000\ 000$ **11. a)** $1, 8, 27$ **b)** $4913; 970\ 299$

Problem Solving: Using the Strategies, p. 423

1. 1, 1, 1, 2, 5; 1, 1, 2, 2, 2; 1, 1, 1, 3, 3 **2. a)** (3, 2, 4), (1, 8, 3), (−3, −4, 2) **b)** (1, −5, −8), (4, −2, −5), (−2, 1, −20) **3.** Divide the coins into groups of three and weigh one group against another. If they balance, the third group contains the counterfeit coin; if they do not balance, the lighter group contains the counterfeit. From the identified group, weigh one coin against another. If they balance, the coin not weighed is the counterfeit; if they do not balance, the lighter coin is the counterfeit.
5. a) 1, 2 **b)** $\frac{3}{2}$ **6.** 60 **7.** 51% **8.** 150 **9.** 7π cm^2

Chapter 6

Getting Started, p. 426

1. a) The next term is the previous term plus 3; 16, 19, 22.
b) The next term is twice the previous term; 48, 96, 192.
c) The next term is the previous term plus 1 more than the difference between the previous two terms; 22, 29, 37.
d) Alternately multiply by 3 and divide by 2 to obtain subsequent terms; 27, 13.5, 40.5. **e)** Alternately add 1 and multiply by 3 to obtain subsequent terms; 31, 93, 94. **f)** The next term is the sum of the previous two terms; 47, 76, 123.
g) The terms are consecutive odd-numbered multiples of 9; 81, 99, 117. **h)** The differences between successive terms are consecutive odd numbers beginning with 3; 37, 50, 65.
i) The next term is the difference between the previous two terms; −9, 16, −25. **2. a)** every third letter in the alphabet; M, P, S **b)** In the alphabet, alternately go forward 3, then back 2, to obtain subsequent letters; G, E, H. **c)** In the alphabet, alternately go forward 2, then forward 1, to obtain subsequent letters; L, M, O. **d)** Letters and numbers alternate. The next letter is the next in reverse alphabetical order; the next number is the previous number plus 1 more than the difference between the previous two numbers; W, 19, V. **3. a)** 26; 32 **b)** Beginning with 8 asterisks, each subsequent diagram has 6 more asterisks than the previous diagram. **c)** $6n + 2$ **d)** 392; 602 **4. a)** 64; 256 **b)** Each subsequent diagram has 4 times the number of triangles as the previous diagram. **c)** 262 144 **d)** 4^{n-1}

Review of Prerequisite Skills, p. 427

1. a) 128 **b)** 81 **c)** 2187 **d)** 64 **e)** 4096 **f)** 96 **g)** −250 **h)** 384 **i)** −8
2. a) $15x^6$ **b)** $8y^2$ **c)** $-4x^4$ **d)** 12 **e)** $28t^8$ **f)** $-44g^8$ **3. a)** 4 **b)** 25
c) −8 **d)** −243 **e)** 16 **f)** 189 **4. a)** 18 **b)** 54 **c)** 9 **d)** −9 **e)** $\frac{1}{8}$ **f)** 36
g) $-\frac{3}{4}$ **h)** 4 **5. a)** $-2x - 22$ **b)** $3y^2 - 15y - 14$ **c)** $z - 16$
d) $15t + 31$ **6. a)** 2 **b)** 35 **c)** 37 **d)** −53 **e)** 1 **f)** −4 **7. a)** $-\frac{1}{2}$ **b)** 9

c) 15 **d)** −5 **e)** −8 **f)** 8 **8. a)** 30 **b)** −13 **c)** 0 **d)** 7.5 **e)** 13 **9. a)** 4
b) 7 **c)** 5 **d)** 7 **e)** 4 **f)** 5 **11. a)** (7, 3) **b)** (−1, −2) **c)** $\left(\frac{1}{2}, 0\right)$
d) (1, −4) **e)** (8, 9) **f)** (6, −10)

Section 6.1, pp. 433–435

1. a) $t_1 = 3$, $t_2 = 6$, $t_3 = 9$, $t_4 = 12$, $t_5 = 15$ **b)** $t_1 = 6$, $t_2 = 8$, $t_3 = 10$, $t_4 = 12$, $t_5 = 14$ **c)** $t_1 = 3$, $t_2 = 1$, $t_3 = -1$, $t_4 = -3$, $t_5 = -5$ **d)** $f(1) = 9$, $f(2) = 8$, $f(3) = 7$, $f(4) = 6$, $f(5) = 5$
e) $t_1 = 2$, $t_2 = 4$, $t_3 = 8$, $t_4 = 16$, $t_5 = 32$ **f)** $f(1) = 0$, $f(2) = 3$, $f(3) = 8$, $f(4) = 15$, $f(5) = 24$ **2. a)** $t_n = 5n$; 25, 30, 35
b) $t_n = n + 1$; 6, 7, 8 **c)** $t_n = 7 - n$; 2, 1, 0 **d)** $t_n = n^2$; 25, 36, 49
e) $t_n = 2n$; 10, 12, 14 **f)** $t_n = -3(2)^{n-1}$; −48, −96, −192
g) $t_n = n - 2$; 3, 4, 5 **h)** $t_n = 0.1n$; 0.5, 0.6, 0.7 **i)** $t_n = \frac{n}{n+1}$; $\frac{5}{6}, \frac{6}{7}, \frac{7}{8}$ **j)** $t_n = nx$; $5x, 6x, 7x$ **k)** $t_n = 1 + (n-1)d$; $1 + 4d$, $1 + 5d$, $1 + 6d$ **3. a)** 0, 3, 6, 9 **b)** 0, 1, 4, 9 **c)** 1, $\frac{1}{2}, \frac{1}{3}, \frac{1}{4}$
d) 2, $\frac{3}{2}, \frac{4}{3}, \frac{5}{4}$ **e)** 0, 3, 8, 15 **f)** −1, 1, −1, 1 **g)** 1, 2, 4, 8
h) 1, 3, 7, 15 **i)** 0, $\frac{1}{3}, \frac{1}{2}, \frac{3}{5}$ **j)** 1, −1, 1, −1 **k)** $\frac{1}{3}, \frac{1}{9}, \frac{1}{27}, \frac{1}{81}$
l) $\frac{3}{2}, \frac{5}{4}, \frac{9}{8}, \frac{17}{16}$ **4. a)** 19, 37 **b)** 67, 83 **c)** 27, 62 **d)** −7, −23
e) 13, 53 **f)** 1, 5 **g)** 1, 144 **5. a)** 0, 1, 2, 3; $t_n = n - 1$
b) −3, −1, 1, 3; $t_n = 2n - 5$ **c)** 3, 6, 11, 18; $t_n = n^2 + 2$
d) 1, 3, 9, 27; $t_n = 3^{n-1}$ **6.** 156; 176 **7. a)** 202.5 t **b)** 207 t
c) 288 t **8. a)** 10.25 cm, 10.5 cm, 11.5 cm **b)** 20
9. a) Beginning with Venus, double the term that is a multiple of 3, add 4, then divide by 10 to obtain subsequent terms in the sequence. **b)** $\dfrac{0+4}{10}, \dfrac{3+4}{10}$,
$\dfrac{6+4}{10}, \dfrac{12+4}{10}, \dfrac{24+4}{10}, \dfrac{48+4}{10}, \dfrac{96+4}{10}, \dfrac{192+4}{10}$,
$\dfrac{384+4}{10}, \dfrac{768+4}{10}$ **c)** Mercury: 0.4 AU, Venus: 0.7 AU,
Earth: 1 AU, Mars: 1.6 AU, Minor Planets: 2.8 AU, Jupiter: 5.2 AU, Saturn: 10 AU, Uranus: 19.6 AU, Neptune: 38.8 AU, Pluto: 77.2 AU **d)** Neptune
10. a) 1020 kJ, 1100 kJ, 1300 kJ **b)** $t_n = 1000 + 20n$
11. a) 30, 32, 34, 36, 38 **b)** $t_n = 28 + 2n$ **c)** 148 **12.** No; many different sequences have these numbers as the first three terms. Three examples: 1, 2, 4, 1, 2, 4, 1, … ; 1, 2, 4, 7, 11, 16, … ; 1, 2, 4, 8, 16, 32, … **13.** Answers will vary.
14. a) $48\,000$, $38\,400$ **b)** $V(n) = 60\,000(0.8)^n$ **c)** 8

Section 6.2, pp. 441–446

1. a) 15, 19, 23 **b)** 15, 9, 3 **c)** −8, −3, 2 **d)** 4, −3, −10 **e)** 10, 11.4, 12.8 **f)** $\frac{9}{4}, \frac{11}{4}, \frac{13}{4}$ **2. a)** 8, 11, 14, 17 **b)** −5, −3, −1, 1
c) 3, 7, 11, 15 **d)** 5, 4, 3, 2 **e)** −7, −12, −17, −22 **f)** 2, $\frac{5}{2}$, 3,
$\frac{7}{2}$ **3. a)** 17 **b)** 49 **c)** −35 **d)** 1.5 **e)** 78.5 **f)** $\frac{23}{3}$ **4. a)** $a = 5$, $d = 4$

b) not arithmetic **c)** not arithmetic **d)** $a = -1$, $d = -3$ **e)** not arithmetic **f)** not arithmetic **g)** $a = -4$, $d = 1.5$ **h)** not arithmetic **i)** $a = x$, $d = x$ **j)** first term: c, common difference: $2d$ **5. a)** 7, 9, 11, 13, 15 **b)** 3, 7, 11, 15, 19 **c)** −4, 2, 8, 14, 20 **d)** 2, −1, −4, −7, −10 **e)** −5, −13, −21, −29, −37 **f)** $\frac{5}{2}$, 3, $\frac{7}{2}$, 4, $\frac{9}{2}$ **g)** 0, −0.25, −0.5, −0.75, −1 **h)** 8, 8 + x, 8 + 2x, 8 + 3x, 8 + 4x **i)** 6, 7 + y, 8 + 2y, 9 + 3y, 10 + 4y **j)** 3m, 2m + 1, m + 2, 3, 4 − m **6. a)** $t_n = 2n + 4$; $t_{10} = 24$, $t_{34} = 72$ **b)** $t_n = 4n + 8$; $t_{18} = 80$, $t_{41} = 172$ **c)** $t_n = 7n + 2$; $t_9 = 65$, $t_{100} = 702$ **d)** $t_n = 3n - 13$; $t_{11} = 20$, $t_{22} = 53$ **e)** $t_n = -5n + 1$; $t_{18} = -89$, $t_{66} = -329$ **f)** $t_n = \frac{2n-1}{2}$; $t_{12} = \frac{23}{2}$, $t_{21} = \frac{41}{2}$ **g)** $t_n = -6n + 11$; $t_8 = -37$, $t_{14} = -73$ **h)** $t_n = 3n + 4$; $t_{15} = 49$, $t_{30} = 94$ **i)** $t_n = -2n + 12$; $t_{13} = -14$, $t_{22} = -32$ **j)** $t_n = x + 4(n - 1)$, $t_{14} = x + 52$, $t_{45} = x + 176$ **7. a)** 49 **b)** 41 **c)** 36 **d)** 42 **e)** 35 **f)** 45 **g)** 44 **8. a)** $a = 4$, $d = 3$; $t_n = 3n + 1$ **b)** $a = -3$, $d = 5$; $t_n = 5n - 8$ **c)** $a = 42$, $d = 2$; $t_n = 2n + 40$ **d)** $a = -19$, $d = 7$; $t_n = 7n - 26$ **e)** $a = 131$, $d = -28$; $t_n = -28n + 159$ **f)** $a = 229$, $d = -9$; $t_n = -9n + 238$ **g)** $a = 1.3$, $d = 0.4$; $t_n = 0.4n + 0.9$ **h)** $a = 5$, $d = -0.5$; $t_n = -0.5n + 5.5$ **9. a)** 14 **b)** $t_n = 5n + 9$ **10. a)** 35, 28, 21, 14 **b)** $t_n = -7n + 42$ **11. a)** 5, 20, 35, 50, 65 **b)** 740; 2990 **12.** 2 **13. a)** 2, 8, 14, 20, 26 **b)** 10, 3, −4, −11, −18 **14. a)** $a = 1896$, $d = 4$ **b)** 1940, 1944; they were cancelled as a result of World War II. **c)** 12, 13 **d)** Answers may vary. **15.** 88 **16.** 7 **17. a)** 140 km **b)** 220 km **c)** $60 + 80t$ km **18.** 2016, 2023, 2030 **19. a)** $45 **b)** $285 **20.** $39 200, $42 600, $45 800 **21. a)** $t_n = 11.54 + 0.83n$ **b)** 22.33 mm **22. a)** 45 **b)** $t_n = 47 - 2n$ **c)** 23 **23. a)** $a = 8$, $d = 2$ **b)** $t_n = 6 + 2n$ **c)** 36; 1440 m **24.** 101 **25. a)** 16 **b)** $t_n = 4 + 3n$ **c)** 79 **d)** 45th **26. a)** 248.55 **b)** 13:12 **27.** 3.8 **28. a)** The first differences all equal d, and so are constant. **b)** The domain is {1, 2, 3, …}. **29. a)** 10:18:48, 10:20:30 **b)** 436 **30. a)** 10:25:24, 10:30:30 **b)** 146 **31. a)** 10:23:45, 10:28:00 **b)** 175 **32.** 6, 8, 10 **33.** 5, 11, 17 **34. a)** Equal; the common value is $2a + 3d$. **b)** Answers may vary; the sum of the second and fifth terms equals the sum of the first and sixth terms. **35.** 319 **36.** $5x + y$ **37. a)** 5 **b)** 4 **c)** −1 **d)** 8 **e)** −2 **38. a)** 2 **b)** 5 **c)** 0 or 4 **39.** $a = 3 - 22x$, $d = 4.5x$, $t_n = 3 - 22x + (n - 1)4.5x$ **40.** $t_n - t_{n-1} = [a + (n - 1)d] - [a + (n - 2)d]$ $= (a - a) + d[(n - 1) - (n - 2)] = d$

Section 6.3, pp. 452–455

1. a) neither; 25, 36 **b)** geometric; 16, 32 **c)** arithmetic; 35, 42 **d)** neither; 16, 22 **e)** arithmetic; 4, 0 **f)** geometric; 2, 1 **g)** neither; $\frac{1}{3}$, $-\frac{4}{3}$ **h)** geometric; 40.5, 121.5 **2. a)** 3; 81, 243, 729 **b)** 2; 80, 160, 320 **c)** −4; 512, −2048, 8192 **d)** −1; 7, −7, 7 **e)** 10; 5000, 50 000, 500 000 **f)** 2; $\frac{16}{3}$, $\frac{32}{3}$, $\frac{64}{3}$ **g)** $\frac{1}{2}$; 4, 2,

1 h) $-\frac{1}{2}$; 50, −25, 12.5 **3. a)** 4, 12, 36, 108, 324 **b)** 20, 80, 320, 1280, 5120 **c)** 1024, 512, 256, 128, 64 **d)** 0.043, 0.43, 4.3, 43, 430 **e)** 8, −8, 8, −8, 8 **f)** −10, 50, −250, 1250, −6250 **4. a)** 4, 8, 16, 32 **b)** 10, 30, 90, 270 **c)** 2, −4, 8, −16 **d)** 5, −15, 45, −135 **e)** −3, −6, −12, −24 **f)** −2, 6, −18, 54 **g)** 0.5, 2, 8, 32 **h)** −1, 1, −1, 1 **i)** 200, 100, 50, 25 **j)** −1000, 100, −10, 1 **5. a)** $t_n = 2(2)^{n-1}$; $t_7 = 128$, $t_{12} = 4096$ **b)** $t_n = 5^{n-1}$; $t_6 = 3125$, $t_9 = 390\ 625$ **c)** $t_n = 4(3)^{n-1}$; $t_8 = 8748$, $t_{10} = 78\ 732$ **d)** $t_n = 64(0.5)^{n-1}$; $t_7 = 1$, $t_{10} = 0.125$ **e)** $t_n = 6(0.1)^{n-1}$; $t_6 = 0.000\ 06$, $t_8 = 0.000\ 000\ 6$ **f)** $t_n = -3(-2)^{n-1}$; $t_7 = -192$, $t_9 = -768$ **g)** $t_n = 729\left(-\frac{1}{3}\right)^{n-1}$; $t_6 = -3$, $t_{10} = -\frac{1}{27}$ **h)** $t_n = 4(-10)^{n-1}$; $t_8 = -40\ 000\ 000$, $t_{12} = -400\ 000\ 000\ 000$ **6. a)** 7 **b)** 10 **c)** 10 **d)** 8 **e)** 10 **f)** 7 **g)** 8 **h)** 8 **7. a)** $t_n = 4(3)^{n-1}$ **b)** $t_n = -3(-2)^{n-1}$ **c)** $t_n = 512(0.5)^{n-1}$ **d)** $t_n = 4^{n-1}$ or $t_n = -(-4)^{n-1}$ **e)** $t_n = 5(2)^{n-1}$ **f)** $t_n = 891\left(\frac{1}{3}\right)^{n-1}$ **8.** 8, $8\sqrt{3}$, 16 **9. a)** 20 000 cm³, 16 000 cm³, 12 800 cm³, 10 240 cm³, 8192 cm³ **b)** 0.8 **c)** 6553.6 cm³, 5242.88 cm³ **10.** 325 779 **11.** 2% **12.** 6 **13.** 4096 **14. a)** 3, 6, 9, 12, 15 **b)** 3, 6, 12, 24, 48 **c)** The graph of the arithmetic sequence lies on a straight line, whereas the graph of the geometric sequence lies on an exponential curve. **15.** 7 **16. a)** $A_n = A_0(0.5)^n$ **b)** 2.5 mg **17.** Yes; the resulting sequence may be obtained from the original by replacing the parameter a by its previous value times this number. **18. a)** $1000 **b)** $1562.50 **19. a)** non-linear **b)** The first differences of the terms of a geometric sequence are not constant. **20. a)** 38.01 cm **b)** 32.64 cm **21.** $t_n \div t_{n-1} = \frac{ar^{n-1}}{ar^{n-2}} = r$ **22. a)** 10 **b)** 4 **23.** Yes, provided the common ratio is not too large. For example, the terms 10, 15, 22.5 are in geometric progression and can be used as side lengths for a triangle. **24.** $y = \frac{x^2}{w}$ **25. a)** 4 **b)** 30 **c)** $\sqrt{mn}$ **26. a)** $a = 3$, $r = 2$, $t_n = 3(2)^{n-1}$ **b)** $a = 6$, $r = -2$; $t_n = 6(-2)^{n-1}$ **27. a)** $a = 5x^2$, $r = x^2$; $f(n) = 5x^{2n}$ **b)** $a = 1$, $r = 2x$; $f(n) = (2x)^{n-1}$ **28. a)** $t_n = (2x)^n$; $t_{10} = 1024x^{10}$ **b)** $t_n = \frac{1}{2}\left(\frac{x}{2}\right)^{n-1}$; $t_6 = \frac{x^5}{64}$ **c)** $t_n = (x^2)^{n-3}$; $t_{25} = x^{44}$ **d)** $t_n = 3x^{10}\left(-\frac{1}{x}\right)^{n-1}$; $t_{20} = -\frac{3}{x^9}$

Career Connection: Accounting, p. 456

1. a) $14 406 **b)** $6174

Section 6.4, pp. 461–464

1. a) 4, 7, 10, 13, 16 **b)** 3, 1, −1, −3, −5 **c)** −1, −2, −4, −8, −16 **d)** 48, 24, 12, 6, 3 **e)** 6, 10, 16, 24, 34 **f)** −2, −4, −7, −12, −23 **g)** 2, −1, −6, −13, −22 **h)** −3, 10, −16, 36, −68 **2. a)** 3, 5, −2, 7, −9 **b)** −2, 3, −1, 5, 3 **c)** 2, 1, −3, −10, −23 **d)** 1, −2, −2, 4, −8 **e)** −1, 2, 4, 8, 24 **f)** 1, 1, 2, 5, 29

3. a) 12, 18, 24, 30, 36, 42 b) 4, 12, 36, 108, 324, 972
c) 1.5, 2.5, 4, 6.5, 10.5, 17 d) −1, 1, −1, −1, 1, −1
4. a) 5, 6, 7, 8, 9, 10 b) 80, 40, 20, 10, 5, 2.5 c) −1, 0, 3, 12, 39, 120 d) 1, 1, 0, −1, −1, 0 5. a) $t_n = -4n + 5$
b) $t_n = 2(3)^{n-1}$ c) $t_n = 20\left(-\frac{1}{2}\right)^{n-1}$ d) $t_n = \left(\frac{1}{2}\right)^n$
6. The explicit formula for the sequence, $t_n = -2n + 12$, shows that the sequence is arithmetic. 7. The explicit formula for the sequence, $t_n = 20(0.5)^{n-1}$, shows that the sequence is geometric. 8. Neither; there is neither a common difference nor a common ratio between consecutive terms. 9. a) 1, 6, 15, 28, 45, 66 c) 91, 120
10. a) 20, 22, 24, 26, 28, 30, 32, 34 b) $t_n = 2n + 18$
11. a) $3000, $1200, $480, $192, $76.80, $30.72 b) 60%
c) $t_n = 3000(0.4)^{n-1}$ 12. a) 1, 4, 9, 16, 25, 36 b) They are perfect squares. c) $t_n = n^2$ 13. a) $t_n = \frac{3}{2} - \frac{1}{n}$ b) $t_n = 2 - \frac{2}{n}$
c) $t_n = 1 + \frac{1}{n}$ 15. a) $x, \frac{x^2}{2}, \frac{x^3}{4}, \frac{x^4}{8}, \frac{x^5}{16}$ b) $t_n = x\left(\frac{x}{2}\right)^{n-1}$ 16. 0
17. a) $t_n = 2^{n-1}$, $t_n = -5(3)^{n-1}$, $t_n = 4(0.5)^{n-1}$ b) The recursion formula $t_1 = a$, $t_n = rt_{n-1}$ yields the geometric sequence $t_n = ar^{n-1}$. c) $t_n = 1000(0.1)^{n-1}$ 18. a) $t_n = 2n + 3$, $t_n = -2n + 8$, $t_n = 5n - 13$ b) The recursion formula $t_1 = a$, $t_n = t_{n-1} + d$ yields the arithmetic sequence $t_n = a + (n-1)d$.
c) $t_n = -6n + 25$ 19. a) 1.4 m, 0.98 m, 0.69 m, 0.48 m, 0.34 m b) $t_1 = 1.4$, $t_n = 0.7t_{n-1}$ c) $t_n = 1.4(0.7)^{n-1}$
20. a) Transfer the top disk to peg B; transfer the next disk to peg C; transfer the disk from peg B to peg C. b) 7, 15, 31
c) $t_1 = 1$, $t_n = 2t_{n-1} + 1$ d) $t_n = 2^n - 1$ e) 255

Section 6.5, pp. 469–471

1. a) 19 900 b) 15 050 c) −8900 d) −14 850 2. a) 110
b) 1150 c) 2550 d) 414 e) 780 f) 180 3. a) 10 098 b) 28 920
c) 400 d) 3275 e) −3960 f) −3738 g) 3382.5 h) −84 4. a) 135
b) 234 c) −40 d) −550 e) 132 f) 2015 g) 270 h) −270.3
5. a) 1275 b) 10 000 c) 8550 d) 10 100 6. 790 7. a) 25 250
b) 4410 8. For the arithmetic series $t_n = a + (n-1)d$, the first term is a and the last term is $a + (n-1)d$. The number of terms is n. Then $\frac{n}{2}[2a + (n-1)d] = n \times \frac{a + [a + (n-1)d]}{2}$.
9. 4950 10. 228 11. a) $1400 b) $53 850 c) fourth
d) $414 000 12. 9, 10, 11; 8, 10, 12; 7, 10, 13; 6, 10, 14
13. 122.5 m 14. 312 15. 5900 m 16. a) 5 b) $60 17. a) 57
b) 105 18. 20 19. 888 20. 295 21. 110°, 114°, 118°, 122°,
126° 22. Answers may vary. 18, 19, 20, 21, 22; 16, 18, 20, 22, 24; 10, 15, 20, 25, 30; 0, 10, 20, 30, 40 23. 8, 11, 14
24. 10, 304 25. −2 or 18 26. 2, 5, 8, 11, 14 27. a) $130x$
b) $10x + 55y$

Section 6.6, pp. 476–478

1. a) 4095 b) 5461 c) 11 718 d) −3280 e) 513 f) 504 g) $\frac{4372}{3}$
h) 0.656 25 2. a) 16 400 b) −59 048 c) 1441 d) 7.9375

e) 90 910 f) −1302 3. a) 511 b) 3280 c) 342 d) 2735 e) 1093
f) 1333.3332 4. 818.4 5. 429 496 730 6. 2046
7. 177 144 cm² 8. a) 40 920 cm b) 34 952 500 cm² 9. 5460
10. $165 984 11. 153 m 12. a) 13 b) 2 c) 160
13. a) $10 485.75 b) 37 days c) Eventually, too much money per day must be set aside. 14. 664.78 cm 15. 508 cm²
16. 63 m 17. 2, 8, 32 or 50, −40, 32 18. 21 845
19. $\frac{3(x^{30} - 1)}{x^2 - 1}$, $x \neq \pm 1$; 45 if $x = \pm 1$. 20. a) $2^{64} - 1$

Review of Key Concepts, pp. 480–485

1. a) 3, 5, 7, 9, 11 b) −2, 1, 6, 13, 22 c) 5, 3, 1, −1, −3
d) 2, 8, 26, 80, 242 2. a) 4 b) 195 3. a) 26, 66 b) −13, −29
c) 44, 95 d) 5, 16 4. a) $t_n = 4n$; 48 b) $t_n = 2n - 1$; 23
c) $t_n = n^2 + 1$; 145 d) $t_n = -5n - 1$; −61 5. a) 15.6 mm
b) 16.2 mm c) 17.4 mm 6. a) 27, 33, 39 b) −9, −14, −19
c) 4, 6.5, 9 d) $-\frac{1}{2}, -1, -\frac{3}{2}$ 7. a) 7, 12, 17, 22 b) 3, 7, 11, 15
c) 3, 0, −3, −6 d) −2, −7, −12, −17 e) $\frac{1}{3}, 1, \frac{5}{3}, \frac{7}{3}$ f) 4.2, 4.4,
4.6, 4.8 8. a) 3, 8, 13, 18, 23 b) −5, −3, −1, 1, 3 c) 4, 1, −2,
−5, −8 d) 0, −2.3, −4.6, −6.9, −9.2 9. a) $t_n = 2n + 1$; 61
b) $t_n = -4n + 2$; −98 c) $t_n = 7n - 11$; 115 10. a) 34 b) 32
11. a) $a = -15$, $d = 4$; $t_n = 4n - 19$ b) $a = 18$, $d = -2$;
$t_n = -2n + 20$ 12. a) 14, 16 b) $t_n = 2n + 6$ c) 56 d) 43
13. a) $t_n = 1987 + 4n$ b) 2127 14. a) neither; 125, 216
b) geometric; 81, 243 c) arithmetic; 30, 36 d) geometric; 4,
−2 e) neither; 0, −6 f) arithmetic; 1.8, 1.5 15. a) 6, 24, 96,
384, 1536 b) 5, −10, 20, −40, 80 c) −3, 15, −75, 375,
−1875 d) 10, 1, 0.1, 0.01, 0.001 16. a) 3, 6, 12, 24, 48 b) 2,
−6, 18, −54, 162 c) 4, −8, 16, −32, 64 d) −1, −4, −16, −64,
−256 e) −2, 4, −8, 16, −32 f) −1000, −500, −250, −125,
−62.5 17. a) $t_n = 3(2)^{n-1}$; 1536 b) $t_n = 2(4)^{n-1}$; 32 768
c) $t_n = 27\left(\frac{1}{3}\right)^{n-1}$; $\frac{1}{9}$ d) $t_n = (-3)^{n-1}$; 729 18. a) 11 b) 8
19. a) $a = 3$, $r = 2$; $t_n = 3(2)^{n-1}$ or $a = -3$, $r = -2$;
$t_n = -3(-2)^{n-1}$ b) $a = -2$, $r = 3$; $t_n = -2(3)^{n-1}$
20. 112 million 21. a) $A_n = A_0(0.5)^n$ b) 12.5 mg
22. a) 19, 11, 3, −5, −13 b) −5, −2, 1, 4, 7 c) −1, 2, −4, 8,
−16 d) 8, 4, 2, 1, 0.5 e) 3, 3, 6, 9, 15 f) −12, −6, 3, 15, 30
g) 11, 18, 27, 38, 51 h) −1, 1, −1, −1, 1 23. a) 2, 8, 32,
128, 512; geometric b) 0, 1, 5, 18, 58; neither c) −3, −7,
−11, −15, −19; arithmetic 24. a) $t_n = 3n - 8$ b) $t_n = 3(4)^{n-1}$
25. a) $t_1 = 3$, $t_n = t_{n-1} + n + 1$ b) 21, 28
c) $t_n = \frac{(n+1)(n+2)}{2}$ 26. a) $2.80 b) $0.20
c) $F = 2.80 + 2n$, where F is the fare in dollars and n is the
number of kilometres travelled. d) $11.80 27. a) 1010
b) 100 c) 1161 28. a) 847 b) −260 c) 808.5 29. a) 1425
b) −190 30. −1275 31. 8 cm, 11 cm, 14 cm 32. 231
33. a) 2046 b) 4921 c) 2016 d) −5460 34. a) 9841 b) 258
c) 15.75 d) 0 35. 671 875 36. a) 2555 b) 6560 c) 728
d) 2735 37. 2044 cm² 38. 29.5 m

1. a) –1, 1, 3, 5, 7 **b)** 4, 7, 12, 19, 28 **2. a)** 40 **b)** 169
3. a) 47, 143 **b)** 116, 1208 **4. a)** 3.5, 4, 4.5, 5, 5.5 **b)** 2, –1,
–4, –7, –10 **c)** 6, 12, 24, 48, 96 **d)** 10, –20, 40, –80, 160
5. a) $t_n = 4n + 2$; 86 **b)** $t_n = -6n + 1$; –125 **6. a)** $t_n = 4^{n-1}$;
1024 **b)** $t_n = 10\,000\left(-\dfrac{1}{2}\right)^{n-1}$; –312.5 **7. a)** 795 **b)** –110

8. 243 **9. a)** 3577 **b)** 1666.56 **10.** –1274 **11.** 23:30 **12.** 2
560 000 **13. a)** no; no **b)** yes; no **c)** The resulting sequence is
neither arithmetic nor geometric. **14. a)** 7, 4, 1, –2 **b)** –2, 0,
3, 7 **c)** 2000, –800, 320, –128 **d)** 2, 3, 1, –2
15. a) $t_n = 2(5)^{n-1}$ **b)** $t_n = 8n - 15$

Challenge Problems, p. 488

1. 34 **2.** n **3.** 9996 **4.** 114 **5.** Wednesday **6.** column C
7. 26 **8.** 108 **9.** The roots of the equation are imaginary.

Problem Solving Strategy: Guess and Check, p. 491

1. Great Bear Lake: 31 792 km^2 **3. a)** 15 000 km **4. a)** 1255
5. top: 9, bottom: 2, middle, from left to right: 8, 4, 1
6. 2660 cm^3 **7.** A = 8, B = 7, C = 4 **8.** Beginning at the top
and moving from left to right, place the numbers in the
order 2, 6, 3, 7, 8, 5, 9, 4, 1. **9.** O = 1, N = 8, E = 2, T = 7
10. three, four, five **11.** 9 **12.** 7, 5, 8; 6, 1, 4; 3, 2, 9
13. Answers may vary. 8 – 7 = 1, 20 ÷ 5 = 4, 9 – 6 = 3
14. A = 12, B = 20, C = 64, D = 4 **15. a)** 219, 438, 657
b) 327, 654, 981; 273, 546, 819

Problem Solving: Using the Strategies, p. 492

1. $\dfrac{2}{3}$ **2.** 3 **3.** 17 **4.** 5 **5.** 78° **6.** 18 **7.** no **8.** 785 m
9. $x = 2y - z$ **10.** D = 1, E = 4, F = 8

Cumulative Review: Chapters 5 and 6, p. 493

Chapter 5

1. a) 20° **b)** 67.5° **c)** 177.6° **2. a)** $\dfrac{4\pi}{9}$ **b)** $\dfrac{13\pi}{9}$ **c)** $\dfrac{19\pi}{6}$

3. $\sin \theta = -\dfrac{3}{\sqrt{13}}$, $\cos \theta = -\dfrac{2}{\sqrt{13}}$, $\tan \theta = \dfrac{3}{2}$ **4. a)** $\dfrac{1}{\sqrt{2}}$

b) $-\dfrac{3}{\sqrt{2}}$ **c)** $\dfrac{1}{\sqrt{3}}$ **5. a)** 13 **b)** 13 **c)** –5 **6. a)** period: π,

amplitude: $\dfrac{1}{2}$ **b)** period: π, amplitude: 3 **7. a)** period: 2π,

amplitude: 1, vertical translation: 3 units downward, phase

shift: $\dfrac{\pi}{3}$ to the right **b)** period: $\dfrac{2\pi}{3}$, amplitude: $\dfrac{1}{2}$, vertical

translation: 5 units downward, phase shift: $\dfrac{\pi}{3}$ to the right

9. a) 210°, 330°

Chapter 6

1. a) 0, 2, 4, 6, 8 **b)** 6, 9, 14, 21, 30 **c)** 0.5, 1, 2, 4, 8

2. a) $t_n = 6n + 3$ **b)** $t_{25} = 153$ **3. a)** 58 **b)** 8 **4.** $t_n = 1(-2)^{n-1}$;
$t_{12} = 2048$ **5.** 240 **6.** 384 **7.** 682 **8.** 855 **9.** 84°, 88°, 92°
10. a) –6, –1, 4, 9, 14 **b)** 800, –200, 50, –12.5, 3.125 **c)** –2,
–1, 2, –2, –4 **11.** $t_n = -3n + 13$

Chapter 7

Answers may vary slightly depending on the application
used to solve them. For example, answers obtained using
tables may differ slightly from those found using the **TVM
Solver**.

Getting Started: Comparing Costs, p. 496

1. Answers may vary. **2. a) i)** $919 **ii)** $1000 **iii)** $1140
b) i) The most expensive option is Option C. This option
may be preferable if Sadie can afford only $95 per month,
and not a lump sum. If Sadie discontinues the lessons, she
may not have continue paying monthly. **ii)** The least
expensive option is Option A. If Sadie can afford the lump
sum and perseveres with the lessons, this option will save her
money. **3.** Answers may vary.

Review of Prerequisite Skills, p. 497

1. a) $100(0.05)^4$, $100(0.05)^5$, $100(0.05)^6$ **b)** 124, 130, 136
c) $(1 + 0.06)^4$, $(1 + 0.06)^5$, $(1 + 0.06)^6$ **2. a)** 1.2 **b)** 0.25

c) $\dfrac{250}{P}$ **d)** $\dfrac{I}{400}$ **e)** $\dfrac{3}{20t}$ **f)** $\dfrac{I}{Pt}$ **3. a)** 5000 **b)** 2000 **c)** 10 000

d) $\dfrac{I}{rt}$ **4. a)** 2 **b)** 5 **c)** 6 **d)** 9 **e)** $\dfrac{I}{Pr}$ **5. a)** 1.1699 **b)** 0.7002

c) 1.3070 **d)** 0.0502 **e)** 0.5835 **f)** 1.9992 **g)** 1.1265 **h)** 2.4117
i) 0.5440 **j)** 0.4803 **k)** 0.8613 **l)** 3.6165 **6. a)** 6146.28
b) 360 149.28 **c)** 71 495.29 **d)** 40 331.81 **e)** 99 250.71
7. a) 5814.07 **b)** 7148.73 **c)** 605 384.93 **8. a)** 0.15 **b)** 0.0613
c) 0.008 **d)** 0.0475 **e)** 0.013 **f)** 0.0025 **g)** 0.07 **h)** 0.0305
9. a) 100 **b)** 75 **c)** 1650 **d)** 941.64 **e)** 551.25 **f)** 281.75 **g)** 90
h) 246.20 **10. a)** $2298.85 **b)** $24 742.25 **c)** $1723.85
d) $2182.70

Section 7.1, pp. 498–500

1. a) $50 **b)** $1050 **c)** $50 **d)** $100 **e)** $1100 **2.**

Number of years	Principal ($)	Interest rate	Interest ($)	Amount ($)
1	1000	0.05	50	1050
2	1000	0.05	100	1100
3	1000	0.05	150	1150
4	1000	0.05	200	1200
5	1000	0.05	250	1250
6	1000	0.05	300	1300
7	1000	0.05	350	1350
8	1000	0.05	400	1400
9	1000	0.05	450	1450
10	1000	0.05	500	1500

4. a) Arithmetic; there is a common difference of 50. **b)** 1050
c) 50, common difference **d)** $t_n = 1050 + 50(n - 1)$

5. a) 1050, 1100, 1150, 1200, 1250, 1300, 1350, 1400,
1450, 1500 **6. a)** Prt **b)** $I = Prt$ **c)** $A = P + I$ **d)** $A = P + Prt$
7. a) 1050, 1100, 1150, 1200, 1250, 1300, 1350, 1400,
1450, 1500 **8. a)** straight line **b)** 1000 **c)** the initial
investment **d)** 50 **e)** the simple interest per year
f) $y = 50x + 1000$ **9. a)** linear; the points lie on a straight
line. **b)** For a fraction of a year, a fraction of the yearly
interest is earned. **e)** 2000 **f)** 2000, equal **10. a)** linear
b) i) $1067.50 **ii)** $1135 **iii)** $1202.50 **iv)** $1270
v) $1337.50 **c)** arithmetic **d)** 1067.50, 67.50
e) $t_n = 1067.50 + 67.50(n - 1)$ **g)** $y = 67.5x + 1000$
h) The equation represents a straight line.

Section 7.2, pp. 508–511

1. a) 3% **b)** 1.5% **c)** 0.5% **d)** 0.016% **2. a)** 4 **b)** 1 **c)** 36 **d)** 5
e) 4 **f)** 24 **3. a)** $578.81 **b)** $75 898.37 **c)** $1208.80
d) $119 268.53 **e)** $269 367.02 **4. a)** $3524.68, $1524.68
b) $35 236.55, $2736.55 **c)** $11 209.55, $1209.55
d) $14 979.78, $6979.78 **5. a)** $3996.73 **b)** $7181.81
c) $16 468.41 **d)** $1 235 035.45 **6. a)** $1060.90 **b)** $1061.36
c) $1061.68 **7. a)** $13 387.77 **b)** $3887.77 **8.** $3526.79
9. $60.76 **10.** $8946.45 **11. a)** Canada Savings Bonds
b) $46.57 **12.** $18 534.54, $10 217.39 **13.** $1658.84
14. 5.87% **17. b)** 5%, compounded annually; 4.95%,
compounded semi-annually; 4.9%, compounded monthly
18. a) 2 years 9 months **b)** 4 months **19.** $12 074.28,
$163.74 **20. a)** 6.95% **21. a)** 4 **b)** 6 **22. a)** true **b)** false
c) true **d)** true **e)** false

Section 7.3, pp. 512–515

1.

Number of years	Principal ($)	Interest rate	Interest ($)	Amount ($)
1	1000.00	0.05	50.00	1050.00
2	1050.00	0.05	102.50	1102.50
3	1102.50	0.05	157.63	1157.63
4	1157.63	0.05	215.51	1215.51
5	1215.51	0.05	276.28	1276.28
6	1276.28	0.05	340.10	1340.10
7	1340.10	0.05	407.10	1407.10
8	1407.10	0.05	477.46	1477.46
9	1477.46	0.05	551.33	1551.33
10	1551.33	0.05	628.89	1628.89

3. a) geometric; common ratio is 1.05 **b)** 1050 **c)** 1.05
d) $t_n = 1000(1.05)^n$ **4. a)** 1050.00, 1102.50, 1157.63,
1215.51, 1276.28, 1340.10, 1407.10, 1477.46, 1551.33,
1628.89 **5. a)** 1050.00, 1102.50, 1157.63, 1215.51,
1276.28, 1340.10, 1407.10, 1477.46, 1551.33, 1628.89
c) $y = 1000(1.05)^x$ **6. a)** exponential **b)** 1000 **c)** the initial
investment **d)** the amount at various points in time **7. a)** Yes,
exponential; the term $(1.05)^x$ grows exponentially. **b)** non-
linear **c)** exponential **8. a)** For a fraction of a compounding
period, a fraction of a compounding period's interest is
earned. **c)** $2653.30 **d)** $2653.30 **e)** equal **9. a)** exponential
growth **b) i)** $1067.50 **ii)** $1139.56 **iii)** $1216.48

iv) $1298.59 **v)** $1386.24 **c)** geometric **d)** 1067.50, 1.0675
e) $t_n = 1000(1.0675)^n$ **g)** The graph represents an exponential
function.

11. Regular Savings Bond

Year	Principal ($)	Interest ($)	Amount ($)
1	500	30	530
2	500	60	560
3	500	90	590
4	500	120	620
5	500	150	650
6	500	180	680
7	500	210	710
8	500	240	740

12. arithmetic; $t_n = 530 + 30(n - 1)$ **13.** 530, 650, 740
14. 650, 680 **15.** $y = 30x + 500$

16. Compound Savings Bond

Year	Principal ($)	Interest ($)	Amount ($)
1	500.00	30	530
2	530.00	61.80	561.80
3	561.80	95.51	595.51
4	595.51	131.24	631.24
5	631.24	169.11	669.11
6	669.11	209.26	709.26
7	709.26	251.82	751.82
8	751.82	296.92	796.92

17. geometric; $t_n = 500(1.06)^n$ **18.** 530.00, 669.11, 796.92
19. 669.11, 709.26 **20.** $y = 500(1.06)^x$ **21.** The graph of the
amount under simple interest exhibits linear growth since it
follows an arithmetic sequence. The graph of the amount
under compound interest exhibits exponential growth since
it follows a geometric sequence. **22.** Answers may vary.

Section 7.4, pp. 523–525

1. a) 2.25% **b)** 1.275% **c)** 8% **d)** 0.75% **2. a)** $7850.16
b) $7799.84 **c)** $7774.04 **d)** $7756.58 **3.** Shorter
compounding periods produce smaller present values.
4. a) $6828.28 **b)** $46 091.89 **c)** $91 403.38 **d)** $51 700.54
e) $229 270.89 **5.** $17 102.75 **6.** $15 301.00 **7.** $6308.77
8. $21 616.27 **9.** $21 359.93 **10.** $4710.92 **11.** $15.81
more must be invested with quarterly compounding.
12. b) 7.2% **14. a)** The formulas are identical, except in the
context of present value PV is calculated from A, rather than
A being calculated from P. **b)** Substitute the value of PV for
P in the compound interest formula. **15.** $34 178.92
16. $12 264.54 **17. a)** The present value is doubled. **b)** In
the formula for present value, the numerator is doubled.
d) no **18.** $4568.46

Section 7.5, pp. 531–533

1. a) 8 **b)** 24 **c)** 5 **2. a)** 0.02 **b)** 0.0325 **c)** 0.0225 **d)** 0.005
3. a) 1 **b)** 12 **c)** 4 **4. a)** $10 757.01 **b)** $9674.76 **c)** $3933.61
5. a) $372.16 **b)** $301.60 **c)** $100.18 **6.** $26 197.40
7. a) $1613.60 **b)** $3227.20 **8.** $2801.07 **9. a)** $617.09
b) $17.09 **10. a)** $2186.98 **12.** $7896.71 **13.** Answers may
vary. **14. a)** $3677.53 **b)** $a = 200$, $r = 1.0025$, $n = 18$ **c)** The
formula for the amount of an ordinary annuity is derived

from the formula for the sum of a geometric series.
15. a) $1624.79 **b)** The interest rate remains at 8% throughout the 4-year period.

Section 7.6, pp. 540–543

1. a) $11 718.83 **b)** monthly **c)** 4.4% **d)** approximately 0.37%
e) $1000 **2. a)** $15 675.34 **b)** $11 102.35 **3. a)** $8187.23
b) $8747.47 **4. a)** $8383.84 **b)** $9954.00 **c)** $10 907.51
d) $11 618.93 **5.** Shorter compounding periods require larger present values, since interest has less time to accumulate. **6. a)** $3317.38 **b)** $2663.80 **c)** $2363.99
d) $2174.71 **7.** As the length of the compounding periods decreases, the payments decrease, since the investment is not gathering as much interest. **8.** $30 056.65 **9. a)** $111 943.89
b) $108 800.92 **10. a)** $149 464.83 **11. a)** $856.37
b) $102 764.40 **12. a)** $6189.85 **13. a)** $134 **b)** $341
14. a) $586 293.98 **b)** The interest rate remains at 5.4% for the 20-year period. **15.** Answers may vary. **16. a)** 16 years
17. $102.83 **19. a)** false **b)** false **c)** false

Section 7.7, pp. 555–558

1. a) 0.493 862 2% **b)** 0.816 484 6% **c)** 0.453 168 2%
d) 1.636 562 4% **2.** Since payments are made monthly, equivalent monthly rates are necessary to determine the interest component of a payment and the outstanding balance. **3. a)** $372.64 **b)** $1488.71 **c)** $2449.00
4. a) 0.797 414 0% **b)** 0.604 491 9% **c)** 0.427 312 8%
d) 0.246 627 0% **e)** 1.388 843 0% **5. a)** 22 months
b) $145.46 **6. a)** 0.796 471 4% **b)** $1176.13 **c)** $127 665.84
7. a) 0.575 003 9%, $1077.03 **c)** $10 827.03
d) 20 years 11 months **8. a)** 0.514 178 4%, $1162.05
b) i) 70.8% **ii)** 70.6% **iii)** 70.5% **c)** The percent of the payments used to pay interest decreases with each payment. This is due to a smaller outstanding principal after each payment. **9.** the 207th payment **10. a)** $5403.88
b) 7 months **11. a)** Answers may vary. **b)** the Garcias, by $3.02 **12. a)** $791.49 **c)** $107.30 **13. a)** $13 440.93
b) $199 140.73 **14.** No; the payments are $311.38
15. $631.3508 **17. a)** Answers may vary; advantage: shorter term, less interest paid in total; disadvantage: higher payments. **b)** Answers may vary; advantage: lower payments; disadvantage: longer term, more interest paid in total.

Section 7.8, pp. 567–570

1. a) $160 500 **b)** $26 912 **c)** $463 920 **d)** $37 220
2. a) $33 915 **b)** $33 000 **c)** $85 000 **d)** $7000
3. a) $8.944 872 **b)** $10.821 941 **c)** $10.916 402 **4.** Only the payment per $1000 is being calculated. **5. a)** $462.38
b) $1036.89 **c)** $1212.75 **d)** $5688.92 **6. a)** $368.40
b) $2062.05 **c)** $723.01 **d)** $2084.32 **7. a)** $110 520
b) $371 169 **c)** $199 550.76 **d)** $400 189.44 **8. a)** $620.38
b) $186 114 **9. a)** $5024.46 **b)** $413 988.48

10. a) i) $992.62 **ii)** $827.13 **iii)** $756.68 **iv)** $722.33 **b)** The payment decreases. **11. a)** $518.24 **b)** $76 354.12
c) $71 691.08 **12.** biweekly; 20 months faster
13. a) $740.50 **b)** $177 720 **14. a)** $1842.02 **b)** $1759.47
c) $1926.14 **d)** The higher the rates, the higher the payments. **15. a)** easy look-up of payment per $1000 **b)** The payment may be found for a wider range of interest rates.
16. a) $1054.73 **b)** $189 848.23 **c)** $803.88 **d)** whether the interest rates are expected to rise or to fall **17. a)** $596.14
b) $496.46 **c)** 14 years 5 months **18. d) i)** $1315.49
ii) $3155.45

Review of Key Concepts, pp. 572–576

1. a) $539.54 **b)** $1774.79 **c)** $4163.53 **d)** $7047.29
2. a) $5361.90 **b)** $1161.90 **c)** No; at 7%, compounded annually, the investment would have yielded $1305.34 in interest. **3. a)** $2703.76 **b)** $2713.48 **4.** Recommend 5.15%, compounded annually; this option yields $27.02 more interest than the next best option. **5. a)** $2351.61
b) $2246.13 **c)** $2060.95 **d)** $5929.60 **6.** $5383.27 **7.** 14%
8. $7295.12 **9. a)** $12 874.45 **b)** $69 482.85 **c)** $29 247.27
d) $34 406.99 **10.** $7450.10 **11.** $75 268.06 **12.** $7446.75
13. a) $52 654.55 **b)** $13 002.51 **c)** $15 855.21
14. a) $2549.73 **b)** $3090.15 **c)** $1469.26 **d)** $1519.84
15. $55 778.78 **16.** $12 235 **17.** $763.38
18. a) 0.493 862 2% **b)** 0.816 484 6% **c)** 0.453 168 2%
d) 1.636 562 4% **19.** 218 months **20.** $45 156
21. a) $23 370.51 **b)** $5723.25 **22. a)** $997.06 **b)** $453.54
c) $1318.77 **d)** $177.94 **23. a)** $1831.49 **b)** $329 668.20
c) $1358.85 **24. a)** $785.69 **b)** $235 707 **25. a)** $678.26
b) $69 482.92

Chapter Test, pp. 577–578

1. $5262.22 **3.** 6.2% per annum, compounded semi-annually, is the better option; this plan yields $14 668.35, whereas 5.75% per annum, compounded quarterly, yields $14 210.05. **4.** $4048.34 **5.** $13 985.34 **6.** $28 279.68
7. $72 225.92 **8.** $245 124.12 **9. a) i)** $843.33 **ii)** $952.83
iii) $1142.78 **b)** $205 700.26 **10.** $48.97

Challenge Problems, pp. 579

1. The company should lease the equipment. **2.** The company should not proceed. **3.** The Andersons' weekly payments are $437.86. **4.** $206 509.61 should be put into the 19-year-old's fund, $156 825.78 should be put into the 15-year-old's fund, and $136 664.61 should be put into the 13-year-old's fund.

Problem Solving Strategy: Use Logic, p. 582

1. Two; turn over the red card and the card with the circle on it. **3.** November = 4, December = 5 **4.** Lions 1, Tigers 0; Tigers 0, Bears 0; Tigers 1, Rams 0; Bears 2, Rams 2; Lions

1, Rams 0; Lions 5, Bears 0. **5.** No; to do so, the car must complete two laps in 2 min. It has already taken 2 min to complete the first lap. **6.** $\frac{1}{5}$; of the 5 possible pairs with at least one white, only one pair has the other ball white. **7.** 3 females, 4 males **8.** (9, 6) and (3, 10) **9.** the odd numbers from 1 to 99 **10.** 8 of each **11.** 26 **12. a)** 125 **b)** 69 375

Problem Solving: Using the Strategies, p. 583

1. $-\frac{1}{40}$ **2.** pattern: 1^9, 2^8, 3^7, 4^6, 5^5, 6^4; next number: 7^3 or 343 **3.** $\frac{9}{2}, \frac{3}{2}$ or $-\frac{9}{2}, -\frac{3}{2}$ **4.** 45 **5.** One possible solution: Fill the 5-L and 11-L containers from the 24-L container, leaving 8 L. Pour the contents of the 5-L container into the 13-L container. Fill the 13-L container from the 11-L container, leaving 3 L. Fill the 5-L container from the 13-L container, leaving 8 L. Pour the contents of the 5-L container into the 11-L container, making 8 L. **6. a)** 17 **b)** The equation is $8(x - 8) = 9(x - 9)$. **7.** $\frac{x^2}{18}$ **8.** 119 (y can be any odd number from 1 to 237.) **9.** Spartans 0, Eagles 0; Ravens 1, Eagles 0; Penguins 3, Eagles 0; Spartans 3, Penguins 2; Spartans 2, Ravens 0; Penguins 1, Ravens 0 **10.** 6 399 999 999 840 000 000 001 **11.** $\frac{1}{9}$ **12.** $\{a, b, c\} = \{2, 3, 4\}$ **13.** If the digit is x, then $\frac{1000x + 100x + 10x + x}{x + x + x + x} = \frac{1111x}{4x} = \frac{1111}{4} = 277.75$. **14.** 648

Chapter 8

Getting Started: Communications Satellites, p. 586

1. 265 465 km **2.** 3073 m/s **3.** 463 m/s **4.** 6.6 **5. a)** $g = \frac{v^2}{r}$ **b)** 0.22 m/s² **c)** 45 **6.** Edmonton

Review of Prerequisite Skills, p. 587

1. a) $-2x^3 + 7x^2 - 15x$ **b)** $-12x^2 - 52x - 16$ **c)** $17x^2 + 26x + 7$ **d)** $107x^2 - 58x + 8$ **2. a)** 5.8 **b)** 7.3 **c)** 5.1 **d)** 7.8 **3. a)** (1, 1) **b)** (5, 4) **c)** (−0.5, −4.5) **f)** (4, −7.5) **5. a)** $y = 4 - 3x$ **b)** $y = \frac{x-2}{4}$ **c)** $y = \pm 5$ **d)** $y = \pm\sqrt{25 - x^2}$ **6. a)** 3, −2 **b)** $\frac{1}{2}$, −2 **c)** 3, $\frac{1}{4}$ **d)** $\frac{1}{2}, \frac{1}{3}$ **e)** $\frac{5}{2}, -\frac{1}{3}$ **f)** 7, $\frac{1}{5}$ **7. a)** 2, −5 **b)** 0.5, −3 **c)** −2, $-\frac{1}{3}$ **d)** 1.2, −1.7 **e)** 0.6, −0.8 **f)** 0.3, −1.6 **8. a)** 36 **b)** 16 **c)** $\frac{9}{4}$ **d)** 9 **e)** $\frac{25}{4}$ **f)** $\frac{1}{4}$ **9. a)** $y = (x + 2)^2 - 9$; minimum of −9 at $x = -2$ **b)** $y = (x - 3)^2 - 19$; minimum of −19 at $x = 3$ **c)** $y = -\left(x + \frac{1}{2}\right)^2 + \frac{121}{4}$; maximum of $\frac{121}{4}$

at $x = -\frac{1}{2}$ **d)** $y = \left(x - \frac{11}{2}\right)^2 - \frac{113}{4}$; minimum of $-\frac{113}{4}$ at $x = \frac{11}{2}$ **e)** $y = -(x + 4)^2 + 16$; maximum of 16 at $x = -4$ **f)** $y = \left(x + \frac{5}{2}\right)^2 - \frac{25}{4}$; minimum of $-\frac{25}{4}$ at $x = -\frac{5}{2}$ **10. a)** $y = 2(x + 2)^2 - 24$; minimum of −24 at $x = -2$ **b)** $y = -3(x - 1)^2 + 9$; maximum of 9 at $x = 1$ **c)** $y = 3(x + 1)^2 - 11$; minimum of −11 at $x = -1$ **d)** $y = 4\left(x - \frac{3}{2}\right)^2 - 9$; minimum of −9 at $x = \frac{3}{2}$ **e)** $y = 0.1(x + 10)^2 - 9$; minimum of −9 at $x = -10$ **f)** $y = -0.2(x + 15)^2 + 45$; maximum of 45 at $x = -15$ **11. a)** (1, 2) **b)** (1, −3) **c)** (5, −2) **d)** (−2, −3) **12. a)** $2x + y - 16 = 0$ **b)** $3x + y + 1 = 0$ **c)** $x - 2y - 8 = 0$ **d)** $2x - 5y - 14 = 0$

Section 8.1, pp. 588–593

8. yes, yes **14.** No, unless E is moved too close to C and no locus is possible. **15.** lengths are equal; yes **16.** right bisector of AB **30.** equal; yes **31.** bisector of $\angle$BAC **32. b)** Yes, provided H and G are adjusted appropriately. **46.** The line parallel to the two given lines and midway between them.

Section 8.2, pp. 598–600

1. a) the angle bisector **b)** the line, parallel to the given lines, at a distance of 3 cm from each **c)** the two lines, parallel to line l, on either side of line l **d)** the circle, concentric with the given circles, with a radius 5 cm **e)** the centre of the square **2.** $x = 1$ **3. a)** parallel **b)** $y = 2x - 1$ **4. a)** $x + y + 1 = 0$ **b)** $2x - 4y - 3 = 0$ **c)** $10x - 12y + 29 = 0$ **5. b)** $x^2 + y^2 = 25$ **6.** $2x - 16y + 41 = 0$ **7.** $10x - 4y - 31 = 0$ **9. b)** $y = 7x$ **c)** The centre of the circle is equidistant from all points on the circle, thus satisfying the equation in b). **10. a)** $x^2 + y^2 = 25$ **b)** $y = x^2$ **c)** $y = \sqrt{x}$ **11. a)** $xy = 0$ **b)** $xy - 2x = 0$ **c)** $x^2 - y^2 = 0$ **12. a)** $y = 0$ **b)** $y = -1$ **14.** 5π m² **15.** (12, 5), (12, −5), (−12, 5), (−12, −5) **16. a)** the strip of width 4 cm, centred on the line **b)** the points outside, or on, the circle of radius 3 cm centred at the given point **c)** the points outside the circle of radius 2 cm, and inside the circle of radius 5 cm, centred at the given point **17. a)** the points on a circle of radius 3 cm, centred at the given point; the points on a sphere of radius 3 cm centred at the given point **b)** the points on the two lines parallel to, and 2 cm from, the given line; the points on the cylinder of radius 2 cm, with the given line as axis of symmetry **c)** the right bisector of the line segment whose endpoints are the given points; the plane that right bisects the line segment whose endpoints are the given points **d)** the line, parallel to the two given lines, and centred between them; the plane that right bisects a line segment perpendicular to, and with endpoints on, the given lines

Section 8.3, pp. 601–607

14. ellipse **15.** F1P1 = CE and F2P1 = ED. Thus, F1P1 + F2P1 = CE + ED = CD. **16.** If F1 and F2 are closer together, the shape becomes wider, or closer to a circle. If F1 and F2 are farther apart, the space becomes narrower or more oval shaped. **17. b)** The distances on either side of the origin are equal and their total distance is more than F1F2. The distance is equal to the length of CD. **c)** As F1 is dragged farther from F2, the ellipse becomes narrower. As F1 is dragged closer to F2, the ellipse becomes wider, and closer to a circle.
29. $|m(AC) - m(BC)| = |m(P1F1) - m(P1F2)|$ and P1F2 = BC and P1F1 = AC, provided C is not between A and B. **33.** A hyperbola with two branches. One branch is created when C is to the left of AB and the other branch is created when C is to the right of AB. **34. a)** Move F1F2 closer together, but still longer than AB. Yes, move A and B farther apart and closer to the length of F1F2. **c)** This only produces half of the hyperbola locus. As F1 moves farther away from F2, the hyperbola becomes wider. As F1 moves closer to F2, the hyperbola becomes narrower, until F1F2 = AB, at which point, the locus becomes an ellipse. The locus disappears as a hyperbola. **47.** Yes; it is a parabola, or the set of points equidistant from point F and line d. **48.** The parabola becomes narrower. The parabola becomes wider. **49. a)** at the origin **c)** The closer F is to the directrix, the narrower the parabola becomes.

Section 8.4, pp. 614–618

1. a) $x^2 + y^2 = 9$ **b)** $x^2 + y^2 = 53.29$ **c)** $x^2 + y^2 = 2$
d) $x^2 + y^2 = 45$ **2. a)** $(x-2)^2 + (y-5)^2 = 9$
b) $(x+1)^2 + (y-3)^2 = 16$ **c)** $(x-3)^2 + (y+2)^2 = 25$
d) $x^2 + (y-2)^2 = 64$ **e)** $(x+3)^2 + (y+4)^2 = 7$
f) $(x+5)^2 + y^2 = 20$ **3. a)** domain: $-10 \le x \le 10$, range: $-10 \le y \le 10$ **b)** domain: $-4\sqrt{5} \le x \le 4\sqrt{5}$, range: $-4\sqrt{5} \le y \le 4\sqrt{5}$ **c)** domain: $-5 \le x \le 11$, range: $-3 \le y \le 13$
d) domain: $-11 \le x \le 3$, range: $-7 \le y \le 7$ **e)** domain: $-11 \le x \le 11$, range: $-14 \le y \le 8$ **f)** domain: $-2 - 5\sqrt{2} \le x \le -2 + 5\sqrt{2}$, range: $7 - 5\sqrt{2} \le y \le 7 + 5\sqrt{2}$
4. a) $x^2 + y^2 = 4$ **b)** $(x-5)^2 + (y-3)^2 = 25$
c) $(x+3)^2 + (y+6)^2 = 16$ **d)** $(x-7)^2 + (y+5)^2 = 16$
5. a) $(x-2)^2 + (y-3)^2 = 25$ **b)** $(x+2)^2 + (y-4)^2 = 41$
c) $(x-3)^2 + (y+2)^2 = 26$ **d)** $(x+4)^2 + (y+1)^2 = 85$
6. a) $(x-4)^2 + (y+5)^2 = 26$ **b)** $(x-2)^2 + (y-4)^2 = 29$
7. a) $x^2 + y^2 = 81$ **b)** $(x-50)^2 + (y-30)^2 = 81$
c) $(x+30)^2 + (y+50)^2 = 81$ **8. b)** The circles are concentric. The circles are different sizes. **9.** $x^2 + y^2 = 1\,785\,062\,500$
10. a) $(x+1)^2 + (y-4)^2 = 5$ **b)** $x^2 + (y-1)^2 = 8$
c) $(x+3.5)^2 + (y+1)^2 = 6.25$
d) $(x+0.5)^2 + (y-0.5)^2 = 26.5$ **11.** $x^2 + y^2 = 2500$
12. a) $x^2 + y^2 = 49$ **b)** yes **13.** $(x-3)^2 + (y+1)^2 = 36$

14. $2 \pm \sqrt{11}$ **16.** Toonie: $x^2 + (y-5)^2 = 1.96$;
Loonie: $(x-3.5)^2 + (y-3.5)^2 = 1.76$;
50¢: $(x+3.5)^2 + (y-3.5)^2 = 1.84$;
25¢: $(x-1.5)^2 + (y+4.5)^2 = 1.43$;
10¢: $(x-3.5)^2 + (y+3)^2 = 0.81$;
5¢: $(x+1.5)^2 + (y+4.5)^2 = 1.12$;
1¢: $(x+3.5)^2 + (y+3)^2 = 0.91$
17. $(x+a)^2 + y^2 = \frac{1}{4}$ and $x^2 + (y+a)^2 = \frac{1}{4}$ for $a = 3, 1$, $-1, -3$ **18.** yes; at two points
19. a) $(x-38)^2 + (y-38)^2 = 1444$ **b)** so they cannot fall into the sewer **20. b)** $(10, -10)$ **c)** $(x-30)^2 + (y-20)^2 = 1296$, $(x+10)^2 + (y-10)^2 = 784$, $x^2 + (y+20)^2 = 196$
21. a) 1 unit to the right, 2 units upward **b)** 3 units to the left, 4 units downward **c)** 1 unit upward **d)** 2 units to the left **e)** h units to the right, if $h > 0$, or $|h|$ units to the left, if $h < 0$, k units upward, if $k > 0$, or $|k|$ units downward, if $k < 0$. **23. a)** No; it is the equation of the point $(3, -4)$.
b) No; no points satisfy the equation. **c)** Yes; it is the equation of a circle with centre $(-1, -3)$ and radius $\sqrt{10}$.
d) Yes; it is the equation of a circle with centre $(0, -3)$ and radius $\sqrt{\sqrt{3}}$. **e)** No; it is the equation of an ellipse. **f)** No; it is the equation of a parabola. **24.** $x^2 + y^2 = 20$ **25. a)** Both points $(0, -3)$ and $(4, 1)$ satisfy the equation of the circle and so are endpoints of a chord of the circle. **b)** $x + y - 1 = 0$
c) The point $(-3, 4)$ satisfies the equation of the perpendicular bisector. **26.** $(x-2)^2 + (y-1)^2 = 25$
27. $(x+1)^2 + (y-3)^2 = 20.25$ **29.** $x^2 + y^2 = 16(\sqrt{2} - 1)^2$
30. $\left(\frac{1}{2}, \frac{\sqrt{3}}{2}\right), \left(\frac{1}{2}, -\frac{\sqrt{3}}{2}\right), \left(-\frac{1}{2}, \frac{\sqrt{3}}{2}\right), \left(-\frac{1}{2}, -\frac{\sqrt{3}}{2}\right),$
$\left(\frac{\sqrt{3}}{2}, \frac{1}{2}\right), \left(\frac{\sqrt{3}}{2}, -\frac{1}{2}\right), \left(-\frac{\sqrt{3}}{2}, \frac{1}{2}\right), \left(-\frac{\sqrt{3}}{2}, -\frac{1}{2}\right)$

Section 8.5, pp. 632–636

1. a) $\frac{x^2}{25} + \frac{y^2}{16} = 1$ **b)** $\frac{x^2}{16} + \frac{y^2}{25} = 1$ **c)** $\frac{x^2}{100} + \frac{y^2}{36} = 1$
d) $\frac{x^2}{36} + \frac{y^2}{100} = 1$ **3. a) i)** $(0, 0)$ **ii)** 8, 4 **iii)** $(4, 0)$, $(-4, 0)$, $(0, 2)$, $(0, -2)$ **iv)** $(2\sqrt{3}, 0)$, $(-2\sqrt{3}, 0)$ **v)** domain: $-4 \le x \le 4$, range: $-2 \le y \le 2$ **vi)** $\frac{x^2}{16} + \frac{y^2}{4} = 1$ **b) i)** $(0, 0)$ **ii)** 6, 4
iii) $(0, 3)$, $(0, -3)$, $(2, 0)$, $(-2, 0)$ **iv)** $(0, \sqrt{5})$, $(0, -\sqrt{5})$
v) domain: $-2 \le x \le 2$, range: $-3 \le y \le 3$ **vi)** $\frac{x^2}{4} + \frac{y^2}{9} = 1$
4. a) $\frac{x^2}{49} + \frac{y^2}{9} = 1$ **b)** $\frac{x^2}{25} + \frac{y^2}{9} = 1$ **c)** $\frac{x^2}{25} + \frac{y^2}{36} = 1$
d) $\frac{x^2}{64} + \frac{y^2}{9} = 1$ **e)** $\frac{x^2}{9} + \frac{y^2}{49} = 1$ **5. a)** $(-2, 3)$; 10, 6; $(-6, 3)$,
$(2, 3)$; $(-7, 3)$, $(3, 3)$; $(-2, 6)$, $(-2, 0)$ **b)** $(3, -1)$; 18, 14;
$(3, 3\sqrt{2} - 1)$, $(3, -3\sqrt{2} - 1)$; $(3, 8)$, $(3, -10)$; $(-4, -1)$,
$(10, -1)$ **c)** $(-1, 3)$; 12, 4; $(-1 + 4\sqrt{2}, 3)$, $(-1 - 4\sqrt{2}, 3)$;

(–7, 3), (5, 3); (–1, 5), (–1, 1) **d)** (3, –2); 8, 2; $(3, -2 + \sqrt{15})$, $(3, -2 - \sqrt{15})$; (3, 2), (3, –6); (2, –2), (4, –2) **6. a) i)** (4, 1)
ii) 8, 2 **iii)** (8, 1), (0, 1), (4, 0), (4, 2) **iv)** $(4 + \sqrt{15}, 1)$,
$(4 - \sqrt{15}, 1)$ **v)** domain: $0 \le x \le 8$, range: $0 \le y \le 2$
vi) $\dfrac{(x-4)^2}{16} + (y-1)^2 = 1$ **b) i)** (–2, –5) **ii)** 6, 4 **iii)** (–2, –2),
(–2, –8), (–4, –5), (0, –5) **iv)** $(-2, -5 + \sqrt{5})$, $(-2, -5 - \sqrt{5})$
v) domain: $-4 \le x \le 0$, range: $-8 \le y \le -2$
vi) $\dfrac{(x+2)^2}{4} + \dfrac{(y+5)^2}{9} = 1$ **c) i)** (3, –1) **ii)** 8, 6 **iii)** (7, –1),
(–1, –1), (3, 2), (3, –4) **iv)** $(3 + \sqrt{7}, -1)$, $(3 - \sqrt{7}, -1)$
v) domain: $-1 \le x \le 7$, range: $-4 \le y \le 2$
vi) $\dfrac{(x-3)^2}{16} + \dfrac{(y+1)^2}{9} = 1$ **d) i)** (–2, –3) 8, 2 **iii)** (–2, 1),
(–2, –7), (–3, –3), (–1, –3) **iv)** $(-2, -3 + \sqrt{15})$,
$(-2, -3 - \sqrt{15})$ **v)** domain: $-3 \le x \le -1$, range: $-7 \le y \le 1$
vi) $(x+2)^2 + \dfrac{(y+3)^2}{16} = 1$ **7. a)** $\dfrac{(x-2)^2}{36} + \dfrac{(y+3)^2}{4} = 1$
b) $\dfrac{(x-3)^2}{49} + \dfrac{(y+2)^2}{9} = 1$ **c)** $\dfrac{(x+1)^2}{16} + \dfrac{(y+2)^2}{36} = 1$
d) $\dfrac{(x-4)^2}{25} + \dfrac{y^2}{9} = 1$ **e)** $\dfrac{(x-4)^2}{169} + \dfrac{(y+1)^2}{144} = 1$

8. a) at the other focus point **b)** $\dfrac{x^2}{3600} + \dfrac{y^2}{400} = 1$

9. $\dfrac{x^2}{2\,235\,025 \times 10^{10}} + \dfrac{y^2}{2\,234\,400 \times 10^{10}} = 1$

10. $\dfrac{x^2}{7\,257\,636 \times 10^{12}} + \dfrac{y^2}{466\,410 \times 10^{12}} = 1$

11. $\dfrac{x^2}{147\,015 \times 10^6} + \dfrac{y^2}{147\,456 \times 10^6} = 1$ **12.** $\dfrac{x^2}{49} + \dfrac{y^2}{49} = 1$

13. a) $\dfrac{x^2}{348\,100} + \dfrac{y^2}{218\,556.25} = 1$ **b)** 230 km **c)** 950 km

14. a) $\dfrac{(x-15.47)^2}{239.3209} + \dfrac{(y-7)^2}{49} = 1$ **b)** 30.94 **c)** 14

15. $\dfrac{x^2}{9} + \dfrac{(y-5)^2}{8} = 1$ **16.** $\dfrac{x^2}{600\,625} + \dfrac{y^2}{599\,400} = 1$

17. 3.73 m **19. c)** The ellipses are concentric. They have different widths. **d)** The ellipses begin to approximate the major axis. **20. b)** 5 **c)** 3 **e)** 2 **f)** 6 **g)** The horizontal stretch factor is the square root of the denominator of the x^2 term, and the vertical stretch factor is the square root of the denominator of the y^2 term. **21. a) i)** 0.75 **ii)** 0.6 **iii)** 0.99
iv) 0 **b)** 1, 0

Section 8.6, pp. 648–652

1. a) $\dfrac{x^2}{9} - \dfrac{y^2}{16} = 1$ **b)** $\dfrac{y^2}{9} - \dfrac{x^2}{16} = 1$ **2. a) i)** (0, 0) **ii)** (–4, 0),

(4, 0), (0, –2), (0, 2) **iii)** 8, 4 **iv)** $\dfrac{x^2}{16} - \dfrac{y^2}{4} = 1$ **v)** $(-2\sqrt{5}, 0)$,

$(2\sqrt{5}, 0)$ **vi)** domain: $|x| \ge 4$, range: all real numbers
b) i) (0, 0) **ii)** (0, –2), (0, 2), (–3, 0), (3, 0) **iii)** 4, 6
iv) $\dfrac{y^2}{4} - \dfrac{x^2}{9} = 1$ **v)** $(0, \sqrt{13})$, $(0, -\sqrt{13})$ **vi)** domain: all real numbers, range: $|y| \ge 2$ **c) i)** (0, 0) **ii)** (0, –5), (0, 5), (–2, 0),
(2, 0) **iii)** 10, 4 **iv)** $\dfrac{y^2}{25} - \dfrac{x^2}{4} = 1$ **v)** $(0, \sqrt{29})$, $(0, -\sqrt{29})$
iv) domain: all real numbers, range: $|y| \ge 5$ **d) i)** (0, 0)
ii) (–2, 0), (2, 0), (0, –3), (0, 3) **iii)** 4, 6 **iv)** $\dfrac{x^2}{4} - \dfrac{y^2}{9} = 1$
v) $(-\sqrt{13}, 0)$, $(\sqrt{13}, 0)$ **vi)** domain: $|x| \ge 2$, range: all real numbers **4. a)** $\dfrac{y^2}{16} - \dfrac{x^2}{25} = 1$ **b)** $\dfrac{x^2}{9} - \dfrac{y^2}{16} = 1$
c) $\dfrac{x^2}{4} - \dfrac{y^2}{21} = 1$ **d)** $\dfrac{y^2}{6.25} - \dfrac{x^2}{2.75} = 1$ **5. a) i)** (2, 1)
ii) (0, 1), (4, 1), (2, –3), (2, 5) **iii)** 4, 8
iv) $\dfrac{(x-2)^2}{4} - \dfrac{(y-1)^2}{16} = 1$ **v)** $(2 - 2\sqrt{5}, 1)$, $(2 + 2\sqrt{5}, 1)$
b) i) (–2, 1) **ii)** (–2, –1), (–2, 3), (–5, 1), (1, 1) **iii)** 4, 6
iv) $\dfrac{(y-1)^2}{4} - \dfrac{(x+2)^2}{9} = 1$ **v)** $(-2, 1 - \sqrt{13})$, $(-2, 1 + \sqrt{13})$
c) i) (2, –1) **ii)** (–1, –1), (5, –1), (2, –2), (2, 0) **iii)** 6, 2
iv) $\dfrac{(x-2)^2}{9} - (y+1)^2 = 1$ **v)** $(2 - \sqrt{10}, -1)$, $(2 + \sqrt{10}, -1)$
d) i) (3, 0) **ii)** (3, –4), (3, 4), (1, 0), (5, 0) **iii)** 8, 4
iv) $\dfrac{y^2}{16} - \dfrac{(x-3)^2}{4} = 1$ **v)** $(3, -2\sqrt{5})$, $(3, 2\sqrt{5})$

7. a) $\dfrac{(x-2)^2}{16} - \dfrac{(y+5)^2}{20} = 1$ **b)** $\dfrac{(y-3)^2}{4} - \dfrac{x^2}{5} = 1$
c) $\dfrac{(x+3)^2}{5} - \dfrac{(y-1)^2}{4} = 1$ **d)** $\dfrac{(y-2)^2}{7} - \dfrac{(x+2)^2}{9} = 1$

8. $\dfrac{(y-2)^2}{9} - \dfrac{(x-2)^2}{16} = 1$ **9.** $\dfrac{x^2}{4} - \dfrac{y^2}{32} = 1$ **10. a)** 10, 6,
$\dfrac{(x-2)^2}{9} - \dfrac{(y-1)^2}{25} = 1$ **b)** $\dfrac{(x-h)^2}{a^2} - \dfrac{(y-k)^2}{b^2} = 1$,
$\dfrac{(y-k)^2}{b^2} - \dfrac{(x-h)^2}{a^2} = 1$ **13. a)** $\dfrac{x^2}{16} - \dfrac{y^2}{20} = 1$ **b)** $\dfrac{y^2}{4} - \dfrac{x^2}{21} = 1$
c) 4 **d)** (–4.8, 2.9) **14. c)** The hyperbolas have the same vertices. Those with larger values of b are flatter. **d)** The hyperbolas close in on the x-axis. **15. a)** The hyperbola is very tight around the line through the transverse axis. **b)** The hyperbola is flat, with the asymptotes approximating the conjugate axis. **16. a)** The asymptotes are $y = \pm x$ for each.
b) $y = x$, $y = -x$ **18. c)** The product of the coordinates of a point is not 0. **d)** The graph of $xy = 0$ is the x- and y-axes.

Section 8.7, pp. 661–664

1. a) $y = -\dfrac{1}{8}x^2$ **b)** $x + 2.5 = \dfrac{1}{6}(y+3)^2$ **2. a)** (4, 3) **b)** (3, 1.5)

c) $(-2.5, 2)$ **d)** $(-3, -3)$ **3. a)** $y = \frac{1}{24}x^2$ **b)** $y = -\frac{1}{16}x^2$

c) $x = -\frac{1}{32}y^2$ **d)** $y = -\frac{1}{8}x^2$ **e)** $x = \frac{1}{4}y^2$ **f)** $y = \frac{1}{12}x^2$ **g)** $x = -\frac{1}{20}y^2$

4. a) $x - 3 = \frac{1}{12}(y+2)^2$ **b)** $y - 4.5 = -\frac{1}{2}x^2$

c) $x + 1.5 = \frac{1}{14}(y-2)^2$ **d)** $y + 1 = -\frac{1}{12}(x+1)^2$

e) $x + 1 = -\frac{1}{8}(y+5)^2$ **5. a)** opens up, V(0, 0), F(0, 4), $y = -4$, domain: all real numbers, range: $y \geq 0$ **b)** opens down, V(0, 0), F(0, −2), $y = 2$, domain: all real numbers, range: $y \leq 0$ **c)** opens right, V(0, 0), F(2, 0), $x = -2$, domain: $x \geq 0$, range: all real numbers **d)** opens left, V(0, 0), F(−4, 0), $x = 4$, domain: $x \leq 0$, range: all real numbers **e)** opens up, V(−2, 3), F(−2, 4), $y = 2$, domain: all real numbers, range: $y \geq 3$ **f)** opens right, V(−2, 5), F(0.5, 5), $x = -4.5$, domain: $x \geq -2$, range: all real numbers **g)** opens right, V(0, −1), F(1.25, −1), $x = -1.25$, domain: $x \geq 0$, range: all real numbers **h)** opens down, V(2, −3), F(2, −8), $y = 2$, domain: all real numbers, range: $y \leq -3$ **i)** opens left, V(2, −6), F(−1, −6), $x = 5$, domain: $x \leq 2$, range: all real numbers

6. $x = \frac{1}{20}y^2$ **7. a)** $x = -\frac{1}{60}y^2$ **b)** 85 cm

8. a) $y - 8 = -\frac{1}{4.5}(x-6)^2$ **b)** 4.4 cm

9. a) $y + 5 = \frac{1}{11.25}(x-7.5)^2$ **b)** 4.4 m **10.** 128 cm

11. $y - 6550 = -\frac{1}{26\ 200}x^2$ **12. a)** 1.52 m **b)** $y = \frac{1}{6.08}x^2$

14. c) The parabolas have the same vertex. The parabolas have different foci. **d)** The parabola becomes narrower. **e)** The foci get closer to the x-axis.

Career Connection: Communications, p. 664

1. a) $y = \frac{1}{80}x^2$ **b)** 80 cm

Section 8.8, pp. 672–674

1. a) hyperbola **b)** ellipse **c)** circle **d)** parabola **e)** hyperbola **f)** hyperbola **g)** parabola **2. a) i)** circle **ii)** $(x-1)^2 + (y-3)^2 = 25$ **b) i)** ellipse **ii)** $\frac{(x+3)^2}{16} + \frac{(y-2)^2}{64} = 1$ **c) i)** parabola **ii)** $y - 2 = \frac{1}{8}(x+3)^2$

d) i) parabola **ii)** $x - 1 = \frac{1}{8}(y-2)^2$ **e) i)** ellipse **ii)** $\frac{(x+4)^2}{16} + (y-3)^2 = 1$ **f) i)** circle **ii)** $(x+2)^2 + (y-3)^2 = 36$

g) i) hyperbola **ii)** $\frac{(y+1)^2}{0.5} - \frac{(x+1)^2}{0.5} = 1$ **h) i)** parabola **ii)** $x + 1 = -\frac{1}{4}(y-2)^2$ **i) i)** hyperbola **ii)** $\frac{(x-3)^2}{9} - \frac{(y+1)^2}{4.5} = 1$ **3. a)** $y + 3 = \frac{1}{8}(x+2)^2$,

$x^2 + 4x - 8y - 20 = 0$ **b)** $\frac{(x-3)^2}{25} + \frac{(y-1)^2}{4} = 1$, $4x^2 + 25y^2 - 24x - 50y - 39 = 0$ **c)** $\frac{(x+3)^2}{9} - (y-2)^2 = 1$, $x^2 - 9y^2 + 6x + 36y - 36 = 0$ **d)** $(x+2)^2 + (y-5)^2 = 16$, $x^2 + y^2 + 4x - 10y + 13 = 0$ **e)** $\frac{(y-4)^2}{4} - \frac{(x+2)^2}{9} = 1$, $4x^2 - 9y^2 + 16x + 72y - 92 = 0$ **f)** $x - 4 = \frac{1}{12}(y-1)^2$, $y^2 - 2y - 12x + 49 = 0$ **4. a) i)** $b = 1$ **ii)** none **iii)** $b > 0$ and $b \neq 1$ **iv)** $b < 0$ **5. a) i)** $a = -1$ **ii)** none **iii)** $a < 0$ and $a \neq 1$ **iv)** $a > 0$ **6. a)** $p^2 - 16p - 32q - 64 = 0$ **7. b)** The plane must be flying at an angle to the ground. **c)** The plane would have to be flying at a 90° angle to the ground.

8. a) i) ellipse **ii)** $(x-1)^2 + \frac{(y+1)^2}{4} = -\frac{1}{4}$

b) i) circle **ii)** $(x-1)^2 + (y-3)^2 = 0$

c) i) ellipse **ii)** $\frac{(x-1)^2}{4} + \frac{(y-3)^2}{3} = 0$

d) i) hyperbola **ii)** $(x+1)^2 - \frac{(y-3)^2}{9} = 0$

Section 8.9, pp. 684–687

1. a) (−3, 9), (3, 9) **b)** (−5, −25), (5, −25) **c)** (0, 0), (2, 4) **d)** (−4, −4), (4, 4) **e)** (2, 8) **f)** (1, −3), (−1, 3) **2. a)** (−1, 4), (−2, 7) **b)** (−6, 0), (−3.6, −4.8) **c)** (3.75, −2.25) **d)** (0, −2), (0.79, 1.96) **e)** no solution **f)** (1.5, 9) **g)** no solution **h)** (−2, 0), (−0.56, −2.88) **i)** (2, 2), (−4, 8) **3. a)** (0, −5), (−4, 3) **b)** no solution **c)** (0, −5), (8, 3) **d)** (10, −6), (8.67, −3.33) **e)** (3.63, −1.69), (−0.83, −3.91) **f)** (−2, 2) **g)** (4, −1), (−3.2, 2.6) **h)** (−6, 0), (12.75, −9.38) **i)** (3, 0), (−2.93, 1.98) **j)** (2, −3), (3.6, 0.2) **k)** no solution **l)** no solution **4. a)** 8.94 **b)** 4 **5. a)** $x^2 + y^2 = 2500$ **b)** 33 min **6.** No; the path does not intersect the circumference of the area being watered. **7.** Yes; the path of the boat takes it within range of the radar. **8. a)** $y = -0.15(x-13)^2 + 7.6$, $y = 7$ **b)** between 11:00 and 15:00 **9.** 23 km/h **10. a)** $y = -0.25(x-12)^2 + 50$, $y = 40$ **b)** from 05:40 through 18:19 **11. a)** (5, 0), (2, 1) **12.** $\sqrt{10}$ **13.** 5.4 m **14.** (−3.5, −2.4) **15. a)** $(0, \sqrt{2})$, (−2, −2) **b)** (0, 1)

Review of Key Concepts, pp. 689–693

2. a) parallel **d)** $y = -2x - 6$ **3. a)** $10x + 12y + 29 = 0$ **b)** $3x - y - 2 = 0$ **4.** $y = 0$ **5. a)** domain: $-9 \leq x \leq 9$, range: $-9 \leq y \leq 9$ **b)** domain: $-2\sqrt{10} \leq x \leq 2\sqrt{10}$, range: $-2\sqrt{10} \leq y \leq 2\sqrt{10}$ **c)** domain: $-2 \leq x \leq 8$, range: $-10 \leq y \leq 0$ **d)** domain: $-1 - 2\sqrt{15} \leq x \leq -1 + 2\sqrt{15}$, range: $4 - 2\sqrt{15} \leq y \leq 4 + 2\sqrt{15}$ **6. a)** $(x-4)^2 + (y+2)^2 = 9$ **b)** $(x-2)^2 + (y+4)^2 = 17$ **c)** $(x-4)^2 + (y+1)^2 = 18$

7. $(x+2)^2 + (y-3)^2 = 2500$ **8. a)** $\dfrac{x^2}{9} + \dfrac{y^2}{5} = 1$

b) $\dfrac{x^2}{7} + \dfrac{y^2}{16} = 1$ **9. a)** $(0, 0)$; $(4, 0)$, $(-4, 0)$; $(0, 2)$, $(0, -2)$;

$(2\sqrt{3}, 0)$, $(-2\sqrt{3}, 0)$; domain: $-4 \le x \le 4$, range: $-2 \le y \le 2$
b) $(0, 0)$; $(0, 5)$, $(0, -5)$; $(3, 0)$, $(-3, 0)$; $(0, 4)$, $(0, -4)$;
domain: $-3 \le x \le 3$, range: $-5 \le y \le 5$ **c)** $(3, -1)$; $(9, -1)$,
$(-3, -1)$; $(3, 2)$, $(3, -4)$; $(3 - 3\sqrt{3}, -1)$, $(3 + 3\sqrt{3}, -1)$;
domain: $-3 \le x \le 9$, range: $-4 \le y \le 2$ **d)** $(-1, 2)$; $(-1, 9)$,
$(-1, -5)$; $(3, 2)$, $(-5, 2)$; $(-1, 2 + \sqrt{33})$, $(-1, 2 - \sqrt{33})$;
domain: $-5 \le x \le 3$, range: $-5 \le y \le 9$

10. a) $\dfrac{x^2}{25} + \dfrac{y^2}{6.25} = 1$ **b)** $\dfrac{x^2}{24} + \dfrac{y^2}{49} = 1$

c) $\dfrac{(x+2)^2}{79.75} + \dfrac{(y-2.5)^2}{100} = 1$ **d)** $\dfrac{(x-2)^2}{25} + \dfrac{(y+1)^2}{4} = 1$

11. $\dfrac{x^2}{33.5241} + \dfrac{y^2}{33.108} = 1$ **12. a)** $\dfrac{x^2}{4} - \dfrac{y^2}{21} = 1$

b) $y^2 - \dfrac{x^2}{15} = 1$ **13. a)** $(0, 0)$; $(-5, 0)$, $(5, 0)$; $(0, 2)$, $(0, -2)$;

$(-\sqrt{29}, 0)$, $(\sqrt{29}, 0)$; domain: $|x| \ge 5$, range: all real numbers
b) $(0, 0)$; $(-2, 0)$, $(2, 0)$; $(0, 6)$, $(0, -6)$; $(-2\sqrt{10}, 0)$,
$(2\sqrt{10}, 0)$; domain: $|x| \ge 2$, range: all real numbers
c) $(-2, 1)$; $(3, 1)$, $(-7, 1)$; $(-2, 5)$, $(-2, -3)$;
$(-2 - \sqrt{41}, 1)$, $(-2 + \sqrt{41}, 1)$; domain: $|x + 2| \ge 5$, range: all
real numbers **d)** $(1, 3)$; $(8, 3)$, $(-6, 3)$; $(1, 9)$, $(1, -3)$;
$(1 + \sqrt{85}, 3)$, $(1 - \sqrt{85}, 3)$; domain: $|x - 1| \ge 7$; range: all
real numbers **14. a)** $\dfrac{x^2}{16} - \dfrac{y^2}{9} = 1$ **b)** $\dfrac{(y-2)^2}{16} - \dfrac{(x+5)^2}{20} = 1$

c) $\dfrac{(x+2)^2}{3} - (y-1)^2 = 1$ **d)** $\dfrac{y^2}{4} - \dfrac{x^2}{32} = 1$

15. a) $\dfrac{(y-35)^2}{225} - \dfrac{x^2}{202.5} = 1$ **b)** 4.7 m **16. a)** $y = \dfrac{1}{12}x^2$

b) $x + 1 = -\dfrac{1}{8}(y - 2)^2$ **c)** $y = -\dfrac{1}{4}(x - 5)^2$ **d)** $y - 3 = -\dfrac{1}{4}(x - 1)^2$

e) $x - 1 = \dfrac{1}{8}(y + 2)^2$ **17. a)** $(0, 0)$, $(0, 2)$, $y = -2$, domain: all
real numbers, range: $y \ge 0$ **b)** $(0, 0)$, $(-3, 0)$, $x = 3$, domain:
$x \le 0$, range: all real numbers **c)** $(3, -1)$, $(3.5, -1)$, $x = 2.5$,
domain: $x \ge 3$, range: all real numbers **d)** $(5, -1)$, $(5, 0)$,
$y = -2$, domain: all real numbers, range: $y \ge -1$
18. a) $x - 3 = \dfrac{1}{4}(y + 2)^2$ **b)** $y + 1 = -\dfrac{1}{8}(x + 2)^2$

c) $x - 2.5 = \dfrac{1}{6}(y - 1)^2$ **19. a)** $y = \dfrac{1}{100}x^2$ **b)** 63.2 cm

20. $y - 10 = -\dfrac{1}{62.5}(x - 25)^2$ **21. a) i)** circle

ii) $(x + 3)^2 + (y - 2)^2 = 1$ **b) i)** ellipse

ii) $\dfrac{(x+1)^2}{25} + \dfrac{(y-2)^2}{4} = 1$ **c) i)** parabola **ii)** $y - 10 = 2(x - 3)^2$

d) i) ellipse **ii)** $\dfrac{(x-4)^2}{4} + \dfrac{(y+1)^2}{9} = 1$ **e) i)** circle

ii) $(x - 3)^2 + (y - 5)^2 = 11$ **f) i)** hyperbola

ii) $\dfrac{(x-2)^2}{4} - \dfrac{(y+4)^2}{36} = 1$ **g) i)** parabola **ii)** $x - 3 = (y + 2)^2$

h) i) hyperbola **ii)** $\dfrac{(x+3)^2}{16} - \dfrac{(y-2)^2}{4} = 1$

22. a) $(x - 2)^2 + (y + 4)^2 = 9$; $x^2 + y^2 - 4x + 8y + 11 = 0$

b) $\dfrac{(x+3)^2}{9} - \dfrac{(y+1)^2}{4} = -1$; $4x^2 - 9y^2 + 24x - 50y + 63 = 0$

c) $\dfrac{(x-2)^2}{25} + \dfrac{(y+2)^2}{16} = 1$; $16x^2 + 25y^2 - 64x + 100y - 236 = 0$

d) $y - 2 = \dfrac{1}{4}(x + 3)^2$; $x^2 + 6x - 4y + 17 = 0$ **23. a)** $(0, -5)$,

$(-4, 3)$ **b)** $(0, -3)$, $(-2, 1)$ **c)** $(0, -3)$, $(8.3, 1.2)$ **d)** no solution
e) $(4.6, 6.6)$, $(-6.6, -4.6)$ **f)** $(4.6, 4.1)$, $(-2.1, -9.1)$ **g)** $(0.5, 1)$
h) no solution **i)** $(-2.5, -4.5)$ **24.** 9.5 km

Chapter Test, p. 694–695

1. $y = 2$ **2.** $yx - x = 0$ **3.** $(x + 1)^2 + (y + 2)^2 = 25$

4. $y - 3 = \dfrac{1}{8}(x + 2)^2$ **5.** domain: $-15 \le x \le 1$, range:

$-6 \le y \le 10$ **6.** $(2, -3)$; $(2 + 3\sqrt{3}, -3)$, $(2 - 3\sqrt{3}, -3)$;
$(8, -3)$, $(-4, -3)$; $(2, 0)$, $(2, -6)$; domain: $-4 \le x \le 8$,
range: $-6 \le y \le 0$ **7.** $\dfrac{x^2}{25} + \dfrac{y^2}{169} = 1$ **8.** $(2, -4)$; $(2, 0)$,

$(2, -8)$; $(5, -4)$, $(-1, -4)$; $(2, 1)$, $(2, -9)$; domain: all real
numbers, range: $|y + 4| \ge 4$ **9.** $(2, 3)$, $(2.5, 3)$, $x = 1.5$,
domain: $x \ge 2$, range: all real numbers

10. a) $(x - 1)^2 + (y - 5)^2 = 5$ **b)** $\dfrac{x^2}{2.25} - \dfrac{y^2}{22.75} = 1$

c) $\dfrac{(x+1)^2}{16} + \dfrac{(y-2)^2}{25} = 1$ **11. a) i)** ellipse

ii) $\dfrac{(x+2)^2}{5} + \dfrac{y^2}{45} = 1$ **b) i)** parabola **ii)** $y + 6 = \dfrac{3}{4}(x + 2)^2$

c) i) circle **ii)** $(x - 1)^2 + (y + 2)^2 = 10$ **d) i)** hyperbola

ii) $\dfrac{(y+1)^2}{9} - \dfrac{x^2}{4} = 1$ **12. a)** $(4, -2)$, $(-0.8, -4.4)$ **b)** $(2, 3)$,

$(3, 2)$ **c)** $(-1.2, 0.6)$, $(-2.8, -2.6)$ **13. a)** $y = \dfrac{1}{20}x^2$ **b)** 28.3 m

14. a) $\dfrac{x^2}{9} + \dfrac{y^2}{4} = 1$ **b)** 3.9 m **15. a)** $y - 2.2 = -\dfrac{11}{20}x^2$

b) $\dfrac{x^2}{4} + \dfrac{y^2}{4.84} = 1$ **c)** $\dfrac{(y-4)^2}{3.24} - \dfrac{x^2}{1.02} = 1$

Challenge Problems, p. 696

1. $y = \pm\dfrac{1}{2}x^2$, $x \ne 0$ **2.** yes; at the point $(3, -1)$ **3.** $2k\sqrt{2}$

4. first ellipse: $\dfrac{x^2}{16} + \dfrac{y^2}{25} = 1$; second ellipse: $\dfrac{x^2}{16} + \dfrac{y^2}{9} = 1$;

third ellipse: $\dfrac{x^2}{4} + \dfrac{y^2}{9} = 1$; fourth ellipse: $\dfrac{x^2}{4} + \dfrac{y^2}{2} = 1$ **5.** 30

Problem Solving Strategy: Use a Data Bank, p.699

1. a) Jupiter, Saturn **2. a)** Newfoundland **b)** Prince Edward Island **3.** (Information from the web site of the Canadian Museum of Nature: http://www.nature.ca/english/eladback.htm) **a)** extinct: a species that no longer exists; extirpated: a species no longer existing in the wild, but existing elsewhere; endangered: a species facing imminent extirpation or extinction; threatened: a species likely to become endangered if limiting factors are not reversed; vulnerable: a species of special concern because of characteristics that make it particularly sensitive to human activities or natural events **b)** all living things, including plants and animals **4.** 70 **5. a)** Answers may vary, depending on the route taken; 4415 km **b)** about 4.5 days **8. a)** Hand geometry takes a three-dimensional image of the hand and measure the shape and length of fingers and knuckles. (Information from the web site http://homepages.go.com/~nuts4pi/edt6030/biometrics.htm) This information is entered into a computer, which can then identify the person when his or her hand is scanned by a security device. **b)** Answers may vary: airports, high-security workplaces, government offices.

Problem Solving: Using the Strategies, p. 700

1. 768 **2.** $\frac{1}{2}$ **3.** 370, 371, 407 **4.** 6 **5.** 2099 **6.** Answers may vary. For example, $123 + 45 - 67 + 8 - 9 = 100$. **7.** $\pi + 3$ or approximately 6.14 m **8.** midnight **9.** $[x - (\sqrt{2} - 1)r]^2 + [y - (\sqrt{2} - 1)r]^2 = (3 - 2\sqrt{2})r^2$ **10.** $(x - 1)^2 + (y - 1)^2 = 2$

Cumulative Review: Chapters 7 and 8, p.

Chapter 7

1. a) The loan at 7%, compounded annually, is the better deal. **b)** $20 201.74 **2. a)** $1169.65 **b)** $232.10 **3.** $129 993.50 **4. a)** $108 900 **b)** $845.08 **c)** $100 558.89

Chapter 8

1 b) $y = 3x + 1$ **2.** domain: $-6 \le x \le 12$, range: $-14 \le x \le 4$ **3.** $\frac{x^2}{169} + \frac{y^2}{25} = 1$ **4. a)** $(5, -11)$; $(5 - \sqrt{23}, -11)$, $(5 + \sqrt{23}, -11)$; $(-7, -11)$, $(17, -11)$; $(5, 0)$, $(5, -22)$; domain: $-7 \le x \le 17$, range: $-22 \le y \le 0$ **b)** $(-1, -3)$; $(-1 - \sqrt{13}, -3)$, $(-1 + \sqrt{13}, -3)$; $(-3, -3)$, $(1, -3)$; $(-1, -6)$, $(-1, 0)$; domain: $-3 \le x \le 1$, range: $-6 \le y \le 0$ **5.** $(-2, 4)$, $(-1.5, 4)$, $x = -2.5$, domain: $x \ge -2$, range: all real numbers **6. a) i)** ellipse **ii)** $(x + 3)^2 + \frac{(y - 2)^2}{4} = 1$ **b) i)** parabola **ii)** $y - 1 = -\frac{1}{8}(x - 4)^2$ **c) i)** circle **ii)** $(x + 2)^2 + (y - 3)^2 = 16$ **d) i)** hyperbola **ii)** $\frac{(y + 3)^2}{4} - \frac{(x - 1)^2}{2} = 1$ **7. a)** $(3.28, -2.28)$, $(-2.28, 3.28)$ **b)** $(0, -1)$, $(-3, 2)$ **8.** 10 cm

Appendix A

Common factors, p. 702

1. a) $2x(2 + 3y)$ **b)** $3xy(1 + 4y + 2x^2y)$ **c)** $5m(m - 6)$ **d)** $2xy(1 - 3x + 4y^2)$ **e)** $2c^2(3c - 2d^2 + d)$ **f)** $2y^2(y^3 - 2y + 4)$ **g)** $5a(x + 2y - z)$ **h)** $pq(3r + 4s - 5t)$ **i)** $9y(3x - 2z + y)$

Evaluating radicals, p. 702

1. a) 4 **b)** 9 **c)** 0.5 **d)** 1.2 **e)** 0.2 **f)** 2.5 **g)** 14 **h)** 0.7 **2. a)** 6 **b)** 9 **c)** 10

Evaluating expressions, p. 702

1. a) -1 **b)** -4 **c)** 0 **d)** 13 **e)** 34 **f)** 31 **g)** 36 **h)** -5 **i)** -20 **2. a)** -3 **b)** 8 **c)** 2 **d)** -51 **e)** 14 **f)** 21 **g)** -8 **h)** -24 **g)** -2

Exponent rules, p. 703

1. a) 3^6 **b)** 2^5 **c)** 5^7 **d)** 4^3 **e)** 3^5 **f)** 4^6 **g)** 2^{10} **h)** x^8 **i)** z^6 **j)** y^{14} **k)** $8x^6$ **l)** $15y^6$ **m)** $16x^{12}$ **n)** $-27z^6$ **o)** $3m^3$

Factoring $ax^2 + bx + c$, $a = 1$, p. 703

1. a) $(x - 5)(x + 4)$ **b)** $(y + 5)(y - 2)$ **c)** $(n - 9)(n + 4)$ **d)** $(m + 6)(m + 3)$ **e)** $(x - 6)(x - 5)$ **f)** $(c - 6)(c + 4)$ **g)** $(8 - y)(2 + y)$ **h)** $(x + 8y)(x + 4y)$ **i)** $(c - 7d)(c + 4d)$

Factoring $ax^2 + bx + c$, $a \ne 1$, p. 703

1. a) $(3x + 4)(x - 2)$ **b)** $(2c - 1)(c + 4)$ **c)** $(m - 2)(4m - 3)$ **d)** $(y + 1)(5y + 6)$ **e)** $(3n - 2)(n + 1)$ **f)** $(x - 3)(6x + 1)$ **g)** $(x - 3y)(3x + 4y)$ **h)** $(x - 2)(5x - 4)$ **i)** $(x + 5)(4x + 3)$ **j)** $(2p - q)(p + q)$

Finding an angle in a right triangle, p. 704

1. a) 22° **b)** 65° **c)** 24° **d)** 51° **e)** 51° **f)** 56°

Finding a side length in a right triangle, p. 704

1. a) 10.5 m **b)** 33.1 m **c)** 11.8 m **d)** 21.3 m **e)** 16.5 m **f)** 31.8 m

Graphing equations, p. 705

1.

a) x	y	b) x	y	c) x	y
-2	8	-2	-5	-2	2
-1	7	-1	-4	-1	3
0	6	0	-3	0	4
1	5	1	-2	1	5
2	4	2	-1	2	6

d) x	y	e) x	y	f) x	y
-2	-4	-2	-8	-2	-5
-1	-3	-1	-5	-1	-1
0	-2	0	-2	0	3
1	-1	1	1	1	7
2	0	2	4	2	11

Graphing quadratic functions, p. 705

1. a) opens up; vertex: $(0, 2)$; axis of symmetry: $x = 0$; domain: all real numbers; range: $y \ge 2$; minimum: 2 **b)** opens down; vertex: $(0, -1)$; axis of symmetry: $x = 0$;

domain: all real numbers; range: $y \le -1$; maximum: -1
c) opens up; vertex: $(0, 0)$; axis of symmetry: $x = 0$; domain: all real numbers; range: $y \ge 0$; minimum: 0 **d)** opens down; vertex: $(0, 1)$; axis of symmetry: $x = 0$; domain: all real numbers; range: $y \le 1$; maximum: 1 **e)** opens down; vertex: $(1, 4)$; axis of symmetry: $x = 1$; domain: all real numbers; range: $y \le 4$; maximum: 4 **f)** opens up; vertex: $(-5, -3)$; axis of symmetry: $x = -5$; domain: all real numbers; range: $y \ge -3$; minimum: -3 **g)** opens up; vertex: $(-2, 3)$; axis of symmetry: $x = -2$; domain: all real numbers; range: $y \ge 3$; minimum: 3 **h)** opens down; vertex: $(-1, -5)$; axis of symmetry: $x = -1$; domain: all real numbers; range: $y \le -5$; maximum: -5

Length of a line segment, p. 706

1. a) $\sqrt{41}$; 6.4 **b)** $2\sqrt{10}$; 6.3 **c)** 5 **d)** $\sqrt{205}$; 14.3 **e)** $\sqrt{11.17}$; 3.3

Midpoint formula, p. 706

1. a) $(-1.5, -2)$ **b)** $(4, 1)$ **c)** $(2.5, -2)$ **d)** $(-4, 0.5)$ **e)** $(3, -0.6)$ **f)** $(0.5, 2)$

Rewriting in the form $y = a(x - h)^2 + k$, $a = 1$, p. 708

1. a) $y = (x + 3)^2 - 5$; minimum of -5 at $x = -3$
b) $y = (x + 2)^2 + 5$; minimum of 5 at $x = -2$
c) $y = (x - 6)^2 - 43$; minimum of -43 at $x = 6$
d) $y = (x - 5)^2 - 20$; minimum of -20 at $x = 5$
e) $y = (x + 4)^2 - 18$; minimum of -18 at $x = -4$
f) $y = (x - 1)^2 - 12$; minimum of -12 at $x = 1$

Rewriting in the form $y = a(x - h)^2 + k$, $a \ne 1$, p. 708

1. a) $y = 3(x + 1)^2 - 9$; minimum of -9 at $x = -1$
b) $y = -2(x + 2)^2 + 8$; maximum of 8 at $x = -2$
c) $y = 2(x - 3)^2 - 15$; minimum of -15 at $x = 3$
d) $y = -4(x - 2)^2 + 7$; maximum of 7 at $x = 2$
e) $y = 3(x - 4)^2 - 48$; minimum of -48 at $x = 4$
f) $y = 4(x + 3)^2 - 41$; minimum of -41 at $x = -3$

Simplifying Expressions, p. 709

1. a) $15x^2 - 10x$ **b)** $2a^2 + 7a$ **c)** $-6c^2 + 13c$ **d)** $-4a + 8$ **e)** $8x^2 - 2x + 2$ **f)** $7x^2 + 12x - 12$ **2. a)** $6x^2 + 6x - 9$ **b)** $-10x^2 - 6x + 3$ **c)** $-5x^2 - 6x - 4$ **d)** $5x^2 - 4x - 5$

Solving linear equations, p. 709

1. a) -2 **b)** -1 **c)** 3 **d)** 4 **e)** 4 **f)** -3 **2. a)** 1 **b)** 17 **c)** $-\dfrac{10}{11}$ **d)** $-\dfrac{3}{5}$

Solving linear systems, p. 711

1. a) $(-2, 1)$ **b)** $(4, -5)$ **c)** $\left(\dfrac{40}{3}, -7\right)$ **d)** $(-5, -3)$ **e)** $(4, -2)$ **f)** $(-1, 3)$

Solving proportions, p. 711

1. a) 8.64 **b)** 1.64 **c)** 8.1 **d)** 2.04 **e)** 0.21 **f)** 1 **g)** 12.4 **h)** 0.275

Solving quadratic equations by factoring, p. 713

1. a) -10, 3 **b)** 1, 2 **c)** -5, -4 **d)** -5, -3 **e)** $\dfrac{5}{2}$ **f)** -1, $\dfrac{2}{3}$ **g)** $-\dfrac{3}{5}$, $\dfrac{3}{5}$ **h)** $-\dfrac{2}{3}$, 4 **i)** 0, $\dfrac{4}{9}$

Solving quadratic equations by graphing, p. 713

1. a) -1, 5 **b)** -4, -1 **c)** -4, 4 **d)** -8, 1 **e)** 1, 5 **f)** -6, -2

Stretches and shrinks, p. 713

1. a) The y-coordinates of $y = 3x^2$ are three times the corresponding y-coordinates of $y = x^2$. **b)** The y-coordinates of $y = \dfrac{1}{4}x^2$ are one quarter of the corresponding y-coordinates of $y = x^2$. **2. a)** The point $(3, 9)$ is transformed to the point $(3, 36)$. **b)** The point $(3, 9)$ is transformed to the point $(3, 3)$.

The quadratic formula, p. 714

1. a) $\dfrac{1}{2}$, 1 **b)** $\dfrac{3}{5}$, $\dfrac{3}{2}$ **c)** 0, $\dfrac{3}{7}$ **d)** -1, $\dfrac{3}{2}$ **e)** $\dfrac{2}{3}$, 1 **2. a)** $\dfrac{-1 - \sqrt{11}}{5}$, $\dfrac{-1 + \sqrt{11}}{5}$ **b)** $\dfrac{1 - \sqrt{7}}{3}$, $\dfrac{1 + \sqrt{7}}{3}$ **c)** $\dfrac{4 - \sqrt{2}}{2}$, $\dfrac{4 + \sqrt{2}}{2}$ **d)** $\dfrac{-1 - \sqrt{15}}{2}$, $\dfrac{-1 + \sqrt{15}}{2}$ **e)** $\dfrac{1 - \sqrt{11}}{5}$, $\dfrac{1 + \sqrt{11}}{5}$

Trigonometric ratios, p. 716

1. a) $\sin A = \dfrac{3}{5}$, $\cos A = \dfrac{4}{5}$, $\tan A = \dfrac{3}{4}$; $\sin C = \dfrac{4}{5}$, $\cos C = \dfrac{3}{5}$, $\tan C = \dfrac{4}{3}$ **b)** $\sin A = \dfrac{5}{13}$, $\cos A = \dfrac{12}{13}$, $\tan A = \dfrac{5}{12}$; $\sin C = \dfrac{12}{13}$, $\cos C = \dfrac{5}{13}$, $\tan C = \dfrac{12}{5}$

GLOSSARY

A

absolute value The distance of a number from zero on a real number line.

acute angle An angle whose measure is less than 90°.

acute triangle A triangle in which each of the three interior angles is acute.

algebraic modelling The process of representing a relationship by an equation or a formula, or representing a pattern of numbers by an algebraic expression.

algorithm A set of instructions for carrying out a procedure.

amortization period The period of time over which a mortgage is paid off with equal payments at regular intervals.

amortization table A table that shows the part of each mortgage payment that is interest and principal, and the principal remaining.

amortize To repay a mortgage over a given period of time in equal payments at regular intervals

amount The sum of the principal and the interest when money is invested or borrowed.

amplitude Half the distance between the maximum and minimum values of a periodic function.

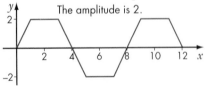

angle bisector A line that divides an angle into two equal parts.

angle of depression The angle, measured downward, between the horizontal and the line of sight from an observer to an object.

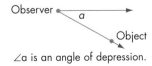

∠a is an angle of depression.

angle of elevation The angle, measured upward, between the horizontal and the line of sight from an observer to an object.

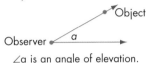

∠a is an angle of elevation.

annuity A sum of money paid as a series of equal payments at regular intervals of time.

arc Part of the circumference of a circle.

arithmetic sequence A sequence where the difference between consecutive terms is a constant.

arithmetic series The sum of the terms of an arithmetic sequence.

astronomical unit The distance from Earth to the sun, that is, approximately 150 000 000 km.

asymptote A line that a curve approaches more and more closely, but never touches.

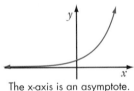

The x-axis is an asymptote.

axis of symmetry The fold line of a symmetrical figure.

B

base (of a power) The number used as a factor for repeated multiplication.

In 6^3, the base is 6.

BEDMAS An acronym that lists the order of operations. BEDMAS stands for **B**rackets, **E**xponents, **D**ivision, **M**ultiplication, **A**ddition, **S**ubtraction.

binomial An algebraic expression with two terms.

$3x + 4$ is a binomial.

C

centre of an ellipse The point of intersection of the major axis and the minor axis.

centre of an hyperbola The point of intersection of the transverse axis and the conjugate axis. The centre is located midway between the vertices and midway between the co-vertices.

chord A line segment joining two points on a curve.

circle The locus of all points in the plane that are equidistant from a fixed point called the centre.

circumference The perimeter of a circle.

coefficient The factor by which a variable is multiplied. For example, in the term $8y$, the coefficient is 8; in the term ax, the coefficient is a.

collecting like terms Simplifying an expression containing like terms by adding their coefficients.

collinear Three or more points that lie on the same straight line.

common difference The constant that is added to a term in an arithmetic sequence to obtain the next term.

> For the sequence 1, 4, 7, 10, ... the common difference is 3.

common ratio The ratio of consecutive terms in a geometric sequence.

> For the sequence 1, 2, 4, 8, 16, ... the common ratio is 2.

complex conjugates Complex numbers of the form $a + bi$ and $a - bi$. Their product is real number.

complex number A number of the form $a + ib$, where a and b are real numbers and i is the square root of -1.

complex plane The complex numbers considered in two dimensions using the real and the imaginary axes.

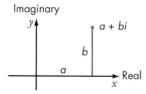

compression A transformation that is a stretch by a factor less than 1.

compound inequality A range of values described by combining two inequalities. $0 < n \le 5$ is a compound inequality that means $n > 0$ and $n \le 5$.

compound interest Interest that is calculated at regular compounding periods and is added to the principal to earn interest for the next compounding period.

compounding period Each period of time for which compound interest is earned or charged in an investment or loan.

conjecture A generalization, or educated guess, made using inductive reasoning.

conjugate axis The axis of symmetry of a hyperbola that is perpendicular to the transverse axis.

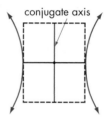

conjugate binomials Binomials of the form $a\sqrt{b} + c\sqrt{d}$ and $a\sqrt{b} - c\sqrt{d}$ where a, b, c, and d are rational numbers. The product of conjugates is a rational number.

conjugate hyperbolas Hyperbolas that share the same asymptotes. The transverse axis of one is the conjugate axis of the other.

constant term A term that does not include a variable.

continuous graph A graph that consists of an unbroken line or curve.

cosine law The relationship between the lengths of the three sides and the cosine of an angle in any triangle.

$$a^2 = b^2 + c^2 - 2bc\cos A$$

cosine ratio In a right triangle, the ratio of the length of the adjacent side to the length of the hypotenuse.

$$\text{cosine} = \frac{\text{adjacent}}{\text{hypotenuse}}$$

coterminal angles Angles in standard position that have the same terminal arm.

counterexample An example that demonstrates that a conjecture is false.

co-vertices of an ellipse The endpoints of the major axis.

co-vertices of a hyperbola The endpoints of the conjugate axis.

cycle One complete pattern of a periodic function.

D

dependent variable In a relation, the variable whose value depends on the value of the independent variable. On a coordinate grid, the values of the dependent variable are on the vertical axis.

In $d = 4.9t^2$, d is the dependent variable.

depreciation The amount by which an item decreases in value.

dilatation A transformation that changes the size of an object. It involves a stretch or a shrink by a scale factor of k.

directrix The line from which each point of a parabola is the same distance as it is from the focus.

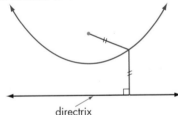

directrix

discontinuity A point at which a function is not defined. The graph has a break at this point.

$f(x) = \dfrac{1}{x}$ has a discontinuity at $x = 0$.

discriminant In the quadratic formula, the quantity under the radical sign, $b^2 - 4ac$.

distance between two points The length of the line segment joining the points. For points (x_1, y_1) and (x_2, y_2), $d = \sqrt{(x_2 - x_1)^2 + (y_2 - y_1)^2}$.

distributive property $a(b + c) = ab + ac$

domain The set of the first elements in a relation.

double root The solution of a quadratic equation where both roots are the same.

E

eccentricity A measure of the shape of an ellipse or hyperbola, given by $e = \dfrac{c}{a}$.

elements The individual members of a set.

ellipse The locus of all points in the plane such that the sum of the distances from two given points in the plane, the foci, is constant.

entire radical A number such as $\sqrt{29}$ or $\sqrt{\dfrac{5}{3}}$.

equilateral triangle A triangle with all three sides equal.

Euclidean distance The square root of the sum of the squares of the differences between the corresponding coordinates of two points.

expansion A transformation that is a stretch by a factor greater than 1.

explicit formula A formula for the nth term of a sequence, from which the terms of the sequence may be obtained by substituting 1, 2, 3, and so on, in turn, for n in the formula.

$t_n = 2n + 1$ is an explicit formula for the arithmetic sequence 3, 5, 7, 9, … .

exponent The use of a raised number to denote repeated multiplication of a base.

In $3x^4$, the exponent is 4.

exponential equation An equation that has a variable in an exponent.

$3^x = 81$ is an exponential equation.

exponential function A function in which a variable is an exponent. It can be defined by an equation of the form $y = ab^x$, where $a \neq 0$, $b > 0$, and $b \neq 1$.

extrapolate Estimate values lying outside the range of given data. To extrapolate from a graph means to estimate coordinates of points beyond those that are plotted.

F

factor Express a number as the product of two or more numbers, or an algebraic expression as the product of two or more other algebraic expressions. Also, the individual numbers or algebraic expressions in such a product.

first-degree inequality An inequality in which the variable has the exponent 1.

$3x + 5 > 2x - 4$ is a first-degree inequality.

focal radii The line segments joining point P on an ellipse or hyperbola to the foci.

foci of a hyperbola The two points in the plane, F_1 and F_2, such that $|PF_1 - PF_2|$ is constant for all points P on the hyperbola.

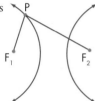

foci of an ellipse The two points in the plane, F_1 and F_2, such that $PF_1 + PF_2$ is constant for all points P on the ellipse.

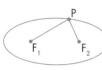

focus of a parabola The point from which each point of a parabola is the same distance as it is from the directrix.

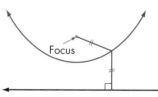

FOIL An acronym used in multiplying binomials where F refers to the product of the first terms, O, to the product of the outside terms, I, to the product of the inside terms, and L, to the product of the last terms.

fractal A curve that generates itself by replacing each side of its original shape with a generator and iterating the process.

function A set of ordered pairs in which, for every value of x, there is only one value of y.

G

generalize Determine a general rule or conclusion from examples. Specifically, determine a general rule to represent a pattern or relationship between variables.

geometric mean If a, x, and b are consecutive terms of a geometric sequence then x is the geometric mean of a and b.

geometric sequence A sequence where the ratio of consecutive terms is a constant. Each term after the first is obtained by multiplying the preceding term by the same number.

geometric series The sum of the terms of a geometric sequence.

gradian An angular measure that has 400 equal parts, called grads, in one revolution.

graphing calculator A hand-held device capable of a wide range of mathematical operations, including graphing from an equation, constructing a scatter plot, determining the equation of a line of best fit for a scatter plot, making statistical calculations, and performing elementary symbolic manipulation. Many graphing calculators will attach to scientific probes that can be used to gather data involving physical measurements, such as position, temperature, or force.

graphing software Computer software that provides features similar to those of a graphing calculator.

greatest common factor (GCF) The monomial with the greatest numerical coefficient and greatest degree, that is a factor of two or more terms.

The GCF of $12ab$ and $8bc$ is $4b$.

greatest integer function $f(x) = [x]$, where $[x]$ denotes the greatest integer that is less than or equal to x.

$[2.5] = 2$, $[-3.6] = -4$, and $[0.0] = 0$

H

half-life The time it takes for half of any sample of a radioactive isotope to decay.

hyperbola The locus of all points in a plane such that the absolute value of the difference of the distances from any point on the hyperbola to two given points in the plane, the foci, is constant.

hypotenuse The longest side of a right triangle.

I

identity An equation that is true for all values of the variable for which the expressions on each side of the equation are defined.

imaginary axis The vertical axis in the complex plane.

imaginary number A complex number, $a + bi$, where $a \neq 0$, $b \neq 0$, and $i^2 = -1$.

imaginary part The part bi in a complex number $a + bi$, where $a \neq 0$, $b \neq 0$, and $i^2 = -1$.

imaginary unit The number i, where $i^2 = -1$ and $i = \sqrt{-1}$.

independent variable In a relation, the variable whose value determines that of the dependent variable. On a coordinate grid, the values of the independent variable are on the horizontal axis.

In $d = 4.9t^2$, t is the independent variable.

inequality A mathematical statement that contains one of the symbols $<$, $\leq$, $>$, $\geq$, or $\neq$.

initial arm The ray, of an angle in standard position, that is on the positive x-axis.

integer Whole numbers and their opposites.

intercept The distance from the origin of the xy-plane to the point at which a line or curve crosses a given axis.

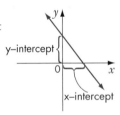

interest rate The percent of the principal that is earned, or paid, as interest.

interpolate To estimate values lying between elements of given data. To interpolate from a graph means to estimate coordinates of points between those that are plotted.

invariant Points that are unaltered by a transformation.

inverse function Two functions that undo each other. A function f has inverse f^{-1} defined by $f^{-1}(b) = a$ if $f(a) = b$.

irrational number A number that cannot be written as the ratio of two integers.

$\sqrt{2}$, $\sqrt{3}$, and π are irrational numbers.

isosceles trapezoid A trapezoid in which the two non-parallel sides are equal in length.

isosceles triangle A triangle with exactly two equal sides.

iteration A method that evaluates a function for an initial value, and uses the output of that calculation as the input for the next calculation.

L

like radicals Numbers that have the same radicand.

$2\sqrt{3}$ and $8\sqrt{3}$ are like radicals.

like terms Terms that have exactly the same variable(s) raised to exactly the same exponent(s).

$3x^2$, $-x^2$, and $2.5x^2$ are like terms.

line of symmetry A line such that a figure coincides with its reflection image over the line.

line segment The part of a line that joins two points.

linear relation A relation between two variables that appears as a straight line when graphed on a coordinate system. May also be referred to as a linear function.

locus A set of points determined by a rule or condition.

lowest common denominator (LCD) The least common multiple of the denominators of two or more rational expressions.

M

magic square A square arrangement of numbers in which the sum of the numbers in any line (horizontal, vertical, and diagonal) is the same.

major axis The longer of the two line segments that form the axes of symmetry for an ellipse.

major axis

mathematical model   A mathematical description of a real situation. The description may include a diagram, a graph, a table of values, an equation, a formula, a physical model, or a computer model.

mathematical modelling   The process of describing a real situation in a mathematical form.

median (geometry)   A line segment that joins a vertex of a triangle to the midpoint of the opposite side.

mean   The sum of a set of values divided by the number of values.

The mean of 2, 4, 6 and 8 is 5.

midpoint   The point that divides a line segment into two equal points. For the line segment joining points (x_1, y_1) and (x_2, y_2), the midpoint is $\left(\dfrac{x_1 + x_2}{2}, \dfrac{y_1 + y_2}{2} \right)$.

minor axis   The shorter of the two line segments that form the axes of symmetry for an ellipse.

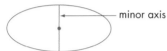

mixed radical   A number such as $3\sqrt{10}$ or $\dfrac{1}{2}\sqrt{7}$.

monomial   An algebraic expression with one term.

$7x$ is a monomial.

mortgage   A special type of borrowing arrangement between a borrower and a lender. The borrower uses property as a guarantee of repayment of the debt.

N

natural number   A number in the sequence 1, 2, 3, 4, … .

non-linear relation   A relationship between two variables that does not fit a straight line when graphed.

O

oblique triangle   A triangle that is not right-angled.

obtuse angle   An angle that measures more than 90° but less than 180°.

obtuse triangle   A triangle containing one obtuse angle.

ordered pair   A pair of numbers used to locate a point in the coordinate plane.

ordinary annuity   A series of equal payments in which a payment is made at the end of each regular payment period.

origin   The point of intersection of the x-axis and the y-axis on a coordinate grid.

P

parabola   The graph of a quadratic function whose domain is the set of real numbers. The locus of all points P in the plane such that the distance from P to a fixed point F, the focus, equals the distance from P to fixed line l, the directrix.

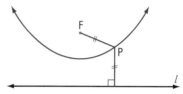

payment interval   The time between successive payments of an annuity.

payment period   Another name for the payment interval.

perfect square trinomial   The trinomial that results from squaring a binomial.

perimeter   The distance around a polygon.

period   The horizontal length of one cycle of a periodic function.

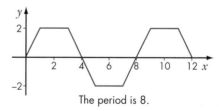

The period is 8.

period (of a pendulum)   The time it takes to complete one back-and-forth swing.

periodic function   A function that has a pattern of y-values that repeats at regular intervals.

phase shift   The horizontal translation of a trigonometric function. Also sometimes called the phase angle.

point of intersection   The point that is common to two non-parallel lines.

point-slope form of a linear equation   The equation for the line through (x_1, y_1) with slope m is given by $y - y_1 = m(x - x_1)$.

polyhedron   A three-dimensional object with faces that are polygons.

polynomial   An algebraic expression formed by adding or subtracting monomials.

polynomial expression   An algebraic expression of the form $a + bx + cx^2 + \dots$, where a, b, and c are numbers.

power   A product obtained by using a base as a factor one or more times.

5^3, x^6, and a^m are powers.

present value   The principal that must be invested today at a given interest rate, compounding frequency, and term, in order to result in a given final amount.

present value of an annuity   The principal that must be invested today at a given interest rate, compounding frequency, and number of payment intervals, in order to provide regular payments over a given term.

prime number   A number with exactly two factors—itself and 1.

2, 5, and 13 are prime numbers.

principal   The original money invested or borrowed.

principal square root   The positive square root of a number.

prism   A three-dimensional figure with two parallel, congruent polygonal bases. A prism is named by the shape of its bases, for example, rectangular prism, triangular prism.

probability   The ratio of the number of favourable outcomes to the number of possible outcomes.

proportion   An equation that states that two ratios are equal.

Pythagorean identity   In trigonometry, $\sin^2 \theta + \cos^2 \theta = 1$ for all values of θ.

Pythagorean theorem   In a right triangle, the square of the length of the longest side is equal to the sum of the squares of the lengths of the other two sides.

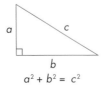

$$a^2 + b^2 = c^2$$

pure imaginary number   A number that can be written as $i\sqrt{x}$, where $i^2 = -1$ and x is a positive real number.

$7i$, $i\sqrt{2}$, and $3i\sqrt{5}$ are pure imaginary numbers.

Q

quadrant   One of the four regions formed by the intersection of the x-axis and the y-axis.

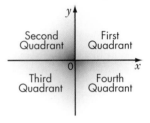

quadratic equation   An equation in the form $ax^2 + bx + c = 0$, where a, b, and c are real numbers and $a \ne 0$.

quadratic formula   The roots of a quadratic function of the form $y = ax^2 + bx + c$, where $a \ne 0$, are given by $\dfrac{-b \pm \sqrt{b^2 - 4ac}}{2a}$.

quadratic function   A function defined by a quadratic equation of the form $y = ax^2 + bx + c$, where a, b, and c are real numbers and $a \ne 0$.

quartic equation   A fourth degree polynomial equation.

quotient identity   In trigonometry, $\dfrac{\sin \theta}{\cos \theta} = \tan \theta$ for all values of θ.

R

radian   The measure of the angle subtended at the centre of a circle by an arc equal in length to the radius of the circle.

radical expression   An expression involving the square root of an unknown.

radical sign   The symbol $\sqrt{}$.

radicand   An expression under a radical sign.

In $\sqrt{ab}$ the radicand is ab.

radius The distance from the centre to any point on a circle.

range of a relation The set of the second elements in a relation.

ratio A comparison of two quantities with the same units.

rational expression The quotient of two polynomials. $\frac{3}{k-1}$, and $\frac{a^2 + b^2}{a + b}$ are rational expressions.

rational number A number that can be expressed as the ratio of two integers, where the divisor is not zero.

0.75, $\frac{3}{8}$, and -2 are rational numbers.

rationalize the denominator The process of multiplying the numerator and the denominator of a rational expression by the same quantity, to make the denominator a rational number.

$$\frac{4}{\sqrt{3}} = \frac{4(\sqrt{3})}{\sqrt{3}(\sqrt{3})}$$
$$= \frac{4\sqrt{3}}{3}$$

real axis The horizontal axis in the complex plane.

real number A member of the set of all rational and irrational numbers.

real part The part a in a complex number $a + bi$, where $a \neq 0$, $b \neq 0$, and $i^2 = -1$.

reciprocals Two numbers that have a product of 1.

x and $\frac{1}{x}$ are reciprocals.

rectangular form A complex number in the form $a + bi$ where (a, b) are its rectangular coordinates in the complex plane.

rectangular prism A three-dimensional figure with two parallel, congruent rectangular bases.

recursion formula A formula consisting of at least two parts. The parts give the value(s) of the first term(s) in the sequence, and an equation that can be used to calculate each of the other terms from the term(s) before it.

$t_1 = 1$, $t_n = t_{n-1} + 3$ is a recursion formula for the arithmetic sequence 1, 4, 7, 10, ...

reduction A dilatation in which the image after the dilatation is smaller than the original image, that is, $0 < k < 1$

reflection A transformation in which a figure is reflected over a reflection line.

reflection image The image of a figure in a plane after a reflection.

reflex angle An angle that measures more than 180° but less than 360°.

relation An identified relationship between two variables that may be expressed as ordered pairs, a table of values, a graph, or an equation.

restriction Any value that must be excluded for a variable.

$\frac{8y}{y + 1}$ has restriction $y \neq -1$.

rhombus A parallelogram in which the lengths of all four sides are equal.

right angle An angle that measures 90°.

right bisector of a line segment A line that is perpendicular to a line segment and divides the line segment into two equal parts.

right prism A three-dimensional figure with two parallel, congruent polygonal bases and lateral faces that are perpendicular to the bases.

right triangle A triangle containing a 90° angle.

roots The solutions of an equation.

S

scalene triangle A triangle with no sides equal.

scientific notation A method of writing large or small numbers that contain many zeros. The number is expressed in the form $a \times 10^n$, where a is greater than or equal to 1 but less than 10, and n is an integer.

second-degree polynomial A polynomial in at least one term of which the variable has the exponent 2, and no term has the exponent of the variable greater than 2.

$x^2 + 5x - 7$ is a second-degree polynomial.

sequence A set of numbers, usually separated by commas, arranged in an order.

series The sum of the terms of a sequence.

simple interest Interest calculated only on the original principal using the formula $I = Prt$.

simplest form of an algebraic expression An expression with no like terms. For example, $2x + 7$ is in simplest form; $5x + 1 + 6 + 3x$ is not.

sine law The relationship between the lengths of the sides and their opposite angles in any triangle.

$$\frac{a}{\sin A} = \frac{b}{\sin B} = \frac{c}{\sin C}$$

sine ratio In a right triangle, the ratio of the length of the opposite side to the length of the hypotenuse.

$$\text{sine} = \frac{\text{opposite}}{\text{hypotenuse}}$$

sinusoidal function A function that is used to model periodic data.

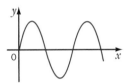

slide A transformation that is comprised of horizontal and vertical translations of an object on the plane.

slope A measure of the steepness of a line. The slope of a line, m, containing the points $P(x_1, y_1)$ and $Q(x_2, y_2)$ is

$$m = \frac{\text{vertical change}}{\text{horizontal change}} \text{ or } \frac{\text{rise}}{\text{run}}$$

$$= \frac{\Delta y}{\Delta x}$$

$$= \frac{y_2 - y_1}{x_2 - x_1}, \ x_2 \neq x_1$$

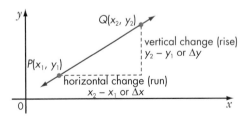

slope and y-intercept form of a linear equation The equation for the line through (x_1, y_1) with slope m and y-intercept b is given by $y = mx + b$.

solving a triangle Finding the values of all the unknown sides and unknown angles of a triangle.

spreadsheet Computer software that allows the entry of formulas for repeated calculations.

square root A number that is multiplied by itself to give another number.

standard position The position of an angle when its vertex is at the origin and its initial ray is on the positive x-axis.

straight angle An angle that measures 180°.

subtended angle An angle for which the arms end on arc of a circle.

supplementary angles Angles whose sum is 180°.

surface area The number of square units needed to cover the surface of a three-dimensional object.

system of equations Two or more equations that are considered together.

symmetry A quality of a plane figure that can be folded along a fold line so that the halves of the figure match exactly.

T

tangent ratio In a right triangle, the ratio of the length of the opposite side to the length of the adjacent side.

$$\text{tangent} = \frac{\text{opposite}}{\text{adjacent}}$$

term A number or a variable, or the product or quotient of numbers and variables.

The expression $x^2 + 5x$ has two terms: x^2 and $5x$.

term of a mortgage The length of time that a mortgage agreement is in effect.

terminal arm The ray, of an angle in standard position, that is not on the positive x-axis.

transformation A move of a plane figure from one position to another position in the same plane.

translation A slide transformation.

translation image The image of a plane figure after a slide transformation.

transversal   A line that crosses or intersects two or more lines.

transverse axis   The line segment between the vertices of an hyperbola that is one of its axis of symmetry.

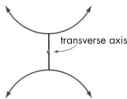
transverse axis

trapezoid   A quadrilateral with one pair of parallel sides.

trigonometric equation   An equation that contains one or more trigonometric functions.

trinomial   A polynomial with three terms.

$x^2 + 3x - 1$ is a trinomial.

V

variable   A letter or symbol, such as x, used to represent an unspecified number. For example, x and y are variables in the expression $2x + 3y$.

vertex   A point at which two sides of a polygon meet.

vertex of a parabola   The point of the parabola at which the graph intersects the axis of symmetry. The vertex is located midway between the focus and the directrix.

vertical line test   A method of determining a function. If any vertical line passes through more than one point on the graph of a relation, then the relation is not a function.

vertices of an hyperbola   The endpoints of the transverse axis.

vertices of an ellipse   The endpoints of the major axis.

volume   The amount of space that an object occupies, measured in cubic units.

W

whole number   A number in the sequence 0, 1, 2, 3, 4, 5, ...

X

x-intercept   The x-coordinate of the point where a line or curve crosses the x-axis.

xy-plane   A coordinate system based on the intersection of two straight lines called axes, which are usually perpendicular. The horizontal axis is the x-axis, and the vertical axis is the y-axis. The point of intersection of the axes is called the origin.

Y

y-intercept   The y-coordinate of the point where a line or curve crosses the y-axis.

Z

zero of a function   Any value of x for which the value of the function $f(x)$ is 0.

zero product property   The property that, if the product of two real numbers is zero, then one or both of the numbers must be zero.

TECHNOLOGY INDEX

Corel® Quattro® Pro
  Activities (Appendix C)
    copying cell content, 769–771
    entering formulas, 767
    entering functions, 768–769
    formatting type and numbers,
      771–772
    getting help, 766
    opening, saving and closing a
      spreadsheet, 766
    printing and copying, 773
    selecting cells, 767

The Geometer's Sketchpad®
  Activities (Appendix C)
    creating a new line segment from
      an existing one, 747
    constructing a circle with a given
      centre and radius, 749
    constructing a line, 747
    constructing a midpoint of a line
      segment, 748
    constructing and measuring a line
      segment, 745–746
    constructing a right bisector of a
      line segment, 748
    construct menu, 745
    display menu, 744
    edit menu, 745
    file menu, 744
    measure menu, 745
    preferences, 743
    tools and menus, 743–744
  conics as the intersection of planes
    with cones, 601–607
  constructing loci, 588–593
  constructing triangles using SSA,
    296–299
  loci and conics, 601–607

Graphing calculator
  changing degree measure to radian
    measure, 333, 374
  changing radian measure to degree
    measure, 332
  converting a decimal to a fraction, 8,
    16, 113, 147
  evaluating expressions with complex
    numbers, 145–147

evaluating expressions with rational
  exponents, 14–15
evaluating expressions with roots,
  123, 128, 347, 625, 642, 644,
  646, 668, 671
evaluating expressions with zero and
  negative exponents, 8
finding the amount compounded,
  503–511
finding the amount of an annuity,
  529–533
finding the future value, 563–570
finding the interest rate, 506
finding the interest rate per month,
  551–558
finding the maximum of a quadratic
  equation, 113
finding the monthly payments of an
  annuity, 529–533
finding the monthly payments of a
  loan/mortgage, 544–558, 562–570
finding the number of payments,
  566–570
finding the present value, 519–525
finding the present value of an
  annuity, 536–543
finding the sum of a series, 467–468,
  474–475
finding the value of payments,
  539–543
finding the zeros of a quadratic
  equation, 131
finding terms of a sequence,
  430–431, 437–439, 448–450,
  459–460
graphing circles, 613, 616
graphing combinations of functions,
  236–243
graphing ellipses, 629, 635
graphing hyperbolas, 646–647, 651
graphing parabolas, 659, 663
graphing functions, 182–183
graphing inverses of functions,
  211–220
graphing periodic functions, 373,
  385, 395
graphing reflections of functions,
  194–207
graphing relations, 173–175

graphing stretches of functions,
  221–232
graphing systems of equations,
  401–402
graphing to find the intersection of
  lines and conics, 675–687
graphing to show equations appear to
  be trigonometric identities, 395,
  399–400
graphing to solve trigonometric
  equations, 402–409
graphing translations of functions,
  184–193
Keystrokes (Appendix B)
  Circle instruction, 717–718
  common denominator function,
    719
  DrawInv instruction, 719–720
  expand function, 720
  factor function, 721
  FINANCE menu, 721
  format settings, 722
  Frac function, 722
  graph styles, 723
  Intersect operation, 724–725
  LIST MATH menu, 725
  LIST OPS menu, 725
  maximum operation, 726
  minimum operation, 727
  mode settings, 728
  sequence function, 728
  Sequence Y= editor, 729
  SinReg (sinusoidal regression)
    instruction, 730
  solve function, 731
  standard viewing window, 731
  STAT EDIT menu, 732
  STATS PLOTS menu, 733–734
  sum function, 734
  TABLE SETUP screens, 735
  TEST menu, 736
  test submenu, 736
  TRACE instruction, 736
  TVM Solver, 737–738
  value operation, 738
  window variables, 739
  Y= editor, 739
  zero operation, 740–741
  Zoom In instruction, 741

ZOOM menu, 741
ZoomStat instruction, 742
Zsquare instruction, 742
modelling restrictions graphically,
83–84
operations with polynomials, 32
solving a linear system, 77
trigonometric ratios, 267–275,
279–282, 285–295, 303–311

Microsoft® Excel
Activities (Appendix C)
copying cell content, 762–763
entering functions, 761
formatting type and numbers,
763–764
getting help, 759
opening, saving and closing a
spreadsheet, 759
printing and copying, 764
selecting cells, 759
using formulas, 760

Spreadsheets. *See also* Microsoft®
Excel *and* Corel® Quattro® Pro
and amortization tables, 544–558,
563–570
and the progress of repaying a
loan/mortgage, 544–558,
563–570

Technology Extension
constructing triangles using SSA
with *The Geometer's
Sketchpad®*, 296–299
radical expressions and graphing
calculators, 143
rational expressions and the
graphing calculator, 70–71
sinusoidal regression, 410
solving exponential equations with
a graphing calculator, 27
solving inequalities with a
graphing calculator, 82
solving quadratic equations, 134

Web Connection, 10, 23, 108, 149,
182, 194, 264, 293, 355, 391, 425,
429, 503, 535, 619, 637, 686

Zap-a-Graph
Activities (Appendix C)
combining transformations, 752
combining transformations of
trigonometric functions, 757
defining a function, 749
finding a maximum or minimum,
750
graphing an ellipse or hyperbola,
753–754
graphing trigonometric functions,
755–756
inverse of a function, 752–753
setting the scale, 750
transforming functions, 750–751
transforming trigonometric
functions, 755

INDEX

A

Absolute value, 637
Achievement Check, 18, 61, 69, 81, 91, 119, 133, 152, 159, 207, 220, 243, 256, 275, 294, 311, 317, 340, 391, 409, 419, 446, 464, 478, 487, 511, 543, 570, 578, 636, 652, 664, 695
Adding polynomials, 28–31
Adding rational expressions, 53–71
Ambiguous case, 300–312
Amortization period, 544
Amortization tables, 544–570
  and spreadsheets, 544–558
Amortize, 544
Amount, 498
  of an ordinary annuity, 526–533
Amplitude, 357
Angle of depression/elevation, 268
Angle of rotation, 351–354
Annuity. *See* Ordinary annuity
Arithmetic sequences, 436–446, 457–464
  and simple interest and linear growth, 498–500
Arithmetic series, 465–471
Asymptote, 183

C

Canada Savings Bonds, 514
Career Connection
  accounting, 456
  communications, 664
  crafts, 420
  microbiology, 26
  publishing, 133
  surveying, 294
  veterinary medicine, 193
Centre of a circle, 609
Centre of a hyperbola, 640
Challenge Problems, 92, 160, 257, 318, 420, 488, 579, 696
Circles, 608–618
Combinations of transformations, 233–243
Common difference, 436
Common factoring, 3
Common ratio, 447

Complex number, 104
  operations with in rectangular form, 144–152
  tools for operating with, 135–142
Complex number system, 100–109
Compound inequality, 2
Compounding period, 501
Compound interest, 501–511
  and geometric sequences and exponential growth, 512–515
Compression, 223
Conics, 601
  circles, 608–618
  ellipses, 619–636
  hyperbolas, 637–652
  intersections with lines, 675–687
  parabolas, 653–664
  with equations in the form $ax^2 + by^2 + 2gx + 2fy + c = 0$, 665–674
Conjugate axis of a hyperbola, 640
Conjugates, 138
Corel® Quattro® Pro. *See* Technology Index
Cosine
  of angles greater than 90°, 276–282
  of angles less than 90°, 266–275
  of any angle, 341–350
  of negative angles, 353–354
Cosine function
  combinations of transformations of, 378–391
  sketching the graph of, 364–365
  stretches of, 367–377
  translations of, 378–391
Cosine law, 283–295, 312
Co-vertices of an ellipse, 622
Co-vertices of a hyperbola, 640
Cumulative Review, 165, 323, 493, 701

D

Dependent variable, 173
Directrix of a parabola, 653
Discontinuous graph, 183
Distinct real zeros, 121
Dividing rational expressions, 44–52, 71
Domain, 174

E

Ellipses, 619–636
Entire radical, 102
Equal real zeros, 121
Equation. *See* Exponential equations, Quadratic equations, Solving linear equations, *and* Trigonometric equations
Expansion, 223
Explicit formula, 457
Exponential equations, 19–27
Exponential function, 513
Exponent laws for integral exponents, 4–11
Exponent rules, 427, 497
Evaluating expressions, 169, 427, 587
Evaluating radicals, 99

F

Factoring $ax^2 + bx + c$, 3
Finding angles in right triangles, 265
Finding sides in right triangles, 265
First-degree inequality, 72
Focal radii of an ellipse, 619
Focal radii of a hyperbola, 637
Focus (foci) of an ellipse, 619
Focus (foci) of a hyperbola, 637
Focus of a parabola, 653
Function, 170. *See also* Exponential function, Functions defined by $f(x) = \dfrac{1}{x}$, Functions defined by $f(x) = \sqrt{x}$, Inverse functions, Periodic function, *and* Quadratic function
Functions defined by $f(x) = \dfrac{1}{x}$
  properties, 182–183
Functions defined by $f(x) = \sqrt{x}$
  properties, 182–183

G

The Geometer's Sketchpad®. *See* Technology Index
Graphing calculator. *See* Technology Index
Graphing equations, 427, 587

Graphing quadratic functions, 99
Geometric sequences, 447–456
  and compound interest and
    exponential growth, 512–515
Geometric series, 472–478
Getting Started
  communications satellites, 586
  comparing costs, 496
  daylight hours, 326
  exploring sequences, 426
  frequency ranges, 2
  human physiology, 168
  parallactic displacement, 264
  store profits, 98

H

Hyperbolas, 637–652

I

Imaginary number, 104
Imaginary unit, 103
Imaginary zeros, 121
Independent variable, 173
Inequalities
  graphing, 3, 74–81
  solving first degree, 72–82
Initial arm, 276
Inquiry process steps, 83–84, 153
Interest. *See* Compound interest *and*
  Simple interest
Interest rate, 498
Intersections of lines and conics,
  675–687
Invariant point, 196
Inverse functions, 208–220
Investigate & Apply
  confocal conics, 584–696
  cosine law and the ambiguous case,
    312
  cost of car ownership, 571
  frieze patterns, 244–245
  interpreting a mathematical model,
    153
  modelling double helixes, 411
  modelling restrictions graphically,
    83–84
  relating sequences and systems of
    equations, 479
Irrational numbers, 105
Iteration, 149

L

Length of a line segment, 587

Linear relation, 500
Loci, 588–593, 601–607
  equations of, 594–600
Logic Power, 143, 262, 446, 674

M

Major axis of an ellipse, 622
Microsoft® Excel. *See* Technology
  Index
Midpoint formula, 587
Minor axis of an ellipse, 622
Mixed radical, 102
Modelling Math
  falling objects, 167, 191, 219, 232,
    242
  making financial decisions, 495, 510,
    533, 542, 569
  measurements of lengths and areas,
    97, 108, 118, 132
  modelling problems algebraically, 1,
    25, 60, 80
  motion in space, 585, 634, 650, 663
  motion of a pendulum, 425, 445,
    455, 478
  ocean waves, 325, 362, 377, 388–390
  ship navigation, 263, 274, 293, 310,
    311
Modelling periodic behaviour, 355–362
Mortgages, 544–570
Multiplying polynomials, 28–31
Multiplying rational expressions, 44–52,
  70

N

Number Power, 26, 435

O

Operations with complex numbers in
  rectangular form, 144–152
Operations with polynomials, 28–31
Operations with radicals, 135–143
Ordinary annuity, 526
  amount of, 526–533
  present value of, 534–543

P

Parabolas, 653–664
Patterns Power, 109, 295, 525
Payment period or interval, 526
Period, 356

Periodic function, 356
  combinations of transformations of,
    378–391
  stretches of, 367–377
  translations of, 378–391
Power, 6
Present value, 516–525
  of an ordinary annuity, 534–543
Principal, 498
Problem Solving
  guess and check, 489–491
  look for a pattern, 161–163
  model and communicate solutions,
    93–94
  solve a simpler problem, 421–422
  solve rich estimation problems,
    258–260
  use a data bank, 697–699
  use a diagram, 319–321
  use logic, 580–582
  using the strategies, 95, 164, 261,
    322, 423, 492, 583, 700
Pure imaginary number, 103

Q

Quadratic equations
  solving, 3, 99, 120–134, 327, 687
Quadratic formula, 99, 587
Quadratic functions
  finding the maximum or minimum
    by completing the square, 110–119
  graphing, 99
  rewriting in the form
    $y = a(x - h)^2 + k$, 99, 587
Quotient identity or relation, 393

R

Radian, 330
  and angle measure, 328–340
Radicals, 99, 100–109, 135–143
Radius, 609
Range, 174
Rational exponents, 11–18
Rational expressions, 35
  adding, 53–71
  dividing, 44–52, 71
  multiplying, 44–52, 70
  simplifying, 35–43, 70
  subtracting, 53–71
Rationalizing the denominator, 138
Rational numbers, 105
Real numbers, 105
Real zeros, 121
Recursive formula, 457–464

Reflections, 169, 194–207
Relation, 170
Review of Prerequisite Skills
    (Appendix A), 702–716

S

Sequences, 428–435
    arithmetic, 436–446, 467–464
    geometric, 447–464
    recursion formulas, 457–464
Series, 465
    arithmetic, 465–471
    geometric, 472–478
Simple interest, 498
    and arithmetic sequences and linear
        growth, 498–500
Simplifying expressions, 3, 427, 587
Simplifying radicals, 100–109
Simplifying rational expressions, 35–43,
    70
Sine
    of angles greater than 90°, 276–282
    of angles less than 90°, 266–275
    of any angle, 341–350
    of negative angles, 353–354
Sine function
    combinations of transformations of,
        378–391
    sketching the graph of, 363–364
    stretches of, 367–377
    translations of, 378–391
Sine law, 283–311
Spreadsheets. *See* Technology Index
Solving exponential equations, 19–27
Solving linear equations, 3, 427, 498
Solving linear systems, 427, 587
Solving proportions, 265
Solving quadratic equations, 3, 99,
    120–134, 327, 587
Solving right triangles, 266–275
Solving triangles, 283–295
Solving trigonometric equations,
    401–410
Standard position, 276
Steps of the inquiry process, 83–84,
    153
Stretches, 169, 221–232
    of parabolas, 169
    of periodic functions, 367–377
Subtended angle, 328
Subtracting polynomials, 28–31
Subtracting rational expressions, 53–71

T

Tangent function
    sketching the graph of, 365–366
Tangent
    of angles less than 90°, 266–275
    of any angle, 341–350
    of negative angles, 353–354
Technology Extension. *See* Technology
    Index
Terminal arm, 276
Term of a mortgage, 544
Transformations. *See* Combinations,
    Reflections, Stretches *and*
    Translations
Translations, 169
    horizontal and vertical, 184–193
Transverse axis of a hyperbola, 640
Trigonometric equations, 401–409
Trigonometric identities, 392–400
Trigonometric ratios, 265, 327
    of any angle, 341–350
Trigonometry
    ambiguous case, 300–312
    cosine law, 283–295
    cosine of angles greater than 90°,
        276–282
    of right triangles, 266–275
    sine law, 283–295
    sine of angles greater than 90°,
        276–282
    special angles/triangles, 341

V

Vertical line test, 172
Vertices of an ellipse, 622
Vertices of a hyperbola, 640

W

Web Connection. *See* Technology
    Index
Word Power, 43

Z

Zap-a-Graph. *See* Technology Index
Zeros, 121

CREDITS

PHOTO CREDITS

x © Robert Holmes/CORBIS/Magma; **xi** © 2001 PhotoDisc *Nature, Wildlife and the Environment 2*, Vol. 44, #44210; **xvi top** Ryan McVay/© PhotoDisc; **xviii top** David Buffington/© PhotoDisc; **1** Ryan McVay/© PhotoDisc; **2** © 2001 PhotoDisc *Nature, Wildlife and the Environment 2*, Vol. 44, #44210; **4** NASA from *Space and Spaceflight* © Digital Stock; **11** Courtesy National Research Council Canada/Conseil national de recherches Canada; **19** Courtesy Hydro One Archives 90-0454-17; **35** Lawrence M. Sawyer/© PhotoDisc; **44** Si-an Deng/Canadian Sport Images; **53** Courtesy Global Skyship Industries; **62** Patricia Fodgen/CORBIS/Magma; **72** Mike Ridewood/Canadian Press Picture Archive; **93** David Buffington/© PhotoDisc; **96** © 2001-Ontario Tourism; **97** COMSTOCK PHOTOFILE LTD./W. Griebeling; **98** PhotoLink/© PhotoDisc; **100** Mike Henson/London Free Press/Canadian Press Picture Archive; **110** Tom Pidgeon/Associated Press AP/Canadian Press Picture Archives; **120** Corel Corporation #149093; **123** NASA; **144** Corel Corporation #424073; **153** Ryan McVay/© PhotoDisc; **161** © Digital Stock *Stormchaser 2* Vol. 188; **167** Courtesy NASA; **168** Kathy Willens/Associated Press AP/Canadian Press Picture Archive; **170** Corel Corporation #0237 002; **184** © Miles Ertman/Masterfile; **194** Corel Corporation #161014; **208** Ryan Remiorz/Canadian Press Picture Archives; **221** © Al Harvey/The Slide Farm; **233** Canadian Tourism Commission 3-H-8-465. Reproduced with the permission of Her Majesty the Queen in Right of Canada; **244** Jules Frazier/© PhotoDisc; **258** Skydome Inc., Toronto, Canada; **262** Canadian Tourism Commission. Reproduced with the permission of Her Majesty the Queen in Right of Canada; **263** J.P. Jerome/Parks Canada/H.06.73.01.10(17); **283** Library of Parliament/Bibliothèque du Parlement CB-051; **300** Corel Corporation #43018; **319** Robert Glusic/© PhotoDisc; **325** A. Sirulnikoff/First Light; **326** Eyewire #e000492r; **328** Cartesia/© PhotoDisc; **341** Martial Colomb/© PhotoDisc; **355** Yokohama Convention and Visitor's Bureau; **367** © 2001 PhotoDisc *Professional Science* Vol. 72 #72148; **378** Don Johnston/Ivy Images; **393** Dave Cannon/Stone; **402** Pascal Crapet/Stone; **421** Robin Nowaacki/Associated Press AP/Canadian Press Picture Archives; **425** © Robert Holmes/CORBIS/Magma; **428** Courtesy NASA GPN-2000-000933; **436** Pierre St Jacques Inc.; **447** © Digital Vision # 074005cm; **457** © Renee DeMartin/Weststock/Image State; **465** © W. Anthony/Weststock/Image State; **472** Tony Freeman/PhotoEdit; **489** Dr. Gary Settles/Science Photo Library; **491** Robin Nowaacki/Associated Press AP/Canadian Press Picture Archives; **495** Spencer Grant/PhotoEdit; **496** Tony Freeman/PhotoEdit; **498** Keith Brofsky/© PhotoDisc; **501** Adam Crowley/© PhotoDisc; **512** © Jeff Greenberg/Visuals Unlimited; **516** © Michael Newman/PhotoEdit; **526** Jack Hollingsworth/© PhotoDisc; **544** Steve Mason/© PhotoDisc; **559** © Digital Vision *Personal Finance*; **585** © NOAO from *Space and Spaceflight* © Digital Stock/ #Img0076; **586** Courtesy Telesat Inc.; **594** Mary Agnes Challoner; **619** Mary Kate Denny/PhotoEdit/PictureQuest; **637** COMSTOCK PHOTOFILE LTD/© E. Masterson/H. Armstrong Roberts; **653** NASA #STS103-726-081;

665 Tom Hanson/Canadian Press Picture Archive; 675 Lori Adamski Peek/Stone; 697 Corel Corporation #283082;

TEXT CREDITS

Calculator templates: Texas Instruments; 296–299 *The Geometer's Sketchpad*®, Key Curriculum Press, 1150 65th Street, Emeryville, CA 94608 1-800-995-MATH; 588–593 *The Geometer's Sketchpad*®, Key Curriculum Press, 1150 65th Street, Emeryville, CA 94608 1-800-995-MATH;717–742 Courtesy of Texas Instruments; 743–749 *The Geometer's Sketchpad*®, Key Curriculum Press, 1150 65th Street, Emeryville, CA 94608 1-800-995-MATH; 749–757 Zap-a-Graph © Brain Waves Software Inc., 2103 Galetta Sideroad, Fitzroy Harbour, Ontario, K0A 1X0 Tel. 613-623-8686; 758–764 Microsoft® Excel are either registered trademarks or trademarks of Microsoft Corporation in the United States and/or other countries; 765–773 Screen shot(s) from Quattro® Pro. Corel and Quattro are trademarks or registered trademarks of Corel Corporation or Corel Corporation Limited, reprinted by permission.

Chapter Expectations: © Queen's Printer for Ontario, 2001.

ILLUSTRATION CREDITS

266 Wesley Bates 270 Michael Herman 274 Michael Herman 424 Michael Herman 534 Michael Herman 580 Bernadette Lau

TECHNICAL ART CREDITS

Tom Dart, Alana Lai, Claire Milne, Greg Duhaney of First Folio Resource Group, Inc.